P9-ECL-608

LAND DEGRADATION IN MEDITERRANEAN ENVIRONMENTS OF THE WORLD

LAND DEGRADATION IN MEDITERRANEAN ENVIRONMENTS OF THE WORLD

NATURE AND EXTENT, CAUSES AND SOLUTIONS

Compiled and Edited by

Arthur J. Conacher
Department of Geography, University of Western Australia

and

Maria Sala
Department of Physical Geography, University of Barcelona

JOHN WILEY & SONS
Chichester • New York • Weinheim • Brisbane • Singapore • Toronto

3 779777
) 1243 779777
and customer service enquiries): cs-books@wiley.co.uk
ge on http://www.wiley.co.uk
or http://www.wiley.com

Other Wiley Editorial Offices

John Wiley & Sons, Inc., 605 Third Avenue,
New York, NY 10158–0012, USA

WILEY-VCH Verlag GmbH, Pappelallee 3,
D-69469 Weinheim, Germany

Jacaranda Wiley Ltd, 33 Park Road, Milton,
Queensland 4064, Australia

John Wiley & Sons (Asia) Pte Ltd, 2 Clementi Loop #02–01,
Jin Xing Distripark, Singapore 129809

John Wiley & Sons (Canada) Ltd, 22 Worcester Road,
Rexdale, Ontario M9W 1L1, Canada

Library of Congress Cataloging-in-Publication Data

Land degradation in Mediterranean environments of the world : nature and extent, causes and solutions / compiled and edited by Arthur J. Conacher and Maria Sala.
p. cm.
Includes bibliographical references and index.
ISBN 0-471-96317-8
1. Environmental degradation. 2. Land degradation.
3. Mediterranean Region—Environmental conditions. I. Conacher, A. J. II. Sala, Maria.
GE160.M47L36 1998
333.73'137—dc21 97–29313
CIP

British Library Cataloguing in Publication Data

A catalogue record for this book is available from the British Library

ISBN 0-471-96317-8

Typeset by Dorwyn Ltd, Rowlands Castle, Hampshire
Printed and bound in Great Britain by Bookcraft (Bath) Ltd
This book is printed on acid-free paper responsibly manufactured from sustainable forestation, for which at least two trees are planted for each one used for paper production.

Contents

List of figures

List of plates

List of tables

Contributors

Jean-Louis Ballais
Institute de Géographie
29 Ave. R. Schuman
Univ. Aix-Marseille
13621 Aix-en-Provence
France

Jela Bilandžija
Ministry of Agriculture and Forestry
Ulica Grada Vukovara 78
10000 Zagreb
Croatia

Mauricio Calderon
Corporacion Nacional Forestal
Alvarez 2760
Vina del Mar
Chile

Consuelo Castro
Universidad Catolica
Ins. de Geografia
Vicuna Macrenna 4860
Santiago de Chile
Chile

Celeste Coelho
Dept Ambiente e Ordenamento
Universidad de Aveiro
3800 Aveiro
Portugal

Arthur Conacher
Department of Geography
University of Western Australia
Nedlands
WA 6907
Australia

Jeanette Conacher
4 Mitchell Rd
Darlington
WA 6070
Australia

N.G. Danalatos
Laboratory of Soils and Agricultural Chemistry
Agricultural University of Athens
75 Iera Odos
Botanikos 118 55
Athens
Greece

Matija Franković
Ministry of Civil Engineering and
Environmental Protection
Avenija Vukovar 78
41000 Zagreb
Croatia

Moshe Inbar
Department of Geography
University of Haifa
Haifa 31999
Israel

Dražen Kaučić
Preradoviceva 39
41000 Zagreb
Croatia

Constantinos S. Kosmas
Laboratory of Soils and Agricultural Chemistry
Agricultural University of Athens
75 Iera Odos
Botanikos 118 55
Athens
Greece

Abdellah Laouina
Comité National de Géographie
Inst. Univ. de la Rech. Scient.
BP 2122 Ryad
10104 Rabat
Morocco

Micheal E. Meadows
Department of Geographical and Environmental Sciences
University of Cape Town
Private Bag X7700
Rondesbosch
6000 Cape Town
South Africa

A. Mizara
Laboratory of Soils and Agricultural Chemistry
Agricultural University of Athens
75 Iera Odos
Botanikos 118 55
Athens
Greece

Antony Orme
Department of Geography
University of California, Los Angeles
1255 Bunche Hall
405 Hilgard Avenue
California 90095-1524
USA

Amalie Jo Orme
Department of Geography
University of California, Los Angeles
1255 Bunche Hall
405 Hilgard Avenue
California 90095-1524
USA

Maria Sala
Department of Physical Geography
University of Barcelona
08028 Barcelona
Spain

Marino Sorriso-Valvo
CNR – Instituto di Ricerca per la Protezione Idrogeologica (IRPI)
Via Verdi 1
87030 Roges de Rende (CS)
Italy

Introduction

Arthur and Jeanette Conacher

About this Book

It was in South Africa, in January 1994, that the International Geographical Union's then fledgling Study Group on Erosion and Desertification in Regions of Mediterranean-type Climate (MED) held its first meeting. The Group's Chair, Dr Maria Sala, who unfortunately was unable to be present, had asked the meeting to consider activities which the Group might undertake. Given the importance of the increasingly severe problem of land degradation, it was suggested that production of a book, written by experts from the various Mediterranean climate regions of the world, would bring together valuable information currently scattered amongst the regions and in various languages. It was further agreed that emphasis should be placed on understanding the cultural and biophysical contexts as well as the nature and extent of the problems, and on discussing solutions. This book is the outcome of that meeting.

It was considered that the terms "erosion and desertification" are too narrow to adequately reflect the scope of the problems, thus the term "land degradation" was used instead. This was defined by the Group to mean:

"alterations to all aspects of the natural (or biophysical) environment by human actions, to the detriment of vegetation, soils, landforms, water (surface and subsurface) and ecosystems."

A further objective of the book was that it should attempt to indicate how, and in what ways (if at all), the seasonal nature of Mediterranean climates affects land degradation: is there something distinctively "Mediterranean" about the problems of land degradation in the world's Mediterranean regions?

Definition of "Mediterranean climate"

A long discussion ensued on this matter at the meeting of the Study Group in Morocco in October 1994, considerably assisted by Dr Moshe Inbar's research. In brief, the schemes of Köppen, Flohn, Trewartha and Horn, Emberger, Nahal, Zohary, UNESCO–FAO, and Aschman were considered and discussed. It was finally agreed to adopt the UNESCO–FAO (1963) definition, since it is based on criteria which are relevant to plant growth. However, it was further agreed to modify that definition (largely but not entirely according to Aschman's criteria) in order to remove some absurdities from the UNESCO–FAO world map. That map delimits some areas as "Mediterranean" even though they have cold, or hot arid and semi-arid climates, and even some areas which do not have a cold season precipitation maximum. Above all, it was agreed that *summer drought* is the distinguishing characteristic of Mediterranean-type climates. It was also recognized that some further modifications may be required for individual regions.

The modifications agreed to are as follows (Conacher 1995):

1. Exclude areas with <275 mm mean annual precipitation for cool, coastal stations and <350 mm mean annual precipitation for warm interior stations;
2. Exclude areas where the mean monthly temperature for any month is <0 C;
3. Exclude areas which receive <65% of their precipitation in the cold half of the year;
4. There should be no maximum mean annual precipitation.

What is "Mediterranean"?

There was a widespread feeling amongst the members of the Study Group that there is something distinctive about regions which are blessed by Mediterranean-type climates, despite their highly dispersed locations (Figure 0.1). The following sections attempt to indicate the nature of this distinctive "Mediterranean" characteristic. It is based on our experience in visiting most (regrettably not all) of the regions, on the work of this book's contributors, and on some of the extensive literature which deals with Mediterranean countries and environments (some useful Mediterranean works include: Arianoutsou and Groves 1994; Brandt and Thornes 1996; Di Castri *et al.* 1981; Di Castri and Mooney 1973; Fox 1991; Girgis 1987; Goldhammer and Jenkins 1990; Groves and Di Castri 1991; Imeson and Sala 1988; Kalin Arroyo *et al.* 1994; McNeill 1992; Meiggs 1982; Mooney 1977; Pate and Beard 1984). At the outset it needs to be stressed that although there is a commonality in climate and therefore a number of other biophysical characteristics amongst the regions, there is also considerable diversity.

The regions discussed in this book are found in five disjunct areas of the world with highly variable geology, soils, history and people but broadly similar climates, vegetation and landscape forms.

A sensual environment

The marked contrasts between hot, desiccating summers and cold, wet winters with intense rain

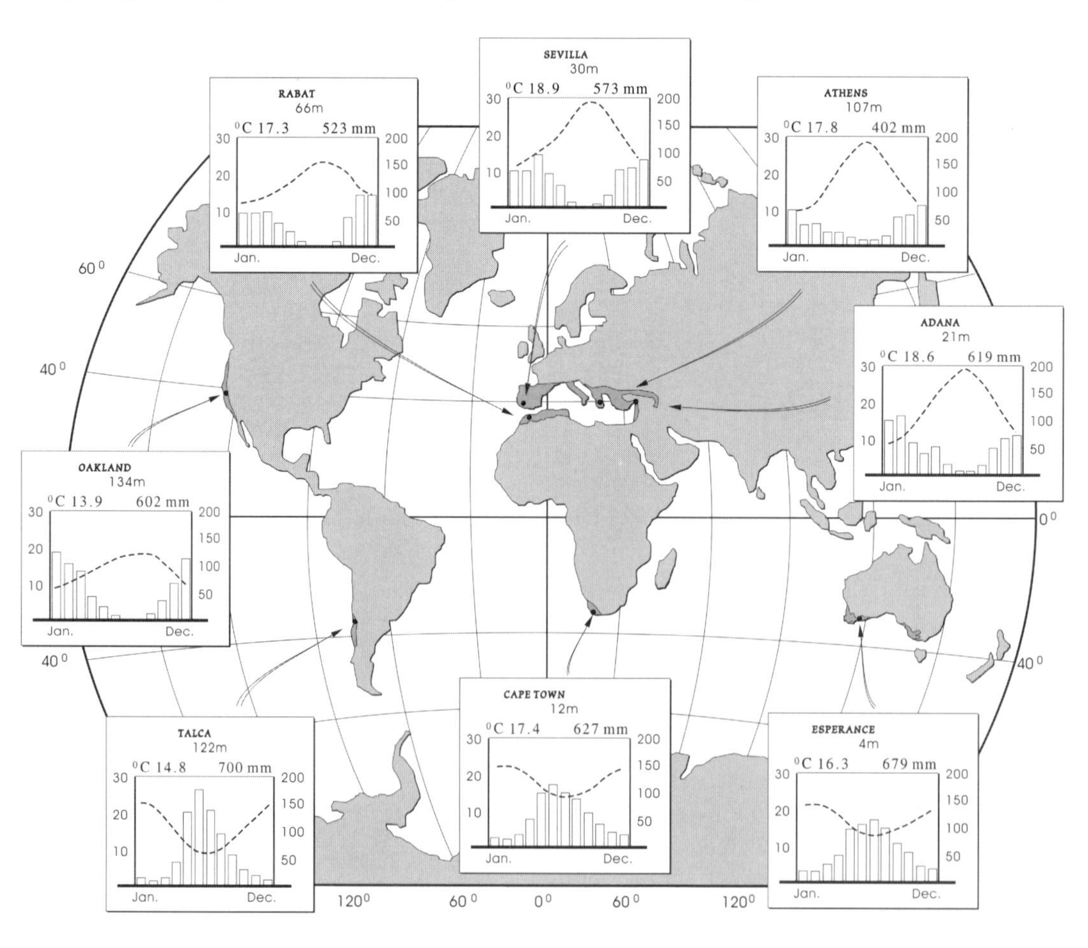

Figure 0.1 *World regions of Mediterranean-type climates* (modified from Archibold 1995, Ch. 5)

and biting winds, are felt directly by the regions' inhabitants. But in addition there are the *visual* characteristics: the bright light and sunshine (even in winter); light colours – of limestone, marble and granite; bright splashes of colour – of geraniums and bougainvillea; olive greens – of olives, *maquis* and *chaparral* – and deep turquoise of the sea; the scenes of terraces, beaches and fire scars, of stark, denuded hills, of high mountains and rural landscapes (vines, olives, fruit trees, grazing lands), of forest and scrub, of crowded towns, cities and fishing ports, and tourists, and of haze (pollution and mists).

There are the heady *aromas* – of aromatics (herbs), sweet perfumes of roses, jasmine, orange blossom and broom, and of cooking (olive oil), and sometimes putrid odours (rotting fish, sewers, rubbish); and *sounds* – seed pods popping in the heat, bleating goats or sheep, wind in the pines, cicadas.

And there is an *intensity* about the Mediterranean regions: of light (white, bright/cool, dark shade); of colour and contrast (bougainvillea reds, plumbago blues and greens, reflected in the regions' art); and passionate natures, revealed in song, dance, language and politics.

People

The Mediterranean Basin is, of course, the cradle of Western civilization, but Mediterranean peoples also colonized the New World Mediterranean regions, particularly California, Chile and southern Australia. The Basin is redolent with its rich history (not forgotten by the inhabitants of the "outer" Mediterranean regions): classical history, the great philosophers and monuments; and the various periods of ascendancy (still dynamic) – Assyrian, Persian, Greek, Roman, Byzantine, Arabic, Moghul, French, Spanish, Portuguese. It is, too, the cradle of agriculture, originating in the Fertile Crescent of Mesopotamia. It is the centre of three major world religions (Judaism, Christianity, Islam), and of many languages and nations. The Basin is distinguished by its art, gardens, architecture and literature, again reflected by more recent imitations or developments in the New World. The climate, history and coastal locations of Mediterranean regions attract more than a third of the world's tourists, who contribute to the life and colour of these regions but also to their environmental pressures.

Heterogeneity

The Old World regions are highly heterogeneous: historically, the Basin was a major crossroads, in trade, invasion and war, and migration; in languages and races; and in fragmented landscapes (both their physical attributes and settlements). There are also dichotomies between rich and poor within and between regions: north–south; urban–rural; eastern–western Mediterranean Basin; Old World–New World. There are shell villages and marginalized indigenes, and contrasts between poverty and opulence, although these are perhaps less distinctively "Mediterranean".

Similarly, various pressures are probably not peculiarly Mediterranean: population growth; rural depopulation; industrialization; loss of traditional ways; political issues; and globalization.

There are, however, a number of distinctive dichotomies between the Old and the New World Mediterranean regions, which in very broad terms could be summarized as: peasant/capitalist economies; village/ranch; long period of settlement/recent; wide gaps in incomes and status/more egalitarian; autocracies/democracies; complex land ownership issues/less so?; twenty nations/single governments; closed sea/open seas; culturally complex/less so; centrally located/peripheral.

The biophysical environments

The world's Mediterranean regions are located very broadly between latitudes 30 and 40 , between temperate and tropical arid zones, and on the west coasts of continents washed by cold ocean currents.

Climate

Climates are characterized by hot, dry summers and cool, wet seasons, but also by marked aperiodicity, strong winds and frontal rain, and winter fogs. Variability results in periods of aseasonal drought (within and between years) and torrential rains. Hot, dry and sometimes adiabatic winds occur in summer, attracting local names (*sirocco*, *meltmi*, *Santa Ana*), and strong, cold winds sweeping from the mountains in winter (*bora*, *mistral*, *tramontana*). In the southern hemisphere the northerly movement of the "roaring forties", and in the north, the southerly movement of Atlantic lows, bring winter gales, while alternating land and sea breezes bring welcome relief to coastal areas in summer. Sea fogs and mists are associated with cold, marine currents off the coasts of California, Chile and southwest Africa, and to a lesser extent southwestern Australia; and heating and cooling effects influence the *marin* in the south of France, Italy and Spain.

Local climates are modified by relief (orographic and rainshadow effects), aspect and altitude; and there are continental influences in the northern and eastern Mediterranean Basin, and arid tropical influences in North Africa (and southern Italy, Greece and the eastern Mediterranean), and in southern California, northern Chile, southwestern Africa and southern Australia. Indeed the location of Mediterranean climate zones in a transition from temperate to semi-arid and arid climates is an important feature which influences many human and biophysical features of the regions.

Geology, landforms and soils

A distinctive feature of the Mediterranean regions is their coastal location, with a coastal plain backed by mountains. The width of the plains varies, of course, as does the height and impressiveness of the mountains; but even in Australia, where the "mountains" are low remnants of long-gone ranges, the pattern of sea, coastal plain and backing escarpment still holds. Mediterranean mountains are often folded and still tectonically active, being at the junctions of continental tectonic plates. Lithologies in the Mediterranean Basin usually have marine origins, with white limestones, dolomites and perhaps marble with ubiquitous karstic landscapes. Coastlines in southern Australia and the eastern Mediterranean have distinctive aeolianites, whilst Chile, California and southwestern Africa have ranges constructed from sedimentary sequences. The latter regions, and North Africa, also have inland troughs with fertile, sedimentary plains. Only in southwestern Australia is the main landmass underlain by ancient (Proterozoic) granitic rocks – elsewhere lithologies are mainly Late Tertiary or even early Quaternary – with their characteristic deeply weathered, duricrusted soils.

Thus the combination of climate and steep relief produces high energy landscapes: there is naturally high erosivity on steep slopes and hence shallow (often calcareous and erodible) soils, particularly where surface cover is poor. Deep valleys, ravines and gullies make travel in the mountains difficult, as do the periodic floods. Landslides, sheetwash, rilling and gullying generate sediments which are deposited as rocky debris fans, with finer materials being deposited on the coastal plains and in the sometimes extensive deltas.

Soils are complex, reflecting parent material, present and past climates and land use, and range from young, azonal soils to palaeosols. They are generally deficient in nutrients, especially nitrogen and phosphorus, but rich in the minerals sodium, iron, calcium and magnesium. Reflecting their origins and age, they seem to be either highly acid or alkaline, rarely neutral. Organic matter contents are low, as are microbial activity and rates of formation. Profiles are thin and poorly developed. Water infiltration rates are often low, due to water-repellent, or stony or compacted soil surfaces; in turn this leads to high rates of surface runoff and erosion. No one soil type dominates; rather, there are several which are common: Terra Rossa (calcareous, iron/clay rich, karstic); siliceous soils (more acidic, leached and generally infertile); "laterites" (supposedly reflecting former tropical climates); alluvial soils (in river valleys, plains, terraces and deltas, and these are amongst the most productive soils); fertile volcanic-derived soils; stony, thin soils and even bare rock, especially in mountainous regions; and sodic or carbonate-rich soils in the more arid margins.

Water

As a result of the physiography and climate of the Mediterranean regions, there are few large, permanent rivers; these generally rise in mountains, sometimes beyond the Mediterranean boundary. Streams are generally short, sometimes deeply incised and characterized by ephemeral and sometimes torrential flows. Limestone lithologies mean that surface water is particularly scarce in many regions. On the other hand groundwater is important, especially in areas underlain by limestone and dolomite, in sedimentary sequences in the plains, and in the deep-weathered Australian soils; this groundwater makes possible summer irrigation for crops and orchards, and summer watering of stock. Springs at the junction between mountain and plain were important in ancient irrigation schemes, such as the Persian *qanats*. A contradiction between water scarcity (with serious associated management and political problems relating to water use and allocation), which is endemic in the Mediterranean regions, and over-supply in the form of seasonal floods, is particularly striking. Schemes to store this floodwater for human use are as old as civilization itself, notably in the eastern Mediterranean, as are the problems of land degradation which are associated with them. Problems of over-extraction, flooding, siltation, pollution, disturbance to natural hydrology, and "ownership" are ubiquitous and are heightened by the strongly seasonal nature of the climate and increasingly heavy demands placed on the resource.

Vegetation

Vegetation is perhaps the most distinctive (with climate) of all the properties of Mediterranean-type regions. It reflects present and past climates, as well as soils, topography, the long history of land use (including the New World), and fire. It is the product of several palaeoclimatic changes of advance and withdrawal of "richer" temperate or "poorer" arid tropical formations which have adapted to environments of fire, alternating wet and arid seasons, and human disturbance.

It is scleromorphic (tough, woody and evergreen) and there are strong morphological (but not taxonomic) similarities amongst the vegetation in the various regions. The *maquis* and *garrigue* Mediterranean Basin types have analogues in the other regions: *macchia* (Italy), *chaparral* and coastal sage (California and Chile), *tomillares* (Spain), *batha*, *goresh* (Israel), *phyrgana* (Greece), *kwongan*, *mallee* (Australia), *fynbos* and *renosterveld* (South Africa). All are adapted to drought, fire and nutrient-poor soils. Fire is endemic in all regions, reflecting the highly combustible nature of the vegetation and a range of natural and human causes. Indeed, some forms of vegetation *require* fire in order to reproduce. These are regions of high fire risk for human populations.

The mesic range of vegetation is taller (1 m +), with small trees and tall shrubs, evergreen stratified, often impenetrable and with deep roots (*maquis*). In contrast, the xeric forms are shrubby (less than 1 m tall), open, aromatic, ericoid, herbaceous, and with shallow roots (*garrigue*). Drought adaptation takes several forms, including sunken stomata, thick leaves, deep tap roots and water storage organs. The plants are seasonally responsive with growth and flowering periods in spring and autumn, and dormancy in winter and summer. They are sensitive to frost. Their highly combustible nature makes them particularly prone to fire, but they respond to fire by seeding and sprouting prolifically (with the important proviso that there are recovery periods of several years between fires).

Mediterranean vegetation is exceptionally rich in species, with some 10 000 in the Mediterranean Basin, 7000 in southwest Africa and 3000 in southwestern Australia. There is also a high degree of endemism, reflecting the "mosaic" nature of the Mediterranean landscapes and past climates. However, plant associations have been altered considerably not only in regions with long periods of human settlement, but also in the more recent colonies (although these areas, too, have long histories of nomadic, hunter/gatherer populations who used fire regularly). This (with climatic fluctuations) makes it difficult if not impossible to determine whether there is such a thing as a "climax" Mediterranean vegetation. Finally, the regions appear to be particularly susceptible to invasions of exotic plants, animals and diseases, again over a long period of time; some of the major "pests" have been introduced from one Mediterranean region to another, rather than from without.

Animals

Animals tend to be ignored in the conventional literature which seems to focus more on vegetation. But a range of native fauna exists in Mediterranean regions, particularly but not solely in the more natural habitat remnants.

Small mammals include badgers, shrews, polecats, mice and bats. Larger, mainly forest animals include pigs, foxes, wolves, boars, bears, mountain lions, large cats and deer. Reptiles are represented by snakes and lizards. Mediterranean regions provide important habitats for a wide range of migratory bird species as well as raptors. Pest birds include sparrows, crows, seagulls and doves. The regions are rich in insect species, with beetles and ants being important.

Soil fauna migrate deep in summer when they are dormant. They have cylindrical bodies and digging structures, heat-resistant eggs, sometimes in the form of cysts, and thick cuticles. Peak activity takes place in spring and autumn. They are particularly important in helping to decompose litter into humus, and they are also an important food source for some birds and animals.

Protection of total habitats, or ecosystems, is essential in order to protect the increasing number of rare and endangered plant and animal species in Mediterranean regions.

The Structure of this Book

Part I provides the context in which land degradation has occurred and continues to occur. It describes the environments – both human and physical – of the 11 Mediterranean regions which have been identified for this book, in order to provide a basis for the main, following Parts II and III. Each region is discussed using a common structure, although the individual contributors have been free to interpret this as appropriate for their regions.

One difficulty concerns the use of soil names. All contributors used "traditional" soil terminology, and this has been allowed to stand. In the absence of detailed data, conversion to the US Soil Taxonomy is impossible; and whilst the FAO classification is less rigorous in this regard, the contributors (supported by the editors) considered that traditional names are still much better known and more widely used by non-specialists such as the probable readers of this book.

Problems of land degradation in these Mediterranean environments are discussed in Part II. Again, the contributors were asked to write to a fairly detailed structure, but it will be apparent that there is a diversity amongst both the regions and the contributors. It is not entirely clear to what extent the varying treatment of the regions, and the recognition of different problems, reflects reality as distinct from the interests of the individual contributors. However, as indicated, a firm attempt was made to persuade each contributor to write to an agreed structure.

Solutions to the problems identified in Part II are discussed in Part III. The overall outline of chapters was agreed to beforehand, but less detail was provided since less was known about this aspect of the work. Contributors were asked to discuss both the solutions which are being implemented and what needs to be done. As will be seen, the responses to this section are particularly uneven; and again, there is uncertainty as to the extent to which this unevenness reflects the actual situation in the different regions, or the research interests of the contributors.

The extent to which the book has achieved its objectives is discussed further in the concluding chapter.

PART I

The nature of the world's Mediterranean-type environments

1

Iberian Peninsula and Balearic Islands

Maria Sala and Celeste Coelho

Introduction

The Iberian Peninsula is located between 35°59' and 43°47' latitude north and extends over 581 535 km², distributed between two countries, Spain with 492 463 km² and Portugal with 88 890 km². The Balearic Islands have an area of 7273 km². Iberia's location between the Atlantic Ocean to the west and the Mediterranean Sea to the east, and between Europe in the north and Africa in the south, is particularly important for the region's environment and culture. The Peninsula is linked to Europe by a narrow and mountainous isthmus 400 km in length, with altitudes exceeding 3000 m, and is separated from Africa by a shallow, narrow strait whose maximum width is only 14 km. Iberia's shape was compared by Strabo with the dissected skin of a bull, and it is often so described.

Despite being surrounded by seas, the extent of the territory (800 km wide and 700 km long) and the coastal mountain fringes give rise to a broad area of continental climate in the centre. The summer drought extends to nearly the entire Peninsula and thus determines the existence of xerophytic and halophytic vegetation species. Destruction of trees seems to have been always very important, not only for timber and agriculture but to a great extent for pasture. Fires are also frequent. Nevertheless, the Peninsula is very rich in species due to its location between Europe and Africa.

Relief and climate determine a high diversity of landscapes, with the diversity being especially marked in a NW–SE direction. The humid-temperate climate of the NW changes more or less gradually until it reaches the arid conditions of the SE.

The Mediterranean Sea facade has been a door open to a diversity of cultures. Apart from the Iberian natives, the three main cultures which have left a marked legacy are the Roman-Christian, the Muslim and the Jewish. Additionally, the Atlantic Sea facade opened the Iberian inhabitants to new worlds, south of Africa towards the Orient, and west to America.

Important land use changes have taken place during the last 30 years in relation to economic development, mainly related to tourism and joining the European Union. The economy has shifted from agro-sylvo-pastoral to a more tertiary-oriented one, although industry and some agricultural products are also important. Mediterranean products such as oranges, wine and olive oil, together with early season vegetables, are the main agricultural exports. On the other hand wheat and sheep from the broad central basins are now in decline. The main development trends such as tourism, intensive agriculture, industry and services are mostly located along the Mediterranean coast.

The Land

Relief

Over half of the territory is more than 600 m above sea level. The relief is strongly contrasted,

Land Degradation in Mediterranean Environments of the World: Nature and Extent, Causes and Solutions.
Edited by A.J. Conacher and M. Sala.

with mountain chains along the coasts and mountainous terrains inland separated by broad depressions (Figure 1.1). Geologically, the Iberian territory comprises two main units (Figure 1.2): Alpine to the east and Hercynian to the west (Julivert and Fontbote 1974; Sala 1984; Sala and García Ruíz 1984). The influence of lithology is important in relation to soils and vegetation. There are three main domains: siliceous (Hercynian areas), calcareous (Alpine areas) and marls and clays (Tertiary depressions).

The Alpine part has been folded and afterwards faulted. It consists of mountain belts along and perpendicular to the coast: the Pyrenees, the Catalan Coastal Ranges, the Iberian Cordillera and the Baetic Cordillera. Adjacent to these major Alpine features are two depressions, the Ebro south of the Pyrenees and the Guadalquivir north of the Baetica, filled with Tertiary and Quaternary sediments. The structure is a result of rift tectonics of Neogene–Quaternary age which affected all Mediterranean coastal areas, giving rise to the present typical structure of horsts and grabens, with the grabens being filled with detrital materials of both continental and marine facies of Miocene and Pliocene age. In the driest parts, basins of internal drainage with salty lakes may be found, normally not exceeding 1 m in depth.

The Iberian Cordillera is the southern limit of the depression and is joined to the northernmost sierras of the Baetic Cordillera creating a complex mosaic of structures in the region of Valencia. The relief forms reflect the presence of both Hercynian and Alpine structures. Faults parallel to the coastline are responsible for block subsidence towards the sea.

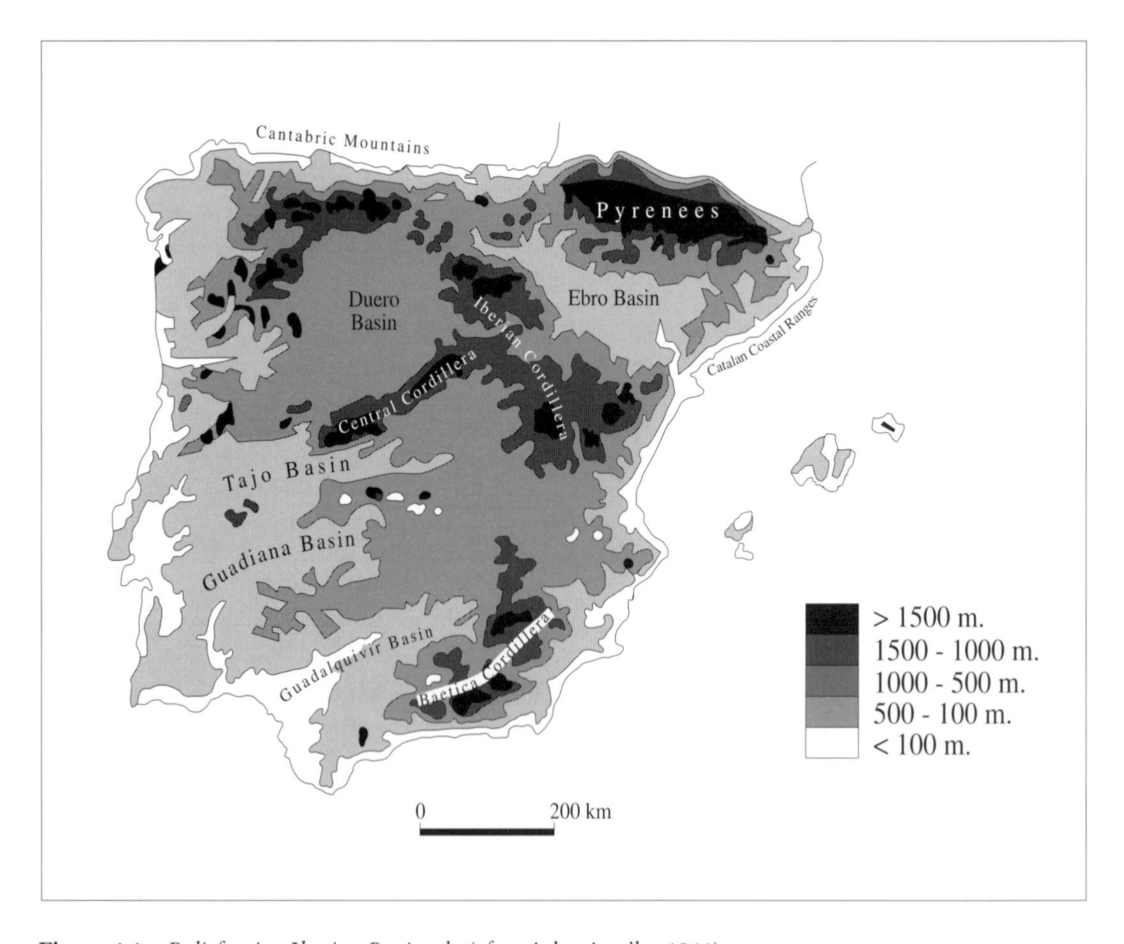

Figure 1.1 *Relief units, Iberian Peninsula* (after Atlas Aguilar 1961)

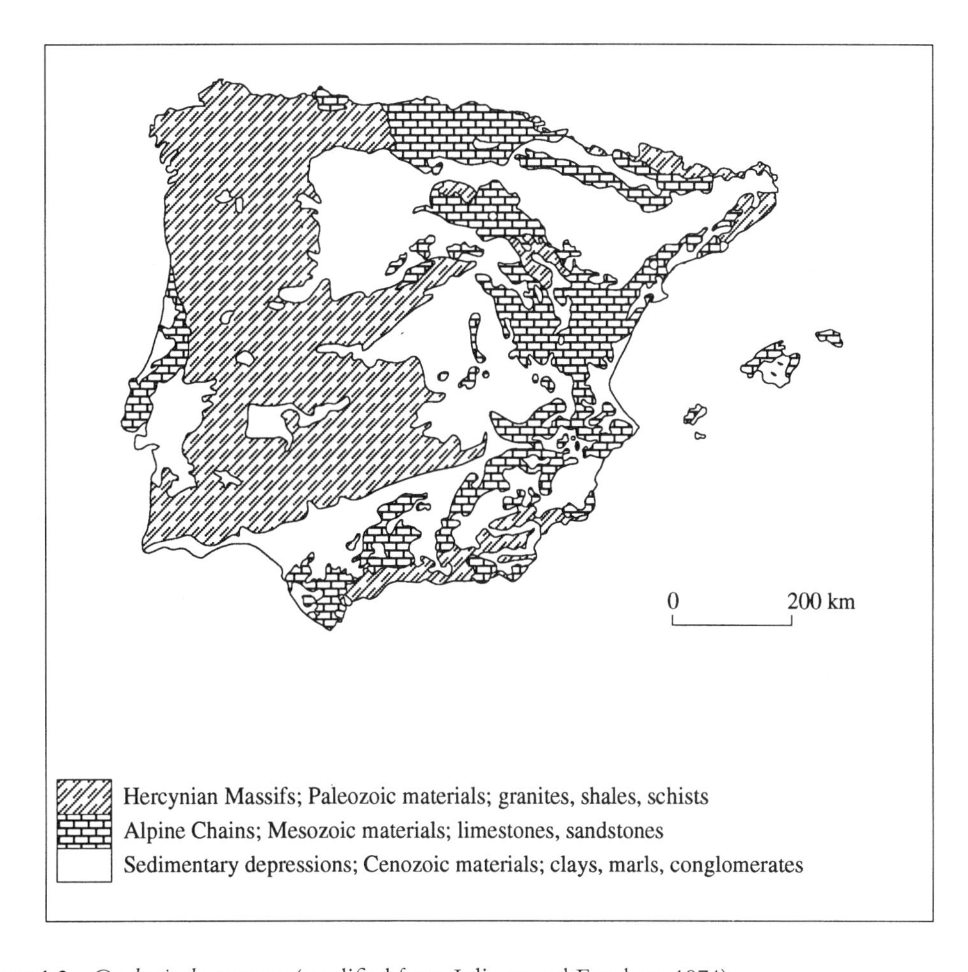

Figure 1.2 *Geological structure* (modified from Julivert and Fontbote 1974)

The Baetic Cordillera extends 600 km from the Strait of Gibraltar to Cabo la Nao. The Balearic Islands mark its prolongation to the east and, past Gibraltar, it is connected to the internal zones of the Rif Atlas in North Africa. Only the scattered limestone masses have resisted erosion, while the marls and shales have been worn down and reduced to hillocks. Former marine levels are numerous and well correlated with the marine levels of other western Mediterranean coastal areas, especially with those of Mallorca.

The Guadalquivir stands at an average altitude of 150 m and ends at the sea in a marshy area, Las Marismas, where accumulation processes are dominant. This plain was still a lagoon during Roman times. Along and close to the right bank of the Guadalquivir, the Hercynian block of Sierra Morena rises steeply to the north as a result of faulting.

The vast remnant of the Hercynian continent, the so-called Iberian massif, suffered repeated folding, granitization and metamorphism, and has been greatly worn down by erosion. It has two distinct elements: a series of massifs built mostly of Palaeozoic materials, and two broad depressions with a Tertiary sedimentary cover of sands, clays, gypsum beds and limestones in a horizontal or subhorizontal position. As a result of these events the Hercynian block now shows the following units:

(a) two basins, the northern one drained by the Duero, with average elevations of 800 m, and the southern one drained by the Tajo, Guadiana and Sado with average altitudes of 600 m to 200 m; and
(b) two massifs aligned from west to east, the Central Massif and the Montes de Toledo mountain.

To the west and south of the Hercynian Massif a narrow belt of secondary rocks, 60 km wide in Portuguese Estremadura, develops as the southwestern extension of the Central Cordillera. It consists of a succession of hills and karstic depressions, the most outstanding being the tectonically formed polje of Mira-Minde. The Algarve, in the south, is a narrow belt of hilly country, bordering the coast from Cape Saeo Vicente to the Guadiana estuary.

Drainage

The general tilting of the Hercynian block to the west and the existence of a vast central plain are responsible for the general tendency of the fluvial net to drain extensive basins towards the Atlantic Ocean. In contrast, on the Mediterranean side only the Ebro has a big drainage area, while the rest are short and steep due to the proximity of the mountains to the coast (Figure 1.3).

The rivers of the Atlantic coast drain 65% of the peninsular land. Their courses are long, about 1000 km, with relatively low energy relief. In contrast, the Mediterranean rivers drain 31% of the peninsular area. Climatic extremes and steep topography combine to produce irregular discharges, with devastating floods (Table 1.1). All the small Mediterranean basins have low yields and are deficient in relation to human demands (Figure 1.4). Subterranean water accounts for approximately 20% of the total runoff.

Basin size is critical: the smaller the basin the larger the relative flood peak and the shorter the lag time. The effects of droughts are also more marked in the small basins due to the insufficient subterranean reservoir of water. Human impacts also have to be taken into account. In the coastal

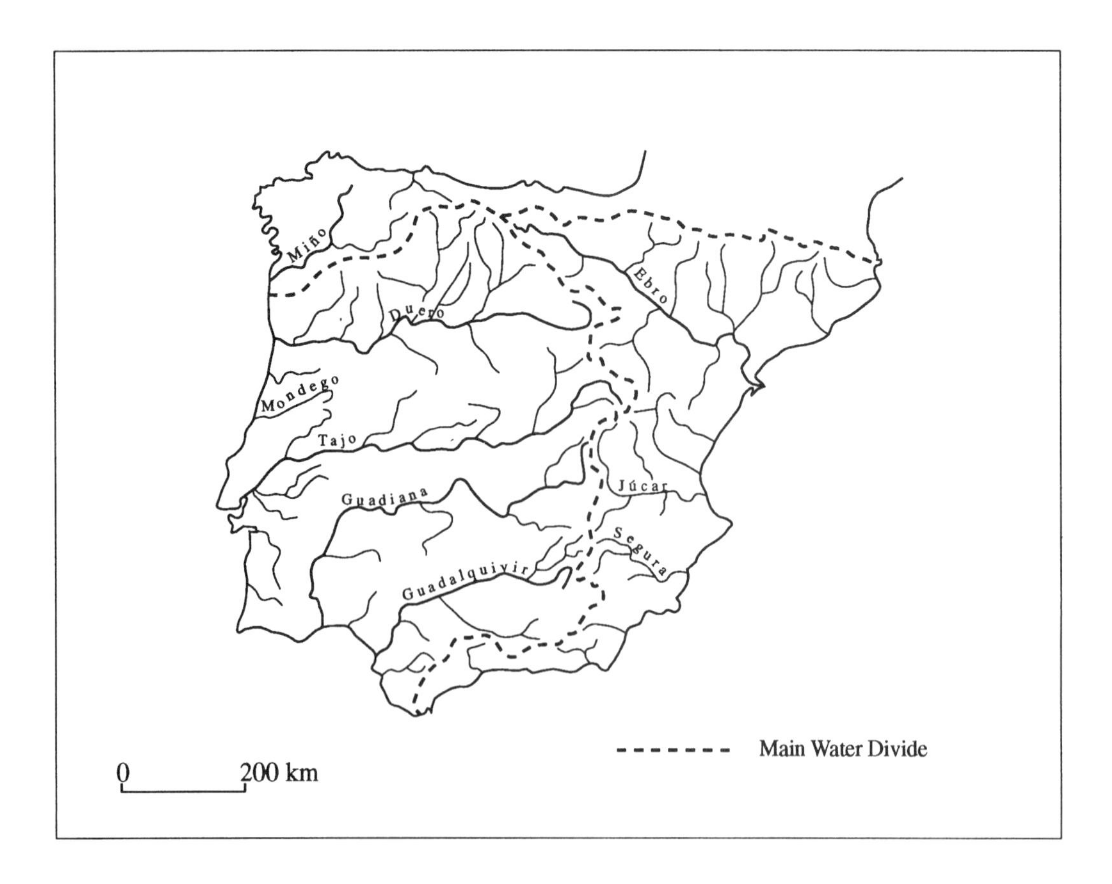

Figure 1.3 *Drainage trends* (after Sala 1989)

Table 1.1 *Iberian river systems: selected data*

River	Gauge st	Area	Runoff							
		(km^2)	(hm^3)	(l/s/km)	coeff. (%)	Qn (m^3/s)	Qc (m^3/s)	Qci (m^3/s)	Qc/Qn	Qci/Qn
Atlantic ocean rivers										
Ebro	Tortosa	84 232	17 258	6.1	29	489	10 000	12 000	21	25
Duero	Puente Pino	63 160	9305	4.2	23	268	7691		29	
Tajo	Alcántara	52 170	8964	5.8	27	301	11 000		37	
Guadiana	Badajoz	48 515	3124	1.6	22	61	2828	3500	46	57
Guadalquivir	Alcalá Río	47 000	5827	4.1	18	186	4675	4800	25	26
Mondego	Coimbra	4929	2500	16.0		79	1154	2457	15	31
Mediterranean sea rivers (from south to north)										
Guadalhorce	Gobantes	966	102	2.3	14	3.2	200		63	
Almanzora	Sta. Bárbara	1850	13	0.3	6	0.3	130		433	
Guadalentín	Puentes	1389	26	0.6	6	0.8	342	2100	417	2625
Segura	Orihuela	13 603	442	0.1	7	14.3	928	971	65	68
Júcar	Mompó	17 876	1640	2.6	15	11.2	632		56	
Turia	La Presa	6294	500	2.5	14	15.6	2674	3700	171	237
Llobregat	Martorell	4561	680	4.7	19	21.4	2420	2785	114	133
Mediterranean *ramblas* (from south to north)										
Guadalfeo		1292				0.3		1142		3805
Albuñol		113				0.3		1518		5060
Nacimiento		616				0.3	55	220	183	733
Almanzora		1100				0.3	38	3100	127	10 333
Albox						0.3		1400		4667
Algeciras	Librilla	52				0.1	46	310	1150	7150
Caldes	La Florida	110				0.3	47	131	157	437
Oñar	Girona	295				1.7	206	552	121	325
Balearic Islands										
Torrent Gros	Palma	215				0.1	31	103	238	792
Major	Soller	50				0.5	16	68	32	136

Qn = mean discharge; Qc = mean maximum discharge; Qci = peak discharge

ramblas, much urbanization has taken place in relation to tourism, adding to the natural imperviousness of the soil.

Neither the climatic conditions, with irregular and small amounts of water and high evaporation, nor the relief, which induces erosion and deposition processes, are favourable to the existence of lakes. Small lakes are present in the mountains and also in relation to karstic features, as in the Guadiana basin. Along the Valencia and Murcia coast and in the Guadalquivir there are littoral lagoons. Endoreism exists in the Duero basin (La Nava de Palencia), in the Guadiana basin (La Mancha) and in the Ebro basin.

Soils

In Iberia soils are highly varied in relation to topographic, climatic and lithologic conditions. Topography is responsible for the shallowness and low degree of development of most soils, and lateral leaching processes are sometimes more important than vertical ones. Climate also influences the slow rate of pedological processes because there is not enough water for solution and leaching. Lithology divides the soils into acid (on siliceous Palaeozoic rocks), alkaline (on limestone Mesozoic rocks), and clayey soils (in the Tertiary depressions).

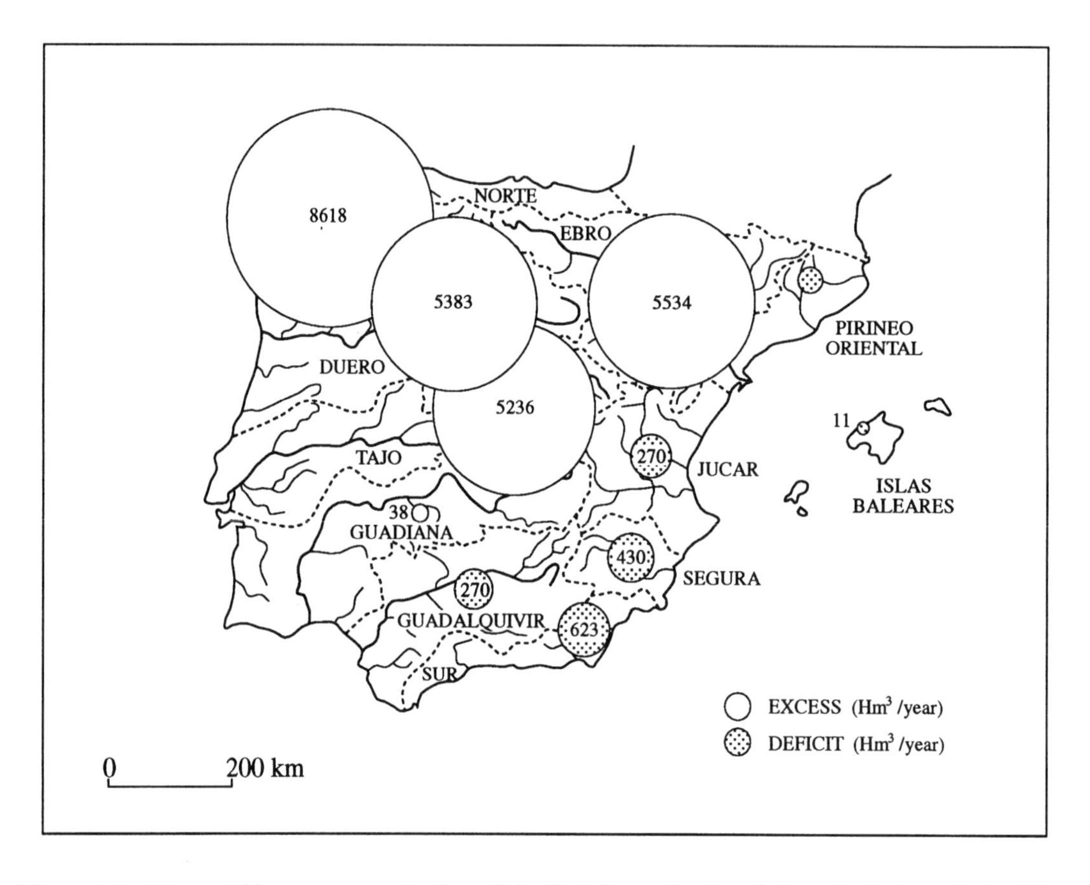

Figure 1.4 *Water yields: mean annual values of the fluvial contribution of the main drainage basins* (after *Atlas Nacional de España* 1996, Hidrología)

In traditional classifications, the Brown Earth soils are described in areas with more than 600 mm of rain. The typical profile is A(B)C, with humus mostly of the moder type, sometimes mull. These Brown Earth soils vary in relation to parent material, being poorer when the rock is acid and the vegetation is *Erica* and *Pinus*, and more fertile when the rock is calcareous and in that case good for agriculture and pasture. In mountain environments soil development is restricted because of erosion. Here the soil has only AC horizons and is called Ranker, supporting *Erica* and *Pinus* vegetation.

Where soils are permeable due to coarse, stony textures, as on colluvial formations at mountain footslopes and on alluvial terraces, leaching takes place and a horizon depleted of fine materials is formed. These are soils with poor physical conditions, difficult to plough. Non-calcic brown soils are used for cereals and legumes in the Duero basin. Fersiallitic, highly rubefied soils are found in Andalucía and the east coast, presenting a thick, dark red argillic horizon and sometimes a calcic horizon. On calcareous rocks Terra Rossa soils are found, with an argillic horizon lying sharply on the parent rock and with frequent rock outcrops. These soils probably reflect past climatic conditions. At present they are used for vineyards and olives.

In more arid areas, such as the Ebro basin, the southeast coast and certain areas in the Duero and Guadiana basins, Sierozems or grey soils are found. Calcareous accumulations (caliche or calcrete) are present in many places, mainly in the Upper Guadiana and along the east coast (Plate 1.1). In recent years an effort has been made to classify the soils following more modern, international classifications (Figure 1.5).

Plate 1.1 *Low mesas near Murcia capped by 1.5 m thick calcrete, formed in fanglomerate: (a) general; (b) detail, showing bedding and fanglomerate pebbles* (cf. Plate 12.8). All photographs are by A. Conacher unless otherwise acknowledged

Climate

The Iberian Peninsula is located in the southern part of the westerly general atmospheric circulation, at the margin of the zone of subtropical high pressures. For that reason it is affected by instability related to the seasonal latitudinal shifts and undulations of the jet stream. In summer, the subtropical high pressure centre of the Azores creates a barrier in front of the fluxes coming from the north and this, together with subsiding air masses, creates a dry and hot season, worsened by the arrival of the hot and dry African air masses. In winter, the Azores high pressure centre is located more to the south and the Atlantic low pressure centres associated with the polar front enter the Peninsula, bringing rains. In certain situations cold, high pressure air masses from the north can reach the Peninsula. The temperate and humid Atlantic air masses are important in the western part of the Peninsula.

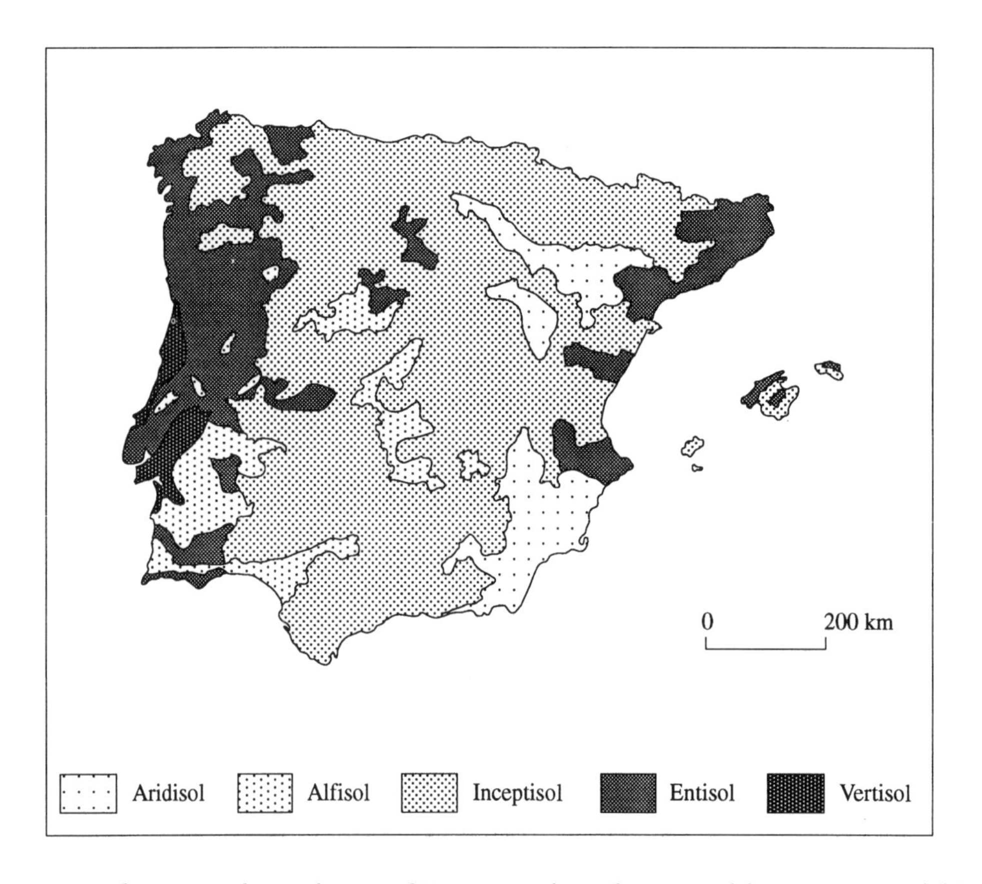

Figure 1.5 *Soil types according to the US Soil Taxonomy* (after *Atlas Nacional de España* 1996, Edafología)

From northwest to southeast, average annual temperatures range from 10°C to 19°C. Temperature ranges are 12 to 14°C in a narrow coastal fringe, and increase progressively inland with the continental conditions. Ranges are at their maximum in the eastern parts of the Tajo and Guadiana basins (20°C) and in the centre of the Ebro basin (more then 21°C). The lowest temperatures are recorded in the Duero and Guadiana basins (Albacete reached –24°C in January 1971) and the highest ones inland of the Guadalquivir (46°C in Córdoba and 45°C in Sevilla) (Figure 1.6). The number of sunny days per year also increases from north (20 days) to south (180 days).

Rainfall diminishes from north to south. The highest rates are registered on the windward slopes of the mountainous systems, while the leeward sides are dry. Mean annual rainfall exceeds 3000 mm in Geres and the Central Cordillera (Sta Estrela 2500 mm) in Portugal. The drier zones, with less than 300 mm, are located in the southeast (Murcia, Almería), in the Duero basin (La Armuña), and in the middle Ebro basin (Los Monegros). Cabo de Gata (Almería) receives only 130 mm and has the pluviometric minimum (Figure 1.7a).

Rainfall variability is remarkable, with dry years such as 1981 when the Peninsula received an average of 495 mm, and wet years such as 1960 when the average was 969 mm. Rainfall intensities are also high, especially in the Mediterranean coastal area where the daily or hourly maximum can be as high as the annual average. Extreme values for a 24 hour rainfall event are 600 mm in Albuñol (Granada) and Zurgen (Almería) in October 1973, and 426 mm in Cofrentes (Valencia) in October 1982. But the most extraordinary value is of 817 mm in Oliva (Valencia) on 3 October 1987, the highest value for one-day rainfall of the century.

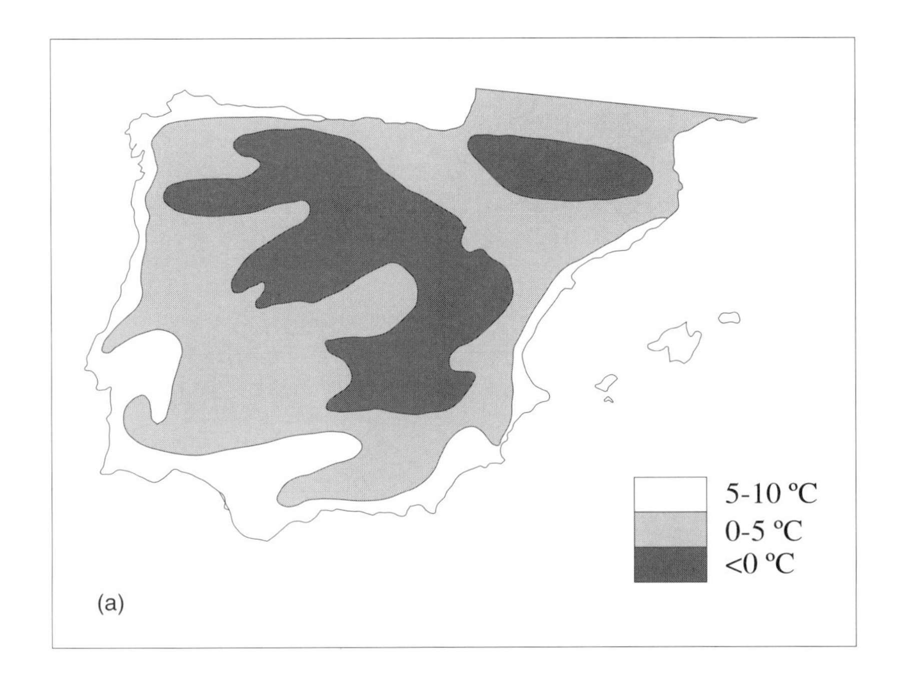

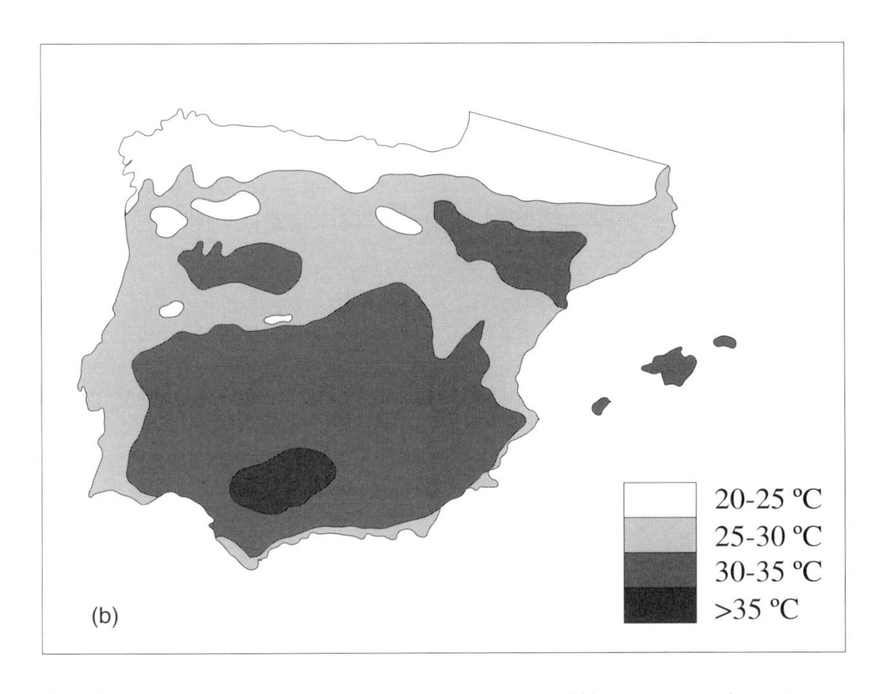

Figure 1.6 *(a) Mean annual minimum temperatures in January; (b) mean annual maximum temperatures in July, 1931–1960* (after Daveau 1982). Temperatures in degrees Celsius

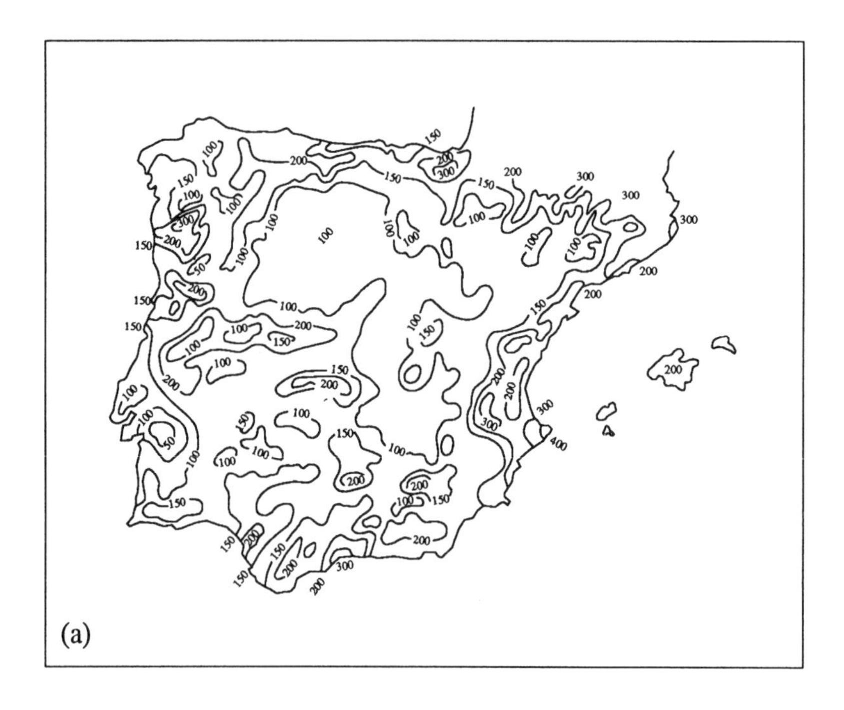

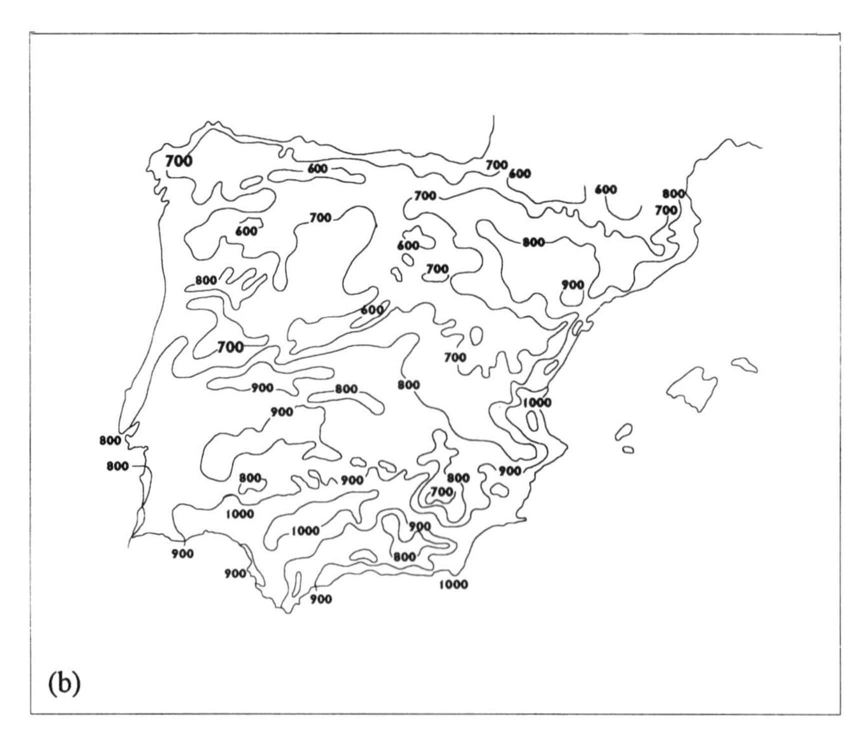

Figure 1.7 *(a) Mean maximum 24 hour rainfall (in mm)* (after Gutierrez-Elorza 1994); *(b) annual potential evapotranspiration (in mm)* (after Font-Tullol 1983)

Average yearly evaporation is above 400 mm in most of the Peninsula, increasing from northwest to southeast, with a maximum (above 500 mm) in the Balearic Islands (Figure 1.7b). Winds are not very strong or persistent except on the east coast (Empordà, Ebro valley, Menorca) and at Cadiz. Velocities of 196 km/h were registered on Montseny mountain (Catalan Ranges) in September 1969, the highest value in the Peninsula. Mist and snow are not frequent.

All six Mediterranean-type climates classified by UNESCO–FAO (1963) occur in the Iberian Peninsula (Figure 1.8).

Plant and animal life

Although relatively poor in vegetation density and coverage, the Iberian Peninsula and Balearic Islands are rich in the diversity of their flora. More than 700 species have been registered. As a whole the predominant vegetation is a xerophytic and heliophytic *matorral*.

Floral and faunal elements come from various sources. Distribution and dispersal of vegetation are related to climatic and tectonic changes that have given rise to changes in the distribution of land and seas. Changes in the configuration and desiccation of the Mediterranean Sea, with opening and closing of the Gibraltar Strait, gave way to the penetration of species coming from south and east. Quaternary climatic changes, with glacial periods, allowed the penetration of northern species. Isolation due to the peninsular character and mountainous topography has also been important. The Balearic Islands have the same characteristics but endemisms are less than in the Peninsula.

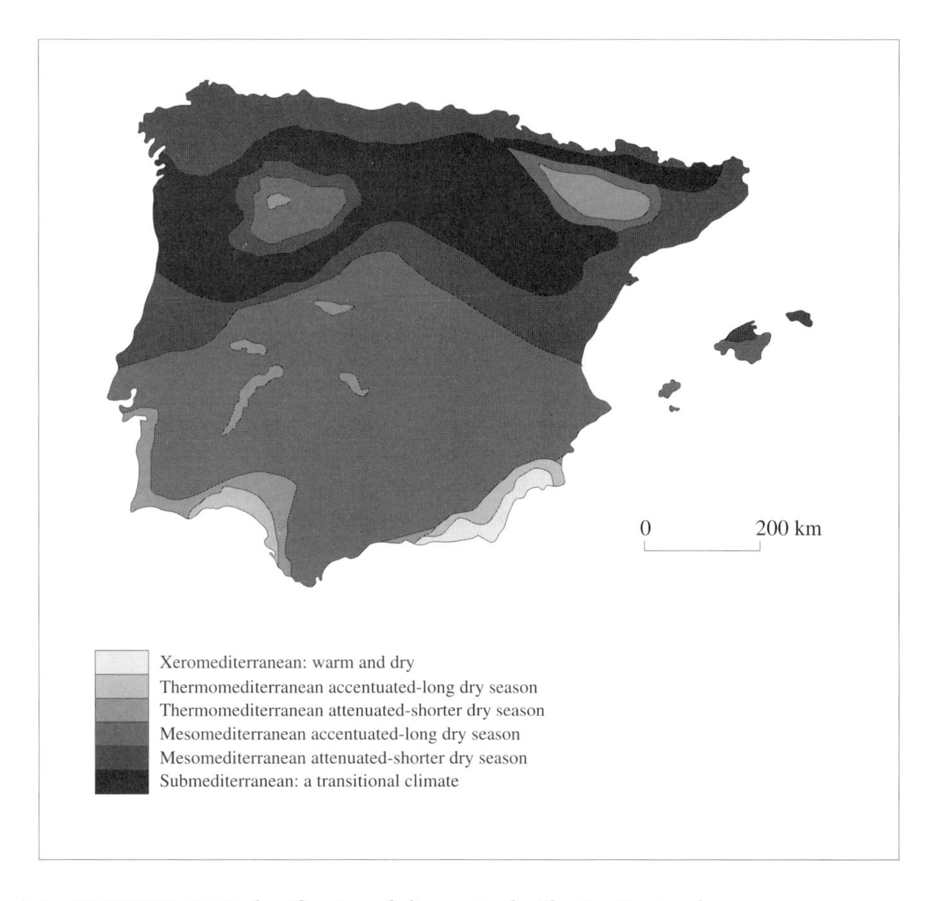

Figure 1.8 *UNESCO–FAO classification of climate in the Iberian Peninsula*

The Mediterranean floristic elements consist of circum-Mediterranean species (*Olea europaea*, *Quercus coccifera*, *Nerium oleander*, *Pistacia lentiscus*, *Mirtus communis*, *Lavandula spica* and *Rosmarinus officinalis*) and western Mediterranean species (*Quercus ilex*, *Q. rotundifolia*, *Q. suber*, *Q. faginea*, *Q. canariensis*, *Thymus polium*). The eastern Mediterranean species are scarce and not significant. Some species are relict from tropical periods (*Chamaerops humilis*, *Ceratonia siliqua*, *Laurus nobilis*), and remnants from colder periods (*Abies pinsapo*, *A. numidica*, *A. cephalonica*). Others are exclusive or endemic, such as *Ulex eriocladus* (SE), *Cistus psilosepalus*, *Genista polyanthos* and *Thymus mastichina*. Endemisms are also found in all mountains that exceed 1500 m above sea level, and in the gypsum soils of La Mancha and the Ebro valley, and in the black soils of Andalucía.

Depending on ecological conditions different community formations are found, ranging from forest to *maquis*, *garrigue* and *matorral*. In its natural state and in subhumid environments, vegetation forms an intricate, stratified ensemble: two tree levels (>3 m, 1.5–3 m), two shrub levels (0.5–1.0 m, 0.2–0.5 m), one herb level (0.–0.2 m), and an abundance of lianas. But in many cases what exists now are extensive areas of scrub (*matorral*), a plant formation type consisting of shrubs.

In the tree landscapes *Quercus* communities are the most widespread in subhumid conditions. There are three species, one maritime (*Quercus ilex*), a continental one (*Quercus rotundifolia*), and a third (*Quercus suber*) which is developed only in sunny, siliceous terrains and with mean annual rainfall over 600 mm (Figure 1.9). *Quercus suber* is distinguished by its thick bark that is removed periodically (every 9–12 years) and used in industry. The *Quercus rotundifolia* community is the most extensive. Poorer in structure, the flora has been deeply altered by humans and at present is not capable of developing a plant community. In many cases it has been substituted by more or less dense *matorral*. *Juniperus phoenicea* and *Juniperous thurifera* are present in arid continental climates.

Pines have been very much favoured by people and thus are present over extensive areas (Figure 1.10). *Pinus halepensis* is adapted to harsh situations; it is often mixed with oaks and especially with shrublands. *Pinus nigra* needs some water in summer and for that reason is located in the coastal mountains. *Pinus pinaster* is used widely in reforestation for its broad ecological tolerance and production of wood and resin (essences). *Pinus pinea* produces pinyons and is also good for wood. *Pinus sylvestris* is present in mountainous continental areas.

In Portugal, pure stands of *Eucalyptus* have been planted extensely. Olive and carob trees have been cultivated and it is difficult to determine their ecologic significance, although the limit of the cultivated olive is often taken as a limit of Mediterranean climatic conditions.

Maquis communities are dense and tall scrub, more than 1 m in height, difficult to penetrate, with a floristic composition similar to that of *Quercus suber*. *Maquis* occurs naturally in situations where *Quercus ilex* cannot compete (the upper parts of mountains and hills, or on very poor stony soils). The *garrigue* formation is less tall and dense and develops mostly on calcareous rock. The most significant species is *Quercus coccifera*, accompanied by several kinds of *Cistus*.

Matorral is a short, often open vegetation. It receives different names in relation to its floristic composition, which indicates different degrees of dryness in the environment. In wetter conditions it is composed of *Cistus* and leguminosae. In drier conditions species are scented (*Thymus*, *Genista*, *Lavandula*, *Rosmarinus*) and thorny (*Chamaerops*, *Rhamnus*, *Asparagus*). Esparto fields (*Lygeum spartum*, *Stipa tenacissima*) are found in the more arid conditions of the southeast.

The zoologic component is more difficult to analyse. Only reptiles, mammals and birds will be mentioned. The cold-blooded reptiles are very well adapted to Mediterranean conditions due to their resistance to the summer drought. Some are endemic (*Podarcis hispanica*). Mallorca is well known for reptile endemism.

Amongst the mammals, small hervibores are very common, such as hedgehogs, moles, shrews and mice. The most common big ones are *Cervus*, *Capreolus*, *Dama* and *Sus scrofa* (wild pig). The wild pig, in contrast to the other species, has extended considerably, especially in the Catalan ranges and eastern Pyrenees, and it is

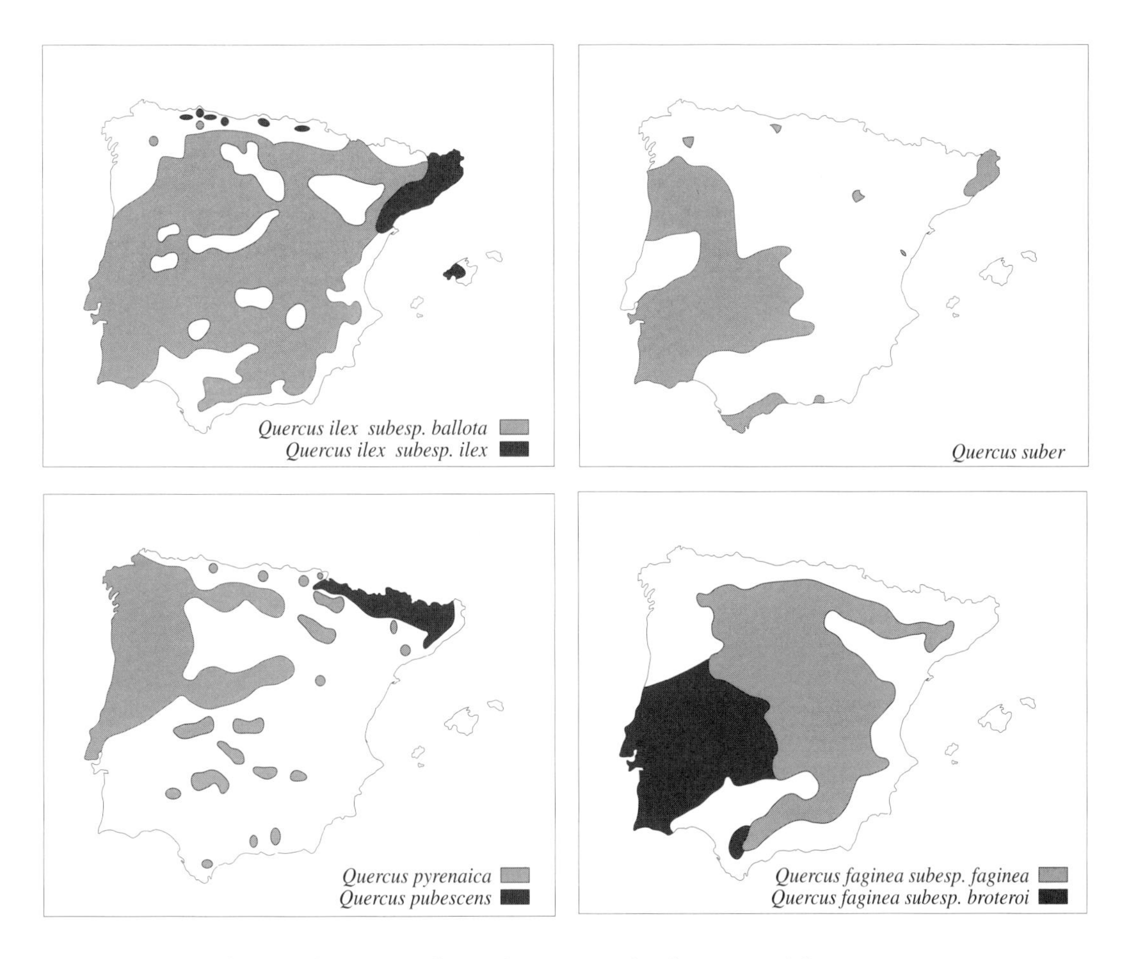

Figure 1.9 *Distribution of* Quercus (after Rubio 1989, and *Atlas Nacional de España* 1996)

actively hunted. There are also more than 25 species of winged mammals. The carnivorous mammals have the largest ecological adaptability, with the most common being weasels, polecats and martens. The mountain cat (*Felis sylvestris*) is also Mediterranean but has extended to the north. Other species are *Lynx pardina* or the Hispanic lynx and *Meles meles* (badgers). Foxes (*Vulpes vulpes*) and wolves (*Canis lupus*) have high adaptability and resistance to anthropic pressures, but have been persecuted for their interference with agricultural and pastoral activities.

Thirty species of birds, 25 of which are migratory, live within Mediterranean vegetation forms. In the southwest are the only species of *Aquila heliaca* (imperial eagle) in the world. There are a few pairs of black vulture (*Aegypius monachus*) and blue elan (*Elanus caeruleus*) living in the west of the Peninsula; other specimens can be found only in the Middle East and in south Asia.

The People

Population composition

Due to the position of Iberia between northern and southern cultures, ethnic and cultural superpositions exist. Traditionally, three main cultural areas have been distinguished: the Celts in the north and west, the Iberians in the south, and the Tertesians in the east. Phoenician, Greek and Carthaginian colonizations were very important along the Mediterranean coast and Balearic Islands.

But the decisive colonization was the Roman, which embraced the entire Iberian territory,

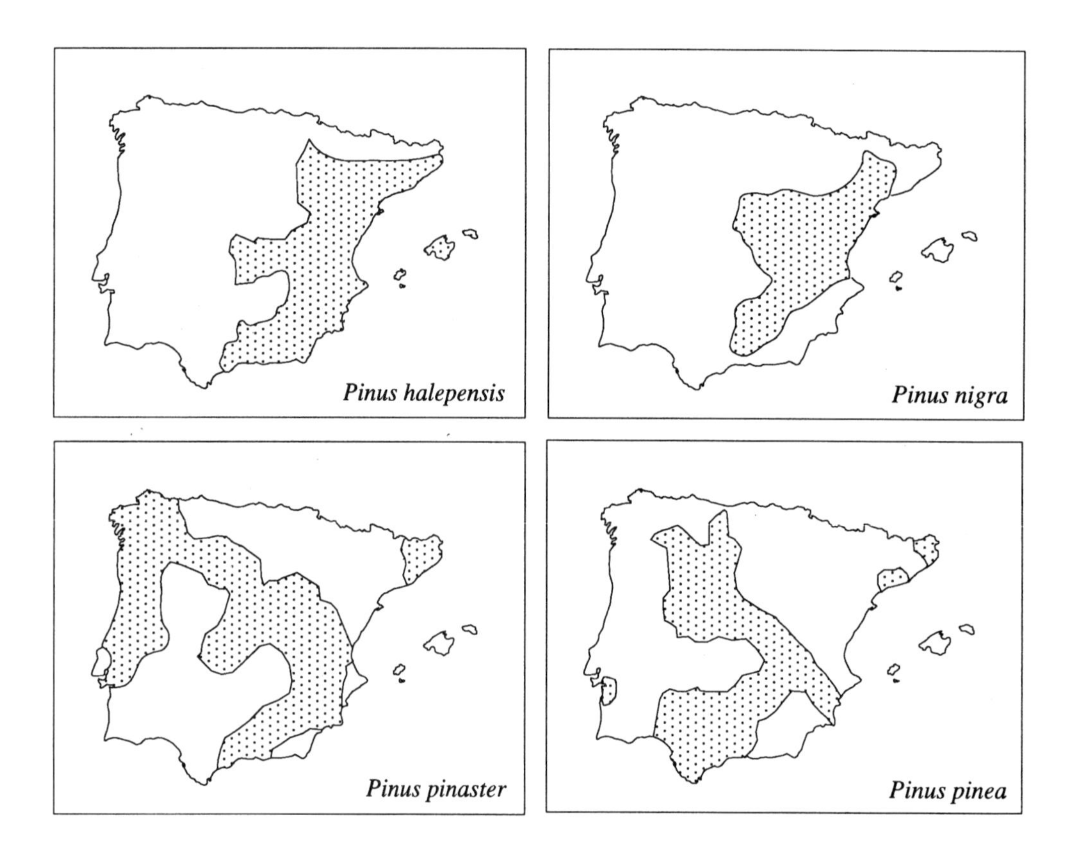

Figure 1.10 *Distribution of* Pinus (after Rubio 1989)

lasted for six centuries and determined a period of cultural, civil and economic progress (Plate 1.2). Economic wealth was related to the exploitation of mines and development of agriculture, with cultivation of the Mediterranean trilogy: wheat, vines and olives. It was centred along the Mediterranean coast and the south of Portugal, and had less effect in the interior lands. Some of the civil engineering constructions have been in use until very recently, for instance drains and sewer systems in many cities, aqueducts and road networks.

The Barbarian invasions from the north destroyed the wealth and civilization created by the Romans without making any fundamental improvement. It was the invasion from the south by Arabs which initiated a new era for Iberia. The division between north and south was maintained because the Arabs became established only in the south and along the coast, as the reconquest war started very early. The north was Christian and rural, dedicated mostly to transhumance pastoralism. The south was Muslim, urban, and dedicated to a highly developed agriculture: orchards, olives, vines, oranges, cotton, sugar cane, rice and mulberries (for silk). During the 10th century Córdoba was the biggest and most cultural city in western Europe. But little by little the differences diminished. The Christian half became related to Europe, to a great extent through Santiago's Pilgrim's Way, and to Mediterranean cities, especially in Italy. Toledo became a cultural centre where the three cultures, Arab, Christian and Jewish, co-existed in peace, being the recoverers and transmitters to Europe of the Greek culture, including the geographical work of Ptolomeo.

Portugal became independent in the 12th century and subsequently developed a strong identity. In the 15th century the great Atlantic Ocean expansion commenced, which had enormous effects on Iberia's population, economy, society

Plate 1.2 *A Roman bridge on the Guadiana River which is still in use today. Photo by Maria Sala*

and politics. Other important changes in that century were the unification of the Castilian and Catalan–Aragonese kingdoms and the final reconquest of Granada in 1492. The historical development of Iberia resulted in the existence of several languages and the official adoption of the Catholic religion.

Demographic trends

The evolution of the population of Iberia reflects its historical and economic development (Table 1.2). During palaeolithic times the population is estimated to have been about 50 000, reaching seven million during the maximum development of the Roman empire. In Spain during the 14th century there was a serious demographic crisis due to the Black Death plague. In the 15th century the Arabs and Jews were expelled by the Catholic kings, causing a fall in population. After a stationary period in the early 17th century, the population reached eight million as a result of economic development related to the discovery of America and the trade of gold and silver. In the next century, and in relation to the emigration to America, the wars of the Spanish empire and the plagues, population decreased again. But since the end of the 18th century there has been a steady growth of about 4% per annum, especially marked during the second half of the 20th century, despite the emigration of workers

Table 1.2 *Demographic trends (source:* Atlas Nacional de España *1996, Informacion Demografica)*

Century/ year	Spain		Portugal	
	population (million)	pop./ km^2	population (million)	pop./ km^2
1st	6.0			
3rd	7.0			
9th	9.0			
14th	6.0			
15th	7.5			
1530	4.8	9.5		
1591	6.8	13.5		
1717	7.5	14.9		
1768	9.3	18.4		
1800	10.5	20.9		
1850	15.5	30.6	3.4	38
1860	16.6	32.9	3.8	43
1890	17.5	34.7	4.7	52
1910	19.9		5.5	62
1920	21.4		5.6	63
1930	23.6			
1940	25.9	51.2	7.2	81
1950	28.1		7.9	88
1960	30.5		8.2	93
1970	33.9		8.0	91
1980	37.7	74.6	9.3	105
1990	39.4	76.4	9.4	105

to Europe. This tendency has changed recently, following the lowering of the birth rate, which is the lowest in Europe. Natality was high until the 1970s (34.5 to 20.0%), related to a dominance of rural populations. By 1984 natality had decreased to 12.2% (11.8% in Portugal in 1991). The reduction of emigration to Europe and a net fall in mortality rates have led to stabilization of the population in the 1990s.

There has been a steady increase in the numbers of people over 65 years of age since the 1950s, ranging from 5% in 1930 to 11% in 1986, especially in the rural areas, but it is still low compared with the average for European countries (13.8% in France and Italy, 15.2% in the UK and Germany).

Settlement patterns

Spain has a low population density, with 77 persons/km^2, compared with 103 persons/km^2 in France, 107 in Portugal, 192 in Italy and 235 in the United Kingdom. Only the provinces of Barcelona, Madrid and Guipuzcoa have population densities higher then 300 persons/km^2, reflecting their urban concentrations. In Iberia the population is irregularly distributed within the territory, mostly concentrated in the coastal fringe and Balearic Islands (62.5%), with densities of 144 people/km^2. Inland lives 37.5% of the population with densities of 44 persons/km^2, except for the Madrid area and the major river valleys. Thus there is a concentric pattern consisting of a highly populated external ring and a "desertified" central area with a population "oasis" in Madrid (Figure 1.11).

Industrialization during the 1960s and 1970s has been responsible for an important migration from rural to urban areas. Catalonia and the Vasc country were the major immigration poles, followed later by Madrid, which was highly industrialized during Franco's time, and the Ebro axis from Rioja to Tarragona. The emigration regions are Andalucía, Extremadura and Castille.

In Portugal, the Lisbon and Porto metropolitan areas contain more than one-third of the population, which has migrated from the dryer areas of Douro, Trás-os-Montes, Beira Baixa and Alentejo. Lisbon and Porto are medium sized cities, the urban network mostly consisting of small towns not exceeding 90 000 inhabitants, such as Coimbra, Braga and Setúbal.

Populations of the larger cities are tending to decline, while the populations of their peripheral areas increase. This is particularly so in Barcelona, which has a small urban area of its own but a wide belt of adjacent cities. For this reason Barcelona is the largest metropolitan area in Spain, with the big adjacent cities of Hospitalet, Badalona, Sabadell, Terrasa and Santa Coloma, which have populations of 294, 227, 184, 155 and 140 thousands respectively. Cities orientated towards tourism have also increased their populations.

The Economy

Although Spain went through a civil war and a period of isolation from Europe and the world during the 20th century, a sharp change was introduced in its economy during the period 1939–1959. Landmarks in the further economic development were, first, the commercial liberalization of 1959, which was introduced by a Stabilization Plan which meant entry to several international organizations (EDCO, IMF, MB, GATT); and second, the Preferential Agreement with the European Community in 1970, followed by entry to the European Community in 1986 together with Portugal.

In 1993 the active population was 39.3% in Spain and 48.4% in Portugal, lower than most European countries. This is related to the still low participation of women in the workforce (38.5% in Spain and 41.3% in Portugal). The distribution in the different sectors has changed considerably (Table 1.3).

Unemployment has been one of the main problems of the last two decades. For the entire Spanish country it has ranged between 23% (1993) and 16% (1990). The worst affected areas are Andalucía (28%) and Extremadura (25%), and the less affected communities are the Balearic Islands (11%), Aragon (12%) and Catalonia (13%). Since the application of the Common Agricultural Policy in 1984, the European Union (EU) has attempted to avoid the problems of unemployment by maintaining rural populations, which at the same time helps to preserve the environment and the landscapes.

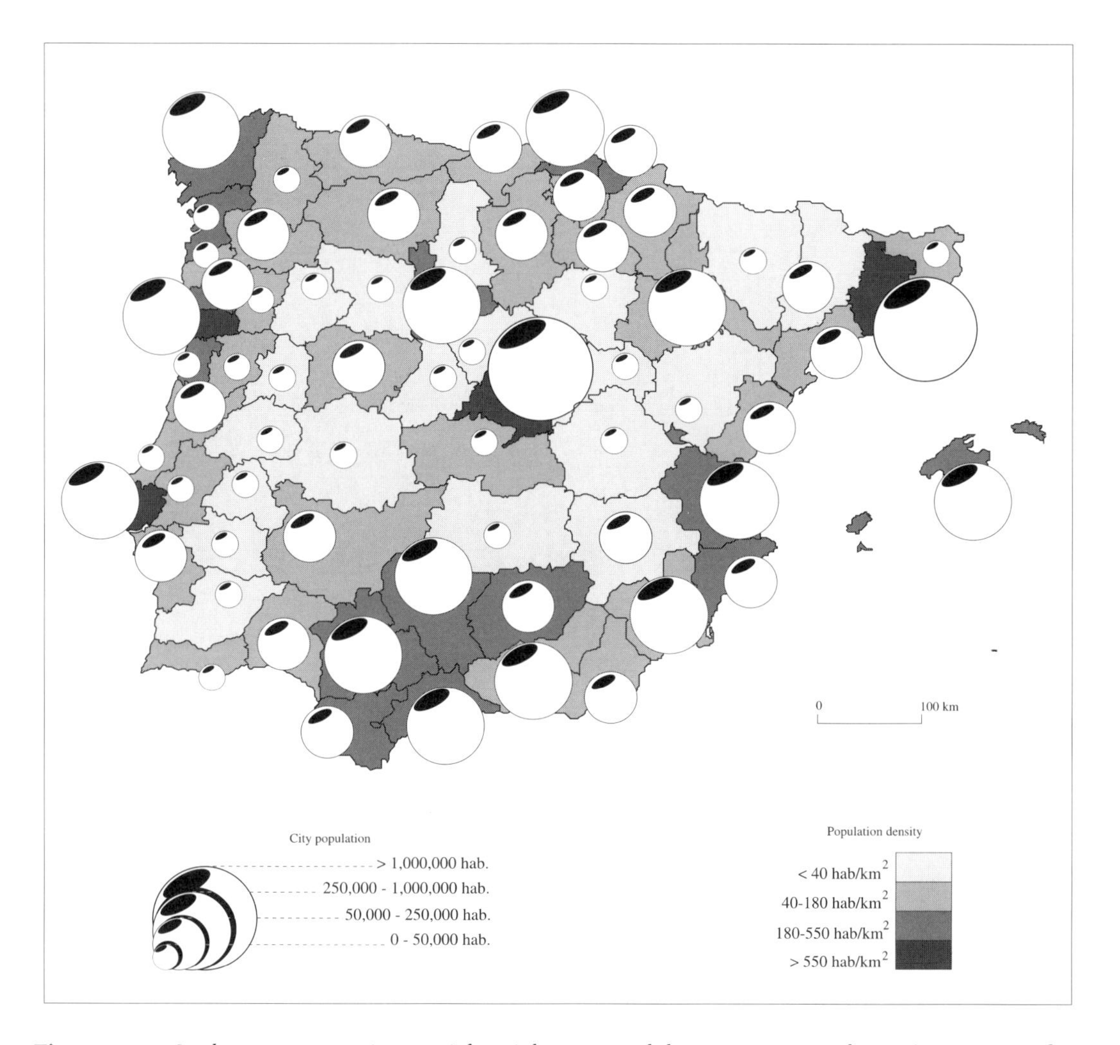

Figure 1.11 *Settlement patterns in 1986* (after *Atlas Nacional de España* 1996, Información Demográfica)

Table 1.3 *Active population by sectors (%)*

Year	Agriculture	Industry	Services	Construction	Country
1900	63.6	16.0	17.8		Spain
1990	10.0	21.2	59.8	8.9	Spain
1990	11.0	39.0	50.0		Portugal

External emigration during the 1960s paralleled Europe's economic development from 1956 with the Treaty of Rome and creation of the European Community. This emigration was positive for the economy because it reduced unemployment and brought money from the workers into the country. The new tendency is the contrary, namely in-migration from North Africa, Spanish and Portuguese America, and Eastern Europe. In Portugal over 1.5 million people emigrated, but the April revolution of 1974 encouraged the independence of African colonies and the return of emigrants.

In parallel with the distribution of population, there is an economic territorial disequilibrium between rich and poor regions. Intensive

Plate 1.3 *Extensive mechanized agriculture on the Spanish plateau*

agriculture, industry and tourism are concentrated along the coast and the Ebro valley. Extensive agriculture and pastoralism are concentrated in the continental areas, except for the Madrid area (Plate 1.3).

Resources and industry

Natural resources in Iberia are not sufficiently important to allow the easy development of industry. One of the pioneering national industries was the textile industry in Catalonia. But the main industrial development came during the period 1939–1959, when development poles and state industries were created in deficient sectors. Liberalization and renovation followed in an effort to merge and compete with the EU countries. Traditional products, especially from mining, have had to be abandoned since 1983.

Mining has taken place in Iberia since Roman times. A great variety of products are mined, with the most important being lead, zinc, pyrite, iron, mercury, potassium, copper, silver, salt and cement. Except for salt, most minerals are located at the margins of the Hercynian block. The present situation is stagnant due to the fact that modern technologies are not applicable or are uneconomic. In Portugal, rock extraction for building and ornamental purposes (marble, limestone, granite, sand, gravel and clay) contributes most to the income from mining activities (75%).

Sources of energy are insufficient and are based on coal and hydro-electricity. External sources are oil, gas and uranium. In Spain, coal is of poor quality and for that reason it is used *in situ* by thermal power stations under state control. A great deal of electricity is generated from hydro-electric power (Figure. 1.12). Seven nuclear plants were constructed in Spain during the 1970s, but controversy is very strong. In Portugal, despite its important uranium reserves, there are no nuclear stations since public opinion has been strongly against it.

Oil is mostly imported in its crude form, refined and distributed through a state company, now privatized. Gas was first introduced through Barcelona by ship from Libya and Algeria and distributed to Zaragoza and Madrid. There is a large expansion plan, with pipe connections from Algeria through Spain to the rest of Europe.

The provincial index of industry shows Barcelona and Madrid as the main poles, followed by Valencia, Tarragona and Zaragoza. The Catalan industry is based on medium and small factories and companies, production diversity and spatial dispersion, in contrast to the Madrid pole. Here, automation, chemicals and agricultural products

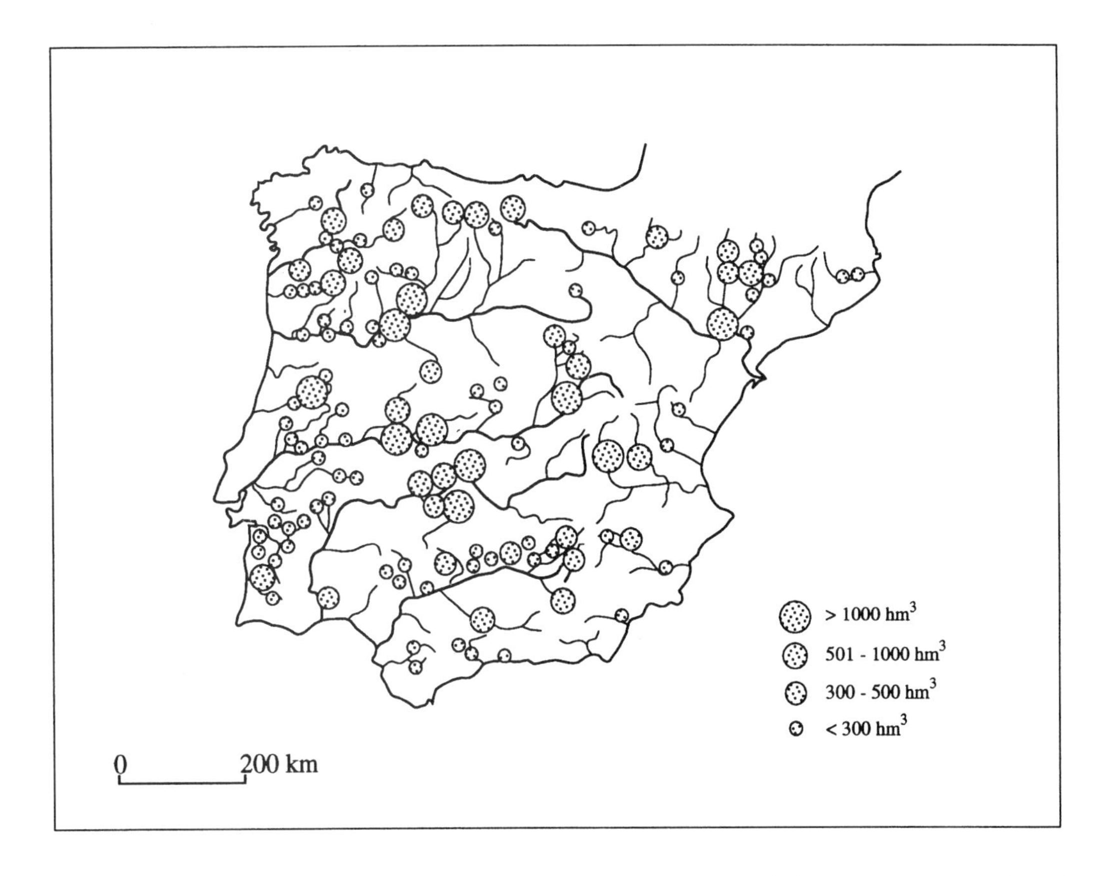

Figure 1.12 *Main reservoirs and their capacity* (after *Atlas Nacional de España* 1996, Actividades Económicas)

are the three pillars of modern industry together with high technology activities. Transportation industries are also important and diversified.

Portuguese industry is located mainly in the littoral, from Setúbal to Braga. In the Lisbon metropolitan area the metallurgic, mechanical, chemical, and shipbuilding and repair industries are dominant. In Porto and Alentejo petrochemicals and textiles are more important. Small clothing and footwear firms represent one-third of the total employment, whilst food processing, wood and cork industries generate most production and employment. Pulp industries are flourishing, related to forestry expansion of eucalyptus.

Agriculture, forestry, fishing

The rural landscapes of Spain are classified as 51% forests and *matorral*, 41% agriculture and 8% unproductive. In Portugal 36% of the land is under forest, 46% under agriculture and agroforestry, and 18% is unproductive.

Agrarian structures are related to both environmental and historical conditions. In the less favoured areas property is *latifundia*, for example in Castilla–La Mancha (the province of Ciudad Real has 33.5% of the Spanish properties of more than 1000 ha), Extremadura (Cáceres has 36.8% of large farms) and Andalucía. These were areas which remained as battlefields or no-man's-land during the centuries of wars between Arabs and Iberians, and were given as war reparations during their reconquest. In areas with better environmental conditions *minifundia* are common, such as the northern half of the Peninsula (Castilla–Leon, Catalunya, Valencia, Murcia and the Balearic Islands). These are the areas which were intensively colonized by the Christian kings (north) and by the Arabs (coastal plains).

Present-day agrarian policy has been orientated to balance this situation by colonization and irrigation in the case of large properties and by concentration of the small ones.

Policies orientated to improving water availability and distribution have been prepared since the 19th century. In particular, irrigation plans have developed intensively during the last 20 years (Figure 1.13).

Two agricultural domains can be distinguished: (a) the narrow coastal and alluvial plains, and the Balearic Islands, mostly irrigated and dedicated to orchard and fruit production, and (b) the continental areas, dominated by extensive dry farming. In addition, extensive cultivation of grapes and olives occurs in both domains, giving rise to typical Mediterranean landscapes (Table 1.4).

In the first domain, Catalonia, Valencia, the Balearic Islands and Murcia account for 20% of the national agrarian production, with 14% of the rural workforce and 26% of the irrigated areas (Plate 1.4). Production is diversified. Orchard products account for more than 20% of total agrarian production, extending over more

Table 1.4 *Areas covered by the most representative crops in Spain, 1993*

Crop	Area (10^3 ha)
Grain cereals	6427
Dried pulses	211
Tuber crops for human consumption	210
Industrial crops	2411
Fodder crops	1236
Vegetables	435
Citrus fruit	270
Non-citrus fruits	940
Vineyards	1281
Olive groves	2011

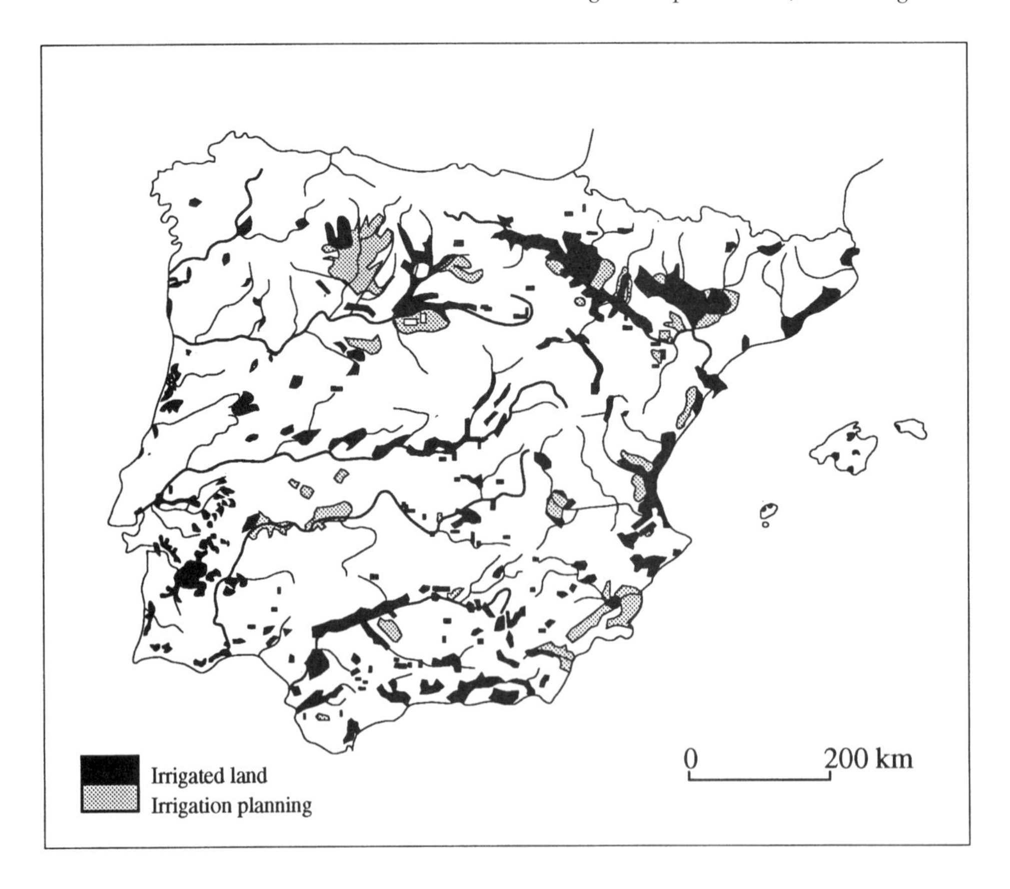

Figure 1.13 *Irrigated land* (after *Atlas Nacional de España* 1996, Actividades Económicas)

Plate 1.4 *Intensive irrigated agriculture and horticulture in the fertile plain around Murcia*

than 2 000 000 ha, and are one of the most important exports. In Almería the development of cultivation under glass houses and on sands has meant a valuable economic development, mainly of export tomatoes, in this traditionally poor province. Citrus fruits are one of the basic pillars of Spain's external commerce. Three-quarters of the national production is located in Valencia and Murcia, while Andalucía's production is increasing. Other fruit trees are located in the Ebro valley and in Catalonia, with more than 30% of the national production. Almonds and hazel nuts are produced in Tarragona (90% of the national production). Rice is produced in the Ebro delta, Valencia marshes and Mondego valley.

The main products of the continental lands are cereals. The dry farming consists of alternating crops with a fallow period, and much of the produce is dedicated to stock feed. Wheat is exported when conditions are favourable. Maize, rye, potatoes and beans are the main products of north and central Portugal. An increase in rental values is expected under the EU policy of setting aside the low productivity lands (the Common Agricultural Policy).

Vineyards in Spain extend over 1 500 000 million ha, the largest area in the world. But the oldest denotative wine region in the world is the Portuguese Douro, where port is produced. In Spain the most important vineyards are mainly located in the Upper Guadiana, Upper Ebro and the dry coastal Mediterranean slopes. Catalonia produces "cava", a sparkling wine similar to Champagne. Productivity is 20.8 hl/ha, lower than in Italy (66.4 hl/ha) and France (59.1 hl/ha). Since 1984 several measures have been taken to introduce better quality grapes.

Olives in Spain extend over an area of nearly 2 000 000 ha. They are concentrated mostly (50%) on the slopes of the Baetic Cordillera, in the Andalucian provinces of Jaén and Córdoba. For environmental, property and input reasons their productivity is low. Emigration and mechanization have increased the costs. Another problem is competition from other oils, mainly soya. In 1972 a restructuring plan was formulated to increase the value of the crop. Consumption by the EU is still less than production, so measures have been introduced to increase it.

Because the EU has a deficit in the production of sheep and goat meat, Spain and Portugal are in a good export situation (Table 1.5). The flocks are mostly concentrated in the drier areas of Alentejo, Salamanca and Extremadura, in the *dehesa* or *montado* association of pasture and forestry. *Dehesa* and *montado* are associations of cork oak and holm oak between which cereals or forage are grown. This association produces a

Table 1.5 *Livestock numbers for the Mediterranean countries of the European Union, 1993 (10^3 head)*

Country	Cattle	Pigs	Sheep	Goats
Portugal	1322	2665	3305	836
Spain	5019	18 234	23 872	2947
France	20 112	12 869	10 401	1029
Italy	7621	8051	10 669	1369
Greece	608	1143	10 069	5821

landscape which resembles a park rather than a forest. In some mountains the animals are still reared in communal lands, with different shepherds in rotation according to the animals under their care. Transhumance is still practised. The production of chickens and eggs is mostly concentrated in Catalonia (one-quarter of the national product), with similar productivity to other EU countries.

Forestry is the more dynamic sector within the agrarian economy of Portugal, contributing 3% of the gross national product. Forest land use trebled this century, from about 1 000 000 to 3 200 000 ha. *Pinus* is the dominant species, mainly distributed north of Tejo, forming the largest contiguous area of pine in Europe (1047×10^6 ha). Portugal is the primary world producer of cork. *Eucalyptus* occupies at present 529 000 ha, its area having doubled since the 1980s, covering extensive areas north of Mondego (Beira Littoral).

Wood production has also increased in Spain since the 1960s thanks to the introduction of various *Pinus* species, although not enough to satisfy the ever-increasing demand. On the other hand, forest values have decreased due to the substitution of natural products by synthetic ones, and by the use of other products for fuel. Cork production in 1984 was 117 625 t, which is double that of previous years and similar to values in the 1960s. Wood production by logging is low, 0.357 m^3/ha, compared with an average of 2.2 m^3/ha in the rest of Europe.

Because the continental shelf is narrow, except in Valencia and Cadiz, the Spanish fleet goes to distant Atlantic fisheries. The percentage of fish coming from the Mediterranean Sea is very low (10% of the total catch). It is worth noting the development of aquaculture of mussels and shrimps in Alicante, the Ebro delta and Mar Menor (Murcia), and the future possibilities of this industry, especially with oysters. Spain is the second country in the world in the per capita consumption of fish, following Japan. In Portugal the coast is rectilinear and exposed to strong winds, so that sheltered harbours are few (Ria de Aveiro, Tejo and Sado). The exploitation of marine resources was important until the 1960s but is now in decline. Portugal occupies seventh place in the EU with regard to fish catches (300 000 t/yr). Today, EU fishing policy is compromising Portuguese fishing and the canned sardine industry due to the free circulation of tinned fish from North Africa.

Transportation

As a result of the centralized model of the Spanish state since the 17th century, the transport net has a radial disposition focused on Madrid (Figure 1.14). Only a few transport corridors connect the Mediterranean developed regions with each other and with Madrid. In Portugal, transport is concentrated in the littoral region while the interior is still badly served.

The construction of the Spanish railroad was initiated in the mid-19th century, with railroad traffic now concentrating on the lines Madrid–Irun, Madrid–Sevilla, Madrid–Barcelona–Cerbère and Lisbon–Porto. The main problems are the ageing structure of the net making high speeds impracticable, and its wider gauge in comparison with other European railways.

Road has become the main method of transport, carrying 80% of the traffic. In Spain there are 25 km of road for each 100 km^2 of territory and 4 km for every 10 000 inhabitants, compared with the European averages of 55 and 9 km respectively. The first toll highway to be constructed connects the French border to Valencia, the most intensely used road of the Peninsula, at present reaching Alicante. The second toll highway links Barcelona with Bilbao along the Ebro valley. Madrid and Lisbon are the only two European capitals not yet connected to a fast highway net (Figure 1.14).

Shipping services are important for goods in Portugal at Lisbon, Leixões and Setúbaland

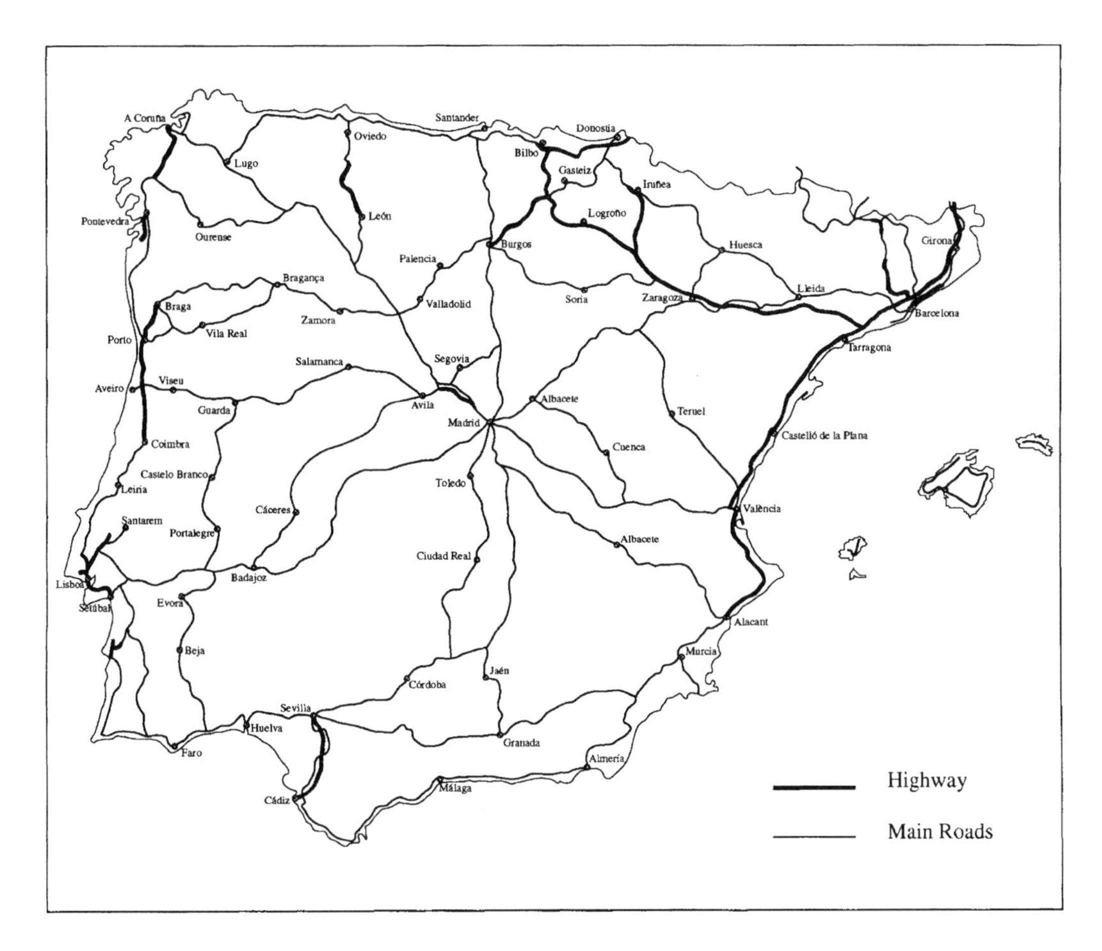

Figure 1.14 *Transport net* (after *Atlas Nacional de España* 1996)

ports. Maritime transport is low in Spain, especially taking into account the length of coasts, the number of ports and the location of a great deal of the industry. Since the 1960s improvements have been taking place in the infrastructure of the ports and in the use of containers, and since 1985 reforms have focused on the renovation of the merchant fleet in relation to EU norms.

The main aerial transport centres are Madrid, Barcelona, Palma de Mallorca and Lisbon. Commercial transport was initiated in 1919 between Madrid and Barcelona. IBERIA and TAP are the state airline companies of Spain and Portugal respectively; at present they have serious economic problems and are suffering competition from other European companies which under EU regulations can operate in Iberia.

Commerce and tourism

Tourism has made possible the development of the Spanish economy. It is also responsible for many of the land use changes and subsequent transformations of the landscapes. The arrival of more than 43 million tourists in 1985 represents the culmination of a development which started in 1951 (Table 1.6). In approximately 10 years a new infrastructure had to be created to cope with the tourist invasion, the biggest investment made in Spain in the last 25 years (Plate 1.5). Tourism provides employment for 11% of the active population, and an input in foreign currency that represents an important percentage of the Spanish payments for imports (31.7% in 1970 and 19.9% in 1982).

Of the tourists who in 1985 spent one night in Spain, 21.7% did so in Catalonia (Costa Brava),

Table 1.6 *Tourist activity in Spain* (source: Tamames 1994)

Year	No. Tourists (10^3)	Income ($ million)	Payments ($ million)
1970	24 105	1681	113
1975	30 122	3404	385
1980	38 027	6968	1227
1985	43 235	8151	1011
1990	52 036	18 593	4253
1993	57 223	19 426	4705
1990			
Germany	6887		
France	11 623		
UK	6286		
Portugal	10 106		
Benelux	3295		
Scandinavia	2073		
Morocco	2132		
USA and Canada	994		
Other	5346		
Spaniards living abroad	3299		

15% in the Balearic Islands, 13.5% in Valencia (Alicantine coast) and 12.4% in Andalucía (Costa del Sol). The basic lodgings are hotels in the Balearic Islands and Costa Brava, apartments in Alicante and Costa del Sol, and camping all along the Catalan coast. The majority of tourists are European (Table 1.6).

In Portugal imports generally outnumber exports giving a negative trade balance of 11.3% (1992). Traditionally, exports were related mainly to agriculture, forestry and fisheries. Since joining the EU in 1986 the exports are mainly textiles, clothes, footwear, cellulosic fibres, paper, timber and cork products. Tourism became an important source of revenue from the 1960s. The Algarve and the coast in Setúbal peninsula around Lisbon became tourist resorts for visitors from European countries. In 1991 Portugal occupied the 13th ranking for world tourism, with about nine million visitors. The full potential is not totally exploited, but there is a need to plan for tourism in the rural areas.

Administrative and Social Conditions

Spanish administrative activities are mostly concentrated in Madrid (47.2% of the political

Plate 1.5 *The coastal resort of Benidorm, Alicante, built entirely for tourists: an example of landscape transformation*

functions, 47.6% of the banking, 49.2% of the companies). It is intended to decentralize part of this power to the newly created Autonomous Communities.

Government

The present-day Spanish Government is a Constitutional Monarchy, with the New Constitution of 1978 declaring the country to be within the capitalist system. The most important political parties are the Socialists (PSOE), Conservatives or Popular Party (PP), the Unified Communists and Ecologists (IU), Catalan Coalition (CIU) and Vasc Nationalists (PNV). Administratively, the country is decentralized in 17 Autonomous Communities.

The Portuguese Constitution of 1976 established a democratic multi-party system. Central government has a prime minister, chosen from the political party with the largest number of votes, elected for four years. Parliament has 230 members representing the Socialists (PS), the Social Democrats (PSD), the Unified Democratic Coalition (CDU, including the Communist Party and the Ecological Party), and the Popular Party (PP, a right wing party).

Education

The law of 1970 restructured education, dividing it into four main levels: pre-school (2–5 years of age), EGB (basic general education, 6–13 years of age), BUP/COU (bachelor and pre-university, 14–17 years of age), universities and polytechnics. Every citizen has the right to education and EGB is compulsory and free. Government and private centres co-exist. Many new universities have been created (in 1996 there were 44 public and six private universities), and given a high degree of autonomy. At this level it is worth mentioning the LRU (Law of University Reform), which considerably changed the medieval structure of the universities and gave all levels of academic staff the freedom to undertake research. At present, serious efforts are being made to place research at an international level.

In Portugal education is compulsory from the ages of 6 to 14 years, although only 65% of the population attains this level. At secondary level (14–17 years old) there are still some marked regional asymmetries, the rural areas having low retention rates to the final year. Portugal has an old tradition in university education, with Coimbra University dating back to the 13th century. Nowadays Lisbon, Porto, Aveiro, Braga, Vila Real, Evora, Covilhâ and Faro are university centres with a high degree of autonomy. Despite the increased number of students since 1976, the rate of participation is still low (0.7% of the student population) compared with other European countries.

Health and welfare

In Spain the development of Social Security had its origins in the National Institute of Prevention created in 1908. In 1919 some aspects became obligatory, such as maternity, labour accidents and professional illnesses. Due to the civil war the obligatory nature of health insurance was not established until 1940. With time, the benefits and range of Social Security have increased substantially. At present it is available to every Spanish citizen even if unemployed. There is also compensation for those without a job, a situation which has occurred to many workers due to industrial and rural reorganization.

In Portugal, infrastructures in health services are well provided; however, not all the regional hospitals are well equipped. Notorious inequities exist related to the number of medical doctors, nurses and hospital beds that are concentrated in Lisbon, Coimbra and Porto regions (almost 60%). The inequality to health rights is not only regional, but is mostly generated by social inequalities. This is particularly apparent in the interior where the infant death rate is still very high (10.8%). The National Security System was developed in the 1960s. Only in the 1980s was a General Regime of Social Security established, covering unemployment subsidies, housing benefits and social care for the underprivileged. A minimum national salary was defined then and is updated yearly according to the inflation rate.

Cultural Life

Culture, as distinct from technology, is one of the main riches of Mediterranean Sea countries, and its influence has extended worldwide. The revival of many Mediterranean cultural traditions is noteworthy, for instance the Mediterranean diet. Both Spain and Portugal are countries with long cultural traditions, where much of the cultural heritage is preserved at all levels and co-exists with modern achievements. In relation to monuments, national and EU funds are dedicated to their conservation. Internationally, Spain is better represented in art (opera singers, painters, architects, etc.) than in industry or commerce. More important than a list of libraries, museums and theatres which can be found in Madrid, Lisbon and Barcelona, and also in all the provincial capitals, are the most remarkable part of the Iberian culture: namely folk activities such as fairs, open-air dances, festivities and pilgrimages. Although Andalucian fairs and religious festivities are well known, similar activities can be found all over Iberia. A more modern but very significant festival of Mediterranean character is Saint George's Day in Catalonia, when gifts are made of a book, as a symbol of culture, and a rose, as a symbol of love.

2
The south of France and Corsica

Jean-Louis Ballais

Introduction

The Mediterranean part of France (the so-called "*Midi méditerranéen*") and the island of Corsica are situated in the south of the national area (between 41° and 45° north latitude). The *Midi méditerranéen* is a narrow fringe, less than 180 km in width, limited by mountains, so that the transition with the oceanic or semi-continental temperate climate is abrupt. Corsica has an exclusively Mediterranean climate. Situated in the north-western corner of the Mediterranean Sea, the *Midi méditerranéen* stretches out for more than 450 km from the Spanish border in the WSW, to the Italian border in the ENE. Corsica (180 km long, 85 km wide) is in the north of the Tyrrhenian Sea, between Sardinia and the Genoa Gulf, about 100 km from the Italian coasts and 220 km from the French ones (Figure 2.1). Marseille is the most important city, followed by Nice and Montpellier.

The Land

Relief

Organized in four sets, the mountains form the main framework of the relief (Figure 2.2). To the west, the eastern extremity of the Pyrenees range which separates France and Spain is orientated W–E. It has a basin and range structure which disappears under the Mediterranean, and reaches its highest point of 2785 m at the Canigou. In the centre, the old Hercynian Massif Central was faulted and tilted during the Tertiary, and especially uplifted in the SE where it exceeds 1500 m (Mont Aigoual, 1567 m; Mont Lozère, 1702 m). To the east, the Alpine chain is made up of two main groups: the W–E secondary ranges and plateaux of Provence (maximum altitude: Mont Ventoux, 1912 m) and, close to the Italian border, N–S orientated folds (altitude lower than 2000 m) which drop rapidly to the sea. Corsica, too, comprises complex middle mountains (Monte Retondo, 2525 m), with the exception of its eastern plain along the coast. The other plains form relatively small areas, close to the sea: the Roussillon basins and Bas-Languedoc plain, with the exception of the Rhône plains, from the north to the south: Comtat Venaissin, Crau and Camargue (the Rhône delta).

Drainage

With the exception of the Rhône – an allogenous river with its sources outside the Mediterranean zone – and its tributaries, the rivers are short coastal drainage systems with small catchments (Figure 2.3). The Rhône comes from the northern Swiss Alps, where the climate is semi-continental temperate. Flowing north–south in an Oligocene graben, the Rhône extends over 512 km in France and has a large discharge (mean rate of flow at Le Theil, 1500 m^3/s), and is well fed by numerous tributaries. The lower course of some of them,

Land Degradation in Mediterranean Environments of the World: Nature and Extent, Causes and Solutions.
Edited by A.J. Conacher and M. Sala.

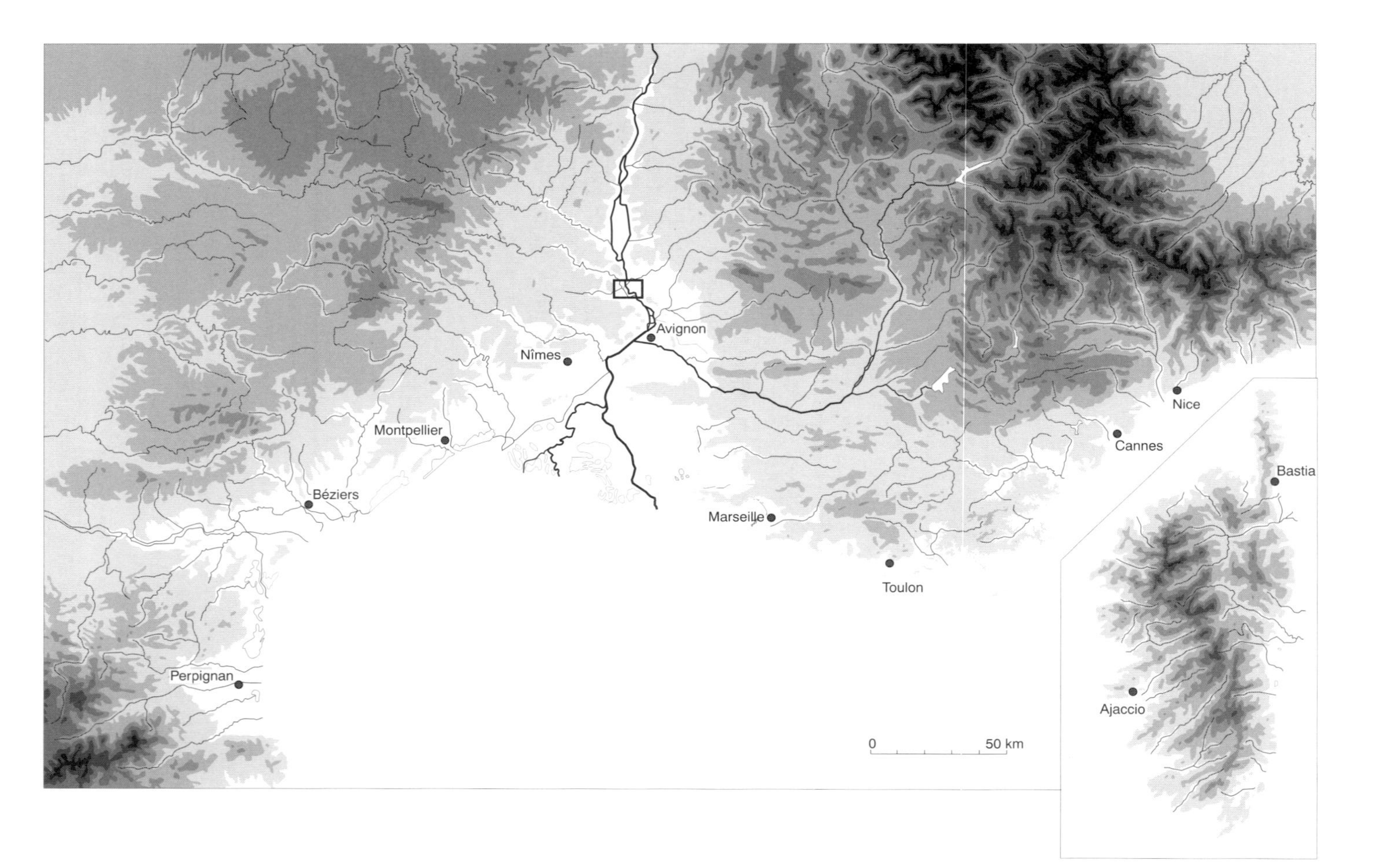

Figure 2.1 *Location of the Mediterranean region of France and Corsica* (modified by cartographer P. Pentsch from general French maps edited by the Institut Géographique National)

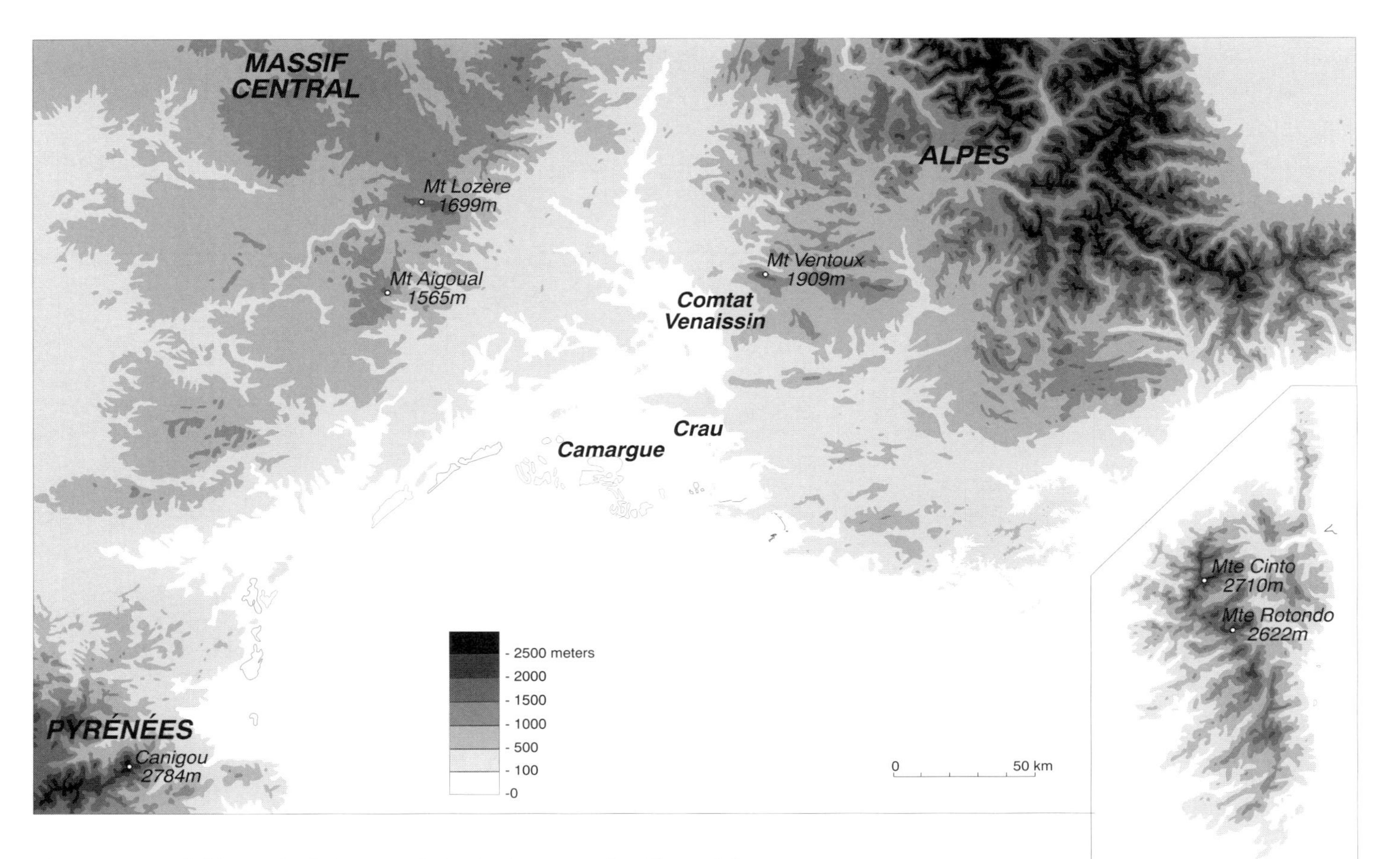

Figure 2.2 *Relief in south of France and Corsica* (source as for Figure 2.1)

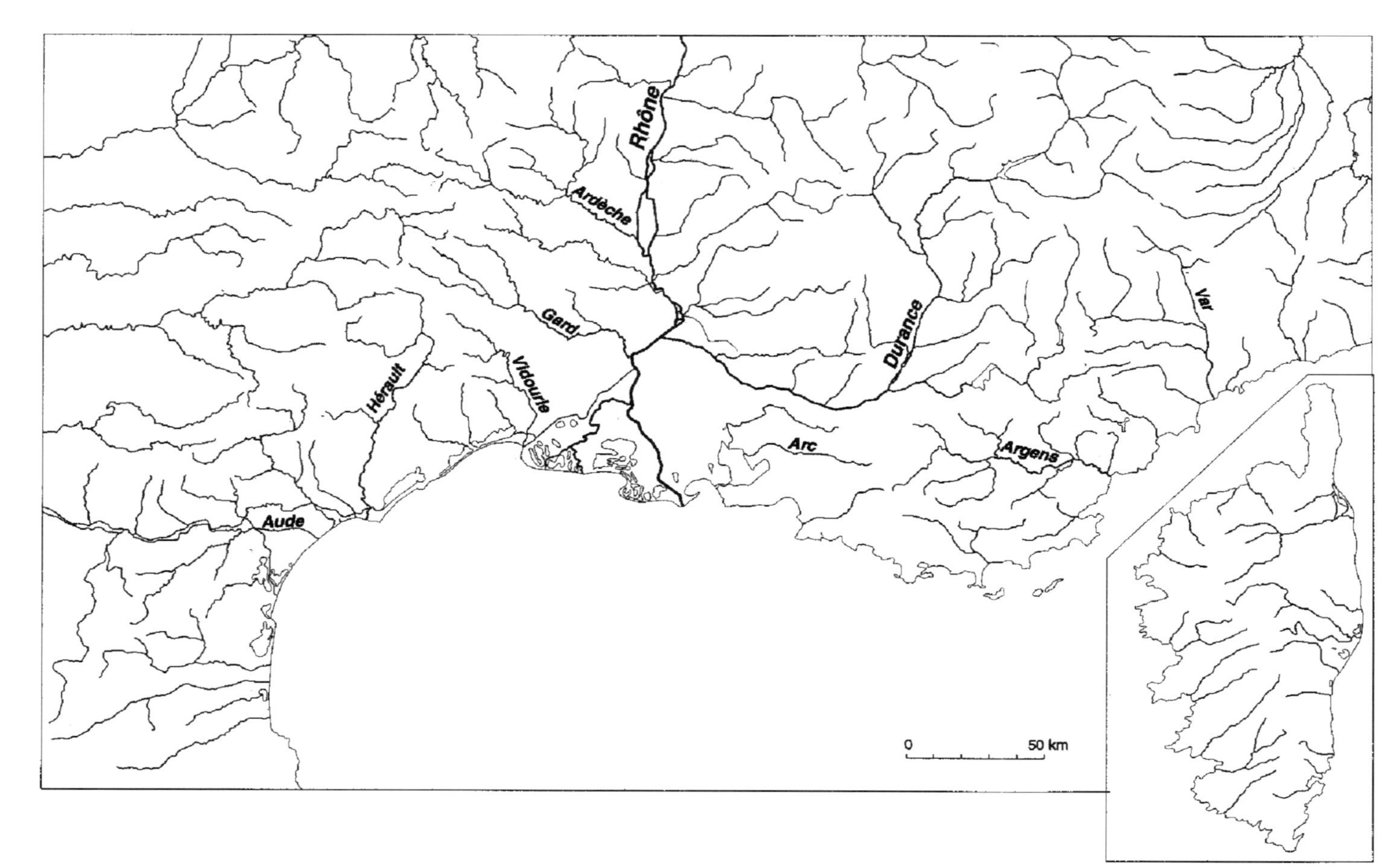

Figure 2.3 *Drainage systems in south of France and Corsica* (source as for Figure 2.1)

including the Durance (length 302 km, with a discharge of 191 m^3/s at Cadarache), the Ardèche (112 km) and the Gard (90 km), is mainly influenced by the Mediterranean climate. The Rhône flows into the Mediterranean across its large delta, known as the Camargue.

The coastal rivers have very irregular regimes and can be grouped into two types. The smaller systems with few tributaries include the more numerous rivers: the Argens (116 km in length, with a discharge of 13.5 m^3/s at Entraigue), Arc (72 km), Hérault (164 km) and Vidourle (100 km), and the Corsican rivers (Plate 2.1). The second type comprises the larger, more complex river systems with important tributaries. They include the Var (135 km, discharge 43 m^3/s at Carros) and the Aude (223 km).

The drainage areas of some French Mediterranean rivers are: the Aude, 5300 km^2; Arc, 780 km^2; Argens, 2800 km^2; Hérault, 3200 km^2; Orb, 1750 km^2; and the Rhône, 95 000 km^2.

Owing to the large areas covered by limestones and well developed karsts (Nicod 1980), important karstic groundwaters exist in the middle mountains, the massifs and the plateaux. They all have important resource possibilities; for example, the Fontaine de Vaucluse, the most powerful spring in France located about 25 km east of Avignon on the edge of the limestone Monts de Vaucluse, has a 29 m^3/s rate of flow and a variable reserve of 80 to 100×10^6 m^3.

Soils

Soil erosion due to cultivation for several thousand years has left behind few of the original soils. Generally, Rankers and Rendzinas occur on the slopes. Some old soils, the so-called "Mediterranean red soils", remain on the Middle or Lower Pleistocene alluvial terraces. Brown Earths have formed on the more recent formations. If the colluvial or alluvial deposits are silty and if the water balance is favourable, the Brown Earths provide good support for crops.

Climate

We have followed the *Bioclimatic map of the Mediterranean zone* published by UNESCO–FAO (1963) to define the limit with the temperate climate (Figure 2.4).

The marked summer rainfall minimum is centred on July: between 10 and 20 mm fall on the coast. The mean annual rainfall is less than 700 to 800 mm on the coast, and falls mainly during the cold season. The secondary minimum is in January and the main maximum, well marked, is in autumn, especially in October. The secondary

Plate 2.1 *The limestone gorge of the Hérault River, Languedoc, with* garrigue *vegetation*

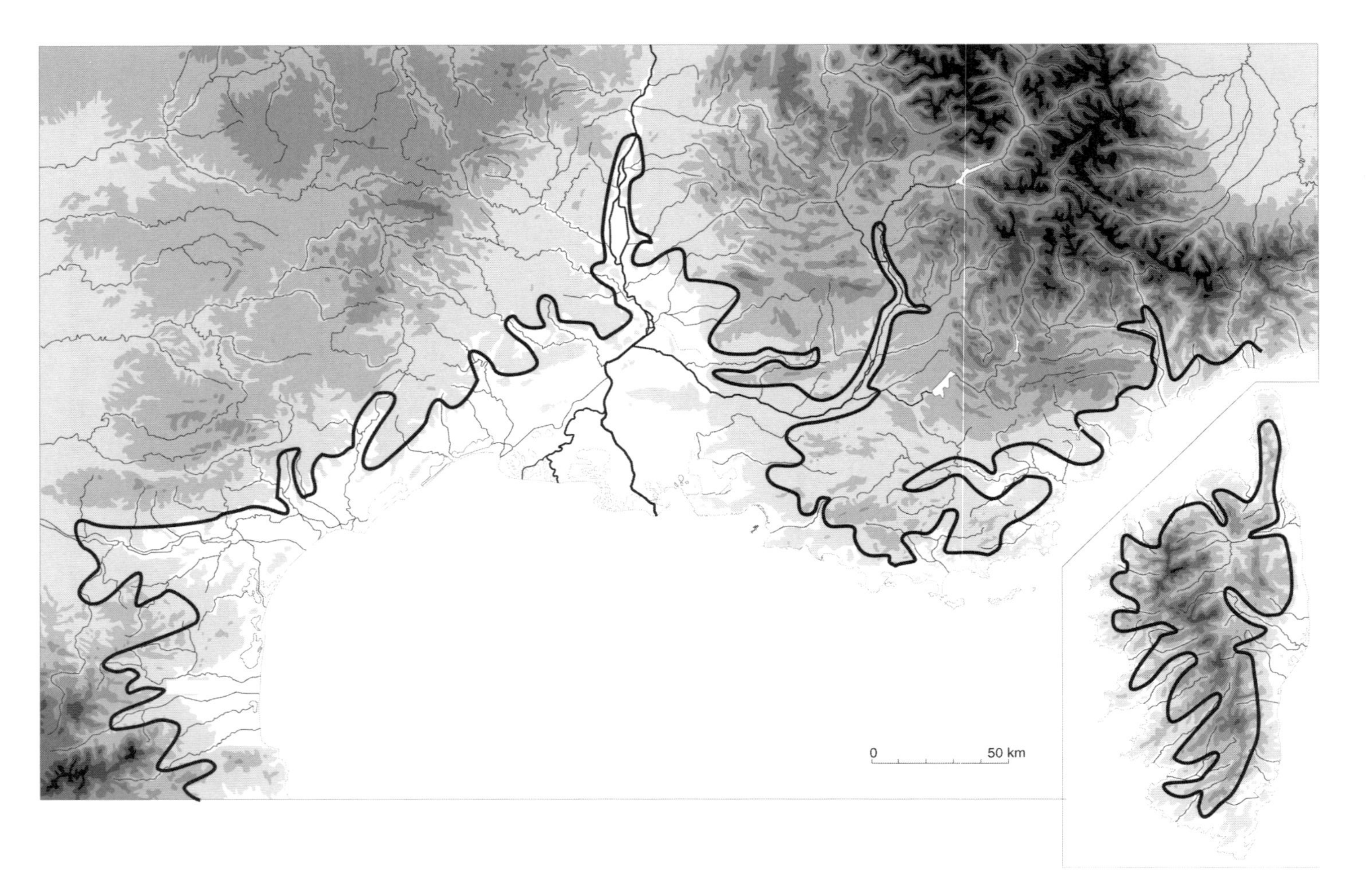

Figure 2.4 *Limits of the Mediterranean climate zone in France and Corsica* (modified by J.-L. Ballais from a map in an old atlas published in the (then) German Democratic Republic)

maximum occurs during spring, which is very characteristic of the French *Midi méditerranéen*. Yearly variations of precipitation are important: in Marseille, between 1864 and 1989 the annual minimum was 283 mm and the annual maximum reached 1088 mm (Douguédroit 1991).

Mean annual temperatures indicate the general mildness: they are always higher than 10°C. The maximum is reached in July or sometimes August, 28°C on the coast and 30°C in the interior, owing to significant irradiation (7525 $Wh/m^2/day$ in Carpentras during July; annual mean, 4294 $Wh/m^2/day$; Wh = watt hour). Temperatures can exceed 40°C, as in 1982 and 1983. The mildness of winters is the basis of the touristic reputation of the *Midi méditerranéen* (the January monthly mean exceeds 6°C and monthly mean minima exceed 0°C). Short periods of very deep cold seldom occur; an exception was during February 1956 (–17°C in Montpellier). But snowfalls are only occasional (Douguédroit and Zimina 1987a).

The continental winds (*cers*, *tramontane*, *mistral*) are dry and often strong. From 1957 to 1981, on average the *mistral* blew 14.6 times each year with a velocity greater than 90 km/h (Douguédroit 1991).

Plant and animal life

Here as elsewhere, the Mediterranean vegetation is characterized by a summer vegetative dormancy due to drought-adapted forms. The xerophytic species have reduced respiratory surfaces or water reserves and their tissues are frequently woody. Here, the pubescent oak (*Quercus pubescens* Willd.) is close to its southern limit. Numerous vegetative species are able to escape from the annual rhythm owing to the edaphic humidity that persists during summer and offsets the air dryness. The "natural" vegetation remains only in very small areas because of the high human impact. Remaining forests on the slopes are marked by the traces of their ancient forestry development, and by reforestation initiatives and fires. Nevertheless, their areal extent will increase: at present they cover 35% of the Provence–Alpes–Côte d'Azur region (Baleste *et al.* 1993) and about 60% of the Var department.

Apart from coastal patches of the *Oleo-lentiscetum* group (Douguédroit and Zimina 1987b), the vegetation is organized into subzones, with the Mediterranean one at the base and the sub-Mediterranean one at higher altitudes. The Mediterranean subzone extends to about 550 m above sea level on the north-facing slopes and 750 m on the south-facing slopes. The groves of holm oaks (*Quercus ilex*) prefer calcic soils and those of cork oaks (*Quercus suber*), acid soils. Many slopes are covered with secondary vegetation formations, with *garrigues* on calcic soils and *maquis* on acid soils (Plate 2.2). Aleppo pine (*Pinus halepensis*) is often associated with them.

The sub-Mediterranean subzone extends from 400 m to 1000 m on the north-facing slopes and from 500 m to 1300 m on the south-facing slopes. In this subzone, there is a continuous vegetative season from spring to autumn. On the other hand, deciduous forests (in particular pubescent oak groves) appear. In the east of Provence are extensive hornbeam forests (a species of *Carpinus*). Everywhere chestnut groves occupy acid soils. The secondary formations are composed of box groves (on a calcareous substratum), broom groves and lavender groves. *Pinus sylvestris* is frequent.

Very large areas have been reforested since the end of the 19th century (about 14 000 ha in the Cévennes according to Faucher (1951)). *Pinus* was the main genus used, either: native *Pinus sylvestris* or imported black pine (*Pinus nigra* sp. *austriaca* L.).

Human activities have reduced wildlife considerably. The great predators like the wolf were totally exterminated during the 19th century and herbivores survive because of drastic limitations on hunting. This is also true of migratory birds, especially pink flamingos.

The People

Composition

The *Midi méditerranéen* is an ancient immigration region, and it has welcomed large groups of people from neighbouring countries, mainly Italy and Spain, during the 20th century. But since the middle of this century, the largest number of immigrants has come from North Africa;

Plate 2.2 Garrigue *in land west of Montpellier, previously grazed by sheep and goats but now abandoned to hunters*

these have been mainly Europeans leaving as a result of the independence of Tunisia, Morocco and, above all, Algeria (300 000 people to the east of the Rhône), but also natives of those countries. The latter are mainly Muslims and Arabic speakers, and they have posed new problems regarding their integration into the national community.

Foreigners constitute between 3% and 9.9% of the population, depending on the region. If the numerous French coming from other regions of the country are added, then 50% of the inhabitants living to the east of the Rhône were not born there. A consequence is the rejection of non-native people, exacerbation of racial conflicts and the growth of "nationalist" parties (Baleste *et al.* 1993).

Demographic trends

The population of the *Midi méditerranéen* is growing twice as fast as the national mean: 1% per year since 1975, mainly owing to immigration, because the rate of natural increase is low (0.08 to 0.22% per year). People under the age of 19 account for 24 to 25% of the population; 51 to 52% are between 20 and 60, and 23 to 24% are over 60 (Baleste *et al.* 1993).

Corsica is a different case: natural increase is negative (–1% per year), but the total population increase (6.4% since 1981) has been due to a positive migratory balance since 1962 which has reversed the trend of population decline (INSEE Corse 1994).

Settlement patterns

The general trend is to an excessive concentration of population in the coastal areas. The inner mountains are almost empty: 2.6% of the total population east of the Rhône live in the Alpes de Haute Provence department versus 43% in the Bouches-du-Rhône department. Densities come down to 20 and even 10 persons/km^2 in the mountains, whereas they reach 347 persons/km^2 in the Bouches-du-Rhône department. Mountains are on the "survival limit", meaning that if population decline continues, the provision of infrastructure (roads, electricity, postal services etc.) will become extraordinarily expensive in relation to the services provided.

In contrast, the coast is over-populated: the environment is degraded, pollution is increasing, xenophobia is latent and the criminality rate is high. East of the Rhône, the spread of urbanization is spectacular on the coastal fringe, with big towns like Marseille (807 000 people), Toulon (170 000), Cannes and Nice (345 000) grouped together in four conurbations of over 300 000 people. West of the Rhône, the main towns of Nîmes, Montpellier, Béziers and Perpignan, are located away from the coast which was marshy and unhealthy for a long time; their origins are ancient or medieval. Corsica has only small towns like Ajaccio and Bastia.

The Economy

The *Midi méditerranéen* economy was for a long time based on agriculture and trade. Today it is dominated by the tertiary sector.

Resources and industry

Mineral resources are poor (lignite at Gardanne and bauxite in Var) and their cost price is high. Water is plentiful, being provided by allogenous flows, permitting the production of hydro-electricity on the Durance and on the Rhône.

The *Midi méditerranéen* and Corsica are under-industrialized: the secondary sector accounts for only 22% of the working population east of the Rhône, 18% in Corsica and 14.2% west of the Rhône (France, 29.6%). Industry is mainly concentrated on the coast and consists of a few big firms and a multitude of small firms. Processing of tropical products, the textile industry, shipyards and iron and steel metallurgy are decreasing, while petrochemicals, chemicals, aeronautics, electronics and data processing replace them. The growth of "high technologies" has been particularly spectacular (Nice and Montpellier regions). The main axis is the Marseille conurbation, around the *étang de Berre* (Berre lagoon) and in Fos-sur-Mer. Toulon has an important arsenal. Everywhere, the building and public works industries advance. But the number of industrial employees is decreasing (employment has declined by 50 000 since 1980 east of the Rhône).

Agriculture

Agriculture occupies only a small proportion of the population (4% east of the Rhône). The area cultivated with cereals has decreased to the benefit of irrigated cultures such as meadows, orchards and, above all, market gardening and flowers. The latter mark the landscape with irrigation canals and sprinkler systems, as well as windbreaks (in the Rhône and tributary valleys, eastern Languedoc and Roussillon). Intensive agriculture is practised on small properties, with the cultivation of flowers being profitable on areas of no more than 1 ha. On the other hand, apples, tomatoes, melons, peaches, apricots and cherries compete with Spanish produce, and over-production crises are frequent.

The area of vineyards has decreased (–14% west of the Rhône between 1979 and 1988), as has wine production, to 28.6×10^6 hl/year. But the Languedoc-Roussillon vineyard remains the world's first co-operative production vineyard; it still covers 35% of the land used for agriculture and groups together 65% of the farming concerns.

In Corsica, only 2% of the surface is cultivated (meadows, citrus fruits, vineyards) and more than one-third of the farming concerns disappeared between 1979 and 1988.

The forest area is increasing rapidly (35% of the surface east of the Rhône, compared with an average of 26% for France as a whole), but it is mainly planted with poor quality and highly combustible *Pinus halepensis*.

The Mediterranean Sea is not as well stocked with fish as the Atlantic Ocean, but its rock fishes are in great demand. The resources are depleting, their economic value is low and production is decreasing.

The tertiary sector

The *Midi méditerranéen* and Corsica have larger proportions of their working populations in the tertiary sector (from 68 to 74%) than France as a whole (64.2%). Tourism is the main economic activity: it accounts for 100 000 permanent employees and welcomes 25 million tourists each year. Coastal tourism dominates (sea bathing, pleasure sailing, golf) but, in the interior, winter skiing sports and ecotourism are increasing.

The communication routes network is one of the densest and most used in France. It comprises railways (soon with the Train à Grande Vitesse), roads and motorways to Spain or Italy. They permit the transit of goods and French tourists, North African, Spanish and Italian people, but also the local population. The port of Marseille is one of the most important in Europe, especially for imported hydrocarbons and exported refined products, steel and containers. Foreign travellers and tourists arrive in Nice and Marseille airports (the second and third most important, respectively, in France) and in Marseille harbour. The Rhône river plays a negligible role although it is dredged.

Despite all these activities, unemployment is high, ranging between 11 and 14% of the working population.

Administrative and Social Conditions

Government

The *Midi méditerranéen* has no administrative unity, and is divided between two administrative regions created in the 1970s, termed Program Regions: Provence–Alpes–Côte d'Azur (east of the Rhône) and Languedoc–Roussillon (west of the Rhône). These Program Regions group together all the Mediterranean zone and some non-Mediterranean departments. Corsica has a special status, being neither a department nor a Program Region: it has been a Territorial Collective since 1991, which means that it has very specific institutions.

The Program Regions have the same powers as all the others in France in that the scope of French decentralization and French law is fully applied. The regional executive councils contribute initiatives in the economic, social and educational sectors. As with most of the other Program Regions, Provence–Alpes–Côte d'Azur and Languedoc–Roussillon are directed by coalitions of right wing political parties, but a "nationalist" extreme right racist and xenophobic party plays an important role.

Education

Education is organized, financed and controlled by the French State. Administrative Regions are responsible for building and upkeep of some school buildings, and they are represented on the boards of directors of the public universities. The main universities are in Aix-Marseille (three universities, with more than 60 000 students), Montpellier (three universities, with 47 000 students) and Nice. Traditionally, the *Midi méditerranéen* has provided a significant number of state servants and middle executives.

Health and welfare

As for education, the health organization (Sécurité Sociale) is the same as in the rest of France.

The Mediterranean climate and the duration of sunshine (sometimes more than 2500 h/yr) attract convalescent people. In the interior, some spas such as Amélie-les-Bains, Digne or Gréoux welcome persons taking the waters.

Culture and Recreation

Cultural life is well developed. The *Midi méditerranéen* and Corsica have a rich and prestigious historic past, going back to the Greek colonization from the 7th century BC in Marseille, Nice, Antibes and Agde. The Roman conquest (2nd century BC) was completed in the *Midi méditerranéen* more than 50 years earlier than in the remainder of France, which meant that the Roman influence was very important. During the Middle Ages, a particular culture was formed which linked with Spanish and Italian as well as other Mediterranean cultures, rather than with northern French cultures. The main part of the *Midi méditerranéen* was conquered or annexed by French kings during the lower Middle Ages, but the Roussillon (north Catalunya) was conquered during the second part of the 17th century, Corsica was purchased from the Genova Republic in 1768 and Nice was given to France by Italy as late as 1860.

The particular culture remains vibrant and is expressed by living languages (such as Catalan, Provençal and Corse) and music. Today, poetry, music and songs are relayed by radio and television, thereby assisting the spread of the distinctive cultures, which differ from the northern French and Parisian ones. They constitute one of the bases of claims for autonomy or even independence (mainly in Corsica).

Summer is the season when most tourist events are organized: theatre (Avignon), opera (Aix-en-Provence) and various concerts.

Football attracts crowds in any season, mainly in Marseille; but all sports are played successfully, on land or in the sea.

Numerous natural parks and reserves, with various status, protect the rare, relatively undisturbed and unpolluted areas, particularly on the coast (Camargue, Port-Cros) and in the mountains (Corsica: 3300 km^2, Mercantour, Lozère).

3
Italy

Marino Sorriso-Valvo

Introduction

The Italian territory with Mediterranean climate is shown in Figure 3.1. Criteria for the identification of this territory are those of UNESCO–FAO (1963), even though data on foggy days may be missing at several gauging stations.

The territorial boundaries of the several islands, the largest of which are those of Sicily and Sardinia, and of the eastern, southern and western sides of the Italian peninsula, are well defined by the Mediterranean Sea itself. The northern boundary corresponds with the Alps–Apennine ridge. However, as climate changes over time, so do climatic/ecological boundaries. For instance, the olive trees, which are one of the typical Mediterranean climate plants, were cultivated north of the Apennine ridges until the 17th century.

The Land

Geology

The geology of Mediterranean Italy is extremely complex, as it is throughout the Mediterranean lands, because the Mediterranean area is the collision zone between the Euro-Asiatic, African and other minor plates and the zone of rifting of the Tyrrhenian Sea. All belong to the Alpine Orogeny. It is also possible to recognize tectonic elements of former orogenic phases (Hercynian, Caledonian).

The following scheme is thus a very simplified description of the geology of the area of interest.

The principal tectono-stratigraphic units outcropping in the study area are:

1. the Alps;
2. the Apennine chain with its bordering Neogene basins and ancient volcanic systems;
3. the Calabro–Peloritan Arc with its internal and external Neogene basins and ancient volcanic systems;
4. the Sardinia micro-craton;
5. the North African domain.

In addition, three active volcanic systems (Aeolian Islands, Mount Etna, Vesuvius) are present in the southern part of Italy.

The first three regions belong to the Maghreb–Apennine–Alps section of the Alpine Orogeny. Sardinia appears to be a part of an ancient continent left behind after the opening of the Tyrrhenian trough. The three volcanic systems belong to completely different magmatic regions and, despite their proximity to one another, there is no link between them: the Etna system takes basaltic lavas from the upper mantle; Vesuvius exhibits lava flows of crustal origin with increasing carbonate (CO_3-rich) contamination; and the Aeolian system has modified the magmatism from basalt to very silica-rich liparite of crustal origin. It is worth mentioning that located here is Vulcano island, from whose name (after the god Vulcano of the Greek and Roman religion) the term corresponding to "volcano" has been derived in all modern western languages.

Land Degradation in Mediterranean Environments of the World: Nature and Extent, Causes and Solutions.
Edited by A.J. Conacher and M. Sala.

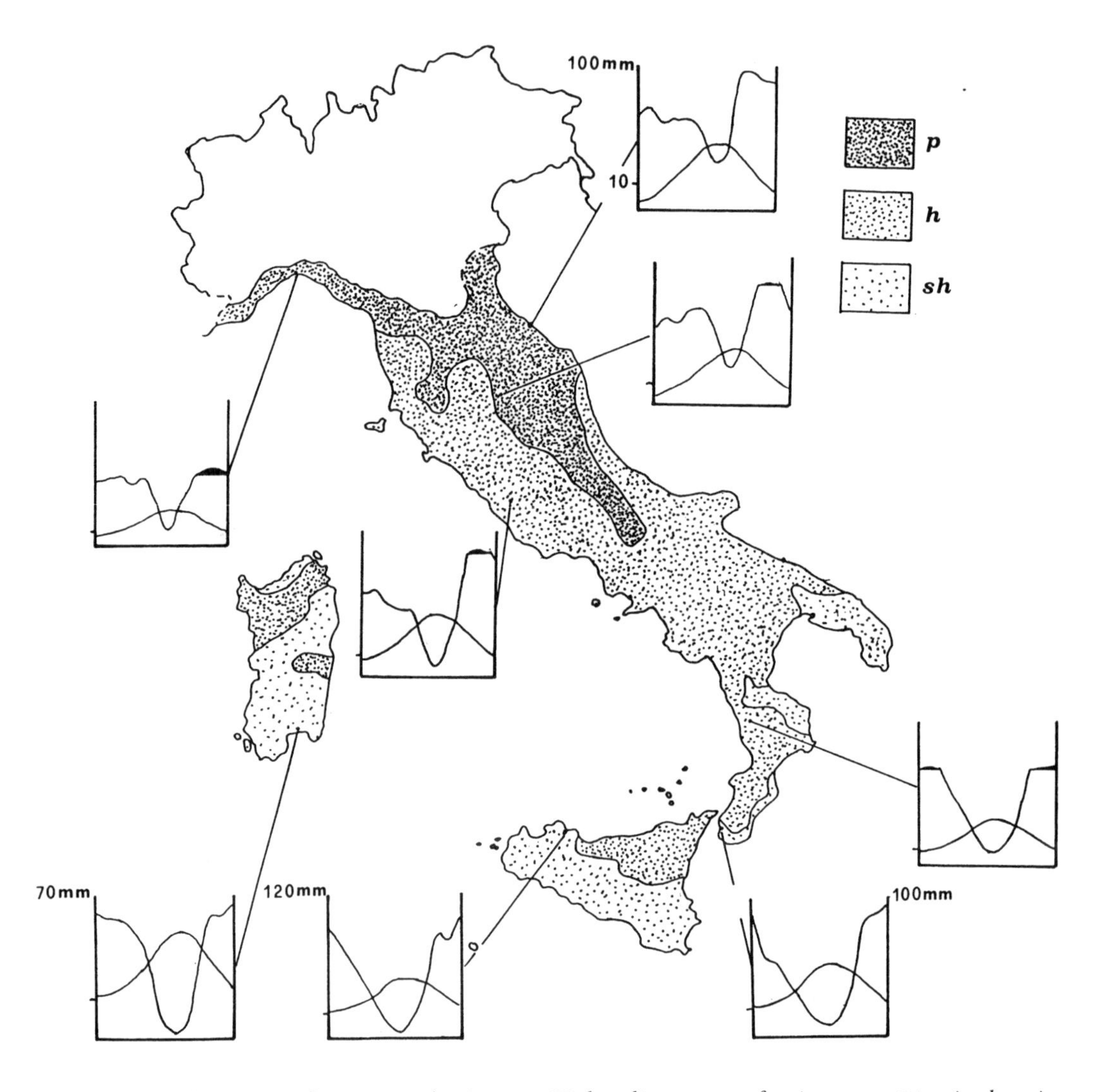

Figure 3.1 *Mediterranean climate in Italy. Gaussen–Walter diagrams are for sites up to 600 m in elevation. Symbols: p = perhumid; h = humid; sh = subhumid* (original diagram by M. Sorriso-Valvo)

Relief

The relief of Italy is strictly controlled by tectonic (Plate 3.1) and volcanic activity and river downcutting and sedimentation (Figure 3.2). The magnificent mountain ranges of the southern Alps, the Apennines and the northern ranges of Sicily, are uplifting at rates of up to 1.4 mm/yr over the last million years, whereas maximum erosion rates, for the same period and at the scale of the mountain range, have been estimated at up to 0.4–0.6 mm/yr (Ergenzinger *et al.* 1978; Sorriso-Valvo 1993). The estimated average erosion rate for all the Italian territory is about 0.1 mm/yr.

The highest mountains in continental Mediterranean Italy have altitudes of more than 2000 m above sea level (Gran Sasso, central Apennines, 2914 m; Monte Pollino, southern Apennines, 2267 m; Mt Terminillo, central Apennines, 2213 m; Mt Saccarello, Liguria, 2200 m). The highest altitude is reached by Mount Etna volcano in Sicily (3323 m). High plateaux of carbonate deposits, dissected by steep-sided canyons, characterize the landscape of Apulia and southeast Sicily

Plate 3.1 *Steep, tectonic coast in western Calabria, with minimal coastal plain. Topography poses a challenge not only for transport routes but for coastal development to accommodate tourists*

(these territories belong to the African Plate). In these lands elevations seldom exceed 600 m above sea level. Highlands carved out of crystalline rocks are typical of northern Calabria (the Sila Massif reaches 1980 m) and eastern Sardinia (the Gennargentu). Altitudes in Sardinia reach a maximum of 1834 m above sea level.

Ranges are characterized by intramontane tectonic depressions forming valleys and lakes. All lakes, except for a few small ones, are now dry, forming intramontane plains. Due to the widespread outcropping of carbonate rocks, karst morphology is developed throughout.

The northernmost part of the eastern side of the Mediterranean land includes the growing delta of the Po river. Other alluvial plains border the peninsula and eastern Sicily, but their extent is limited to a few tens of square kilometres at most.

Drainage

As a result of the high relief and short distances from divides to the coasts, Mediterranean Italy does not permit the development of large rivers. Major rivers are the Tiber (452 km long, area of basin 17 679 km^2), Arno (241 km long, area of basin 8247 km^2), Volturno (175 km long, area of basin 5455 km^2) and the Ombrone (161 km long, area of basin 3480 km^2), just to mention rivers longer than 150 km.

In areas with rapid rates of uplift, such as Lucania, Calabria and eastern Sicily, erosion processes are intense. The debris budget thus provided cannot be evacuated to the sea, so that thick, coarse-grained and steep riverbeds have developed (Plate 3.2). These streams, which are also characterized by an extremely high seasonal variability of their discharge, are called *fiumara* (in Calabria) or *jumara* (in Sicily). They appear to be aggraded, but at present they are degrading. An aggradational phase was observed following major storms in 1973 (Ergenzinger *et al.* 1978; Sorriso-Valvo 1994; Gabriele *et al.* 1994). This suggests that the present conditions are in equilibrium with minor oscillation towards aggradation or degradation, according to the temporary availability of debris coming from slopes during recurrent, extreme meteorological events, and its exhaustion by erosion. In north Calabria, the recent discovery of two lacustrine deposits due to landslide damming, both at about 1 m above the present level of the riverbeds, but dated about 200 and 1800 years BP, suggests that this system has been in substantial equilibrium over the last 2000 years.

Silting of ports and widespread decay was initiated throughout the Mediterranean countries from the 8th century BC, but not at the same time (Vita-Finzi 1969; Bruckner 1986). In some places it was continuing in late medieval times, as at the ports of Rome (Ostia) and Pisa, one of the four Italian Marine Republics (with Genova, Amalfi and Venice).

Soils

Mediterranean Italy is one of the areas characterized by Alfisols, suborder Xeralfs (USDA Soil Taxonomy), but in fact virtually all types of soils

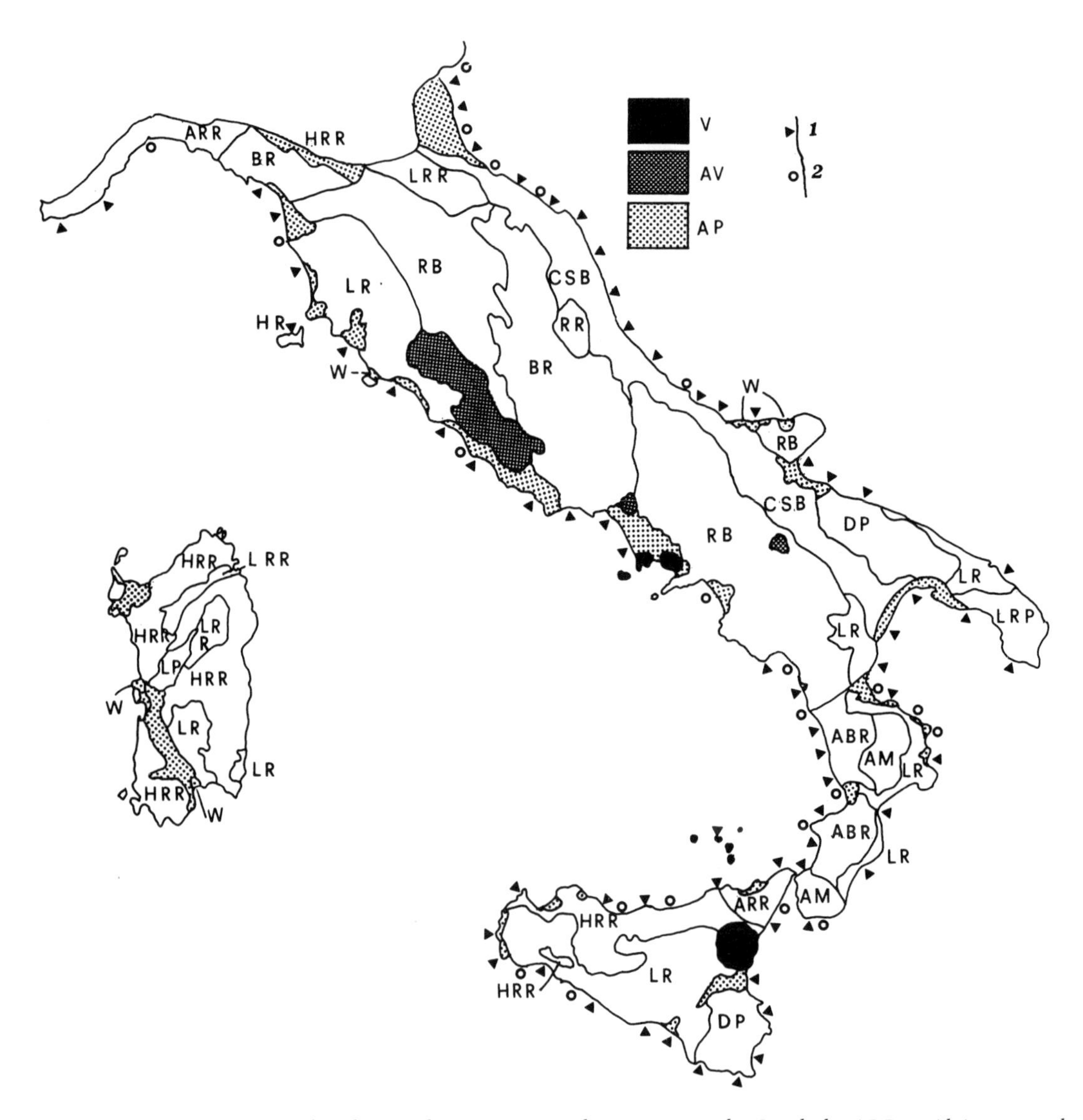

Figure 3.2 *Zoning of relief and coast dynamics in Mediterranean Italy. Symbols: ARR = Alpine rugged range; AM = Alpine units massif; ABR = Alpine units basin and range; AP = alluvial plain; AV = ancient volcanic; BR = basin and range; CSB = consequent sub-parallel basins; DP = dissected plateau; HRR = high relief range; LR = low relief hills; LRP = low relief plateau; LRR = low relief range; RB = range and basin; V = volcano; W = wetland; 1 = eroding coast; 2 = accreting coast* (original elaboration by M. Sorriso-Valvo from information in Rechembach *et al.* (1992) for DEM, Ogniben *et al.* (1975) for geology, and Biasini *et al.* (1993) for coast dynamics)

occur there, except those of very cold climates. In his synoptic work on soils in Italy, Mancini (1966) identified a large number of soil associations. In northeast Calabria, Dimase and Iovino (1988) produced a soil map for an area of about 700 km², with elevations ranging from sea level to about 1600 m. With a wide range of lithologies, except volcanic rocks, the area is representative of Mediterranean Italy except volcanic areas. The authors distinguished 25 soil associations, mostly comprising soils that are normally found on lands currently or formerly covered by deciduous forest or prairies. There are also hydrophilic soils on lacustrine deposits

Plate 3.2 *Coarse gravel riverbed, eastern Calabria. Bedload is largely supplied by landslides in steep, upper parts of the catchment. Note bare, grazed ground beneath olives*

and alluvial riverbeds, and a high percentage of bare rocks (for example *fiumara* riverbeds and rocky free-faces of some fault scarps and cuestas).

Most relevant for this work is the degraded condition of the soils, which is extremely important for the nation's economy. However, there is no permanent programme for even the detailed mapping of soil degradation in Italy. Erosion studies refer to badlands, but these are already abandoned, soil-deprived lands, sometimes with a very low possibility of reclamation. The important problem with soil erosion is on cultivated lands, where nutrients and fines – that is, the mineral-releasing portion – are eroded every year because of inadequate ploughing techniques. In general, all arable lands, which have been ploughed for many centuries, are now in a critical condition. Soils are very deprived of nutrients, so that the massive use of mineral additives and organic manure is necessary. As is well known, problems arise with chemical additives, mostly in terms of groundwater pollution. This occurs mostly in association with the highly productive fields of north and central Italy, but is spreading to south Italy, as in Puglie, where the karst circulation accelerates the pollution of groundwater.

Climate

Precipitation is represented by rain, normally falling with high hourly intensities (tens of millimetres per hour, but sometimes over 100 mm/h). Hail may occur with destructive effects on crops in summer, due to convective storms. Snow is normally confined to high mountains, but if the winter frontal system approaches from the northeast or northwest, snowfalls may occur at sea level, especially along the eastern coasts (Adriatic Sea). This is a rare occurrence, as tropospheric air circulation is essentially zonal, so warm frontal systems from the southwest are more frequent in Sicily, Sardinia, and the southern Italian peninsula.

Rainfall–temperature diagrams of some Mediterranean stations are shown in Figure 3.1. To date, detailed climatic maps have been produced only for Tuscany (Rapetti and Vittorini 1994a). In northern Tuscany, annual precipitation may reach over 3000 mm, while in the central part of the region precipitation is around 900 mm. In the southern part, rainfall averages about 1500 mm in the mountains, 800 mm in the hills and lowlands, and drops to 406 mm/yr along the coast, where up to four dry months may occur. The climate on the islands is even drier, with five dry months.

Variability of precipitation and temperature is characteristically high.

Exposure strongly influences local climatic conditions, and it may be expected that Mediterranean-type climate may reach the highest elevations in southern areas.

Storms reach maximum intensity in Calabria, especially in the southeastern territories (Ferrari *et al.* 1990), where rains due to frontal systems are intensified by orographic conditions. In Calabria and Basilicata, according to Caloiero and Mercuri (1982), from 1921 to 1980 ten major destructive events occurred in Basilicata and 56 in Calabria, resulting in flooding and landsliding (Giangrossi 1973; Cotecchia and Melidoro 1974; Sorriso-Valvo *et al.* 1992). In Calabria there has been an average of more than one extreme event each year.

In 1951, 100 mm of precipitation in three days involved 28% of the region (about 4600 km^2), while over an area of about 330 km^2 rainfall reached 1000 mm and over 80 km^2, 1400 mm, with 1496 mm in southern Calabria. Here, a single day's rainfall yielded more than 50% of the annual average. These figures are exceptional in the time series of all Calabrian rain gauging stations from 1921 to 1995. The return period has been assessed at about 50 years (Ferrari *et al.* 1990). Hourly rainfall intensities of 138 mm/h were recorded in the same region in October 1953.

Conditions are similar in other parts of Mediterranean Italy, but generally with lower intensities and with montane modifications (more humid summers, snowfall in winter) in the high mountains.

Drought should also be considered. In some parts of southern Italy, it is a common experience to have one or more summer months practically dry. Droughts are also experienced when winter months receive very low rainfalls. This occurred on several occasions during the 1980s and 1990s, not only in Mediterranean Italy. The most dramatic drought occurred in 1988–1990. The months with greatest rain loss were January 1989 and February 1990 (Conte 1994), due to persistent conditions of high atmospheric pressure all over the Mediterranean basin, centred over the northern Mediterranean Sea.

It has been suggested that in south Calabria, daily rains constituting more than 5% of the average yearly precipitation are increasing, whereas rainy periods longer than three days are decreasing (Sorriso-Valvo *et al.* 1992). This tendency to more extreme climatic conditions may be due to the enhanced greenhouse effect, and might be a precursor to desertification in Mediterranean countries.

Plant and animal life

The biosphere of Mediterranean Italy is all but natural, due to the long-term influence of human activities. However, there are some very sparsely populated areas where plant and animal life has been retained or has recovered most of its natural character (Naveh and Liebermann 1984).

Plant life

The distribution of Mediterranean-type vegetation in Italy and its islands is shown in Figure 3.3. By comparing Figure 3.3 with Figure 3.1, it can be seen that the vegetation with Mediterranean characteristics does not cover exactly the same areas as the zones with Mediterranean climatic characteristics. This might be due to the different criteria for ecobotanic and climatic zoning, or the influence of medium-term climatic changes which are reflected in the climatic data but not yet in vegetation characteristics.

Figure 3.4 illustrates the influence of altitude on vegetation in Italy, Sicily and Sardinia. The main characteristics of the Mediterranean vegetative zones are (Pignatti 1982):

1. *Mediterraneo-arida*, with *macchia* vegetation and palms (mean annual temperature 18°C);
2. *Mediterranea*, with evergreen forest of warm climate (15°C);
3. *Sannitica*, with mixed deciduous forest with oak domain (11–13°C);
4. *Colchica*, with mixed deciduous forest of temperate climate (8°C);
5. *Subatlantica*, in mountain zones, with deciduous forest with beech domain (8°C);
6. *Irano-nevandese*, spherical, spiny scrubs (around 5°C);
7. *Mediterraneo-altomontana*, only grass (1°C).

It is clear that altitude influences the distribution of vegetation types. Note that the *Irano-*

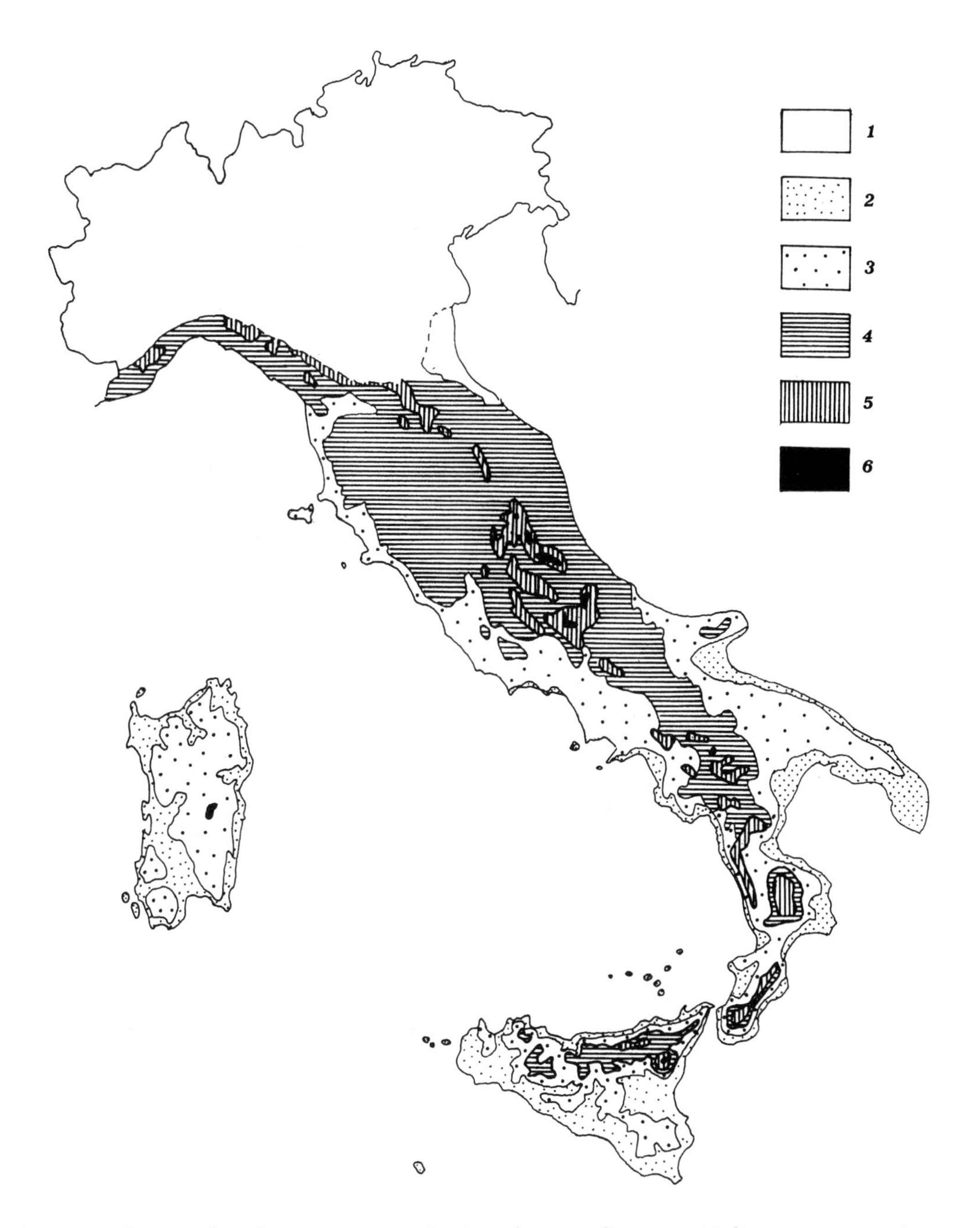

Figure 3.3 *Zoning of Mediterranean vegetation in Italy. Legend: 1 = non-Mediterranean vegetation; 2 =* Mediterraneo arida*; 3 =* Mediterranea*; 4 =* Sannitica*; 5 =* Subatlantica *and* Colchica*; 6 =* Mediterranea altomontana *and* Irano-nevandese. *For characteristics of the vegetation associations refer to the text* (simplified from Pagnatti 1982)

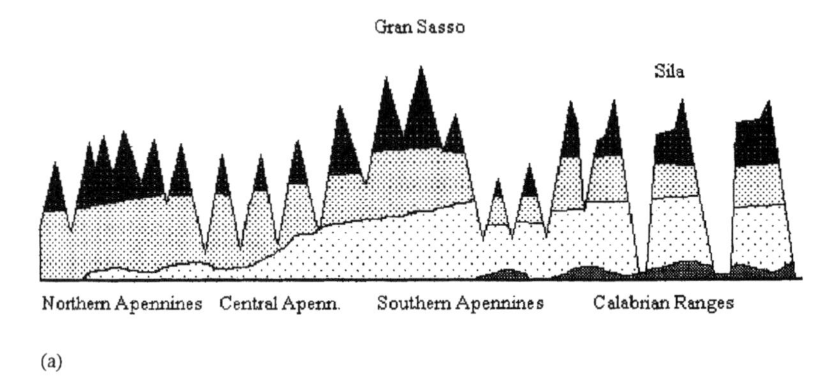

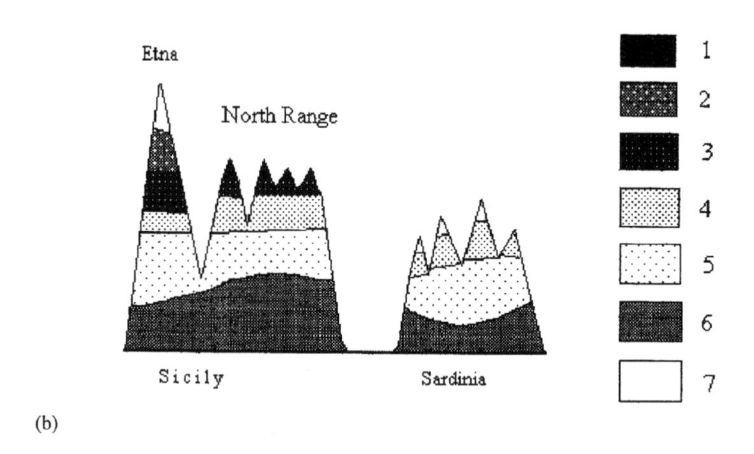

Figure 3.4 *The influence of altitude on vegetation in (a) Italy, and (b) Sicily and Sardinia. The key is the same for both figures and is explained in the text* (simplified and modified from Pignatti 1982)

nevandese zone is found only on the slopes of Mt Etna, in Sicily, and that the *Boreale* "extrazonal" type of mountain areas should also be included, with evergreen forest (conifers and ericacea) and a mean annual temperature of 4°C.

The extent of the different types of vegetation can be assessed from Figure 3.3. The national average of forested land, expressed as a percentage of potentially productive territory, is 28.8%. Some Mediterranean regions have a higher forest-cover rate: Liguria 60.6%, Tuscany 43% and Calabria 35.6%. Table 3.1 gives details for Calabria.

The typical Mediterranean bush, the *macchia*, is confined to an area of only 0.74% of the regional territory. This percentage is probably an under-estimate, because some territories have been recolonized recently by *macchia* as a result of land abandonment. The extent of these areas is difficult to assess.

Table 3.1 *Extent of forest types in Calabria, Italy* (source: Maiolo 1993a)

Forest type	Extent (ha)
Evergreen	100 224
Pure deciduous	152 355
Mixed	301 743
Coppice	31 990
Macchia	11 094
Total	479 286

Pine forests (*Pinus nigra* var. *calabrica*) constitute some 80% of evergreen forests. Nearly 63% of pine forests are in natural conditions, where "natural" is used in the sense that human intervention is older than the remaining records. Perhaps the term "renaturalized" conditions should be used. With regard to grasslands, the typical formation of degraded areas is a steppe-type association dominated by a variety of legume, *Hedisarum coronarium*, which is typical of badlands in non-cultivated fields. Some varieties of this grass, however, produce good cattle feed, and are widely seeded in south Italy.

For Calabria, the vegetation zonation according to altitude can be reduced to three main levels: Lauretum, Castanetum and Fagetum. Of these, the Lauretum zone is the most typically "Mediterranean", extending up to about 800 m above sea level, with local variations according to aspect. It can be subdvided into two subzones: the Warm and the Cold Lauretum.

The Warm Lauretum corresponds to the *Mediterraneo-arida* type (Pignatti 1982) and extends up to 400 m above sea level. It consists mainly of thermo-xerophilic varieties of the *macchia* and steppe. Dominant shrubs characterize the different types of *macchia*: laurel (*Laurus nobilis*), lentisk (*Pistacia lentiscus*), oleander (*Nerium oleander*), myrtle (*Myrtus communis*), several species of broom (*Genista*), arbutus (*Arbutus unedo*), and several others. The Indian fig cactus is present throughout "picturesque South Italy", but actually this plant, together with another typical circum-Mediterranean plant, the aloe, comes from America. Orange, lemon, mandarin, jasmine, and several other fruit trees were introduced to southern Italy by Arabs (7th century AD).

In the higher Cold Lauretum, corresponding to the *Mediterranea* vegetative zone, trees are dominant, with several varieties of oak (typically the evergreen oak and the cork tree) and pine. This zone was the one most subject to clear-cutting for agricultural reclamation in medieval and subsequent times. The olive trees there were introduced by Greek colonizers (8th century BC), but vines were probably already known by local inhabitants. Modern cultural practices have led to the loss of several varieties of cereals (such as wheat) and fruit trees, and at present there is a strong biological genetic impoverishment in the Mediterranean zone.

Some special environments, of high scientific interest, are present. Some of them have been declared zones of national interest and national or regional parks have been created. It is worth mentioning the Maremma, a zone between Lazio and Tuscany, where horses are raised semi-wild in a typical *macchia* environment characterized by swamps. Another zone of interest is the Circeo Park, with typical maritime pines and swampy vegetation.

Animal life

Animal life in Italy is rather degraded. While vegetation can recover and recolonization is possible, animal life seldom recovers where mammals and birds are concerned, with the exception of readily adaptable species. In fact, from 1889 to 1976, seven species of birds and one species of mammal have become extinct in Mediterranean Italy (Massa and Mingozzi 1991) (Table 3.2).

Other animals live in very limited areas: the brown bear, wolf, otter, wild goat, seal and eagle. In recent decades, cities and waste disposal areas are being visited increasingly by wild animals. As for bears in North America, wolves have been seen at night close to a waste disposal zone in a tourist resort area in north Calabria. Snakes, including vipers, approach waste disposal areas hunting rats. Hawks hunt birds in Rome. Nocturnal predatory birds and starlings leave the

Table 3.2 *Extinctions of animal life in Italy since 1880* (source: Massa and Mingozzi 1991)

Species	Zone	Year of last sighting
Francolinus francolinus	Lazio	1889
Turnix sylvatica	Sicily	1920
Haliaëtus albicilla (sea eagle)	Sardinia	ca. 1956
Aegypius monachus (vulture)	Sardinia	ca. 1961
Gypaetus barbatus	Sardinia	ca. 1965
Pandion haliaetus (hawk)	Sardinia	1968
Oxyura leucocephala	Sardinia	1976
Dama dama (deer)	Sardinia	1968

forest for the cities. Seagulls reach cities more than 100 km inland.

Besides the banal European fauna, the few wild animals of biological interest and typical of Mediterranean Italy that remain in their normal habitat are the following (Massa and Mingozzi 1991).

1. Mammals: brown bear (*Ursus actos*, in the Apennines), wolf (*Canis lupus*, besides the few already mentioned for Calabria, in the Apennines), fox (*Vulpes vulpes*), marten (*Martes martes*), stone marten (*Martes foina*, not present in the islands), polecat (*Mustela putorius*, not in the islands), weasel (*Mustela nivalis*), badger (*Meles meles*, not in the islands), otter (*Luntra luntra*), wild cat (*Felis sylvestris*), wild boar, chamois (*Rupicarpa pyrenaica ornata* in the Apennines), wild goat (*Capra hircus* in the Montecristo island and *Ovis musiform* in Sardinia), different squirrels (including the distinctive black squirrel of Mount Sila, Calabria) and several types of rabbit, hare, porcupine and hedgehog. Deer have been reintroduced artificially, while the lynx is re-entering Italian territory from the northeast, and may reach Mediterranean territories in the next few years. In Sardinia, there is probably still a variety of seal (*Monachus monachus*), which is very difficult to see. Other marine mammals are common, including a variety of whale, recently discovered as a typical species of the Mediterranean Sea.
2. Birds: eagle (*Aquila chrysaetus*), vulture, several species of hawk (either migrating or permanent), and some species of nocturnal predatory birds. Recently herons have been more frequent in lakes and swamps (throughout the peninsula and islands, and also in high mountain lakes). Vultures, in particular, are reappearing, probably because of the return of semi-wild sheep breeding in some regions like Sardinia. Other typical species of birds are: *Alectoris graeca, Alectoris barbata* (Sardinia only), *Perdix perdix, Coturnix coturnix, Phasianus colchicus* (not in the islands), *Colinus virginianus*, some varieties of woodcock and rare hoopoe. All these species are under pressure by hunters. Other birds, belonging to the Anseriform group, are migratory, even if a limited percentage of them pass the winter season in the Mediterranean basin: very few swans (*Cygnus olor*), more geese (*Tardona tardona*) and a variety of duck (several species of *Anas*, and also *Netta rufina, Aytya ferina, A. nyroca* and *A. fuligula*).
3. Reptiles: four varieties of viper (except in Sardinia), other colubrides, several peculiar subspecies of lizard – gecko, triton and salamander.

The People

Evidence of human settlements dates back to the Late Palaeolithic (mostly in the Grimaldi Cave, in Liguria, and in Sicily), with evidence of the spread of agriculture along the Italian peninsula from 8200 to 7500 years BP. The first appearance of the true Mediterranean population, that is, the anthropological type with brown hair and eyes, a light brownish skin, short height and thin body, that we find today as typical of inner parts of southern Italy and limited parts of Sicily, joined the Mediterranean Basin from the east in the Early Neolithic. They spread rapidly throughout, bringing the cultivation of wheat and barley, and the breeding of sheep and goats.

Besides the rich tradition of Greek mythology and its adaptations in local myths, written chronicles and historical reports were initiated with Greek invasions in the 8th century BC, in southern and central Italy. Meanwhile, a population which had come from the Middle East in the 11th century BC and settled in central Italy, began the expansion of its domain over the entire peninsula and then over the circum-Mediterranean, Middle East and north European territories – the Romans.

By that time, a different people whose origin is uncertain, the *Italioti*, were living in Italy, basing their economy essentially on agriculture and pasture. Several Italian regions still bear the names of these ancient people, whose cultural heritage was to some extent regulated but never completely eliminated by Rome. These populations are known to us by Roman names: *Sanniti*, *Tusci*, *Lucani*, *Sardi*, *Apuli*, *Brezi*, *Siculi*, *Sicani*, and so on. Romans themselves mixed with local populations of *Etruschi*. Some Greek colonies remained in southern Italy after the complete affirmation of Roman domination. Some of these still remain in the Greek-speaking enclaves of south Calabria and Sicily. In the north, recurrent invasions of people coming from eastern Europe and Asia, the *Galli* (Gauls), resulted in wars with fluctuating fortunes for Rome, which in some circumstances was subject to temporary invasion. Some of these people remained in the Alpine regions, such as the *Galli Orobici* who give their name to the Orobic Alps. Some others arrived in south Italy, such as the Goths in the 5th century AD.

After the decay of the Roman Empire (5th century AD), several populations coming from the Mediterranean Basin (Arabs, Castilians, Catalans, French) or from northern Europe (Normans, Swedes, Austrians), settled as conquerors at different times in different regions of Italy. Recent immigration from North and Central Africa, and from Eastern Europe as well, will in the future increasingly complicate this racial mixing.

The present population in the Italian Mediterranean climate territories is about 40 million, that is, 70% of Italy's total population. More than eight million are concentrated in cities with more than 500 000 inhabitants. Uncertain numbers of immigrants from Africa (mostly from Morocco, Tunisia, Algeria, Ethiopia, Somalia and Nigeria), Eastern Europe (the former Yugoslavia, Albania, and a few from Poland) and Asia (Philippines and Sri Lanka) are estimated at over one million, and are still increasing. Conversely, some 40 million people emigrated from Mediterranean Italy to worldwide destinations (especially to northern Europe, North and South America and Australia) during the second half of the 19th and the first half of the 20th centuries. Some return home after retirement, especially from northern Europe.

The Economy

Resources and industry

Italy is one of the seven most industrialized countries of the world. Its primary (agricultural), secondary (industrial) and tertiary (services, mostly commercial and tourism) systems are strongly advantaged by the imaginative character of commercial operators, but sometimes suffer from a certain incapacity for rigorous organization or from the heavy threat of gangsters, and from the lack of natural resources. Imports of oil are a major cost, followed by iron and other metallic and non-metallic minerals.

Most of the industrial resources of Italy are concentrated in the north, that is, outside the Mediterranean climate area; in the latter, however, the most important industrial areas are those around Rome, Naples, Bari, Taranto, Palermo, Catania and the oil refineries of south Sicily.

Power consumption reflects the degree of development of a country. In Italy, most of the demand is for electrical power. Other forms represent a fraction of one percent of the total power consumption. This is true also for Mediterranean Italy, and notwithstanding the existence of a research agency devoted to so-called "alternative energy sources", little of real importance has been done so far to encourage the use of other sources of energy for heating or other small-power plants.

The different sources of electrical power for Italy as a whole in 1963 and 1993 are shown in Table 3.3 (Corte 1995). In 1994, total demand was 240×10^9 kWh, of which imported electricity from other European countries accounted for 17.7%. Of the total amount, 32.6% is used by factories, 25.6% by transport systems, 33.2% by domestic users, 3% by agriculture, and 5.6% for other uses. Sicily is self-sufficient in its use of electricity, while Calabria exports more than 50% of the power it produces to the industrial settlements of Taranto. Power plants were installed in the 1970s following the great plan of development of the southern regions, but with the oil crisis and the subsequent economic problems, factories were never located in these regions. Structures like these power plants are now known as "cathedrals in the desert".

Table 3.3 *Sources of electricity for Italy in 1963 and 1993* (source: Corte 1995)

Source	% in 1963	% in 1993
Hydro-electric	64	20
Thermal (oil, gas)	31	78
Geothermal	4	2
Nuclear	1 (5% in 1986)	0

Table 3.4 *Current and future water demands in Italy* (source: Corte 1995)

Use	Year	Demand ($m^3 \times 10^6$/yr)	Year	Demand ($m^3 \times 10^6$/yr)
Home	1981	5800	2015	7600
Agriculture	1977	22 859	2000	26 195
Industry	1981	6800	2015	13 300
Power plants	1985	5400	1995	6400

Table 3.5 *Comparison of agricultural productivity in Sicily and the north of Italy (wheat production as a percentage of the total for Italy)* (source: Corte 1995)

	No. of farms	Used land (ha)	Production of wheat (%)
Sicily	404 204	1 598 501	10.4
Emilia-Romagna	150 736	1 232 220	18.6

Water consumption is another parameter for assessing the economic conditions and potentialities of a nation. Table 3.4 shows the demand for water by the main uses at present and, as estimates, in the future. Industry is expected to have the greatest relative increase in demand for water (Bilello 1984).

Industrial production from Mediterranean Italy represents about 30% of total products. It is represented mostly by raw materials and the processing of agricultural and fishing products, and low-tech mechanical products, with a few exceptions such as high-tech electronics factories near Rome, and an aircraft factory near Naples.

Handcrafts are one of the most appreciated of Italy's products. Florentine handcrafts are perhaps the best known, but high-class tailoring and fashion are the most important sources of income for Italian handcrafts. Large-scale fashion production is also well known, but it belongs to the industrial sector rather than to handcrafts. Low-quality handcrafts are suffering from the low-cost products of eastern Asia, so that in the souvenir shops it is easy to find Italian souvenirs made in Hong Kong.

Agriculture and fishing

Of the approximately 22 700 000 ha of agriculturally useful land, 63.8% is located in the Mediterranean regions, and 74.2% of land is actually used for agriculture. However, agricultural technology is less advanced in the Mediterranean regions. A measure of that is given from the comparison of the number of farms in Sicily and Emilia-Romagna, with the area of land in use and the production of wheat as percentages of the nation's total production (Table 3.5).

Major agricultural products of Mediterranean Italy are wheat (5 861 600 t), grapes (7 102 200 t), citrus fruits (3 206 000 t), olives, orchard products, and the derivatives wine and olive oil (Plate 3.3). They were, until the 1960s, the most important economic resource for Mediterranean Italy, but today tourism is taking the lead in income-producing activities. Emigration reduced the farming population at the end of the 1800s and the beginning of the 1900s. Another strong decrease occurred in the 1950s and '60s, when domestic emigration towards the industrial cities of the north began. The recent increase of tourism in southern Italy, and the decrease of employment opportunities in the north, have recently encouraged a return migration towards the Mediterranean lands. These returned people, however, as well as the young ones looking for a job, dislike humble tasks which are left to (mostly illegal) immigrants from Africa and Eastern Europe.

Fishing in Italy produces some 356 000 t/yr. The largest part of this production (about 70% as a gross estimate) comes from fishing companies based in Mediterranean Italy. The Italian seas are not very productive, due to over-fishing and seawater pollution caused by agricultural and industrial waste disposal; thus these companies fish in the Atlantic Ocean or on the North African banks, with agreements with North African countries. This is a source of problems, because sometimes fishing boats stray into unauthorized waters.

Tourism

Tourism is an important source of income, due to the cultural heritage and favourable climatic conditions of Italy. Three of the four major "art cities" of Italy are Mediterranean: Rome,

Plate 3.3 *Intensive agriculture and horticulture west of Florence*

Florence and Naples, not to mention Pisa, Siena, Palermo and the ancient Greek cities in Sicily. With regard to landscape beauty, it is sufficient to refer to Capri, Amalfi, Taormina, the Mount Etna volcano, most of Sardinia and the Aeolian Islands, and the Tuscan countryside. However, most of these resources are being heavily exploited, and in some cases (Calabria and parts of Sicily, for example) the lack of urban planning has caused environmental and landscape devastation of the coastal areas, from which it is now very difficult to recover.

Transportation

Italian transportation systems are limited by the topography of the territory and the activities of certain lobby groups. Topography makes high-speed methods very expensive, as they have to be constructed on bridges and in tunnels for a large portion of their routes (Plate 3.4). It has been said that since both truck and tyre manufacturers are responsible for most of the mechanical industry in Italy, their lobby has no interest in the development of other transport systems. Whatever the reason for the failure to develop an adequate railroad-based transport system, this is a big problem, because it has significant consequences for road traffic safety. In addition, wider and wider roads are needed, noise pollution near cities is severe, and energy consumption is higher than desirable.

Ship transport is limited to tankers for oil, and cargo from overseas. Passenger transport to islands is by ferries. Trains to and from Sicily cross the Messina Straits on ferries. A 50-year-old dream bridge should link Sicily to the continent, but it seems that there may be a long wait.

The number of passengers is relevant: in 1992, nearly 25 million passengers left or reached Italian ports. The great majority is domestic traffic. More than 16 million passengers fly every year in Mediterranean Italy through 26 major Italian airports. Fiumicino (Rome) is the largest one as regards both domestic and international traffic (36.7% of passengers).

There are more than 19 682 km of railways in Italy (57% electrified), of which 62% is in Mediterranean Italy. The road network extends over 303 518 km, of which 58% is in the Mediterranean regions. Highways cover only 6301 km (55% in the Mediterranean territory).

Balance of trade

The import–export balance for Italy in 1993, for the different product categories, is shown in Table 3.6 (Corte 1995). In 1992 the balance was

Plate 3.4 *Road construction in the tectonically active and still naturally vegetated, though landslide-prone mountains of southern Italy poses a challenge which is being met in impressive fashion by civil engineers*

negative. The import–export balance is rather unsteady for Italy, as it is strongly influenced by the monetary market and other very changeable variables. In 1992, the exchange rate of the Italian lira was depreciated by a Government statement. Depreciation of the currency occurred again in 1994/95, this time as a result of speculation on the stock markets. The drawback of depreciation, for Italy, was the negative effect on trade, principally petroleum, which resulted in surging inflation. This situation has now been overcome.

Employment

In January 1995, the rate of unemployment was about 12% in Italy, but it reached 55% for young people in the poorest provinces in the south. Although this seems to be a very difficult situation, it appears that Italians do not want to do hard work, which is being taken by immigrants. In Mediterranean Italy, where new factories are being installed taking advantage of European Union financial assistance, problems in finding employees result from the lack of specialized workers. But fishermen and field workers in Mediterranean Italy are increasingly non-Italians. Besides the refusal of Italians to take these jobs, the problem is enhanced by the illegal employment of immigrants whose labour costs are less than half those of regular workers. This fact sometimes leads to police raids or strikes by Italian workers, but no one seems really interested in solving the question. On the other hand, a large number of unemployed people have one or more unofficial jobs, the so-called *lavoro nero* (black job), and a small but very active minority, which has been estimated at a few hundred thousand throughout the country, works for the *mafia*. Others, finally, make their living day-by-day, finding irregular, short-term or seasonal work. All this unofficial work is part of the so-called "underground economy" which principally sustains people living in large cities like Naples, Rome and Palermo.

Table 3.6 *The balance of imports against exports in Italy for 1993, for the main product categories (source: Corte 1995)*

Category of product	Balance (million US$)
Agriculture, logging, livestock, fishing	–6765
Mineral industry	–14 420
Transformation industry	41 750
(mechanical industry only)	(24 074)
Total	20 565

Administration and Social Conditions

The Corte dei Conti, the national organization which checks the financial administration of public and government agencies, has argued that in 1993 the Ministry for the Environment not only made inadequate use of its budget but was unable to spend all its budget. Unused funds increased by 21% with respect to 1992, reaching about $2300 million. This should be sufficient to indicate that the main problem in Italy is the inadequate bureaucracy.

The recent crisis of policy in Italy reflects the inadequate mentality of the political class, with few exceptions. But much worse, the recently discovered corruption at every level has opened a frightening view of Italy's future. Reasons for widespread corruption are similar to those underlying the spread of the *mafia*. Connections between the political world and the *mafia* are finally being discovered, even although they have always been suspected.

Social programmes in Italy were like a fairy tale, with the nation being close to bankruptcy until a few years ago. Retirement conditions were the best in the world, medical care and assistance were free for everybody, with a few exceptions. Again, it was discovered that the generosity of medical assistance enabled undue financial advantage to be obtained for a few people and nearly all political parties. This was also the case for highway network programmes, some branches of which were financed but never constructed. Now, medical assistance is minimal. Retired persons' salaries will be computed on the basis of much more strict criteria. Taxes increase every year.

The situation in private firms and schools is much better, but it cannot compensate for the damage caused by the failure of the national system. However, the situation is steadily improving now.

Education

The education system is far from perfect. Elementary school is fair, but study programmes for the intermediate and high-school degrees are obsolete. At university, the percentage of students who abandon their courses before graduation is near 70%. The University Diploma (a degree intermediate between bachelor and graduation) was introduced only five years ago, and PhD courses about 10 years ago, while postgraduate specialization courses have always been active for medicine. Industry and tertiary firms require a large number of engineers and economics specialists who are not available on the labour market. Young professionals do not apply for such positions because they wish to be freelance consultants rather than employees.

Only 25% of candidates pass the exam for professional geologists. The reason is that university still educates science-orientated rather than profession-orientated graduates, missing the requirements of the work market and creating a category of frustrated, highly educated but unemployed people.

The student population is decreasing because of the reduction in birth rates over recent decades. Elementary and low-intermediate school is compulsory; 62.9% continue to high school, and there is a high mobility between regions for university and post-doctoral courses, so the provenance of students at those levels is unknown. The total population of students in 1993 was 1 074 330 and there were 54 750 teachers, with a 1:19.6 teacher:student ratio, which is too low. The worst teacher:student ratios are for law (1:84.1), economics (1:46.5) and policy (1:46.4). The best is for agriculture (1:9.5).

Culture and Recreation

Culture was, for Italians, a matter for rich people until World War II. Only recently has "mass culture" spread throughout the population, and cultural activity has increased enormously in terms of the performing arts. In practically all cities there is a "traditional theatre" and at least one musical association. At present, 56% of the 843 Italian theatres are in the Mediterranean zone, with opera theatres being the most important. "Serious" shows such as plays are also performed here, but seldom musicals. Concert halls are less common. A great deal of classical music is performed in any place at any time, with a large number of young people in the audience, but still disco dancing is the favourite Saturday night activity of young people.

Italian TV programmes still preserve some cultural content: science documentaries, news reports, opera, a few plays and tragedies, and good quality movies can be found among lots of trash programmes and advertisements.

Soccer is the most popular sport in Italy, and it is played by millions of people, 99% male, throughout the country, and at every level. Other sports have varying numbers of supporters, depending on the presence of good performers, and include bicycle races, basketball, athletics, car racing (the Ferrari!), motorcycling, skiing, horse races, swimming, rugby, regatta, boat races and gymnastics.

With regard to the publishing world, the most ancient Italian publishing house is the Accademia dei Lincei, founded in Rome in 1603. At present, however, the major publishers are in northern Italy, although Rome and Florence are sites of important printing activities.

Several reliable sources state that 50 to 70% of the world's cultural objects (depending on what is included) are in Italy. Only about 30% of these are shown in museums and archaeological sites. Actually, very little is done to preserve this huge treasure of mankind from decay. Only recently, due to the sponsorship of some famous firms, several important monuments and masterpieces (one example for all: the Sistina Chapel fresco paintings by Michelangelo, in Rome) are being restored to their original splendour. But it is too little, so far. UNESCO has included among the 411 sites of the World Patrimony, only seven Italian sites, four of which are in Mediterranean Italy (downtown Florence, Piazza del Duomo in Pisa, downtown Gimignano (Tuscany), the Sassi (an amazing cave city) of Matera, in Basilicata). In November 1997, ten more Italian sites were included in the World Patrimony. Six of them are in Mediterranean Italy.

4
The Croatian Adriatic coast

Jela Bilandžija, Matija Franković and Dražen Kaučić

Introduction

Croatia's Mediterranean environment occurs in a long zone between the Adriatic Sea and the inland region. In the northeast it is bordered by a belt of mountains and in the southwest by the Adriatic Sea. The region extends from the Slovenian border over the peninsula of Istria, to the northern Adriatic coast and the islands, Dalmatia and its islands, and the eastern border with Montenegro. The area characterized by a Mediterranean-type climate coincides with the Adriatic karst region (Figure 4.1).

The very indented coast (the most indented in the European Mediterranean region and, after the Norwegian coastline, the second in the world), a precious cultural heritage, relief diversity and the clean sea, make the Croatian Mediterranean area a particularly attractive tourist destination and an important economic resource for Croatia. The total length of the coast is 5790 km, of which 4012 km belong to the islands and 1778 km to the continent. There are 718 islands in Croatia, of which 66 are permanently inhabited, and 467 cliffs and crests (Statistical Year Book 1993).

Traditional economic activities of the region are agriculture (mainly vegetable and fruit growing) and fishing. The well known vineyards are situated in Istria and in Dalmatia (especially in its southern parts). Tourist development has intensified in recent decades. Its main basis is the variety of attractive beaches, especially during summer, and a climate which is convenient for outdoor recreation all year round.

The Land

Relief

There are few countries where the depth of calciferous rock layers reaches more than 1000 m, as in the case of the Dinaric karst, and this has resulted in the richness and diversity of surface and underground karst phenomena and forms (scarps, funnel-shaped holes, inlets, karst fields and caves). The main geological foundations of the region are sedimentary Mesozoic calcareous rocks (Roglić 1975).

The region's landforms, characterized by coastal indentations, coastal slopes, peninsulas and islands mainly of limestone origin with layers of dolomite and impermeable flysch, are known worldwide as the "Dalmatian type". The landforms have resulted from the sinking of the folded zone of the Adriatic littoral; the lower synclinal parts were transformed into bays and creeks and the anticlines into islands and peninsulas (Jovanović 1986).

Drainage

The region's drainage is distinguished by a lack of surface water during most of the year and by very specific and complex underground water regimes. Surface waters appear occasionally during the rainy season. On the karst, only a few surface rivers reach the sea: the Mirna in Istria (53 km), the Cetina (105 km), Krka (72.5 km), Zrmanja (69 km) and the Neretva which flows through Dalmatia in its lower reaches (20 km). These rivers are the main source of drinking water for the local population. The purity of the rivers

Land Degradation in Mediterranean Environments of the World: Nature and Extent, Causes and Solutions.
Edited by A.J. Conacher and M. Sala.

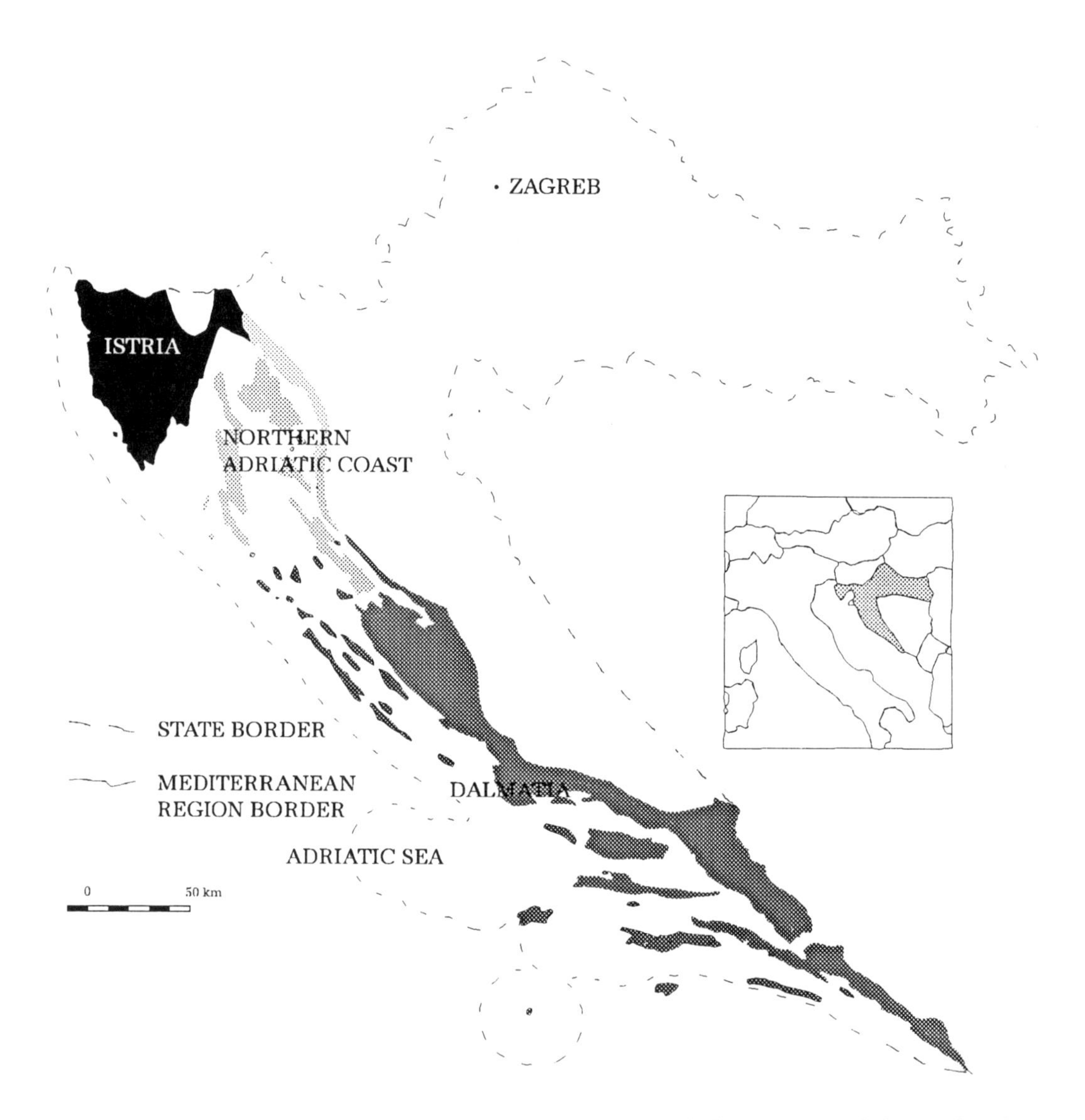

Figure 4.1 *The zone of Mediterranean-type climate in Croatia, including regions and places referred to in the text* (compiled from *Statistical Year Book* (1993) and meteorological reports for the period 1973–1992; State Meteorological and Hydrological Service)

Cetina, Krka and Zrmanja is still acceptable for human use as well as for breeding sensitive fish species. These rivers are much more polluted downstream and near the river mouths (State Directorate for Environment 1994). The Neretva river is used to irrigate fields near Opuzen and Metković, where excellent vegetable and fruit crops are grown. Occasional surface waters and wetlands are important as supplies of drinking water for cattle and wildlife.

A major problem concerns untreated industrial wastes and sewage which reach the groundwaters due to the permeable geological formations, thereby polluting the sources of drinking water. However, the greatest difficulties with drinking water supplies occur on the islands. Only a few of them (Krk, Brač and Cres) have water supply systems. People living on the other islands use traditional rain collectors.

Soils

Most soils in the region are old. They are classified in the humus-accumulative soil group (A(B)C profile form) and cover 16 400 km². The common characteristic of all soils is that they almost never occur in a continuous spatial pattern of uniform depth, owing to the specific relief forms (Figure 4.2.). This is one of the reasons why an intensive agriculture is not possible. The other limiting factor is insufficient water supply.

The main soil type in Istria is Terra Rossa, with Istria being the only region where this type of soil occurs in a relatively continuous pattern. It originates from the continuous weathering of the Triassic, Jurassic and Cretaceous limestones and dolomites. It is a non-carbonate type of soil with a mostly neutral pH reaction. The humus content in the littoral zone of cultivated and eroded red soils is 1–2%, but in the areas covered by native vegetation, it can reach 4% (Škorić 1977). Because of its high degree of erodibility, the humus-accumulative layer of red soil frequently has been removed by erosion.

The other main soil type is a Calco-cambisol. It covers the limestone and dolomite mountains, often in succession with Terra Rossas. Calco-cambisols and Terra Rossas differ in colour and in the properties of the humus-accumulative layer (the content of humus in Calco-cambisols under native vegetation is 8–10%; Škorić 1977; Ćirić 1984). By its characteristics it is analogous to Terra Rossas because they both belong to the same stage of development. Intensive cultivation of Calco-cambisol areas is not possible due to the altitude at which they occur (up to 1700 m) and

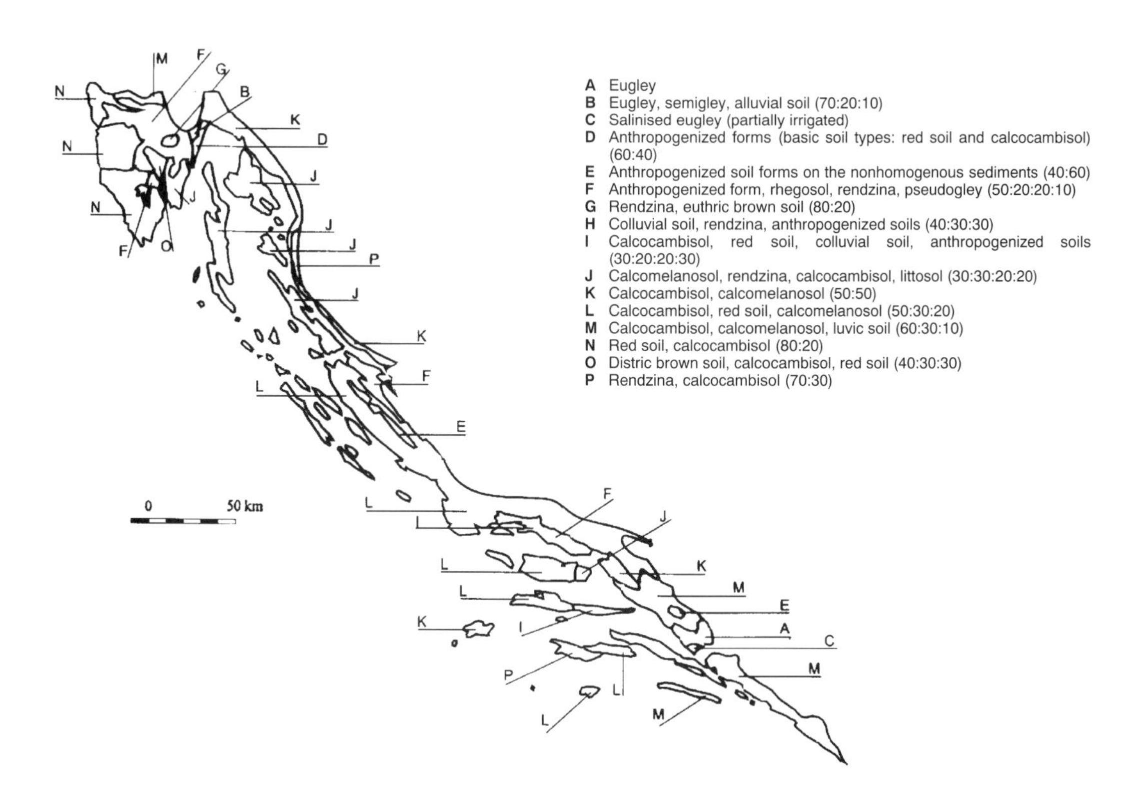

Figure 4.2 *Soil types and their spatial pattern in the zone of Croatia's Mediterranean-type climate* (source: J. Martinović and A. Valanović 1990)

their spatial discontinuity. The largest areas of Calco-cambisol are in Dalmatia.

The main soil type on dolomite is a Rendzina (AC profile), which contains more than 10% of carbonates ($CaCO_3$); but its proportion of the total soil area is not high. Rendzinas originate on strewn and erodible carbonate parent rock and have alkaline properties (pH 7–8; Škorić 1977). They are well supplied by nutrients. Rendzinas on convenient relief forms and at lower altitudes are covered mostly by native forest vegetation.

Besides these main soil types, fragmentary hydromorphic soils can be found (alluvial, eugley, pseudogley), as well as Luvisols and Rankers. These soils occupy 3363 km^2.

Climate

The mountain belt of Dinaridi forms a climatic partition between the karst region and the inland region (Figure 4.3). The main factors responsible for the climate of the karst region can be divided into two groups: those which influence climate generally, and those which cause contrasts in the climates of the coastal and inland parts of the karst region. Included in the first group of factors are the Azores anticyclone during summer and winter, and the Siberian anticyclone during winter. The differences between continental and sea temperatures constitute the second group of climate factors. Owing to the relief forms, the boundaries between climate subtypes should be considered as broad, transitory regions (Seletković and Katušin 1994). During the cold season (October–March), huge air masses warmed by the sea meet and interact with cold air masses formed in the area of the winter continental anticyclone above the Dinara mountain chain. Cold air reaches the coast and is soon warmed, while its upper parts remain cold. This stratification is unstable and frequently causes winter thunderstorms.

The highest mean annual rainfalls occur in the northern Adriatic zone (Rijeka), exceeding 1500 mm, whereas in the southernmost parts of Dalmatia (the island of Lastovo), annual rainfall never exceeds 700 mm. July is the hottest month, as is true for the whole Mediterranean region. Maximum summer temperatures range from 31.8°C (in Zadar) to 39.2°C (Šibenik). The mean air temperature in July ranges from 23.0 to 24.9°C except in Pazin (20.5°C, but Pazin is located in the zone of climate transition). January is the coldest month. Mean air temperatures in January never fall below 5°C, except in Pazin (2.6°C). Mean monthly air temperatures increase from Istria to Dalmatia.

Winter frosts are frequent in Istria and occasional in Dalmatia. Snow is rare, especially in the southern part of the region.

The characteristic cold season winds are a severe, northeastern wind called *bura* and a southern wind called *jugo*. A westerly wind called *maestral* is the most frequent during the hot season, and it moderates the summer heat.

Plant and animal life

The geographic position of Croatia is very distinctive within central Europe. It includes Alpine, Dinaric, Mediterranean and Pannonian geographic, climatic and other influences. This complex combination of environmental conditions has led to a great variety of flora and fauna species (Horvatić 1967; Horvat 1962). The best example of biotope, species and genetic richness is the Adriatic karst region of the Croatian Mediterranean territory, with the Croatian Dinaric region being defined as the *locus typicus* of this famous natural heritage. Because of many endemic and relict plants, arthropods, freshwater fishes and lizards, it is one of the richest endemic centres of European flora and fauna (Jalžić and Pretner 1977; Tvrtković, pers. comm.).

The southern slopes of Velebit mountain, together with the islands of Prvić and Goli Otok, contain 47 steno-endemic species. Biokovo mountain contains 17, Konavli 10, and the island of Vis with the surrounding isles 16 endemic plant species (Borzan *et al.* 1994).

The main natural vegetation types comprise steno- and eu-evergreen and sub-Mediterranean broad-leaved vegetation. The steno-Mediterranean zone is characterized by the dominance of olives (*Olea silvestris*), mastic tree (*Pistacia lentiscus*), carob (*Ceratonia siliqua*), *Euphorbia dendroides*, myrtle (*Myrtus communis*), *Juniperus phenicea* and Aleppo pine (*Pinus halepensis*) (Plate 4.1). The eu-Mediterranean zone is characterized by the dominance of holly oak (*Quercus*

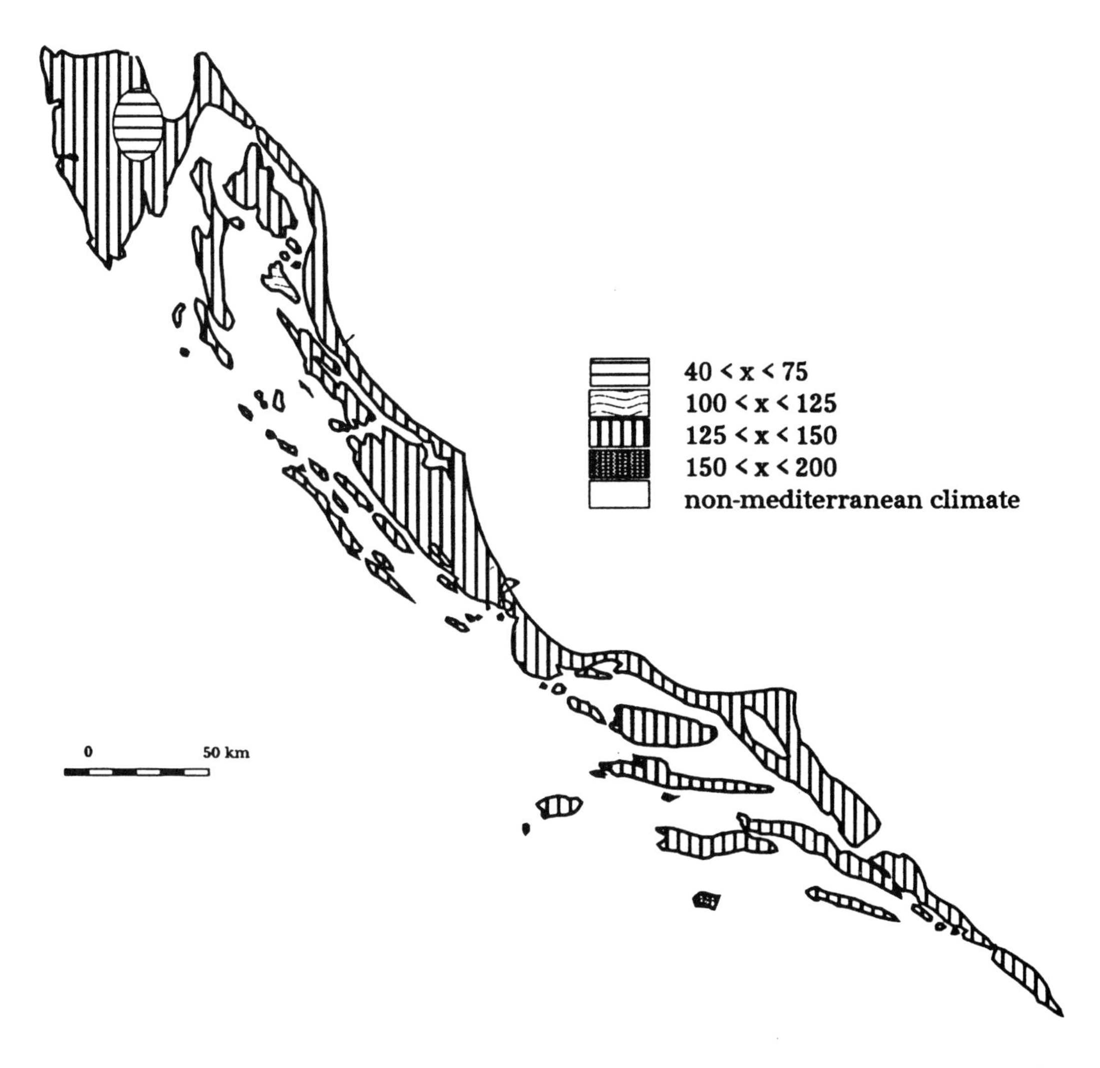

Figure 4.3 *Mediterranean types of climate after Xm values (UNESCO–FAO 1963)* (compiled from meteorological reports for the period 1973–1992; State Meteorological and Hydrological Service)

ilex), kermes oak (*Q. coccifera*), flowering ash (*Fraxinus ornus*), mastic tree (*Pistacia lentiscus*), interlaced honeysuckle (*Lonicera implexa*), strawberry tree (*Arbutus unedo*) and knee-holly (*Ruscus aculeatus*). The sub-Mediterranean zone is characterized by a dominance of pubescent oak (*Quercus pubescens*), *Q. virgiliana*, oriental hornbeam (*Carpinus orientalis*), hop hornbeam (*Ostrya carpinifolia*), black pine (*Pinus nigra*), red-berried juniper (*Junperus oxicedrus*), *Coronilla emeroides* and *Sesleria autumnalis* (Horvatić 1958; Marković-Gospodarić 1966; Gaži-Baskova and Bedalov 1983; Trinajstić 1986). The original properties of the vegetation have been preserved only in inaccessible regions and protected areas (two strict nature reserves, five national parks and three nature reserves). In accessible areas, the native forest vegetation has been changed mainly to anthropogenically degraded forest forms such as *maquis*, brushwood, *garrigue*, grassland and man-made pine forests. In some areas, the degradation has culminated in barren, stony surfaces.

Unfortunately, there are almost no data on the diversity of taxonomically lower plants (lichens and mosses) in Croatia. Exact numbers of taxa of

Plate 4.1 *A remnant of the originally rich vegetation of the Croatian region. The oldest olive tree in Croatia is in Kaštel Štafilić in Dalmatia, with an estimated age of more than 1500 years. Photo by S. Kovač*

ferns and taxonomically higher plants are not well known either. There is an estimate of 3600–3800 species (around 4500 taxa, including subspecies, varieties and forms) of higher plants in Croatia.

The real status of insects in the Mediterranean region in Croatia is difficult to evaluate at the moment (except for butterflies and some particular small groups and/or species, such as dragonflies), because the basic research on them is lacking (Jakšić 1988; Novak 1970; St Quentin 1944).

Freshwater fish fauna of the Mediterranean region in Croatia are very important and interesting. From 50 autochthonous freshwater fish species (that number is probably not the final one), five are considered extinct: *Acipenser nacarii, A. sturio, Salmo dentex, Leuciscus turskyi* and *L. ukliva*. Some of them, such as *Salmo marmoratus, Salmothymus obtusirostris* and 10 other species, are extremely endangered as well. The exact status of most freshwater fish species is unknown (Mrakovčić *et al.* 1995).

There are 19 species of amphibians known from Croatian territory, one of them being extremely important in the Dinaric cave system – *Proteus anguineus* (Tvrtković, pers. comm.).

The state of data on reptiles is very similar to that for amphibians. There are 34 species which are known on Croatian territory. Some of them are very endangered, such as three species of marine (*Caretta caretta, Chelonia mydas* and *Dermochelis coriacea*) and other reptiles (*Hemidactilus turcicus, Tarentola mauritanica, Natrix tessellata* and *Vipera berus, V. ursinii*), all requiring urgent protective measures (Tvrtković, pers. comm.).

As a group, birds are almost completely protected (except five species – *Corvus corone cornix, C. frugilegus, C. monedula, Pica pica* and *Garrulus glandarius*). In fact the sparrow (*Passer domesticus*), probably the most numerous bird species in Croatia which is not endangered at all, is protected equally by law with, for example, *Hieraetus fasciatus*, one of the most endangered eagles of which less then five pairs remain (Floriani 1991).

There are 103 mammal species in Croatia (including introduced species), of which 43 are endangered. Some of them, such as the Mediterranean monk seal (*Monachus monachus* – one of the world's 10 most endangered mammals) and otter (*Lutra lutra* – which lives in the delta of the Neretva river) are on the verge of extinction and there is no knowledge of their present status (Anon 1994).

The People

The Croats inhabited the territory of Illyricum during great migrations in the 6th and 7th centuries (Sabadi 1994). Illyricum was bordered by today's Vienna and Budapest to the north, Greece to the south and the Morava and Vardar rivers (Macedonia) to the east (Jakić 1993). From that time they lived either in an independent state (between the 7th and 12th centuries, in 1941–1945, and since 1990) or annexed to other countries (the Austrian empire, Austrian–Hungarian monarchy, the kingdom of Serbs, Croats and Slovenians and two former Yugoslavias). The conquering raids influenced the national composition in the region (Croats 87%, Italians 7% and other national minorities). The number of inhabitants in the region is 1 360 000 (28% of the total population of Croatia; Statistical Year Book 1993).

Most of the population (86%) declare themselves as Roman Catholics; the rest are Muslims, Greek Catholics, Orthodox and atheists (Statistical Year Book 1993). At the beginning of the 20th century, the region experienced great overseas emigrations, and one-third of the Croat population now lives outside Croatia. After World War II there was a massive shift of population from villages into towns. Before the war, only 28% of the population lived in towns (Korenčić 1979), whilst the comparable figure today is 68%. Village populations are characterized by declining numbers and increasing age. Declining birth rates are a serious demographic problem (Statistical Year Book 1993).

The Economy

Tourism has become the main economic activity in the region, and directly employs about 8% of the workforce. Tourism is responsible for 6.5% of the gross national product (GNP), accounting for 30% of Croatia's total exports (State Directorate for Environment 1994). Agricultural production, especially vegetable and fruit growing, is extensive, and contributes significantly to the income of families on small-holdings. Agriculture (including fisheries) from the Mediterranean region contributes 2% of Croatia's GNP. Olive growing and wine production have intensified in recent years.

There are 10 factories manufacturing fish products. The factories are socially significant in the region because they employ the local population and thus prevent them leaving.

Resources and industry

Energy production in Croatia is based mainly on fossil fuels (86.6%) such as liquid fuels, gas and coal (State Directorate for Environment 1994). A considerable amount of fossil fuels is imported. Most of the oil-bearing fields are located in other Croatian regions. There are some coal mines in the Mediterranean region, but the coal structure is very unfavourable. There are large quantities of lignite with low caloric value and of coal with higher caloric value, but also containing a high percentage of sulphur. This coal is used only by a thermo-electric power plant in Plomin in Istria. During the current transition to a market-based economy, it is planned to close most of the mines.

Industrial production in the Mediterranean region is located within or adjacent to towns, mostly in the territory of Rijeka, which has 15% of Croatia's industrial enterprises, and in Pula and Split (State Directorate for Environment 1994). Shipbuilding is the main industry in the region. All industrial activities influence the environment significantly, despite protective measures which apparently have not been sufficient. For this reason the coking coal plant in Bakar was closed in 1994.

Regional electricity requirements are met by national thermo- and hydro-electric power plants and by importing certain amounts of electricity from the nuclear power plant in Krško (Slovenia).

Agriculture, forestry and fishing

Agricultural activities in the region are related mainly to production from small family holdings. The resulting income is not sufficient to cover expenses, and additional income is obtained from non-agricultural activities. Most rural communities have dual economies.

The most common means of production are from non-irrigated fields, and are restricted to an annual crop. Intensive agricultural production (mainly of citrus fruits and vegetables) on irrigated fields is organized in the valley of the Neretva river. Recently, there has been increasing interest in vineyard revitalization and olive growing.

Forests and forest land of the Mediterranean region cover 972 655 ha (39.56% of the total forest area in Croatia), but only 57 066 ha are occupied by highly productive forests, mainly pines (*Pinus halepensis, P. nigra* and *P. maritima*). The native forest vegetation, in different stages of degradation, covers 518 005 ha; the remaining 397 584 ha consist of barren, stony surfaces (Topić in press). The ecological functions of forests are much more valuable in the region than wood production. Forest vegetation serves the local population as a source of by-products – seeds, leaves, fodder, fuelwood and roundwood – and by protecting waters and soils. On the other hand, wood production provides very low or no economic benefit. Pinewood, not highly valued, was sold in Bosnian markets but the trade has stopped as a result of the war.

Croatia's Mediterranean forest vegetation, although indigenous in origin, suffers from air and water pollution. Between 7.5 and 10% of all trees experience visible crown damage (Prpić *et al.* 1991). Fires pose the greatest threat to forest vegetation. The average area burnt each year in the region (9440 ha in the period 1986–1990) is almost twice the area planned for annual reforestation in Croatia.

Fishing in the region, except in Lake Vrana in Dalmatia, and the delta of the Neretva river, is mainly related to the sea. There are 422 fish species in Croatian sea territory, of which about 100 are used as human food, and 30 are caught as a permanent economic resource (Homen, pers. comm.). The average catch of 26 470 t/yr consists mostly of sardines (60%), with tuna comprising 2.6% and other fish species even smaller proportions (Statistical Year Book of Croatian Counties 1993). Some estimates indicate that the sea fishing possibilities are not used optimally, and that the annual sardine catch could be doubled or even trebled, as well as the catch of some other species. In Croatia, average annual fish consumption is 9 kg per capita. The fish factories' production capacity is 30 000 tonnes annually (State Directorate for Environment 1994).

Recently commenced aquacultural projects of fish, shell-fish and crab breeding are very important in the process of regional revitalization which is especially focused on the islands.

Transportation

The shape of Croatia causes difficulties in establishing a network of railways and roads which would connect more effectively the Mediterranean and the inland regions. Ships and road vehicles are mostly used in the region. Krk, Pag, Murter and Čiovo are the only islands which are connected with the continent by bridges; the rest can be reached only by ships. Pula, Krk, Mali Lošinj, Zadar, Brač and Dubrovnik have small but well equipped airports for national and international airlines. Railway transport (except for Rijeka) was completely stopped in 1990 due to the war and reactivated in 1995. The number of roads and their technical characteristics are not adequate for the transport requirements either in the region or in Croatia as a whole (48.4 km per 100 km^2; Statistical Year Book of Croatian Counties 1993).

Administrative and Social Conditions

Government

The Republic of Croatia was recognized in 1992. It was re-established after half a century by democratic elections which elected representatives of several parties which constitute today's Parliament. Local authorities function through municipal, town and county councils. Rights of national minorities are secured by participating in county councils, proportionally according to their share of the county's population. The Parliament consists of two parts, the Chamber of Counties and the Chamber of Representatives. The Government and its Prime Minister represent the executive authority. The system of justice is being established according to the models of traditionally democratic countries. It functions independently of state authorities.

Education

Primary school is obligatory in Croatia for all children of 7–15 years old. Secondary schooling is the subject of choice of about 45% of pupils in the region (the state average is 42%). This is an improvement on 1969, when only 36.8% of pupils continued secondary schooling (Statistical Year Book of SFR Yugoslavia 1970; Statistical Year Book, 1993). For the time being, primary and secondary education in the region is possible only in public schools. Secondary schools are located in towns and this causes financial obstacles for many children from the rural areas. There is a scholarship system but it is insufficient to cover the expenses of non-resident pupils.

Graduate and post-graduate education can be obtained at the public universities (in Rijeka and Split) and faculties (in Opatija, Zadar and Dubrovnik) in the region. There are two universities outside the region (in Osijek and Zagreb). The University of Zagreb was established in 1669 and has the highest reputation. Of the population living in Croatia's Mediterranean region, 8.6% has tertiary education compared with the state average of 7.5% (Statistical Year Book of Croatian Counties 1993).

Health and welfare

The Ministry of Health, together with local governments, is establishing a new system of health and welfare insurance. Basic medical treatment in public hospitals is free of charge for certain population categories, whose income is below a minimal amount. Other people contribute to the costs of medical treatments. The basis of private medical institutions has been developed since 1990, whilst the state Medicare programme is financed by income taxes.

Pensions and subsidies are provided by state funds and are guaranteed by Government to people older than 65 years. Loss of income due to work-related accidents or diseases is compensated by the employing company in the case of 1–30 days of absence from work. In the case of more than 30 days of absence, the loss is compensated by state funds.

The number of governmental or non-governmental organizations for helping people with special needs is insufficient. The voluntary Catholic organization CARITAS is the best organized one. It was indispensable in supporting several hundred thousand refugees during Serbian aggression.

Culture and Recreation

The cultural heritage consists of the achievements of different civilizations which have inhabited the region. There are several historical urban entities of protected architectural heritage, such as Dubrovnik and Split (both on the list of World Natural and Cultural Heritage), archaeological areas such as Brijuni, Issa, Narona and Nesactium, and cathedrals and palaces.

This region is characterized by the diversity and dynamics of its cultural activities. The annual theatre open-air festival *Dubrovačke ljetne igre* in Dubrovnik attracts many artists from all over the world. Some of them consider that Dubrovnik has the best stage for Shakespeare's *Hamlet*. The Children's Festival in Šibenik and the Film Festival in Pula contribute a great deal to the public's cultural development.

Croatian theatre has a long tradition. One of the first public theatres in Europe was opened on the island of Hvar in 1612 (Batušić 1978). Theatres in Rijeka, Split and Dubrovnik perform classical and modern repertoires of national and world playwrights.

In most of the towns, artistic achievements can be seen in museums (51), several hundred galleries and private collections (Statistical Year Book of Croatian Counties 1993). Artistic workshops are organized every year, with a scale of topics and professional levels, for graduate artists as well as for amateurs.

Public information is provided by the state radio and TV network and by a few private channels, and there is a wide range of state and private newspapers, magazines and periodicals.

The most popular sports in the region are soccer and basketball. The most famous Croatian basketball players come from Dalmatia (for example Dražen Petrović and Toni Kukoč). Sailing (the ACY match-race cup in Rovinj) and water polo are also popular.

5
Greece

C.S. Kosmas, N.G. Danalatos and A. Mizara

Introduction

Greece is a rugged, mountainous country with large variations in altitude within relatively short distances. Owing to its predominantly steep terrain and adverse climatic and bioclimatic conditions, the country is facing considerable soil erosion problems. However, on undisturbed soil surfaces, soil formation has proceeded at rates faster than geological erosion, so that untruncated soil profiles classified in a variety of soil orders are found even on steep slopes where the natural vegetative cover remains intact. Thus, the natural environmental factors, adverse as they are, would not have caused the present extensive land degradation in Greece without the synergetic action of people. Important recent work on land degradation and management is included in Bonazountas and Katsaiti (1995), Ellenic Soil Science Society (1996) and GEOTE (1996).

The country is located between the latitudes 34° 48' 02" and 41° 44' 58" north and longitudes 19° 22' 41" and 29° 38' 27" east. Its borders are defined to the north by Albania, the former Yugoslavia and Bulgaria, to the northeast by Turkey and to the east, south and west by the Aegean and Ionian seas. It includes over 2000 islands, 75 of which are inhabited, and covers a total area of 131 191 km^2. The maximum length from north (Rodopi) to south (Crete) is 727 km, while the maximum width is 570 km. The capital of Greece is Athens with other major cities being Pireus and Thessaloniki.

The Land

Relief

The greatest part of Greece is mountainous with steep slopes (Figure 5.1). About 49% of the surface area has slopes greater than 10% (Table 5.1), with only 36% comprising lowlands with slopes less than 5%. Areas at elevations above 800 m above sea level occupy 28.6% of the country. The highest elevation is 2917 m, on Mount Olympus.

Greece forms part of Alpine Europe. It can be divided into several physiographic zones based on lithology, folding and homogeneity of phases, and tectonism. The zones correspond to a succession of submarine depressions and ridges, originally formed within the Alpine geosyncline, and with a dominant NNW–SSE orientation. The main geological zones in Greece are, from west to east: the Apulian zone, the pre-Apulian zone, the Ionian zone, the zone of Gavrovo–Tripolis, the zone of Olonos–Pindos, the Pelagonian zone, the zone of Vardar, the Servo–Macedonian zone and the zone of Rhodopi. These zones have greatly affected soil formation and vegetation cover.

Drainage

The main rivers of Greece originate from mountains located in Bulgaria and the former Yugoslavia and debouch into the Aegean Sea. Examples include the Evros (550 km long),

Land Degradation in Mediterranean Environments of the World: Nature and Extent, Causes and Solutions.
Edited by A.J. Conacher and M. Sala.

Figure 5.1 *Simplified physiographic map of Greece*

Table 5.1 *Distribution of land by categories of slope and altitude* (source: Yassoglou and Kosmas 1988)

Slope (%)	Total land (%)	Altitude (m a.s.l.)	Total land (%)
<5	35.8	<400	51.3
5–10	15.3	400–800	20.1
>10	48.9	>800	28.6

originating from Aemos mountain, the Nestos (234 km), originating from Rodopi and Orvilos mountains, the Strimon (115 km) and the Aliakmon rivers (285 km), originating from Grammos mountain.

Major rivers originating within Greece include the Pinios (185 km), with its source in Erymanthus mountain, and the Eurotas (80 km), originating from Taygetus mountain. A number of rivers originate from Pindos mountain and debouch into the Ionian Sea, such as the Kalamas (99 km), Louros (75 km), Arachthos (120 km) and the Acheloos rivers (215 km).

The main lakes are Pamvotis in Gianenna (with a maximum depth of 23 m), Trichonis in Akarnania (97 m), Iliki in Viotia (22 m), Prespa (with a maximum depth of 37 m in the Greek part), Kastoria (30 m, all in western Macedonia), Doirani (Greek part 22 m) and Tachinou in eastern and central Macedonia. There are three major lagoons, namely Messologiou, Vistonis and Agoulinitsa.

Groundwaters are mainly present beneath alluvial plains located along the coast or in the valleys. This water is intensively exploited for irrigation of summer crops, with adverse consequences on soils due to intrusions of seawater.

Soils

Because of the steep slopes, the soils on mountains are extremely eroded, shallow and poor (Lithosols; FAO–UNESCO 1989), and useless for any agricultural activity. On the uplands, the steep slopes combined with destruction of natural vegetation (by fire, cultivation and overgrazing) have caused soil erosion, resulting in Cambisols, Luvisols and Regosols (FAO–UNESCO 1989), often to the extent that parent rock is exposed on the surface (Lithosols). In view of their low fertility and productivity it is questionable whether these soils should be used for pasture or for forests.

Lowland soils are more productive, and can be subdivided into three main groups. The first group includes soils formed after reclaiming former lakes such as Gianitsa, Tenagi Filipon, Kopais and Karla. These are the most fertile soils, well structured and rich in organic matter (mainly Mollisols and Vertisols; FAO–UNESCO 1989), but they cover only a small part of the lowlands. The second group includes cultivated soils of moderate to high fertility; they are temporarily flooded and are characterized by shallow groundwater tables and high organic matter content (hydromorphic counterparts of Fluvisols, Cambisols, Luvisols and Vertisols). The third group includes lowland soils with low organic matter content and with moderate to poor fertility. Therefore their fertility is mostly determined by their texture, the type of clay, and soil thickness. About 150 000 ha in the lowlands contain appreciable amounts of soluble salts to such a degree that they need reclamation before use.

Besides their unfavourable topography, Greek soils are characterized by low organic matter content. About two-thirds of the cultivated soils contain only 1% organic matter, whereas less than 14% of the soils contain 3% or more organic matter. Reduction of organic matter content causes structural degradation and soil erosion as well as lack of nitrogen, which characterize 87% of the cultivated soils (Fassoulas and Fotiades 1966). Many soils in Greece, both on uplands and in the lowlands, are formed in calcareous deposits and are rich in $CaCO_3$, whereas soils with a positive reaction to Ca-fertilization are less abundant (Figure 5.2). About 70% of the soils have an alkaline or very alkaline reaction, 12% neutral and 18% acid (Fassoulas and Fotiades 1966). Fixation of P as well as Zn, B and other elements is common in the alkaline soils.

Based on soil, climate and topography, land of high potential quality constitutes 19% or 24 919 km^2 of the total land surface, with 18% or 23 394 km^2 being of moderate quality and 57% or 75 775 km^2 of low potential quality (CORINE 1990). Much of the low quality land is used for traditional, low capital intensity

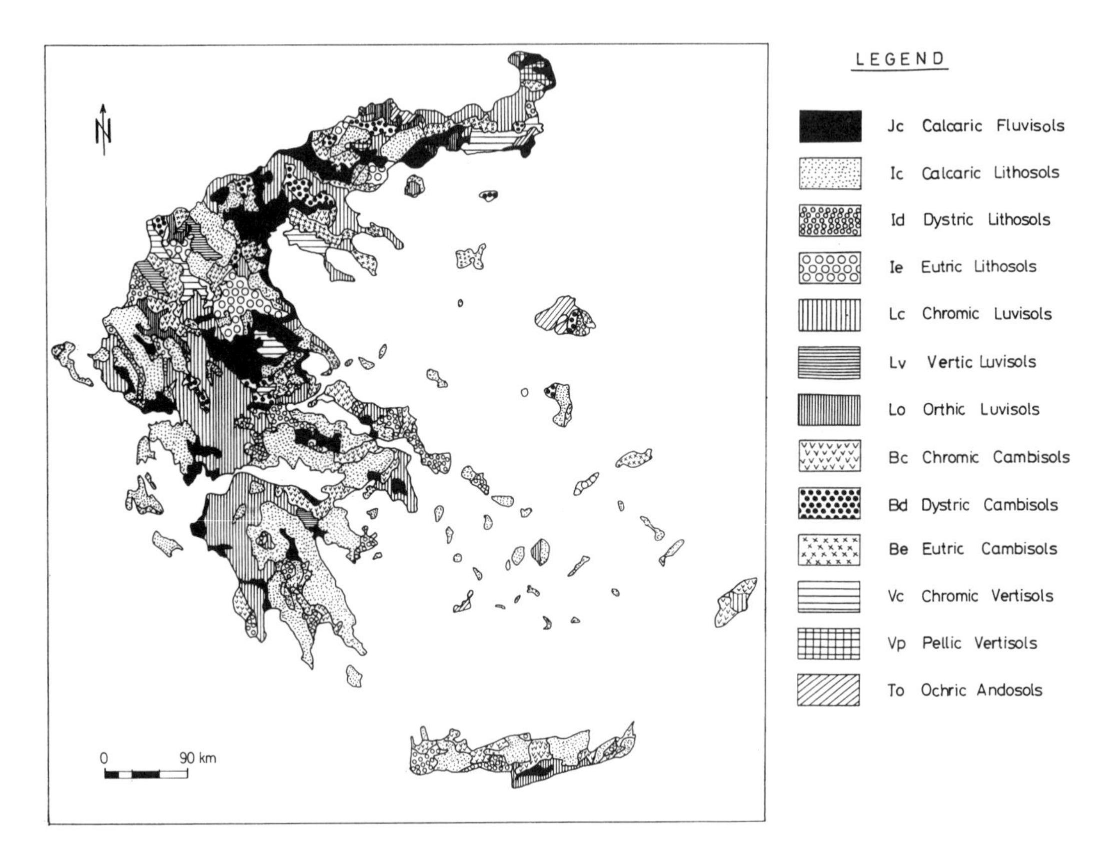

Figure 5.2 *Soil map of Greece* (source: Soil Geographical Database of Europe at scale of 1:1 000 000, version 3.1B, 1995, Commission of the European Communities)

farming systems which are important in maintaining the characteristic Mediterranean landscapes.

Climate

The climate of Greece belongs to the Mediterranean type, according to which most rains fall in the cold period of October–March whereas the summer months of July and August are almost without any precipitation. A more detailed investigation of the country's climate, however, shows some significant climatic variations over short distances. Generally, the following zones may be distinguished (Figures 5.3 and 5.4):

- *the mountainous zone*, having the characteristics of an Alpine climate. This zone includes the Pindos mountain chain which, running in a NNW–SSE direction, separates the country into two parts with different climatic characteristics, especially in rainfall;
- *the continental zone* of northern Greece including the mainland of Epirus, Macedonia and Thrace, and most of Thessaly, having a climate changing gradually from pure Mediterranean to the colder climates of central Europe;
- *the marine Mediterranean zone* of Iona, including the coastal regions of western Greece and the Ionian islands;
- *the Mediterranean mainland zone*, including the southeastern part of Greece (Aegean) up to Thessaly and the Aegean islands. The climate of this region is similar to the marine Mediterranean, but with lower winter temperatures and longer summer droughts.

Such climatological differences are due to the complex vertical and horizontal distribution of

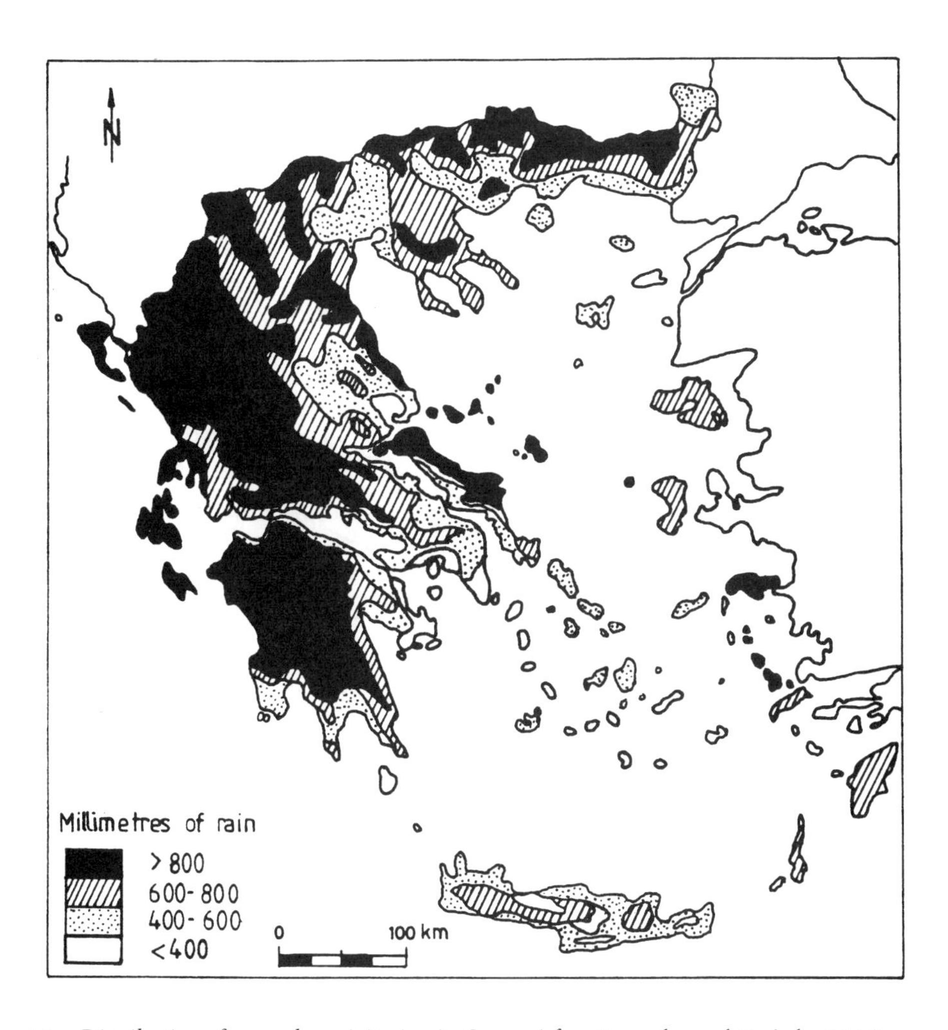

Figure 5.3 *Distribution of annual precipitation in Greece* (after Fassoulas and Fotiades 1966)

Greece, the great number of islands and the atmospheric disturbances which control the weather and more generally the atmosphere above the eastern Mediterranean. Thus, western Greece is influenced by low pressures in the western Mediterranean, whereas eastern Greece is under the influence of the Siberian anticyclone. During winter, which starts in November in the north and December in the south, the weather is unstable due to frequent changes of low and high pressures. The amount of rainfall ranges from 780 to 1280 mm per year in the western part of Greece, reducing by about half in the eastern part, ranging from 380 to 640 mm per year.

Spring is short because cold fronts may affect the region frequently in March, whereas May is rather warm especially due to the appearance of the first *etesian* winds and diminishing low pressure systems. *Etesian* winds are strong, north winds which blow in the period May–August, and are particularly strong in July and August. They were given their name by the ancient Greeks due to their periodicity (Greek: *etos* = year). The *etesians* create drought, which is interrupted only by some local rainfalls of tropical origin. Heatwaves in the plains may last for some time, in contrast to the islands and the coastal areas where the coastal winds and strong *etesians* ameliorate the climate.

The first rains start to fall with the onset of low pressure systems in autumn, especially in western Greece, whereas in the eastern part, autumn may be prolonged until December.

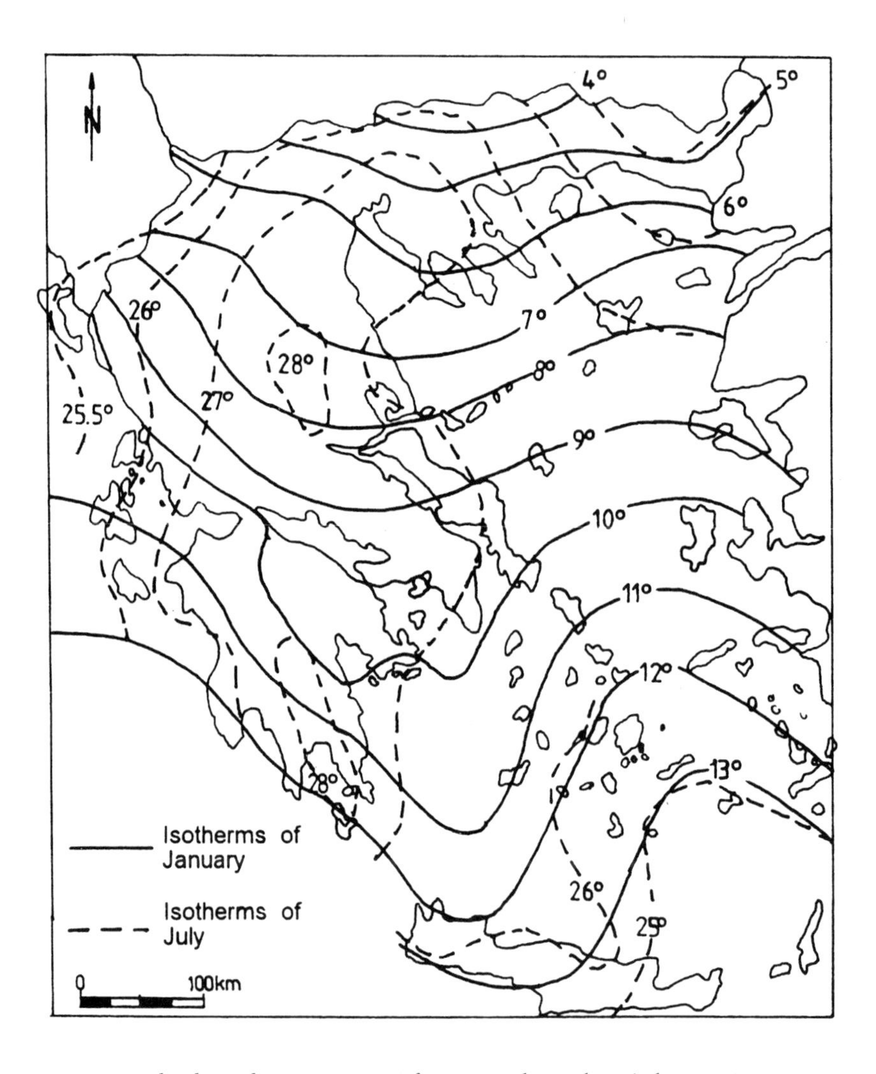

Figure 5.4 *January and July isotherms, Greece* (after Fassoulas and Fotiades 1966)

As far as temperature is concerned, there is considerable variation. Greece lies between the isotherms of 14.5°C and 19.5°C. During the cold period, temperature increases with decreasing latitude, whereas in the warm period, and especially between May and August, temperature increases from the coast to the mainland and particularly the plains. In winter, the lowest temperatures occur in northern Greece, occasionally reaching –20°C, whereas in the southern parts and Aegean islands, temperature rarely falls below 0°C. In summer, temperatures greater than 40°C may occur in the lowlands of the Greek mainland, whereas in the islands and coastal areas temperatures rarely reach 40°C, due to the *etesians* and local winds.

With regard to sunshine duration, certain Greek regions rank among the first in southern Europe. The western coast of Peloponnesus, together with the Ionian coast and the Aegean islands, have more than 3000 hours of sunshine per year. The remaining coasts have about 2500 hours and inland this figure is about 2300 hours.

Vegetation

Large areas of the country are not adequately protected by vegetative cover. About 50% of

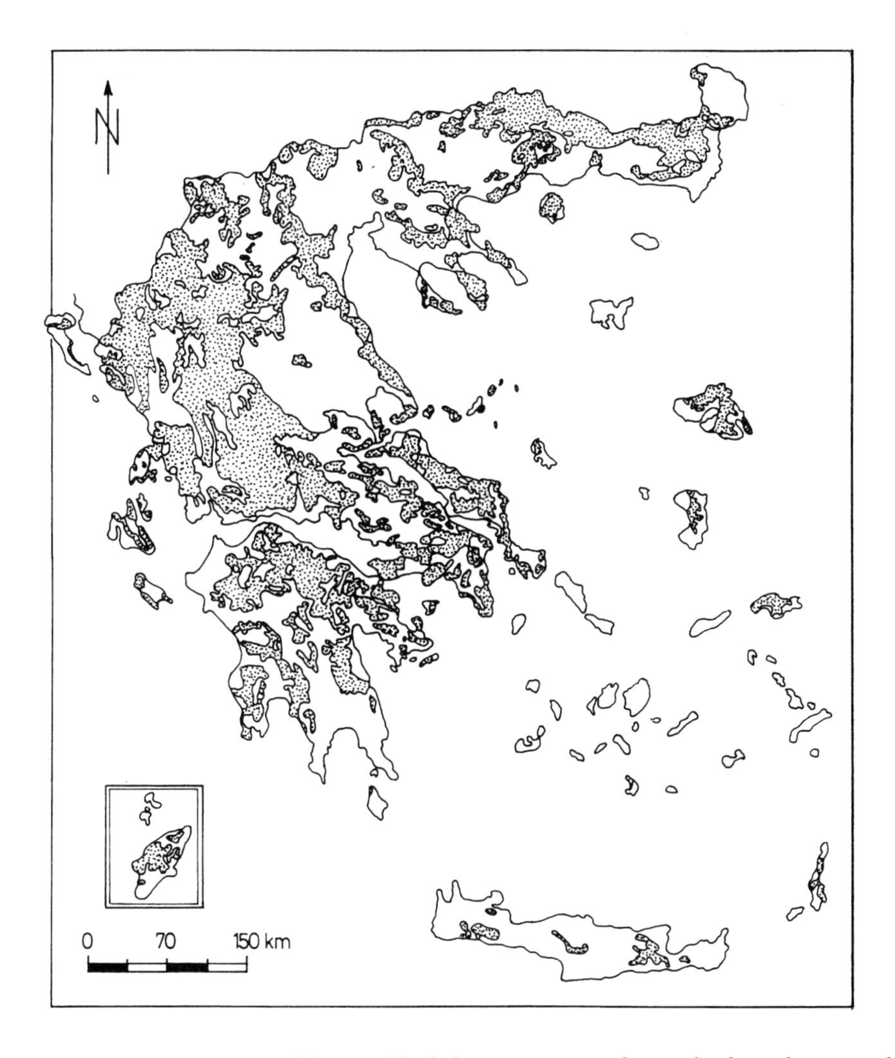

Figure 5.5 *Vegetation cover map of Greece. Shaded areas represent forest, shrubs and vegetated pastures: the remaining areas are cultivated and bare lands* (based on the forest map of Greece, Greek Forest Service)

the country is bare or cultivated land (Figure 5.5).

Greek vegetation is typical of the eastern Mediterranean, being adapted to the dry climatic conditions of this region. The lower zone is characterized by *maquis* vegetation and particularly by evergreen shrub species such as *Quercus coccifera, Quercus ilex, Arbutus unedo, Arbutus andrachne, Phillyrea media, Pistacia lentiscus* and *Pistacia terebinthus*. Apart from *maquis*, this zone is characterized by shrubs among which the most important species are *Erica arborea, Erica verticillata, Genista acanthalados, Poterium spinosum, Euphorbia acanthothamnos, Thymus capitatus, Anthylis hermaniae, Calycotome villose, Phlomis fruticosa* and various species of *Cistus*.

Conifera are represented by two main species, *Pinus* and *Abies*, which form extensive forests of great economic value throughout the country. Of similar importance but of smaller extent are the conifera species *Cuperus* and *Picea*, the first in the Aegean islands (Samos, Crete, Dodecanesse) and the second in northern Greece where the southernmost extent of conifers occurs. Common pine species are *Pinus peuse, Pinus leucodermis* and *Pinus halepensis*. Other conifera such as *Juniperus* and *Taxus* are found but they do not form individual plant associations. Filices are represented in Greece mainly by *Pteris aquilina*, which dominates the mountain regions and many forests, forming dense populations occasionally 2 m high, retarding the natural regeneration of the forest.

Annual plants commonly grown in Greece belong to the families of Graminae, Compositae, Leguminosae, Cruciferae, Caryophyllaceae, Labiatae, Umbelliferae and Liliaceae.

The People

Population composition

The population of Greece is 9.9 million, almost exclusively of Greek nationality. The average population density is 75 persons/km^2. Greek is the official language of the country, having evolved from the ancient Greek language. Greeks are predominantly (96%) of the Christian Orthodox faith, which is the country's official religion. The active workforce comprises 39.2% of the population, with 28.9% employed in agriculture, 27.4% in industry and 43.7% in other sectors.

Demographic trends

The population of Greece is increasing very slowly, with growth rates being lower than the global average of about 2%, and also much lower than the low population growth rates of western and central European countries. Birth rates have followed a negative trend to reach the very low rates of other European countries, especially France, rather than the rates observed in the bordering countries such as the former Yugoslavia and Bulgaria. Less than 22% of the population is under the age of 15, whereas 13.3% is above the age of 65. Women represent 50.5% of the population, and of those, 33.7% are in the active workforce.

Settlement patterns

About half of the Greek population is concentrated in the three biggest cities, namely Athens, Pireus and Thessaloniki. Athens (population 2.6 million) is located in the southern part and is the capital of the country. Pireus (population about one million) is located 8 km from Athens and has the biggest harbour in Greece. Thessaloniki is the third biggest city (population about 800 000) and is located in northern Greece. Other cities include Patras (400 000), Heraclion and Larissa.

Population movement to and concentration in the big cities started many decades ago, especially after the 1950s, and is still continuing, although at much slower rates (Plate 5.1).

The Economy

Fourteen years after the entrance of Greece to the European Economic Community, the Greek economy might be expected to be among the modern western economies. But despite rapid development during recent years, the country belongs to the poorer partners within the European Union (EU) with a gross income of about 4500 ECU per capita (equivalent to about US$5100 at the January 1996 exchange rate). Per capita incomes (ECU) in 1988 in some other EU countries were: Germany 13 543, Italy 9600, Spain 5602, and Portugal 2851 (Commission of the European Communities 1988).

Resources and industry

Geological substrata in Greece include a variety of minerals, of which many are of economic significance. Minerals found in exploitative quantities include bauxite and the mixed sulphur deposits (galinitis, sphaleritis and ironstone). Iron-bearing minerals generally are smiris (Naxos), varitini, chromite (Vourinos) and molybdenite. Perlite, which is mined in some Aegean islands, has an important application in agriculture, especially in hydroponic cultures.

Lignite constitutes the main energy resource of the country and occurs mainly in western Macedonia (Ptolemais), Evia (Aliveri) and Peloponnesus (Megalopoli). Bauxite, mined mainly in the central part of the mainland, constitutes an important export product. Lead mining activities in Laurion, a well known argentiferous mine since ancient times, have declined in the last decade due to the international mining crisis.

Plate 5.1 *Part of ancient Athens, viewed from the Acropolis, is represented here by the treed areas in the foreground, with modern Athens expanding over agricultural land. The population of Athens is about 2.6 million; as little as 100 years ago it was only 30 000*

Marble is mined in several areas throughout the country. The famous Penteli marble, from which the Acropolis was constructed, is well known throughout Europe. Gypsum constitutes an important resource which is used extensively in industry and agriculture.

Oil and gas deposits have been found recently in the north Aegean Sea (Prinos field) and account for about 8% of the national energy needs for oil. The Prinos field produces 9000 barrels per day. Other fields are the W Katakolo (to be developed) and a new reservoir discovered only recently in N Prinos (estimated production 3000 barrels per day).

The Greek manufacturing industry, the second important sector of the economy, has developed rapidly since the end of the 1960s, and today industrial products constitute an increasing sector of the country's exports. The contribution of industry to the gross national income is 28.5% (Commission of the European Communities 1988).

Shipping is another important sector. With about 3700 commercial ships, Greece is among the leading nations in this sector, despite the serious difficulties which face further development of shipping due to the general crisis over the past few years.

Agriculture, forestry and fishing

Agriculture plays a major role in the economy. Together with forestry and fisheries, its contribution to the country's economy is about 18.5%. Greek agriculture has modernized rapidly since the late 1950s when self-sufficiency in wheat production was attained. Since then, soil amelioration and mechanization, application of fertilizers, pest and disease control, introduction of improved varieties, and expansion of the total irrigated area have led to a dramatic increase in agricultural output.

The country's food situation had already improved substantially before entry to the European Union in 1981. After that, agricultural development focused on maximization of fodder and cash crop production which resulted in intensive arable cropping on all fertile, irrigable lands. Further mechanization and expansion of the irrigated area to 1 000 000 ha were realized soon after the country became a full member of the European Union (Boyatzoglou 1983); the national production targets of major crops (such as maize, cotton and sugar beet) were achieved as early as 1985.

Rapid agricultural development, however, resulted in great surpluses of some (Mediterranean) products particularly after Spain and Portugal entered the European Union. Considering the present rate of expansion of irrigated land (40 000 ha/yr) towards a potential of 1 900 000 ha, it may be expected that surpluses of some commodities will increase even further, while other agricultural sectors lag behind in development. Examples of surpluses are oranges, grapes, olive oil, cotton, tobacco and vegetables such as watermelons and tomatoes. Some of these products have to be destroyed in some years, depending on total production and market requirements. For example, the proportion of oranges destroyed ranges from 10 to 25%, except for some years of very high market demands. The products are concentrated in selected waste sites by the farmers, who receive money from their local government or the European Union. As another example, in 1988 about 35% of the production of watermelons remained in the fields. More recently, the over-produced fruit (oranges, pears and apples), or a big proportion of the over-production, is not destroyed but is offered free to public organizations such as schools and hospitals.

At the same time as large quantities of fruit and vegetables are being destroyed each year, imports of meat and milk products (exports of which are virtually nil) total a staggering 1.5 billion ECU.

Tourism and commerce

Tourism has become an important income source since the mid-1980s, and contributes about 10% of the national income (there were 9.8 million arrivals in 1993 according to unpublished data supplied by staff of the Greek Tourism Organization).

The main exports of Greece are as follows (with the contribution to the export value in brackets): industrial products (up to 45%) – clothing, chemical products (3.3%), minerals and metals (5.8%); and agricultural products (28.7%) – tobacco, cotton, resins, vine, olive oil and citrus. The main exports are to the European Union (63.5%), the United States of America (7.1%), Egypt (2.4%), the former Soviet Union (1.4%), and the former Yugoslavia (1.3%). Energy supplies are 64.5% dependent on imports. Unemployment has increased recently, and in 1995 reached about 10% of the active population.

Administration and Social Conditions

Government

Greece is a parliamentary presidential democracy. Legislation is conducted by the Parliament and the President of the State who approves and publishes the laws. Twelve of the 300 members of the Parliament are nominated by the political parties in relation to their percentages in the Parliament. The remaining members are elected directly by the people, whilst the President is elected from the Parliament for a period of five years. The President appoints the Prime Minister and after that appointment, the Ministers.

Greece is administratively divided into 52 Prefectures and the following 13 Peripheries: East Macedonia and Thrace, Central Macedonia, West Macedonia, Epirus, Thessaly, Ionian Islands, West Greece, Sterea Hellas, Attica, Peloponnesus, North Aegean, South Aegean and Crete.

Every citizen older than 18 years is obliged to vote but candidates for Parliament should be older than 21 years. There are no restrictions for political parties. Military service is compulsory for males and voluntary for females. The main national holidays are 25 March (commemorating the beginning of the revolution against the Ottoman Empire in 1821) and 28 October (the beginning of the war against the Third Reich, 1940).

Education

Education is offered at three levels: primary (six years of education), secondary (gymnasium three years, lyceum three years) and university levels (four to six years). The first nine years of education are compulsory and free of charge, a basic personal and social right of every Greek citizen. After completing the lyceum and subject to examination, about 30% are admitted to the universities.

There are 14 universities spread around the country, of which six are located in Athens. Agricultural education at the highest level is offered by three agricultural universities (four or five years of study). At a lower level, agriculture is also offered at higher technological schools (three years of study).

Health and welfare

The commonwealth system, which is controlled by different public or private agencies, provides pensions and benefits after the age of 65 and a wide range of welfare services for people with special needs. Hospital services are covered by health insurance which is compulsory for every working citizen. Unemployed family members are covered by the insurance of their parents until a certain age. Insurance for farmers is provided by taxes paid by all Greek citizens.

Government health authorities (Ministry of Public Health) provide health services both at the first aid (free of charge) and hospital levels, paid by insurance.

6
The eastern Mediterranean

Moshe Inbar

Introduction

There is no unequivocal definition of the area of Mediterranean-type climate in the eastern Mediterranean, as is true for other similar areas in the world. Although the FAO map identifies extensive areas in this region as "Mediterranean" (UNESCO–FAO 1963), the area of Mediterranean-type climate *sensu stricto* is restricted to the coastal areas of Anatolia, the eastern Mediterranean coastal zone and adjacent mountain ranges, and most of the island of Cyprus (Figure 6.1). The region is commonly identified as the Fertile Crescent, wedged between the Mediterranean Sea to the west and the Syrian and Arabian deserts to the east and southeast.

To the north and inland in Anatolia, precipitation increases over the ranges of the Taurus mountains. In the central coastal part, in Syria, Lebanon and Israel, precipitation decreases sharply towards the Syrian desert. Toward the south, there is a clear border with the Negev and Sinai deserts. The physiographic character of the area has a pronounced effect on precipitation: above 1000 m, snow is more frequent and landscapes change from Mediterranean to subalpine environments, such as the Troodos mountains in Cyprus and the Lebanon and Anti-Lebanon ranges. However, high mountains do not cover extensive areas. The eastern Mediterranean coastline has no islands and the coastal area is characterized by a parallel mountain range, bordered to the east by the main Syrian–African Rift, creating a series of north–south coastal and rift valleys, similar to the topographical features of California and Chile (di Castri 1991).

In the Anatolia region of Turkey, the length of the coastal area, including the Black Sea and Mediterranean Sea coasts, is about 5000 km. Along the eastern Mediterranean, the coastal area extends over 600 km in length and between 200 km and 100 km in width. Dry, continental areas to the east are not included, but mountain ranges in Jordan have typical Mediterranean landscapes (lands partly covered by dense thickets of evergreen woody shrubs and plants, with dry summers and wet and mild winters, called *maquis*, *chaparral*, *matorral*, *mallee*, etc. (di Castri *et al.* 1981, p. 1)), and are included in the area.

Located between the Mediterranean Sea and the deserts, the region exhibits sharp climatic gradients ranging from a typical Mediterranean maritime climate to fully desert environments over short distances (hundreds of metres). The region's mosaic-like landscapes are important as a geographic background to its history and human development.

For several millennia, the area was densely populated. The human influence on the landscape started with Early Man, some 1.5 million years ago. Throughout historical times, the area served as a bridge between adjacent African and Asian civilizations. In historical times the Mediterranean area was probably one of the most human-impacted landscapes in the world. By the late 20th century, the total population was estimated to be around 30 million.

Land Degradation in Mediterranean Environments of the World: Nature and Extent, Causes and Solutions.
Edited by A.J. Conacher and M. Sala.

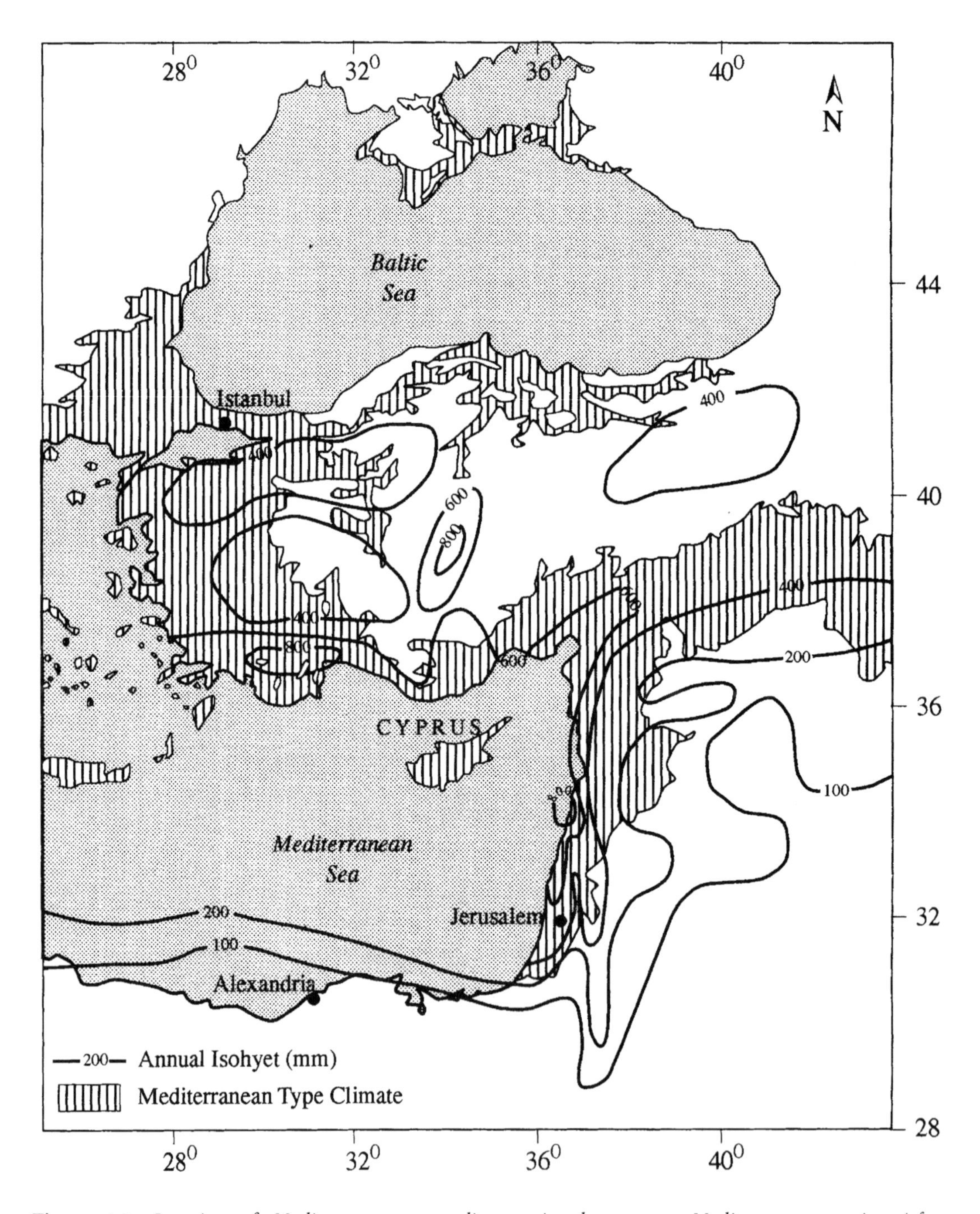

Figure 6.1 *Location of Mediterranean-type climates in the eastern Mediterranean region* (after UNESCO–FAO 1973)

The Land

Relief

To the north, in Anatolia, the relief is determined by the Pontic and Taurus mountains which are part of the Alpine chain, folded and lifted in Tertiary times. The Kyrenia and Troodos mountains in Cyprus have similar characteristics. In the Levant, the physiographic features are determined by the north–south Syrian–African Rift, and four distinct physiographic regions can be distinguished.

1. The coastal plain, a narrow undulating area with its elevation rising gradually from 10–20 m at the coastal cliffs to about 100 m above sea level, extends from the sea to the foothills, with Nile sand accumulated in its southern beaches.
2. The central mountain belt, including the Lebanon mountains to the north and the Galilee, Samaria and Judea mountains to the south, has elevations of 2000 m and 3000 m in north Lebanon, but elevations are generally lower at between 500 m and 1000 m above sea level.
3. The Rift Valley area (Plate 6.1) is a linear depression trending north–south, extending 1000 km from Anatolia to the Red Sea. In its northern part it is a fertile and water-rich area, which feeds the Orontes, Litani and Jordan, the main perennial rivers in the area. The Rift includes the freshwater Kinneret and saline Dead Sea lakes, the latter being the lowest point on earth, 400 m below sea level.
4. To the east, in its northern part, are the Anti-Lebanon mountains, reaching an elevation of 2800 m above sea level (Plate 6.2), and the volcanic Golan Heights and Jordanian mountain belt in its southern part. Fresh fault scarps several hundreds of metres long occur in the rift, along with *en echelon* strike slip faults, tectonic basins and compressional structures (Garfunkel 1981).

Drainage

The main water resources of the "Near East" – the Nile river in the south and the Tigris and Euphrates along its northern edge – are outside the Mediterranean area, flowing through desert areas. Most of the rivers have ephemeral flow regimes and are characterized by flash floods. Increasing demand for water has diminished drastically the base level discharge of most streams, and they are becoming increasingly polluted from sewage and other sources (see Part II).

Plate 6.1 *The Hula valley in northern Israel, part of the Rift Valley. This former swamp and lake area flooded after heavy rains in February 1992. Photo by M. Inbar*

Plate 6.2 *Mount Hermon, south of the Anti-Lebanon mountains, reaching 2814 m above sea level. Snow cover remains only during winter. Photo by M. Inbar*

Water resources are relatively abundant in Turkey and Cyprus, while in the Levant area where rainfall is sparse and water the mainstay for survival, disputes over water repeatedly have become a *casus belli* and the focus of political and military struggles (Caelleigh 1983). In Cyprus, the rivers originate in the Troodos mountains; the Pedieos river flows eastward, the Karyoti river west and the Kouris river to the south. In Turkey the water divide between the main Mesopotamian streams flowing to the Indian Ocean and the Mediterranean-flowing streams, is located in Anatolia. Most of the Mediterranean-orientated basins have a Mediterranean-type climate. The principal rivers are the Goksu, Seyan and the Orontes, which flows from the Syrian Rift Valley. The main rivers of Lebanon and Israel – the Litani in Lebanon and the Jordan in Israel – flow in the Rift Valley. More than 90% of their flow is regulated. All of the flow of the coastal streams is regulated and the water is used for domestic and irrigation purposes. Groundwater is a major water source, but over-exploitation leads to salinization in the coastal area.

Soils

In the proximity of the desert areas, the soils are composed predominantly of wind-borne loess, which originates in the Syrian and Sinai deserts. Part of it is redistributed by the rivers. In the coastal plains, red-sand soils intercalate with carbonate-cemented sandstone. The most common mountain and hilly topography soils are the Terra Rossa or red Mediterranean soils, which develop on hard limestone rocks. Calcareous soils are found on soft limestones and marls. All of the soils developed in the mountain and hilly areas are thin – usually less than one metre thick – and underlain by bedrock. Deep clay soils on alluvium are found in the inner valleys and are most favourable for agricultural purposes. Most soils are severely eroded, with erosion rates reaching several hundred tonnes per kilometre squared per year (Inbar 1992). In the lower parts, soils have been affected by salinization problems due to continuous irrigation.

Agriculture is limited to 10–30% of the total area and the main crops are orchards and vegetables. Cereals (mainly wheat and barley) are grown in large fields in alluvial valleys. The mountainous area is characterized by small, terraced fields.

Climate

The long, dry Mediterranean summer is dominated by a dry, high pressure cell, which in

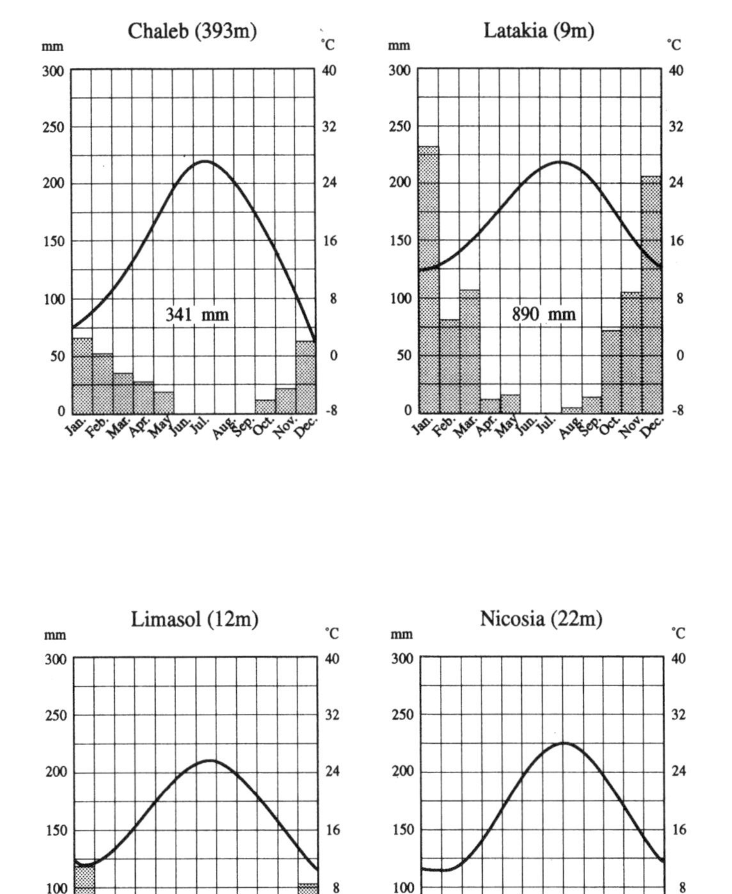

Figure 6.2 *Climographs for selected sites in the eastern Mediterranean region* (compiled by M. Inbar from various sources)

winter retreats to allow a series of cold fronts to cross the eastern Mediterranean from the west through Cyprus to Lebanon and northern Israel. The cyclonic depressions in the eastern Mediterranean are small in area and length with 10–13 mbar of difference between periphery and centre. The trajectory of the depression determines the boundary of the humid Mediterranean area; the arid and semi-arid areas are found to its south and east, on the edge of the 300 mm isohyet. Unlike the western Mediterranean areas, the dry season is very pronounced and lasts for six months (Figure 6.2). The dry pattern increases toward the south and east. Rainfall intensities are about 50–100 mm/day. In the Levant coastal areas intensities of 20 mm/h have a 50% annual frequency and are the major factor in soil erosional processes.

A breeze from the Mediterranean Sea is common during the summer months. Dry, easterly desert winds, called *hamsin* in Arabic and *sharav* in Hebrew, occur in the transition seasons. Air temperature depends on elevation and distance from the sea. Mean annual temperatures in the coastal Levant area are 18 to 20°C, and 3 to 4°C

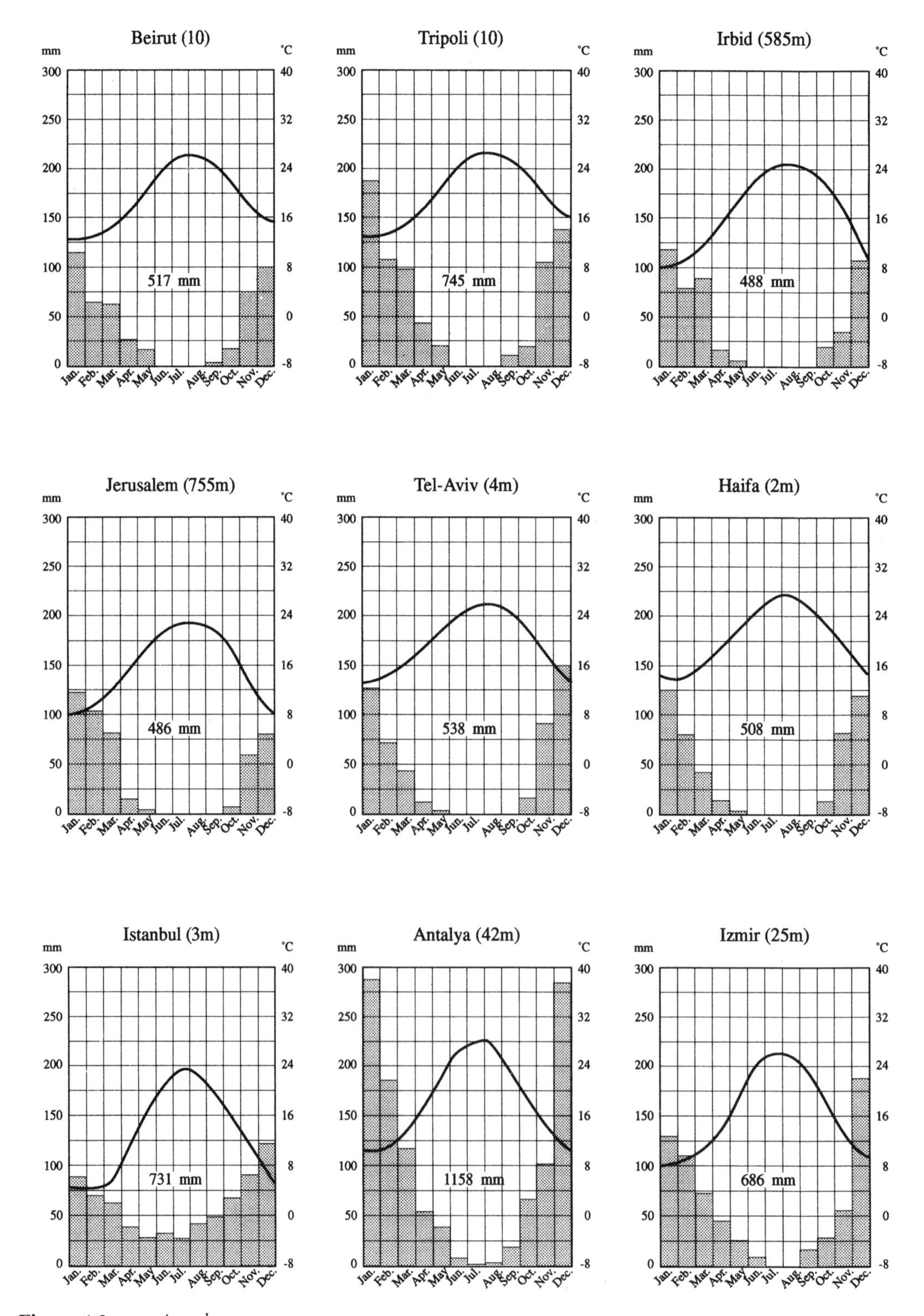

Figure 6.2 *continued*

less in the mountain areas. Diurnal temperature amplitudes are up to 20°C at inland locations, whereas in coastal areas they are generally less than 10°C. Maximum temperatures recorded in the Rift Valley have reached an extreme of 54°C in the Beth-Shean valley. Cold waves flowing from Siberia in winter bring air temperatures of –5 and –10°C in the inner valleys and high mountains. Frost is not common in the coastal areas, but below-zero winter temperatures occur frequently in coastal areas of Cyprus and Anatolia. Snow is important in the high mountains as a feeder of permanent springs and flow in rivers. Snow is rare at lower altitudes and is a minor component of the water budget.

High radiation and high summer temperatures determine a negative annual water budget for most of the Mediterranean areas. The aim of water management is to store winter water surplus for summer consumption. The same strategy is used by the natural vegetation in order to survive the long, dry summer period.

Plant and animal life

The present composition and distribution of the flora and fauna have been influenced by three main processes:

1. the transitional location of the area between humid and arid environments;
2. climatic changes since the last glaciation; and
3. human impacts, with the extinction of many faunal species and the introduction of cultural flora and fauna (Aschmann 1973).

Matorral landscapes are the dominant natural vegetation in the Mediterranean areas. The sclerophyllous, evergreen shrublands are widespread on the Aegean coast of Turkey, Cyprus and the Levant countries (Plate 6.3). They include the arborescent *matorral* with *Pinus halepensis* and *Quercus calliprinos* (Quezel 1981; Zohary 1973), and *maquis* and *garrigue* communities with *Quercus ilex*, *Pistacia palestina* and the low, thorny formations of *phrygana* which are very common in all areas which have suffered from over-grazing and burning for centuries. The eastern Mediterranean region has a more xerothermic nature and merges on its borders with the semi-arid, Irano-Turanic steppe formations (Naveh 1984). In addition, many natural (*P. halepensis*) and planted (*Pinus brutia*) pine forests are found in all the Mediterranean areas. Sclerophyll communities are replaced after large fires or intensive forest clearing by low dwarf shrubs, called *phrygana* in Greece and *batha* in Israel.

Plate 6.3 *Typical Mediterranean forest with* Quercus calliprinos *in the hilly, western Galilee region. Photo by M. Inbar*

The most interesting and impressive wildlife fauna is the bird community and its migratory character. Along the eastern Mediterranean, millions of birds cross from Europe and Asia to the equatorial areas of Africa in winter, returning to the north in spring. Goats and donkeys are the main domestic animals and *Sus scrofa* (wild pig) and gazelle the main wild animals.

The People, Economy, Administration and Social Conditions

The east Mediterranean region includes the island of Cyprus, with its two political entities, a large part of Turkey (both in Anatolia and in its European part), a small part of Syria, and most of the settled regions of Lebanon, Israel and Jordan. The diversity of population background among the countries is too complex for a general regional overview, therefore the description of the human characteristics is given by country. Demographic data are from the charts published by the Population Reference Bureau (Haub and Yanagishita 1994).

Cyprus

The people

Cyprus is the third largest island (9200 km^2) in the Mediterranean, with a population of 750 000. Politically, it is divided between the Republic of Cyprus with a Greek majority – constituting 80% of the total population, belonging to the Greek Orthodox church – and the state of Turkish Cyprus on the northern third of the island, with a Muslim population. Nicosia (population 250 000) is the capital for both entities.

The economy

The gross national product (GNP) for the Greek part is $9820 per capita (1994 value), and is larger than in the Turkish part. The economy is based on tourism, shipping, construction and business, with less reliance on agriculture. Agriculture is more prominent in the Turkish sector. About one-quarter of the agricultural land is irrigated, with citrus as one of the main crops, together with grapes, potatoes and cereals. Grazing by sheep and goats is traditional and is found in the mountain areas. Water projects – like the Paphos and Southern Conveyor Project – have enlarged the irrigated areas by 10 000 ha, increasing crops of export fruits such as citrus and avocado. Further agricultural development requires major water projects to transport water across the island.

Industry was severely affected by the 1974 war. Principal sectors are clothing, footwear, refined petroleum, cement, asbestos and construction. Exports cover only half of the imports, and tourism is a main source of foreign exchange. Electricity is supplied to the Turkish zone by the southern Greek zone in exchange for water supply. Generally, the economy is growing and the recovery of the southern part is quicker than in the northern part.

Administration and social conditions

The new Republic of Cyprus was formed in 1960, with an agreement that the elected Parliament would comprise 70% Greeks and 30% Turks. Since 1974 there has been a clear political division, and minorities from both parts – about 180 000 Greek Cypriots and 45 000 Turkish Cypriots – migrated and became war refugees. The Turkish political entity is not recognized internationally except by Turkey, and reunification talks have continued since 1988.

Annual population growth is 1.3% and life expectancy is 73 years for men and 76 for women. Infant mortality is 16 per thousand in the Greek part and 19 per thousand in the Turkish part. Literacy in the Greek part is more than 90%.

Lebanon

The people

The estimated population for 1994 was 3.5 million, living in an area of 10 400 km^2, the smallest country in the area but one of the most densely populated and with a high rate of population increase. About one million people live in the capital, Beirut, and surrounding areas, and more than 80% of the country's population is urban. Ethnically the population is a mixture of Muslims (both Sunnite and Shiite; 57%), Christians (36%) and Druze (7%). Over the past 20 years, ethnic, religious and political factions have created a situation of political instability which affects all aspects of human life.

The economy

Until recent years Lebanon acted as a banking and trade service centre for the Middle East. Political and military instability has severely disrupted this role.

About 23% of the country is cultivated, mainly by olives, citrus and vineyards. The main cereal-growing district is the Bekaa, in the Rift Valley. Grazing by goats and sheep is practised in the hilly and mountain areas. Forests cover only 7% of the country. Cultivation of drugs and related manufacturing are a significant component of the country's economy. There is increasing domestic demand for water resources, with the total available resource for Lebanon being about 1500×10^6 m^3/yr.

There are no major mining resources, and industry is restricted to manufacturing enterprises. The main trans-Arabian oil pipeline, from the Persian Gulf to the Mediterranean, is not functioning and the refinery at its outlet is only partly operational.

Administration and social conditions

In 1946 Lebanon gained independence from the French administration and was ruled by an alliance of the main religious communities. This power sharing collapsed during the civil war which started in 1975, and the warring factions try to control the national political and administrative institutions in a rapidly changing situation.

Syria

The people

The Mediterranean-type climate area of Syria covers the coastal Syrian strip (200 km in length) and all its western part to the east of the Anti-Lebanon mountains until it reaches the arid Syrian desert. It covers about 20% of the country, but contains about 80% of the total population of 14 million people. Syria's Mediterranean zone includes the cities of Homs and Alep and is the most densely populated area of the country, except for the area around the capital, Damascus. There are more than 100 people/km^2 in the coastal strip and more than 50 people/km^2 in the mountain area and inner valleys.

The average annual rate of population increase is 3.7%. Ethnically, the vast majority (72%) are Muslim Sunnites, with ethnic minorities including the Kurds and Turkish-speaking communities in the north, Armenians in the cities, and Druzes in the southeast parts of the country.

The economy

The economy is part publicly owned and part private, based on agriculture and manufacturing. The GNP per capita is only $1170. The heavy burden of military expenditure affects economic growth, and the decline of economic aid from the Gulf Arab countries brought the country to a state of crisis. The main export products are petroleum, cotton and phosphates. Syria is an oil-exporting country and in 1986 oil products accounted for 42% of the country's total exports (Fisher 1993a).

Agriculture's position has declined in the Syrian economy. In 1965 the agricultural sector employed 50% of the labour force, declining by 1984 to 25% (and 17% of the GNP). The Euphrates Dam project added an area of 240 000 ha of land for irrigation. Syria has large water resources – 2800 m^3/person/yr – mainly from the Euphrates basin, and their development will increase the extent of irrigated land, mainly in the Jezira Valley. However, this area between the Euphrates and Tigris is not located in the Mediterranean-type climate region.

Administration and social conditions

Syria became independent from French rule in 1946. Hafez Assad has been the Head of State since 1971, and has been re-elected every seven years. The government is appointed by the President. Assad's power relies on the army and his ethnic Alawi minority. Political and military conflicts with its neighbours – Iraq, Turkey,

Lebanon, Israel and Jordan – create an unstable situation. The trend towards a general peace in the Middle East may release the prolonged political tension in the region.

Life expectancy is 65 years for men and 67 for women. Population growth is 3.7% and the infant mortality rate is 56 per thousand. Illiteracy in 1983 was 50% of the total population.

Turkey

The people

Turkey is the largest (780 000 km^2) and most populous country (62 million people) of the eastern Mediterranean. Ninety-five percent of the country lies in Asia, the Anatolia region between the Black Sea and the Mediterranean Sea, and 5% in Europe. The Mediterranean area covers a wide fringe along the coasts of Anatolia and includes all of the European part of Turkey. Except for the capital, Ankara, which lies on the continental inner part of Anatolia, most of the population and large cities, including Istanbul (5.5 million people), are in the Mediterranean-type climate area. The annual rate of population increase is 3.2%; the population density is 70 people/km^2 – and is much higher in the Mediterranean areas. The large majority are Muslims from the Sunnite sect, while Kurds are the largest minority in Eastern Anatolia. Turkish is the spoken language.

The economy

Turkey has been in an intensive process of industrialization and modernization since Ataturk's (the first President of the modern republic) radical reforms in the 1920s. The GNP is $1950 per capita and annual growth rates in the 1980s fluctuated between 2.5% and 8%. Several decades ago Turkey relied mainly on agriculture, but recently the agricultural sector has contributed about 15% of GNP (Fisher 1993b). The total cultivated land is 250 000 km^2, about one-third of the total land area; but only 20% of cultivated lands are irrigated. The development of water resources in the Euphrates basin by dam construction, including the large Ataturk Dam, is part of the GAP (Guneydogu Anadolu Proves) programme, a large development scheme for southeast Anatolia. Principal crops are fruits, cotton, tobacco and cereals. Grazing by sheep, goats and cattle is found in all the regions and especially on the Anatolian plateau.

Rich mineral resources, including iron, bauxite, copper and chromium, provide the raw materials for the growing manufacturing industry. Thermal power plants supply two-thirds of electricity with the remaining third being generated by hydroelectric sources, which are increasing with the completion of the dams on the Euphrates and Tigris rivers. Total demand by the end of the century will be 170×10^9 kWh. Major developing industrial sectors are textiles, petrochemicals and engineering. Tourism is growing fast, mainly in the Mediterranean coast areas.

Administration and social conditions

Turkey became a republic after World War I with the abolition of the Ottoman Empire. A multi-party parliamentary system developed, and political conditions have been relatively stable in recent years. Disputes with Greece concern the Cyprus issue and jurisdiction of the Aegean Sea, and disputes with Syria and Iraq, water rights and borders. The Kurdish minority – 18% of the population – and their struggle are the main internal conflict.

Life expectancy is 67 years for women and 64 for men. Annual population growth is 2.2% and is declining. About two-thirds of the population is urban.

Israel

The people

Over 90% of the population of Israel (5.4 million) and administrated territories (2.1 million people) is concentrated in the Mediterranean region, especially in the coastal strip with the main urban districts of Tel Aviv (one million people) and Haifa (600 000). A further million live in the central district and 600 000 in the Jerusalem district. Eighty-two percent of the State of Israel population are Jews, and the others mainly Arab Muslims, Arab Christians and Druzes.

The natural annual rate of population increase is 1.5%, but the main reason for the growth of the Jewish population has been in-migration. The most recent wave of in-migration – over 500 000 people – arrived from the former Soviet Union countries between 1990 and 1995.

The economy

GNP per capita of $15 000, is the highest of the eastern Mediterranean countries. The economy is similar to developed European countries, based on high-tech products, financial and business services, tourism and manufacturing. The agriculture sector is relatively small, about 5% of domestic product, but provides most of the local foodstuffs except for grains; the main crops are citrus fruits, avocado and flowers. About half of the cultivated area, 500 000 ha, is under irrigation, consuming 75% of the annual water resources (about 1800×10^6 m^3). Further economic and demographic growth may warrant a decline in available water for the agricultural sector. Disputes over water have become the focus of political and military struggles between Israel and neighbouring countries, but recent peace agreements have achieved a growing degree of co-ordination for managing the region's water resources (Inbar and Maos 1984).

Lacking resources of fuel and power, almost all of the energy requirements are imported, mainly coal for electricity production and petroleum for other needs. Mineral resources are mainly from the Dead Sea; these are primarily potash, bromide and magnesium products. Phosphates are also mined in the Negev area.

Administration and social conditions

Israel became independent in 1948 but was not recognized by its neighbouring Arab countries. Several wars followed until a first peace agreement was reached with Egypt in 1978. A peace agreement has been reached subsequently with Jordan and a peace process commenced with the Palestinians. The political system is based on an elected government by the Parliament.

Life expectancy is 74 years for men and 78 for women. The infant mortality rate is 13 per thousand and illiteracy 5% in 1992.

Jordan

The people

Only 8% of the total area of Jordan (89 206 km^2) has a Mediterranean-type climate, but that region contains over 90% of the country's population of about 4.2 million people. The Mediterranean region includes the capital, Amman (1.5 million people), and the cities of Zarka (400 000) and Irbid (280 000). The population has grown very rapidly from 350 000 in 1948 to three million in 1992, mainly as a result of immigration of Palestinians from the West Bank (Soffer 1992a). The annual rate of population increase is 3.3%.

The economy

The GNP is $1120 per capita. The economy is based on agriculture, manufacturing and trade with the neighbouring oil-rich countries. Water resources are about 1000×10^6 m^3/yr, or 250 m^3/person/yr, which are the lowest values for the region. The main irrigated areas are in the Jordan Valley, using the Yarmouk waters (130×10^6 m^3) and dams on the eastern drainage system to the Jordan valley. The main crops are vegetables, olives and grains in the higher plateau areas. Grazing by sheep and goats is practised intensively. Small industry and manufacturing are developing rapidly. The port of Akkaba is the only sea outlet, and it also serves other countries such as Syria and Iraq. The main mineral resources are similar to those of Israel, and include potash from the Dead Sea and phosphate.

Administration and social conditions

Jordan became independent from British rule in 1946, after being part of the Ottoman Empire. The country is ruled by King Hussein from the Haschemite dynasty, and a government appointed by the King. The large majority of the population are Muslim Sunnites; the former nomadic tribes, Bedouins, are in a transition period of settling in permanent villages. Life expectancy is 64 years for men and 70 for women and infant mortality is 49 per thousand. Illiteracy is about 25%.

7 North Africa

Abdellah Laouina

Introduction

The northern limit of the Mediterranean region of North Africa is located at 35–37°N, and the region has often been compared with an island between the sea and the steppe transition towards the Sahara. However, its outline is irregular because of the influence of relief (Figure 7.1).

Whereas the three countries Morocco, Algeria and Tunisia, extend over some 3 000 000 km², the Mediterranean domain represents less than one-third of that area (the Tell alone covers 415 000 km²). The Maghreb environment is characterized by its variety which is derived from the geographical and socio-economic characteristics of the territory (Troin *et al.* 1985; Despois and Raynal 1967):

– it is situated on the southern shores of the *Mediterranean Sea*, bordering on the steppe and the desert (Birot 1964). This accounts for the regular accentuation of drought towards the south. It is a transition zone, but only on the Atlantic coast is the transition smooth;
– it belongs to the *Mediterranean and Alpine worlds*, characterized by mountainous relief and recent orogenesis;
– it is one of the *Mediterranean countries*, marked by ancient civilizations and cities, and a system of agro-sylvo-pastoral production (Birot 1964);
– it is on the *African continent*, previously colonized but now within the group of southern developing countries, and hence familiar with the fundamental problems of being under-

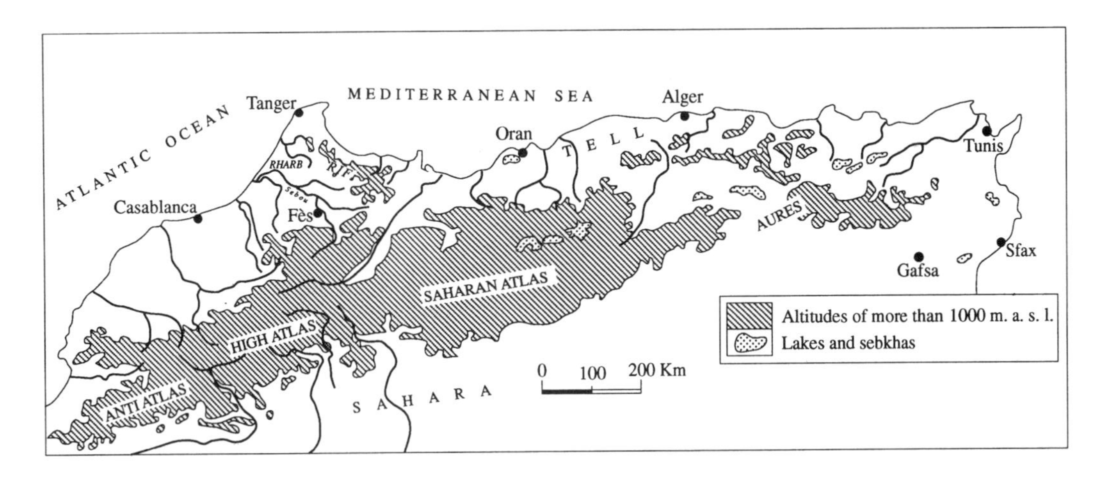

Figure 7.1 *The Mediterranean zone in the Maghreb; relief of the North African Mediterranean regions*

Land Degradation in Mediterranean Environments of the World: Nature and Extent, Causes and Solutions.
Edited by A.J. Conacher and M. Sala.

equipped, the non-mastery of technology, indebtedness, the demographic boom, swelling cities, and the abnormally large tertiary and informal sectors of the economy.

Each of these Mediterranean countries is influenced both by the physical environment and political tendencies:

- *Morocco*, with its wide open spaces and development focused on the Atlantic fringe. Arrangements aimed at mastering water and transforming the semi-arid plains in productive regions have been particularly important, and have been directed at increasing the value of agriculture;
- *Algeria*, a country with limited windows on the Mediterranean, with plains subdivided into many small basins. Agriculture is the weak point of the country's economy;
- *Tunisia*, with an extensive opening to the Mediterranean but with a more restricted area of Mediterranean-type climate, a relatively large population, limited resources and a tendency to over-crowd the coastline, leaving the hinterland abandoned and lacking in infrastructure.

The variety of conditions accounts for the basic imbalance between the well endowed regions and those with few resources. Owing to the relief and the climate, penetration was difficult and mountains provided a barrier to invaders. This context causes problems for modern communications and the spread of settlements.

Within the Mediterranean part of North Africa, it is possible to differentiate two types of sectors according to their economic potential.

(1) In the favourable environments, well watered agriculture produces favourable yields, and rainfall and relief allow rotations of cultivation. Within this domain, the region which receives over 600 mm of rainfall annually exhibits a wide range of winter and summer cultivation, without being constrained by summer evaporation or winter temperatures. Within that domain is a distinction between the extensive plains where mechanization and other innovations have been introduced (Plate 7.1), and the densely populated hills with more traditional structures (Plate 7.2).
(2) The difficult environments extend more or less over the high and dissected mountains (Plate 7.3). These conditions impose particular constraints and populations are often too large in relation to the local resources. In other respects, given their inaccessibility, the mountains have poor infrastructures and little innovation in land management.

Plate 7.1 *Extensive mechanized agriculture in a broad valley draining the Rif mountains. Stubble burning is commonly practised*

Plate 7.2 *Intensive agriculture: irrigated fruit and vegetables being grown for local markets in flatter parts of the densely populated mountains*

Plate 7.3 *Example of agriculturally poorer, more difficult terrain in the Rif mountains. The trees are cork oaks, and the land is grazed by goats and also cultivated in small patches*

The Land

Relief

Two-thirds of the North African Mediterranean region is mountainous, extending over an area 2000 km long and 300–400 km wide (Figures 7.1 and 7.2). There are many individual massifs with rounded summits and high planation surfaces. Sharp-edged peaks are rare. The mountains are composed of hard rocks which, with recent tectonics, account for the high altitudes of several thousand metres, and steep slopes (Despois and Raynal 1967).

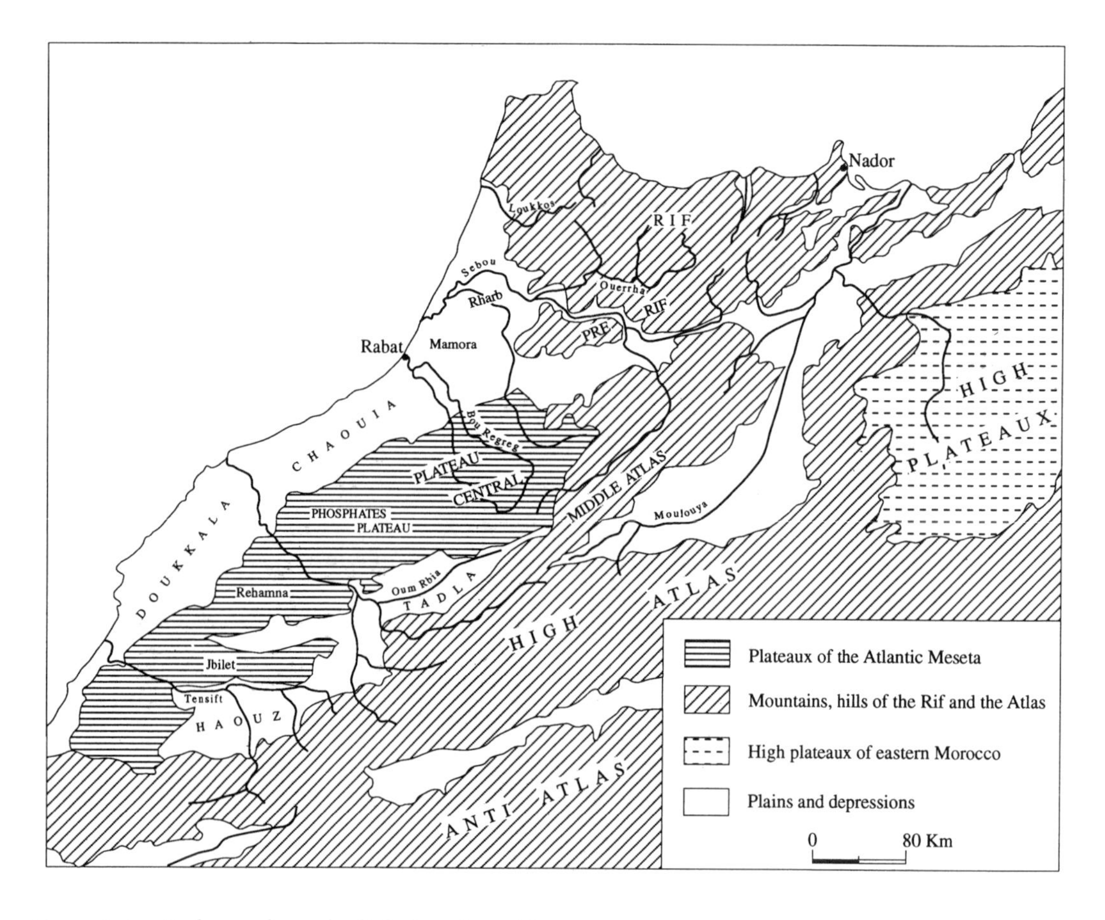

Figure 7.2 *Hydrography and relief of northern Morocco*

The North African relief corresponds to a continental and Mediterranean fringe which is intertwined with a Precambrian African domain. This domain is bordered, in the Mediterranean region, by planed Hercynian chains which evolved in furrows and troughs of sedimentation in the Mesozoic and at the beginning of the Tertiary. Orogenesis ended with the formation of two mountain alignments, roughly parallel, the Rif–Tell in the north and the Atlas in the south.

The Alpine Orogeny in the north accounts for the upthrust schists and sandstones, whereas in the Atlas mountains, the sedimentary strata originated in an epicontinental sea with carbonates dominant. In the Atlas, the folded sedimentary strata are often faulted. The High Western Atlas is the highest mountain in North Africa, formed in old basement. The relief units are as follows (Michard 1976; Martin 1981; Marre 1987).

A young, mountainous relief, the Rif–Tell, corresponds to the most recently uplifted maritime strip along the African continent. It is a domain characterized by its structural complexity and numerous contact zones (Wildi 1983). The vigour of the upheaval accounts for the intense, recent dissection by streams flowing north to the Mediterranean Sea. Tectonic activity has created broad depressions, especially in Algeria (Figure. 7.3). Within the Rif–Tell are four sectors, from west to east: the Rif, the most massive and most continuous yet also the most representative sector of the variety of the chain; the western Tell, a simple structure with two ridges, the Dahra and the Ouarsenis, separated by the Chelif depression; the eastern Tell, a massive and complex domain; and the Tunisian Tell, where highly folded, small chains pass to Jurassic folds.

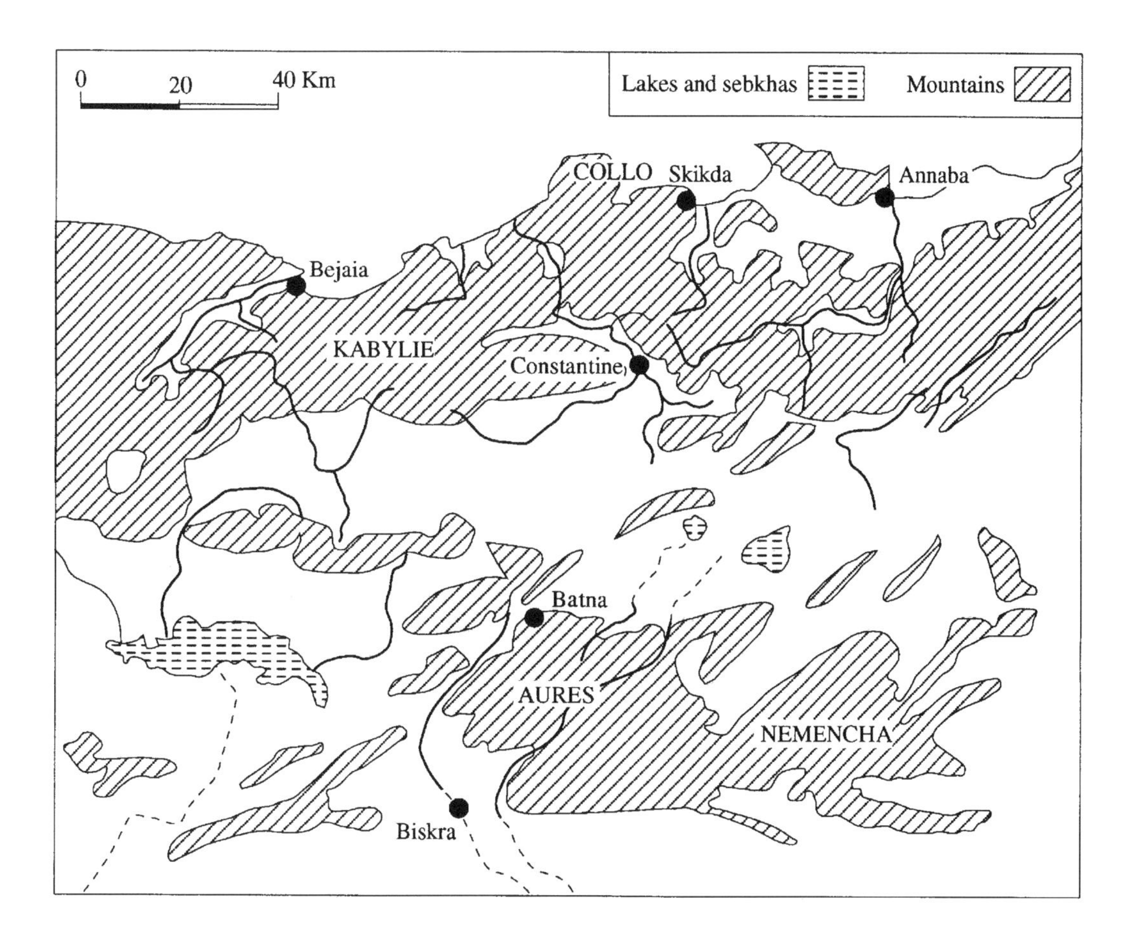

Figure 7.3 *The relief of eastern Algeria*

The Atlas is broader than the Tell. It is here that the highest chains of North Africa are to be found (4165 m above sea level at Toubkal Peak in the High Atlas, 3300 m in the Middle Atlas and 2324 m in the Aures). The Palaeozoic base, with its numerous faults, explains the alignment of structural units (Despois and Raynal 1967; Michard 1976) which in turn influence the "heaviness" of the relief. The mountain chain is composed of massive units with steep slopes. Strong, recent upheavals account for the high altitudes and intense dissection. The valleys make poor communication routes because the cols are elevated and the high valleys isolated.

Atlantic Morocco is representative of dissected relief affecting large areas with rejuvenated drainage, but there are remnant planation surfaces (Plate 7.4), large basins and broad valleys containing Tertiary and Quaternary sediments (Figure 7.2).

The Mediterranean and Atlantic coastlines extend over more than 3000 km and are highly irregular in the north, with open bays framed by promontories. The sea is often directly dominated by mountains. In Morocco, the Atlantic coast has both beaches and cliffed coastlines (Laouina and Watfeh 1993).

Drainage

North African drainage lines are often short and seldom exceed 500 km (Figures 7.2 and 7.3). Channels are often not graded, and have sections with contrasted orientations. High evaporation rates account for water deficits and local endoreism, even in the vicinity of the coastline. Owing to the climatic conditions, especially the concentrated nature of the rains and high evaporation, runoff coefficients are often low; thus water

Plate 7.4 *The planation surface on folded Palaeozoic beds (but with a sedimentary veneer), near Rabat. Relative relief here is 150 m and mean annual rainfall about 550 mm*

shortages are common, even in the well watered regions. Droughts in the early 1980s and 1990s resulted in many springs drying up, and also many wells, demonstrating the precariousness of the resource. But drought is not the sole cause of this problem: over-exploitation of groundwater by over-pumping is probably also a cause of the exhaustion of the resource.

Only the Moroccan drainage network is really organized, because it radiates from a high domain represented by the High Atlas and the Middle Atlas. In Algeria, several drainage lines originate in the steppe zone and cross the Tellian domain. Indeed, there is a wide range of variables in relation to fluvial supply (Troin *et al.* 1985).

First, nival zones produce abundant and regular supplies of water. They are located at altitudes higher than 1500 m above sea level in the eastern Tell, at more than 1900 m in the Middle Atlas, and above 2500 in the High Atlas. Second, stony covers in the mountains favour water infiltration. And third, the calcareous massifs have the most interesting conditions for producing regular supplies of water; this is particularly the case in the Middle Atlas (Martin 1981).

The following classification of North African drainage lines can be used:

– regularly abundant streams but which have low flows in summer on the plains. An example is the Oum Rbia which has a minimum discharge of 34 m^3/s (Loup 1962);
– streams with a small, but regular flow. This type of regime applies to most rivers of the Tell and to the rivers of Atlantic Morocco;
– *oueds* with intermittent flow. These are mainly the local streams on the semi-arid plains.

Soils

The climate, topography and lithology account for the spatial differentiation of the soils and their great variety. Different combinations of soils allow the introduction of complementary land use. But the soils are often vulnerable and degraded, especially those with bare surfaces and on calcareous parent materials. As a result, the soils are often stony.

The main problem is the conservation of a sufficiently deep soil to allow sustained production, with an upper organic-rich horizon, well structured and sufficiently porous to allow good water penetration. The vulnerability of soils is linked to slope, on the one hand, and plant cover on the other. Soil fertility is enhanced by manures and fertilizers, but it is often reduced by the effects of soil loss (erosion) and desertification.

In the wetter, subhumid, maritime environments, the red soils have relatively high clay contents and sometimes contain nodules and ferruginous coatings (Maurer 1968a, b). In the higher rainfall regions, leaching is important, resulting in the impoverishment of the upper soil horizons. In continental environments, owing to the alternation of rubefaction, leaching and crusting, soils are thinner. In the mountains, red, brown and forest soils rich in humus occur locally, but more generally soils are stony with rocky outcrops.

The semi-arid environments are the main region where soils develop calcareous crusts (Laouina 1984, 1990; Ruellan 1971). Both stony and less developed soils occur on terraces and *glacis*, red soils on calcareous parent materials, and Vertisols in the plains, with high contents of clay and silt-sized materials having good water-retention capacities (Plate 7.5). Water reserves are

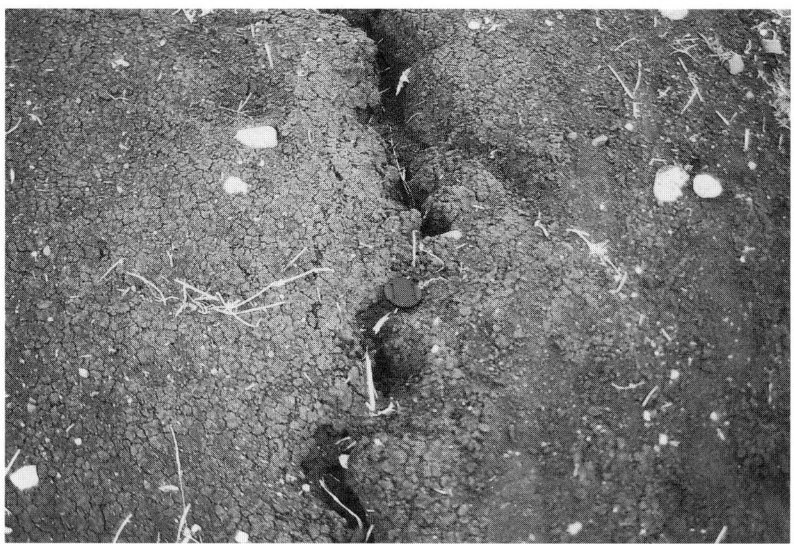

Plate 7.5 *Extensively cultivated Vertisols (a) with fissuring upon drying (b): Pre-rif hills*

assured from one season to another in the deep soils on the plains, but this is not the case in the thin soils of the hilly or mountainous regions. Maintaining soil water reserves by fallow systems, however, poses the problem of erosion of the surface horizons.

Climate

For two reasons – first the presence of the Atlantic Ocean, then the influence of the high barrier of the Atlas mountains – the desert is relatively far to the south. In other respects, climate deteriorates progressively from north to south, with an increase of irregularity and of marked spatial variations linked to the orographic effect (Figure 7.4). This explains the distant thrust southwards of the Mediterranean zone in Morocco along the coast. Inland, the reinforced aridity starts only to the south of the High Atlas, although dry zones appear in the Haouz, on the northern side of the Atlas (Delannoy 1971). Further to the east, in Orania and the western and eastern parts of the Maghreb, the barriers that the Middle Atlas and the Rif constitute cause the steppe to extend well to the north, practically to the Mediterranean coastline.

North Africa is affected by both polar and tropical air masses; this accounts for the diversity of winter situations, which can be more or less cold or humid. In other respects, the disposition of fronts is fundamental and accounts for the differences between the western and eastern parts of the Maghreb (Despois and Raynal 1967).

Maritime influences are also fundamental; the Mediterranean Sea is warm (20°C in Oran), whereas the Atlantic Ocean is colder. On the Tunisian coast, the Syrtes Sea is warmer again, and generates local cyclogeneses. The influence of the W–E orientated relief accounts for the variation between the two major facades and the *foehn* situations which can reach the Mediterranean Sea.

Temperatures undergo considerable variation, with short periods of frost during incursions of Arctic air. Likewise, intermittent arrivals of heatwaves from the Sahara are recorded up to the coastal domain, with temperatures up to 45°C. Insolation is strong and the humidity moderate, except on the coast. This accounts for the prominent diurnal amplitudes (Birot 1964; Despois and Raynal 1967).

Two kinds of thermal climates can be distinguished. The first is a mild coastal climate, with low amplitudes; the isotherms are parallel to the coast in Morocco and in Tunisia (Figure 7.5). The absolute minima there are greater than 0°C. Winter averages are much more differentiated (19.9°C in Essaouira, 23°C in Tanger and 27°C in Sfax). Humid maritime and often mild winds (the *estival* sea breeze which penetrates up to 40–50 km inland) are responsible for the temperate character of the coastal fringe, especially along the Atlantic (Troin *et al.* 1985).

The second type of thermal climate is an interior, continental climate, cold in winter and very hot in summer; in the Constantine region the average temperature in winter is 4–5°C, with 30 to 50 days of frost a year. In summer, temperatures increase rapidly; averages for July are everywhere greater than 27°C, and average maxima reach 38–40°C. Average minima in summer are 16–20°C. Diurnal amplitudes are 10°C in winter and 16–18°C in summer. This is linked to the influence of relief; indeed, the thermal altitudinal gradient is low in winter (0.4°C/100 m) but stronger in summer (0.7°C). This accounts for the exaggeration of the continentality of intramontane depressions (Troin *et al.* 1985; Despois and Raynal 1967).

The seasons are more distinct than in Europe: summer is longer and drier with some rainstorms; winter is often humid but has dry, anticyclonic phases. Rain maxima occur in different periods of the season: to the west, in Morocco and in Orania, there is a first maximum in November and a second in February/March, with rainstorms in the mountains; to the east, the rainfall maximum is in December, with a secondary peak in May. Evaporation rates are very high. Everywhere, evaporation exceeds rainfall by more than 1 m.

The climate is characterized by a strong irregularity (Figure 7.6). Deviations in relation to annual averages are 15–25% in the Tell. In extreme cases, they are up to 100%. The irregularity also concerns the commencement of the rainy season and the duration of the dry one (Bensaad 1993; Benzarti 1987; Amar 1965; Couvreur-Laraichi 1972). The seasonal concentration of rainfall is

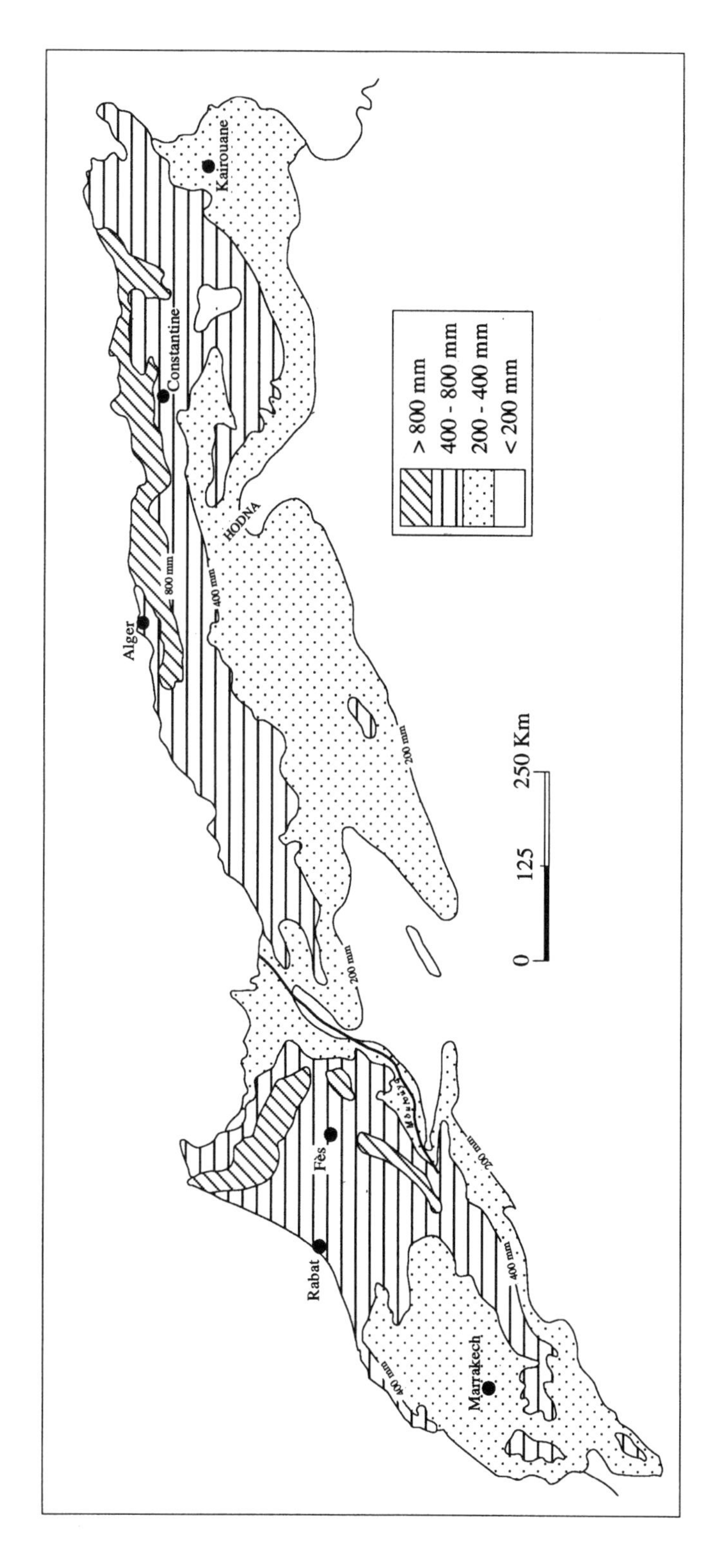

Figure 7.4 *Distribution of mean annual rainfall in the Mediterranean regions of North Africa* (reproduced from *Atlas du Maroc* (1962, sheet 4a) by permission of Comité National de Géographie du Maroc)

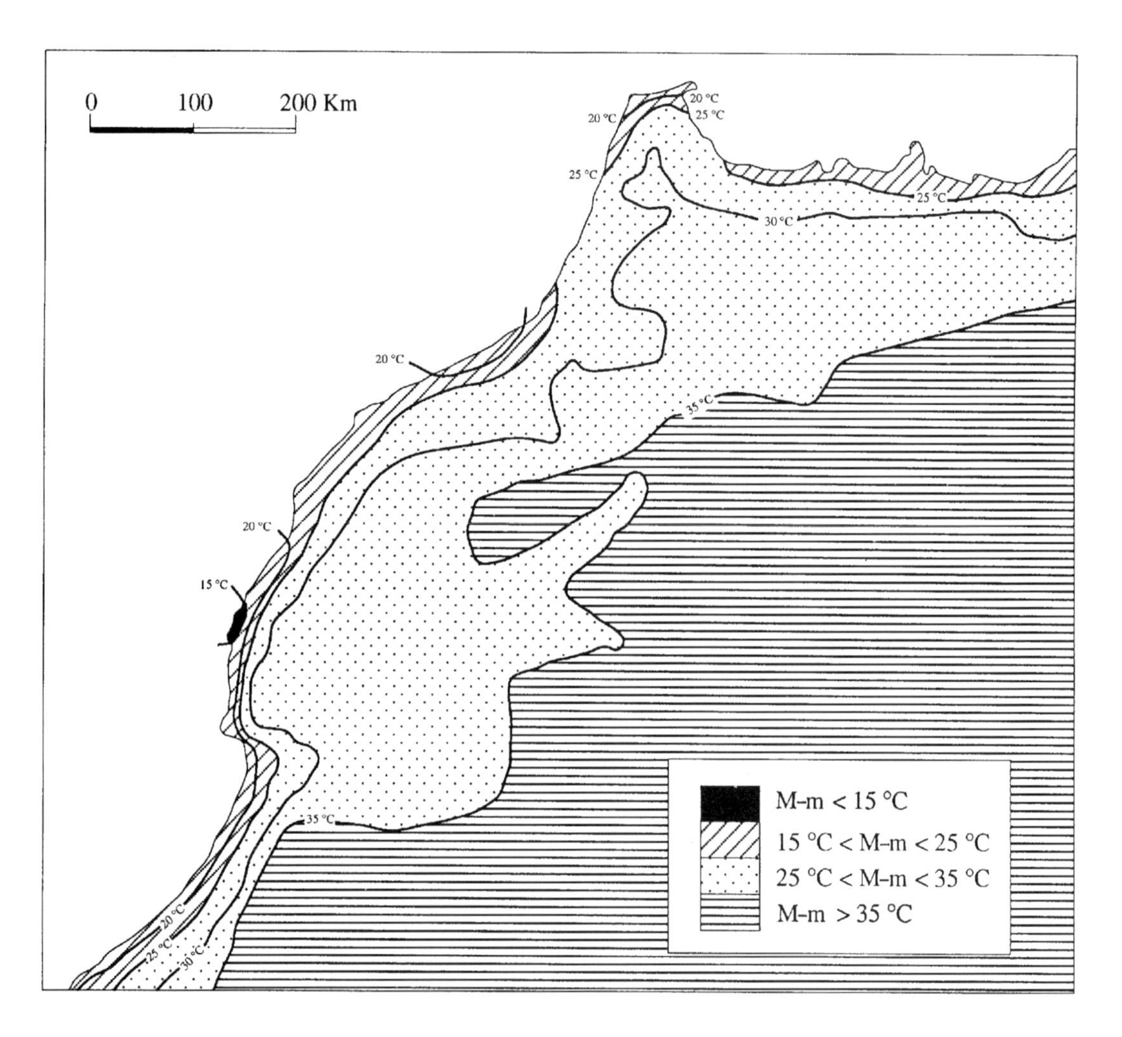

Figure 7.5 *Range of temperature extremes, Morocco: M = average maximum of the hottest month; m = average minimum of the coldest month* (reproduced from *Atlas du Maroc* (1963, sheet 6b) by permission of Comité National de Géographie du Maroc)

another basic characteristic of the climate. There are 120 rainy days in the eastern Tell, declining to 75 days in the rest of the humid zone, particularly in the Rif (Despois and Raynal 1967).

The fundamental role of relief and its orientation must be stressed, especially the contrast between windward and leeward slopes. Isohyets generally run parallel to the contours (Gaussen 1958). Humid zones, which receive more than 600 mm of rain, have islets receiving more than 1 m and semi-arid zones receiving between 300 and 600 mm of rain, with the latter being more extensive. The higher rainfall regions can be further subdivided into two kinds of regions: first, zones with heavy rain at low altitudes, namely in Atlantic Morocco and in the eastern Tell; and second, high rainfall zones at high altitudes in the High Tell and Atlas (Figure 7.7).

Semi-arid regions include: plains and nether plateaux, in particular coastal plains which are relatively meridional, such as the Doukkala as well as sheltered maritime regions such as the Oranian Tell and the eastern Rif; semi-continental basins such as the Sais of Fes; or the Constantinian and meridionally positioned mountains such as the southern slopes of the High Atlas and some summits of the Anti-Atlas (Despois and Raynal 1967; Troin *et al.* 1985).

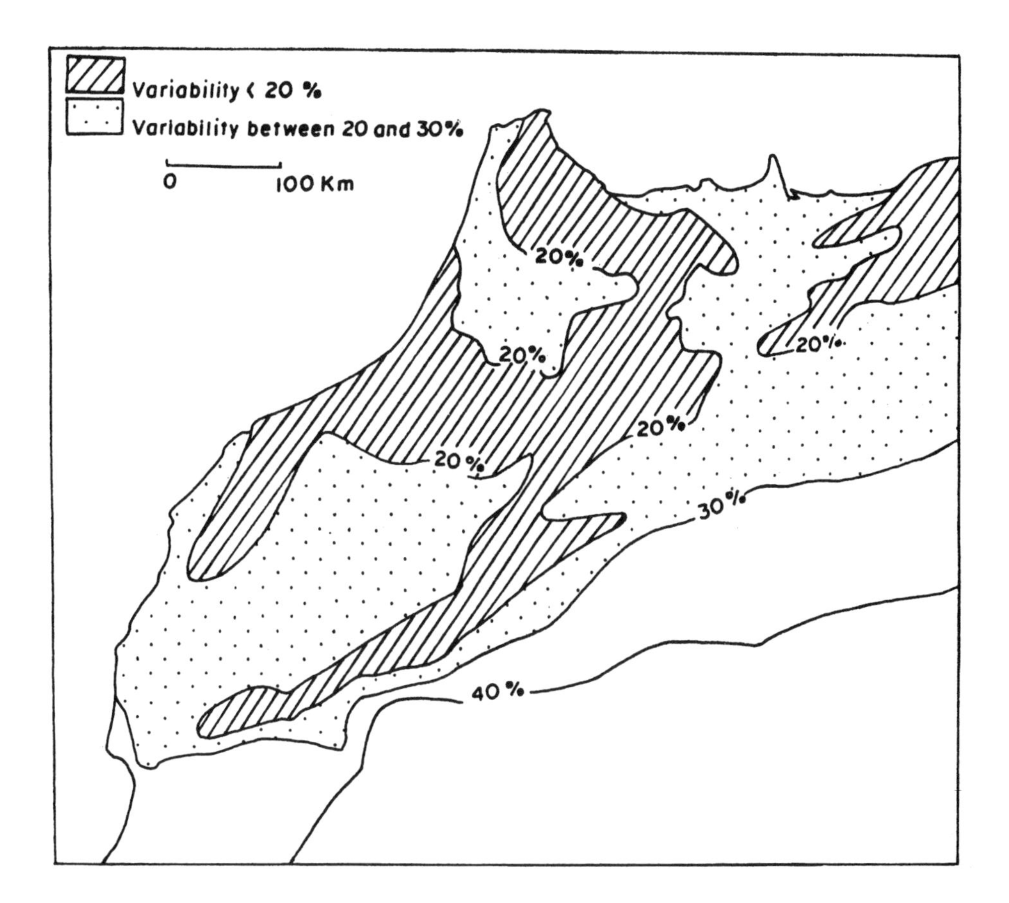

Figure 7.6 *Inter-annual variability of rain in Morocco* (reproduced from Cote and Legras (1966) by permission of Association Nationale des Géographes Marocains)

Plant and animal life

In general, the Maghrebian environment is characterized by biological diversity and a great wealth of flora and fauna, despite the poverty of many soils. There is a strong endemism of fauna and a large number of bird species, many of which are endangered.

The North African plant cover is characterized by its variety and can be classified into different formations, from the Mediterranean forest to the steppe, with an increasing degradation from north to south. The complexity of the influences accounts for that physiognomic and phytosociologic variety. Criteria are climate (zonal subdivisions and continentality), relief (namely altitude and aspect), lithology (siliceous, calcareous and schistose lands), and the human context (demographic load and land use).

Subhumid Mediterranean forest is located in environments receiving more than 600 mm of rainfall annually. These forests include cork oak with dense and shrubby understorey, the most typical forest occurring in littoral areas with as little as 500 mm rainfall and atmospheric humidity, such as the Mamora and the East Tell. Green oaks cover 2 000 000 ha and extend into the semi-arid environment with rainfalls of 400 mm. These forests are often degraded to *matorrals*, and grow on the southern slopes of the eastern Tell, on the northern slopes of the western Tell and in the higher rainfall parts of the Atlas, up to 2800 m (Benabid 1982; Al Ifriqui 1993; Deil 1987, 1988; Tatar 1995; Zitam 1987) (Plate 7.6).

Subalpine formations occur from 1800 m to 3100 m in the continental Atlas, and comprise

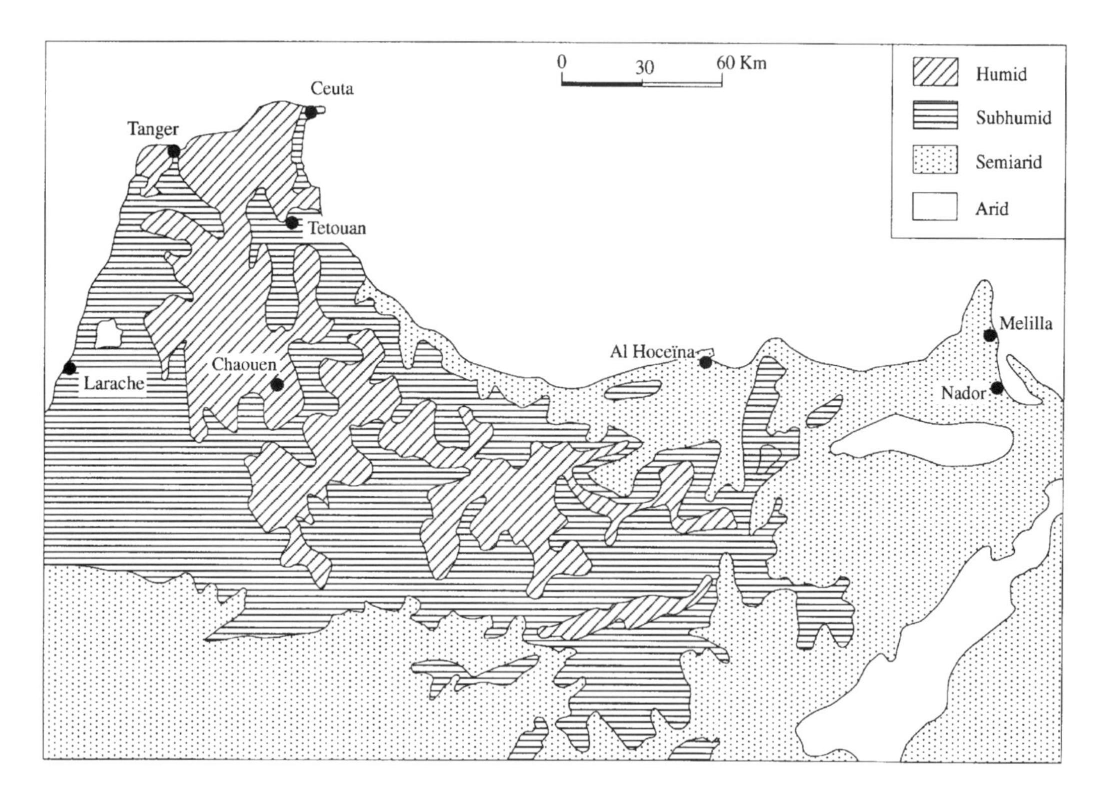

Figure 7.7 *Bioclimatic zones of the Rif* (reproduced from *Atlas du Maroc* (1963, sheet 6b) by permission of Comité National de Géographie du Maroc)

Plate 7.6 *Degraded subhumid "Mediterranean forest": burnt* matorral *in centre background, and* Pinus *on higher ground, on the fringe of the Middle Atlas*

Plate 7.7 *Cedars at a Middle Atlas ski resort*

conifers, firs (*Abies pinsapo*), cypress (*Cypressus atlantica*) and cedar (*Cedrus atlantica*) (Plate 7.7). Juniper (*Juniperus thurifera*) appears at the top, and beyond, the vegetation is composed of cushion bushes.

In environments receiving less than 600 mm of rain annually, several formations represent the transition towards the steppe: the green oak in *matorral*; resinous trees like *Pinus halepensis* which cover 1 000 000 ha, especially in Algeria and in Tunisia; the "thuya" (*Tetraclinus articulata*), which also spreads over 1 000 000 ha and is the most representative of these environments; and on sunny slopes, junipers (*Juniperus phoenicea*), the oleasters (*Olea eropaea*) *matorral*, lentiscuses (*Pistacia lentiscus*), the dwarf palm tree, and asphodels which are more particularly related to the hills and the depressions.

Contrasts are also noticed between the maritime and continental domains. For example, in the Moroccan southwest region, the argan tree (*Argania spinosa*) and euphorbia represent formations with tropical affinities; in the Tunisian Sahel, elements of Saharan flora can be found; and in many semi-arid mountains, the thurifer juniper, the thuya and the green oak are the most common species.

The North African fauna result from many successive upheavals in this isolated space between the Mediterranean and the desert. There is a high rate of endemism, with more than 50% of species being found only in the Maghreb. The fauna are more linked to a European origin, with only some of the mammals having African and tropical origins.

The species are increasingly endangered as a result of human pressures. More than 100 species of mammals are endemic to the Maghreb and are currently under threat: examples are the "Mouflon à manchettes" (*Ammotragus lervia*); the gazelles, particularly *Gazella dammah* and *Gazella dorcas*; many kinds of monkeys, such as *Macaca sylvanus*, the hyena (*Hayena barbarus*); and "panthers" (*Panthera pardus*). On the coast, the most famous mammal is *Monachus monachus* (Phoque moine) which is now well protected.

The People

The most recent censuses were conducted in 1987 for Algeria, and 1994 for Tunisia and Morocco. Around 1990, there were 25 million inhabitants in Morocco, 25 million in Algeria, and eight million in Tunisia. Since 1950, the population of the three countries has practically trebled. This has been due to the high rate of natural increase (2.7% in Morocco, 3.1% in Algeria and 2.6% in

Tunisia). But North Africa has already begun its demographic transition. The growth rate, therefore, is tending to decrease strongly, particularly in Tunisia and Morocco. For example, in Tunisia the annual rate of natural increase was 2.3% for the decade 1984–1994; but in 1984 the rate was 3.27%, declining to 1.82% in 1993. Birth rates declined strongly, from 4.85% in 1965 to 3.36% in 1990. Fecundity also decreased, from 7.2% to 4.7%, but it still remains higher than the world average. The death rate, which was 1.81% in 1965, has fallen to 0.72%, although infant mortality remains high (6.4%). Nevertheless, the population remains very young (about half the population is less than 19 years of age). Population densities (Figure 7.8) are 10–15 inhabitants/km² in the pastoral mountains and 20–30 inhabitants/km² in the populated ranges, such as the High Atlas and the Aures (Kassab and Sethom 1980; Cote 1983, 1996; Troin *et al.* 1985).

Population growth varies between rural and urban environments. From 1980 to 1990, growth was about 4% in cities whereas in rural areas it was only 1.5%. This is linked to the important population migrations which particularly concern the inhabitants of the countryside. It is in the mountains, in the most disfavoured zones, where population structure is the most imbalanced, with a lack of young adults, particularly males, owing to temporary work migration towards the cities

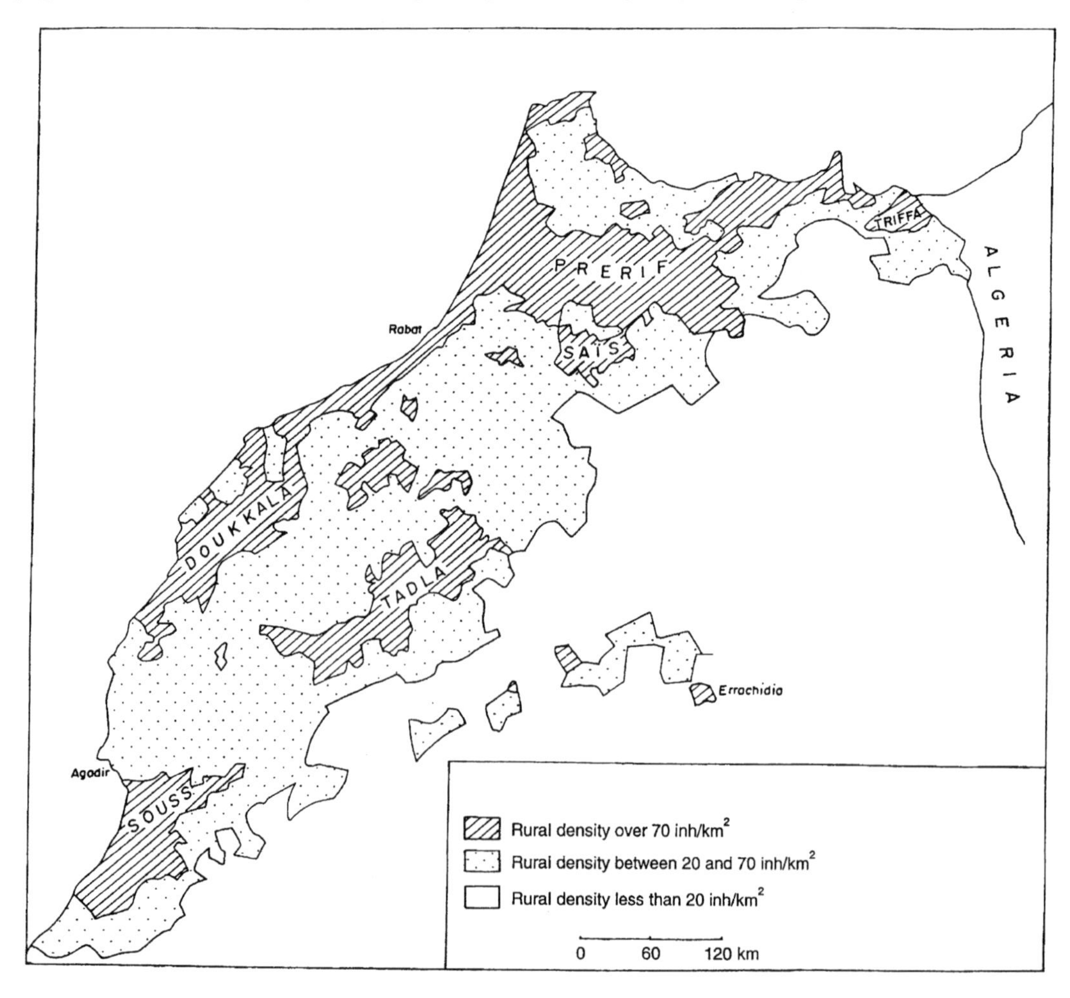

Figure 7.8 *Rural population density in northern Morocco, 1982* (from *Atlas du Maroc* (1984, sheet 31.3) by permission of Comité National de Géographie du Maroc)

and abroad. Great migratory movements have been recorded. Most mountains, as well as parts of the high plateaux and plains, have experienced a relative decrease of population.

Socio-economic analyses shows a low active population rate (24% in Algeria), which is related to the youthful population and the imbalance in the activity of both genders. Yet, only 16% of the population really have an occupation, and unemployment reaches 23% to 25%. The imbalance between rural and urban populations is marked. Rural populations, which represented 58% to 60% of the population in 1980, had only 47% of incomes. School attendance rates also show an inequality between rural and urban dwellers, as does the female activity rate (Troin *et al.* 1985).

Locally, human pressures on the environment are increasing, as is the case in the Rif mountains. The population of the Central Rif alone, in the zone of the massive chains of the Sanhaja and Rhomara, increased from 77 000 to 163 000 inhabitants between 1950 and 1980 (Maurer 1990). The rate of growth seems, in fact, to have increased, from 1.6% between 1930 and 1950, to more than 2.5% subsequently. Such rates are not found in the other Maghrebian mountains, and occur only in rural areas which are being fully developed, as in irrigated areas.

Several factors account for such a particular growth in the Rif: first, the reunification of Morocco and the opening of the Rif to the exterior world; second, the setting up of a basic infrastructure and the formation of a number of urban centres having commercial and administrative activities, thereby providing incomes; and finally, the development of illicit and highly remunerative activities such as the cultivation and trading of cannabis on the one hand, and illegal trade in relation to the two cities occupied by Spain (Ceuta and Melilla) on the other.

The Economy

Resources and industry

Data for the Mediterranean region of North Africa are not available separately, thus the discussion of the economy concerns the three countries as a whole.

The economic potentials of the three countries are unequal, and are not based exclusively on the resources of the Mediterranean regions. The steppe and Saharan environments have a very important potential in minerals, pastures and fish resources. But the essential part of agricultural and industrial resources, of infrastructure and equipment, is concentrated in northern regions. In Morocco, for example, a line from Agadir to Oujda delimits that part of Morocco known as being "useful" and roughly corresponds to the limit of regions with favourable climates.

The gross domestic product (GDP) of the three countries in billions of US dollars is 46 for Algeria, 13.3 for Morocco and 7.76 for Tunisia (Troin *et al.* 1985). On a dollars per capita basis in 1982, these data convert to 2266 for Algeria, 1395 for Tunisia and 864 for Morocco. However, the Tunisian and Algerian GDPs are inflated by the production of hydrocarbons from the Saharan regions of those two countries. The structure by economic sector shows a clear differentiation between the three countries: industry holds the first rank in the Algerian economy with 58%, whilst services are most important in the economies of both Morocco and Tunisia (45%). Agriculture has a secondary role in the economies of the three countries (18% in Morocco, 15% in Tunisia and 6% in Algeria).

In Morocco, the largest population and most important economic activities are concentrated in the northern provinces representing half the area of the country. Agriculture is a developing sector which has not yet reached its full potential, despite big efforts and especially the policy of irrigation. Phosphates, for which Morocco is in the first rank of the world's exporting countries, are the most important mining resource, but their importance in the economy of the country has tended to decline because of the fall in prices. Tourism represents a sector in full development, yet it experiences many difficulties. Energy dependence is the main economic weakness, and partly accounts for the high indebtedness of the country. However, the policy of restructuring has reduced the rate of increase in the demand for energy.

In Algeria, the Mediterranean environment represents a small part of the national territory (Cote 1983). The vast Sahara produces hydrocarbons which are partly value-added in the northern provinces. The choice was for heavy industrialization and the creation of several coastal nodes (Spiga 1993). This accounts for the spatial imbalance between four littoral provinces which experience higher than average demographic growth and important economic development, and the remaining areas. One of the consequences has been the disorganization of the agricultural economy through the loss of rich soils in the urban and industrial growth zones, and also the competition for water. Additionally, there has been a general agricultural decline in many regions affected by rural depopulation (Pérennes 1993).

In Tunisia, the current industrial development is important but it does not compete with the agricultural sector, which has retained its vitality. The choice was to develop the tertiary sector, especially tourism, which is growing rapidly (Kassab and Sethom 1980).

Agriculture and forestry

With regard to the agricultural sector, three systems of organization of space in the Tellian system can be distinguished.

(1) There is the system of plains and low plateaux with rich potential, where a modern agriculture is developing and undergoing an increasingly diversified commercialization.
(2) The Atlas system associates agriculture and grazing in varying proportions. Pastures from bordering steppes are used as well as undergrowth pasturing and the use of high altitude meadows (Plate 7.8). The piedmont lands and the cultivated terraces are in the middle mountains, in addition to the sloping, temporary fields. These latter have the tendency to extend at the expense of the degraded pastures. Current developments include: abandoning lands which are difficult to till; limiting summer displacements; in the Atlas, the settlement of summer pastures bordering the Middle Atlas; and migration towards the cities and abroad.
(3) The third agricultural system is the Tellian one, which is also based on an agro-sylvo-pastoral economy. But it concerns regions where rural demographic growth is still important, despite big problems of resource degradation. Very often, the active population working in the agricultural sector exceeds 50% and can even reach 75% in some

Plate 7.8 *Degraded (over-grazed) steppe pasture with High Atlas in background. Tussocks are* Stipa tenacissima, *which is also common in southern Spain. Early October (1994) snow is unusual in the High Atlas, even though these peaks rise to 3800 m above sea level*

sectors, such as the Province of Taounate (Rif mountains). In many regions, agriculture is no longer of primary importance, and in some areas, abandoned sectors denote situations of uneasiness (Maurer 1991). As a matter of fact, the abandonment seems to be temporary and concerns only isolated sectors. Throughout history, episodes of abandonment and of reconquest have followed one another (Pascon and Van der Wusten 1983). Today, migration does not entail a systematic abandonment of the lands, because association of the land owners, living in the towns, with local rural populations allows the pursuit of cultivation.

In the Tellian regions, agricultural resources are evolving rapidly (Maurer 1991). Cereal cropping prevails not only on the good soils but also on the margins. It is increasingly accompanied by leguminous plants, accounting for the limitation of the fallow period to less than 30% of the cultivated areas. An important extension of market gardening on irrigated parcels of land is noticeable; their value can be up to 20 times that of a good harvest of cereals. These cultivations receive important contributions of manure. Arboriculture is an old tradition, namely with the olive tree that progresses in many sectors. Finally, grazing plays an important part in the incomes of rural population, with sheep essentially in the dry mountains and goats as well as cattle in the humid mountains.

Within the Tellian mountains there is, then, a certain evolution of the agricultural sector in the sense of diversification and improvement in yields. But the evolution remains weak for several reasons: competition from the products of the plains whose operating costs are lower; competition from other economic sectors such as industry; and the growth of less demanding and highly remunerative activities, such as the cannabis trade and smuggling.

Yet, in North Africa, agriculture no longer represents a secondary economic sector if one examines its share of GDP. In other respects, the importance of the rural world is tending to decline.

With regard to forestry, there are 6 000 000 ha of forests in Morocco, of which only 800 000 ha are well delimited and protected. Reforestation covers 500 000 ha. However, the rate of reforestation of the whole of Morocco is very low (7.8%), although it reaches higher figures in the most humid regions (31.5% for the central north region, 25.6% for the northwest and only 17% for the Tensift region). Productivity is interesting only in the subhumid and humid environments.

At present in Morocco, forestry provides a basic livelihood for some communities (Direction de l'Aménagement du Territoire 1992). Yet forests are a resource whose management is suffering many problems (over-consumption, illicit cutting and maladjustment of legislation to the varied social conditions). Forest profitability is becoming increasingly precarious. The most important species, like the cedar, the fir and the pine, are threatened in the short term. Potential annual production of the forest is as follows: 351 000 m^3 of timber, 440 000 m^3 of plywood, 826 000 m^3 of industrial wood, 2 700 000 m^3 of firewood and 447 000 m^3 of cork. For firewood, annual cutting rates are estimated at about 10×10^6 m^3, which represents three times the rate of renewal. That means a reduction of the forest area of about 20 000 to 25 000 ha/yr. Direct clearing for cultivation is about 4500 ha/yr. Forest pastures also play an important role in the grazing of cattle. Overgrazing may be having an impact on forests, especially in dry years. Its effect is added to excessive cutting, forest clearing and fire.

Administration and Social Conditions

The three countries have a common history as much through the essential feature of their settlement by the Berber people, as through their belonging, by cultural and political links, to the Arab and Muslim world as a whole. All of them are seeking to establish close economic links with Europe. However, the political, economic and social orientations are clearly different.

With a constitutional monarchy, recognizing the multiplicity of parties and unions while praising liberal and economic options, the Kingdom of Morocco has long maintained strong central control of the economy. But it is now in the process of rapid privatization of many important sectors.

Algeria has long been a socialist republic basing its system on one very strong party, state control and rigorous planning. Yet recently, the tendency to economic liberalization has had some success, questioning again the agrarian revolution, and a certain opening to the private sector; whereas political democratization has been a failure. Nowadays, the political system is a temporary presidential system requiring strong military support.

The Tunisian system is a parliamentary republic where the influence of the formerly single party, having become the official party, remains predominant. It is a presidential regime which chose the liberal option and a wide opening towards the exterior while favouring foreign investments and tourism development.

In the three countries there exists a strong middle class having sundry origins, from the traditional and urban middle class in Morocco and Tunisia and from the former single party in Algeria. The disfavoured classes represent the majority of the population which is composed of the destitutes from the countrysides and the uprooted people of the city. The middle class has the potential to develop further, mainly in Tunisia.

Culture

The North African Mediterranean regions conserve the remnants of old Mediterranean cultures with old rural civilizations, especially in the mountains. This old culture associated Mediterranean agriculture with the nomadic grazing of sheep. The most important characteristics of this way of life are the strength of familial communities and a close attachment to the land.

Successive events have contributed to the present-day culture:

- an old Berber background;
- Phoenecian and Roman colonization of coastal areas, with Roman occupation of the land and Christianization of some of the population, particularly urban traders;
- the Islamization of North Africa, followed by the introduction of some Arabic tribes and the Arabization of the language of plains dwellers, while the mountain people retained their Berber languages; and
- the French and Spanish colonizations of the 19th and 20th centuries, which partly transformed the way of life of the people, bringing modern techniques and a new culture, transplanted on to the existing varied and heterogeneous culture.

The North African Mediterranean regions have in common their membership of the Arab world, the Islamic religion and the use of two languages, Arabic and French. The Islamic religion (Sunnite Islam) marks the landscape and the society and also the legal, social and political systems.

8

Greater California

Antony Orme and Amalie Jo Orme

Introduction

The Mediterranean region of western North America covers approximately 350 000 km^2, extending from southeast to northwest as a narrow, elongated zone between latitudes 30°N and 43°N and between longitudes 115°W and 125°W (Figure 8.1). This region thus lies primarily within the southwest United States, mostly in California, but extends northward into southern Oregon and southward into Mexico along the western margin of Baja California Norte. It is bordered westward by the Pacific Ocean and eastward by the subtropical Mojave and Sonoran deserts and, further north, by the temperate Nevada desert. Although the region stretches latitudinally over 1600 km, it never exceeds 300 km in width and in the south is little more than 100 km wide.

This region, here termed Greater California, contains about 35 million people. Most are concentrated in three large metropolitan areas: around San Francisco Bay in northern California, around Los Angeles in southern California, and astride the United States–Mexico border around San Diego and Tijuana. Other important centres lie within California's Central Valley, including the State capital of Sacramento, and along the coast between the major metropolitan areas. Beyond the cities, lowland economies are primarily agricultural whereas the extensive and rugged mountain ranges are mostly forested, their cover ranging upwards from *chaparral*, through deciduous woodland to coniferous forest and locally to alpine meadows.

Unlike the classic Mediterranean region of southern Europe and North Africa, Greater California has no extensive east–west seaway, no large islands, and no internal peninsulas or semi-enclosed seas. Instead, because the dominant tectonic grain extends from northwest to southeast, the principal mountain ranges parallel the Pacific coast, thereby restricting oceanic influences and imparting a NW–SE trend to climatic and ecologic regionalism. Thus the Coast Ranges impede the eastward penetration of ocean influences, while the higher Sierra Nevada, Transverse Ranges and Peninsular Ranges form an effective eastern margin to the region. Their considerable height ensures that rain-bearing winds off the Pacific Ocean release most of their moisture against west-facing slopes, leaving leeward slopes in a significant rainshadow that results in mostly arid conditions. The crestline of these mountains is thus a significant parting between the humid temperate ecological domain inland from the Pacific Ocean, and the dry domain that extends across much of western North America eastward to beyond 100°W.

The Land

Tectonics and relief

n understanding of Late Cenozoic tectonism is fundamental to any interpretation of Greater California's present physical and ecological character. Tectonism dictates the primary relief, which in turn constrains the atmospheric circulation, and the resulting climate patterns are

Land Degradation in Mediterranean Environments of the World: Nature and Extent, Causes and Solutions.
Edited by A.J. Conacher and M. Sala.

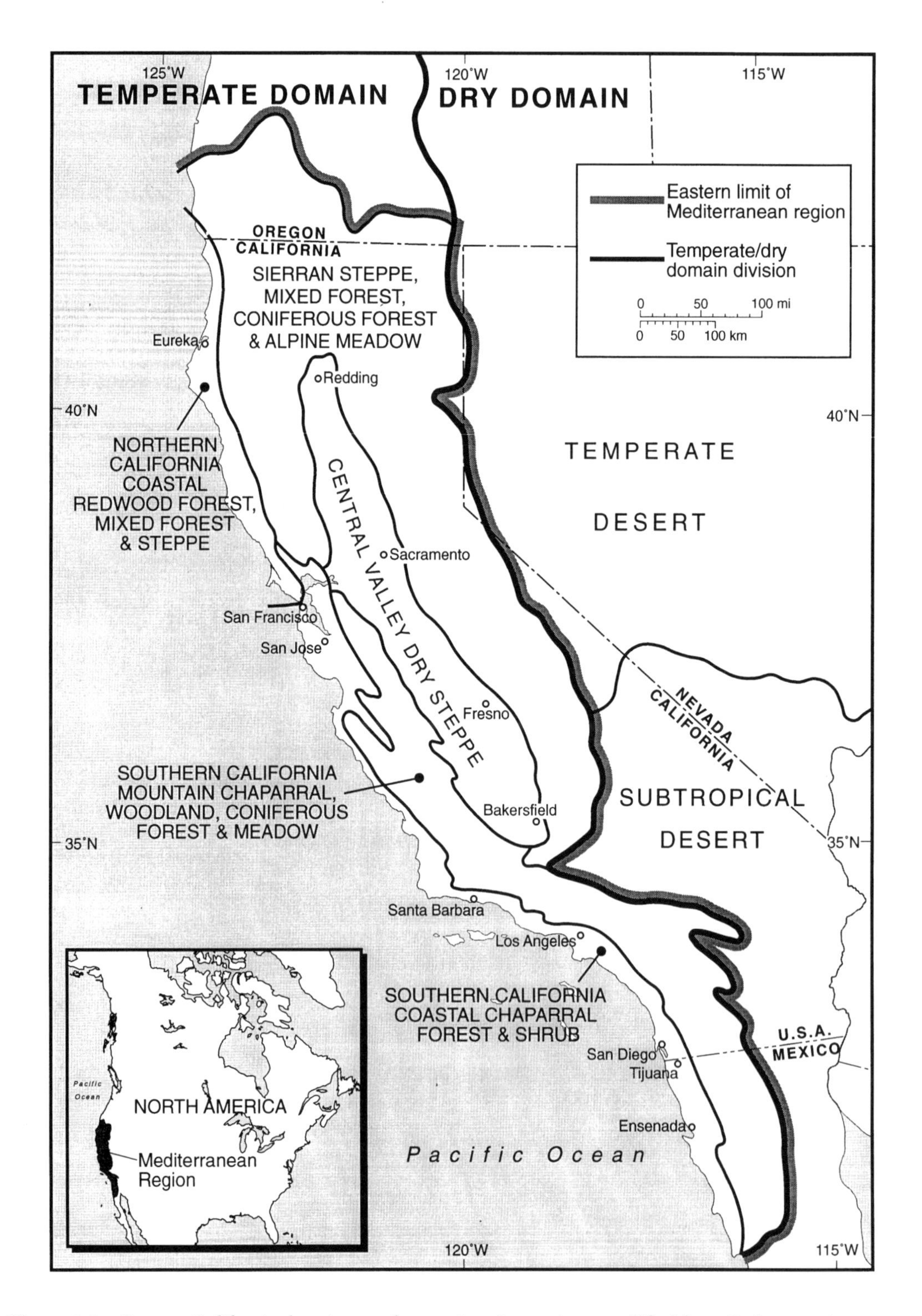

Figure 8.1 *Greater California: locations and ecoregions* (ecoregions modified from Bailey 1995)

reflected in runoff, erosion and vegetation. The region's changing tectonic geometry, a complex response to Late Cenozoic plate motions, also helps to explain the present distribution of plants and animals. More recently, the principal distinction in human endeavour, between well populated lowlands and sparsely occupied mountains, also reflects this tectonic imprint. Recurrent earthquakes, the expression of continuing tectonism, are capable of causing significant loss of life, livelihood and property throughout the region, notably in cities straddling the San Andreas Fault zone (for example, the M7.1 Loma Prieta earthquake near San Francisco, October 1989) and in the compressive Transverse Ranges around Los Angeles (for example, the M6.7 Northridge earthquake, January 1994) (Figure 8.2).

In essence, Greater California developed from differential uplift during Neogene times as the westward-moving and westward-growing North American Plate over-rode various subducting plate segments lying east of the East Pacific Rise spreading centre. Then, some five million years ago (5 Ma), part of this region transferred across the East Pacific Rise on to the Pacific Plate and began accelerating away from the main North American Plate. Continuing lateral shear saw the opening of the Gulf of California along major transform structures to the southeast and, as these propagated upwards into the continental crust further north, separation along a series of *en echelon* strike-slip structures, of which the San Andreas Fault is now the most significant (Figure 8.2). Massive crustal deformation involving both compression and extension preceded, accompanied and followed plate separation.

Meanwhile, the Sierra Nevada, which had been little more than 1000 m high at the onset of the Neogene around 24 Ma, reached 2500 m by 10 Ma, 3500 m by the close of the Neogene around 2 Ma, and rose a further 500 m during the Quaternary. Further north, massive volcanism associated with continuing plate subduction near the coast characterized the northern Sierra Nevada and Cascade Range. West of these mountains, the subsidence which had formed a shallow sea along the Central Valley was offset by the influx of Neogene terrigenous sediments from the Sierra Nevada, while along the coast differential uplift, subsidence and faulting raised and then dislocated the Coast Ranges.

Thus, in general terms, Greater California comprises two distinctly mobile and seismically active structural units separated by the San Andreas Fault system (Orme 1992a). To the east, at the leading edge of the North American Plate, lie the northern Coast Ranges, Central Valley, Cascade Range and Sierra Nevada. To the west, and moving northwestward at a rate of 6 cm/yr on the western limbs of the East Pacific Rise, lie the southern Coast Ranges, the Transverse Ranges of southern California, and the Peninsular Ranges forming the mountainous spine of Baja California.

Climate

The climate of Greater California is in many respects typically Mediterranean (NOAA 1973). Cool wet winters alternate with warm dry summers, with temperatures and length of dry season increasing towards the south while precipitation increases towards the north and with elevation. In summer, the Hawaiian high pressure cell dominates, winds flow into the region from the northwest, drought prevails, and temperatures in interior valleys may exceed 40°C. In winter, this high pressure cell contracts, mid-latitude cyclonic systems predominate, and storms from the westerly quadrant bring rain and, at higher elevations, snow.

In other respects, the regional climate is less typical. The geometry of the Gulf of Alaska and western North America ensures that eastward-moving cyclonic systems strike hard against west-facing mountains but, with warm fronts commonly deflected northward, cold fronts are the principal source of moisture for the Mediterranean region. Further, when cyclonic systems stall against the coast, secondary circulations import abundant moisture from the warmer Pacific to the southwest. In either case, precipitation during storm events is frequently intense and persistent, leading to rapid runoff from mountain slopes and severe flooding in nearby lowlands (Figure 8.3). Offshore, the cool California Current, part of the North Pacific surface gyre which has intensified over the past 24 Ma, is particularly

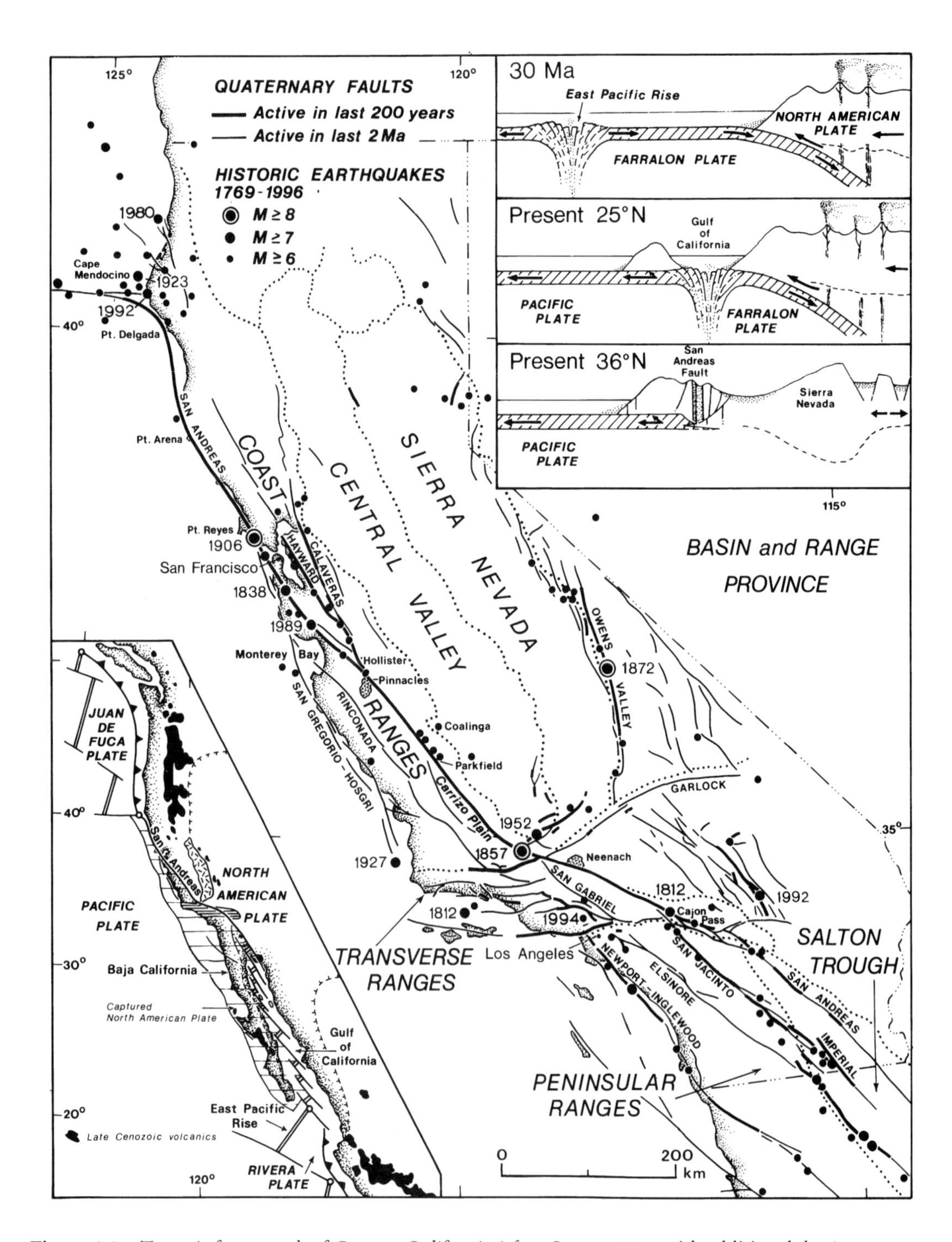

Figure 8.2 *Tectonic framework of Greater California* (after Orme 1992a, with additional data)

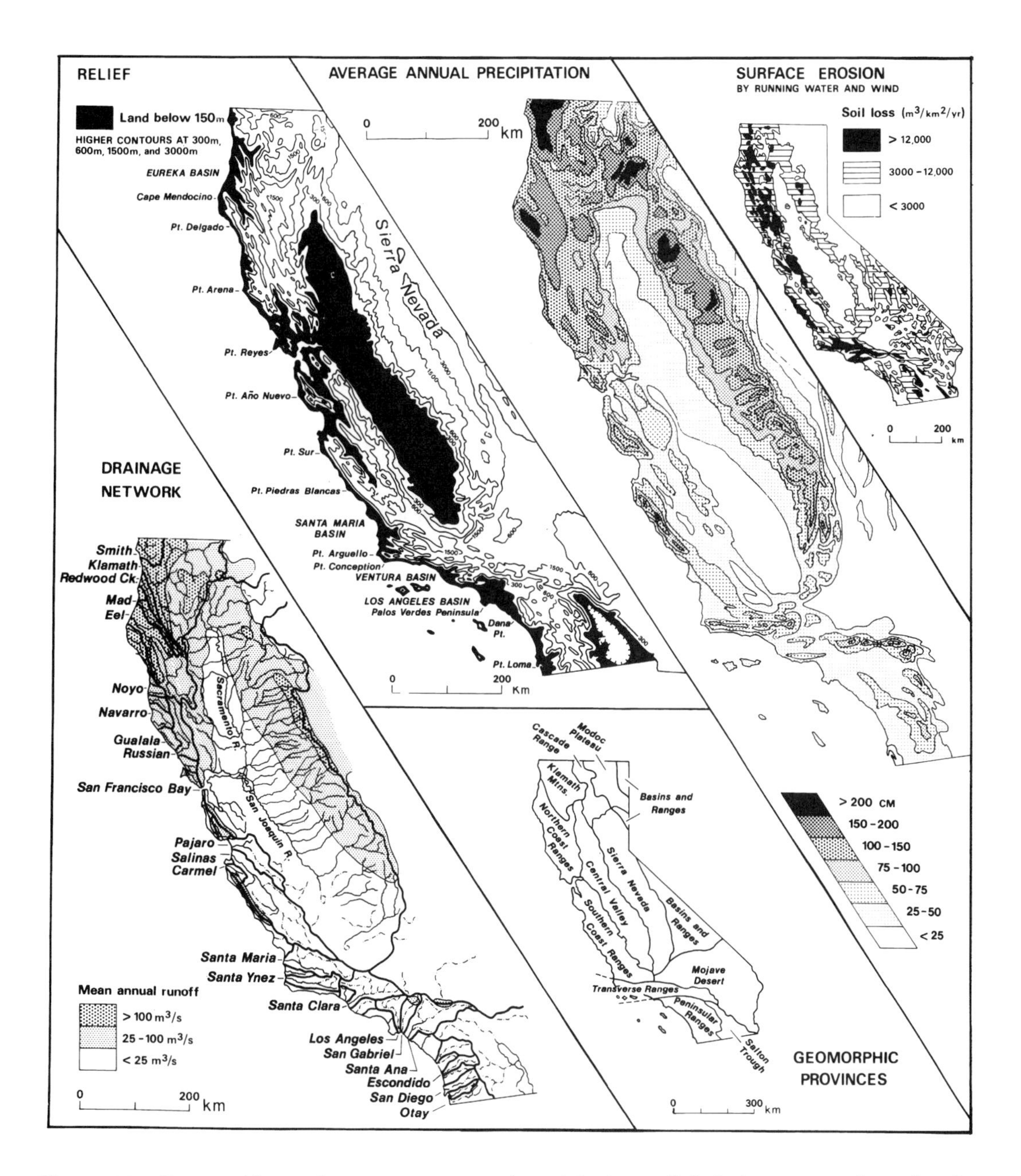

Figure 8.3 *Geomorphic provinces, average annual precipitation, relief, drainage network, and surface erosion* (after Orme 1985)

effective in moderating coastal climates, leading to extensive advection fogs and low-level inversions, especially in summer when warm subtropical air from the Hawaiian cell reasserts its onshore flow but is cooled on nearing the coast (Plate 8.1).

Drainage and water resources

The region's drainage and water resources reflect both tectonic and climatic imprints (Kahrl 1979). The Coast Ranges are drained by many poorly integrated streams whose swift, erratic courses reflect recent uplift; long streams often flow

Plate 8.1 *Advection fog impinging on the central California coast. The fog forms as warm moist air approaches the coast over the cool California Current*

along synclinal or faulted troughs, other streams are often antecedent to the ranges they cross. In the north, where the rainy season is longer, major rivers are perennial. Further south, as the dry season lengthens and precipitation decreases, summer flows diminish and many rivers become intermittent, although surface flows are still flashy and dangerous during winter storms and ample groundwater persists beneath most floodplains. Further inland, rivers draining the tilted western slopes of the Sierra Nevada formed large lakes in the Central Valley at intervals during the Quaternary. As these lakes evaporated, drainage became integrated along the NW–SE axis of the valley, the more northerly streams joining the 600 km Sacramento River, the more southerly draining to the 560 km San Joaquin River, with both systems joining to exit through the Carquinez Strait into San Francisco Bay and from there through the Golden Gate to the Pacific Ocean (Figure 8.3).

The Central Valley contains vast groundwater resources but, as discussed in Parts II and III, both the quality and quantity of these resources have come under threat in recent decades. Discharge from California rivers approximates 90×10^9 m^3/yr, 41% of this by rivers draining from the Northern Coast Ranges directly to the Pacific, a further 32% from the Sacramento River basin, with much of the rest from the San Joaquin River basin (Figure 8.3).

Ecosystems

Reflecting the above generalities, the Mediterranean region of Greater California can be divided into five reasonably distinct ecosystems (Bailey 1995). These ecosystems reflect well the linkages that exist between geomorphic provinces, climatic regions and major units of the California Floristic Province (Hickman 1993). Although one may debate the appropriateness of this division, and indeed the inclusion of such elements as alpine meadow and semi-arid steppe, these ecosystems as a whole all respond to a Mediterranean-type climate regime characterized by summer drought, modified in response to such relief factors as elevation and rainshadow (Figure 8.1).

Northern California coastal redwood forest, mixed forest and steppe (12 000 km²)

Under the influence of the cool California Current, a narrow strip of marine terraces, low hills and coastal valleys along the California coast north of the Golden Gate is characterized by cool, foggy summers and mild wet winters. Annual temperatures average 10° to 14°C and frosts are rare. Annual precipitation ranges from 1000 to 2500 mm, essentially in winter and, although two or three months in summer are rainless, fog precipitation continues to play an important ecological role. This region, essentially below 1000 m and between 30 and 50 km wide, has more days of dense fog than elsewhere in the United States (Figure 8.1).

Vegetation in this fog belt is dominated by coast redwood (*Sequoia sempervirens*), a hygrophyllic, warm temperate species which often exceeds 100 m in height and may live from 500 to 1800 years. Redwood forests also include Douglas fir (*Pseudotsuga menziesii*), western hemlock (*Tsuga heterophylla*), western red cedar (*Thuja plicata*), and Port Orford cedar (*Chamaecyparis lawsonia*), together with a well developed understorey of California rosebay (*Rhododendron macrophyllum*), western azalea (*Rhododendron occidentale*), salal (*Gaultheria shallon*) and California huckleberry (*Vaccinium ovatum*), with sword fern (*Polystichum* spp.) and redwood sorrel (*Oxalis oregana*) among the ground cover. A pine–cypress forest association occurs along the immediate coast and dry headlands support grassy steppe. Drier, south-facing slopes inland are covered by mixed forest comprising coast live oak (*Quercus agrifolia*), tanoak (*Lithocarpus densiflorus*) and Pacific madrone (*Arbutus menziesii*). Forest soils are typically Ultisols, with Mollisols beneath grassland. Mule deer (*Odocoileus hemionis*) and Roosevelt elk (*Cervus canadensis roosevelti*) are common larger mammals, while among other fauna the habitats of the northern spotted owl (*Strix occidentalis*) and various anadromous fish are considered at risk.

Southern California coastal chaparral *forest and shrub (34 000 km²)*

This ecosystem extends southward across deformed marine terraces, rugged coastal hills, and mostly narrow valleys from the Golden Gate in central California to Bahia San Quintín in Baja California Norte. It lies mainly below 1000 m but reaches 50 to 100 km inland in the Santa Maria and Los Angeles basins. Compared with the northern California coast, the climate is hotter and drier in summer, milder and less rainy in winter. Annual temperatures average 10 to 18°C and annual precipitation ranges from 250 to 1250 mm, with summer drought becoming more pronounced southward. Ecologically important advection fogs are common during morning hours in early summer, but usually burn off as the sun rises.

The vegetation is dominated by *chaparral*, a dense sclerophyllous forest of low trees and shrubs whose composition varies with elevation and exposure. Along the immediate coast, the *chaparral* is dominated by coastal sage (*Salvia* spp.), further inland by chamise (*Adenostoma fasciculatum*), manzanita (*Arctostaphylos* spp.), sugar bush (*Rhus ovata*), scrub oak (*Quercus berberidifolia*) and, rarely, Nuttall's scrub oak (*Quercus dumosa*), and in moister, often north-facing habitats by additional oaks and lemonadeberry (*Rhus integrifolia*). Towards the north and in protected habitats elsewhere, several endemic conifers survive, notably Monterey cypress (*Cupressus macrocarpa*), Monterey pine (*Pinus radiata*), Bishop pine (*P. muricata*) and Torrey pine (*P. torreyana*). Broadleaf deciduous trees such as cottonwood (*Populus* spp.) and western sycamore (*Platanus racemosa*) grow well in riparian habitats.

Originally, the larger valleys and coastal basins probably carried extensive sagebrush (*Artemisia* spp.) and grassland communities, but most such areas have been converted to agricultural and urban use. With prolonged summer drought, lightning fires are common and indeed under natural conditions the *chaparral* may be considered a fire-climax vegetation, although the natural regenerative cycle has been disrupted in recent decades by arson, prescribed burning and fire suppression. Soils in this region are mostly Alfisols and Mollisols, high in bases and relatively fertile with adequate soil water. Few large mammals have survived human incursion but the opossum (*Didelphis marsupialis*), North America's only marsupial, is noteworthy, while coastal wetlands and offshore waters provide important habitat for migratory and resident birds, sea lions, sea otters, seals and whales. However, more than 80% of the coastal wetlands that existed along the California coast in 1850 have been lost to reclamation.

Central Valley dry steppe (48 000 km^2)

California's Central Valley, a Neogene sea transformed by fluvial and lacustrine sedimentation into a relatively flat lowland, is some 750 km long, 50 to 100 km wide, and extends from sea level in the Sacramento–San Joaquin delta to around 150 m (Figures 8.1 and 8.3). The valley floor is flat where lakes once existed, gently sloping where alluvial fans and river terraces extend beyond the mountain front, but more dissected along the margins. Summers are hot and dry, winters mild and wet with rainfall varying from 150 mm in the sheltered south to 750 mm near gaps in the Coast Ranges further north. Advection fogs penetrate inland from San Francisco Bay and over the lower Coast Ranges; radiation fogs are common along the valley floor. Annual temperatures average 15 to 20°C, lower in the sheltered south.

Prior to significant human impact, the Central Valley was characterized by a dry steppe vegetation dominated by native bunch grasses (*Festuca* and *Poa* spp.) and needlegrass (*Achnatherum* spp.), and by extensive wetlands. Fire, grazing and tillage have eradicated all but a few native grass stands over the past 150 years and introduced many alien species of *Avena*, *Bromus* and *Festuca*. More than 90% of the wetlands have been eliminated by reclamation. Rivers flow through alkaline flats characterized by greasewood (*Sarcobatus vermiculatus*), pickleweed (*Salicornia* spp.), salt grass (*Distichlis spicata*) and shad-scale (*Atriplex canescens*), while tule (*Scirpus acutus*) marshes border the lower Sacramento and San Joaquin rivers. Soils are mostly Entisols at lower elevation, with Alfisols higher above the valley floor and Aridisols in the more arid south. Human impacts have also caused most large mammals such as the Tule elk (*Cervus nannodes*) to disappear or retreat into the hills. However, bobcat (*Felis rufus*), coyote (*Canis latrans*) and the endangered San Joaquin kit fox (*Vulpes macrotis*) sortie forth from woody cover, mule deer live in shrubby areas, and about 3000 Tule elk survive in scattered locations. Small mammals, birds and reptiles are common, with golden eagle and various hawks as the main avian predators and rattlesnakes as important predators of rodents.

Sierran steppe, mixed forest, coniferous forest and alpine meadow (177 000 km^2)

The Sierran ecosystem covers the southernmost Cascade Range, the Klamath Mountains, the Northern Coast Ranges, and the massive Sierra Nevada. The region is characterized by steep mountain slopes and deep valleys, accentuated in the Sierra Nevada and locally elsewhere by the effects of repeated Pleistocene glaciation. The long western slope of the asymmetric Sierra Nevada rises to 4418 m in Mount Whitney, from which the eastern slope descends precipitously to around 1000 m in the Owens Valley. Further north, the volcanic pile of Mount Shasta reaches 4317 m while all ranges have extensive country above 2000 m. Mean annual temperatures range from 15°C at lower elevations to 2°C along the Sierra crest. Although average temperatures along the crest may fall below 0°C for some winter months, and thus are untypical of Mediterranean regimes, the areas involved are insignificant at the scale discussed here.

The lower western slopes of the Sierra Nevada experience 250 to 400 mm of rainfall annually but the summer season is long and dry. Precipitation rises to around 1800 mm between 1000 and 2000 m, with more falling as snow. Between 2000 and 3000 m, the dry season is much shorter but precipitation is less, around 1000 to 1300 mm, and mostly falls as snow (Figure 8.3).

There are four vegetation zones. In the foothill zone below 1000 m, a parkland of coniferous, deciduous, *chaparral* and grassland associations occurs. Conifers include foothill pine (*Pinus sabiniana*); deciduous trees include blue oak (*Quercus douglasii*) and other oaks. *Chaparral* is dominated by manzanita (*Arctostaphyllos* spp.) and buckthorn (*Rhamnus* spp.); and grasslands include both native and introduced species. The montane zone lies between 600 and 1800 m in the Cascade Range, between 1200 and 2100 in the central Sierra, and between 1500 and 2500 or more in the southern Sierra. This zone is characterized by Pacific ponderosa pine (*Pinus ponderosa*), Jeffrey pine (*P. jeffreyi*), sugar pine

(*P. lambertiana*), Douglas fir (*Pseudotsuga menziesii*), white fir (*Abies concolor*), California red fir (*A. magnifica*), incense cedar (*Calocedrus decurrens*), and locally by the unusually large and long-lived giant sequoia (*Sequoiadendron giganteum*), 50–80 m high, 5–8 m in diameter, and typically more than 1500 years old. The subalpine zone, rising 300–500 m upslope from between 2000 and 3000 m depending on latitude and exposure, is dominated by mountain hemlock (*Tsuga mertensiana*), lodgepole pine (*Pinus contorta*), western white pine (*P. monticola*) and whitebark pine (*P. albicaulis*). The alpine zone above the timberline is characterized by low shrubs and arctic–alpine flowers.

Alfisols predominate at lower elevations, Ultisols on moister, higher slopes, and Entisols along alluvial corridors. Common large mammals include black bear (*Ursus americanus*), mountain lion (*Felis concolor*), coyote and mule deer. There are numerous smaller mammals and birds, including the porcupine (*Erethizon dorsatum*) and predators such as owl and hawk.

Southern California mountain chaparral, *woodland, coniferous forest and meadow (79 000 km²)*

This ecosystem embraces the mountainous terrain of the Southern Coast Ranges, the Transverse Ranges of southern California, and the Peninsular Ranges that extend southward from the Los Angeles Basin into Baja California (Figure 8.1). The Southern Coast Ranges, which extend south from the Golden Gate to the San Rafael Mountains, are low rugged mountains which, though frequently rising above 1000 m, are insufficiently high to form a major climatic barrier. The Transverse Ranges extend 350 km eastward from Point Arguello and are unique in departing from the characteristic NW–SE trend of Greater California, due to massive tectonic rotation during Neogene times. Towards the east, subsequent compression and thrusting have forced the San Gabriel Mountains up to 3068 m and the San Bernardino Mountains to 3505 m. The Peninsular Ranges are a series of parallel ranges which reach 3293 m in the San Jacinto Mountains 150 km east of Los Angeles, and in Baja California form the Sierra Juarez and reach 3086 m in the Sierra San Pedro Martir. The eastern Transverse and Peninsular Ranges thus form a considerable barrier to weather systems approaching from the west, resulting in a rapid transition to desert conditions in their rainshadows further east.

The climate of this mountainous region is characterized by hot dry summers and mild wet winters. Annual temperatures average 12 to 18°C in the lower mountains but decrease with elevation to around 0°C on the higher crests. Precipitation ranges from 300 to 1000 mm annually, mostly as rain but with some snow at higher elevations (Figure 8.3). Crestal areas and leeward slopes are influenced by summer thunderstorms imported from the south and east under monsoonal conditions.

Much of the area is dominated by *chaparral* forest and scrub, similar in composition to vegetation in the neighbouring coastal zone, but with less sage and more oak woodland. The Transverse Ranges in particular show a distinct contrast between scrub-covered south-facing slopes and forested north-facing slopes. The *chaparral* forest is dominated by sclerophyllous species such as California live oak (*Quercus agrifolia*), canyon live oak (*Q. chrysolepis*), interior live oak (*Q. wizlizenii*), California laurel (*Umbellularia californica*), Pacific madrone (*Arbutus menziesii*), golden chinquapin (*Castanopsis chrysophylla*) and Pacific bayberry (*Myrica californica*). The *chaparral* scrub contains at least 40 species of fire-adapted evergreen shrubs dominated by chamise (*Adenostoma fasciculatum*) and manzanita (*Arctostaphyllos* spp.), but also including Christmas berry or toyon (*Heteromeles arbutifolia*), Nuttall's scrub oak (*Quercus dumosa*), mountain mahogany (*Cercocarpus betuloides*), sugar bush (*Rhus ovata*) and California lilac (*Ceanothus* spp.). Riparian woodlands of cottonwood (*Populus* spp.) and sycamore (*Platanus racemosa*) line most valleys while coniferous forest dominates higher elevations.

Soils are complex and relatively young. Alfisols are dominant in the north, Entisols in the south, and Mollisols towards the coast. Large mammals include mule deer, mountain lion, bobcat, coyote and grey fox (*Urocyon cinereoargenteus*). Of the avian fauna, the endangered California condor (*Gymnogyps californianus*) with its 3 m wing span is noteworthy, saved recently from extinction by careful breeding.

Natural hazards

Natural hazards are expressions of the physical environment, and often have direct impacts on the region's peoples. Earthquakes are a recurrent reminder of California's tectonic framework; floods and landslides occur in response to high magnitude, high intensity winter rainstorms; and firestorms driven by outflowing *Santa Ana* winds are a frequent autumn hazard. Though fire incidence has been modified by fire-suppression programmes, leading to questionable fuel build-up and distorted fire ecologies, this has been offset by an increase in arson fires. Similarly, whereas flood potential has been countered by extensive flood-control devices, particularly around the Los Angeles Basin, the environmental cost is reflected in natural streams that now flow in concrete troughs, in lowered water tables, and in beaches starved of sediment trapped in foothill debris basins. Damage estimates provided by California's State Office of Emergency Services for the period 1993–1995 include $20 billion for the 1994 Northridge earthquake, $2 billion for the storms of winter 1995, $1 billion for the firestorms of autumn 1993, and $50 million for the Mediterranean Fruit Fly infestation of 1994 (which is not wholly controlled).

The Human Context

Before European contact, Greater California was occupied by around 100 reasonably distinct groups of Native Americans speaking dialects of six linguistic families. These peoples survived on hunting, fishing and gathering, their livelihood being most favoured along the coast and in the Central Valley (Heizer 1978). Although Spain claimed the region in the 16th century, little effort was made to consolidate this claim before the late 18th century (Bryant 1848, pp. 278–285). Then, the Portola expedition of 1769 led to the establishment of four *presidios* (forts), three *pueblos* (towns), and a chain of 21 Franciscan missions from Baja California northward to the San Francisco Bay area, many of which formed the basis for subsequent growth. The last and most northerly mission, San Francisco Solano, was established north of San Francisco in 1823.

Spain exercised precarious rule over the region and in 1822 it became part of independent Mexico. Continued neglect by Mexican authorities, combined with secularization of Franciscan mission lands in the 1830s and the arrival of foreign trappers, traders and settlers, mostly American but including Russians from Alaska, led in 1848 to the transfer of Alta California to the United States, the boundary with Mexico being drawn just south of San Diego. This acquisition was part of a broader territorial settlement that concluded a significant period of American expansion, given political if not ethical dignity by the term "manifest destiny", which ended in war with Mexico and the Treaty of Guadalupe Hidalgo of 1848. To the north, though coveted by Spain, Britain and Russia, United States hegemony over the Oregon territory between latitudes 42° and 49°N was recognized in 1846 and part of this territory became the State of Oregon in 1859.

Thereafter both California and Oregon began to prosper from immigration, exploitation of such natural resources as timber, fish and minerals, and then agriculture (Brewer 1864; Newell 1894; Orse 1974). From 1850 to 1880, agriculture was dominated by livestock ranching, but grain farming rose to prominence between 1880 and 1900 by which time irrigation farming was expanding. Nineteenth century urbanization was largely limited to expansion around former *presidios*, *pueblos* and missions, but the 20th century saw the rapid growth of major metropolitan regions and major restructuring of the economy towards industrial and commercial interests. To the south, development of Mexico's largely arid peninsula of Baja California was slow and was to occur mostly along the United States border.

The people

In 1994, the State of California officially had 31 431 000 inhabitants (US Bureau of the Census 1994), together with an uncertain number of illegal immigrants. Subtracting the small population along the state's non-Mediterranean eastern margin, but adding the largely rural population of southern Oregon and some 800 000 people in the Mediterranean part of Baja California Norte, and allowing for a 5% increase since 1994, probably yields a present total population of around 35 million for Greater California.

In California, white persons of non-Hispanic European stock make up about 54% of the population, Hispanic peoples (mostly Mexican) 28%, African Americans 7%, peoples from eastern Asia 10%, and Native Americans less than 1%. Roman Catholics form the single largest religious group, followed in size by several Protestant groups and a significant Jewish minority. The population of southern Oregon is mostly white; that of Baja California Norte almost entirely Hispanic, mostly *mestizo* (mixed Spanish–Native American ancestry), with a sizeable Native American minority. English is the common language throughout, even in Baja California Norte, but Spanish is also important, especially in the south, and most other immigrant groups maintain some linguistic and cultural identity.

Total population has grown rapidly throughout the region since the mid-19th century, even as the rural–urban emphasis has changed. In 1850, the State of California contained 93 000 people, 8% of whom were classified as urban. Of the official population of 29 760 021 in 1990, 93% were urban and this value has probably increased since. This rapid growth has been due partly to natural increase and partly to high immigration because, despite occasional floods, fires, earthquakes and riots, California continues to be perceived elsewhere as a sunny and relaxed "lotus land" – a land of opportunity. For example, of the 260 000 persons admitted legally to California in 1993, 25% came from Mexico, 11% from the Philippines, 10% from Vietnam, 7% from the former Soviet Union and 5% from mainland China. Much of Baja California's recent population growth is due to the arrival of border industries directed towards the California market and of people seeking entry to California.

The principal metropolitan areas are the San Francisco Bay area, containing seven million people in 1990; the Greater Los Angeles area (10 000 km^2) containing 15 million people, and the San Diego–Tijuana area with around three million persons. The Central Valley contains Sacramento (370 000), State capital and manufacturing centre, Fresno (354 000), and many smaller towns whose growth has long been linked with agriculture. Only 7% of the population is now classified as rural, reflecting a historic absence of traditional villages and the rapid change within the past century from smaller, labour-intensive farms to large, mechanized enterprises.

The economy

In terms of the value of goods produced, manufacturing now constitutes 78% of California's economy, agriculture 11%, mining 10% and fishing less than 1% (US Bureau of the Census 1990). Some aspects of the economy, such as citrus and wine production, reflect Mediterranean conditions, while the aerospace and motion-picture industries are both favoured by the usually clear sunny weather of southern California. Whereas manufacturing payrolls amount to about $50 billion annually, the payroll of those employed in government, education, construction, business, trade, tourism and other service industries has risen to over $150 billion in recent years. In terms of gross domestic product, California's $800 billion in 1993 was 15% of the United States total and ranked eighth in the world (behind the United States, Japan, Russia, Germany, France, Britain and Italy). About 60% of California's total trade is with Asia.

Petroleum, natural gas

From the Gold Rush of the mid-19th century, mining has been important to the economy, but petroleum and natural gas now account for more than half of all mineral value. Petroleum production is concentrated in southern California, where the state's first oil refinery was built in the Santa Clara Valley in 1877, but as older fields onshore have declined in output the focus has shifted to the continental borderland offshore, with heightened public awareness of environmental issues. Natural gas is produced mostly from the northern Central Valley.

Agriculture and forestry

Agriculture, initiated during the Spanish and Mexican periods, remains important though its contribution to the economy has declined relative to other sectors. With about 78 000 farms averaging 170 ha in size, California contains only 3% of farms in the United States but accounts for 10% of national farm income. Some 200 commercial crops account for about 63% of this income, with production of 50 crops leading the nation. Leading crops include cotton, sugar beet, rice, barley, hay, citrus, grapes and other fruits, and a wide range of vegetables. California accounts for 85% of United States wine production, mostly from the Napa Valley and the Southern Coast Ranges. Livestock and their products account for the remaining farm income. The strength of the agricultural sector reflects the coincidence of a Mediterranean climate, especially the long growing season of lowland areas, with water resources for irrigation, particularly in the Central Valley and coastal basins, though the continued use of irrigation water has led to major environmental and political problems within the state. Ocean fishing accounts for 8% of the national catch.

Forestry accounts for about 10% of national timber production, mainly from Douglas fir, redwood and other softwoods in the Northern Coast Ranges and the Sierra Nevada. About one-third of the 18 million hectares of forest land in California is devoted to commercial logging, but depletion of old-growth resources has forced this industry into sustained-yield timber harvesting, while concerns about the impact of logging on slope stability and habitat have led to major restrictions in recent decades. Much of the remaining forest land is reserved for parks and wilderness, notably Yosemite (established 1890, 308 000 ha), Sequoia (1890, 163 000 ha) and Kings Canyon (187 000 ha) National Parks in the Sierra Nevada, and Redwood National Park (45 000 ha) in the Northern Coast Ranges.

Manufacturing

Some 40 000 manufacturing firms, employing two million workers and centred on the major metropolitan areas, account for almost four-fifths of California production. Notable among these industries are electronics in the San Francisco Bay area (including "Silicon Valley" south of San Jose), aerospace in southern California (now contracting in the post-Cold War era), and food processing in the Central Valley, Salinas Valley, Santa Maria Basin, Oxnard Plain, and many cities. The Los Angeles metropolitan area contains half the state's manufacturing employees. Some 32% of the state's electrical energy is derived from hydro-electric plants, about half of which is imported from beyond the Mediterranean region. A further 58% is provided by nuclear and fossil fuel steam generators, mostly along the coast, and the rest from solar and wind-powered facilities and geothermal sources.

Military

Since before World War II, California has served as an important focus for military training, research and manufacturing, and during the Cold War between 1945 and 1990 military expenditures accounted for a large portion of the tax base. Some military facilities, such as Vandenberg Air Force Base in northern Santa Barbara County, cover large areas. However, with the alleged "outbreak of peace" following the Cold War, and the ensuing realignment of military needs and priorities, many such facilities are being closed, including the large Fort Ord army base on Monterey Bay, various air force bases around Sacramento and Los Angeles, and two naval shipyards. The ultimate impact of such closures on environment and economy is presently uncertain, although part of Fort Ord is reverting to a coastal dune reserve.

Communications

Until 1869, California remained somewhat isolated, accessible either by challenging overland stage, by ship around Cape Horn or across the Pacific, or by ship and land over the Panama Isthmus. When the Central Pacific Railroad, built eastward from Sacramento with mostly Chinese labour, met the Union Pacific Railroad, built westward from Omaha with mostly Irish labour, at Promontory Point, Utah, in 1869, the first intercontinental rail link was completed and travel time from New York to Sacramento was suddenly reduced to seven days. Thereafter the expansion of railroad facilities provided wider markets for the state's growing agricultural and mineral products, while land grants to competing railroads sparked a land and population boom. San Francisco became the largest city and commercial centre on the Pacific coast while Los Angeles remained a modest town. Opening of the Panama Canal in 1914 further improved access to California.

The advent of cars and trucks, particularly after 1920, led to the slow decline of railroads and railroad interests, and the growth of paved highways. Today, only 10 000 km of operating railroad track survive in the state, mostly orientated north–south but with important links eastward from major metropolitan areas. From the 1950s onward, the creation of the federal interstate highway system and improvement of state and local highways paved the way, together with federal housing subsidies and tax incentives, for the urban population explosion of the later 20th century. Then, the conversion of international shipping and surviving railroads to container traffic led to further growth in ports serving major metropolitan areas. Air traffic also grew rapidly, with both the San Francisco and Los Angeles areas developing several major airports while over 500 miscellaneous airports emerged statewide. However, the rise of private car ownership after 1950 and the demise of railroads as passenger carriers led to increased highway congestion and air pollution, partly and belatedly offset by the provision of new light rail systems, first in the San Francisco Bay area after 1972, then in Sacramento and San Diego in the 1980s, and more recently with a long overdue system to serve the sprawling Los Angeles area.

Administration, society and culture

Government

Under a constitution adopted in 1879, California is governed by a popularly elected executive headed by a governor and a bicameral legislature composed of a 40-member senate and an 80-member assembly. The judiciary is headed by a seven-member supreme court, five district courts of appeal, and various lower courts. California has 58 counties, relatively small units in the north, much larger in the south, most of them governed by five-member elected boards of supervisors. Most cities function with mayors and/or a council-manager system. Law enforcement falls to state police, primarily the California Highway Patrol, to county sheriffs and their staff, and in the larger cities to urban police departments.

As part of a 50-state federal system of national government, California elects two senators to the 100-member United States Senate and, based on the 1990 census, 52 members to the 435-member House of Representatives. As the most populous state in the Union, California's congressional delegation can thus be expected to carry significant weight in national affairs, with 52 representatives compared with 31 from New York, 30 from Texas and 23 from Florida. Political emphasis focuses mostly on two parties and in recent years the popular vote has tended to split nearly evenly between Democrats, the slightly more liberal party, and Republicans, now the somewhat more conservative party. Northern California residents, non-white minorities and recent immigrants tend to favour Democrats; southern California, businesses and many whites favour Republicans, but voter preference is strongly influenced by local issues such as immigration and candidate personalities.

That part of Oregon which falls within the Mediterranean region is similar in government structure to California but, though more rural and conservative, tends to be politically more independent and often liberal on social issues. The Mediterranean part of Baja California is politically and culturally part of Mexico, and the State of Baja California Norte is headed by a governor and chamber of deputies. The Partido Revolucionario Institucional has maintained control in one form or another since the 1920s, but distance from Mexico City and widespread contacts with California have created a more independent spirit. Tijuana is very much a border city – lively, enterprising, hopeful and squalid – a city of some opulent mansions but many cardboard shacks. Were it not for the attraction of "lotus land" to the north, Tijuana would not be a large city.

Education

California's highly respected educational system comprises over 7000 public primary and secondary schools, and a three-tier tertiary system comprising junior colleges, 19 state colleges and universities, and the University of California. The University, founded in 1868, remains the only public PhD-granting institution in the state, now embracing 150 000 students on nine campuses of which those at Berkeley and Los Angeles and the Scripps Institution of Oceanography in San Diego are prominent. There are also many quality private schools and colleges. Among the large private universities, Stanford University (1885) and the California Institute of Technology (1891) in Pasadena are notable for research, the latter associated with the Jet Propulsion Laboratory and the Keck and Palomar observatories which play a major role in space exploration.

Health care and social services

Health care and social services are provided by a combination of federal and state assistance, private organizations and personal insurance. For example, the federal government provides social security assistance to wage earners and their families for retirement, disability and death. Health insurance for senior citizens is provided by Medicare and for the poor by Medicaid. Government assistance is also provided to the blind, the elderly poor, and to poor families with dependent children, including such aids as school lunches and food stamps. California and Oregon have many good facilities for health care but gaining access to them remains a major problem, particularly among the moderately poor and new immigrants who do not qualify for Medicare or Medicaid and have insufficient private insurance. Indeed, the tremendous influx of immigrants over recent decades continues to place great stress on both the individuals concerned and government's ability to provide for them. The situation is significantly worse in Mexico, such that emigration to the United States becomes an attractive option. Furthermore, the presence on both sides of the international border of people who barely survive within or are disaffected by the system poses the potential for unlawful activities and social unrest, as shown by high urban crime rates and the Los Angeles riots of 1992.

Social issues

California is essentially a multi-cultural society, more so in the south than the north, more so in the cities than the rural areas, and more so in recent decades than in the state's first hundred years. Even as Native American and colonial Spanish traditions withered under Anglo-Saxon influence after 1850, so the dilution of Anglo-Saxon traditions by later immigrants, including the re-emergence of an Hispanic American culture, has continued to change the fabric of society. Meanwhile, many African American people in the inner cities continue to feel disadvantaged by the system. Furthermore, large-scale immigration from eastern Asia has given rise to ethnic neighbourhoods such as Little Tokyo and Koreatown in Los Angeles. The "melting pot" so nurtured in concept by the United States has become more of a stew in which the individual components maintain their cultural identity and traditions. The potential for inter-racial conflict has risen accordingly.

9

Chile

Consuelo Castro and Mauricio Calderon

Introduction

Chile is a country with a distinctive geography, extending over a length of 4200 km but with an average width of only 200 km. To the east, the country is bordered by the western slopes of the Andes, which were formed by the subduction of the Nazca lamina plate beneath the South American one. To the west, Chile is bordered by a marine trench almost 8000 m deep.

The Chilean Mediterranean environment extends between 31° and 37°S (Figure 9.1). To the north, the rainfall regime changes to semi-arid and in the south, to a more humid environment.

The north–south disposition of the Andes mountains contributes to the great diversity of Chile's bioclimatology. Major relief patterns are orientated north–south, introducing a significant spatial variety in Chile's central climatological patterns, but there are also intermontane, fluvial basins which have continental features because of the rainshadow effect of the coastal range (Figure 9.2). The fast-flowing rivers are used intensively for a wide range of productive activities, as well as to generate electricity and for household consumption. In this environment, the seasonal and orographic variety in rainfall and humidity, as well as differences of gradients and soils, result in vegetation assemblages ranging from semi-desert formations to hygrophyllic, temperate woodlands.

Human settlements are concentrated in the Mediterranean part of the country, with most of the population located in the central tectonic depression, and also on the coast (Figure 9.3). The agriculturally favourable soils and littoral terraces encourage human settlement, leaving the Andes mountains almost unpopulated. The cultivated areas are preferentially dedicated to fruit trees, vineyards, cereals and vegetables. Much of Chile's agriculture, and the biggest cities and ports such as Valparaíso and San Antonio, as well as the national capital, Santiago, are located in the Mediterranean zone.

The Land

Relief

The mountains and highlands cover 80% of the Chilean continental surface. Tertiary tectonic activity generated the Andes mountains and the coastal range, separated by a central depression (Figure 9.2). In this way, there are four different longitudinal features which extend along almost the entire country: the coastal terraces, the coastal range, the central tectonic depression, and the Andes mountains.

The coastal terraces are well developed, especially where the rivers flow out to the sea, distinguishing coastal and fluvio-marine stepped terraces. Near the Aconcagua River, the coastal terraces generally have an extent of 15 to 20 km, an altitude of 200 m, and have been profoundly dissected by deep streamlets. There are important dune fields in the Mediterranean region, especially to the north of the outlets of the big rivers, notably the Aconcagua and Maule rivers.

Between latitudes 33° and 37° south are very distinctive coastal ranges. At the latitude of

Land Degradation in Mediterranean Environments of the World: Nature and Extent, Causes and Solutions.
Edited by A.J. Conacher and M. Sala.

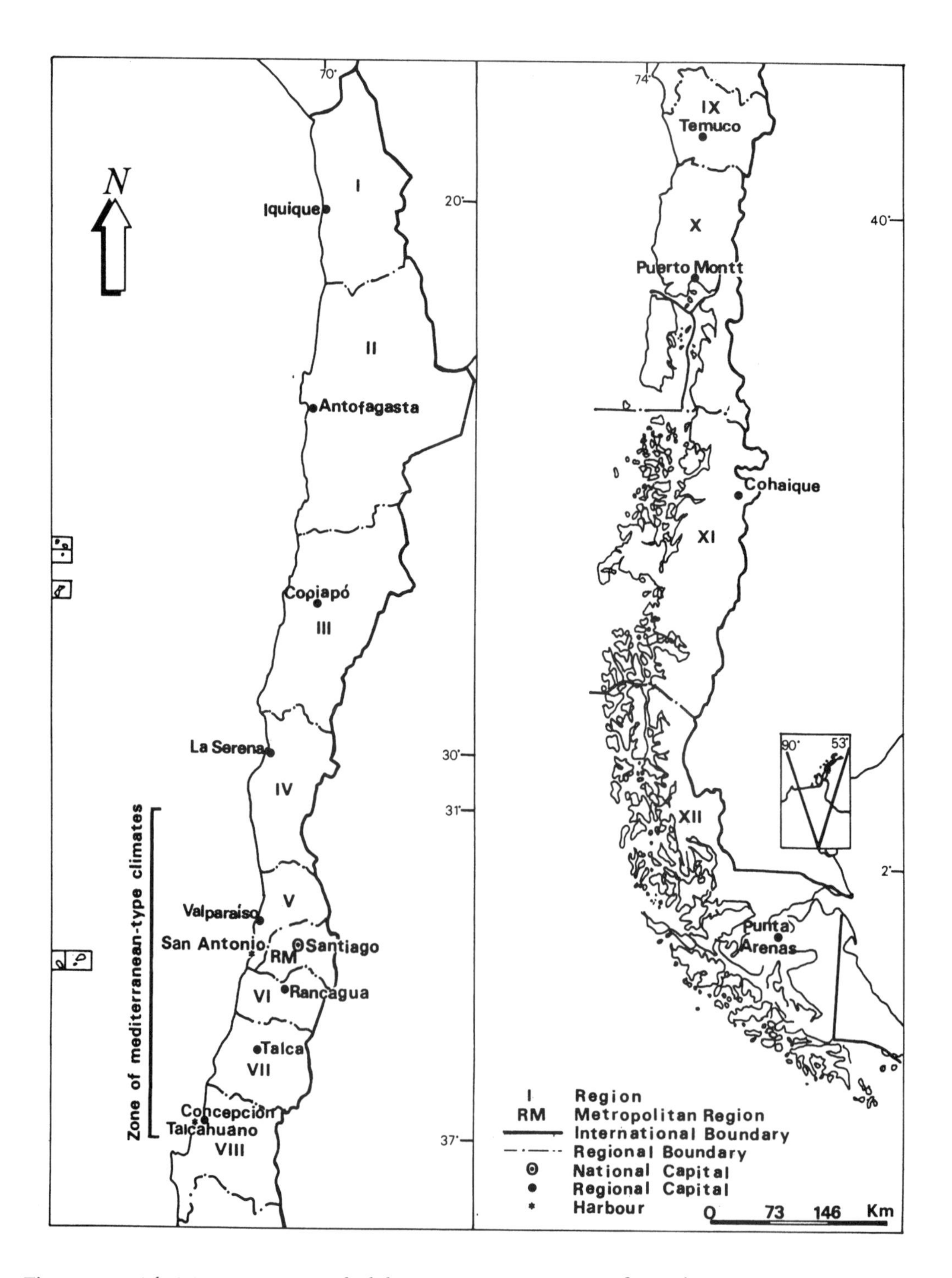

Figure 9.1 *Administrative regions of Chile* (source: Instituto Geográfico Militar 1993)

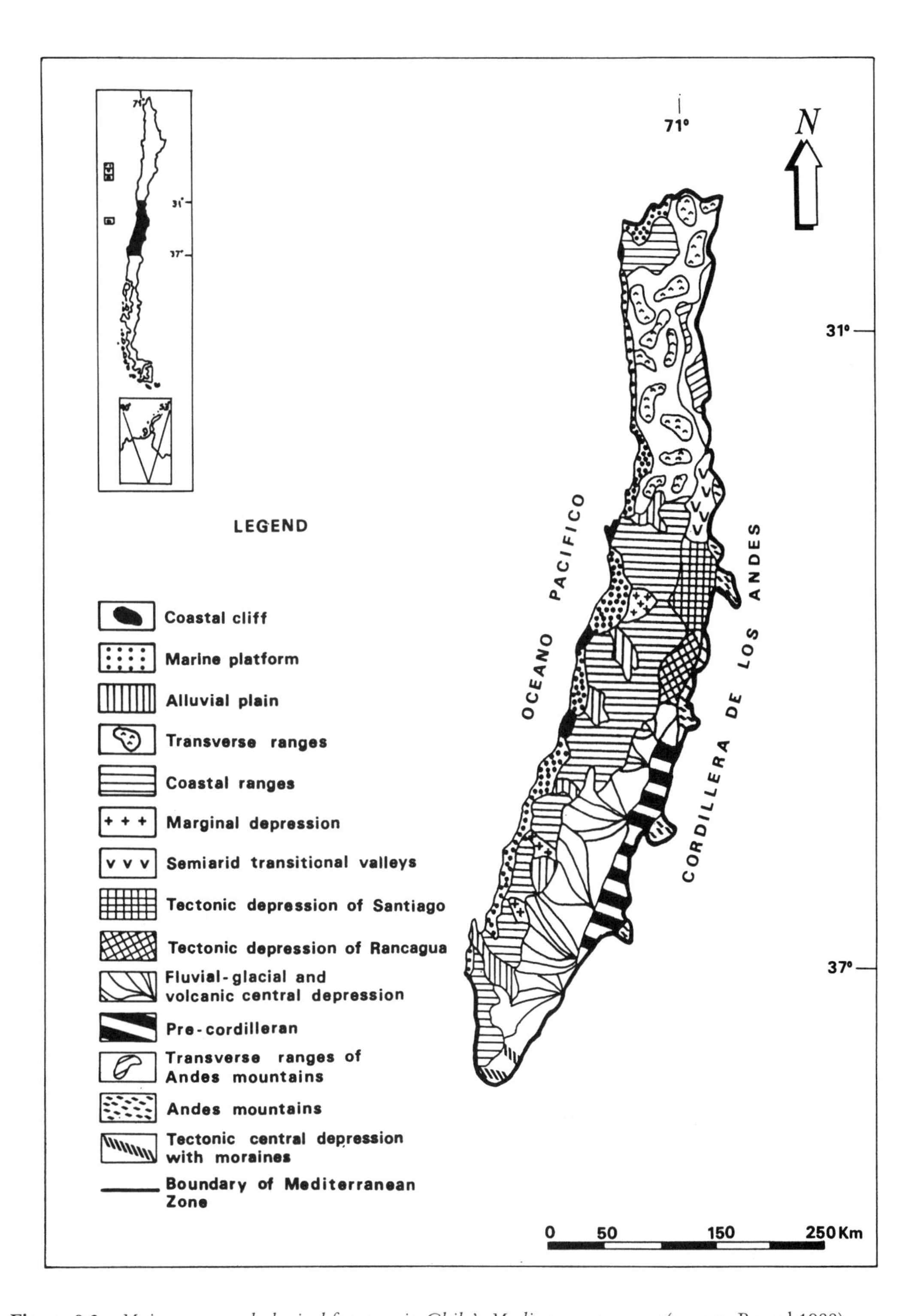

Figure 9.2 *Major geomorphological features in Chile's Mediterranean zone* (source: Borgel 1988)

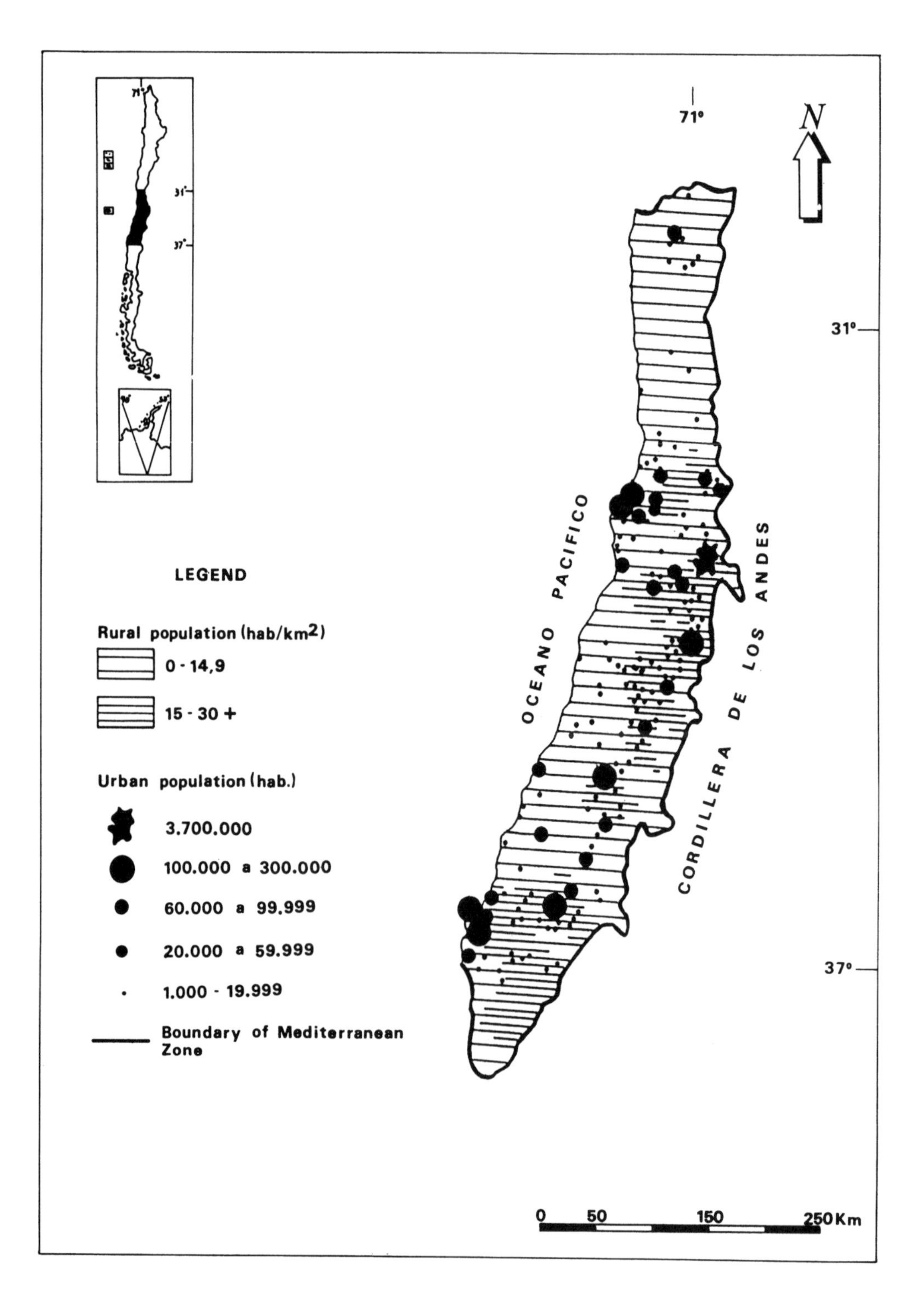

Figure 9.3 *Distribution of population in the Mediterranean zone of Chile* (source of data: Instituto Geográfico Militar 1993)

Valparaíso and the Metropolitan Region are the important Roble hills, with an altitude of 2242 m, and Vizcachas which has a height of 2220 m. Altitudes diminish south of Santiago to about 1000 m above sea level.

The width of the central tectonic depression increases to the south, from 12 km at latitude 34°S to 20 km at 35°S, 51 km at 36°S and 74 km at 37°S. It has been aggraded by alluvial, glacial and volcanic sediments. Here are located the big basins surrounded by mountains: the Santiago and Rancagua, in the central-eastern part of the Mediterranean region (Figure 9.2). The first basin extends 80 km from north to south and has an average width from east to west of 35 km, and the second extends for 60 km from north to south in its centre, and has an average width of 30 km. The Santiago and Rancagua basins are important reserves of groundwater in the central tectonic depression, which is irrigated by an improving network of hydrographic basins surrounded by mountains (Figure 9.4).

The Andes reach their maximum elevation above sea level between 31° and 33°S, at "Olivares" (6252 m) and "El Plomo" (5430 m). Transverse hills predominate in this part, linking the Andes mountains to the coastal ones, resulting in a confused topography. The transverse ranges of the Andes foothills have an altitude ranging between 600 and 1000 m, and their W–E orientation is caused directly by the erosive action of the rivers. Thus, the relief is characterized by transverse valleys (Petorca, La Ligua and Aconcagua) which are limited by transverse ranges (Figure 9.2).

Drainage

Channel sizes of the Andean rivers increase towards the south of the Mediterranean region (Table 9.1). The rivers have torrential discharges and mixed regimes, with increased quantities of rain-fed discharge in winter and increased quantities of snowmelt discharge in spring and the beginning of summer. The combined sources of runoff secure a permanent contribution of water and sediments to the Central Depression.

The altitude of about 5000 m of the Andes mountains to the east of the Mediterranean region and the relative narrowness of the territory condition the torrential form of stream runoff. Tributaries of the larger rivers have a predominantly snow-fed regime, whereas the tributaries of the intermediate and smaller rivers have mainly pluvial regimes.

Table 9.1 *Principal characteristics of mixed regime rivers of the Chilean Mediterranean zone* (source: Niemeyer and Cereceda 1984)

River basin	Area (km^2)	Length (km)	Discharge (m^3/s)
Choapa	8124	160	8.73
Petorca	2669	112	0.63
La Ligua	1900	106	1.10
Aconcagua	7163	177	39.00
Maipo	15 380	250	92.30
Rapel	14 177	230	162.00
Mataquito	6190	221	153.00
Maule	20 295	240	467.00
Itata	11 090	180	186.00
Biobío	24 029	407	899.00

The hydrographic systems are torrential from the high ridge of mountains, with great variability, to the central depression, allowing the generation of hydro-electricity. Impoundments also provide water for irrigation and potable water for cities and industries. The water of the Maipo River, regulated by dams in the coastal range, provides almost 70% of the demand for water by Santiago city and almost 90% of the irrigation requirements of the basin (Toledo and Zapater 1989). Another intensive use of the water of the Maipo basin is the generation of hydro-electricity, with the Volcan (capacity 13 000 kW), Queltehues (43 500 kW), Maitenes (24 500 kW) and La Florida (13 500 kW) power stations.

Soils

In the northern part of the Mediterranean area, which has a tendency towards a semi-arid climate, there is moderate pedogenetic activity related to weathering, producing Mollisols and Alfisols (Rovira 1984).

From the Aconcagua River to the south, favourable climatological conditions associated with a mesophyte vegetation allow the recognition of five pedogenetic environments. Mollisols are formed in the littoral area, and are related to subhumid and semi-arid conditions. On the

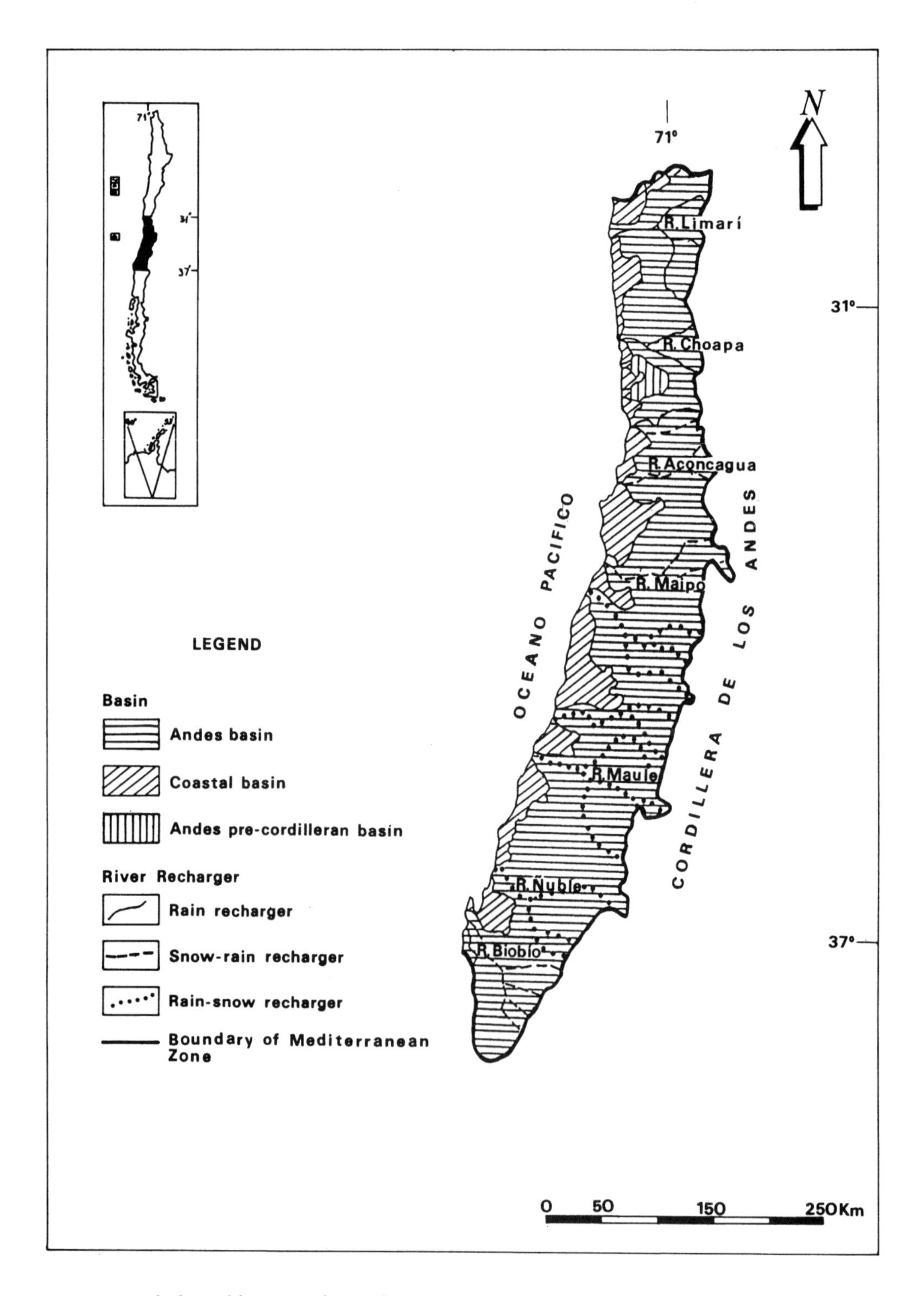

Figure 9.4 *Hydrological basins in the Mediterranean zone of Chile* (source: Instituto Geográfico Militar 1993)

coastal range there are Alfisols over granitic rocks. In the central tectonic depression, which from Santiago is a sedimentary basin, there are Inceptisols and Brown Soils. The most fertile soils are those which are established on alluvium in the central tectonic depression, characterized by good drainage and light texture. Subjected to irrigation, these soils are suitable for the cultivation of cereals, fruits, vegetables and other horticultural products. At the base of both the Andes mountains and the coastal range, as on the alluvial fans in contact with the central depression, there are Entisols.

Important deposits of volcanic ash associated with Inceptisols occur in the southern part of the Mediterranean area, whilst the Mollisols on the sedimentary marine terraces in the littoral area are related to subhumid and semi-arid conditions, with prairies (pastures). Surface soil materials have a medium texture, with large quantities of ferromagnesium concretions. Erosion of the reddish soils is severe, in the form of sheet erosion, ravines and gullies. The sedimentary parent materials are derived mainly from granitic and metamorphic rocks, and also Quaternary marine sediments. Fertility is moderate, and the soils are able to sustain cultivation of cereals and other grains and leguminous forage plants.

Climate

The South Pacific Ocean anticyclone is located between latitudes 20° and 40°S, to the west of the Chilean coast, with the 100°W meridian as an axis. This anticyclone shows an annual displacement of 10° from north to south, taking its most southerly position in summer, and its most northerly position in winter. The arrival of atmospheric perturbations in winter generates rains in the Chilean Mediterranean zone. Regionally dominant winds come from southern and southwesterly directions, and are related to the South Pacific anticyclone.

Strong latitudinal differences result in a temperature gradient, with temperatures decreasing from north to south by 0.3°C every 100 km. At the same time, there is an annual rainfall increment of 132 mm for every 100 km. At 36°S there are the same number of dry and humid months.

In Chile the characteristic Mediterranean climates occur between 0 and 1500 m above sea level. Rainfall is low and fluctuates between 200 and 900 mm/yr. To the north of 31°S there is interpenetration of desert and Mediterranean influences. In contrast, to the south of 37°S the reciprocal influences are Mediterranean and oceanic.

Rains caused by the action of the active polar front predominate during the cold season, from May to September; the warm season, which is present during the remainder of the year, is dry. In the northern part of the Mediterranean area the rains are less frequent and more irregular, whereas to the south the rains become more abundant and more regular. The duration of the dry season is very variable; nevertheless, its duration diminishes from north to south (Table 9.2). There are variations caused by latitude and relief.

The coastal range separates the western, humid side, with few thermal fluctuations, from the eastern side, which is drier and has continental thermal influences. Because of the physiographic conditions of the Mediterranean Chilean area, there are three kinds of bioclimatological Mediterranean zones: littoral, interior and mountainous (Pre-Andes), according to Di Castri and Hajek (1977).

On the littoral border between 31° and 34°S a cloudy climate is recognized which shows the typical oscillations of the anticyclonic regime. Because of the proximity of the sea, the temperature is moderate. For example, the mean annual temperature of Valparaíso is 14.4°C. The warmest month is January, which has an average temperature of 17°C, and the coldest one is July, with an average temperature of 11.4°C. Mean annual rainfall is 462 mm, most of which falls between May and August, with a dry period during the other seven to eight months.

The variability of rains during and between years influences human activities, because droughts and floods have been recurrent stimuli which have impacted on the economy of the Mediterranean zone of the country.

Plant and animal life

In the Chilean Mediterranean zone there is a dominance of the spiny steppe of *Acacia cavenia* in the central tectonic depression, the coastal ranges and the Andes mountains piedmont and

Table 9.2 *Climatic data* (source: Oficina Meteorologica de Chile 1995)

Meteorological station [altitude, m] (latitude)	Mean annual temp. (°C)	Mean Jan temp. (°C)	Mean July temp. (°C)	Annual range (°C)	Mean annual rainfall (mm)	No. driest months
Zapallar [30] (32°01'S)	14.1	17.6	11.4	6.2	384.3	8
Valparaíso [41] (33°01'S)	14.4	17.8	11.4	6.4	462.2	7
Santiago [520] (33°27'S)	14.5	20.9	8.1	12.8	291.3	8
Talca [97] (35°26'S)	14.7	21.9	8.3	13.6	734.0	6
Constitución [2] (35°20'S)	13.9	18.2	10.1	8.1	959.8	4
Concepción [10] (36°50'S)	13.1	18.0	9.1	8.9	1320.1	3

soft hills. This steppe is formed by shrubs which can reach heights of 2 to 6 m; it forms an association of annual grass and shrubs such as the huañil (*Proustia pungeus*), trevo (*Dasyphyllum diacanthoides*), chilca (*Baccharis glutinosa*) and others. The thorn (*Acacia cavenia*) has an economic value because it is used to make charcoal, and it has been extended in zones where the sclerophyll native forest has been felled.

Animal life is variable but in danger of extinction because of human persecution, for example the American lion (*Felis concolor*), nutria (*Lutra provocax*), the coipo (*Myocastor coypus*) and others. There is a great variety of birds such as the tenca (*Mimus thenca*), the chercán (*Troglodytes aedon*), the diuca (*Diuca diuca*) and the loica (*Sturnella loyca*).

On the coastal ranges, in zones where rainfall fluctuates between 400 and 1000 mm/yr, there is a forest whose principal species include quillay (*Quillaja saponaria*), litre (*Lithraea caustica*) and boldo (*Peumus boldus*). An azonal forest occurs in association with this vegetation, which includes amongst its more characteristic species one of the more southerly palms in the world, located in association with small streams of the coastal range. This is the Chilean palm (*Jubaea chilensis*). It is nearly 15 m tall, and it is developed between latitudes 33° and 34°S, where there are suitable microclimatic conditions, near Viña del Mar and Valparaíso. The palm has been widely exploited for the production of honey, for which it was felled. Today the palm is nearly extinct.

Because the Chilean Mediterranean region is the most populated part of the country, it is intensely affected by humans, who have felled the sclerophyll forest for firewood and charcoal, and for medical purposes. On the coastal range, where the forest is degraded, there is the coligüe (*Chusquea coleu*), which is very combustible in summer. The more common native grasses in this formation include yuyo (*Brassica campestris*) and cardo (*Cynera cardunculus*). This plant cover has encouraged people of the coastal range and Andes piedmont to practise permanent pasturing of goats and sheep, contributing in this way to erosion of the slopes.

Table 9.3 shows the more abundant native trees of the Mediterranean zone, from the Choapa to the Laja rivers.

The People

The pre-Columbus settlement of the Mediterranean zone of Chile was by the indigenous people, who established different means of domination over the natural territories where they settled. The Changos people were nomads, involved mainly in hunting and fishing, who settled on the

Table 9.3 *Native trees of the Mediterranean zone of Chile* (source: Quintanilla 1983)

Common name	Scientific name	Sector	Distribution*
Thorn	*Acacia cavenia*	Central Tectonic Depression	Regions IV–VIII
Cypress tree of the mountain	*Austrocedrus chilensis*	Pre-Andes mountains	Regions V–VIII
Coigüe, Raulí	*Nothofagus dombeyi, N. alpinal*	Pre-Andes mountains	Region VIII
Lenga	*Nothofagus pumilio*	Pre-Andes mountains	Regions VII and VIII
Oak tree, Raulí, Coigüe	*Nothofagus oblicua, N. alpino, N. dombeyi*	Pre-Andes mountains	Regions VII and VIII
Oak tree, Hualo	*Nothofagus oblicua, N. leoni*	Pre-Andes mountains	Regions VI–VIII
Chilean Palm	*Jubea chilensis*	Coastal range	Regions IV–VI

* For locations of regions, refer to Figure 9.1

littoral terraces, while the central tectonic depression was populated by the Picunches, who were sedentary farmers and shepherds (Toledo and Zapater 1989).

During the 16th century the European discovery and conquest of the American Chilean territory by Spanish soldiers, farmers and tradesmen took place. They were solely males, and this fact caused a profound mixture between the people. The Chilean Mediterranean zone became the area in which the indigenous culture united with the Spanish people, and this produced the Chilean nationality. This process happened mainly during the 17th and 18th centuries. So, the Chilean nationality is the result of the intermingling of the Picunches, Diaguitas and Mapuches indigenous peoples with Indo-Europeans from the Iberian Peninsula, especially from Andalucía, Castilla and the Vasc country (Toledo and Zapater 1989).

Today, 75% of the Chilean people (total population nearly 10 million) live in the Mediterranean region, the traditional home of the nationality. Santiago has 38% of the total Chilean population and it is the centre of all national activities. The administrative centralization of the country has caused a serious imbalance in population distribution (Figure 9.3). The official language is Spanish, and 83% of the inhabitants of the Mediterranean area are Catholics.

Rapid population growth has been one of the constant aspects of the demographic situation of Chile for nearly 50 years, although the rate of growth has slowed in recent decades. Since 1980, a large proportion of the population is in the 20 to 64 year age group, whereas previously it was in the under-20 group (Figure 9.5).

The most important migratory movement has been from rural locations and smaller cities to Santiago. Between 1920 and 1960 this migration was explained by the attraction of the industrial, commercial and administrative development of Santiago. From 1960 to 1980 the country was subjected to a considerable amount of reform of property (co-operative, collectivization, individual property), as well as modernization of collective transportation systems. This situation encouraged many people to leave the rural zones and small cities to go to the Metropolitan Region of Santiago, incrementing in this way the accumulation of people and the urban importance of this zone.

The Economy

The Chilean Mediterranean region accounts for 70% of the country's total production. The most important sector is services, which account for 31% of gross national product, followed by manufacturing industry (21%), commerce (16%), agriculture, pastoral farming and forestry (9%), and mining (8.5%).

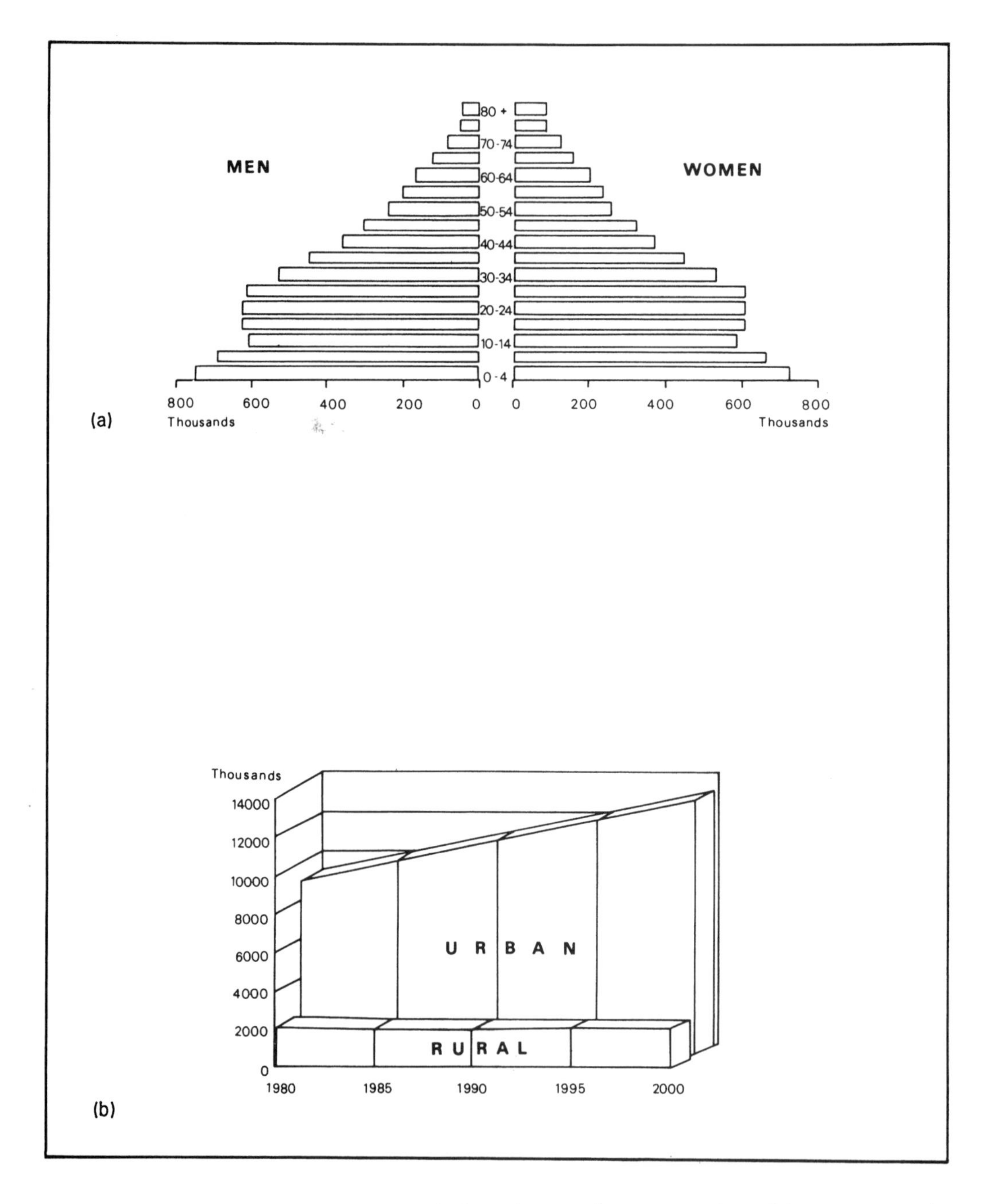

Figure 9.5 *(a) Population composition and (b) urban–rural projection for Chile* (source: Instituto Geográfico Militar 1993)

Resources and industry

Chile traditionally has been one of the world's major producers of copper. In the Mediterranean area the most important copper mine is El Teniente in the Andes mountains, located at an altitude of less than 2500 m, in Region VI (Figure 9.1). Another important copper mine is Saladillo (Minera Andina), located in the Aconcagua valley in Region V. Chuquicamata (Region II) and

El Salvador (Region III) also have major copper mines. El Teniente and Chuquicamata account for 66% of national copper production. There are two medium-sized centres of copper mining in the Metropolitan Region: at La Diputada de Las Condes, in the pre-Cordilleran zone at an altitude of 2000 m, and the deposits of La Africana-Lo Aguirre, located 20 and 30 km west of Santiago.

The Metropolitan Region accounts for 37% of the national production of limestone and calcium carbonate. Region V of Valparaíso is also an important producer, with its contribution of 25% destined for cement production.

Manufacturing activities in Chile constitute nearly 30% of gross national product. Most of the industry in Chile is in the Mediterranean area, especially in the Metropolitan Region of Santiago (56.6%), Region V of Valparaíso (10%) and Region VIII of the Biobío which is characterized by a large number of small and medium-sized establishments. The most important industry is the processing of metals (mainly copper) and steel. The major steel industry of the country is located in the industrial centre of Huachipato in Region VIII. Motor vehicle parts are manufactured in areas where the principal markets for cars exist, in Regions V and VI.

The oil industry is also located in the Mediterranean area, mainly in the industrial centres (Con cón and Quintero in Region V, and Talcahuano in Region VIII) where factories are based on oil by-products. Food industries are distributed throughout the Mediterranean regions. Wood industries are associated with the forest areas from the Metropolitan Region to the south. Cellulose, paper, pasteboard and furniture are the principal products of this important industry.

All this industry is supported by electrical energy, mainly hydro-electricity (53%). The principal hydro-electric power stations are located in Region VI (Rapel), Region VII (Colbún-Machicura) and Region VIII (El Abanico and El Toro) with 1900 MW of installed power.

Agriculture, forestry and fishing

The best soils of the country are in the Mediterranean zone. The Metropolitan Region has 7% of class 1 and 2 soils; Region VI has 10% of these soils. Here is also located the biggest proportion of irrigated lands (Figure 9.6). Most of the agricultural production of the country is concentrated in the Mediterranean region (Table 9.4), which accounts for 97% of the country's fruit orchards, 34.4% of the cattle and 72% of the tractors. The principal crops, according to the areas which they occupy, are: cereals (wheat and maize); vegetables (legumes and potatoes); and cash crops (marigold, sugar beet, leaf tobacco).

Fruit orchards have increased considerably since 1970, due to the strength of external markets. Between 1972 and 1990, the area assigned to fruit trees nearly quadrupled (Errazuriz *et al.* 1992). Rice production from irrigated lands during warm summers is important in the southern part of the Mediterranean area (Regions VI, VII and VIII), south of Talca. The most important farm exports are apples, grapes, nectar fruits and lemons.

In the northern part of the Mediterranean area, cattle are replaced by goat-rearing, one of the fundamental economic activities of the agricultural communities to the north of the Aconcagua River. This activity is often the only source of wealth for the *cabreros* (rural small-holders).

At the present time the principal plantations and timber-producing areas are in the southern half of the Mediterranean zone (Figure 9.6). The natural woodlands were more extensive, but their over-use provoked their almost total extinction in many sectors of the pre-Cordilleran and the coastal ranges. In the Mediterranean region the depredation occurred over several centuries, because the forest provided timber for construction and heating. Forest clearing in order to sow wheat was stopped in mid-1994 in the area of natural forest on the rising ground of the coastal range.

It has been estimated that in Chile there are more than 225 species of fish, of which more than 56 have commercial importance. The principal species in the marine water of the Mediterranean zone are found in relation to the Humboldt Current. Among the species of commercial importance are the Spanish sardine, the spiny sea-fish (*Trachurus murphy*) and the common merluce (*Merluccius gayi*). Prawns (*Pleuroncodes*) are important among the commercial crustaceans (Morales and Cañon 1985).

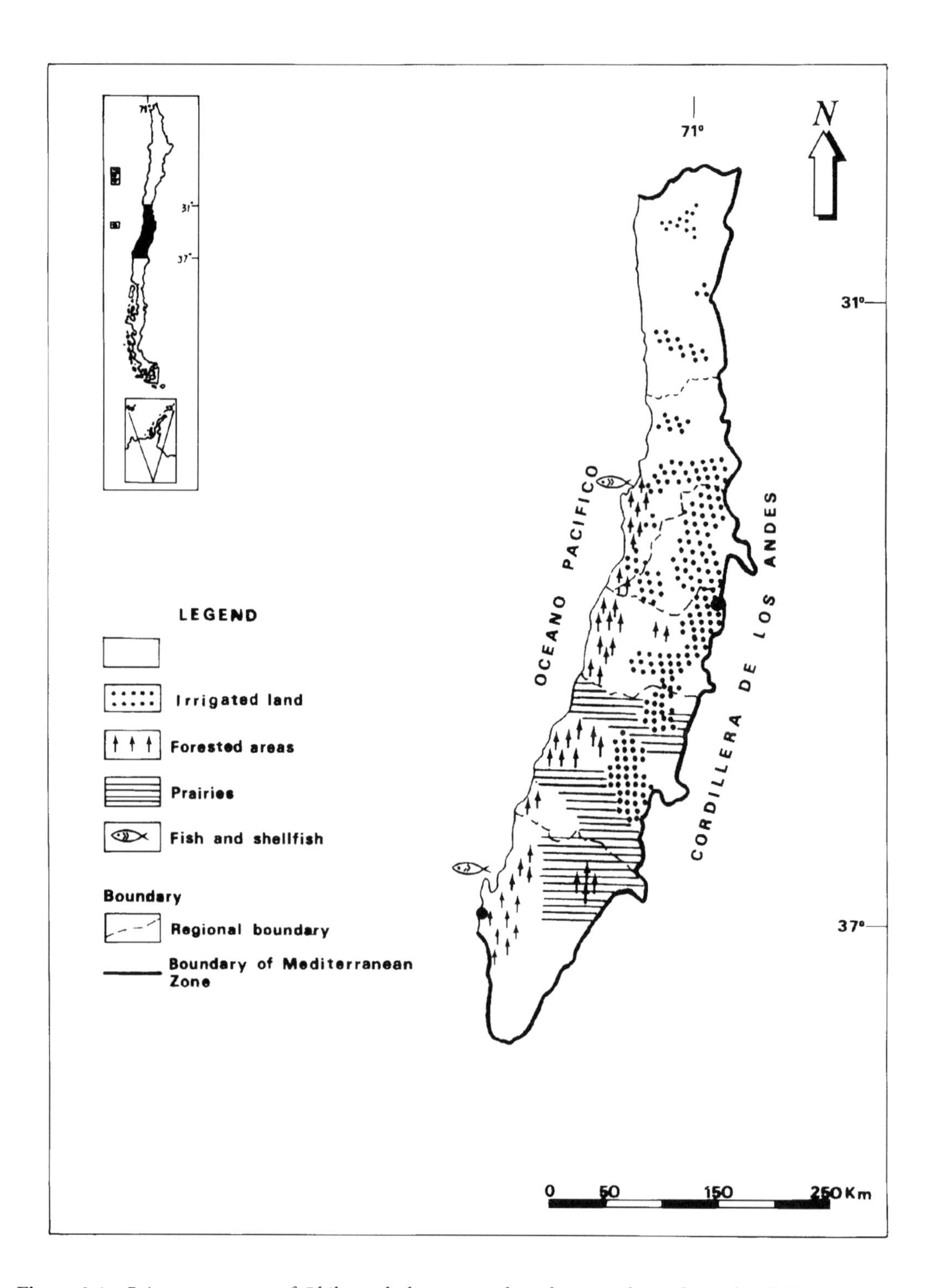

Figure 9.6 *Primary resources of Chile, excluding minerals and energy; large dot is the El Teniente copper mine* (compiled by the authors)

Table 9.4 *Land use (ha) in the Mediterranean area of Chile* (source: Instituto Nacional de Estadísticas 1994)

Region	Crops	Pastures	Other lands	Total
IV Coquimbo	11 240	32 930	57 170	101 340
V Valparaíso	20 240	223 680	185 740	429 660
Metropolitan Santiago	39 330	235 180	226 650	501 160
VI O'Higgins	109 370	460 380	254 150	823 900
VII Maule	139 780	817 830	649 740	1 607 350
VIII Biobío	182 070	945 460	808 870	1 936 400
Total Mediterranean area	502 030	2 715 460	2 182 320	5 399 810
Relation to the country	63%	59%	65%	62%

Most of the fishing is carried out for industrial purposes, such as the production of fishmeal, oils, pastes and tinned fish. This production profoundly exploits the fishing grounds, using factory boats which process the fish onboard. The remainder is processed in land-based factories. Most of the factories of the Mediterranean area are in Valparaíso. Line fishing is used mainly to obtain fish and crustaceans for fresh consumption, but it also contributes to fish processing.

Transportation

The principal mode of transport is by road, representing more than the 60% of passenger movements. The country is connected from north to south by the Panamerican Highway, which links up with a series of transverse routes of national, regional or county character (Figure 9.7). There is a good highway infrastructure in the Mediterranean zone, mainly because of the concentration of the principal urban centres of the country. Table 9.5 shows the proportion of roads which are paved (13%), covered with asphalt (41%) and unpaved (46%).

Terrestrial transportation routes run over long distances from north to south, avoiding accentuated relief, but having to cross a number of transverse rivers. Thus, considerable engineering works to avoid or overcome natural obstacles have been required. Rail transportation, as with roads, extends its routes from north to south with transverse branches. Chile has nearly 17 000 km of railways which are used mainly for the transport of goods.

Despite the long littoral, there are few protected natural ports for maritime transportation; this fact has forced the construction of special protective structures. The principal maritime ports are Valparaíso, San Antonio and Talcahuano. Through them are exported industrial products, the mining production of El Teniente and Saladillo, steel and related products, and fruit products. Additionally, other ports of chemical/oil importance are Quintero-Ventana, Las Salinas and San Vicente.

In relation to air transport, the country has two international airports located in the Metropolitan Region, with the principal airport being Comodoro Arturo Merino Benítez, in Santiago. In relation to national air transportation, the important airport is Carriel Sur in the Mediterranean area (in Region VIII) which, with the airports of Cerrillos and Pudahuel in Santiago, have the greatest national movement of passengers in the zone, although their numbers do not exceed 250 000 a year.

Administrative and Social Conditions

Government and administration system

Since 1974, Chile has been divided into 13 political/administrative regions as part of a governmental decentralization process, including an attempt to reduce the concentration of economic activity. The Mediterranean area of Chile extends from Region IV of Coquimbo to Region VIII of BíoBío, including the Metropolitan Region, where Santiago, the national capital, is located (Figure 9.1).

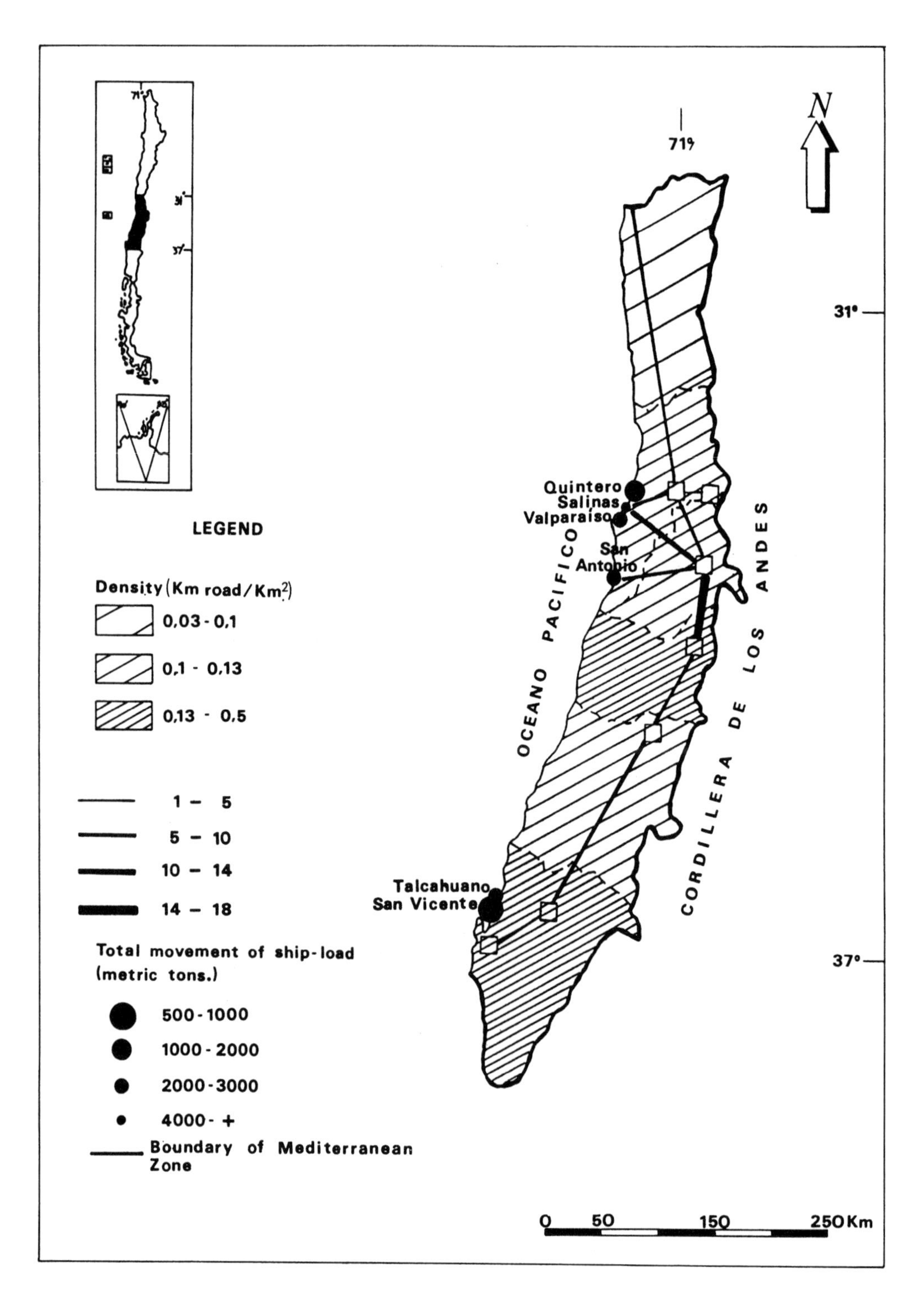

Figure 9.7 *Transportation in Chile; line thicknesses are proportional to the number of cars paying tolls per hour in different parts of the various routes* (source of data: Ministero de Obras Publicas (1989), compiled by the authors)

Table 9.5 *Road network in Mediterranean Chile* (source: Ministerio de Obras Publicas 1989)

Region	km	Paved roads (%)	Covered with asphalt (%)	Not established (%)
IV Coquimbo	5 046	15.1	10.3	74.5
V Valparaíso	3 481	32.6	15.7	51.5
Metropolitan Santiago	2 544	39.1	27.8	33.1
VI O'Higgins	4 205	12.8	46.5	41.0
VII Maule	7 426	7.6	41.8	50.5
VIII Biobío	11 579	10.1	36.4	53.5
Total Mediterranean area	34 281	15.3	32.3	52.5
National total	79 222	13.0	41.0	46.0

Each region is governed by a Regional Governor (Intendente Regional), who directly represents the Republic's President, who is the chief of State and Government. Many Ministries (Estate Office) are territorially organized into Regional Ministerial Offices (SEREMI), helping the Regional Governor. The regions do not have their own legislative system but they are administered by a participant subsystem (Regional Development Assembly) which integrates the public, private and worker sectors and the Provincial Governments of every region. The legislative system in Chile is represented in the National Parliament, located in Valparaíso, which combines the Hall of Deputies and the Senate House. Judicial power is exercised by the magistrates of the Supreme Court in Santiago, who are appointed by the President, and the judges of the courts of appeals and other tribunals.

The regions are divided into provinces which in turn are divided into counties, which are governed by the mayor of the city and represent the jurisdictional territory of the local government or municipalities.

After an interruption of 16 years to the political life of the country, today there is a Government Coalition (Coalition for the Democracy), which integrates the Christian Democracy Party, Socialist Party, Party for the Democracy, and others. The opposition is a coalition of right wing parties represented by National Renovation, the Democratic Independent Union and the Centre of Centre Union.

The present political/administrative organization of Chile is based on the Political Constitution of 1980 (amended in 1989), and applied through the regionalization process and the political issues exercised by the government and the opposition sectors.

Education

Since 1965 the educational system has been structured into three levels: (a) pre-basic education, until five years old, which is regulated by the National Junta of Childhood Schools; (b) basic education, from six until 13 years of age, which is obligatory; and (c) high school, which is distinguished into universities, professional institutes and technical centres.

Between 1980 and 1981 profound transformations were made. These consisted of the transfer of most basic education to municipal control, and the creation of a new university system which permits the creation of private universities and the reunion of the traditional universities belonging to the Rectors Assembly.

Students of traditional universities receive state benefits to pay their education costs. The most important universities are located in the big cities. Because of that, the principal universities are in the Mediterranean zone: the University of Chile (with 16 230 students) and the Pontificy Catholic University (with 16 230 students), in Santiago; the Catholic University of Valparaíso (with 7020 students) in Valparaíso; and the University of Concepción (with 11 460 students), in Concepción.

The rate of illiteracy has been reduced from 11% in 1970, to 5.7% in 1992, which is one of the lowest in Latin America. Public spending on education accounted for 3.1% of gross national product in 1993.

Health and welfare

In Chile there is a system of free medicine and a private health system. The state system funds come from State Health Securities and the private system requires the payment of health insurance premiums. With respect to the health infrastructure, there is a National System of Health Service with 180 hospitals, 413 consultative health centres, 1076 emergency centres and 1214 rural health establishments, with a total of 32 579 hospital beds. The infrastructure of the private service has 31 hospitals, 178 clinics, 392 polyclinics and 467 medical centres, with a total number of 11 063 hospital beds.

The consultative health centres and state emergency centres are free and attract more than 18 million medical visits a year, distributing doses against diseases and foods for pregnant women and infants. The number of inhabitants per doctor in 1989 was 930, while expenditure of the public sector on health accounted for 5.9% of gross national product in 1992.

Social security in Chile has a system of private administrative centres of pensions, which are funded by more than 70% of the active population who contribute to the funds. The pensions are provided to old people who draw on the social security service and public charities.

Culture

Chile has one of the strongest cultural infrastructures in Latin America.

Today the country has 294 public libraries, including the National Library (with 3.5 million volumes), the Library System of the Pontifice Catholic University of Chile (1.3 million volumes), the Central Library of the University of Chile (one million volumes), the Library of the National Congress (800 000 volumes) and the Library of the University of Concepción (300 000 volumes). Chile also counts on important archives such as the National Archive (583 000 volumes).

There are many Academies (of Sciences, Arts, Social, Political and Moral Sciences, and Medicine), created in 1964. There are also the old Chilean Languages Academy (founded in 1885) and the Chilean History Academy (founded in 1933). There are several excellent museums, which are devoted to natural history, archaeology, arts, mineralogy, popular art and history. The National Art Museum (founded in 1880), the National Museum of Natural History (founded in 1830), the Precolumbus Art Museum (founded in 1981) and the National Historical Museum (founded in 1911) are all in Santiago. The Office of Arts of Representation is the principal dramatic centre (teaching and acting) of the country; in Chile there are many theatre rooms devoted to dramatic acting and to the exhibition of popular and musical works. The Chilean National Ballet, the Symphony Orchestra of the University of Chile, the Symphony Orchestra of Santiago and other philharmonic orchestras offer ballet performances and concerts, and there is an opera season between March and November in the municipal theatres of the big cities or in the theatres of the principal universities.

A series of artistic and cultural events is also held during the year. Important ones include: the National Folklore Festival in San Bernardo; Movies Festival of Viña del Mar; International Song Festival, in the same city; the Architecture Biennal, in Santiago; The National Book Fair, in Santiago; and the International Handcrafts Fair, in Santiago.

Much of Chile's colonial architectural heritage has been destroyed by catastrophic seismic events. Important buildings include the Church of San Francisco (1586–1628), the Casa Colorada (1769), the Cathedral Church (1780–1789), the Palace of la Moneda (1799), the palace of la Real Audiencia (1804–1807) and the Real Palace Casa de Aduana (1805–1807), all located in Santiago.

10
The southwestern cape of South Africa

Michael E. Meadows

Introduction

At the southwestern tip of Africa, in the Western Cape Province of South Africa, a Mediterranean-type climate, coupled with the combination of geology, topography, soils and environmental and cultural history, has resulted in a remarkable landscape with a unique vegetation formation. Perhaps more than any other feature of the natural environment, this highly species-rich vegetation type, known as *fynbos*, is what distinguishes the southwestern Cape from other regions of Mediterranean-type climate. Its diversity, in terms of numbers of plant species and genera, rivals even that of the tropical rain forests and suggests that the long-term evolution and development of the southwestern Cape environment has followed a distinctive and complex course. Although there are generally close relationships between climate and vegetation, the distributions of *fynbos* and of the "Mediterranean" climate in this part of southern Africa do not exactly coincide; *fynbos* vegetation is more widely distributed, which suggests that factors other than climate are the main ecological determinants in the region. The area of Mediterranean-type climate (Figure 10.1) approximates the administrative region known as the Western Cape Province (the terms southwestern Cape and Western Cape Province are used more or less interchangeably), although the characteristic vegetation is actually distributed considerably further to the east and reaches beyond Port Elizabeth (see Cowling 1992).

In many ways, this region represents an "island" at the southwestern corner of Africa, isolated by a series of folded mountain ranges trending parallel to the coast. It is a curious feature that the region is so limited in its extent, especially when considering the greater area occupied by this type of climate in Australia, which is similarly exposed to a broad, open ocean. The limited area is most likely a consequence of the mountain ranges acting together with a strongly upwelling west coast current (Benguela), which intensifies summer aridity. The area of Mediterranean-type climate extends from Cape Columbine in the west (18°E), to Cape Agulhas in the south and east (20°E, 35°S), and to just west and north of Vredendal in the north (32°S), an area of some 40 000 km^2. The limited and distinctive nature of this type of environment lends a unique character to the region which is also mirrored in its cultural geography. Despite its location on the African mainland, the insular nature and unique cultural history make it a region which has a much more "European" atmosphere than the rest of South Africa. The major city in the region is the spectacularly located Cape Town, nestling at the foot of Table Mountain. Other important centres include Stellenbosch, Paarl, Worcester, Swellendam, Malmesbury, Clanwilliam and Piketberg.

Land Degradation in Mediterranean Environments of the World: Nature and Extent, Causes and Solutions.
Edited by A.J. Conacher and M. Sala.

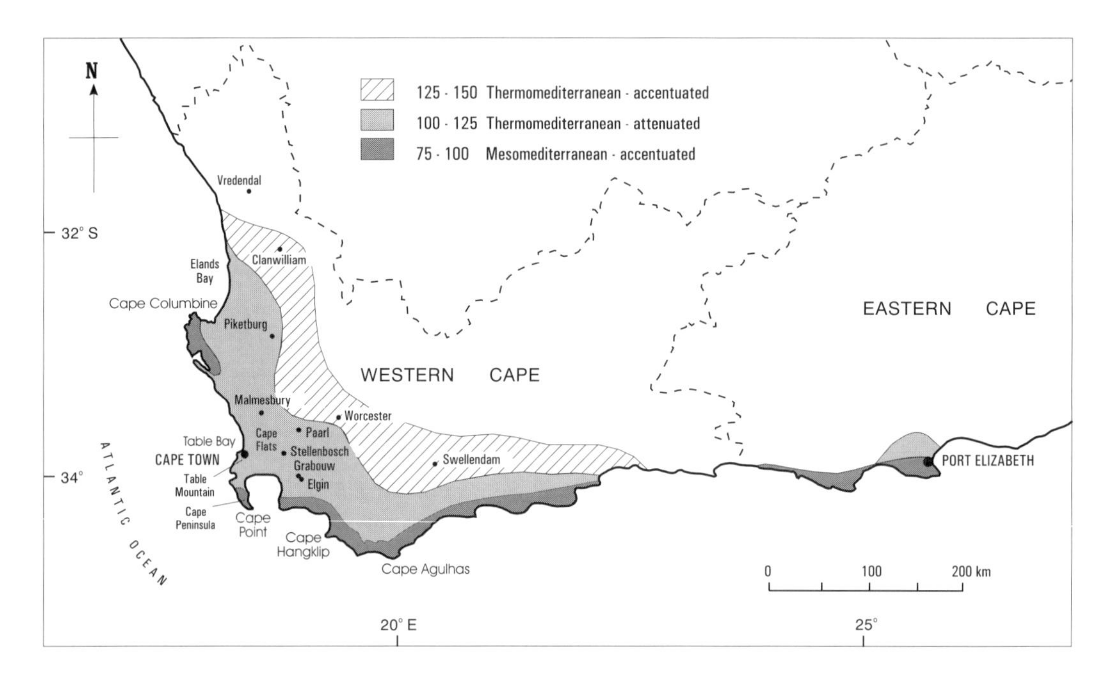

Figure 10.1 *The southwestern Cape, showing the extent of "Mediterranean-type" climates and location of places mentioned in the text.* Original map by the author

The Land

Relief

Wellington (1955) divided the region into two physiographic elements, namely the mountains of the Cape Folded Belt and the plains of the coastal foreland (Figure 10.2). Underlying geology is essentially responsible for the major topographical features, as softer parent materials (for example, the shales of the Malmesbury Group) are associated with more subdued relief, while more resistant rocks produce well developed uplands with steep slopes.

The Cape Folded Belt consists of a series of pronounced anticlinal folded mountain ranges, constructed of quartzitic rocks (Plate 10.1). Two zones of folding are distinguishable (Lambrechts 1979): a western zone in which the folds trend north–south and form a wide arc concave to the west; and an eastern zone running generally west–east. At the meeting of the fold axes, at the so-called syntaxis, folding is variously oriented, but it is here that the altitude of the Cape fold mountains reaches its maximum (du Toit's Peak, 1900 m above sea level).

The coastal foreland, between the folded belt and the coast, is generally a smoothly undulating plain with a maximum elevation of 200 m above sea level, most often underlain by shales, phyllites and, occasionally, granite. Along the west coast down to the south of Cape Town, a former coastal plain with altitudes below 50 m is completely mantled in drift sands of Quaternary to recent origin which extend up to 50 km inland in some areas. The landscape is one of low dunes, mainly vegetated except near the coast. To the south of the western segment of the folded mountains, the gently sloping foreland is covered by Tertiary limestones and marine marls, a situation which reflects the importance of sea-level changes in determining the geomorphology of the region.

Drainage and hydrology

The steep slope gradients, seasonality of rainfall, confinement of the Mediterranean-type climate by the Cape Folded Belt mountains, shallowness or porosity of soils in the catchments – all combine to produce characteristically small, short

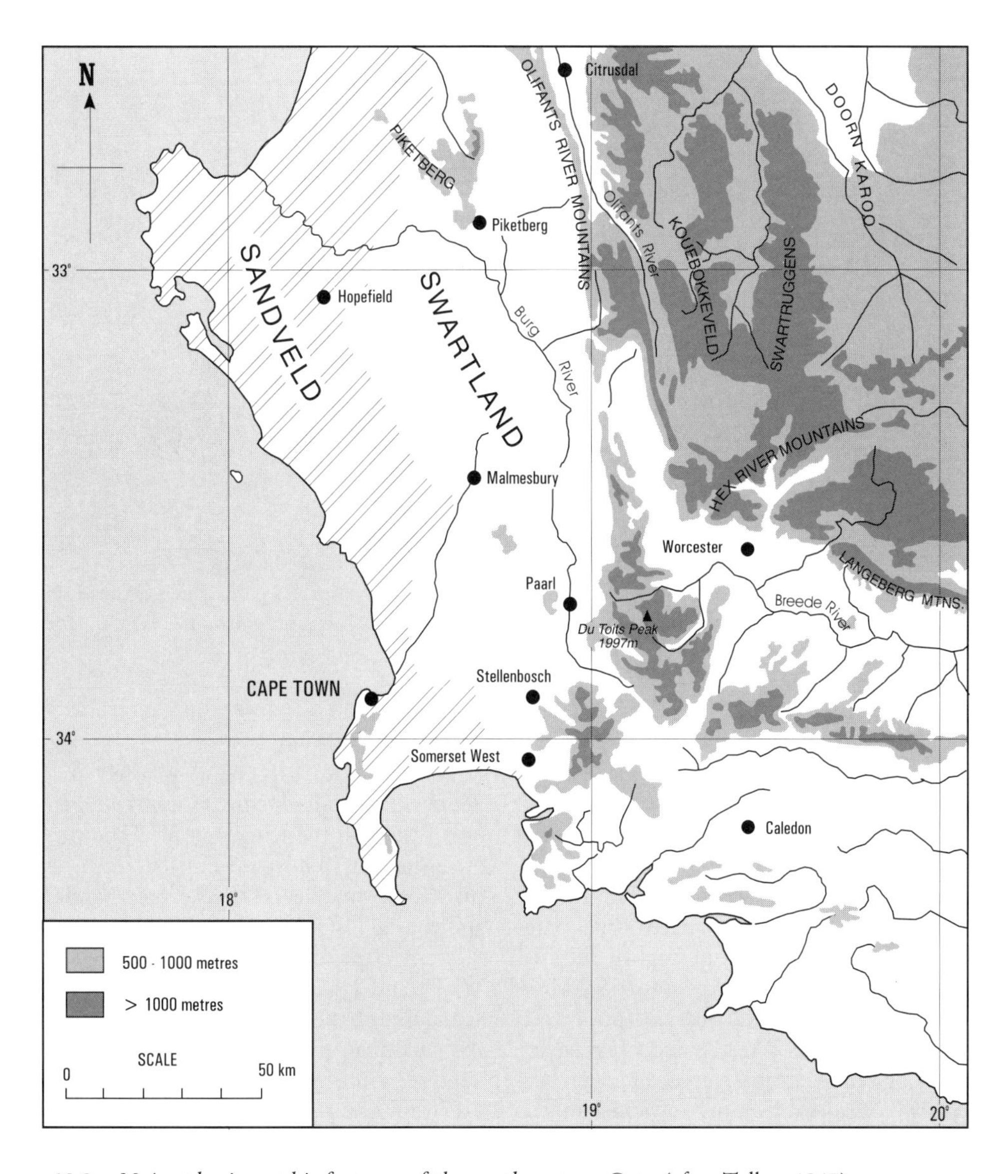

Figure 10.2 *Major physiographic features of the southwestern Cape* (after Talbot 1947)

and often ephemeral rivers in the southwestern Cape. As a consequence, there are relatively few larger catchments and associated rivers, and even fewer naturally occurring standing bodies of water (all the larger lentic waters are associated with impoundments). The major catchments are shown in Figure 10.3, the three largest being those of the Olifants, Berg and Breede rivers. Southwestern Cape rivers contribute around 10% of South Africa's mean annual runoff, representing approximately what could be expected proportionately from the geographical area occupied by its catchments. The Olifants River, draining the largest catchment in the region at more that 49 063 km^2, receives on average less than 400 mm precipitation annually, has a mean annual runoff of 1 008 106 m^3 and a coefficient of variation for runoff of greater than 50% (Midgley *et al.* 1994). Major groundwater resources, mainly because of the nature of the geological substrates, are confined to the larger coastal dunefields and the coastal limestones of the Agulhas region.

Plate 10.1 *Tightly folded and steeply dipping quartzitic beds in the Langeberg Mountains near Montagu in the southwestern Cape*

Soils

The diversity of the southwestern Cape landscapes results in a great variety of soil types and associations in the region, and parent material characteristics frequently dominate as a soil-forming factor. The quartzitic rocks of the folded ranges are usually associated with pale-coloured, shallow, sandy and highly leached oligotrophic lithosols (Lambrechts 1979). Morphologically immature and nutrient-poor soils are also found on the Tertiary and Quaternary driftsands of the west coast area. Heavier textured soils, considerably more enriched in nutrients, are developed on the shales and phyllites of the coastal foreland where Malmesbury Group rocks prevail. The isolated granitic outcrops of the region are characterized by deep, red, kaolinitic soils (Lambrechts 1979), which are also richer in nutrients (and therefore agriculturally more valuable) than the quartzitic soils of the mountains. As in any other polygenetic landscape situation, not all soil and weathering products are necessarily contemporary in that there is widespread evidence, in the form of either ferric or siliceous duricrusts, of long-term weathering under climates which may have differed markedly from those which currently prevail (Lambrechts 1979).

Climate

Southern Africa in general is affected by dry, anticyclonic high pressure systems which migrate northwards in the winter months and permit the extension of transient, rain-bearing westerlies (namely frontal depressions, ridging anticyclones and coastal low pressure systems) into the southwestern Cape at that time (Cole 1961; Preston-Whyte and Tyson 1988; Deacon *et al.* 1992). This accounts for the seasonality of precipitation and the fact that the region experiences predominantly winter rainfall and, accordingly, a Mediterranean-type climate.

Climate diagrams for three meteorological stations in the region (Ceres, Cape Town and Grabouw) show rainfall, associated with the passage of frontal systems, occurring mainly during the period June to September, although the pattern of rainfall distribution is less strongly seasonal towards the east and south (Figure 10.4). North of Cape Town, the cold Benguela Current exerts a strong influence on the climate and there is marked aridity. Thus, despite the occurrence of dominant winter rainfall, most areas north of Clanwilliam do not classify as having a Mediterranean-type climate because the mean annual totals are well below the 275 mm threshold adopted in this volume.

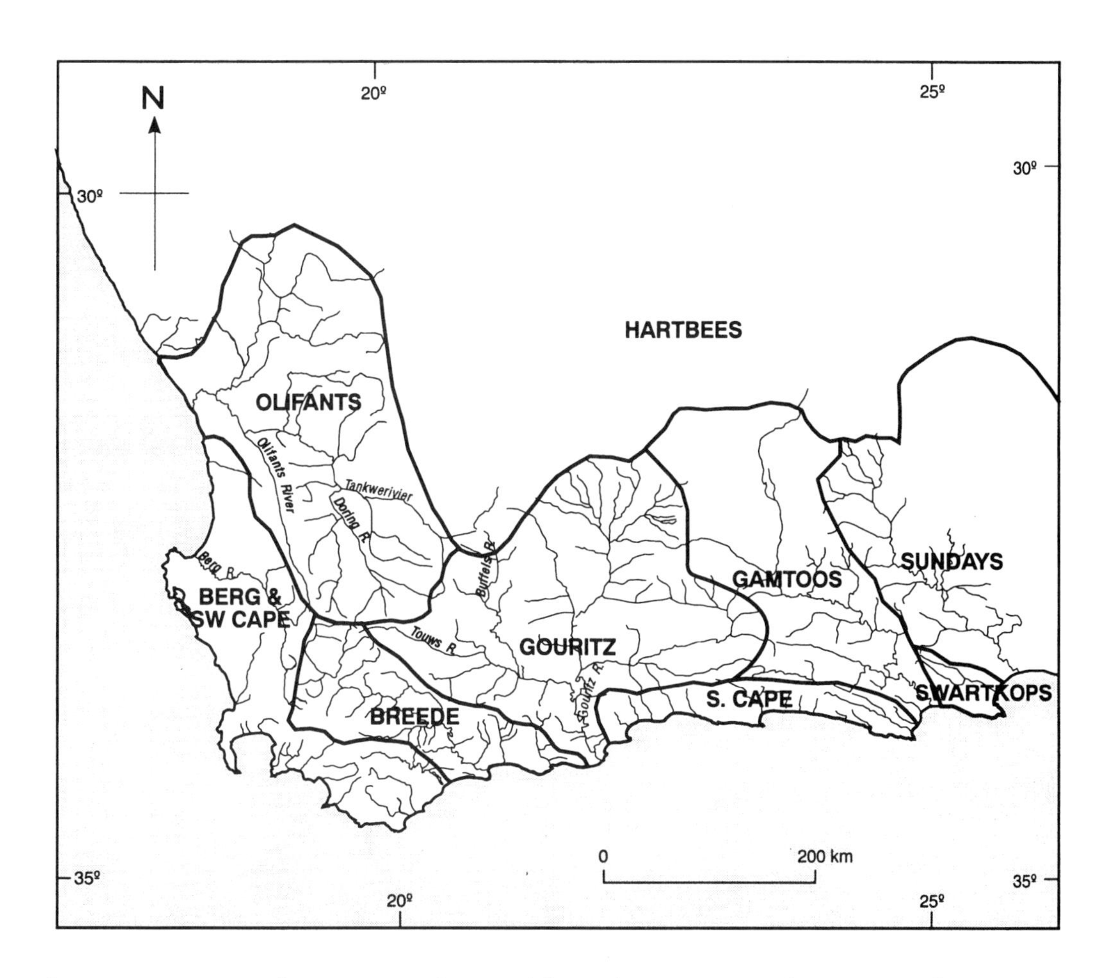

Figure 10.3 *Major catchment areas and rivers of the southwestern Cape* (after Davies *et al.* 1993)

Within the region, mainly as a consequence of the variability of relief, there is enormous variation in annual amounts of rainfall. Altitude and aspect (the winter rains come predominantly from the northwest) ensure that some mountain peaks receive more than 2500 mm per annum (Fuggle and Ashton 1979), while areas of the coastal foreland record less than 400 mm. Rainfall gradients are often very steep over short distances. For example, mean annual rainfall at Maclear's Beacon (the highest point on Table Mountain) is 1900 mm whereas at the hospital, some 700 m lower in altitude and less than 3 km away, it is less than 600 mm (Wellington 1955).

During the summer months, ridging high pressure systems frequently bring moist air onshore in strong southeasterly winds, producing a characteristic cloudiness adhering to many of the mountain peaks nearer the coast (most famous of all is the "tablecloth" of Table Mountain; Plate 10.2). Fog precipitation from such clouds results in prodigious amounts of precipitation, much of which is probably available to plant communities at the higher altitudes. Fog precipitation occurred on Table Mountain over 123 days during one calendar year, and on a further 89 of 126 rain days (Nagel 1956). In this way, total annual precipitation is substantially augmented (total fog precipitation on Table Mountain exceeds 3000 mm according to Nagel (1956)). The summer drought is completely absent in this type of habitat circumstance.

Maximum 30 minute rainfall intensities are generally less than 50 mm/h and 24 hour values

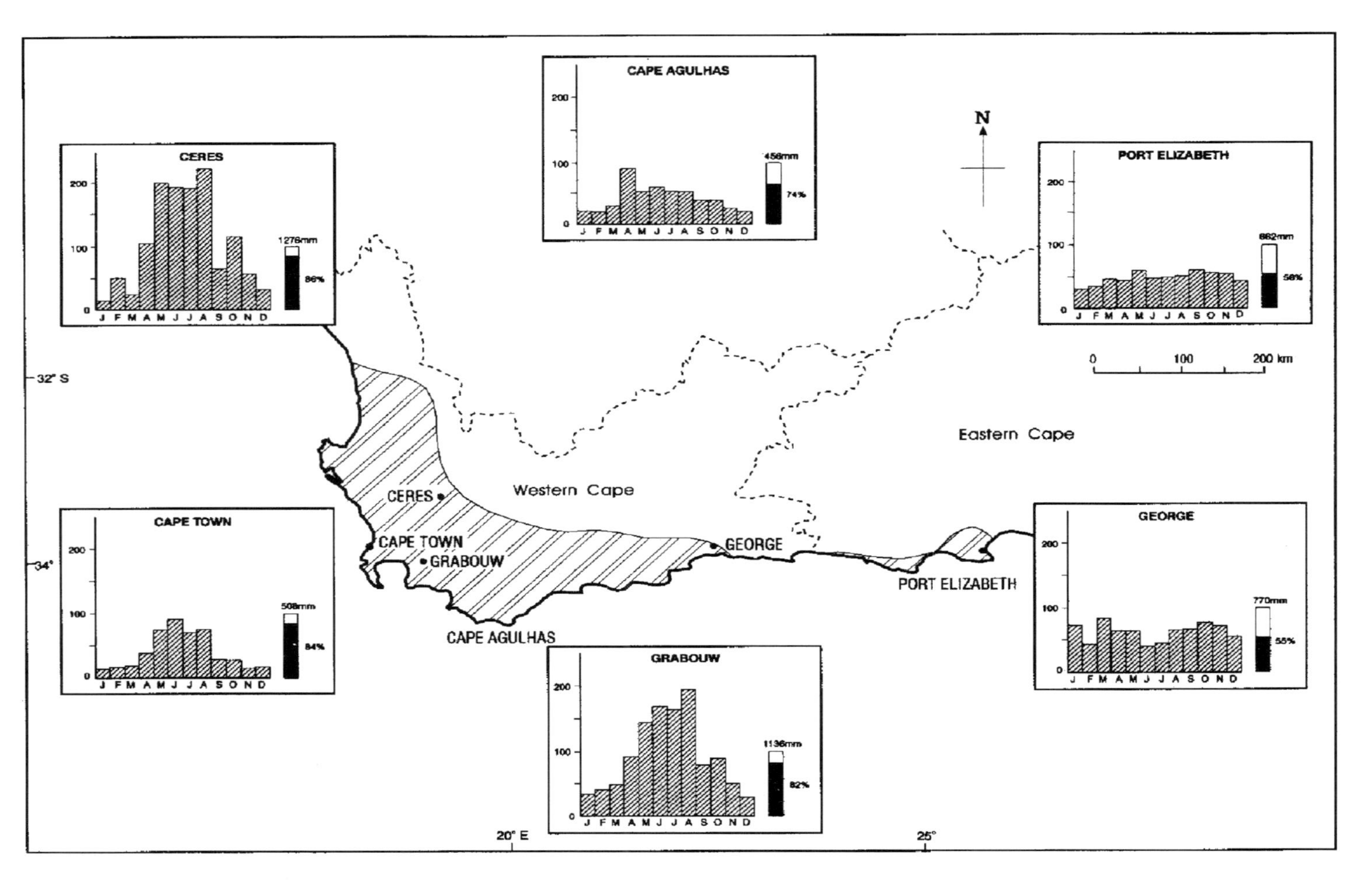

Figure 10.4 *Climate diagrams for the southwestern Cape, showing mean annual rainfall distribution* (after Cowling 1992)

Plate 10.2 *Table Mountain and its "tablecloth". As in California and Chile, fog accounts for a significant proportion of moisture received by the vegetation*

rarely exceed 100 mm (Schulze 1965). Indeed, Nagel (1956) points out that effective intensities for mist and fog generally exceed those of rainfall on Table Mountain. Thus, the southwestern Cape is relatively less susceptible to extreme flooding events and associated soil erosion than most Mediterranean regions. Countering this tendency, however, are the steep slopes which dominate much of the landscape, and the role of fire in periodically removing vegetation cover.

Temperatures are strongly influenced by the proximity of a cool South Atlantic Ocean and the Benguela Current, and are generally around 17°C with an annual range of only 8°C at the coast (Fuggle and Ashton 1979). Owing to the well developed relief, however, there is considerable local variation; maximum temperatures inland may exceed 30°C in summer months when the northwesterly *berg* (mountain) winds blow, representing a significant fire hazard. Frost is unusual near the coast but more common in July and August in the interior or at altitude (Fuggle and Ashton 1979), where its economic importance (in terms of the deciduous fruit industry, for example) is enormous.

Given the region's location at the tip of a continent which reaches out into the southern oceans, it is not surprising that this is a windy region. The turbulence of the atmosphere is perhaps epitomized by the southeasterlies (the "Cape Doctor", so named for its legendary reputation for blowing away the winter's ills) which, especially on the Cape Peninsula, may blow hard continuously for more than a week during November to February.

Plant and animal communities

The vegetation of the southwestern Cape is remarkable in respect of its species diversity. Good (1964) identified only six phytogeographical "kingdoms" globally, assigning an entire division to a minute (certainly in comparison to the other five) geographical area at the southwestern tip of Africa – the Cape Floristic Region, or Capensis (Taylor 1978).

The most remarkable feature of the flora is its species richness and degree of endemism (Table 10.1; Cowling *et al.* 1992). Bond and Goldblatt (1984) estimate the total flora to be in the region of 8500 species which, in terms of species density, for a total area of just 90 000 km^2 is richer than the flora of Greece and Portugal respectively by a factor of approximately three, richer than the Californian flora by a factor of seven, and richer than the flora of southwestern Australia by a factor of eight (Cowling *et al.* 1992). The diversity is constructed at several levels; that is, within-habitat (alpha diversity), between-habitat (beta diversity) and between-area (gamma diversity) values are all moderate to very high when compared with other regions of similar climate. Levels of endemism are also impressively high: 68% of the species are found nowhere else, a degree of uniqueness which is similar to that of large islands (for example, New Zealand), and all the more interesting because the region is contiguous with mainland Africa and of course not at all an island in the geographical sense (Goldblatt 1978). More than 200 genera are restricted to Capensis, possibly the highest rate of generic endemism in the world (Good 1964), and seven plant families are entirely confined to the Cape Floristic Region.

Taxa indicative of the Cape Floristic Region include numerous members of the Restionaceae, Proteaceae and Ericaceae families, although there are many other families and genera which may be regarded as "typical". Several plant families are especially diversified in this flora, in particular the Ericaceae, Asteraceae and Orchidaceae. Single-species dominance is unusual, and it is the variety of different species which best characterizes the vegetation.

Structurally, there is a measure of uniformity. Cody and Mooney (1978) concluded that, in terms of structure and physiognomy, the communities of the Mediterranean Basin, California, Chile, southwestern Australia and southwestern South Africa are strikingly similar. All areas are dominated by evergreen shrubs, frequently dwarf in character, with sclerophyllous leaves. The so-called *fynbos* communities of the southwestern Cape are no exception; typical *fynbos* vegetation types prevail across much of the region, although they occur in a mosaic with other, non-*fynbos*, plant communities (Figure 10.5).

The term *fynbos* was originally coined by European settlers (the Dutch word is *fijnbosch*) and was used to describe the vegetation, equated with *macchia* in Europe, characteristic of the area around Cape Town, dominated by woody, evergreen plants with small hard leaves. The *fynbos* biome consists of several phytogeographically and ecologically distinctive groups of plant communities which occur in relation to particular combinations of climate, substrate, topography and fire frequency.

Oligotrophic soils on uplands derived from the Table Mountain Group sandstones, together with soils developed on the granites, support mainly mountain *fynbos* vegetation. Typically a shrubland, the vegetation is characterized by ericoid, proteoid and restioid leaf forms. Trees are absent, with the exception of some of the larger proteas, for example *Protea nitida* and *Leucadendron argenteum*, and the Clanwilliam cedar, *Widdringtonia cedarbergensis*, which is a gymnosperm restricted to the Cederberg range.

There has been much speculation concerning the scarcity of trees (Moll *et al.* 1980), but it is probably related to the fire regime and seasonal aridity in conjunction with low nutrient status soils. Restioid (reed-like) elements are practically ubiquitous in these communities, although more common on deep sands and in seasonally wet habitats. An important variant of *fynbos* occurs on the acidic sands of the coastal lowlands, particularly in the Greater Cape Town area, where acid sand plain *fynbos* is especially endangered due to the onslaught of urbanization.

Calcareous dunes at the coast support a xerophytic community known as *strandveld*, which may be regarded as transitional between true *fynbos* and the karroid communities of the drier

Table 10.1 *Area and plant diversity in some Mediterranean climate regions* (source: Cowling *et al.* 1992)

Region	Area ($10^6 km^2$)	No. species	Species density (10^6 spp./10^6 km^2)
Cape floristic region	0.09	8550	94.4
Californian floristic province	0.32	4452	13.9
Southwest botanical province	0.31	3611	11.6
Greece	0.13	*c.*4000	30.8
Portugal	0.09	*c.*2500	27.8

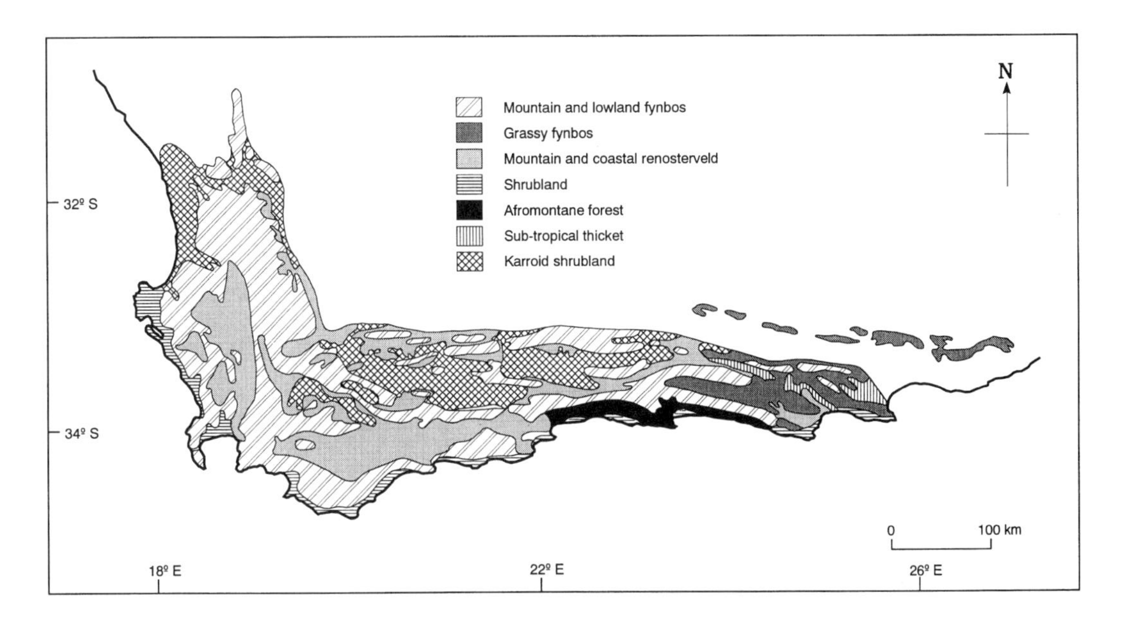

Figure 10.5 *Vegetation of the southwestern Cape* (after Cowling 1992)

interior. Here, the ericoid and proteoid elements are less abundant and the vegetation is more succulent in nature (for example, *Zygophyllum morgsana*), and may even form low scrub forest with small trees such as milkwood (*Sideroxylon inerme*) and *Rhus glauca*.

Renosterveld is an interesting plant community associated mainly with the more nutrient-rich shales of the Malmesbury Group, although there is little of it remaining. The vegetation is dominated by *Elytropappus rhinocerotis*, so-called *renosterbos* (from the Dutch, meaning rhinoceros bush). The community is again rich in species, although the true Cape Floristic Region elements are much less important and, phytogeographically, this vegetation has more in common with the interior of South Africa.

Another non-*fynbos* community is Afromontane forest, mainly found on southeasterly facing slopes and in ravines on sandstones of the mountains and coastal uplands. These forests are similar to those found across the rest of the African continent, more usually at higher altitudes, although at the Cape they extend virtually down to sea level. The most important trees are the yellow-woods, *Podocarpus falcatus* and *P. latifolius*, and numerous other species which were much in demand for commercial purposes in a landscape which was otherwise, as noted above, largely devoid of trees.

By comparison, animal communities in the *fynbos* are less well known and considerably less conspicuous. Unlike the savannas of South Africa, *fynbos* supports few large animals, and most of these have been rendered extinct, or nearly so, by successive waves of human occupation. Many of the mammal herbivores and carnivores once found in *fynbos* still survive in other biomes, but for the endemic blue antelope, Cape lion and quagga, local extinction around the turn of the 19th century meant global extinction.

In any event, the nutrient deficiency of *fynbos* soils, combined with a relatively uniform vegetation structure and frequent fires, ensured that diversity and population numbers of larger animals were low by comparison with other parts of the subcontinent. Antelope, such as the common duiker, grysbok and klipspringer, survive in small but viable numbers in the Cape mountains and provide the main source of food for the only remaining large carnivore of the region, the leopard, although this species is now extremely rare. There are also many varieties of smaller mammals, birds, reptiles and amphibians which thrive in *fynbos* and which play an intricate and important role in the ecology of the various plant communities. Small herbivores, such as the vlei rat and striped mouse, along with numerous varieties of insect, in particular ants, are crucial to the seed dispersal and survival mechanisms of many *fynbos* shrubs (Cowling and Richardson 1995).

The People

The region, together perhaps with the circum-Mediterranean region itself, has arguably the longest record of human habitation of all the Mediterranean climate regions (Deacon 1983) and has been occupied since the dawn of the genus *Homo* more than one million years ago. Deacon (1983) believes that even relatively small numbers of people would have impacted considerably on the fire regime, such that anthropogenic landscape alteration was possible even 100 000 years ago.

By the end of the Pleistocene, so-called Later Stone Age hunter-gatherers (known as *Khoisan*) prevailed in the southwestern Cape. The arrival of *Khoikhoi* pastoralists around 2000 years or so ago, with sheep and cattle, facilitated a more settled lifestyle. They co-existed with the *strandloper* (literally, beach-walker) populations at the coast and the *San* in the mountains. There would also have been economic contact with the Bantu-speaking Africans of the summer rainfall grasslands and savannas to the east.

The arrival of European colonists was confirmed from 1652 onwards. A new social and economic order emerged as they systematically, through the use of political force, and insidiously, through the introduction of endemic diseases, completely subjugated the indigenous population and rapidly rendered it extinct as an identifiable separate culture.

It is estimated that population numbers were 18 000 *Khoikhoi* in the Cape Town area at the time of European occupation (Thom 1958, in Deacon 1983). Since then, numbers have risen to more than four million people in the Western Cape Province, of which almost 90% are urbanized (Wesgro 1994).

Despite the unacceptability of ethnic labelling, it remains necessary in order to fully appreciate the degree to which "race" has determined the geography and demography of South Africa. Around 54% of the region's population is classified as Coloured (2.2 million), with the remaining people divided approximately equally between Africans (one million) and Whites (0.9 million). In total, this represents some 10% of the population of South Africa, which means that population densities are about average for the country as a whole, at 28.0 people/km^2. Currently, the average annual population growth rate in the region is high, marginally exceeding the national mean at 2.83% (Wesgro 1994). The increase is made up of both intrinsic growth and a substantial rural–urban migrational influx, mainly from the eastern Cape former "homelands". Greater Cape Town dominates the region demographically and constitutes some 72% of the population; almost three million people live within the metropolitan area. Linguistically, there are three dominant languages: mother-tongue is still essentially determined by race, that is English (white – 21% of the population), Afrikaans (coloured and white – 62%) and Xhosa (African – 16%).

The region was subjected during the apartheid era to the segregation of its towns and this led to developmental dualism (Development Bank of Southern Africa 1994). The initial grand plan would have been to retain the region's historical character as mainly coloured and white. Socio-economic imperatives, however, have led to some dramatic population developments during the past decade or so. Thus, while the privileged white community has remained largely economically active, educated and demographically stable, between 1990 and 1993 the African population increased four-fold from less than 300 000 to more than one million as a consequence of the relaxation of influx control. A substantial proportion remains marginalized, despite the exciting political developments of the 1990s. Socio-economically and demographically, the conditions of the large coloured population of the region are transitional between the affluent whites and the disadvantaged Africans.

Population projections (Table 10.2) reflect these differences. Thirty percent of the population is younger than 15 years of age, a little lower than the national average (Wesgro 1994). Nonetheless, the population of the region is expected to exceed six million people by the year 2010, with a markedly increased proportion of Africans.

Table 10.2 *Population projections (thousands) for the western Cape* (source: Wesgro 1994)

Year	Total	Coloured	(%)	African	(%)	White	(%)
1994	4198	2258	53.8	1001	23.8	939	22.4
1997	4570	2410	52.7	1193	26.1	967	21.2
2000	4920	2545	51.7	1380	28.1	995	20.2
2005	5475	2780	50.8	1650	30.1	1045	19.1
2010	6045	3020	50.0	1925	31.8	1100	18.2

The Economy

The Western Cape ranks second among the nine South African provinces in per capita gross domestic product (GDP) and, although its share of the national population is only around 10%, the region contributes almost 15% of the GDP. This confirms the region's relatively high economic performance, and the Cape metropolitan area currently represents South Africa's fastest growing economy.

Agriculture, forestry and fisheries

Agriculture represents a relatively small, but highly significant sector of the region's economy. It accounts for just 6% of the gross regional product and employs around 8% of the workforce (Wesgro 1994), but is highly significant because it indirectly supports a substantially greater proportion of the economy through associated food and beverage industries, and because it is one of the most widespread land uses (approximately 25% of the land area of the region; Plate 10.3). All the typical crops of a Mediterranean climate region are grown in abundance, in particular citrus and deciduous fruit, grapes for table and wine, wheat and various vegetables, in addition to mixed stock farming.

Particular combinations of climate, geology and soil favour different agricultural activities. For example, winter wheat is established on the nutrient-rich soils developed on Palaeozoic shales of the Malmesbury area, deciduous fruit dominates on the high-lying quartzitic sandstone plateau around Elgin, and vineyards are

Plate 10.3 *Intensive agriculture and horitculture near Paarl, viewed from du Toit's Kloof. The steeper country is not farmed or grazed, which accounts for the generally low sedimentation rates in the rivers and streams of the southwestern Cape*

established on the colluvial and alluvial soils of the Stellenbosch and Paarl districts. Deciduous fruit production (Plate 10.4), which accounts for more than half of the region's earnings from agriculture, is particularly profitable, representing over 40% of the European import market and in 1993 earning the country almost US$750 million in foreign exchange (Table 10.3).

The wine industry has grown since the lifting of sanctions during 1992, and contributes a further 25% of the earnings from agriculture. In 1993 production topped one billion litres, and exports at 2.3 million cases were 124% up on the previous year (Regional Development Advisory Committee 1994). There are now more than 100 000 ha under vines and this is likely to increase further (Odendaal and Horwood 1995). Wheat yields continue to expand, and more than 1 500 000 t were harvested in 1993, well over half the domestic demand.

Future climate change and land reform are unknown factors which may affect the situation. The issue of land ownership, historically concentrated in the hands of relatively few white farmers, is currently on the discussion table and already 30 000 new black (coloured and African) small farmers are to be settled in the region.

The southwestern and southern Cape are significant forestry areas, although they are secondary to the major wood-producing areas of eastern and northeastern South Africa. Since indigenous forests are not abundant in the region, the forestry sector is based almost entirely on state-owned and private plantations of fast-growing

Plate 10.4 *Irrigated orchards and vineyards in the Elgin district*

Table 10.3 *Fruit industry in the western Cape, 1994 (source: Wesgro 1995)*

Industry sector	No. of producers	Production (t)	Gross earnings (US$ million)	Employment
Fresh	1800	752 000	471	178 000
Canning	1400	220 000	99	48 000
Dried	2600	250 000	31	30 000
Processing	–	290 000	43	4000
Total deciduous	5800	1 512 665	644	260 000
Wine	4800	1 111 000	62	50 000
Total	10 600	2 623 665	707	310 000

species of *Pinus* and *Eucalyptus*. Of the Western Cape Province's 13 000 000 ha, 0.588% are under pine plantations, 0.026% under eucalypts and 0.002% under wattle plantations (Le Maitre *et al.* 1995). These plantations supply wood for a variety of industrial purposes, with high quality furniture at one end of the spectrum and crates for agricultural produce at the other. The forests also represent a significant element in the region's tourist attraction, although the negative impact of the plantations on water supply and the conservation of indigenous species may be considered a disadvantage.

Fishing boasts a very long history in the region and continues to play an important role in its economy. The 1000 km of coastline to the north and southeast of Cape Town produces 85% of South Africa's fishing harvest annually (Wesgro 1994). The rich fishing grounds are based on an equally rich nutrient source fuelled by the upwelling Benguela Current. In 1991, more than 1 000 000 tonnes of fish were brought ashore in Western Cape harbours (Wesgro 1994), although the industry has been subject in recent years to the vagaries of world markets and prices, and the number of 10 000 employed appears to be in decline. The most important species include crayfish, pelagic fish (anchovies and sardines) and hake, all of which are subject to quota, together with the uncontrolled sector, which includes tuna.

Industry

The structure of the various industrial sectors in the region is shown in Table 10.4 in comparison to the national situation. In sharp contrast to the rest of South Africa, the mineral wealth of the region is limited and the most important primary component of the economy is agriculture. The most significant mining activities are based on the extraction of heavy metals from dune sands some 350 km north of Cape Town. However, the relatively low profile of mining has probably contributed to the diversification of the manufacturing and service sectors (Wesgro 1994). Moderate growth is anticipated in several of the region's industrial clusters, including food processing and beverages, the clothing and textile sector, petrochemical products, metal working, printing and publishing, and the furniture and wood processing sector (Odendaal and Horwood 1995). Industry in the region benefits from the fact that apartheid policies offered the coloured population a reasonable, by comparison to African, standard of education and social standing and the labour force now offers skill levels significantly above the national average.

The backbone of industry in the southwestern Cape is the processing of its agricultural products. Food and beverages together contribute more than 25% of the region's industrial output and employ more than 60 000 people, second in importance only to clothing, which employs some 150 000 people but has a smaller share of the gross regional product (Wesgro 1994). The closely related textile industry is also important, although external competition has put this sector under very heavy pressure in recent years. Other significant elements of industry include a buoyant financial and insurance sector (the head offices of five of the country's six major insurance companies are based in Cape Town), a vibrant

Table 10.4 *Industrial sector structure: western Cape and South Africa compared, 1990* (source: Wesgro 1994)

Sector	Western Cape (%)	South Africa (%)
Agriculture, forestry, fisheries	7.8	5.6
Mining	0.2	12.9
Manufacturing and processing	24.6	24.2
Electricity, gas, water	2.7	4.1
Construction	4.2	3.6
Trade, accommodation	15.7	12.8
Transport, communications	9.9	8.9
Finance, insurance	15.1	11.1
Community/social services	19.8	16.8
Total	100.0	100.0
Gross domestic product (US$ million)	10 000	73 800
Western Cape share of RSA	13.1%	

retail, wholesale and import–export division and a strongly developed public service element.

Special mention must be made of tourism, for this is rapidly developing into the major growth industry. The lifting of sanctions also appears to have paved the way for a boom in this industry, although it is difficult to separate this from expansion in other industrial sectors. If multiplier effects are taken into account, tourism, both international and national, contributes more than 6% of the gross regional product (Wesgro 1994). In 1992, there were more than one million visitors to the region, and the number is thought to be expanding at a rate of 10% per annum (Odendaal and Horwood 1995). Indeed, such are the attractions of the scenic and cultural attributes of the Western Cape, that more than 80% of the tourists who visit South Africa spend at least some time in the region. A prime example of the development of this sector is the Victoria and Alfred Waterfront area of Cape Town which attracted more than 14 million visitors in 1994 and continues to grow in terms of size, facilities and popularity.

Transportation

The legacy of apartheid is still obvious in the context of transportation within the region as a whole, and within metropolitan Cape Town in particular. The marginalized poor, mainly African and coloured people, were forced to live on the periphery of the major centres. This situation still prevails such that their journeys to work are generally longer than those of the white population. There is a heavily subsidized (and heavily subscribed) public transport system in the form of train services which operate throughout the Western Cape. The commuter rail network carried an estimated 178 million passengers in 1994, representing almost one-third of the national total (Wesgro 1994). The last decade, however, has seen the extensive development, as elsewhere in South Africa, of private transport based on the minibus taxi. All told, the transportation industry employs some 60 000 people.

There is a remarkably high rate of car ownership in the region, with 240 registered vehicles for every 1000 people; this compares with 300 in the United Kingdom and 140 in Greece (Wesgro 1994). Road conditions are highly dependent on locality and are certainly deficient in many of the townships. Road connections within the region and with the rest of South Africa, on the other hand, are excellent and a very high proportion of the roads are tarred. Air transportation of both people and goods has been increasing in recent years in line with national and international trends. Sea transport and the use of the major harbour in the region, Cape Town, has declined over the same period, although a substantial proportion of the region's international trade is still routed this way.

Administrative and Social Conditions

Government

South Africa, after almost four decades of apartheid and many more decades of colonial rule, finally acceded to democracy in 1994. In the country's first non-racial elections, a transitional Government of National Unity, headed by the State President, was returned to power to see through the first five years of democratic politics. However, in 1996 the predominantly white National Party withdrew from the governing coalition.

At national level there is an upper house, the Senate, with 90 members, and a lower house, the National Assembly, with 400 members, half of whom are coupled to particular provinces in proportion to their size. The provincial legislatures, of which there are nine, represent the next tier of government. The Western Cape Provincial Government has 41 members belonging to parties represented proportionally and headed by a Premier. The Western Cape was the only one of the nine provinces to receive a majority of seats held by the National Party (the former white minority government's leading party). This apparently surprising situation, given the reputation the region has for its more liberal attitudes, arose because it is the only province with a non-African majority and the coloured vote favoured a more conservative approach than that offered by the African National Congress, which gained a massive majority nationally.

The provincial legislature holds jurisdiction over some administrative areas of considerable importance in the context of land degradation, including, agriculture, environment and nature conservation. The next tier down is the local authority; the situation remains in a state of transition, since elections for this sector were due to take place in most provinces (but not the Western Cape, where boundary and demarcation disputes delayed the process) at the end of 1995.

Education

The effects of apartheid planning policies still dominate the educational structure of South Africa, and the southwestern Cape region is no exception. The new constitution is not yet fully established, but it is likely that this will make schooling compulsory. At the moment, access to educational facilities is highly dependent upon geographical location which, because of the effectiveness of apartheid planning, means that former white group areas are relatively well served by public and private schools, whereas the situation in former coloured and black group areas falls well short of desirable goals, especially in the burgeoning townships and informal settlements. Despite this, the Western Cape Province has the highest rate of adult literacy in the country, although at 72% this would be considered low by developed world standards (Development Bank of Southern Africa 1994). Schools in the middle and upper income sectors certainly offer a quality of education comparable with those of developed countries and it is one of the key goals of the Government of National Unity's Reconstruction and Development Programme (RDP) to address inequality in educational standards between residential areas. At tertiary level there are several technical (including agricultural) and teacher-training institutions and three universities.

Health and welfare

While private health facilities in the region are excellent, the situation in the public sector is once again highly skewed and dependent upon the "logic" of apartheid. Access to health care for most of the population in the peripheral townships is limited at best. Even though the first heart transplant operation was performed in 1967 in a provincial hospital, the health and welfare situation in 1994 is a major cause for concern. There are three major provincial hospitals in the region, with a total of just under nine beds per 1000 population (Wesgro 1994). This compares favourably with the country as a whole (four per 1000), but is low by developed world standards, especially given that access to these beds is far from even.

If infant mortality can be taken as an indication of general health standards, then the regional average (44 per 1000) is rather better than the national average (62 per 1000; Wesgro, 1994). The inequalities are stark: the situation for white infants is probably more comparable with Switzerland; that for the coloured population compares with Poland; and that for black children is more like the situation in Mexico (Wesgro 1994). Welfare generally is in a state of transition and it remains to be seen whether the Reconstruction and Development Programme is able to address the plight of the region's unemployed, disabled and elderly people.

Cultural Life

The distinctive nature of the southwestern Cape extends well beyond its physical environmental features and there is a definite and unique, if somewhat intangible, character to this region which is a product of its landscape, its history and its people. Table Mountain perhaps embodies the spirit of the region. It is at once a powerful illustration of the complexity and beauty of nature, a beacon to travellers, an icon of hope in the emerging democracy, and an internationally recognized symbol of the nation.

The culture of the region is rich with the influences of its history and the "new" South Africa is at last giving expression to this diversity. Thus, there is a provincial orchestra and ballet, the State Theatre and numerous other theatres and cinemas, museums and art galleries and all the other typical accoutrements of "western" art. There is, in addition, an emerging and vibrant "alternative" art culture, epitomized by the music scene, which is rapidly revealing a creativity and flair that appeared to be suppressed during the apartheid era. Surely one of the crucial environmental challenges of the 1990s for this region is to find ways of channelling some of its cultural diversity and vigour into fostering and protecting the diversity and vigour of its landscapes, fauna and flora?

11
Southern Australia

Arthur and Jeanette Conacher

Introduction

Australia's Mediterranean environments occur in two discrete zones (Figure 11.1). Western Australia's Mediterranean zone is separated from that in South Australia and Victoria by the arid, Nullarbor Plain. The UNESCO–FAO (1963) map shows the Mediterranean region as extending from the west coast as far north as Geraldton, through the southern, populated portion of South Australia into a broad belt across Victoria and almost as far as Wagga Wagga in New South Wales. However, Wagga Wagga does not have a markedly seasonal rainfall regime, and we have drawn the eastern boundary of the region along the Murray River. The west–east straight line distance is about 3100 km, and the north–south distance, from Geraldton to Portland in Victoria, is the equivalent of some 1000 km.

To the north, the Mediterranean zone passes quickly into semi-arid and then arid environments, and in the extreme east, into more seasonally equable rainfall areas. The Southern Ocean borders the region to the south. Cooler and moister areas along the south coast in the west and east are not classified as Mediterranean.

In general, the land is gently undulating and relatively featureless, the main exceptions being the dissected Darling Plateau in the west and the Mt Lofty Ranges in South Australia. The Mediterranean region incorporates Australia's major, extensive, grain-cropping area (especially wheat), associated with sheep raising on improved pastures. Well known wine-producing areas are

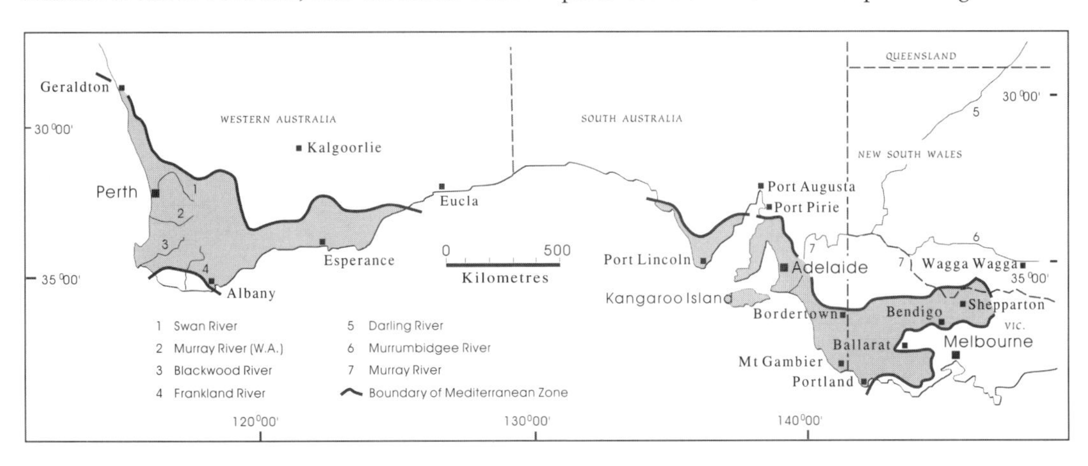

Figure 11.1 *The zones of Mediterranean-type climates in Australia, including major rivers and places referred to in the text. Mediterranean boundary modified from UNESCO–FAO (1963) as discussed in the text*

Land Degradation in Mediterranean Environments of the World: Nature and Extent, Causes and Solutions.
Edited by A.J. Conacher and M. Sala.

located in the region, especially in South Australia, and large tonnages of bauxite are mined in the Darling Plateau. The major cities are Perth and Adelaide, located on the coast; they are the capital cities of their respective States of Western Australia and South Australia. At least 80% of the population of these states lives in the Mediterranean climatic zone. Although Victoria's capital, Melbourne, is located to the south of the Mediterranean region, approximately 40% of Victoria's agricultural land lies within the Mediterranean region, where agriculture is dominated by dryland crop/pasture farming. There are also significant areas of irrigated pasture and horticulture in the north of the region. Nevertheless, the region of Mediterranean-type climates is generally less significant for the State of Victoria than the other two states.

The Land

Relief

Unlike Mediterranean regions in other parts of the world, described in the previous chapters, Australia's Mediterranean regions are low relief and low energy landscapes. In the west, there is a rise of little more than 300 m from the sandy, coastal plain of the Perth Basin to the Western Australian plateau – the topographic expression of the Archaean granitic craton known as the Yilgarn Block (Figure 11.2). The highest point on the dissected Darling Plateau is only 582 m above sea level. To the south lies the Stirling Ranges National Park, a naturally vegetated, uplifted horst which rises abruptly from the surrounding farmlands to 1110 m above sea level.

The Eucla Basin separates the Yilgarn Block from the Archaean–Proterozoic Gawler Block which constitutes most of the Eyre Peninsula in South Australia (Figure 11.2). Only the southern half of the peninsula is included in the Mediterranean region. Holocene dunes and swales, vulnerable to wind erosion, are distinctive features. East across Spencer Gulf and Gulf St Vincent are the Proterozoic Adelaide geosyncline and the Palaeozoic Kanmantoo fold belt which extends south to Kangaroo Island (Figure 11.2). The two gulfs and the Mt Lofty Ranges separating Gulf St Vincent from the Murray River in the south, reflect these folds, and also coincide with Australia's most active seismic zone.

In the southeast of South Australia, and in western Victoria, are extensive Quaternary coastal dune ridges, with centres of Pliocene/Pleistocene volcanics in the Mt Gambier–Millicent area. These are western extensions of the western volcanic plains of Victoria which were probably active as recently as 1000 years ago. Volcanic ash covers 110 km^2, producing fertile soils. Most of this area is less than 100 m above sea level, although the land rises to 1167 m above sea level in the Grampians. The land also rises gently to the eastern extremity of the

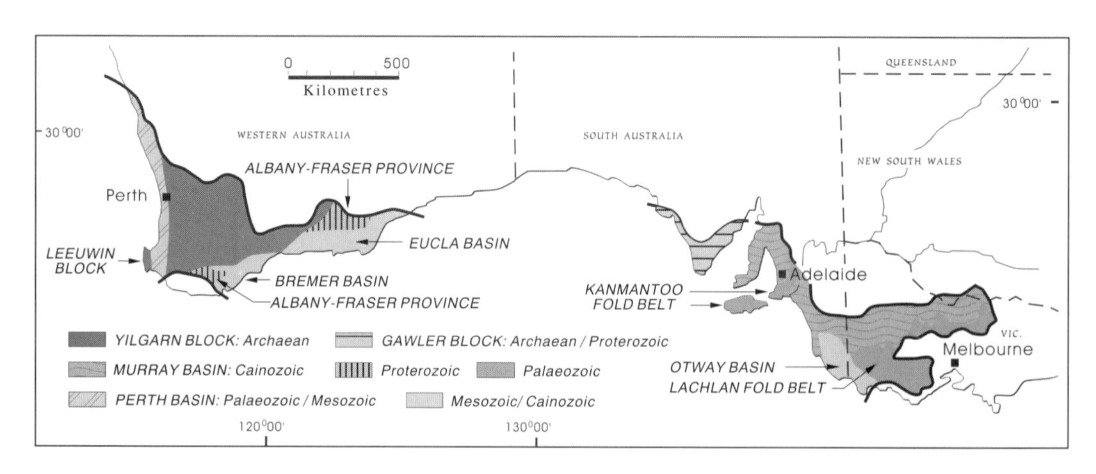

Figure 11.2 *Major structural elements of Australia's Mediterranean region* (source: *Atlas of Australian Resources*, vol. 5, Geology and Minerals 1988, p. 11)

Mediterranean region, but is still below 500 m above sea level.

Drainage

With a few exceptions, drainage in Australia's Mediterranean areas is characterized by: a lack of surface water; ephemeral streams; remnants of ancient drainage systems; and extensive groundwater resources. The main river is the Murray (Figure 11.1), which drains an area of more than 1 000 000 km^2, one-seventh of Australia. But only short portions of the river in the Shepparton area of Victoria, and at the mouth in South Australia, are included in the Mediterranean region. The main significance of the river to the region lies in its provision of water for the city of Adelaide and for irrigation. These uses are under threat as the concentration of soluble salts in the river water is rising inexorably.

The other rivers of note in Australia's Mediterranean region are in the southwest of Western Australia. They include the Blackwood and the Swan–Avon (Figure 11.1), together draining most of the "wheatbelt" (Figure 11.3) which lies between Esperance and Geraldton (Figure 11.1). Here, too, salinization is a major problem. However, unlike Adelaide, Perth is not dependent for its drinking water on the major rivers, but on shorter streams which rise in the uncleared Darling Ranges, and on groundwater beneath the coastal plain.

Groundwater provides an important source of water for both rural and urban areas in the Mediterranean regions, but high salt concentrations often limit its use, especially for human consumption. Thus, reliance is placed on various freshwater collection methods including earth dams on farms, storage tanks which collect rainwater from roofs, and piped scheme water from metropolitan dams (especially in the Western Australian wheatbelt).

Lake Alexandrina, at the Murray River mouth, is a major freshwater lake, but its water quality is maintained artificially by barrages which prevent the incursion of seawater. Also at the Murray mouth is the increasingly saline Lake Albert, still used for irrigation. Other freshwater lakes include the polluted Lake Bonney and the crater lakes of Mt Gambier in South Australia.

The Coorong, southeast of the Murray River mouth, is a long (140 km), narrow, shallow and saline lagoon renowned as a wildlife sanctuary. Inland, both South Australia and Western Australia have numerous, seasonally wet saline playas. However, the major playas are located outside the Mediterranean region. In Western Australia, during occasional periods of extensive regional rain, some of the saline playa systems overflow into palaeodrainage systems and contribute water to the Swan–Avon and Blackwood rivers. Many smaller, seasonal lakes occur in the Western Australian wheatbelt: some were fresh

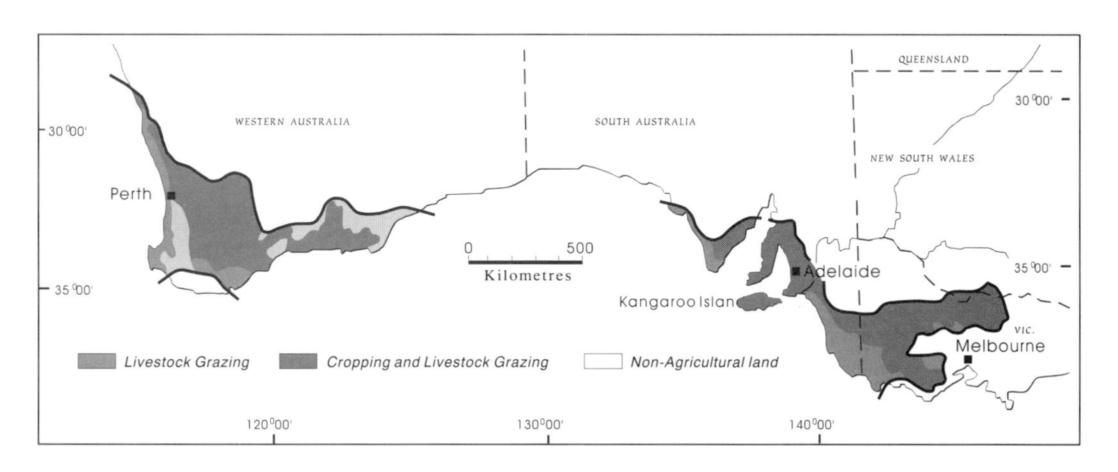

Figure 11.3 *Main rural land uses* (compiled from Division of National Mapping 1982, pp. 18–19 and 1:10 000 000 Map of Farm Types)

in the early days of European settlement but most are now saline. These lakes, and numerous wetlands on the coastal plains of the west coast of Western Australia and the southeast portion of South Australia, were particularly important for wildlife habitats. A large proportion of those wetlands has now been severely altered or destroyed.

Soils

In general terms, Australia's soils have been forming for a long time. Contemporary soils often overlie deep (to 40 m or more), intensely weathered regoliths. The thin, sand-textured surface soil horizons are leached and low in organic matter (generally <1%) and soil nutrients, and have poor structure. Locally, soils may also be acidic, solodic, saline or alkaline, calcareous, hard-setting or water repellent. These characteristics, together with hot, summer droughts and inappropriate land uses, lead to serious management problems.

In southwestern Australia, the southern Eyre Peninsula, parts of Kangaroo Island, and in the southeast (in both South Australia and Victoria), the dominant soils are classified as yellow, and also brown or red, duplex (texture-contrast) soils with conspicuously bleached subsurface horizons (Figure 11.4). Due to impermeable subsoils, soil conditions are periodically too wet and too dry. Inherent fertility is very low to moderate (Division of National Mapping 1980).

Relatively small areas of deep sands occur along the west and south coasts, on the Eyre Peninsula and, more extensively, in the eastern parts of South Australia, extending into Victoria (Figure 11.4). Shallow, sandy and gravelly soils overlie bauxitic duricrusts on Western Australia's Darling Plateau, with more extensive areas adjacent to the bauxites described as yellow duplex soils containing much ironstone gravel and naturally low in nutrients. Similar soils are also present east and south of Adelaide, and on parts of Kangaroo Island.

In the eastern parts of the Western Australian wheatbelt, saline soils are associated with the palaeodrainage systems and saline playa chains. There are also some relatively minor areas of calcareous soils. These are more extensive in South Australia – on the Eyre and Yorke Peninsulas, and north of Adelaide – and they also cover extensive areas of the flat Murray valley (mostly to the north of the Mediterranean climatic zone). The Eyre and Yorke Peninsulas also have areas of shallow soils underlain by rock, and shallow loam soils.

All of the above soil types have limitations for agriculture, with varying degrees of severity. However, there are areas of relatively fertile soils, identified as soils generally without limiting chemical or physical properties (Figure 11.4). These occur in relatively small areas of Western Australia, with the most extensive portion associated with the metamorphic zone along the Avon valley, and in more extensive areas in South Australia to the east and north of Gulf St Vincent, again associated with metamorphic rocks. The soils are red and also brown or yellow duplex soils without subsurface bleaching. They are "mostly favourable for a wide range of uses except that phosphorus and nitrogen are commonly required. Their physical properties allow plant roots to penetrate deeply and thus to take advantage of their high water-holding capacity" (Division of National Mapping 1980, key to the *Map of Soil Resources*).

Climate

The major control over weather in the Mediterranean region is the movement of an anticyclonic belt which lies in an east–west direction across the continent for about half the year. In winter this system moves to the north, allowing a procession of cold fronts to bring cool, cloudy weather, strong westerly gales and heavy rain (and sometimes hailstorms) to southern areas of the continent. In summer, the anticyclonic belt has its axis off the south coast. Hot, dry and dusty easterly and northerly winds prevail.

Several tropical cyclones develop off the northwest coast of Australia each summer, from November to April. They usually move south and then inland, where they become intense, rain-bearing depressions. With southward incursions of deep barometric troughs and local convectional disturbances, the cyclonic depressions are responsible for the occasional, widespread, regional summer rain which can cause

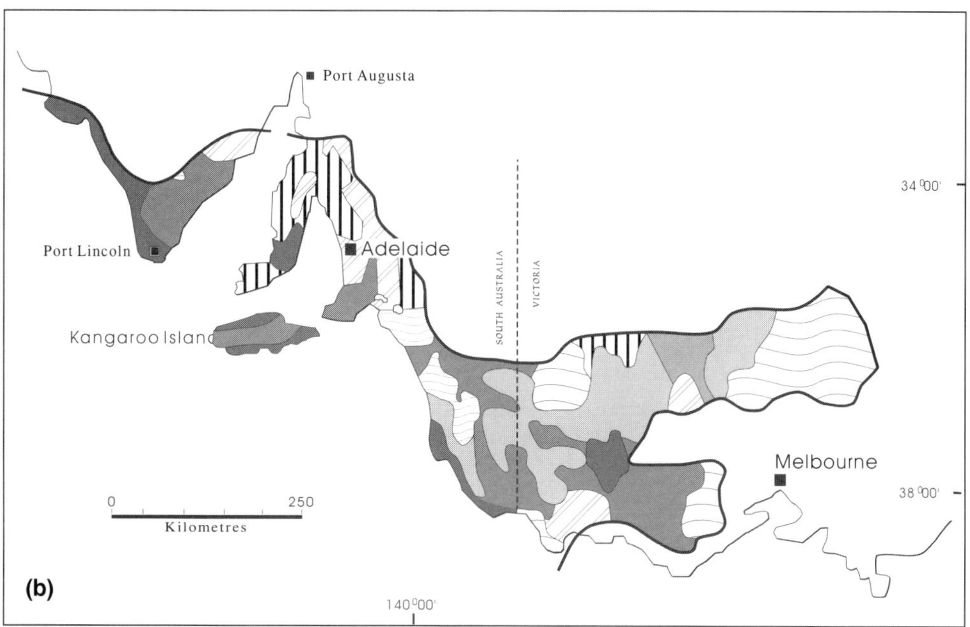

Figure 11.4 *Main soil associations of Australia's Mediterranean region: (a) southwestern Australia; (b) southeastern Australia* (adapted from Division of National Mapping 1980, 1:5 000 000 *Map of Soil Resources*, based on soil properties that affect land management). Legend is the same for both maps

considerable damage, as well as short and intense summer storms which cause localized soil erosion and flooding. Rainfall intensities of > 100 mm/h (over a 5 min period) have been recorded in Adelaide, and 228 mm in 4 h near Truro (*South Australian Year Book* 1994, p. 6).

As is characteristic of Mediterranean regions, more than two-thirds of the year's rain falls during winter (May–October). The highest mean annual rainfalls occur in the southwest of Western Australia, exceeding 1400 mm at several locations in the Darling Ranges, but decreasing to <300 mm/yr at the eastern margins of the wheatbelt. In contrast, mean annual rainfall barely exceeds 600 mm in the Mt Lofty Ranges east of Adelaide, and most of South Australia receives <500 mm/yr. Unlike the wetter southwest of Western Australia, nowhere in South Australia does rainfall exceed evaporation. Rain decreases rapidly to the east and the north in both regions, with increasing distance from the ranges and the coast. In most of the Victorian Mediterranean region, mean annual rainfall is <700 mm. However, mean annual rainfall approaches 1000 mm along the coast and a little higher in the Grampians.

The hottest months are January and February; July is the coldest. Although temperatures are generally higher in the north and centre of the continent, Western Australia's highest recorded temperature of 50.7°C was recorded at Eucla on the south coast (Figure 11.1). Welcome afternoon sea breezes generally moderate uncomfortable summer heat in coastal areas, and may extend 100 km inland.

Minimum temperatures drop below –1°C on occasion, and frosts may be widespread, especially in July and August, but they are not generally troublesome. Snow is very rare, though winter hailstorms are sometimes damaging. Overall, it is climatic variability which causes the greatest difficulties for Australian farmers in the Mediterranean regions.

Plant and animal life

Australia has a rich genetic diversity with a high degree of endemism. It is estimated that 88% of its reptiles, 94% of frogs and 70% of birds occur nowhere else (Table 11.1; *Year Book Australia* 1994, p. 433). In Australia's Mediterranean regions, plant life, especially, is characterized by its species richness and adaptation to fire, drought and low soil fertility. Vegetation types are generally mesic and range from forest and woodland through to heath and scrub associations, having their analogues in the French *garrigue*, Spanish *matorral*, South African *fynbos* and North American *chaparral*. The hardwood, evergreen and aromatic properties of this vegetation are often associated with a range of commercially

Table 11.1 *Australia's biodiversity* (source: *Year Book Australia* 1994, Table 14.1)

	Number of species
Flora	
Vascular plants	22 000 species, >90% occur naturally in Australia, 83 spp. presumed extinct, 840 spp. threatened with extinction within 10–15 years
Non-vascular plants	About 28 000 species of algae. About 3500 spp. of mosses, liverworts and lichens. About 10 000–20 000 spp. of large fungi. About 250 000 spp. of microfungi
Fauna	
Birds	850 spp. of which 70% occur naturally only in Australia, 16 spp. extinct, 15 spp. endangered or vulnerable
Reptiles	700 species, 88% occur only in Australia
Insects	54 000 known insects with at least as many to be identified
Mammals	276 native land mammals, 20 spp. extinct, 38 spp. endangered or vulnerable
Amphibians	About 180 spp., all of which are frogs, 94% of frogs occur nowhere else
Fish/molluscs	3600 spp. of fish and tens of thousands of spp. of molluscs

useful products, including timber, oils, scents, resins, seeds, flowers and honey.

Figure 11.5 shows the general distribution of contemporary vegetation in the region. By far the greatest portion of the previously naturally vegetated region now has a cover of introduced pastures and crops. The associated land uses are mainly extensive mixed grains and sheep farming, and in the southern parts of South Australia and Victoria, primarily livestock grazing.

Western Australia has the only significant area of closed native forests. These *Eucalyptus* forests have an extremely rich understorey: many of the more than 6000 plant species found in Western Australia grow in these forests. They are also associated with the only permanently flowing, fresh streams in the entire region. Dominant trees on the deeply weathered, "lateritized" soils are jarrah (*Eucalyptus marginata*) and marri (*Eucalyptus calophylla*) and, on deep red earths, the majestic karri (*Eucalyptus diversicolor*) (Plate 11.1). The latter grows to 90 m and is associated with the wettest areas of the southwest. Unfortunately, the forests' ecological viability is threatened by extensive clear-felling for an export woodchip industry, fire, and the spread of a severe disease known as "jarrah dieback" (Plate 11.2).

Remnant open forests and woodlands, dominated by stringybark (*E. obliqua* and *E. baxteri*) occur in South Australia, east and south of Adelaide and in the southeast, and in Victoria.

There are important, species-rich heath communities on the coastal sandplains north of Perth, around the south coast of Western Australia, and on deep sands in Western Australia's wheatbelt and in Victoria (Plate 11.3). Remnants of *Eucalyptus* woodlands are present throughout the Western Australian wheatbelt and (with regard to the mallee form) more extensively in several parts of South Australia and Victoria (Figure 11.5). Natural tussock grasslands north of Adelaide have nearly all been put under the plough.

As a result of the extensive clearing of the natural vegetation, wildlife is now restricted largely to the forests, the woodland remnants and the heath communities. The wildlife is under continuing threat by a range of human activities.

The People

Australia's original inhabitants, the Aborigines, occupied the land for at least 40 000 years. Many were located in better-watered, coastal areas. It is not possible to provide an estimate of their previous numbers for the Mediterranean region. Indicatively, the early 19th century Aboriginal population of Western Australia has been estimated at between 50 000 and 60 000. The population was then decimated by shooting, poisoning, starvation and disease following the arrival in 1829 of the first permanent European settlers at what is now Perth (Stannage 1981, Part

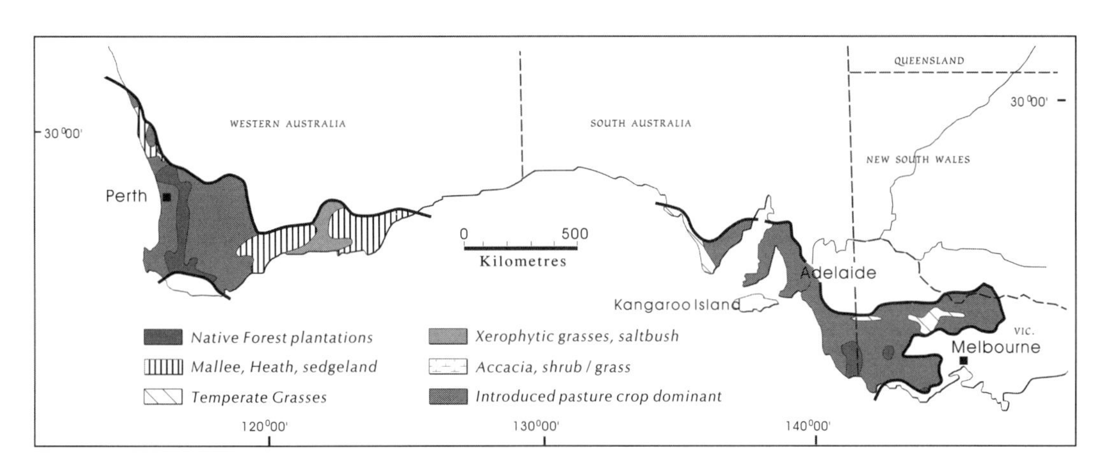

Figure 11.5 *Contemporary vegetation cover of Australia's Mediterranean region* (based on Division of National Mapping 1982: 1:10 000 000 *Map of Australia Native Pastures*)

Plate 11.1 *(a) Jarrah (dry sclerophyll) and (b) karri (wet sclerophyll) forest. Virtually all jarrah (*Eucalyptus marginata*) trees of this size have been logged. Jarrah (known as "Swan River mahogany" in the 19th century) is often associated with marri (*Eucalyptus calophylla*) and has a rich understorey. Most "old growth" karri (*Eucalyptus diversicolour*) now occurs only in areas reserved from logging*

1). Subsequently, the population recovered, and in 1996 Aborigines constituted almost 3% of the state's population of 1.8 million (*Western Australian Year Books*).

The 2.45% average annual rate of increase in Western Australia's total population during the 20th century exceeded that for Australia as a whole (1.71%) and indeed all other states (*Western Australian Year Book* 1992, p. 6–1). That growth reflected gains from immigration and the prosperous minerals and agricultural resource base of the state. On the other hand Victoria, and to a lesser extent South Australia, have lost population, relatively, to Queensland.

Population composition in the three Mediterranean areas broadly reflects that of Australia as a whole, although there are some marked regional differences. Predominantly Anglo-Saxon and Celtic in origin, the European population is now much more diverse and, as a result, culturally enriched. By the 1990s a quarter of Australia's population continued to be overseas-born. Although British, New Zealand and Western European immigrants still predominate, other main source areas include Europe's Mediterranean countries and Asia. English remains the language of the country, and in 1986 over 70% of the population regarded themselves as Christians (*Year Book Australia* 1994, pp. 123–124).

Unlike most of the other Mediterranean regions of the world, Australia's population is relatively young. In 1992 the median age was 32.6 years, but increasing; the proportion of those

Plate 11.2 *Dieback-affected jarrah forest near Mt Cook, Darling Ranges, Western Australia*

Plate 11.3 *Heath-type vegetation on deep sands in the central wheatbelt of Western Australia. Most of these associations and woodlands have been cleared for extensive agriculture*

over 65 years of age increased from 9.9% in 1982 to 11.4% in 1992. Population densities, too, are much lower than in the other Mediterranean countries, averaging less than 2 persons/km^2 in 1990. However, this low density masks the fact that more than 85% of the population lives in urban areas (*Year Book Australia* 1994, pp. 115–118).

Perth, Adelaide (both over one million people) and Melbourne (over three million) are primate cities, containing >75% of their states' populations. The few, larger country towns barely exceed 30 000 people, and towns of less than 2000 people have been losing populations in recent years. Farm populations are dispersed, with

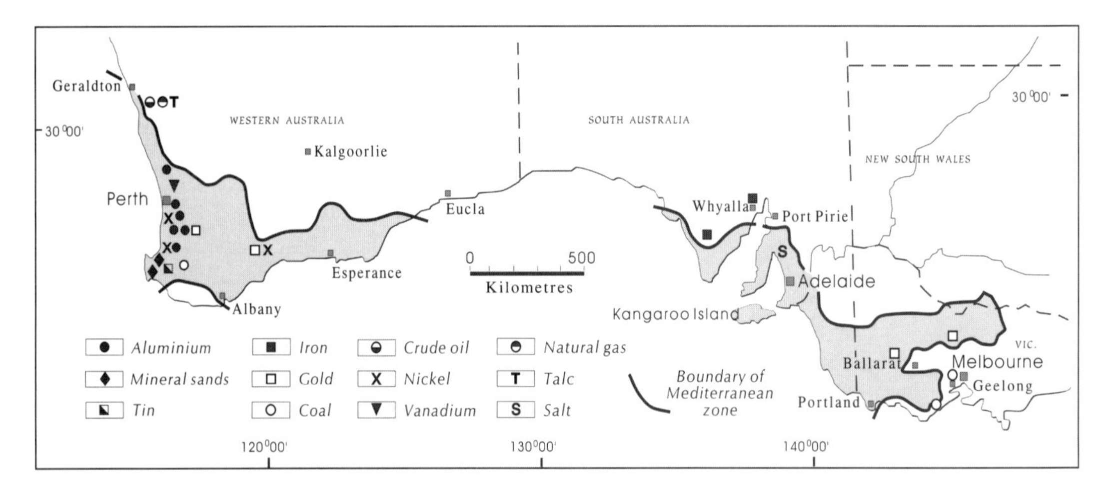

Figure 11.6 *Mining and industrial areas in Australia's Mediterranean regions* (source: *Atlas of Australian Resources*, vol. 5, Geology and Minerals 1988, p. 26)

homesteads being located on the farms rather than in nucleated villages. Many rural areas are characterized by a loss of younger people, leaving behind an ageing population and a suite of social and welfare problems. Thus the Mediterranean region is dominated economically and culturally by three coastal cities, only two of which (Perth and Adelaide) are actually located within the region.

The Economy

The economy of the Australian Mediterranean region is based to a major extent on primary industries. Agriculture is pre-eminent, especially extensive cropping of grains and grazing by sheep for meat and wool. The region also includes a high proportion of Australia's major wine-producing areas. Additionally, there are more intensive agricultural activities, including fruit and vegetable production and dairying, some of which is based on irrigation. The Western Australia zone has significant forestry and maritime fishing industries, and the world's largest tonnages of bauxite are mined in the Darling Ranges southeast of Perth. Tourism, especially from Asia, is of increasing importance throughout the region, with ecotourism, based on forests, "outback" and farm-stay experiences becoming popular.

Resources and industry

The annual turnover from mining in Australia was valued at almost US$24 billion in 1990/91, employing 5% of the labour force. Western Australia is Australia's leading mining state, accounting for about one-third (US$8 billion) of the country's production in 1990/91, well ahead of the next state, Queensland (US$4.8 billion). Gold, iron ore, oil and gas are its leading commodities, but they are located outside the Mediterranean zone. However, alumina refined from bauxite mined in the Darling Plateau near Perth, as well as production from the Northern Territory, is Australia's fourth leading export commodity by value (*Year Book Australia* 1994, 504). Other significant minerals include mineral sands at Eneabba and Capel, gold at Boddington and coal at Collie (Figure 11.6).

Victoria's important oil, gas and brown coal deposits are outside the Mediterranean region. However, the state's US$56.8 million (1990/91) gold industry centres on Stawell (*Victorian Year Book* 1993, 158). Ballarat and Bendigo, the sites of a major Australian gold rush in the 19th century, are foci of a lively tourist industry.

South Australia's most important mining activities are also located outside the Mediterranean region's boundary. Nevertheless, those mining activities (copper, gold and uranium at Roxby Downs, and iron ore and other metals in the

Middleback ranges to the north of Eyre Peninsula) exert some economic and environmental influence over parts of the Mediterranean zone. Gypsum is mined from Kangaroo Island, and there are large solar salt works north of Adelaide.

Several major industrial areas based on the mineral resources are located in the Mediterranean regions of Australia. Western Australia's industrial area is located along the shores of Cockburn Sound, within the Perth metropolitan region. The area has a number of major industries including oil, alumina and nickel (with cobalt as a by-product) refineries, a fertilizer works, a major grain-exporting facility, a growing shipbuilding and repair industry, and a naval base on Garden Island, on the opposite side of Cockburn Sound. Bauxite is also refined at four refineries south of Perth, with some of the alumina being shipped to Portland in Victoria (within the Mediterranean region) for smelting. Tin, silica and gold are also smelted in the region. Fluoride is an important industrial emission which has caused concern both to grape growers in the Swan valley (primarily from brickworks), and to conservationists, concerned over the adverse effects from the silica smelter on the state forests on the Darling Plateau. Mineral sands are processed in the region and concerns have been expressed at radioactive fallout from these plants.

South Australia has a number of important industries. The state is a major centre for the manufacture of automobiles, at Elizabeth to the north of Adelaide. It is also an important manufacturer of electrical appliances. Some industries, particularly the lead smelter/refinery at Port Pirie (the world's largest, with an annual production capacity of 235 000 t; AUSLIG 1988, p. 43), have significant environmental implications due to the fallout of lead, cadmium and sulphur on surrounding soils and marine sediments. Iron ore from the Middleback ranges is processed at Whyalla iron and steel works (1 100 000 t annual capacity; AUSLIG 1988, p. 44) just to the north of the Mediterranean region. Natural gas is piped from Moomba to Port Stanvac, near Adelaide. Two pulp and paper industries in the southeast have a significant adverse impact on Lake Bonney.

Apart from an aluminium smelter at Portland, Victoria's main industries are not located in the Mediterranean region. However, it is important to note that Victoria and New South Wales have the highest state populations, contain Australia's two largest cities (Sydney and Melbourne), are responsible for two-thirds of Australia's total manufacturing turnover, dominate the national retail trade sector, and have the major political influence over Australia's federal governmental structure.

There are no nuclear power stations in Australia. Electricity for the Mediterranean region is generated from coal (at Collie in Western Australia), natural gas and oil (Perth, Adelaide and Melbourne).

Agriculture, forestry and fishing

Agriculture in the Mediterranean region is limited by rainfall seasonality and variability, the lack of fresh water and infertile soils. Australia's major grain/sheep area stretches from the Western Australian wheatbelt (15 500 000 ha), through most of South Australia's 7 000 000 ha and a big proportion of Victoria's 9 000 000 ha devoted to crops and pasture, although the western state is the major wheat producer (Figure 11.3). Other important grains are oats and barley, and crops such as lupins and oilseeds are increasing. Wheat/sheep farms in southern Australia average 1000 ha in size, and 2000 ha in Western Australia. Between a quarter and half of the farm area goes under crop during winter, with the remainder under pasture.

Dairy cattle for metropolitan fresh milk markets are important on the irrigated Swan coastal plain in Western Australia, south of Adelaide and around Mt Gambier in South Australia, and in the southwest coast and Shepparton areas of Victoria. Beef cattle are relatively more important in Victoria.

Australia is enjoying rapid growth of its wine industry. The Swan valley adjacent to Perth has long produced fortified and table wines, but the newer grape-growing areas in the Margaret River and Mount Barker districts of the southwest are assuming greater importance. It is South Australia, however, which is Australia's leading wine and brandy producer, with nearly half the national turnover, valued at nearly US$400 million in 1990/91 (*South Australian Year*

Book 1994, p. 225). Some of the most famous wine-growing areas are located in the Coonawarra area and the Barossa valley near Adelaide.

Southwestern Australia has declining beef, apple and citrus industries, while pome fruits are grown near Adelaide and irrigated fruit crops around Shepparton in northern Victoria. Manjimup in the southwest of Western Australia and the Ballarat region of Victoria are associated with potato growing. Other vegetables are grown in the far southeast of South Australia, in the Horsham–Mt Gambier area, and in southwestern Victoria and northern Victoria's irrigated areas.

Logging of the 1 800 000 ha of state-owned, indigenous hardwood forests in the southwest of Western Australia has been a significant industry since European settlement. Jarrah has outstanding hardwood qualities and is exported worldwide. Karri is a particularly prized timber from the region. In 1976, a controversial woodchip industry was introduced, involving the milling each year of almost 1 000 000 t of unprocessed marri, karri and jarrah woodchips, which are exported through the port of Bunbury to Japan. Unfortunately, the forests coincide with the major bauxite ore bodies as well as freshwater catchments and wildlife habitats, generating important land use conflicts.

Plantation forestry is growing in importance, partly in response to pressures on the indigenous forests. Western Australia has >100 000 ha, mostly in its Mediterranean area, under softwood and hardwood plantations. South Australia also has about 100 000 ha in plantations, mainly in the southeast. Victoria has >230 000 ha, but much of it is outside the Mediterranean zone (*Year Book Australia* 1994, p. 488).

Although Australian marine waters contain >3000 fish species, fewer than 100 are exploited commercially. More than 60% of the catch is exported, mainly to southeast Asia (*Year Book Australia* 1994, p. 496). Rock lobsters, western king prawns (shrimp) and scallops off the west coast, and Australian salmon and southern bluefin tuna off the southwest and south coasts, form the basis of the industry. It is centred mainly on Geraldton and Fremantle (the port of Perth) in Western Australia, and Port Lincoln in South Australia. As is true elsewhere, over-fishing is a controversial problem.

Transportation

The most important mode of transport throughout the region is by road, although Perth's relative isolation places considerable importance on its air links. Except for Perth's electrified suburban rail system, the transcontinental Indian–Pacific line from Sydney through Adelaide to Perth, and the "Ghan" from Adelaide to Alice Springs (the latter two services are particularly important for tourists), passenger rail transport has generally declined. Many branch lines in the agricultural areas have been closed, both as a result of and contributing to rural depopulation.

Internal and international air transportation is competitive, relatively cheap and growing rapidly, for both people and goods. In 1991/92, Perth and Adelaide airports each handled more than two million domestic passengers, compared with Sydney's 10 million and Melbourne's eight million (*Year Book Australia* 1994, p. 663). State shipping services, on the other hand, perform a useful function but appear to be uneconomic. Perth's (especially) and Adelaide's hinterlands benefit from busy ports.

Administrative and Social Conditions

Government

The people living in the Mediterranean region of Australia are represented at the federal level in Canberra (the Australian capital) and have their own State Parliaments in Perth, Adelaide and Melbourne. Australia has six federated sovereign states which, with the Northern Territory and the Australian Capital Territory, comprise the Commonwealth of Australia. The British monarch is constitutionally the Head of State. The Crown, represented by State Governors, and the Parliaments, comprising Upper (Legislative Councils) and Lower (Legislative Assemblies) Houses, constitute the legislature of the three Mediterranean states. There is an overdue and increasingly strong republican movement. Executive government is based on the "Cabinet" system, led by a State "Premier" as distinct from the Federal "Prime Minister". A third tier of government, that of local government, functions through city, town and shire (or county) councils. A party political system operates at federal and state levels, with the conservative side represented by the Liberal and National (previously Country) Parties, often in coalition, and the "left wing" represented by the Labor Party.

The two major factors in the development of the Australian legal system have been its British origin and the Australian Commonwealth constitution of 1900. The judiciary is independent of parliament although judges of the Supreme Courts (state) and High Court (federal – the ultimate court of appeal) are appointed by the state and federal governments, respectively.

Education

Education is compulsory from the age of six years in Western Australia, five in the other states, to the age of 15 years. An increasing number of children continue beyond the upper age. In 1991, retention rates to the final year of high school were 71.3% nationally (*Year Book Australia* 1994, p. 310), but with some marked regional differences. Most children are educated in government (public) schools, but more than 20% go to private schools, mostly operated by religious institutions. For children of high school age (12 years plus) in rural areas, there is often no choice but to be sent to expensive, private boarding schools in the city.

At tertiary level, Western Australia has five universities, all located in Perth. The University of Western Australia (UWA) is the oldest and most prestigious, although not the largest. South Australia has three universities, located in the city of Adelaide. The University of Adelaide is that state's equivalent to UWA. Victoria's universities are located outside the Mediterranean region, mostly in Melbourne, but there are several regional campuses including one at Ballarat.

The Office of Technical and Further Education (TAFE) co-ordinates the provision of vocational and training programmes at a number of locations throughout the Mediterranean region. It offers important alternative courses, geared towards trade and business qualifications, to children of upper school and adult ages. There are also a number of agricultural colleges, some at high school level, others affiliated to universities, throughout the region.

Health and welfare

The Commonwealth and state government health authorities, together with boards of health under local governments, maintain health services. Basic public hospital services are provided "free". The federal Medicare programme, supported by a levy on all wage and salary earners, is supplemented by private health insurance schemes.

The Commonwealth government provides pensions and benefits, repatriation services and, with state and numerous voluntary agencies, a wide range of welfare services for people with special needs. State agencies operate in the field of child welfare and distribute emergency relief in situations where federal assistance is not available. State Workers' Compensation legislation provides for compulsory insurance by employers to compensate employees for loss of income caused by work-related accidents or diseases.

Culture and Recreation

The arts thrive in South and Western Australia, perhaps in response to the relative isolation of those states, although Melbourne lays claim to being the nation's cultural centre. The bi-annual Adelaide Festival, centred on its Festival Theatre complex, is a major tourist attraction. It attracts worldwide contributions in drama, music, ballet, films and art, as does the annual (and older) Festival of Perth. The German origins of much of the Barossa valley in South Australia are recognized in its annual harvest festival. Many buildings of historical or architectural importance throughout the region are protected through listing on a Register of the National Estate.

Western Australia, South Australia and Victoria have symphony and youth orchestras, and opera and ballet companies. State and local libraries are numerous and well patronized. Perth and Adelaide have several museums and art galleries, and a casino each. Country areas are serviced by a range of arts boards and benefit from visits by touring groups and workshops. Australia is particularly fortunate in its Special Broadcasting Service (which broadcasts a wide range of foreign language, international television programmes, especially films), and the national Australian Broadcasting Commission, which provides excellent non-commercial radio and television services to both city and country areas, as well as overseas.

The climate encourages outdoor recreation, with beaches and sheltered waters being major attractions in summer. Australian Rules football is the major winter sport and cricket the summer activity, but the people engage in virtually every sport known. Other popular recreational activities include wilderness experiences, bushwalking and camping, driving offroad vehicles, trout fishing, gliding, prospecting for gold and gemstones, snorkelling, and deep-sea fishing.

PART II
Problems of land degradation

Introduction

Arthur Conacher

Part I has provided brief introductory descriptions of the 11 Mediterranean regions identified for this book. Part II, the major section, now discusses the problem of land degradation, how it has developed historically, the causes of the problem, and some of its broader implications. As discussed in the main Introduction, for the purposes of this book land degradation is defined as:

"alterations to all aspects of the natural (or biophysical) environment by human actions, to the detriment of vegetation, soils, landforms, water (surface and subsurface) and ecosystems".

Although the sequence of description, history, causes and implications is logical, it is in fact difficult if not impossible to so categorize much of the material. "Causes", for example, obviously incorporates historical developments, although the attempt here is to place the more process-orientated material in the chapter dealing with causes. Some "implications" are themselves problems of land degradation; some problems are causes of other problems, and so on. Nevertheless, everything cannot be dealt with simultaneously, so the attempt is made to follow the sequence outlined.

In the world's Mediterranean regions, rural land degradation generally is associated more with agricultural and pastoral activities than with other forms of land use. However, urban, industrial, mining and forestry land uses also have important impacts on the biophysical environment and are discussed where relevant.

Some forms of degradation are visibly evident, such as gully erosion or vegetation loss. Other forms are much less obvious, such as loss of soil structure or deteriorating water quality. Generally, no one form exists on its own, which often makes it difficult to isolate the most serious problems and assess their causes and effects. There is also a distinction between direct, on-site degradation, where the damage occurs at the same place where the causal processes are operating, and indirect, off-site or regional degradation, where the damage occurs away from the causative actions or processes.

Chapters 12 and 13 describe the main problems of land degradation in terms of their nature, extent and severity in the various Mediterranean regions. Soil erosion, flooding, vegetation loss and degradation are dealt with in Chapter 12, while Chapter 13 discusses drought and water shortages, water and air pollution, the effects of industrialization and urbanization, and problems in the coastal zone.

At a world level, Spain is perhaps the only European nation that experiences the risk of desertification due to erosion over nearly half of its territory. Bennett (1960) wrote that 60–80% of the arable land in the central and southern Spanish provinces was being seriously degraded. He also stated that on slopes of more than 5% the edaphic film and parent rock were being eroded, gullying was preventing land from being used, and soils seldom had an ABC profile.

Serious erosion is taking place in the eastern Spanish basins of the Andalucía, Ebro and Guadiana rivers. In Extremadura and eastern Andalucía more than half the surface is under

Land Degradation in Mediterranean Environments of the World: Nature and Extent, Causes and Solutions.
Edited by A.J. Conacher and M. Sala.

serious erosion risk. Other degradation problems are related to the availability and quality of water resources, whilst pollution associated with industrial, tourist and intensive agricultural developments is an important problem in certain areas.

Up to the end of the 19th century, as in many parts of the Mediterranean Basin, deforestation and exploitation of the residual forest constituted the main forms of degradation of the geosystems in the south of France and in Corsica. These problems have since been replaced by forest fires, floods, soil erosion and air and soil pollution.

Mass movement, flooding and accelerated erosion are the most impressive aspects of land degradation in Italy. Soil salinization exists and may be relevant in some places, but does not appear to be an important problem so far, even though irrigation has been practised since Greek times (8th century BC). Subsidence has been recognized recently in some areas, such as in the Romagna and northeast Calabria alluvial plains. Its relevance is mainly with regard to cultural aspects, such as the recent subsidence of the city of Ravenna or the subsidence of archaeological sites, which may be deceiving planners as to the long-term consequences on land use. Subsidence following the extraction of groundwater and oil is a problem in parts of California.

Soil erosion has long been identified as a major problem in Greece and the eastern Mediterranean region. Clearing of land by fire and deforestation was one of the first farming methods that enabled humankind to develop agricultural activities, triggering processes of soil erosion. More broadly, soil degradation is now recognized as a major problem: soils are degraded not only by removal as in erosion, but by the addition of toxic materials such as salts (the salinization problem), residues from fertilizers and herbicides, the deposition of solid and liquid wastes, or by the depletion of nutrients or essential elements.

Many of the degradation processes are not new in the Levant. Accelerated soil erosion transformed the forested hills and mountains of Lebanon and Galilee into bare rock surfaces, and the fertile valleys of Mesopotamia into saline flats. However, soil degradation has also increased due to other factors, and today it is a serious problem.

Erosion, flooding and sedimentation are serious land degradation problems in North Africa and Chile, associated with population movements and periods of colonization. In California, rapid population growth and over-exploitation of the land for minerals, oil, timber and agriculture resulted in major problems, many of which have been ameliorated to some extent in recent years. In contrast, vegetation decline and species invasions are the main concerns in the southwest Cape, where good land management appears to have ameliorated earlier problems of soil erosion by water and wind.

The main forms of degradation in the Mediterranean regions of Australia are loss of natural vegetation, secondary salinization, accelerated erosion and loss of water quality. In 1975, it was estimated that 51% of Australia's agricultural and pastoral land needed treatment for erosion and vegetation degradation. Of this land, 56% required treatment with engineering or conservation works, especially in the non-arid zone, with the balance needing only better management practices. In the extensive cropping zone, much of which lies within the Mediterranean regions, about one-third of the area required treatment for wind and water erosion. Subsequently, the problems recognized by the 1975 survey have probably become worse – certainly in the case of secondary salinization – and additional land degradation problems have been identified.

Many land degradation processes have been going on for millennia. In Australia, for example, human impacts on vegetation through Aboriginal use of fire took place for at least 40 000 years. Here and in the other Mediterranean regions, fire and seasonal drought have resulted in vegetation species and associations with very distinctive characteristics. But even though land degradation has long been recorded in most of the Mediterranean regions, and was certainly recognized by classical scholars, in general it is only in recent decades that concerns over the problem have received serious attention. This reflects an intensification of agricultural practices and worsening of land degradation in association with population pressures and technological change, amongst other causes. But it has also been due to: an increased global awareness of broader environmental problems; a movement of issues into the political arena, and an improved understanding of the physical, economic, social and political processes which affect the production base of contemporary agriculture, community well-being and natural ecosystems. Chapter 14 considers the historical development of land degradation in the Mediterranean regions.

There are both direct and indirect or underlying causes of land degradation – with some being more readily identified than others – and these are evaluated in Chapters 15, 16 and 17. The nature of the biophysical environment itself, which is also what distinguishes "Mediterranean"-type environments from others, is an underlying contributor to the problems of degradation and is discussed in Chapter 15. Direct causes through human influences are assessed in the following two chapters: the clearing of vegetation and agricultural practices in Chapter 16, and a range of other human actions in Chapter 17. The latter chapter also considers underlying social and policy factors, including human attitudes and perceptions, as being partly responsible for land degradation.

Land degradation directly affects the viability of natural ecosystems as well as agricultural productivity in terms of yield losses and loss of income. But it also imposes a range of indirect costs both on and off the farm. Some of these include the costs of: repair and prevention; damage to ecosystems; adverse effects on human well-being and health; and rural depopulation. These and other implications of land degradation are discussed in Chapter 18.

12

The main problems of land degradation: their nature, extent and severity

1: Erosion and soil deterioration, flooding, vegetation loss and degradation

---

Compiled by Arthur Conacher and Maria Sala from contributions on the following regions:

Iberian Peninsula and Balearic Islands	**Maria Sala** and **Celeste Coelho**
South of France and Corsica	**Jean-Louis Ballais**
Italy, Sicily and Sardinia	**Marino Sorriso-Valvo**
Greece	**Constantinos Kosmas, N.G. Danalatos** and **A. Mizara**
Croatian Adriatic coast	**Jela Bilandžija, Matija Franković, Dražen Kaučić**
Eastern Mediterranean	**Moshe Inbar**
North Africa	**Abdellah Laouina**
California	**Antony** and **Amalie Jo Orme**
Chile	**Consuelo Castro** and **Mauricio Calderon**
Southwestern Cape	**Mike Meadows**
Southern Australia	**Arthur** and **Jeanette Conacher**

Problems of accelerated erosion (including flooding), other aspects of soil degradation (referring to changes in chemical, biological and physical soil properties), and vegetation loss and degradation (including fire) are discussed in this chapter. Broader issues including drought, water shortages, pollution of air and water, urban and industrial waste disposal, and degradation of coastal areas are discussed in Chapter 13. A difficulty is that many problems of land degradation can and do extend well beyond the place where the causal actions and processes are taking place (such as eutrophication of estuaries), and that some forms of land degradation, such as species extinctions, are only applicable for large regions and not at a local or farm scale. A fundamental difficulty in presenting this material is the fact that most aspects of land degradation are interrelated, including cause and effect. The loss of vegetation, and fire, are particularly good examples. Both are problems in their own right, and both cause a range of other problems, particularly accelerated soil erosion. Whilst an attempt has been made to discuss the problems of land degradation in a logical sequence, the nature of the problems makes it very difficult to do so.

Accelerated Erosion

Some Mediterranean environments experience rates of soil loss which are unsustainable. As discussed in Chapters 15 and 16, such losses are mainly associated with the nature of land use, loss of vegetative cover, drought, fire and torrential rains.

Soil losses from Spain's main drainage basins have been calculated by the Water Authorities in relation to the silting of dams, and the results show very high degradation rates. Comparison with 16 major world rivers indicates that only two (Ching and Lo in China) have specific degradation rates higher (>7000 t/km^2/yr) than some of the Spanish ones (Lopez-Ontiveros 1984). Rates of more than 4000 t/km^2/year are found in 19 of the dams in Spain's Mediterranean coastal rivers, in seven dams of the Guadalquivir, in five of the Tajo and four of the Duero (MOPU 1984–1987). The Guadalquivir basin has the highest degradation rates (Table 12.1).

Annual soil loss was estimated in 1982 at 1 000 000 t over an area of 5 000 000 ha in Spain, being "high" over 2 300 000 ha, and "very high" over 800 000 ha (MOPU 1984–7). According to the hydrology branch of the Forest Service, a quarter of the territory suffers from serious erosion, 38% has moderate or low erosion rates and only one-third is not affected by erosion. The southern half of the Iberian Peninsula is the worst affected (Figure 12.1). In relation to agriculture, the highest erosion rates occur in association with tree crops, followed by herbaceous dry farming.

Local and plot studies being undertaken in many areas of Spain (Sala *et al.* 1991) indicate the great variability of runoff and erosion processes within the territory in relation to particular environmental and land use conditions.

Table 12.1 *Specific degradation of Spanish drainage basins estimated from silting of dams* (MOPU 1989a)

Basin	No. of dams >40 t/ha/yr	No. of dams <20–40 t/ha/yr	No. of dams 10–20 t/ha/yr	No. of dams <10 t/ha/yr
Duero	4	5	20	5
Tajo	5	29	14	0
Guadiana	0	12	2	2
Guadalquivir	7	39	10	0
South	8	9	0	0
Segura	0	3	11	1
Levant	9	8	13	2
Ebro	0	20	39	12
East Pyrenees	2	3	5	0

Land Degradation in Mediterranean Environments of the World: Nature and Extent, Causes and Solutions,
Edited by A.J. Conacher and M. Sala.

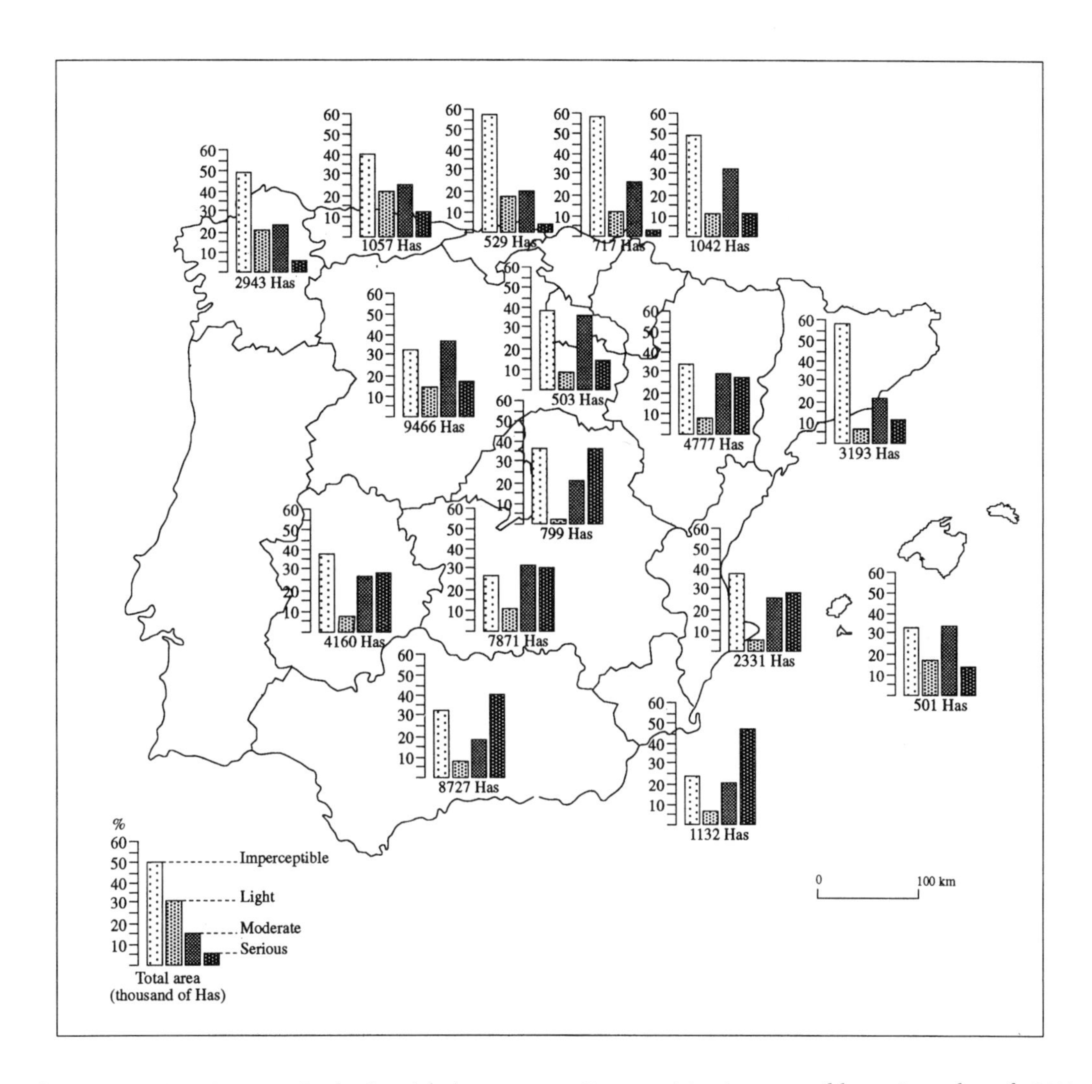

Figure 12.1 *Erosion rates in the Spanish Autonomous Communities: imperceptible erosion – loss of <25% of A horizon; light erosion – loss of up to 75% of A horizon; moderate erosion – the entire A horizon has been eroded and also up to 25% of B horizon; serious erosion – occurrence of gullying interfering with agricultural activities* (after MOPU 1989a)

Quirantes *et al.* (1991) have provided water erosion values ranging from 0.3 to 1.4 mm/yr and about 0.3 mm/yr for wind erosion in the Andalucian basins of Gualchos, Albuñol and Verde. Evidence from rainfall simulation experiments shows that sediment concentration in runoff depends mostly on the amount of fine size carbonates and the salinity of soils.

In the badland areas of the Ebro basin, measurements with erosion pins in selected plots by Benito *et al.* (1991) have shown an average ground lowering of 3–9 mm/yr in an infilled valley and 8–17 mm/yr on a slope. These erosion rates are directly related to cumulative rainfall during each measurement period. Erosion records indicate that 87–92% of the total sediment production in the infilled valley and 73–82% of that produced from the slope was generated by runoff processes operating in inter-rill areas. In the badland areas of Valencia (Calvo, pers. comm.), erosion pin measurements have given erosion rates of 50 mm/yr in certain sectors, but spatial variability is very high.

Erosive processes by piping have an important effect on the development of gully systems and badlands in the southeast of Iberia, as in the Almanzora basin (Martín-Penela 1994). Gullies develop in two different topographic contexts: on steep slopes and on subhorizontal surfaces of abandoned agricultural land (Plate 12.1). These pipe networks have developed during the last few centuries.

In gypsiferous soils of the Ebro basin, Navas (1991) has found, by means of simulated rainfall, that an average of 50% of total rainfall is transformed to runoff. The greatest runoff yield was produced from plots with stony soils and steep slopes and thus with the lowest percentages of plant cover. Suspended gypsum yields ranged from 0 to 3.9 g/m^2/h. Erosion of these gypsiferous soils has a clear effect on runoff salinity as shown by the increase in solute release when sediment concentrations increase.

Plate 12.1 *(a) Largely abandoned, gently sloping terraced land at Mula (southeast Spain); (b) now being eroded by subsurface-initiated gullies. F. Gallart kindly provides scale*

Abandoned fields are becoming a very important part of mountain landscapes due to current land use changes. The work of García-Ruíz *et al.* (1991) shows that when plant colonization is periodically disturbed, sheetwash and erosion are important, leading to rilling and the development of stone pavements at the surface.

Road building in mountain areas has a notable impact because it involves the removal of vegetation, soils and thousands of cubic metres of rock. Subsequently, hill-roads are affected by accelerated runoff, erosion and mass movements. In many areas of Spain the abandonment of forestry and the recreational use of forests is leading to an increase in road construction and with it an increase of erosion. Studies of the impact of roads in the Iberian Cordillera by Arnáez-Vadillo *et al.* (1991) show that sheetwash, rockfall and slides are the main erosion processes.

In Portugal, intensive use and recent changes in land use are responsible for 30% of soils being under high erosion risk, 54% under medium erosion risk and 15% under low erosion risk (Commission of the European Communities 1992).

Few quantitative studies of soil erosion are available in Portugal. In 1960 in Baixo Alentejo, a station was installed by the Ministry of Agriculture. Altogether 15 erosion plots (20 × 8 m) were installed in thin soils under different crop rotations, and a control plot under fallow-wheat in schist slopes ranging between 10° and 20°. Results from a 10 year period show a 2.38 t/ha/yr loss in the control plot, reducing to 1.16 t/ha/yr when a rotation of wheat plus leguminosae was used. If the soil is covered with pasture and kept undisturbed, soil loss is reduced to 0.27 t/ha/yr (Rosa 1980).

Determinations based on the silting of reservoirs and small dams by Rocha (1993) show a high production of sediment load in the mountain areas of northwest Portugal, with values ranging from 80 to 40 t/ha/yr. In the drier areas of Alentejo and Algarve, sediment production is also high. In Alentejo values for dams silted after only a few years in use demonstrate high sediment production (Table 12.2).

In the lower Tejo valley a measured reduction in sediment production during floods was observed following the construction of dams upstream (1 200 000 m^3 over the period 1899/1900 to 1952/53, compared with 350 000 m^3 in 1990).

Table 12.2 *Siltation of dams in Alentejo, Portugal*

Dam	Lifetime (yr)	Sediment production (m^3/km^2/yr)	Area
Montargil	2	21	Alentejo
Campilhas	11	289	Alentejo
Arade	10	762	Algarve

Soil erosion is severe on cultivated plots of land in France, particularly in vineyards (Durbiano 1988). Annual soil losses of 2 mm have been recorded. This kind of land degradation is more rapid in the Mediterranean domain than in the temperate zone.

In order to understand the potential of erosion in the Mediterranean territory of Italy, the results of a recent study in the Campania Region are useful (Buondonno *et al.* 1993). Maximum potential erodibility was assessed over 460 km^2 using a modified Universal Soil Loss Equation. Results were grouped in five classes, corresponding to five ranges of calculated maximum erodibility (Table 12.3). These values may appear rather over-estimated, but it must be considered that they are maximum potential erodibility values. In fact, the real erosion rate calculated for the whole study area was 16.03 t/ha/yr. This value compares closely with values obtained through suspended load measurements in some watersheds within and near the study area: 13.09 and 16.34 t/ha/yr respectively.

In a badlands zone of south Calabria, the erosion rate in a small bare catchment has been assessed at about 40 mm/yr (Sorriso-Valvo *et al.* 1992). In northern Calabria, erosion measurements from simulated rainfall experiments in badlands, for an intense (>25 mm/h) storm with a duration of 30–40 min, yielded rates of nearly

Table 12.3 *Five classes of calculated maximum erodibility, Italy* (source: Buondonno *et al.* 1993)

	Calculated erodibility	
	(mm/yr)	(t/ha/yr)
Weak	<2	<26
Moderate	2–4	26–52
Intense	4–12	52–156
High	12–24	156–312
Severe	>24	>312

zero from grass-covered plots, 0.05 mm from grass-devoid, forested plots, and 0.14 mm from bare plots (Sorriso-Valvo *et al.* 1995). Storms with intensities similar to those of the experiments occurred only once in a 10 year measurement period. Thus, values over 10 mm/yr are also exceptional if based on data obtained from small experimental plots.

Landslides are an important form of accelerated soil erosion in Italy, and the distribution of landslide-prone areas in Mediterranean Italy is shown in Figure 12.2 (see also Chapter 18).

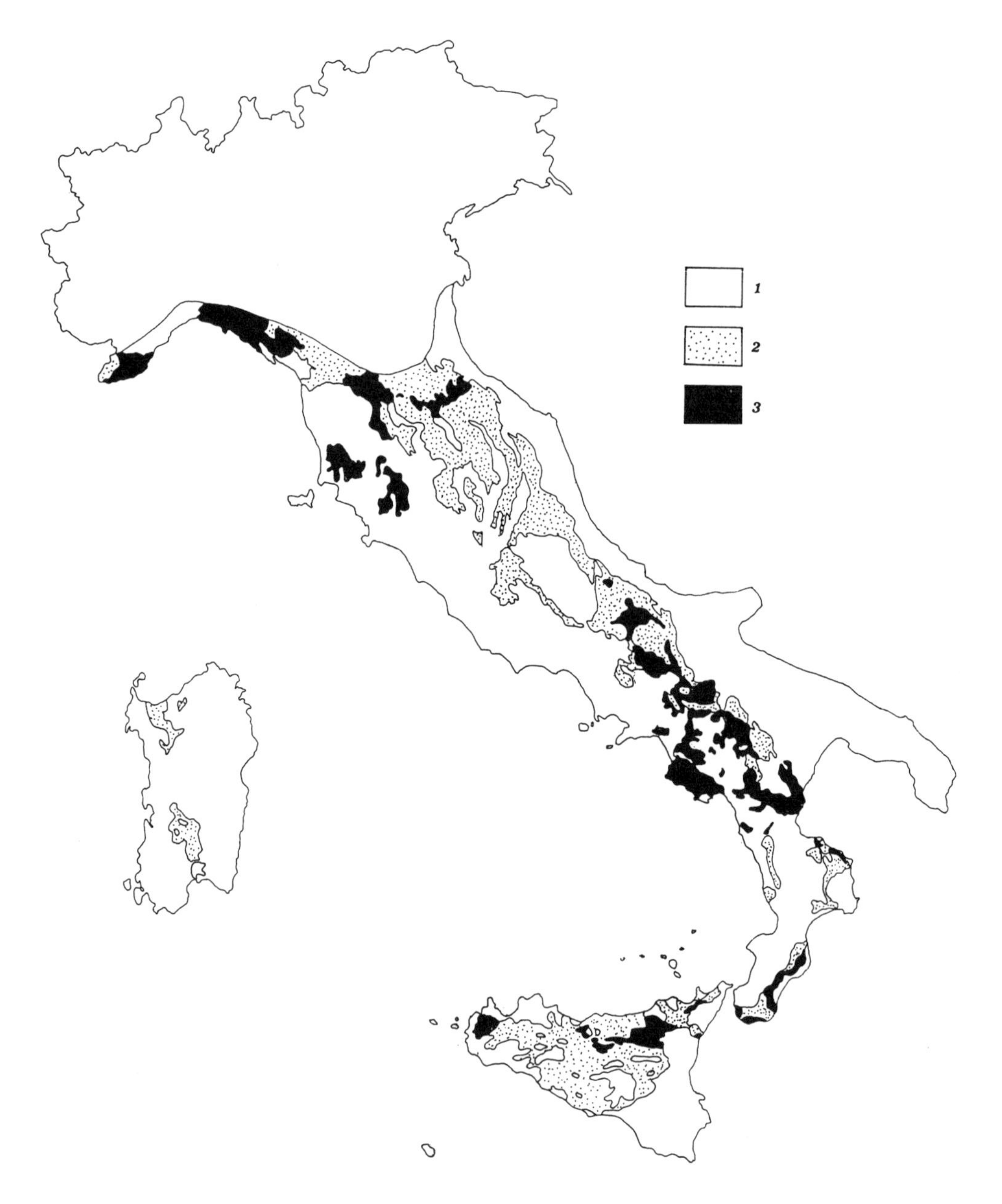

Figure 12.2 *Landslide-prone areas of Mediterranean Italy. Legend: 1 = least incidence of landsliding (e.g. less than 5% of the area); 2 = intermediate incidence (<20%); 3 = high incidence (>20%). Original compilation by M. Sorriso-Valvo of information from several sources*

Another distinctive form of "erosion" in Italy is subsidence. It is generally due to the extraction of water, or gas plus water, from unconsolidated sediments underlying alluvial plains. In the Romagna coastal plain, for example, subsidence of 1.3 m in 40 years has affected the industrial zone, the city centre and the countryside of Ravenna, with problems for field drainage and monuments. After the exploitation of artesian water was stopped, there was a recovery of 0.35 m over five years. The phenomenon seems to have stabilized, but many problems remain for farmers. A similar situation occurs in the proximity of Pisa, where increments to the inclination of the "Leaning Tower" have been enhanced by overpumping from water wells. The alluvial plain of the River Crati in northeast Calabria is subsiding for the same reason, although by much smaller amounts (Guerrichio *et al.* 1976). Along the Appia road in Latium, in an area extending over several tens of square kilometres, a subsidence rate of 4–5 mm/yr is variously attributed to tectonic activity, decomposition of the organic content of the sediments, exploitation of groundwater (in areas where the sediments are predominantly sand), and reclamation drainage (Serva and Brunamonte 1994).

Subsidence induced by human activity also affects 16 000 km^2 of California, mostly in the Central Valley where maximum subsidence approaches 10 m. Subsidence is most extensive in the western and southern San Joaquin valley and in the Santa Clara valley south of San Francisco, where it results mostly from excessive pumping of groundwater. Other subsidence problems are linked to the hydro-compaction of moisture-deficient deposits, the oxidation of organic soils, and the withdrawal of fluids from oil and natural gas fields.

In Croatia, 26% of agricultural and forest land in the area of Mediterranean climate has been eroded, with the problems of topsoil erosion being most evident on burnt areas. For example, 5–11 mm of topsoil at the Aleppo pine habitat (Rendzina on dolomite parent rock) on Peljesac peninsula were eroded from September 1979 to January 1980. This is equivalent to 100 t/ha and is considered to be intolerable (Martinović 1987).

The main problem of land degradation in Greece is the loss of soil volume capable of supporting vegetative cover. Lands of high quality are scarce due to irregular terrain with steep slopes, soil limitations, high rainfall variation and long periods of misuse. According to the results of the CORINE project (1992), 43% of the land is classified as having a high potential erosion risk, mainly in the south and west, where soils, terrain and climate combine to create suitable conditions (Figure 12.3). In contrast, about 20% of the country has a low erosion risk, mainly in the broad belt through Macedonia, Peloponnesus and Thessaly.

Wind erosion represents a serious hazard for land degradation especially in the Aegean islands. Strong north winds in combination with weak vegetative protection create favourable conditions for wind erosion, particularly during the summer or autumn period when the soils are dry. Erosion of the topsoil in unprotected areas may account for several hundreds of tonnes per hectare each year.

Erosion rates and trends in the eastern Mediterranean are similar to those of Spain, and are discussed in some detail in Chapters 15 and 16.

Soil erosion by water is spatially variable in North Africa. In some sectors, it totally transforms the area while in others it seems inoperative. The latter is particularly the case on extensive outcrops of coherent and slightly weathered rocks. Very often, overland flow erodes only superficial formations and susceptible rocks. In other respects, it is on steep gradients that the action of concentrated overland flow is most effective (Plate 12.2). But on the piedmonts, shifting channels account for the importance of sheetwash (Joly 1952).

Regionally, erosion is most severe in the Rif–Tellian world, for structural, lithologic and climatic reasons. In the Atlas mountains, there are slight differences between Atlantic Morocco and the rest (Despois and Raynal 1967). Depressions in soft rocks have been hollowed out (Triassic formations). The hydrographic network is disorganized and ends at the extensive basins. In Morocco, for example, the Rif, which covers but 4% of the surface, represents 60% of the soil loss of the country (Table 12.4). In other respects, erosion is mostly concentrated in zones of reduced economic value.

Figure 12.3 *Potential soil erosion risk map of Greece* (source: Yassoglou and Kosmos 1988)

According to measurements on the slopes of the Pre-Rif and the Rif (Heusch 1970), between 0 and 300 mm of rainfall are distributed as runoff each year; that is, the runoff coefficient varies from 0% in the driest years to 25–30% in the wettest ones. However, it is usually less than 10%. Soil loss is even more variable, ranging from insignificant in many cases to the record

Plate 12.2 *Rilling and gullying in colluvium and shales near Rabat: (a) general view; (b) detail. Processes are a combination of concentrated flow, piping and mass movement, following a complex sequence of events resulting from changing infiltration rates caused by over-grazing*

figure of 54 t/ha/yr, practically comparable to sediment loss from gullies and badlands (Heusch 1970; Laouina 1993b).

In Tunisia, four mapped and gauged watersheds have yielded the following data (Heusch 1990): 4% of the surface undergoes serious erosion (150 m^3/ha/yr); 16% of the surface undergoes a moderate rate of erosion (50 m^3/ha/yr); and 80% of the surface undergoes a weak erosion rate (of 11 m^3/ha/yr).

Considerable areas are affected by mass movements (Maurer 1975). These are either shallow features, which uncover bedrock in headward scars with a chaotic accumulation of materials downslope, or landslides related to deep infiltration of water. Mudflows are a response to saturation of materials and constitute a transition to torrential flows. In the most severely affected part of the Rif, in a small area of about 450 km^2,

Table 12.4 *Specific erosion in North Africa by region based on Fournier's formula* (source: Erosion Project, Project PNUD/FAO, 1981)

Region	Area (% of whole country)	Ds*
Oum Rbia–Tensift–Sebou	17	50–200
Eastern Rif–Middle Atlas, Western High Atlas	11	200–500
Eastern Middle Atlas, Atlantic Rif	2	500–1000
Northern Rif–Pre-Rif	2	1000–2000
Central and Western Rif	2	2000–5000

Ds = soil loss in t/km²/yr

mass movements cover more than half the surface (Figure 12.4). In the Pre-Rif, the proportion is smaller, but the forms are still impressive. The sectors in question have both soft rock outcrops and thick detrital covers.

Erosion and mass movement are areally significant in California, where conversion of chaparral to grassland in the San Gabriel Mountains led to an eight-fold increase in soil slips and to a commensurate increase in channel widening downstream (Rice *et al.* 1969; Orme and Bailey 1970, 1971). The infrequent but often intense, high magnitude rainstorms so typical of California's Mediterranean climate readily strip shallow O and A soil horizons from hillslopes (Plate 12.3).

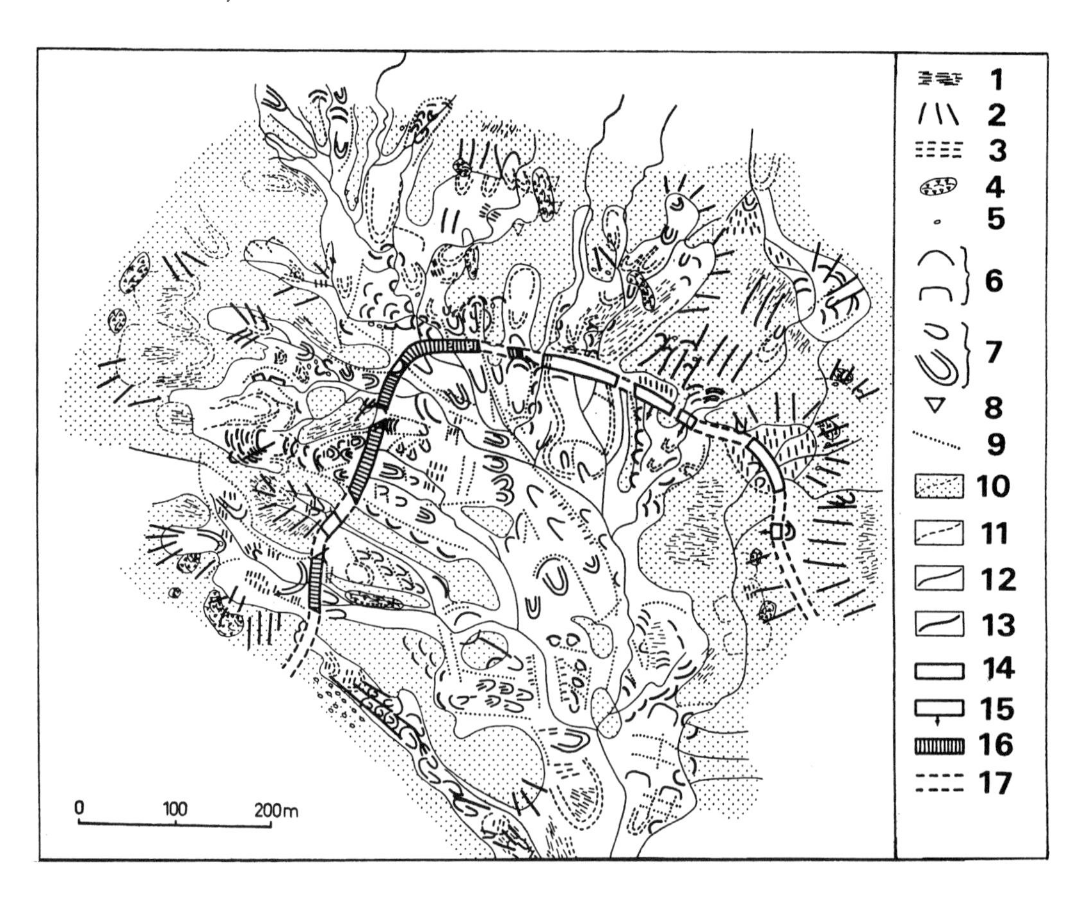

Figure 12.4 *Slope evolution in the Senuhan watershed, western Rif* (modified from Maurer's original work in 1962). *Legend: 1, sheet wash; 2, gullies; 3, spreading materials; 4, landslides; 5, small scars; 6, large scars; 7, solifluction and mudflows; 8, rock slides; 9, slip plane; 10, old forms without recent changes; 11, old forms, not stabilized; 12, current changes; 13, changes in March 1992; 14, road section threatened; 15, defect in roadside drainage; 16, destabilized section of road; 17; stable road section*

Plate 12.3 *Soil slippage on steep hillslopes in the San Gabriel Mountains, southern California, after misguided conversion of fire-ravaged chaparral to grassland in the early 1960s, followed by heavy winter rains in January and February 1969. Photo by A.R. Orme*

In Chile, soil deterioration and erosion are considered the most severe problem of the degradation processes of natural resources. Details of erosion in the Mediterranean regions can be found in Table 12.5.

Field measurements of erosion by surface runoff indicate losses of 25 to 100 t/ha/yr, with an average loss of 60 t/ha/yr being the equivalent of 4–5 mm of surface lowering per year.

The only study to cover most of continental Chile was undertaken in 1979 by the Institute of Natural Resources Research (IREN). It consisted of the analysis of Landsat images. The most important degradation problems by erosion are located in the high plateaux, in the Andes chains of Norte Chico, the Coastal Cordillera (particularly between Aconcagua and Cautín), in part of the Pre-Andes Cordillera (in the central and south zones), in the central region of the longitudinal valley south of Nuble river, and in the Patagonian *estepes* (CONAMA–MINAGRI 1994).

Soil and biological potential losses by wind erosion are of some importance due to the accumulation of fine sands in the form of dunes. From the inventory of rural lands since the 1960s, in the Mediterranean sector about 67 000 ha of littoral dunes were formed between the

Table 12.5 *Soil erosion (10^3 ha) in the Mediterranean regions of Chile* (source: CONAMA–MINAGRI 1994)

Region	Area	Area studied	Erosion high	Erosion serious	Erosion moderate	Erosion low	% Eroded
IV Coquino	3964	3459	–	654	1425	1379	85
V Valparaíso	1638	894	51	232	147	464	55
Metropolitan	1578	559	95	379	59	17	36
VI O'Higgins	1595	973	188	554	211	20	61
VII Maule	3052	1538	152	662	687	37	51
VIII Biobío	3601	2362	176	819	1168	200	66
Total	15 428	9786	729	3309	3696	2117	59

regions of Coquimbo and Arauco. There are also about 56 000 ha of continental dunes in the Bío-bío region. The distribution of soils covered by dunes is presented in Table 12.6.

Soil erosion has been regarded as one of South Africa's most significant environmental problems. In the southwestern Cape, it has been taken for granted that the seasonality of the climate, together with the pressure of agriculture upon the land, has promoted soil erosion on a large scale. This opinion is illustrated with reference to the words of H.H. Bennett, who pronounced, in respect of the wheatbelt of the southwestern Cape (the so-called Swartland), that soil erosion there was "violently in progress" Bennett (1945, 8). This theme was followed by Talbot (1947), who stated that soil erosion by gullying was a blight on more than 1600 km^2 of the Swartland, representing 25% of the area of this agriculturally important natural region (Figure 12.5; Plate 12.4).

More contemporary quantitative assessments appear to be based on individual catchments (for example, Scott 1993) and therefore only give a general indication of the regional picture. Scott's (1993) work on a sandstone catchment with both pine plantation and *fynbos* vegetation (Plate 12.5) points to remarkably low rates of erosion given the steep slopes and seasonal climate, with annual erosion rates of less than 100 t/km^2. However, the degree to which such conditions are comparable with catchments with different geological substrates, slope characteristics and soils is debatable. A more regional approach has been adopted by Meadows and Meades (in preparation, see Chapter 14), who have observed significantly less gullying in the recent Swartland landscape than that documented in Talbot's (1947) study. This is perhaps suggestive of the efficacy of soil conservation practices, although the improvement does not seem to have been paralleled by increases in yield (H. Germishuys, pers. comm.). In general, the problem of soil erosion by water appears to be less marked than in other Mediterranean regions, although the data base is small.

Talbot (1947) interpreted widespread drift-sands in the western lowlands of the south-western Cape as indicative of a major wind erosion problem (Figure 12.5), assumed to be driven by a combination of severe fires, overgrazing and an energetic wind regime. Even in inland areas, there were some 6000 ha of devegetated and mobile sands in the Sandveld in 1938. Ironically, perhaps, one form of land degradation seems to have replaced another in this instance, as invasion of exotic tree species has markedly reduced the extent of wind erosion in the region, even in the coastal areas (Holmes and Luger, in press).

Table 12.6 *Distribution of soils covered by dunes* (source: CONAMA–MINAGRI 1994)

Province	Littoral area (10^3 ha)	% Dunes	Interior area (10^3 ha)	% Dunes	Total area (10^3 ha)
Coquimbo	4.2	5.7			4.2
Aconcagua	0.9	1.2			0.9
Valparaíso	2.5	3.3			2.5
Santiago	4.4	5.8	0.1	0.2	4.5
Colchagua	2.0	2.6			2.0
Curico	0.8	1.0			0.8
Talca	1.6	2.1			1.6
Linares			0.04	0.0	0.04
Maule	15.5	20.0			15.5
Nuble	0.6	0.8	7.3	12.0	7.9
Concepción	4.1	5.5	25.4	45.0	29.5
Arauco	30.7	41.0	1.0	1.8	31.7
BíoBío			22.6	39.0	22.6
Malleco			0.1	0.2	0.1
Total	67.3	89.0	56.5	10.0	123.8

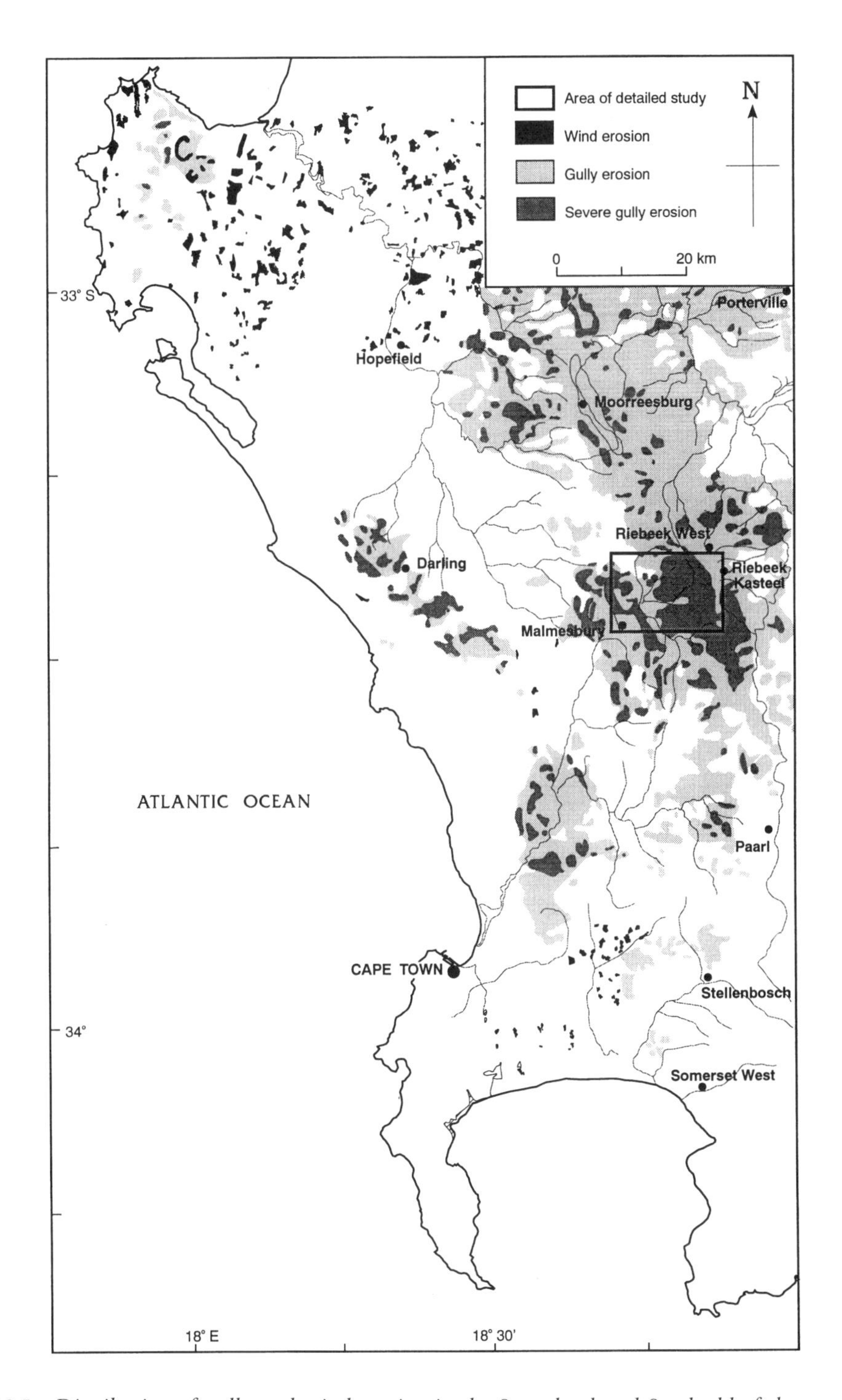

Figure 12.5 *Distribution of gully and wind erosion in the Swartland and Sandveld of the southwestern Cape* (after Talbot 1947). *The rectangle shows the location of Figure 14.7*

Plate 12.4 *Gullying in shales, yielding dispersable clays: northern parts of southwest Cape, mean annual rainfall about 370 mm. Most of this gullying has now been controlled (Chapter 20)*

There is very little information on the soil erosion situation elsewhere in the region. The steep slopes and strongly seasonal climates of the sandstone uplands might be expected to produce high rates of erosion. However, the generally shallow nature of the soils mitigates against agricultural use of these areas, and the only real transformation problems are produced by inappropriate burning regimes and wildfires. In the absence of specific data on the issue, coarse approximations of sediment yield data suggest that, contrary to expectations, current erosion rates from the southwestern Cape mountains are low by South African standards, varying between 50 and 200 t/km²/yr (Pitman *et al.* 1982). There is evidence of mass movement in the form of recent and fossil landslip scars on steeper slopes in the region, but the process has never been investigated or quantified.

Wind and water erosion affect around 800 000 ha of Western Australia's 15 500 000 ha area of extensive cropping and livestock, most of which is in the region of Mediterranean-type climate (Table 12.7). Another 2 000 000 ha are susceptible to erosion (Table 12.8). Wind erosion is particularly variable from year to year and season to season (Plate 12.6). Erosion of topsoil from unprotected, cultivated land on susceptible soils may amount to several hundreds of tonnes per hectare each year. In an environment where natural soil formation rates are estimated in millimetres per hundreds of years, such rates are completely unsustainable (Edwards 1988, 1993).

In South Australia, Matheson (1986, p. 133) estimated that 250 000 ha (4% of the 7 000 000 agricultural ha in the state, most of which are in the Mediterranean zone) require treatment to control water erosion. This area was in addition to the 220 000 ha (approximately 3% of the agricultural area) which had already been treated to June 1984. The soils most affected are duplex red brown earths and hard setting sandy loams. Some 2 900 000 ha of mainly duplex soils have the potential to be affected by water erosion (Table 12.8). For South Australia's Eyre Peninsula, Heathcote (1992) has discussed changes in available biomass over time, accelerated soil erosion, increased soil salinity, declining agricultural productivity, and declining population-carrying capacities. There is considerable variability in erosion amongst different periods, largely coinciding with droughts but also with periods of agricultural intensification. In one storm event in 1988, 234 000 ha of soils on Eyre Peninsula were affected by wind erosion (Community Education and Policy Development Group 1993, p. 84).

Floods

By definition, Mediterranean environments are characterized by strong, seasonal climatic contrasts. It is therefore not surprising that these regions experience particularly severe floods – although the flood season is not always in winter – which in turn affect a wide range of environmental properties and human activities. Floods accelerate erosion processes, cause the loss of

Plate 12.5 *Part of the Jonkershoek mountains, illustrating* fynbos *vegetation, in which the 265 ha catchment studied by Scott is located. Erosion rates are as low as 0.15 t/ha/yr, despite the steep slopes*

arable soil and the aggradation of gravel-bed rivers. At the same time, solid discharge contributes to the silting of alluvial plains and reservoirs.

Spain has a particularly grave problem as far as floods are concerned. Although floods occur in all the river systems, the main areas affected are the Mediterranean coastal streams. Floods have been a common feature in relation to torrential rains (Gil Olcina 1983), both on coastal floodplains and in urban areas, especially in intermittent or ephemeral, coastal streams (*ramblas*).

The most spectacular floods occur on the south Mediterranean coast, in the streams draining the Baetic Cordillera (Sala 1989). Data from the October 1973 floods indicate discharges of more than 1000 m^3/s in *rambla*-type streams, with lag times of 2–6 h, a few minutes peak and a flood duration of only 3–6 h. Further north, the Segura, Jucar and Turia rivers also experience floods, in these cases with big economic losses in relation to the intensive agricultural activity on their alluvial plains.

The streams draining the Catalan Coastal Ranges suffer from increased urban use of their watersheds and river beds (for housing, parking and driving), as a result of which human and economic losses caused by flooding are often high. Studies of the annual discharge of Catalan coastal rivers before and after urbanization and industrialization within their watersheds show marked increases in total runoff, peak discharges and lag times (Sala and Inbar 1992). This phenomenon affects the entire Spanish Mediterranean coast, and is bound to cause channel changes associated with erosion and sedimentation.

Floods constitute the second form of land degradation in the south of France and Corsica. The *Midi méditerranéen* and Corsica have very sudden, violent floods: for example, the Ouvèze river flood of 22 September 1992 (Arnaud-Fassetta *et al.* 1993), which caused extensive damage. Other serious floods occurred in the south of France in 1940 and 1986, causing the loss of many lives and serious economic damage (Chapter 18).

Several Italian cities are located along rivers: Rome on the Tiber; Florence and Pisa on the Arno; Pescara on the Pescara; Genova and Reggio Calabria on several minor but very active streams, and so on. The most vivid floods in living memory, in Mediterranean Italy, were those in 1951 and 1953 in Calabria, and in 1966 in Florence.

In Calabria, an unprecedented rainfall event was recorded in 1951, mainly on the Ionian side

Table 12.7 *Status of the soil resource in Western Australia's agricultural areas: total area 16 000 000 ha, c. 6.5% of the state* (source: Grant 1992, p. 88)

Condition of soils	Estimated annual cost (US$ million/ yr)	Trends	Status of knowledge	Comments
433 090 ha recognized as saline, up to 1 000 000 ha may be affected*	50	On average 11 000 ha each year turning saline	Improving	Up to 2 477 000 ha could be affected if no action taken
500 000 ha waterlogged in average year	760	Minor increases each year	Good	Up to 700 000 ha affected in wet year
50 000 ha affected by wind erosion	17	Stable or decreasing	Fair	Improved management is reducing area affected
750 000 ha affected by water erosion	17	Stable to decreasing with appropriate management	Good	Improved management in past decade is reducing eroded area
Area affected by water repellence varies greatly in response to soil characteristics and weather	120	Area affected is increasing in response to use of legumes in farming systems	Fair	Up to 5 000 000 ha are susceptible
3 500 000 ha with soil structure decline	56	Area affected is reducing	Good	Minimum tillage and direct seeding are reducing problem
8 500 000 ha affected by subsoil compaction	122	Area affected stable	Fair	All susceptible soils already affected
250 000 to 500 000 ha affected by soil acidification	4	Area affected increasing	Fair	Up to 11 000 000 ha susceptible
12 700 ha have organo-chlorine pesticide residue notices	n/a	Area affected stable	Good	Agricultural use of organochlorine pesticide banned in 1990

*1994 estimates by the Western Australian Department of Agriculture indicated that 1 804 450 ha of agricultural soils were affected by secondary salinity, with a potential to increase to 6 108 980 ha (Natural Resources Management Services Unit 1996, table 1)

of the region. As a consequence, peak stream discharges reached 1000 m³/s, compared with normal values of 10–20 m³/s. Serious loss of life and property occurred (Chapter 18). In Florence, there was a major flood on 4 November 1966, which inundated a large part of the historical city. This flood was not the only one to affect the city and its countryside in historical times: there have been very heavy floods on several occasions in most centuries, with most major events occurring in November/December, coinciding with the wettest part of the year. Nevertheless, there were summer floods in 1547 and 1557 (Principe and Sica 1967).

In North Africa, rainstorms can result in concentrated, sudden floods (for example, an instantaneous flow of more than 10 000 m³/s in the Sebou; more than 8000 m³/s in the Moulouya;

Table 12.8 *Some comparative data on land degradation in Australia's three "Mediterranean" states* (sources: *State of the Environment* Reports and Natural Resources Management Services Unit 1996)

Problem	WA	SA	Vic
Forests cleared (%)	50		
Woodlands, shrublands cleared in agric. areas (%)	90	>80	>65
Native grasslands destroyed in agric. areas (%)	n.a.		99.5
Secondary soil salinity area affected (ha)	1 800 000	>225 000	275 000
potential area (ha)	6 100 000		
Waterlogging, crop (ha)	>500 000		
Waterlogging, pasture (ha)	1 300 000		
Wind and water erosion (ha)	1 000 000	470 000	3 100 000 (sheet & rill)
susceptible area (ha)	2 000 000	2 900 000	
Soil acidity (ha)		1 000 000	
potential area (ha)	375 000	2 500 000	6 400 000
Water repellence – susceptible area (ha)	5 000 000	2 200 000	
Soil structural decline (ha)	>8 000 000		9 800 000
Organochlorine residues (ha)	13 000		

Plate 12.6 *Wind blowing nutrient-rich fines from a wheat field in the northern wheatbelt of Western Australia during the drought of 1976. The photo was taken in August, when the wheat crop would normally be knee-high*

4200 m^3/s in the Chelif, and 2500 m^3/s in the Medjerda; Troin *et al.* 1985). The Medjerda can often discharge 20% of its annual flow in one day. In small streams, the values are still more impressive (for example, the Lao which discharged 1530 m^3/s from a watershed of 939 km^2 in 1951, and the El Hammam river in Algeria, which reached 1500 m^3/s on 13 May 1948). However, it is in the steppe regions where floods are the most sudden and devastating (for example, the flood of the Zeroud river in the Tunisian steppes; Aït Hamza *et al.* 1995). Erosion is less important in the karstic regions (Hakim 1985; Nicod 1993). In the purely Mediterranean domain, floods are restricted to the lower plains, such as the Gharb, or to the small Mediterranean rivers.

The plain of the Gharb has experienced 50 floods in 50 years. The most famous one was that of 1963 which covered 200 000 ha. The phenomenon is linked to the position of the base level downstream from the Rif and the Middle Atlas. Two rivers, the Sebou and the Ouerrha, meet at the commencement of the plain, with the combined river then meandering below alluvial levees. Floodwaters emerge through open breaches in the levees. Since 1953, the frequency of the phenomenon seems to have accelerated, which poses the question of the extent of human influence and the effect of the degradation of the plant cover. Since then, the state has arranged the construction of a number of dams which will soon completely regulate the waters of the Sebou and its tributaries (Troin *et al.* 1985).

Within the Tellian or Atlas mountains, violent storms and persistent rains may give rise to very violent floods. The Beni Chougran, in Algeria, may receive half the annual rainfall in three days, and very rapid floods occur shortly afterwards. In 1927, the Algiers–Mostaganem road was cut over 200 km: all the bridges were damaged, an entire quarter of the city of Mostaganem was swept away, and the dam on the Fergoug river was breached.

Sedimentation and River Changes

As described in preceding paragraphs, land degradation involves erosion and local landsliding in some areas. The debris produced is transported by floods downstream and deposited as ridges within braiding channels or as fans at the confluence with major valleys or at mountain fronts.

In southern Italy, *fiumara*-type streams (Chapter 3) result from the aggradation of riverbeds as a consequence of increased coarse debris budgets from slopes which are not balanced by a corresponding increase in the transport capability of the streams. This situation has been present fairly continuously for about the last 2500 years. The order of magnitude of the debris budget is illustrated by the results of studies conducted in the rugged basins of Fiumara Buonamico and Fiumara Laverde (Ergenzinger 1988) and in the relatively smooth basin of an unnamed small creek in south Calabria (Mandaglio pers. comm.).

In the Laverde basin (area about 160 km^2), the average erosion rate has been measured at 10 mm/yr, while in the Buonamico basin (about 60 km^2), the erosion rate has been assessed at about 2.5 mm/yr for the last 50 years. In the 23 km^2 smooth basin, the erosion rate has been assessed at 1 mm/yr over the last three years. On a long-term basis, the average erosion rate for the two larger basins decreases to about 0.14–0.25 mm/yr. This indicates that the sediment mass transport may vary over time, and it depends on the occurrence of extreme events which can produce a large amount of debris. After one extreme event has caused the mobilization of materials on the slopes and deposition at their base and in the riverbed, then it takes 20–25 years for the stream to recover from the additional aggradation and to return to "normal" stream conditions.

There is thus a pulsating behaviour of material transport, and a long sequence of debris waves moves downstream every winter towards the terminal fan and the sea. For the Buonamico basin, the magnitude of sediment yield varies from 1000 to 10×10^6 m^3/yr, but the low frequency–high magnitude events are those which do most of the physical work in moving the debris either along the river or down the slopes. When such events occur, debris transport exceeds normal values by several orders of magnitude.

This represents a condition of high hazard for fields and orchards cultivated close to the "normal" river channel, as in these circumstances the deposition of the coarse debris occurs during the extreme flood, and with a high probability in the following 10–15 years. Such a situation, illustrated also by Gabriele *et al.* (1994), is to be found for all *fiumara* streams of Calabria (Sorriso-Valvo 1993, 1994) and northeast Sicily (Plate 3.2). The problem is more hazardous on the steep fans of the Tyrrhenian mountain front (Figure 12.6), where the last flood, producing debris flows rather than true floods, occurred some 250 years BP, depositing layers more than 4 m thick (Plate 12.7) (Sorriso-Valvo and Sylvester 1993). In this area, the resources at risk are not merely the land, but also the villages which have been built there recently as tourist resorts. In this case, however, it is not clear whether the return period of extreme events capable of mobilizing the debris along the slopes and the stream bed is longer than in south Calabria, or whether the present conditions are different (for example due to reforestation), and such events will not recur until conditions have changed. Also, the climatic character of south Calabria differs from that of the northern Tyrrhenian coast, where rainstorms are more frequent but less intense (Versace *et al.* 1989).

Considering all the active or recently active fans as zones of potential invasion by debris brought by flood or stream-channel debris flows, some 5% of the region, that is about 750 km^2, is under the threat of such an event. An area of 1 to 50 km^2 is estimated to be affected by debris invasion every year. The phenomenon is present with similar intensity in Liguria and parts of Lucania, Campania and Tuscany, but it is much less frequent or totally absent in the remaining parts of Mediterranean Italy.

In Spain, changes in the coastal streams (*ramblas*) have been well documented in the Valencia area (Segura 1990). Gravel extractions from the main channels, undertaken systematically since the 1950s, and the construction of dams have diminished sediment availability and thus favoured river incision in many reaches. For instance, the remains of a Roman bridge have been undercut by 0.5 m during this period and similar incision has been observed at modern bridges. Segura (1990) has indicated that the erosion layer in the *rambla* channel ranges from 30 to 90 mm in depth, the most important being in the main channel, whilst Pardo Pascual (1991) found an increase in channel depth of more than 3 m along several kilometres in the Carraixet *rambla* (Plate 12.8).

Pardo Pascual (1991), using data from construction agencies, has indicated that the volume of fluvial sediments removed from the main basins which drain to the Gulf of Valencia increased from 214 030 m^3 in 1980 to 501 147 m^3 in 1986, with an average rate of 294 708 m^3/yr. The total volume of fluvial sediment removed during this period was 2 658 694 m^3. Sediment retention by dams is illustrated by studies carried out after the collapse of the Tous dam in 1982, where a total accumulation of 400 000 m^3 of mud and sands was found.

Work in Morocco has shown the effects of floods and sediment loads on the rivers themselves. Downstream effects are discussed in Chapter 18. Turbidity may be very high during floods (in the Medjerba, ranging from 10 to 15 g/l for small floods and 80 to 100 g/l for the biggest ones, which gives an average load of 14×10^6 m^3/yr). Discharge/load ratios vary considerably during floods, according to the bedload materials and the resistance of the banks and slopes that are close to the channel. The river is often at the limit of its transport capacity and continually shifts its bed while eroding the banks, which increases its load (Plate 12.9). In widened sections, it spreads its coarse load on the convex bank. In the gorges, it drags its load. It may also behave like a mudflow, for example after crossing soft rock outcrops. The profile is thus endlessly in a precarious balance.

The case of the Sbiba river (Boujarra 1993) is significant: over a distance of 7 km, the surface area of the channel increased from 26 ha to 72 ha in 20 years. That means average losses of land of 2.2 ha/yr. In fact, the erosion was very rapid and occurred in 1969. In places, the El Hammam river has eroded its bank by as much as 15 m in one day (Benchetrit 1972). Meanders develop very rapidly and the river thereby increases its actual length, and hence its range of action. Sediment accumulation is important during these floods. The most serious aspect, however, is the silting up of the dams, which is discussed further in Chapter 18.

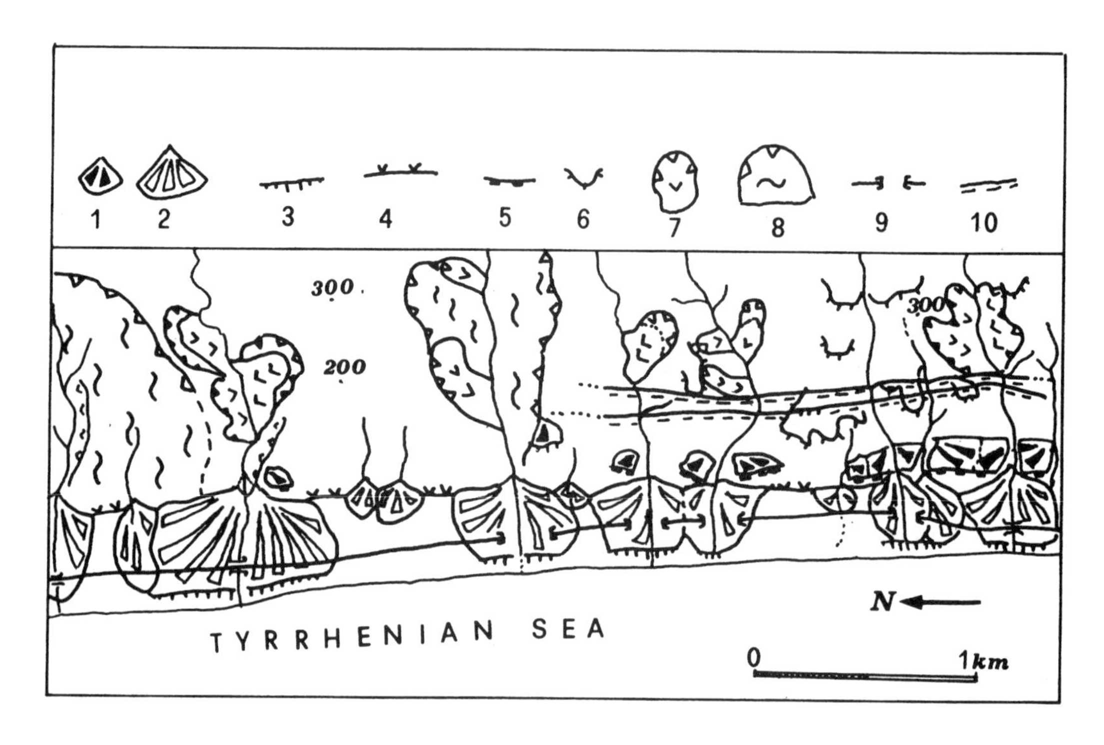

Figure 12.6 *Alluvial fans on the seaward side of the Tyrrhenian mountains in north Calabria. Legend: 1 = Tyrrhenian age fan; 2 = Holocene fan (inactive at present); 3 = active marine-cut cliff (length of "teeth" proportional to cliff height); 4 = foot of mountain front; 5 = upper edge of ancient cliff; 6 = outer edge of marine terraces (from 70 000 to 100 000 years BP); 7 = landslide; 8 = deep-seated rock creep; 9 = highway tunnel (under fans); 10 = active normal faults* (data from Sorriso-Valvo 1993)

Plate 12.7 *The partial burial and recovery of the Palazzo Bardano. This building is located on one of the alluvial fans on the seaward Tyrrhenian mountain front (Figure 12.6). Photo by M. Sorriso-Valvo*

Plate 12.8 *One of six knickpoints in the* rambla *Cervera, here cut into caliche (CO_3-cemented gravels). The entire feature is incised into a large Pleistocene fan. Knickpoints here supposedly reflect base level changes, but the* rambla *has all the characteristics of a (large) discontinuous gully. It has also been disturbed by gravel quarries and bridge construction, as well as by land degradation in the catchment. The* rambla *has an ephemeral flow regime, averaging 2.7 flows per year, in a 500 mm rainfall area (with recorded rainfall intensities of up to 200 mm in 2 h)*

Removal of native vegetation has also disrupted the hydrological system in California, leading to increased runoff which is reflected in increased flooding, erosion, sediment delivery and habitat changes downstream. Logging, mining and agriculture have each generated such responses, most dramatically in the vast sediment yields that were flushed from placer gold workings in the Sierra Nevada foothills during the later 19th century (Chapter 17).

Other Forms of Soil Degradation

Soils exhibit many forms of degradation other than their physical loss by erosion and redeposition elsewhere as sediment. Types of soil degradation discussed here include salinization, acidification, structural decline, waterlogging and water repellency, declining fertility and pollution by chemical residues. In some regions, the consequences of these forms of soil degradation probably exceed the effects of physical soil loss.

The Iberian Peninsula suffers from a wide range of soil degradation problems other than erosion, particularly increased salinization and calcrete accumulation (Figure 12.7). Saline soils cover an area of 1900 km^2 and gypsiferous ones an area of 6425 km^2. In the cultivated alluvial plains, especially in Guadalquivir, soil degradation problems are related to insufficient organic matter and to an excess of fertilizers. The hilly areas with sylvo-pastoral land use (*dehesas*) suffer from compaction by stock.

Locally, and sometimes temporarily, soil contamination is becoming a serious problem. The main contamination sources are industrial and mining wastes and agricultural fertilizers and pesticides. The industrial impact is not well known and only in Catalonia have some inventories been undertaken in order to determine the magnitude and the characteristics of the problem (MOPU 1989b). In agriculture the use of chemicals has notably increased and with it the contamination of the soil. Nevertheless, the use of fertilizers and pesticides is much lower in Spain and Portugal than in north European countries (Table 12.9). The high-risk zones are the Mediterranean coastal plains and the alluvial plains of the main rivers.

Plate 12.9 *A broad, gravel- and sand-bed river with constantly shifting channel and highly erodible banks in semi-arid (250 mm/yr) eastern Morocco. Increased channel instability reflects increased peakedness of the hydrograph, following land degradation in the catchment. Compare the Spanish* rambla *(Plate 12.8) and the Italian* fiumara *(Plate 3.2)*

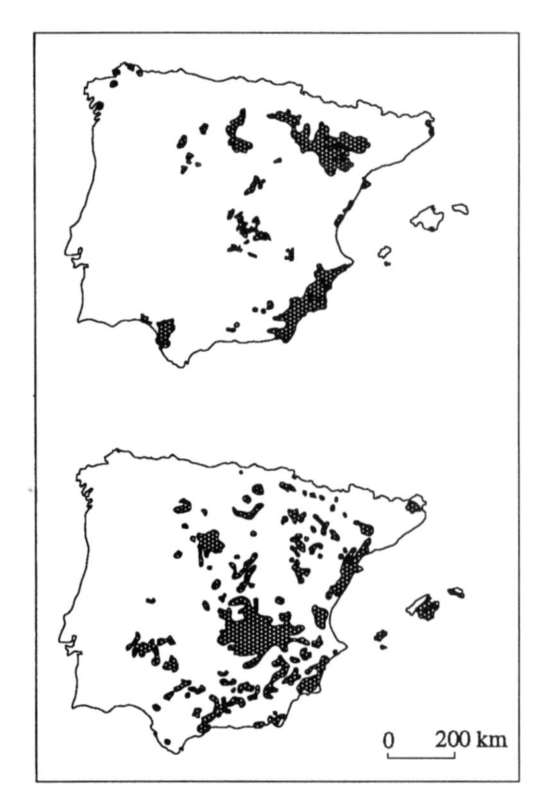

Figure 12.7 *Salinization (top) and calcrete accumulations, Spain* (after Atlas Nacional de España 1996)

In Portugal, the uncontrolled use of chemical fertilizers and pesticides is causing soil pollution in agricultural areas, for example high copper concentrations in vineyard soils. The disposal of liquid and solid effluent from intensive pig and cattle farms is also contaminating the soil. Untreated domestic solid waste disposal creates point-source pollution problems, particularly in coastal dune systems subject to pressure for tourism development. Industrial pollution affects soils in areas of the littoral, from Setúbal, Lisbon, Aveiro and Oporto, since effluent discharge has been uncontrolled for decades, causing enormous damage to soils and also to groundwater. In the interior, mining activities at Beira Baixa do not treat effluent discharge and are also contaminating soils and water. The production of toxic wastes has been estimated at 1 043 000 t/yr (DGQA 1992), of which 75% are directly disposed at the soil surface or subsoil in municipal rubbish tips without any treatment.

Salinization due to irrigation is a long-term problem which has been studied for several years in Italy. There are no data at the national level, since the problem is probably overwhelmed by other, more urgent forms of degradation.

One interesting case of soil salinization was caused by rapid subsidence following brine extraction. The subsidence occurred at the base of a

Table 12.9 *Use of chemicals in agriculture in the Spanish Autonomous Communities in 1988, compared with EU countries* (source: MOPU 1989b)

	Cultivated area (ha)	Pesticides (kg/ha)	Fertilizers (kg/ha)
Mediterranean communities			
Andalucía	4 167 300	9.2	98.1
Aragon	1 897 900	2.4	95.0
Baleares	257 200	4.6	30.1
Castilla-La Mancha	4 252 700	1.3	49.8
Castilla y Leon	4 097 700	1.4	112.7
Catalunya	1 044 100	11.8	153.2
Valencia	933 400	29.9	192.2
Extremadura	1 484 700	3.7	89.9
Madrid	276 100	4.9	82.8
Murcia	596 000	18.1	69.0
La Rioja	179 600	11.8	303.4
Atlantic communities			
Asturias	31 200	16.9	228.6
Cantabria	19 400	16.0	500.9
Galicía	558 600	7.0	95.9
Navarra	373 400	1.4	136.9
Pais Vasco	104 000	7.3	158.6
Total Spain	20 415 400	6.2	99.3
EU Communities in 1986			
France	18 928 000		300.9
Germany	7 453 000		427.3
Greece	3 940 000		173.9
Ireland	800 000		787.5
Italy	12 200 000		172.2
Netherlands	892 000		783.6
Portugal	2 760 000		87.3
UK	7 077 000		355.5

slope, triggering a landslide which filled the subsidence crater. As a consequence, some 800 000 m^3 of brine was squeezed out of the crater, flooding more than 10 ha of good olive groves, killing all the trees and rendering the land useless for some 10 years. In this case the economic loss was in the order of millions of US dollars, not to mention the fact that mining was stopped for a while. As mining restarted, other brine collapses occurred.

A different condition occurs in soils forming from materials which are already salt-rich, such as the salty Messinian clays and the gypsum beds in central Sicily, northern Calabria and the northeastern Apennines. Periods of drought or very hot summers may result in the formation of salt crusts, which may exhibit small gypsum lenses and "desert roses". An estimated 8000 km^2 of land in Mediterranean Italy is affected by this problem.

A recent, unpublished graduate thesis (Viscomi 1994) reports that the salinization hazard in Calabria due to climatic factors is low to moderate, and is limited to the Ionian coastal strip. This area might extend to the low hills adjacent to the coastal plains if the inferred rise in global temperatures becomes a reality.

Forms of soil degradation in Greece include soil structure decline, loss of organic matter, reduced aggregate stability and salinization, which are discussed further in Chapters 15 and 16. About 150 000 ha in the lowlands, especially along the coast, contain appreciable amounts of soluble salts to such a degree that they need reclamation before any use.

Soil acidification affects a great portion of the cultivated land in Greece. Old alluvial terraces or areas with soils formed on acid parent materials show a rapid deterioration of soil fertility and depressed productivity. It is estimated that more than 450 000 ha present serious toxicity problems. The use of large amounts of ammonium fertilizers over the previous few decades has resulted in a rapid increase of soil acidity. Comparing unpublished data from the early work by Yassoglou and his colleagues in 1965 from their soil survey of the Pinios plain with contemporary data, shows that the acidity of soils of the old alluvial terraces of the Pinios river (Peloponnesus) has increased over the last 30 years by about 2 pH units.

In the eastern Mediterranean and North Africa, there is progressive salinization of soils mainly in irrigated areas and low-lying areas which are subject to strong evaporation and rising groundwater tables. The problem concerns all the irrigated lands but it is particularly serious in semi-arid regions. In sum, it is a difficult problem to solve, especially in the plains of the Gharb and the Triffas (Laouina 1987; refer also to the section on water pollution in Chapter 13, and the discussion on irrigation agriculture in Chapter 16).

Soil degradation, involving changes in the physical, chemical and biological structure of the soil, has been an inevitable consequence of vegetation change in California and is exacerbated by irrigation agriculture. Soil degradation related to agriculture commonly occurs as salinization, the hyper-concentration of naturally occurring substances such as sodium, selenium, boron and molybdenum, and the accumulation of artificial nutrients associated with fertilizers and pesticides.

An important portion of the more productive areas in central and south Chile have soils with waterlogging problems (Table 12.10). Although the general distribution of soils with this problem is known, available data are not sufficient to evaluate properly the characteristics and intensity of the problem. Soils at risk of salinization and the presence of phytotoxic components are located in the districts of la Serena and Coquimbo, in the lower course of the Limarí River in Region IV, and in the districts of Colina, Lampa and Pudahuel of the Metropolitan Region (Table 12.10). Saline and sodic soils of the desert and semi-desert zones occupy vast areas and are used for forestry.

There are few data on other aspects of soil deterioration for the southwestern Cape. Scotney and Dijkhuis (1990) have documented significant recent changes in the fertility status of South African soils, including lowered organic matter content, declining nitrogen levels, increases in soil acidity, a fall in micro-nutrient status and expansion of the area subject to saline and alkaline conditions, but the data are not regionally specific. Features such as the decline in soil structure, the formation of plough pans and consequent reductions in infiltration capacity are probably widespread in the agricultural areas of the southwestern Cape, particularly in the Swartland, but evidence remains circumstantial. The regular occurrence of fire in the natural vegetation of the southwestern Cape suggests that water repellency could be a problem, at least periodically. The vegetation itself may produce waxes and other organic compounds which could favour repellency, but fire is commonly regarded as the most important factor (see Chapter 18).

Soil salinization and acidification have been the subject of national surveys (du Plessis 1986; Fey

Table 12.10 *Soils in Chile with waterlogging and salinization problems* (source: MINAGRI 1994)

Region	Area (10^3 ha)	Waterlogging (10^3 ha)	Salinization (10^3 ha)	% Region with problem
IV Coquimo	3964.7	12.7	5.0	0.5
V Valparaiso	1637.8	29.8		1.8
Metropolitan	1578.2	56.9	13.9	4.5
VI O'Higgins	1595.0	139.5		8.7
VII Maule	3051.8	141.3		4.6
VIII Biobío	3600.7	166.1		4.6

et al. 1990), but again data are not available for separate regions. Conditions strongly favour salinization on the shale-derived soils of the Swartland. Here, a combination of elevated nutrient loadings from the parent material and high evaporation rates in summer promote accumulation of soluble salts in the soil profile. Farmers of the Swartland regularly plant and harvest saltbush (*Atriplex* sp.) along drainage lines in an effort to remove these salts from the system.

In 1994, 1 800 000 ha, or about 9% of cleared land in the agricultural areas of the southwest of Western Australia, were affected by secondary, dryland soil salinity (Natural Resources Management Services Unit 1996), defined as the salinization of soils which were previously agriculturally productive and which are (or were) farmed without irrigation (Plate 12.10). Importantly, a further 4 300 000 ha were considered as being susceptible to salinization (Table 12.8). Some farms already have more than 30% of their land so affected, rendering the property economically unviable. In the state of South Australia, more than 225 000 ha of land throughout the agricultural areas are affected by secondary salinization, a considerable increase on previous estimates (Community Education and Policy Development Group 1993). In addition, some 275 000 ha are affected by secondary salinization in the state of Victoria. Unlike Western Australia, about half of Victoria's salt-affected area is irrigation-induced, and some of the Victorian areas lie outside the Mediterranean climatic zone (Scott 1991).

Another form of soil degradation in the southwest of Western Australia is waterlogging (which is often closely associated with salinity problems, soil structural decline and duplex soils), which affects at least 500 000 ha (more in wet years). Soil structure decline and compaction are also significant forms of land degradation in terms of areas affected (more than 8 000 000 ha) and production losses caused. Manifestations include reduced aggregate stability, porosity and permeability, increased surface sealing and soil bulk densities, and the development of subsoil hardpans, which are common at depths of 200–400 mm.

Table 12.11 indicates the extent and severity of soil structural decline in Victoria. Two-thirds of the state's soils suffer from moderate to severe structural decline, especially across the northern and south-central regions. The huge dust storm from the Victorian mallee which enveloped Melbourne in February 1983 was symptomatic of something seriously wrong in the agricultural areas (Middleton 1984) (Plate 12.11).

Although soil acidification affects a relatively small proportion of Western Australia's

Plate 12.10 *Secondary, salt-affected soil in the rain-fed (dryland) Western Australian wheatbelt*

Table 12.11 *Land degradation on Victorian farmland 1985–1991: soil structural decline* (source: Scott 1991, table 18.4)

	Total area (10^3 ha)	Insignificant/low (10^3 ha (%))	Moderate (10^3 ha (%))	Severe (10^3 ha (%))
Broad-acre crop land	5807	2547.9 (44)	906.8 (16)	2334.1 (40)
Horticulture	199	58.3 (29)	71.5 (36)	66.8 (34)
Dryland pasture	7199	1240.5 (17)	4385.9 (61)	1502.1 (21)
Irrigated pasture	531	7.5 (1)	47.7 (9)	474.4 (89)
Plantations	50	12.8 (26)	31.7 (65)	3.7 (8)
Remnant vegetation	936			
Total agricultural land	14 722	3868 (26)	5445 (37)	4381 (30)

Plate 12.11 *Dust storm engulfing Melbourne on the afternoon of 2 February 1983. Photographer not known*

wheatbelt soils, it has the potential to affect more than half of all agricultural land in the state (Table 12.7). It is associated with other forms of soil degradation, notably the availability of nutrients to plant roots. Water repellency is another extensive problem which affects wheatbelt soils (Table 12.8). Its relevance to land degradation is that it not only starves plant roots of moisture but also leads to excessive overland flow and hence waterlogging, soil erosion and possibly salinization.

The extent of areas affected by chemical residues of synthetic fertilizers, pesticides and veterinary chemicals is more difficult to assess, given the limited amount of environmental monitoring in the past, although their use is widespread. In Western Australia, there are two main areas of concern with pesticides. Residues from past uses of organochlorine pesticides contaminate an estimated 13 000 ha of soils (Table 12.7), especially in orchards and potato-growing areas (Grant 1992). The second concern is the huge increase in herbicide use associated with minimum tillage of croplands. The associated on-farm problems have included non-target damage and increasing resistance of weeds to the herbicides (Conacher and Conacher 1986).

In South Australia, Matheson (1986) has discussed soil fertility decline, especially nitrogen deficiency, soil acidification and salinization, other forms of soil pollution (pesticide residues and heavy metal pollution), the decline of soil structure, and the loss of prime agricultural land to urbanization, conservation and other uses. There are 2 500 000 ha of land prone to acidification (with a pH of <5.5) in the root zone, and about 1 000 000 ha already adversely affected (Table 12.8). A further 2 200 000 ha are susceptible to water repellency (Community Education and Policy Development Group 1993, p. 86). The best documented case in South Australia of heavy metal pollution is at Port Pirie, where above-normal concentrations of lead, cadmium and zinc have contaminated wheat and vegetable crops (Table 12.12). High levels of cadmium in root vegetables were still causing concern in 1993 (Community Education and Policy Development Group 1993).

Table 12.12 *Maximum heavy metal concentrations (in ppm) found in vegetables surveyed in Port Pirie* (source: Environmental Protection Council of South Australia 1988, p. 25)

Vegetable	Lead	Cadmium
Silver beet	2.20	0.49
Rhubarb	0.75	0.19
Tomatoes	0.07	0.21
Beetroot	1.90	0.35
Carrots	1.60	2.60
Lettuce	0.22	0.18

NHMRC Standards: lead 4.0 ppm, cadmium 5.5 ppm

Vegetation Loss and Degradation

Most of the above problems of land degradation have been partly caused by the removal, degradation or replacement of the vegetative cover. However, vegetation loss and degradation are also forms of land degradation in their own right, and are so treated in this section.

Little of the indigenous vegetation remains in many parts of the Mediterranean Basin, due to its long period of human settlement; indeed, there is considerable room for debate as to what constitutes truly "indigenous" vegetation in that region. In contrast, removal of indigenous vegetation – usually to make way for agriculture – is recent (and indeed ongoing) in the New World. Not only does this make possible a range of studies comparing disturbed with "undisturbed" environments (although no environment is wholly undisturbed by human actions), but it also means that the sense of loss as native vegetation is removed, degraded and invaded, is far more acute.

Vegetation and fauna have been intensely altered by human actions in the Iberian Peninsula. But despite this fact, ecosystems and species have been maintained in a far better conservation state than those of northern Europe. The so-called natural heritage is still impressive. The problem is, can economic development continue without destroying this heritage? The equilibrium reached through the agro-sylvo-pastoral system of *dehesa* and *montado* (Chapter 1) was partly lost during the 1960s due to more intensive exploitation, aiming at a self-sufficient production of cereals, importing agricultural techniques from northern Europe.

In common with many Mediterranean or seasonally arid areas in Portugal, the indigenous mixed oak forest in Spain has been replaced almost entirely by *Cistus*-dominated *matorral* on hillslopes and by cultivated dryland farming on the plateaux. The *Cistus* plants are often 1.5 m high and, where not degraded, the ground surface is completely covered. Limited grazing continues on these areas and occasionally it has been cleared to provide young, less woody vegetation for grazing.

The effects of over-grazing by small ruminants and pigs became apparent in the 1960s due to overloading the carrying capacity of *montado* and *dehesa*. Recently, the intensive rearing of cattle has led to soil and cork degradation. Holm oak *montado* has been intensively cleared for the production of charcoal, in particular in the Guadiana River basin during the 1980s. Severe cork oak death is occurring in the Alentejo littoral, especially during dry years. Factors considered to accelerate cork oak death are leaving wounds when extracting the bark, intensive exploitation, keeping cork waste in the soil, and increasing pests.

To the end of the 19th century, deforestation and exploitation of the residual forest constituted the main forms of degradation of the geosystems of southern France and Corsica, and probably of most other regions around the Mediterranean Sea. Today, these problems have been replaced by forest fires, floods, soil erosion and air and soil pollution.

The whole Mediterranean region in Croatia, for example, has experienced severe degradation and the loss of native forest and grassland vegetation. These human impacts are obvious over 397 584 ha (26% of the total agricultural and forest land in the area of Mediterranean climate) of lost native vegetation and 518 005 ha (34%) of degraded vegetation (coppice, *maquis* or brushwood). The remaining 37% (excluding pine plantations) is assigned to agricultural use (mainly extensive grazing).

Original vegetation forms are present only as fragments in inaccessible and protected areas, and the vegetation of the region is neither protected nor evaluated properly, with only 3.2% of the total forest and agricultural area being protected in national parks and strict nature reserves (Table 12.13). The indigenous, non-degraded vegetation of *Quercus ilex* is best preserved on the island of Rab in the northern Adriatic (Raus 1976).

Agricultural areas in the Croatian Mediterranean region represent degraded forest and grassland vegetation areas, because they originated in human activities such as burning and rooting out. Only 33.7% of agricultural land is being intensively cultivated (190 890 ha; Topič in press).

Forest land in Greece covers about 8 200 000 ha, of which less than 2 500 000 ha are really productive forests (Papamichos 1985). In 1825 the total forest area covered about 48% of the country, compared with about 20% today. Natural vegetation continues to be destroyed at an annual rate of 0.076% of the total area of Greece (Alexandris 1989). The rest of the area is covered by shrubs and *prygana*, sparse forest trees or bare land resulting from forest degradation. Extensive deforestation and intensive cultivation of the sloping lands since ancient times have led to soil erosion and degradation through the progressive inability of the vegetation and soils to regenerate themselves (Chapter 14) (Plate 12.12).

Degradation of the plant cover takes two main forms in North Africa. On the one hand, there is a rapid physiognomic defacement of the vegetal

Table 12.13 *Main protected areas in Croatia's Mediterranean-type climate region* (source: Floriani 1991)

Croatian category	Protected area (ha)	Locality	Protected since	IUCN category
National parks				
Brijuni	3635	Istria	1983	II
Paklenica	3617	North Adria	1949	II
Komati	22 375	Dalmatia	1980	II
Krka	14 200	Dalmatia	1983	II
Mljet	3100	Dalmatia	1960	II
Strict nature reserves				
Hajdučki i Rožanski ku	1220	North Adria	1969	III
Bijele i Samarske stije	1175	North Adria	1985	III
Nature reserves				
Telašćica	6706	Dalmatia	1988	VIII
Velebit (Biosphere res.)	200 000	North Adria	1981	III, VIII, IX
Biokovo	19 550	Dalmatia	1981	VIII

Plate 12.12 *According to C. Kosmas, these hills north of Athens have been grazed and burnt repeatedly, until the soils are so thin that trees will no longer regenerate; even planted seedlings fail to establish*

formations through excessive wood cutting and by over-grazing. Dense forests are transformed into open woodlands, and woods can even take on a steppe character (Aalounne 1995; Labhar 1995).

On the other hand, there is an effective recession of the forest due to the encroachment of cultivation. Thus, forests extend over no more than two-thirds of their initial extent. The shortage of land accounts for the clearing of even steep slopes, often without consideration of soil conservation. Cultivated lands are increasing rapidly, namely in the piedmonts, the plateau slopes of the *meseta* and on the low relief of the middle mountains, with the cultivation of the most humid pasture lands (for example that of the Middle Atlas) progressing at a particularly rapid rate. Annually, 6000 ha of forest are lost in the provinces of Al-Hoceima, Azilal and Taza through clearing (Benabid 1992).

In Algeria, two vegetation species have receded considerably in comparison with their natural distribution: the green oak has contracted from about 1 800 000 to 600 000 ha, and the thuya (*Tetraclinis articulata*) from 520 000 to 130 000 ha (Plit 1983).

In the Tell, the *maraboutic* forests (remnants of the original vegetation around sacred cemeteries) represent the only samples of the primary plant cover. Cedar, for example, is no longer present in the form of closed forests, but occurs on north-facing slopes over 1600 m above sea level (Deil 1987) (Plate 7.7). Elsewhere, it is very sparse and gives way to secondary cystic and broom-infested groupings. Young plants are confined to glades and the edges of aged stands. Regeneration really succeeds only in sites which are protected, for example in the vicinity of forest houses. Thus, it is on the south-facing slopes that cedar has been most widely degraded, partly for climatic reasons, because this aspect is hotter and has less snow, and partly because human pressures are stronger in these positions.

The example of the high Tunisian steppes (Boujarra 1993) is significant. Since 1952, cultivation has spread considerably at the expense of pastures and forests; rangelands have been reduced by as much as 70%, according to the district, and forest by 10–15%. But the forest of Aleppo pine (*Pinus halepensis*) has been cleared extensively (the dense forest has undergone a 68% reduction in area and has been replaced by open forests). Locally, the density of pines per hectare has decreased from 28 to 21 from 1963 to 1983, while the rate of cutting has increased (from 1.2 tree/ha/10 years in 1963–1973, to 3.7 trees/ha/10 years in 1973–1983).

The loss of native vegetation is of prime importance in California, notably through the conversion of primeval forest, *chaparral* and grassland to secondary forest, degraded scrub, pasture, tilled fields, orchards and built landscapes. Clearance of native vegetation occurred in several ways – through mining, logging, grazing, cropping, reclamation, railroad and highway construction, urban growth, industrial development and, of course, fire (Plate 12.13). In some instances, notably through urban development, the effect was dramatic and complete. In others, such as logging, native species returned but the character of the old-growth forest was lost to even-age stands of second-growth timber. In yet other instances, notably in the grasslands, removal of native bunch grasses led to their purposeful or incidental replacement by exotic annual species such that the character of the primeval grasslands was lost forever.

Wood resources are very important in Chile. Forests are the second source of energy in the country (25%), producing $6 \cdot 10^6$ to $7 \cdot 10^6$ m^3 of wood. In addition, in recent years the pressure on the native forests to produce woodchips has grown probably to unsustainable rates (as is also the case in southeastern and southwestern Australia). One of the reasons is that this product has been internationally commercialized since 1987 (1970s in Australia).

Another pressure on vegetation is caused by over-grazing, which is coupled with the loss of biodiversity. The Coastal Cordillera in central Chile is one of the most affected areas due to the combined action of dry farming and grazing, complemented by tree felling (MINAGRI 1994). In addition, wood resources are being threatened by endemic pests. The exotic plantations, such as the monoculture of *Pinus radiata* introduced from Argentina, are especially susceptible to insects, for example *Rhyacimia buoliana*, a lepidoptera which covers extensive areas. Rabbits are also damaging wood plantations and their impact can reach 75%. The *Eucalyptus* plantations also suffer from pests, whose impact is effective after two or three years, especially in plantations growing in dry areas. In the arid and semi-arid Regions I–V, 8 500 00 ha of forests have been lost to logging; whilst in the semi-arid and subhumid Regions V–X, 7 700 000 ha have been lost to a combination of logging, fires and agriculture (MINAGRI 1994).

Arguably the most important and certainly the most visually obvious manifestation of land degradation in the southwestern Cape has been the progressive and on-going degradation of natural vegetation. This takes various forms, including fragmentation in response to wholesale transformation, structural alteration involving reductions in vegetative cover, and biogeographical adjustments involving changes in biodiversity due to selective removal or invasion by exotic invasives.

There are probably no plant communities in the region which do not bear at least some signs of human activity. The most recent regional-scale assessment of the extent of remaining natural vegetation was that by Moll and Bossi (1984), although there is a more recent analysis which deals only with the area immediately around Greater Cape Town (McDowell *et al.* 1991). Moll and Bossi's (1984) survey under-estimated current levels of transformation for three reasons: first, their aim was to map vegetation which had been completely transformed, meaning that partly transformed or simply degraded vegetation communities were mapped as "natural"; second, the analysis was based on satellite imagery and this sometimes made it difficult to distinguish natural communities from those significantly invaded by alien tree species; and third, the degree of transformation has probably increased substantially for at least some of these vegetation types during the period since the data were published (McDowell *et al.* 1991).

The data, shown in Table 12.14, indicate that, overall, 67% of the original natural vegetation of the region remains intact. This compares with an approximate national average of 78% (Macdonald 1989). The mountain vegetation types of the southwestern Cape are relatively unaffected by anthropogenic transformation as a significant proportion, 93%, of the montane vegetation of the region appears to be in a largely natural state. However, there are indications that such a picture is far too optimistic. The data do not agree, for example, with the analyses of Richardson *et al.* (1992), who documented the areas of the region infested with exotic tree species (see below). It can be concluded that the Moll and Bossi (1984) assessment of the remaining areas of

Plate 12.13 *(a) Firestorm in the Santa Monica Mountains near Los Angeles, October 1978. Destruction of native and exotic vegetation increases flood and debris-flow hazards during rainy winters, at which time flood-control structures such as the Sepulveda Dam (foreground) serve to mitigate urban flooding. Photo by L. Loether. (b) Fire-ravaged suburban area in the Oakland Hills east of San Francisco Bay, following the 728 ha firestorm of October 1991 which resulted in the loss of 25 lives and 2903 houses*

natural vegetation in the region (Table 12.14) are absolute maximum values in the case of all the plant communities. Some communities are clearly very seriously impacted, in particular those *renosterveld* communities developed mainly on the nutrient-rich Malmesbury Shale-derived soils which have proved to be so valuable agriculturally. McDowell *et al.* (1991), in their survey of the Cape Flats, noted that very substantial proportions (up to 57%) of the lowland *fynbos* and non-*fynbos* communities have succumbed to the advances of urbanization alone.

The loss of native vegetation cover in the Australian Mediterranean regions, too, has probably been the most visible and potentially damaging form of land degradation, particularly in relation to changes in soil properties and land surface hydrology, and loss of water quality and of plant and animal species. In the agricultural areas, more than two-thirds of the original vegetation has been removed, mostly very recently (since 1920).

About half of the original 4 000 000 ha of eucalypt forests in the southwest of Western Australia has been cleared since European settlement commenced in 1829. More significantly, over 90% of the original woodland and heath vegetation has gone from parts of the wheatbelt areas of that state, mostly since 1920. The thorough nature of the clearing is indicated in Plate 12.6. Much of the remainder is in a scattered, degraded and poorly protected form (Figure 12.8).

More than 80% of the native vegetation in South Australia's agricultural region has been cleared, with less than 1% remaining in some districts (Community Education and Policy Development Group 1993, p. 81). The largest uncleared blocks are found in conservation reserves or national parks in the southeast, on Kangaroo Island and on Eyre Peninsula (Figure 12.9). However, 26% of South Australia's plant communities are poorly or not at all conserved.

Victoria, too, has experienced heavy losses of its original vegetation, with more than 65% of the state cleared (Figure 12.10). That which remains in the Mediterranean region, mainly in the southwest areas and in the Grampian Mountains, is in a semi-natural condition and subject to a range of pressures. In rural areas, trees continue to be lost at an estimated rate of 1% each year. Native grasslands, which once covered more than a third of the state, have been almost completely destroyed (Scott 1991). Indeed, serious regional losses of Australia's grasslands have often been overlooked due to preoccupation with the loss of forests (Kirkpatrick 1994).

Fire

The current number and extent of forest fires are amongst the most serious environmental problems in all the world's Mediterranean regions. In addition to the loss of vegetation, forest fires induce changes to the physico-chemical properties of soils, often leading to water repellency, and thus to increased runoff and erosion. They also destroy wildlife habitat, cause loss of human life and damage infrastructure (Plate 12.13) (Chapter 18).

The areas affected by forest fires are increasing throughout the Mediterranean Basin: 200 000 ha/yr from 1960 to 1971, 470 000 ha/yr from 1975 to 1980, and 660 000 ha/yr from 1981 to 1985 (Le Houerou 1987): a trebling in 30 years.

Table 12.14 *The extent of remaining natural vegetation in the southwestern Cape. Values are for 1981 and losses of natural vegetation have continued to increase* (source: Moll and Bossi 1984)

Plant community	Area (km²)	Remaining area (km²)	Natural vegetation lost (%)
Afromontane forest	3844	2930	24
Strandveld	4453	2072	24
Mountain *renosterveld*	4754	3448	27
Coastal *renosterveld*	15 285	2256	85
Coastal *fynbos*	8770	4627	47
Mountain *fynbos*	37 310	34 652	7
Total	74 416	49 985	33

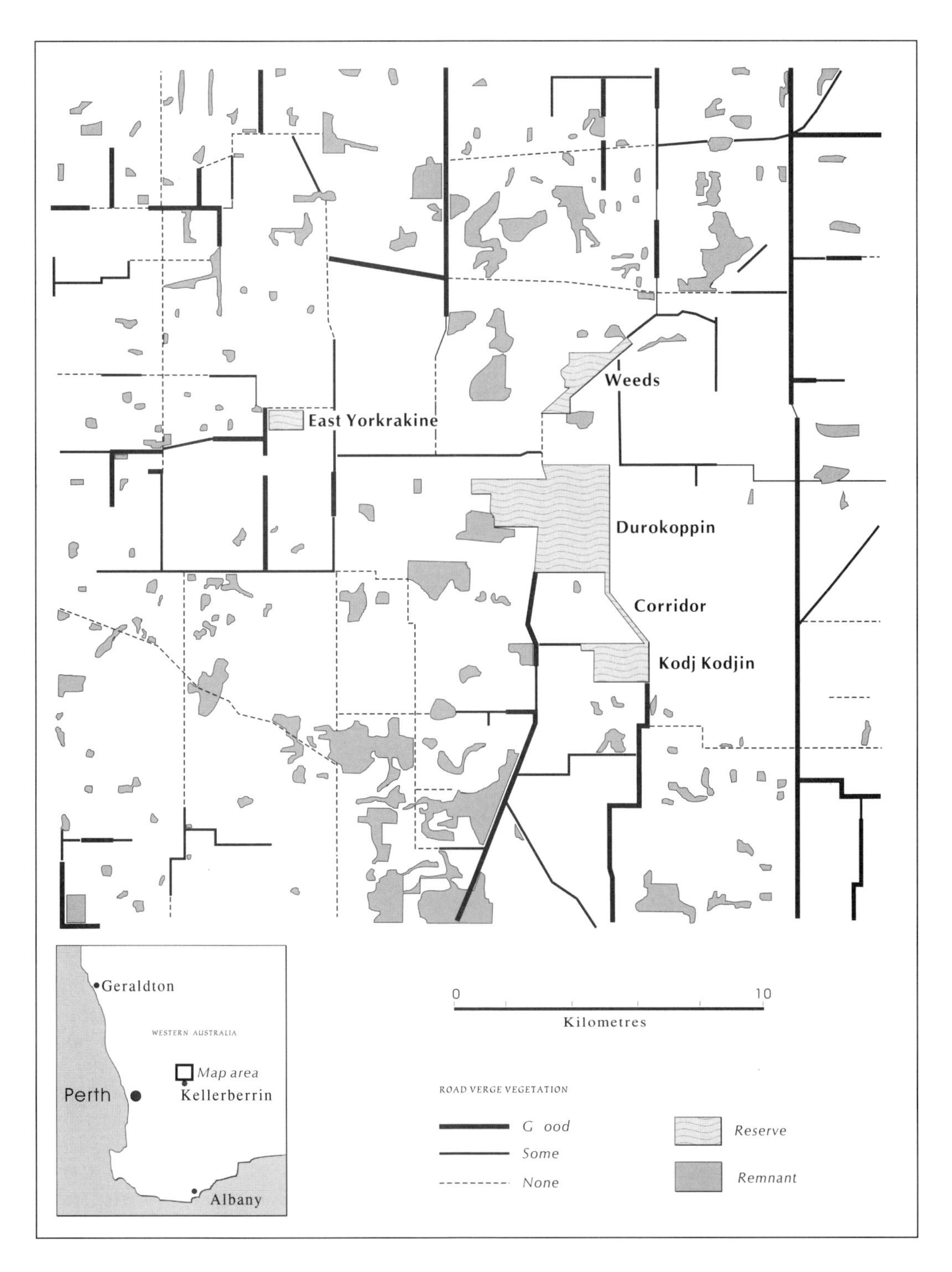

Figure 12.8 *Remnant native vegetation in the central Western Australian wheatbelt* (source: Hobbs and Saunders 1993). *The reserves shown are some of the largest in the central wheatbelt: most remnants are small and unprotected. Only about 7% of the native vegetation in Kellerberrin Shire (in which the map area is located) still remains*

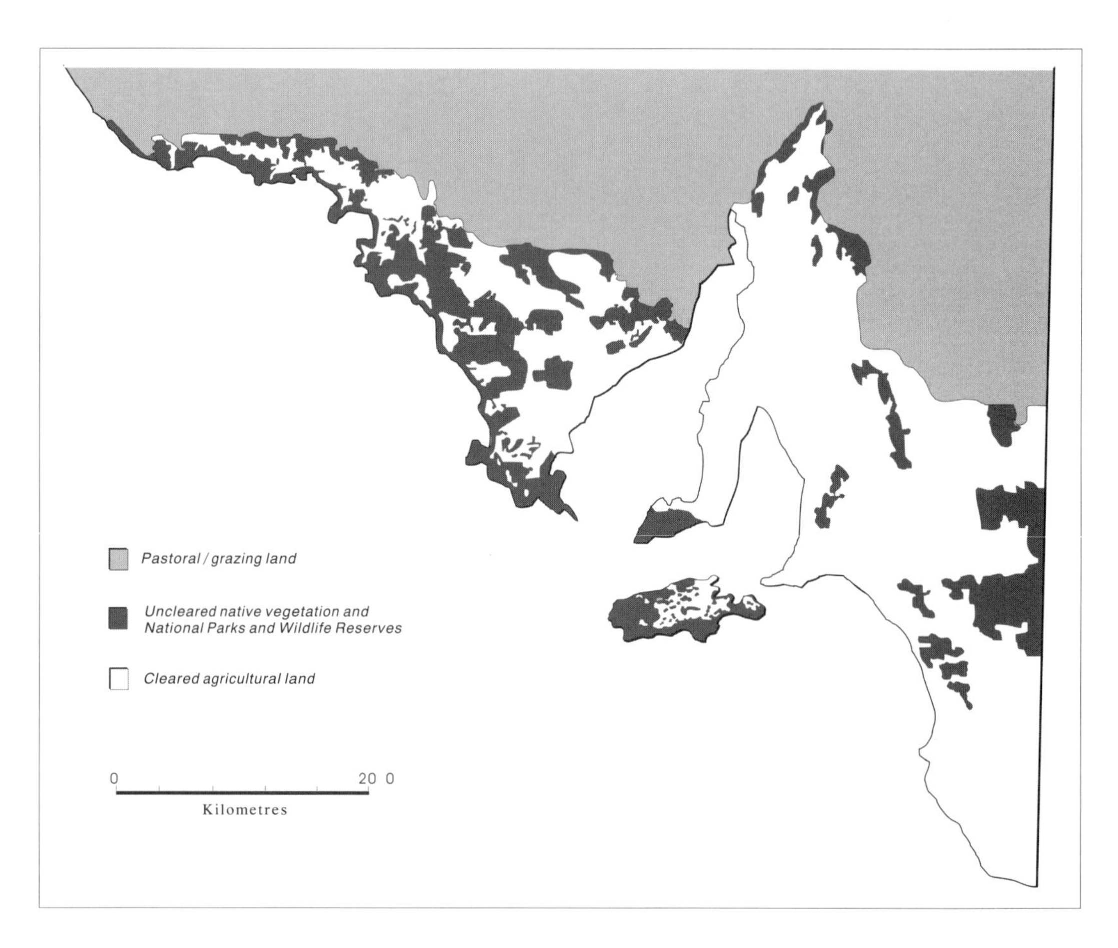

Figure 12.9 *Extent of natural vegetation clearance in the agricultural regions of South Australia* (source: Harris 1986, figure 2.15)

Since the mid-1980s, more than 94 000 forest fires have occurred and nearly 2 500 000 ha have been burned in Spain. As stated by Campos Palacín (1992), during the decade 1976–1985 the total area affected by fires (972 790 ha) exceeded that subject to afforestation (830 222 ha). Although not all the burned areas are forests, the economic losses are important (Table 12.15; Plate 12.14).

It must be borne in mind that 50% of the area of Spain is forested (woodland and scrubland), and that in the last 12 years fires have affected 9.4% of that area. It is difficult to quantify the loss to the environment that this represents. The region of Galicia, 67% of which is forest, has the greatest problem, especially in the provinces of Pontevedra and La Coruña. In contrast with the areas nearer the Mediterranean, in Galicia fires do not last a long time and scrub is the most affected area. On the Mediterranean strip, the most affected areas are the regions of Valencia and Catalonia, these being precisely the areas where there are still large forests (Plate 12.15). In Valencia, the number of fires and the affected area rose dramatically after 1977, going from a yearly average of 5–10 ha to 40 ha in 1978, and to 60 ha in 1979. Catalonia, with a forested area of 1 164 200 ha (9.87% of the Spanish total), also saw an increase in forest fires, reaching annual figures of between 4000 and 17 000 ha in 1978 and a maximum of 42 417 ha in 1986 (data from Departament d'Agricultura, Ramaderia i Pesca, DARP). However, safety measures taken since

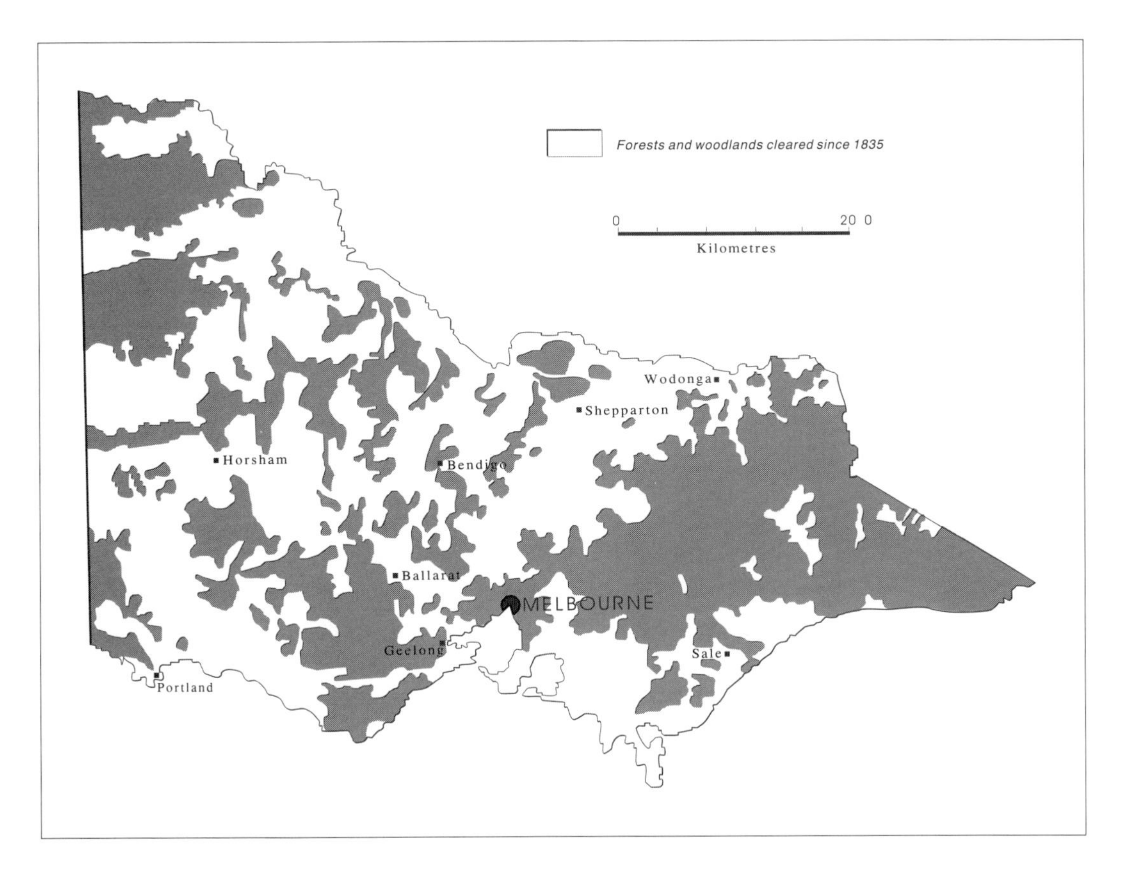

Figure 12.10 *Vegetation clearance in Victoria* (source: Scott 1991)

Table 12.15 *Forest fires: type of vegetation affected and economic losses, Spain (selected data from Annual Reports of ICONA (National Institute for the Conservation of Nature) and MAPA (Ministerio de Agricultura, Pesca y Alimentacion))*

Year	No. of fires	Area affected (ha)			Loss of primary products (US$ million)
		Timbered	Untimbered	Total	
1963	1302	13 279	9400	22 679	2.4
1968	2109	20 547	36 081	56 628	4.2
1973	3765	40 559	54 698	95 257	8.6
1978	8324	159 264	275 603	434 867	70.8
1983	4880	57 832	59 767	117 599	32.3
1988	9262	40 484	83 484	123 968	41.5

then have brought the average down to around 1000 ha a year. Un fortunately, 1994 was a bad year in relation to forest fires.

Fires have become increasingly common over the last 20 years in the pine and eucalyptus forests of north-central and south-central Portugal. In the wetter areas of central Portugal (Agueda basin), recovery is rapid under pines due to the development of a thick layer of *maquis* which completely covers the soil surface, inhibiting the detachment of soil particles by splash.

In France, an average of 42 000 ha were burnt each year from 1981 to 1985, that is to say 2.3% of the entire forests, *maquis* and *garrigues* in the

Plate 12.14 *House partly destroyed by wildfire on the Spanish plateau in 1994*

Plate 12.15 *Part of the upper Llobregat valley, Catalonia, affected by an extensive and severe forest fire in 1994. Photo by J. Conacher*

Midi méditerranéen. As Chapter 14 shows, this was a sharp increase over earlier periods. If this rate continues, all the French Mediterranean forests will be burnt twice in a century. That seems to be confirmed by the great fires in the Sainte Victoire Mountain (5000 ha in 1989) and in the Maures massif (about 30 000 ha in 1990).

Corsica represents a very important proportion of the annual total of about 35 000 ha burnt each year in the Mediterranean part of France and Corsica, according to the Entente Interdepartementale 1988 (Figure 12.11). Thus, for example, 83% of the whole burnt area in the south of France was located in Corsica in 1977 (Husson

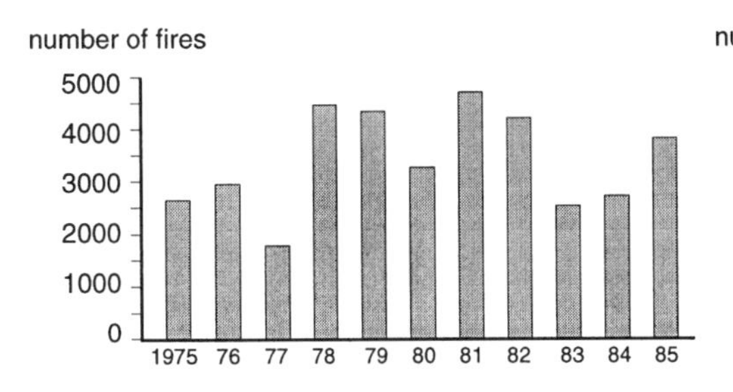

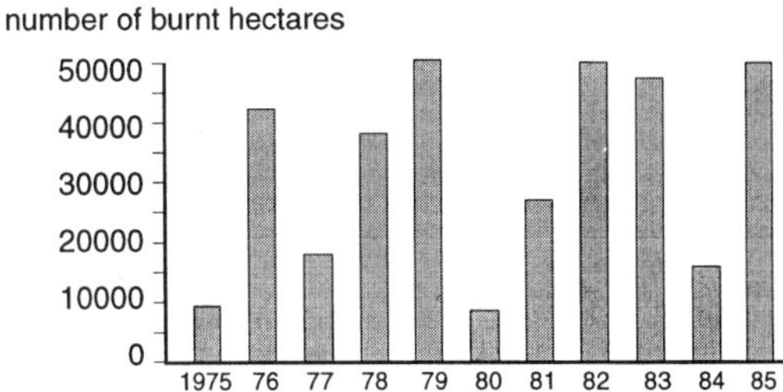

Figure 12.11 *Forest fires in the* Midi meditérranéen *and Corsica* (modified by J.-L. Ballais from Battesti 1991)

1985). More generally, from 1975 to 1985, the burnt areas in Corsica covered about 10 000 ha/yr (Battesti 1991). Nevertheless, in France, forest fires are not restricted to the Mediterranean part of the territory: all the French forests are subject to fire.

In the 25 years since 1970, about 3 000 000 ha of land of every type has been burnt in Italy (Bourriau 1993). The number of fires and affected areas from 1970 to 1988 display a positive trend. In 1993, 11 932 fires destroyed 104 385 ha. Most of such fires (about 75%) strike the Mediterranean regions, where there is a long, dry and hot season, during which most of the annual plants of the steppe and understorey die and dry out, making conditions extremely prone to fire. Sardinia, however, is the region where fire strikes most: from 1970 to 1990 there were 53 940 fires which destroyed 163 347 ha of forest, pasture and *macchia*.

Forest fires are also amongst the most important land degradation factors in Greece. They have become particularly frequent in the pine-dominated forests during the last 50 years. Of course, extensive fires also occurred previously. In 1888, for example, a fire in eastern Attica burned 75 000 ha (Alexandris 1989). Part of the vegetation was regenerated during the following years. More recently, in the period 1964–1975, the average extent of areas burnt each year was 12 900 ha (Alexandris 1989). The rate of burning increased dramatically in the following period 1976–1986, reaching an average of 37 800 ha per year.

Forest fires have increased in recent decades in Israel, with the increase of reforestation, abandonment of marginal mountain areas and the increase in natural reserves; on average, 5% of the forests in Israel are burnt each year.

In Morocco, fire destroys 3000 ha of forest and alfa (*Stipa* sp.) every year (Zitan 1987). In Algeria, fires also devastate big areas every year: 7202 ha in 1975; 25 927 in 1985; and 15 000 in 1990.

Fire has long been both a form and a primary cause of land degradation in Greater California, but its evaluation in a human context is problematic and uncertain. The fire landscapes of southern California and northern Baja California offer an informative contrast in ecosystem dynamics under protectionist management and a more traditional economy respectively (Minnich 1983). As public concern for watershed protection and wildfire control grew in southern California, so protectionist policies led unintentionally to undesirable changes in the ecosystem. *Chaparral*, woodland and coniferous forests thickened such that less frequent fires increased in magnitude and intensity, leading perversely to greater mass movement and flood impacts in denuded watersheds. In contrast, as a result of isolation and a small population, similar ecosystems in northern Baja California have experienced neither such intense exploitation nor protectionist management. Because there has been little fire control, burns are frequent, small and low in intensity. Thus the modern Baja California landscape probably resembles more closely the late 18th and 19th century landscape of southern California.

Since the 1970s, 68 300 forest fires have occurred in Chile, burning an area of 775 473 ha, of which 48 876 ha were tree plantations and 655 930 ha native vegetation. Of the natural areas burned, 39.4% were shrubs (*matorral*), 33.7%

pastures and 26.8% forest. The most affected regions have been Valparaíso (193 800 ha) and Bío-Bío (127 087 ha). Recovery of these terrains has been mostly nil in Coquimo and O'Higgins, while Maule and BíoBío have been afforested with exotic species, mostly pines and eucalyptus (MINAGRI 1994). In the burnt plantations the most affected species are *Pinus radiata* and *Eucalyptus*.

In many parts of Australia, bushfires, whether accidentally or deliberately lit, are a major concern, and the Mediterranean zones are particularly vulnerable with their hot, dry summers and combustible vegetation types. For example, Table 12.16 lists the damage caused by major bushfires in the Mt Lofty Ranges of South Australia, and Figure 12.12 shows the disturbing increase in the areas burnt by bushfires in the non-pastoral areas of South Australia. Webster (1986) has described areas the size of some European countries being burnt out during major conflagrations in South Australia since 1939.

Large bushfires have raged every 3–5 years in southern Australia over the past 50 years (Department of Primary Industries and Energy 1990). One huge fire in February 1983 ("Ash Wednesday") claimed 76 lives and destroyed 2463 houses from Victoria through to South Australia (W. McCarthy, Fire and Rescue Service WA, pers. comm.). The fire danger had been high to extreme for 30 days, with the fires starting on a day with temperatures exceeding 40 C, and burning out of control over an area of tens of thousands of hectares on a broad front across southern Victoria and southeastern South Australia. Each year in Australia, an average of 15 000 bushfires burn 23 000 km² of forest, grass, crops and townships, many of them in the southern Mediterranean region. Prior to the Ash Wednesday fire, 850 bushfires had burnt in Victoria and on that fatal day alone, 180 fires burned out of control (Webster 1986).

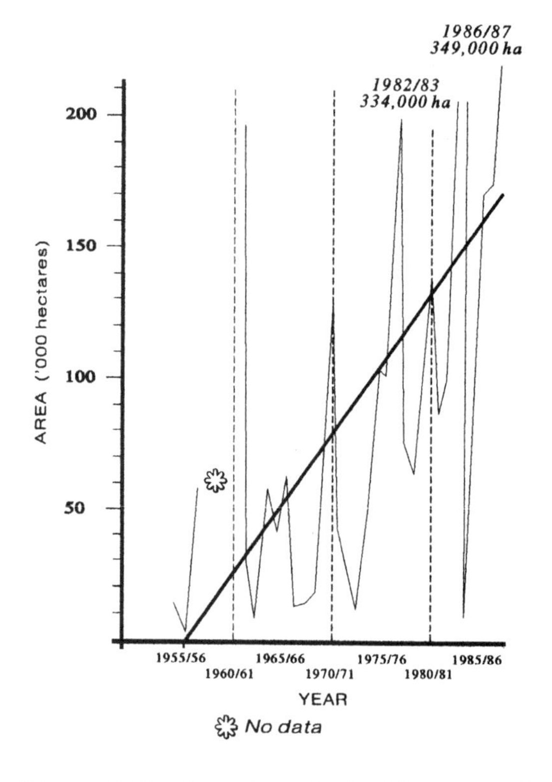

Figure 12.12 *Areas burnt by bushfires in South Australia, 1949/50 to 1986/87, excluding pastoral areas* (source: Environment Protection Council of South Australia 1988, figure 5.7)

However, the most extensive fires in Australia occur in the non-Mediterranean regions in the north of the continent. Here, fires burn most years over broad fronts of tens of kilometres, with durations of several weeks and even months.

Table 12.16 *Damage caused by major bushfires in the Mt Lofty Ranges* (source: Environmental Protection Council of South Australia 1988, table 5.6)

Year	Duration (days)	Area (ha)	Deaths	Houses burnt	Cost (US$ million)
1939	14	61 000	0	?	0.4
1955 (Black Sunday)	1	40 000	2	?	2.4–3.2
1980	1	3700	0	51	5.2
1983 (Ash Wednesday)	1	33 000	14	300	160

Biological invasions

The Mediterranean environments have been characterized by the translocation of plants from one region to another over many millenia, to the extent that it is difficult to state what is truly "indigenous" Mediterranean vegetation in any particular region (see, for example, Groves and di Castri 1991; di Castri *et al.* 1990).

Israel illustrates the situation nicely. Plant invasions probably started as a result of anthropic influences and there is evidence that oak (*Quercus calliprinos*) gave way to pine (*Pinus halepensis*) (Naveh and Vernet 1991). Pollens in Lake Kinneret sediments indicate a shift to cultural cultivars (olives, grapes) after the Bronze period, about 5000 years BP (Baruch 1987). The present major plant invasion is by *Eucalyptus*, a typical tree in most Mediterranean landscapes. Introduced from Australia about a century ago, mainly for swamp desiccation purposes, it has thrived in all kinds of environments, and more recently is also being used as a commercial timber. Over-grazing has wrought a strong reduction of primary productivity and standing crop, developing new habitats: *garrigue*, *maquis*, cultivated sites, and an evolutionary process characterized by a reduction of the vegetative apparatus towards a transition to human-modified habitats (Pignatti 1983).

These changes are perhaps being more acutely felt in the New World, where the proliferation of introduced plants and animals at the expense of native species is an inevitable consequence of human colonization from overseas. Among the more notable exotic plants in California, for example, blue gum (*Eucalyptus globulus*) and related species were widely introduced from Australia after 1850, mostly to provide windbreaks on farms. *Eucalyptus* grows readily on unstable sites and has been widely planted on shifting dunes and eroding hillslopes near the coast. Among the weeds, various members of the plantain family (*Plantago* spp.) have spread extensively in the wake of introduced domestic animals. Some plants such as cardoon (*Cynara cardunculus*), sweet fennel (*Foeniculum vulgare*) and Johnsongrass (*Sorghum halepense*) were introduced as food items or forage by persons of Mediterranean ancestry, only to escape into the wild as troublesome weeds. Many other weeds were introduced accidentally at an early date. For example, adobe bricks from Spanish buildings of the pre-Mission period contain seeds of filaree (*Erodium cicutarium*), curly dock (*Rumex crispus*) and sow thistle (*Sonchus asper*), all from Europe (Ornduff 1974). Notable arrivals during the Mission period included black mustard (*Brassica nigra*), bull thistle (*Cirsium vulgare*) and wild carrot (*Daucus carota*).

In the southwest Cape, invasion of the indigenous *fynbos* communities by exotic trees is widely regarded as one of the most serious forms of land degradation in the region (Richardson *et al.* 1992). The 12 most widespread invasives are all trees, many of which come from Australia; they have all spread within a relatively short time period across large areas (Table 12.17).

Richardson *et al.* (1992) concluded that certain invasive trees, especially *Hakea sericea* and *Pinus pinaster*, have caused major modifications to the montane communities in the region. The silky *Hakea*, for example, covers 4800 km^2, representing 14% of the area of mountain *fynbos*; and 19% of the quarter-degree grid squares of the *fynbos* biome as a whole have dense stands of the pinaster pine (Figure 12.13; Richardson *et al.* 1992). The situation in the lowlands is at least as severe, although the successful invading species in this case tend to be *Acacia saligna* and *Acacia cyclops*.

The introduction of exotic fauna also produces degradation problems; for example, in the southwest Cape, the Argentine ant *Iridomyrmex humilis* (origin, Brazil) has proved especially problematic, as it fails to bury the seeds of myrmecochorous (ant-dispersed) *Protea* species and hence renders them susceptible to granivory (seed predation), with consequences for species richness and plant community structure. Other introduced species have proved problematic in interfering with ecosystem processes and causing plant and animal extinctions, for example the North American grey squirrel (*Sciurus carolinensis*), the European starling (*Sturnus vulgaris*) and several species of alien mollusc (Macdonald and Richardson 1986). Much of the research attention, however, has been focused on alien trees and shrubs.

Chapters on Australia in di Castri *et al.* (1990) and Groves and di Castri (1991) provide details

Table 12.17 *Alien tree infestation in the southwestern Cape: the 12 most widespread exotic species* (source: Richardson *et al.* 1992)

Species	Region of origin	Year introduced	Reason introduced	Quarter degreee grids invaded (%)
Acacia saligna	Australia	1848	Botanic garden	60
Acacia longifolia	Australia	1827	Horticulture	42
Acacia cyclops	Australia	1857	Botanic garden	65
Pinus pinaster	Mediterranean	*c.*1680	Timber	30
Leptospermum laevigatum	Australia	1850	Dune stabilization	10
Hakea sericea	Australia	1858	Botanic garden	30
Pinus radiata	California	1865	Timber	17
Pinus halepensis	Mediterranean	1830	Timber	?
Paraserianthus lophantha	Australia	1835	Botanic garden	14
Acacia melanoxylon	Australia	1848	Botanic garden	26
Acacia mearnsii	Australia	1858	Botanic garden	47
Hakea gibbosa	Australia	1835	Botanic garden	4

of biological invasions in Australia's Mediterranean-type regions. In these regions, there are several hundred species of introduced weeds (many of which originate from other Mediterranean regions, especially the Mediterranean Basin; Scott 1991, p. 308–311), insect pests and diseases which adversely affect agricultural areas and native vegetation (Conacher and Conacher 1995). Feral cats and foxes are highly efficient predators of small native animals (Chapter 17). Other problem pests include introduced rabbits, rats, mice and feral cattle, horses, goats and pigs. Rabbit plagues have drastic effects on vegetation cover and aggravate soil erosion (Chapter 17), especially in the drier areas. And there have been at least 14 mouse plagues in Victorian cereal areas since 1900, causing multi-million dollar crop losses (Scott 1991, p. 316).

Some Australian species have also caused problems. In drier areas, growing kangaroo populations are increasing grazing pressure on the land. Some species of parrot cause severe damage to horticultural and cereal crops and are serious pests for farmers attempting to establish trees on their properties. These problems are symptoms of imbalanced ecosystems.

Pests are also problems in the Old World. In Croatia, for example, forest vegetation on karst is endangered by fungi (*Dothiostroma pini, Phellinus pini, Cenangium ferruginosum, Sphaeropsis sapinea, Cryphonectria parasitica*), insects (*Thaumatopoea pityocampa, Ryacionia buoliana, Diprion pini, Neodiprion sertifer, Corabeus bifasciatus, Lymantria dispar*) and other less important pests, as well as (especially) by fires (Halambek and Novak-Agbaba pers. comm.; Glavaš *et al.* 1994).

Loss of wetlands

The cumulative effect of these kinds of changes is the disruption of total ecosystems, with wetlands being a particular example. Many wetland areas of the Mediterranean regions are important overwintering or nesting habitats for migratory birds, and the loss or degradation of these areas seriously threatens many bird species.

Wetland loss is an important aspect of vegetation change in California. Though never extensive beyond San Francisco Bay, some 80% of the coastal salt marshes which existed in 1850 have been reclaimed for urban and industrial use. Between 1945 and 1965 alone, California lost 67% of its coastal wetlands to development, mostly around San Francisco Bay and between Los Angeles and San Diego. Some 96% of the interior freshwater wetlands in the Central Valley have been lost to agriculture.

In South Africa, Cowan (1995) has reviewed the situation nationally. Wetlands have clearly

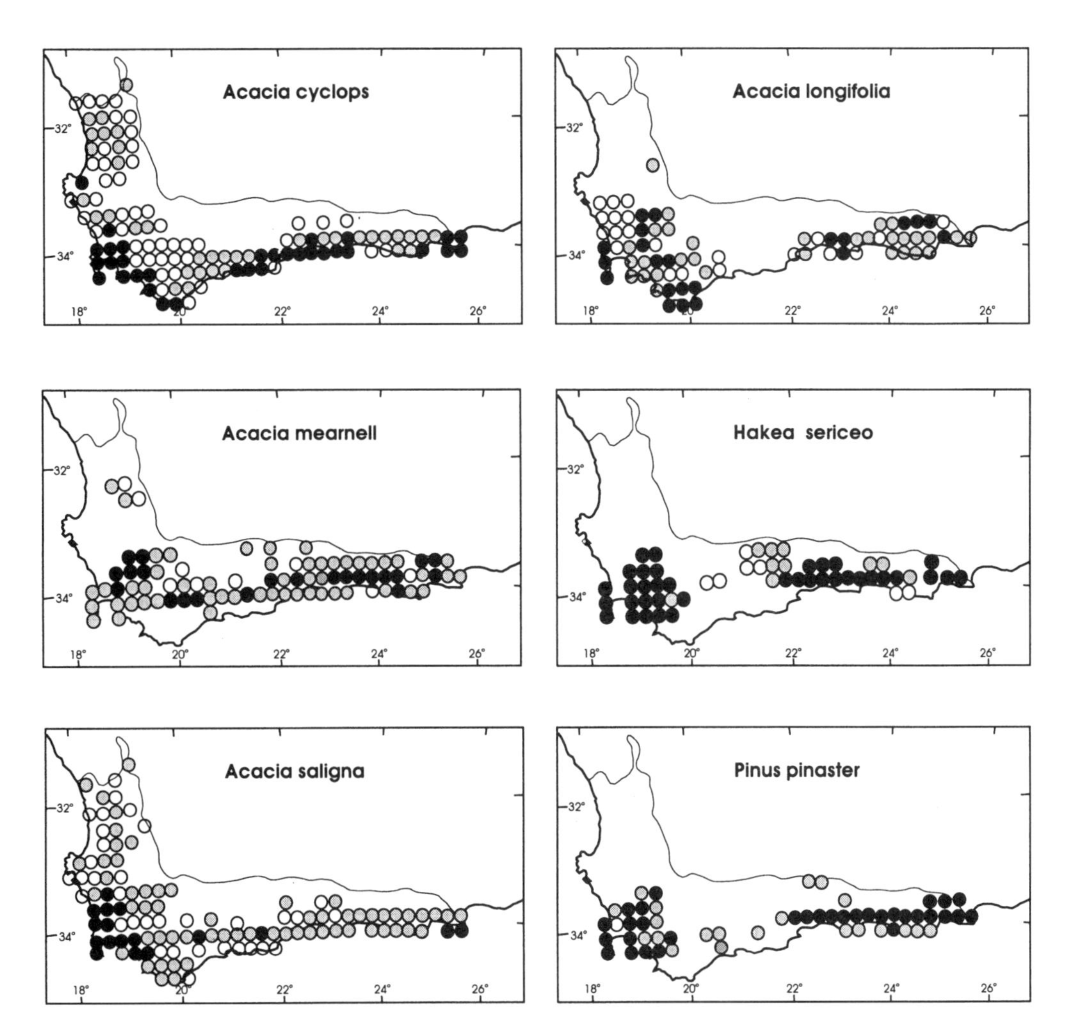

Figure 12.13 *The distribution of important invasive plant species in the southwestern Cape. Solid circles = quarter degree squares with dense stands greater than 1 ha in extent; grey circles = quarter-degree squares with moderately dense stands less than 1 ha in extent; open circles = quarter-degree squares with scattered plants only* (after Richardson *et al.* 1992)

been lost at a rapid rate under the various impacts of, for example, agricultural development (including irrigation and addition of fertilizers), invasion of catchments by exotic invasives, erosion (incorporating increased sediment supply), the construction of dams, and urban development. The losses are difficult to quantify, because many areas that were once wetlands were impacted without ever having been recorded as such in the first place. The square in front of the Cape Town City Hall was once a wetland with, among other wildlife, a thriving hippopotamus population! (Begg 1986, cited in Cowan 1995). Cowan (1995) estimates that more than 50% of the southwestern Cape wetlands and riverine ecosystems have been seriously degraded or lost altogether and that the condition of the region's estuaries, at the receiving end of negative catchment developments, is a particular cause for concern.

More than 70% of the original wetlands on the Swan coastal plain in Western Australia have been drained, filled or polluted. In South Australia, Jensen (1986) has discussed degradation problems of drainage of wetlands and loss of wetland habitat, flooding and pollution. In the southeast of the state, approximately 53% of formerly flood-prone land is now available for agriculture. But this extensive drainage programme has also resulted in the loss of all except 7% of former, permanent wetland areas. And between a third and half of Victoria's original wetlands have been destroyed or altered by drainage, infill, use as evaporation basins or sinks for chemical pollutants from industry, mining or agriculture (Scott 1991).

13

The main problems of land degradation: their nature, extent and severity

2: Drought, water shortages and water quality, and other forms of degradation

---

Compiled by Arthur Conacher and Maria Sala from contributions on the following regions:

Iberian Peninsula and Balearic Islands	**Maria Sala** and **Celeste Coelho**
South of France and Corsica	**Jean-Louis Ballais**
Italy, Sicily and Sardinia	**Marino Sorriso-Valvo**
Greece	**Constantinos Kosmas, N.G. Danalatos** and **A. Mizara**
Croatian Adriatic coast	**Jela Bilandžija, Matija Franković, Dražen Kaučić**
Eastern Mediterranean	**Moshe Inbar**
North Africa	**Abdellah Laouina**
California	**Antony** and **Amalie Jo Orme**
Southwestern Cape	**Mike Meadows**
Southern Australia	**Arthur** and **Jeanette Conacher**

Drought

Annual summer drought is characteristic of Mediterranean environments, and is both a problem in its own right and a cause of others, notably water shortages (discussed in the following section). Drought severity varies, however, and can reach serious proportions.

The North African situation is illustrative of a number of Mediterranean regions, particularly those which border on more arid regions (Spain, the eastern Mediterranean, southern California, the northern parts of Mediterranean Chile, the northern parts of the southwest Cape, and southern Australia).

Drought is present everywhere in North Africa, at least temporarily. It introduces severe constraints and can be catastrophic. This is explained by the fact that the Mediterranean Maghreb is on the fringe of the desert. Drought lasts everywhere for at least three months, during which rains are negligible (apart from the late summer storms); at the same time temperatures are excessive. The *chergui* (the east wind) and the south wind bring Saharan temperatures to the coast. Evapotranspiration consumes more than 70% of the rains and 90% on the south side of the Tell (Despois and Raynal 1967). Hence there is a general lack of water in the spring/summer period, when evapotranspiration increases: the landscape becomes uniform and dusty winds become the rule (Laouina 1982).

Even the rainy season consists of a succession of more or less long periods of fine weather with rainy passages. Stable conditions may occur in winter throughout the Maghreb, but a real contrast between western and eastern North Africa is often apparent.

The annual drought affects the whole economy. These may be cases of global deficit representing real catastrophes, or cases of inter-seasonal irregularity (a lag in the onset of the rains, a winter deficit of soil moisture storage, delayed spring rain, or low snowfalls, resulting in a deficit in the growth of meadow herbs and reduced replenishment of aquifers).

Since the beginning of the 20th century, Morocco, for example, has experienced practically balanced phases of precipitation surpluses and deficits. This balance seems to have broken since 1975, as the number of dry years has a tendency to exceed the humid ones. In some cases, drought extends over several years. For example, the 1980–1984 drought in Morocco was perhaps the most severe experienced in the country over a period of 1000 years (Stockton 1988). In those cases, water scarcity becomes obvious with falling water tables, exhaustion of springs and declining river flows. The soils then lack water reserves and a high mortality of trees can be recorded. Dam storage deficits reveal the severity of the drought (Table 13.1). The average water deficit experienced by the agricultural regions was 28% in comparison with the normal average. The effect was also important for groundwater, because many springs dried up.

In the Algerian Tell, the frequency of the years of high drought is 10% on the coast and 20% in the Tell (Bensaad 1993). The 110 year old Tunis–Manoubia station has recorded 50% of deficit years (Benzarti 1987), with 11 drought sequences of more than one year, the longest of which lasted for eight consecutive years (1940–1947). Among the deficit years, 36% have a deficit of more than 10%. Spatially, the northwest of Tunisia experiences the lowest variation coefficient (standard deviation/average). The coefficient is 0.186 in the region which receives a mean annual rainfall of 663 mm, whereas it is 0.235 for 525 mm in the northeast, 0.360 for 292 mm in the Sahel and 0.358 for 285 mm in the centre of the country.

Table 13.1 *Deficits in Moroccan dam storages during the 1980–1984 drought* (source: Stockton 1985)

Region	*Oued*	Deficit (%)
Middle Atlas/Tadla	El Abid	55
Eastern Morocco	Moulouya	46
Pre-Riffian hills	Inaouen	80

Land Degradation in Mediterranean Environments of the World: Nature and Extent, Causes and Solutions.
Edited by A.J. Conacher and M. Sala. 1998 John Wiley & Sons Ltd.

Water Shortages

Water shortages and droughts are not, of course, restricted to North Africa. All Mediterranean regions experience them to greater or lesser degrees. In the Iberian Peninsula, for example, there is an increasing demand for water from agricultural, industrial and urban uses. Except for the north and northwest fringe and some parts of the main drainage basins, economic development is not being matched by the development of water resources.

The problem is the uneven seasonal and spatial distribution of water resources within the country. For example, regulated water resources amount to 1233 m^3/person/yr (4174 m^3/person/yr in the Ebro basin, 3798 m^3/person/yr in the Duero, but only 363 on the SE coast and 301 on the NE coast). Present demand averages 3 m^3/person/day, of which two-thirds is for agricultural uses and one-third for industrial and urban uses. The demand estimated for the year 2000 is 45 042 Hm^3/yr, similar to the total available resources (45 185 Hm^3/yr). However, while the resources of the Duero, Tajo and Ebro basins are regulated and exceed demand, elsewhere (except for the Segura basin which is already in deficit) resources are sufficient for the expected demand only to the year 2000. Development problems can be expected in the future for the Mediterranean coastal areas and the Balearic Islands, and over-exploitation of aquifers is already taking place (Figure 13.1).

In Greece, too, water consumption will be further increased following the continuous increase in tourism. Combined with the substantial increase in water requirements associated with high-input agriculture, this will bring about a significant water allocation problem with further degradation of the plains through salinization.

If one were to define a single resource that lies at the roots of Greater California's rapid development and population growth, that resource would be water. Despite summer drought, early farm towns such as Los Angeles could be established where mountain streams descended steeply to the lowlands, replenishing lowland aquifers with water that with modest effort could be used throughout the year. However, there were limits to these resources, not only in terms of discharge but because flashy runoff often shed water and debris quickly seawards, causing floods but allowing little time for groundwater replenishment. Thus, if agriculture was to prosper and towns were to grow to meet the visions of their developers, larger and more reliable water resources had to be found.

Los Angeles was among the first communities to recognize the value of seeking water closer to its source and, accordingly, between 1907 and 1913 completed a 375 km aqueduct to bring Sierra Nevada waters to Los Angeles via the Owens Valley (Figure 13.2). In 1940 this was extended northward to the Mono basin for a total aqueduct length of 544 km, and a second aqueduct was completed in 1969. Meanwhile, between 1931 and 1941, the Colorado River began to be tapped by Los Angeles and other communities in southern California, including San Diego in the late 1940s. Further north, shortly after 1900, cities in the San Francisco Bay area also began seeking ways to expand their water supplies, mostly through reservoirs on the western slopes of the Sierra Nevada and along the margins of the Central Valley. For example, after a long political struggle, authorization for the Hetch Hetchy reservoir in the Sierra Nevada was given in 1915 but water supplies to San Francisco only began in the 1930s.

Shortly afterwards, farms and towns in the Central Valley began to benefit from an ambitious project that incorporated the state's largest reservoir at Lake Shasta, completed in 1945 with a capacity of $5.6 \cdot 10^9$ m^3. These waters supplied important needs but growth in turn created a thirst for more water and the desire for a more co-ordinated approach to water supply problems. This led in 1957 to authorization of the California Water Project, designed to move water through a system of natural streams and aqueducts from the Sacramento River basin into the San Joaquin Valley and southern California. Oroville Dam on the Feather River, at 230 m the highest dam in the United States, was completed in 1968 as part of this project.

Such developments did not occur without conflict, at first between competing users and later between water interests and environmental concerns. Northern Californians began to perceive that their water was being diverted to support

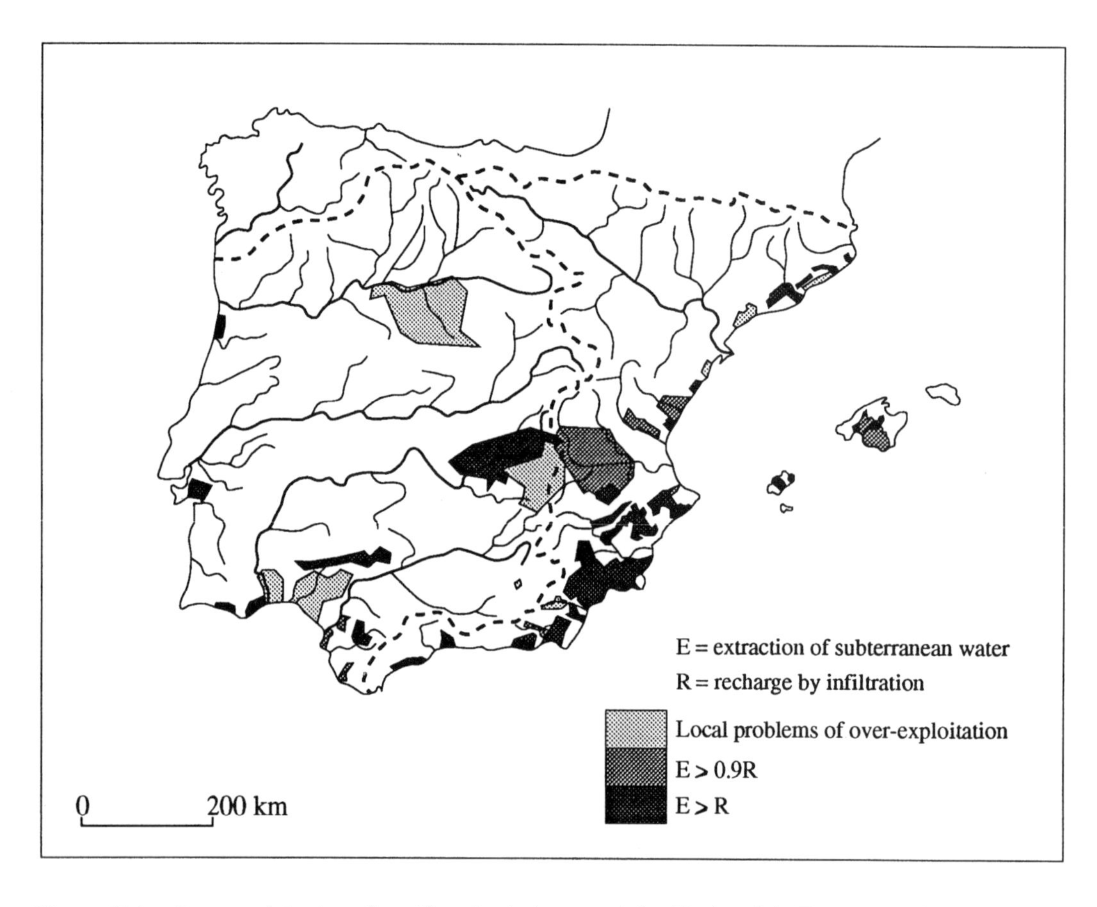

Figure 13.1 *Over-exploitation of aquifers, Spain* (source: Atlas Nacional de España 1996)

southern California's irrational growth, a belief that even led to suggestions for partition of the state. In reality, much of the captured water was actually being used by agricultural interests in the Central Valley, leading to progressive soil salinization and diminishing crop yields. Nevertheless, the vast Los Angeles and San Diego metropolitan areas could not have reached their present size without importing water for agriculture, industry and people. Water use is a contentious issue, highlighted during prolonged drought by water rationing and calls for improved recycling, desalinization of ocean water, and more rational consumption.

The water problem: regional and environmental conflict

Many Mediterranean countries experience natural water shortages during summer drought periods and in some years this deficit is acute. However, a number of other factors exacerbate water scarcity or availability to the point of crisis. These factors include: prolonged periods of drought; increasing population demands (agricultural, industrial and urban needs) which reduce supply and adversely affect quality; issues of sovereignty where water supply transcends regional or national boundaries; problems of upstream interception (dams, diversion, drainage, pumping) causing gross interruptions to hydrological regimes, or water pollution (see following section). In extreme cases such problems may give rise to international conflict and, as pressures increase, could forseeably witness countries go to war.

It is in the eastern Mediterranean where water shortages are perhaps most serious, particularly given the international implications of water resource developments in a politically unstable

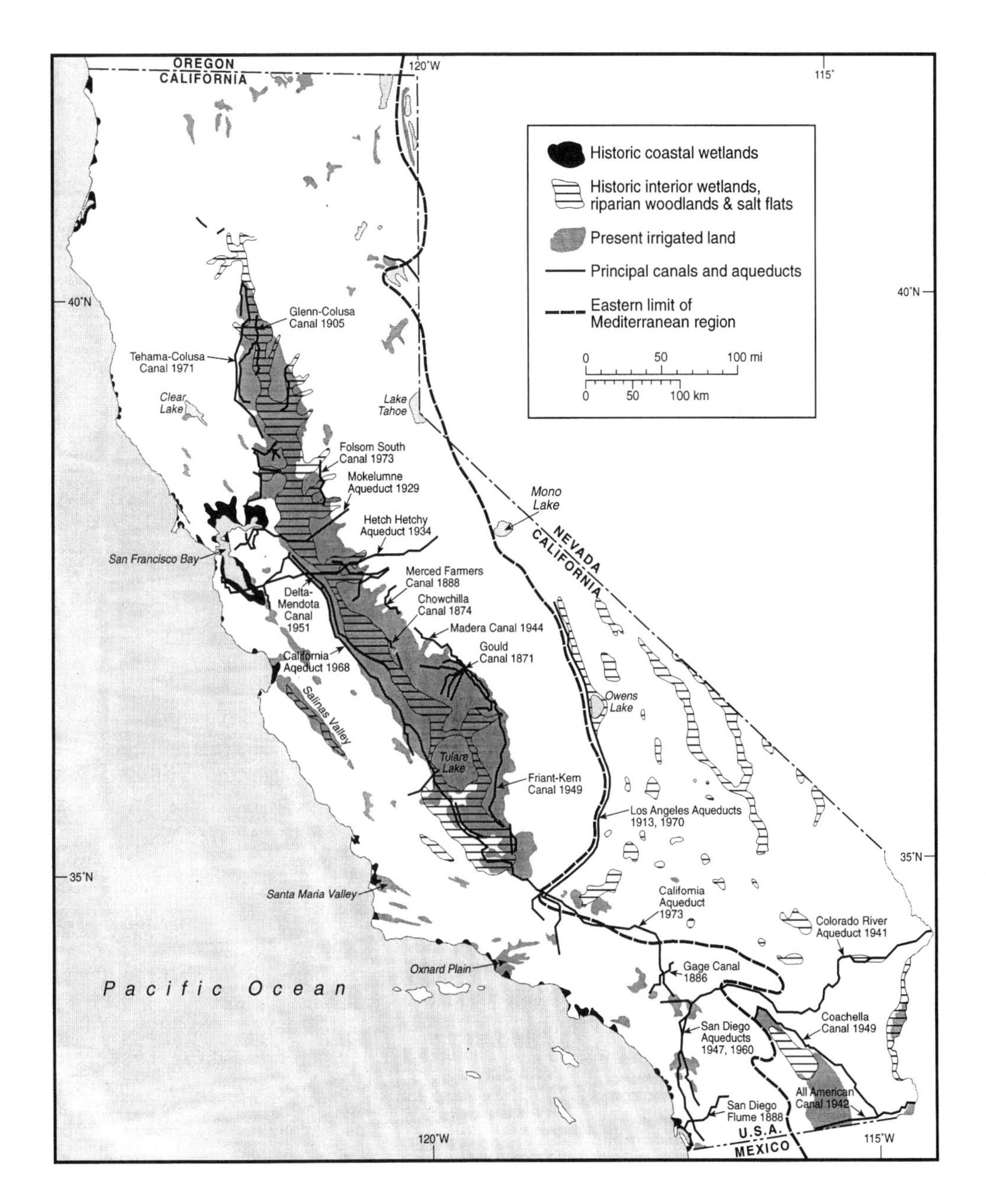

Figure 13.2 *California water use showing selected aqueducts and canals, historic wetlands and present irrigated areas* (based in part on Kahrl (1979), with additions)

region. The depletion or pollution of water resources is probably the major problem of land degradation in the countries of the old Fertile Crescent. In an area where rainfall is sparse and there is a shortage of water, natural processes are exacerbated by population demand, political tensions, national conflicts and disputes that have repeatedly become the focus of political and military struggles. Water sources and catchment areas, river basin boundaries and river channels often transcend political borders, but countries usually consider water as a national resource at their sovereign disposal. A probable global climate change in the near future may reduce rainfall and bring the situation to a political or even military crisis.

The average annual growth rate of population in the Middle East is 3.2%, but the rate is higher for Jordan (3.6%) and for the Palestinian population (5%) which is one of the most rapid in the world. By 2005, the population of Jordan, Israel and the Palestinians will be more than 18 million, compared with 10 million in 1995 (Naff 1993). Water resources are limited and will not increase to match population growth. In this area water resources are already used to almost 100% of their potential. As in other developing countries, the main problem lies in social structures rather than in the limited economic resources.

Syria is illustrative of the problem. More than 80% of the country is in the arid or semi-arid climate belt with less than 200 mm of rainfall per year. Water resources are scarce, except for the northeastern area crossed by the Euphrates River, and a small border region where the Tigris River flows (Figure 13.3). Cultivated land in Syria extends over about 6 100 000 ha, of which 50% is in the Euphrates basin; in 1991, 12% or 700 000 ha were irrigated (Table 13.2; Wakil 1993). The Syrian claims are for 15 · 10^9 m^3 of the Euphrates River flow, not including the Khabur tributary with another 2.2 · 10^9 m^3. Available water resources total 23 · 10^9 m^3 (Table 13.3), two-thirds of which are from the Euphrates basin. Demands during initial negotiations by Turkey, Syria and Iraq from the Euphrates waters were respectively 14 · 10^9, 18 · 10^9 and 13 · 10^9 m^3, amounting to 45 · 10^9 m^3. Since the average flow is only 32 · 10^9 m^3, an eventual compromise from all parties is clearly required.

Table 13.2 *Actual (1989) and projected irrigation areas, Syria: 10^6 m^3* (source: Wakil 1993)

Basin	Actual (10^3 ha)	Projected (10^3 ha)
Khabur	130	205
Euphrates	144	594
Aleppo	98	98
Orontes	155	237.5
Coastal	50	67.5
Damascus	75	82.3
Upper Jordan	18	30.0
Desert	–	20.0
Total	670	1350

The Euphrates and Tigris river waters account for 80% of the country's water resources. The natural inflow into Syria of the Euphrates is 30 · 10^9 m^3, but Syria has to leave most of the flow to the downstream neighbour, Iraq, in order to prevent a conflict. A large dam, the Tabqa, with a storage capacity of 12 · 10^9 m^3 was built by 1974 to irrigate about 400 000 ha; by raising the dam crest by 20 m the dam will supply water to irrigate 600 000 ha. A second regulating dam, Al Baat, was also completed and a third, the Tishrin dam, is under construction. Other projects are being developed in the tributary to the Euphrates, the Khabur River, that will irrigate 150 000 ha by the building of three dams to be completed in the 1990s.

During dry years, the decrease of flow affects both downstream countries in the system, Syria and Iraq, aggravating the problem. Syria plans a further dam, upstream of the Tabqa dam and Lake Assad. Other Syrian development plans include the Orontes, flowing to the Turkish Mediterranean coast, and tributaries of the Yarmouk

Table 13.3 *Available water resources in Syria: 10^6 m^3* (source: Wakil 1993)

Basin	Surface water and springs	Groundwater	Total
Khabur	1695	500	2195
Orontes	2609	356	2865
Coastal	2386	236	2622
Desert	125	100	225
Euphrates*	13 000	300	13 300
Total	21 575	2038	23 613

* Assumed Syrian share

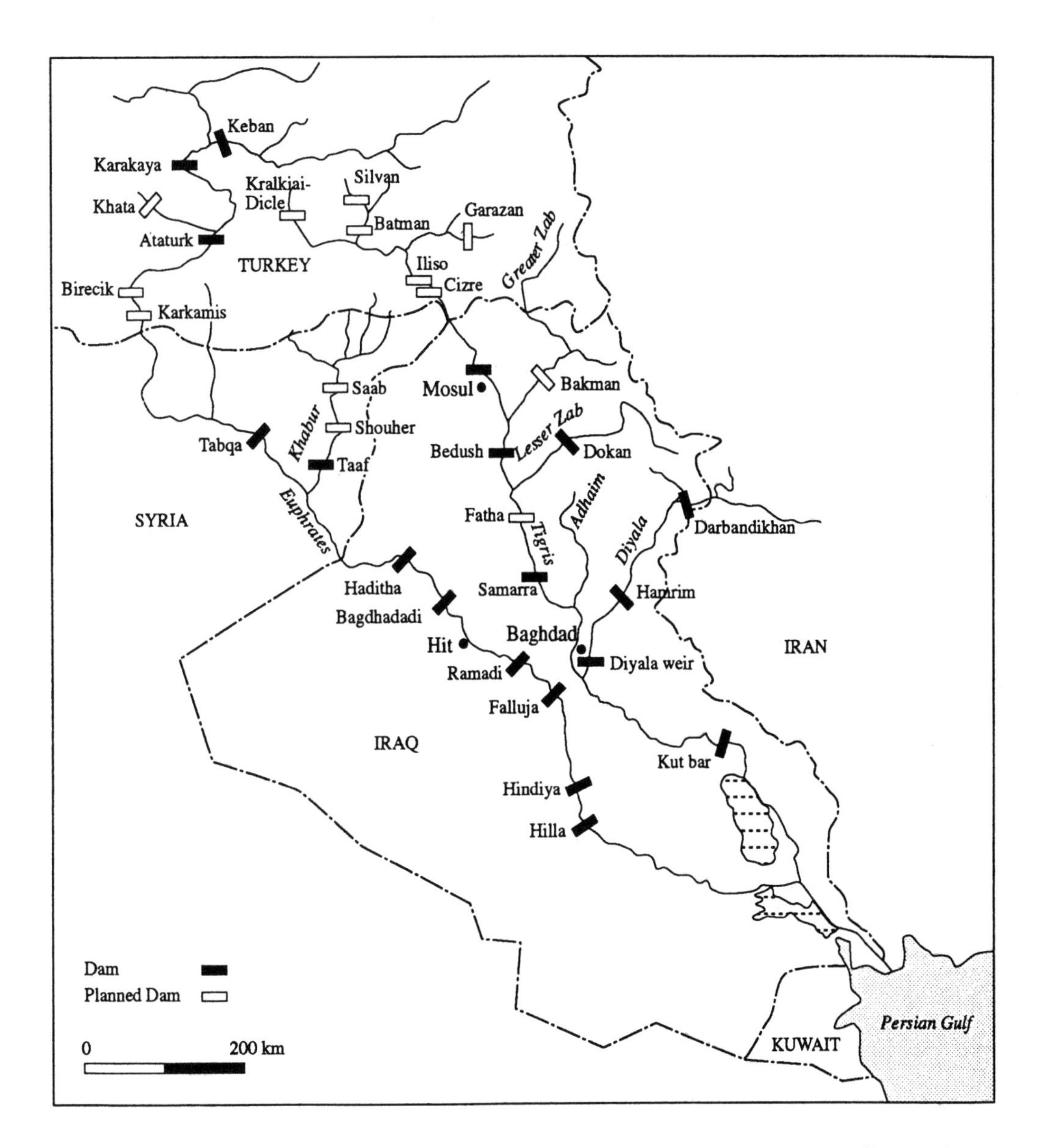

Figure 13.3 *Major dams in the Euphrates and Tigris drainage basins* (modified from Soffer 1992b)

River in the south, which flows into the Lower Jordan River and affects primarily the water needs of Jordan. In this case Syria is in the basin headwaters and less willing to compromise with Israel and Jordan, the downstream countries.

The political water problem between Turkey, Syria and Iraq exemplifies many of the problems of water resources management between riparian states. Without a comprehensive agreement, the allocation of water and the deterioration of water quality through increasing amounts of salts, fertilizers and other residues, may create dangerous situations.

Further south, the Lower Jordan River flows in a meandering pattern along a straight distance of 100 km between Lake Kinneret (the Sea of Galilee) and the Dead Sea. The natural flow, including the inflow of the Yarmouk River, was $1.2 \cdot 10^9$ m^3. By using Lake Kinneret as a natural dam and diverting or using almost all of the Jordan flow, as well as a large part of the Yarmouk and other tributaries being diverted by the East Ghor Canal, the total Jordan River inflow into the Dead Sea has decreased to only 300–400 $\cdot$ 10^6 m^3 yearly. Moreover, the quality of the river water has deteriorated as water from saline springs and domestic and agricultural effluents have transformed the once sacred river into a channel with brackish water, containing up to 3000 ppm of salts.

Due to the reduced flow of water into the endoreic system of the Dead Sea, its water level drops by an average of 0.4 m/yr, reaching 409 m below sea level in 1995, about 19 m below the average level during the first half of the century. Its southern shallow part, with an elevation of –405 m, is used as artificial evaporating ponds for potash mining, pumping water from the deep northern basin of the sea. The planned Mediterranean–Dead Sea canal, or the alternative Red-Dead Sea canal, proposed to generate hydro-electric power, may bring the Dead Sea back to its historical level.

Both Lebanon and Cyprus have relatively large water resources with values of 1200 m^3/person/yr and 2000 m^3/person/yr respectively, being located in a region with over 600 mm average annual precipitation. However, population increase and rapid rates of economic development are creating problems relating to the impact of dam construction on the environment downstream from the dams.

The most important water development scheme in Lebanon was on the Litani River, which has an annual flow of 700 $\cdot$ 10^6 m^3 (Naff and Matson 1984), by the construction of the Qir'un Dam, completed in 1966 with a storage capacity of 220 $\cdot$ 10^6 m^3, mainly for hydro-electric power generation. Seventy-five percent of the water is diverted by tunnel to the Awali River, reducing irrigation possibilities in the Bekaa area and on the coastal plain, but increasing hydro-electric power production (Naff and Matson 1984). The use and diversion of the water altered the natural hydrological regime and changed the downstream channel environment. Irrigation development schemes in the Upper Bekaa valley and along the narrow coastal strip may increase the demand for water and degrade water quality by irrigation return water.

Water Pollution

Problems relating to water shortages are compounded by deteriorating water quality, which is a global problem. However, the seasonality of Mediterranean climates means that pollutants may accumulate during the high temperature months with low or negligible flow, exacerbating problems of both quality and quantity. A heavy reliance on irrigation in some areas compounds the problem.

In the Iberian Peninsula, river pollution occurs in industrial and urban areas, mainly in relation to chemical industries, especially paper plants, the heating of water in nuclear plants, industrial waste (grape pulp, heavy metals) and urban waste. It is particularly serious in the surroundings of Barcelona (Besos and Llobregat rivers). The other sources of water pollution are agricultural inputs of fertilizers and pesticides (Table 12.9). The Llobregat River, north of Barcelona, is one of the most polluted rivers of the Peninsula, followed by the Guadalquivir and Guadiana, while the rivers with the best water quality are the Ebro and Duero. Other problems of water pollution are the eutrophication of dams.

Aquifers are more protected than surficial waters but when pollution occurs it is generally irreversible. The main problems of groundwater pollution have the same origin as river pollution. The most affected areas are the coastal Mediterranean, especially the NE (Catalonia) and SE areas (Andalucía), the Balearic Islands, the Jucar River (Valencia) and parts of the Guadiana (Campos de Montiel, Badajoz), Tajo (Madrid–Toledo), and Duero (Zamora–Salamanca) rivers. In the Balearic Islands and in Catalonia, over-exploitation leads to marine intrusions (Figure 13.4).

There are problems of surface water quality resulting from industry and urbanization in littoral areas of Portugal and from intensive agriculture in Algarve. North of Douro, rivers and streams are heavily polluted by textile and dye industries. Tannery, abattoir wastes and untreated sewage pollute small streams north of Lisbon, whilst pulp and paper factories pollute the rivers Vouga and Tejo. The Guadiana waters are polluted by urban development, food industry and agriculture, partly coming from Spain, causing eutrophication (*azolla* blooms) in river

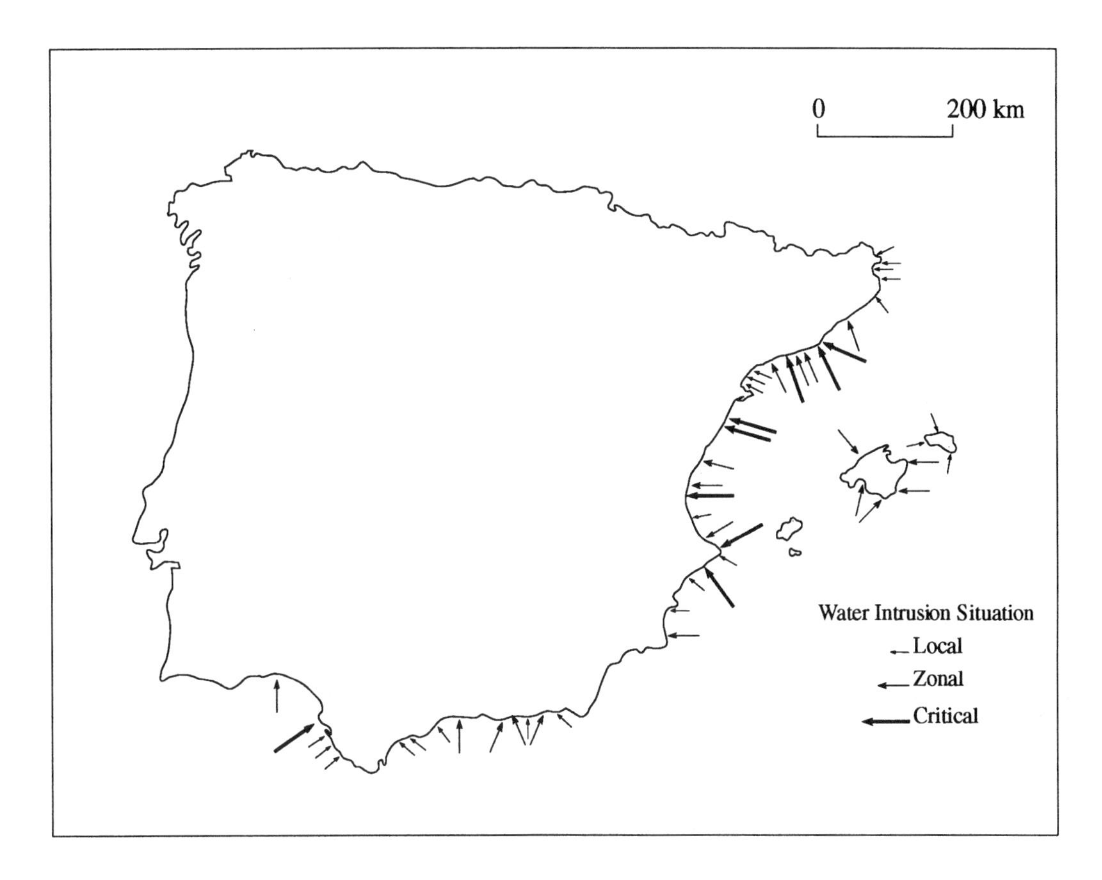

Figure 13.4 *Marine intrusions into groundwater, Spain* (source: Atlas Nacional de España 1996)

segments, especially during drought years. In Alentejo pollution comes from abandoned mining activities. Some groundwaters have been over-exploited and polluted in Lisbon, Setúbal and Algarve (limestone and karst aquifers). The coastal waters and estuaries around Oporto, Lisbon and Setúbal are also affected by untreated sewage and industrial wastes.

In the south of France, at the beginning of the 1980s, water pollution mainly concerned coastal areas (Figures 13.5 and 13.6), with the exceptions of the Durance (locally), the Colostre and the Coulon rivers and some very small rivers such as the Sorgues, Meyne and Argens. In particular, the Huveaune, Arc, Touloubre, Loup, Var and the Paillon rivers were locally very polluted (Pelissier 1980).

The area of Istria and the northern Adriatic have 50 sampling points where water quality is monitored. Water purity is still satisfactory. The upper parts of these streams are mostly clean, but become moderate to highly polluted downstream as a result of discharges of untreated or inadequately treated industrial and household waters, as well as runoff from agricultural land (there are no data on non-point pollution sources).

In the territory of Dalmatia, the water quality of streams and lakes has been monitored periodically since 1961 and continuously since 1972. As in Istria, the upper parts of streams are less polluted than downstream or near effluent outlets (State Directorate for Environment 1994).

The purity of groundwater in Croatia is currently acceptable, but it is endangered by uncontrolled water discharges and/or waste disposal because of permeable ground layers, especially during the rainy season.

In contrast, groundwater pollution is a major aspect of land degradation in Greece. Measurements of nitrate concentrations in 146 wells

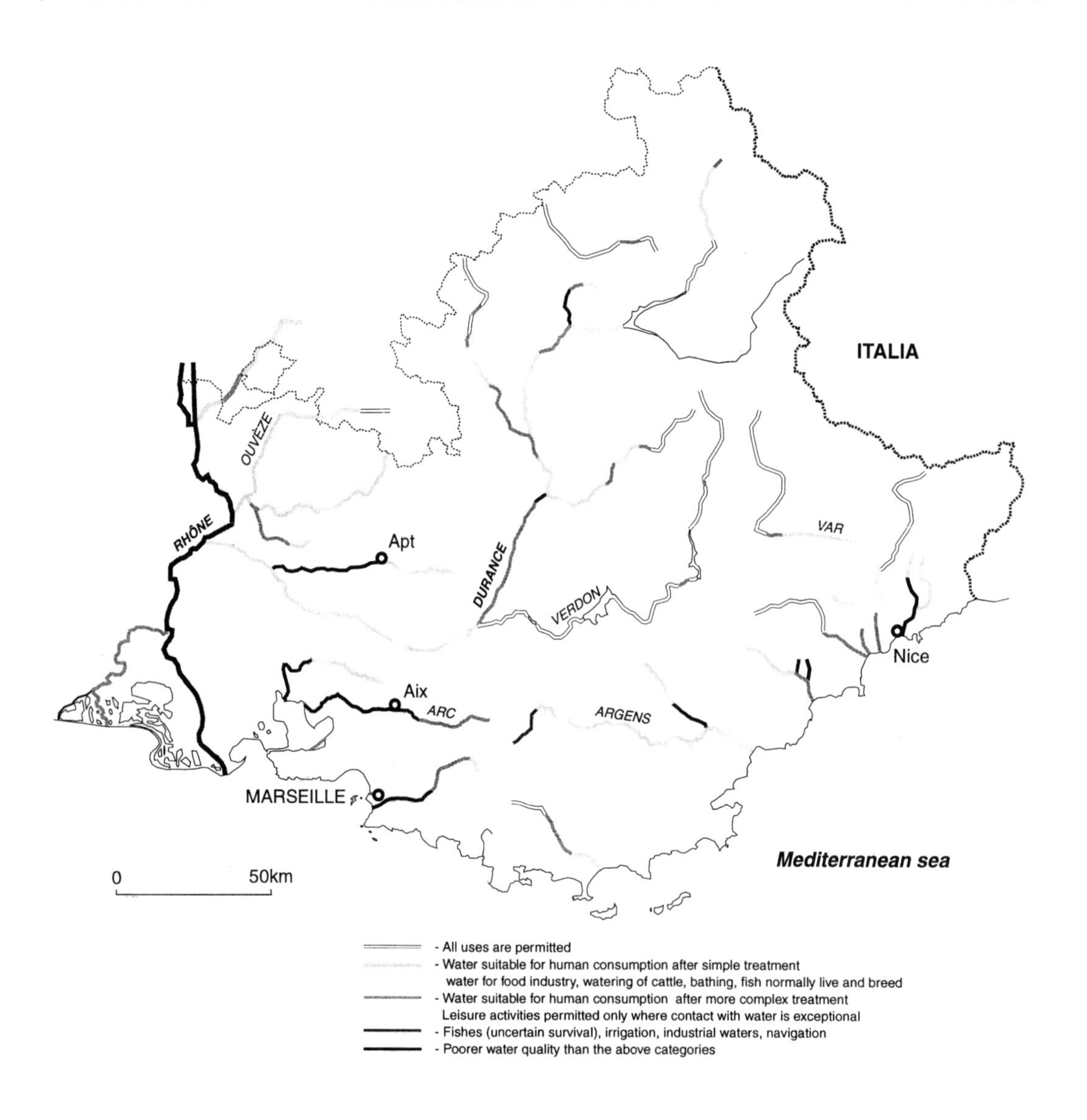

Figure 13.5 *Physico-chemical quality of the water of the river systems east of the Rhône* (modified from Pelissier 1980)

across the country revealed that 50% of wells had concentrations less than 50 mg/l, 26% had concentrations ranging from 50 to 100 mg/l, and nitrate concentrations exceeding 100 mg/l were found in 24% of the wells (Kosmas and Moustakas, unpublished data). On the other hand, phosphates were found in only trace amounts.

Salinization of water has probably been the most complex problem encountered in the various water development projects of the eastern Mediterranean countries. For example, in 1995 the salinity of Lake Kinneret was approximately 220 ppm (parts per million chlorides), compared with 20 ppm for the Jordan water's inflow into the lake (Plate 13.1). Although the 220 ppm level can be tolerated for domestic and some agricultural uses, it is considered high for the irrigation of citrus groves and subtropical fruits. The diversion of saline springs from the shores of Lake Kinneret reduced a yearly inflow of 50 000 t of salts into the lake. On the other hand a gradual

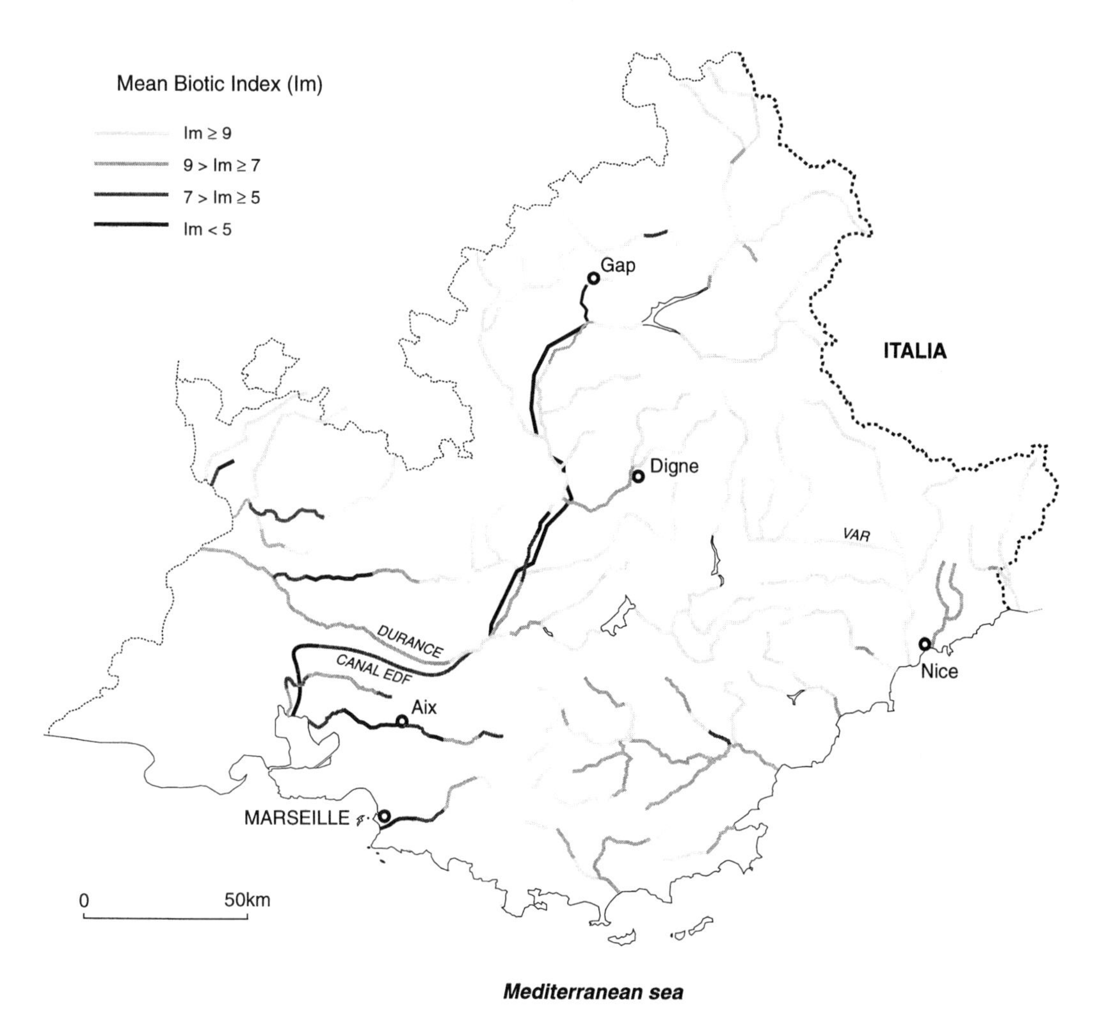

Figure 13.6 *Biological quality of the water of the river systems east of the Rhône* (modified from Pelissier 1980)

deterioration of water quality occurred along the Upper Jordan River: as a result of the drainage of the Hula Valley and the use of fertilizers, more nutrients, especially nitrates and phosphates, are reaching Lake Kinneret. Persistent pesticides flush into the lake, constituting another health hazard (Schick and Inbar 1972). Increasing urbanization in the catchment area is particularly critical. Some quantities of untreated sewage from uncontrolled sources may occasionally reach the lake.

Another problem is the rapidly expanding role of the Upper Jordan basin and Lake Kinneret as an inland recreation area. The environmental impact of recreational activities is expected to compound the problems of water quality produced by agricultural and urban effluents. Last but not least are the changes wrought by the retention of water which may affect the natural and historical landscapes for which the Jordan River is famous.

Nitrogen is a major pollutant of groundwater in Morocco. Between 10 000 and 13 000 t/yr of nitrogen migrate towards the water table and the rivers, mainly in irrigated lands (Conseil Superieur de l'Eau 1991c). In Tadla, groundwater is seriously polluted by fertilizers and insecticides through the leaching of irrigation waters. Pollution by nitrates varies from well to well, with concentrations ranging from 10 to 50 mg/l. In

Plate 13.1 *Inflow of the River Jordan to Lake Kinneret (Sea of Galilee)*

some sectors, the concentration of nitrates tends to increase by 5 mg/l each year. The origins of this surplus of nitrates are 20 kg/ha/yr from rain and irrigation waters, 80 kg from the mineralization of organic nitrogen, and 100 kg from fertilizers. In many sites, the tolerable threshold has already been exceeded.

In California, water degradation generally occurs when the natural system is changed by vegetation clearance and subsequent agricultural, industrial and urban land uses. Its expression in the landscape may be found in changes in stream discharge and load, water temperature, and contamination by pollutants including bacteria and trace metals, leading to nutrient enrichment and increased organic carbon content. The influx of pesticides, high total dissolved solids (TDS), and elevated levels of PCBs in surface waters may also accompany the introduction of agriculture, ranching and urbanization. For example, though banned some 20 years ago, residual pesticides such as DDT, toxaphene, dieldrin and chlordane are found in tributaries to Mugu Lagoon in southern California. Table 13.4 lists TDS, salts,

Table 13.4 *Recommended amounts and measured concentrations of chemicals in the Calleguas Creek watershed* (source: USDA 1995)

Components	Recommended (mg/l)	Calleguas Creek (mg/l)	Revolon Slough (mg/l)	Arroyo Las Posas (mg/l)	Arroyo Simi (mg/l)
Sodium	100	10–126	327–600	180	9–270
Calcium	50–150	>56	225–466	130	24–300
Magnesium	50–200	2–33	73–180	43	4–100
Chloride	250	20–200	131–185	5–190	9–205
Sulphate	250	54–1550	1083–2325	20–440	18–1100
Nitrate	45	0–35	0.4–248	0–50	6–20
TDS	500	118–702	2160–4623	1180	156–2275
DDT, DDD, DDE	1.0	2.0	2.0		
Dieldrin	1.9	2.0	2.0		
PCBs	14	30	30		

DDT and related pesticides, and PCBs for several watersheds in Ventura County.

Degradation linked to water use also results from the discharge of hot water by thermal power plants, mostly powered by petroleum and natural gas but including six nuclear plants which produce 23.3% of the state's electricity. Most of these are along the coast where hot-water discharge has caused significant ecological change.

The main water pollution problem in Chile is related to the lack of treatment of used waters. Of particular importance is the contamination of the waters of the Metropolitan Region, where an important percentage of the country's orchards are located. Table 13.5 shows the contents of faecal coliforms in the surface waters.

The quality of surface and groundwater resources in the southwest Cape is initially highly dependent on geology, although in the case of surface waters, many streams in the region are seriously polluted, particularly in their lower reaches. The rivers have also been subject to the spread of alien aquatic plants such as *Myriophyllum aquaticum* (parrot's feather), *Eichornia crassipes* (water hyacinth) and *Salvinia molesta* (Kariba weed), as well as marked reductions in flow through the spread of alien trees, and through the establishment of impoundments and irrigation (Davies *et al.* 1993). Eutrophication, due to inputs of both agricultural fertilizer and domestic sewage, is a problem in nearly all of the numerous small coastal lakes of the Cape Flats.

Total dissolved solids in groundwater aquifers based on quartzitic sandstones rarely exceed 50 mg/l, whereas in the case of the Malmesbury Shales, concentrations in excess of 6000 mg/l are possible (Department of Water Affairs 1994). This is, of course, the underlying reason why salinization problems commonly result from use of such water for irrigation.

Developments in irrigation and fertilization practices in recent years have affected groundwater quality in certain areas. Maclear (1994) has described a situation for an aquifer on the western coastal lowlands of the region, where extracted groundwater failed to meet South African potable water standards at least in part because of high nitrate and phosphate concentrations, the source of which appears to be fertilizers added via the centre-pivot irrigation method. Maclear (1994) reported that seven of 19 borehole water samples in the vicinity of Elands Bay exceeded South African Bureau of Standards recommendations for maximum allowable limits (for human consumption) in, for example, sodium (400 mg/l), nitrate (10 mg/l) and total dissolved solids (2000 mg/l).

There are numerous, increasing problems of deteriorating water quality throughout the Mediterranean regions of Australia, exacerbated by the long summer droughts which reduce the flushing regime of the water bodies. Problems include nutrient enrichment, causing eutrophication and accumulation of nitrates, the accumulation of heavy metals and other toxic wastes, salinization, and sedimentation.

Less than half the major rivers in the southwest of Western Australia are now potable (WAWRC 1992) (Figure 13.7). All were previously fresh. Rivers and streams which have their headwaters in forested catchments are the freshest but are under increasing threat from a range of pressures on the forests. The larger drainage systems, which have their headwaters in the wheatbelt, are the more seriously degraded from the effects of salinity, loss of fringing vegetation, river "improvement" projects, siltation, eutrophication and chemical pollution. Additionally, pesticide residues in a number of southwestern rivers, and in aquatic organisms, are often at unacceptable concentrations (Olsen and Skitmore 1991).

Table 13.5 *Contents of faecal coliforms (NMP/100 ml) in river flow, Chile* (source: MINAGRI 1994)

Region	<1000	%	1001-10 000	%	>10 000	%
IV Coquimo	33 200	80	7250	17.5	1050	2.5
V Valparaiso	25 850	40	27 150	42	11 600	18
Metropolitan	10 100	8.5	31 600	26.5	76 900	65
VI O'Higgins	25 200	29	17 500	20	44 400	51
VII Maule	82 100	32	71 800	28	101 900	40
VIII Biobio	82 700	48	23 700	14	64 750	38

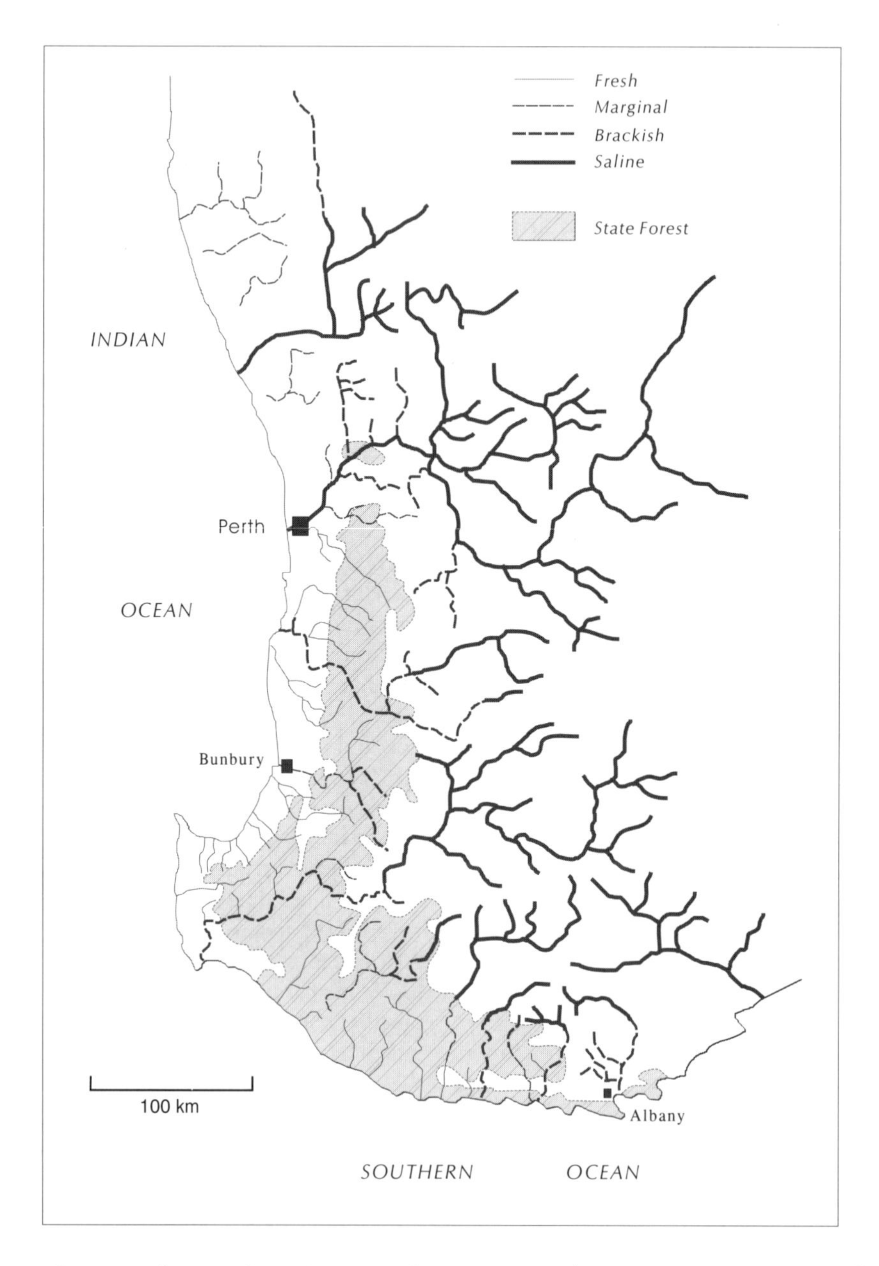

Figure 13.7 *Stream and river salinization in southwestern Australia* (source: WAWRC 1992, figure 9)

The terminating estuaries and lagoons of the southwestern rivers have also been degraded by the same processes: 22 of 50 estuaries studied are considered to be in a poor condition (WAWRC 1992). Several estuaries (the Peel/Harvey in particular) and rivers (Swan and Canning) have become severely eutrophic as a result of nutrient enrichment, and experienced serious algal blooms in the summers of 1994–1996. Two major harbours, near Perth and at Albany, have been severely degraded as a result of nutrient enrichment and toxic chemical pollution. Radioactive substances from mineral sands treatment plants have caused concern, and there have been two major oil spills offshore.

Some groundwaters have become polluted, a few seriously: in the Perth Basin, 1112 sites have contaminated soil and groundwater. Sixty-three percent of these sites are in the Perth metropolitan area, where groundwater is an important resource: it provides 40% of Perth's drinking water (Grant 1992, p. 79). Over-extraction of groundwater in some areas has resulted in seawater incursions.

In the state of South Australia, increasing salinity and the changed flow regime of the Murray River are probably the major concerns with regard to water quality. In an average season the river provides about half of the state's total supply of water for stock, irrigation, domestic and industrial consumption. Regulation of the Murray by constructing weirs upriver and extracting water for irrigation has reduced mean annual flow from $>10 \cdot 10^9$ to $6.5 \cdot 10^9$ l m^3. This has increased the frequency of moderately elevated salt concentrations, changed dissolved oxygen, temperature and sedimentation regimes, and affected wetland habitats as well as water supply (Environmental Protection Council of South Australia 1988).

At the mouth of the Murray, Lake Albert has a rising salt concentration, causing concern to irrigators of lucerne and pasture crops. The salinity of streams has also increased elsewhere: for example the Tod River on Eyre Peninsula, the Middle River on Kangaroo Island, and streams in the Mount Lofty Ranges, where salt concentrations have increased by a factor of four since European settlement. In addition, there is evidence of rising saline groundwater tables in many parts of South Australia's agricultural areas, with serious implications for further salinization and waterlogging (Community Education and Policy Development Group 1993, pp. 34–37).

Perhaps the best known example of water pollution in South Australia is Lake Bonney, near Millicent in the southeast. At one time this was the largest freshwater lake in the state, 23 km long by 2–5 km wide. However, as a result of pollution by effluent from two pulp mills, the water today "is a murky red brown, the lowered lake level is edged by mud flats, and few birds or fish are seen" (Jensen 1986, p. 173).

An example of compound pollution is the Bremer River, on the eastern slopes of the Mount Lofty Ranges, draining into Lake Alexandrina. It is polluted by soluble salts due to clearing; sulphuric acid, iron and aluminium from an abandoned pyrites mine; and high organic and bacteriological pollution from intensive agriculture, a (now abandoned) tannery, abattoir wastes, and overflows from sewage treatment works.

The Mount Bold Reservoir on the Onkaparinga River becomes severely eutrophic on occasions, and there have been blue/green algal blooms in the Murray River and lakes Alexandrina and Albert. Nitrate pollution of groundwater has caused concern for drinking water supplies in the southeast and parts of Eyre Peninsula.

The coastal waters of South Australia, notably around Port Lincoln, are also affected by pollution – untreated sewage, effluent from meat and fish processing works, and industrial wastes (Shepherd 1986). Serious contamination of marine sediments with toxic, heavy metals from heavy industry at Port Pirie and Whyalla occurs in the upper Spencer Gulf (Figure 13.8). Although just to the north of the Mediterranean zone, some of the contaminants undoubtedly affect marine organisms which are eaten by the Mediterranean region's population. Thermal discharges from power stations and oil spill incidents and volumes are also cause for concern. The number of oil spills increased significantly from the decade of the 1970s to the 1980s and now averages about six each year, although most are small (Environmental Protection Council of South Australia 1988; Community Education and Policy Development Group 1993).

In the state of Victoria, between 25% and 80% of surface waters tested in freehold agricultural areas are in a "poor to degraded" condition with regard to nutrient enrichment, sedimentation, turbidity and salinity. Various pesticides have been detected in waterways in a number of locations, especially where there is intensive agriculture. Residue concentrations have occasionally exceeded thresholds for the health of humans and aquatic organisms. In addition, groundwaters in the major northern irrigated regions have risen to within 2 m of the surface, posing a significant salinization and waterlogging threat.

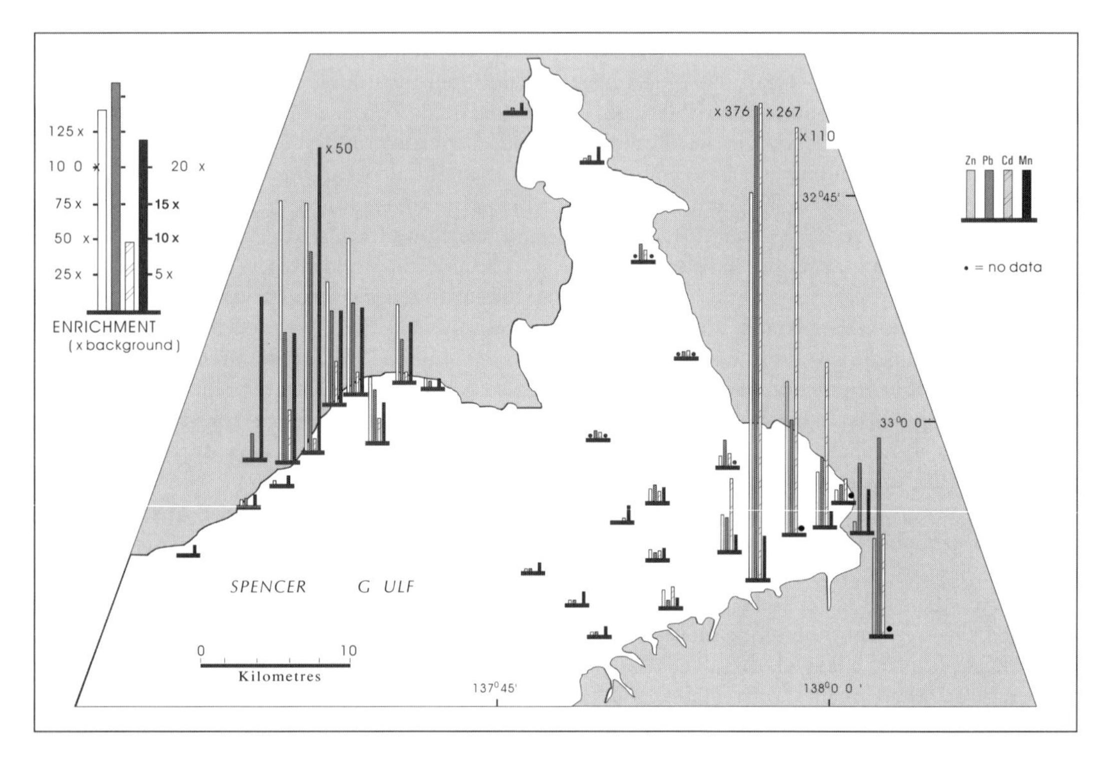

Figure 13.8 *Heavy metal contamination of sediments in the upper Spencer Gulf* (source: Shepherd 1986, figure 7.12)

Air Pollution

Air pollution (including photochemical smog from vehicles) is an important degrading element in land degradation, affecting plants and animals, soil and water quality, and human health. In some respects human activities in rural areas are also partly responsible for the problem.

Table 2.1 of South Australia's *State of the Environment Report* for 1988 lists 24 major air pollutants and their potential impacts (Environmental Protection Council of South Australia 1998). The concentrations of many of those pollutants have decreased since the early 1980s, but with some significant exceptions, including carbon monoxide, ozone and particulate matter.

Industrial pollutants such as sulphur dioxide, fluoride, dioxins, hydrocarbons and heavy metals impact on vegetation, soils and animals around the industrial areas of Kwinana (Perth), Port Pirie and Whyalla (South Australia) and Portland (Victoria) (refer to the respective *State of the Environment Reports*). Increasing ultraviolet radiation due to the "hole in the ozone layer" has known effects on plants and ecosystems (Tolba *et al.* 1992, 34–5; refer also to the special issue of *Ambio* 15(3), May 1995, on the environmental effects of ozone depletion). Ironically, at lower levels in the atmosphere, *increased* ozone concentrations due to urban pollution have adverse effects on crops and people (OECD 1989, table 6.3; Emsley 1992). Dust from mining and quarrying operations can deposit toxic materials on vegetation, soils and water bodies and adversely affect quality and productivity.

Air pollution has become a major problem in southern France due to the concentrated development of some big factories, especially chemicals and petrochemicals, and is contaminating soils in Lower Provence. For example, the Sainte Victoire mountain soils contain abnormal concentrations of sulphur (Vaudour *et al.* 1991). In contrast, acid rain does not affect agricultural or forest land in Spain, partly due to that country's lower industrialization compared with northern European countries, but also due to the calcareous substrate of much of the land, and to the long dry periods.

It needs to be recognized that agricultural activities are a direct contributor of greenhouse gases, and thereby land degradation. Victoria's *State of the Environment Report*, for example, lists eight agricultural sources of greenhouse gases. They are: biomass burning (crop waste); enteric fermentation (methane from animals); soil processes, including cultivation; plant respiration; fertilizers, especially nitrogen; legumes and rhizobia; anaerobic decomposition (methane from wetlands); and fossil fuel use (combustion engines). The report notes that current global estimates for the combined contribution of agricultural and associated land uses to greenhouse gas emissions vary, but are as high as 33% (Scott 1991, p. 57).

In Western Australia, agricultural activities are responsible for nearly 25% of greenhouse gases: 10% from clearing (CO_2), 13% from animal digestion (methane) and 0.4% from fertilizers (NO_2). Quantities are not given for rural burn-off, including stubble burning and forestry practices, or for soil microbial actions (WA Greenhouse 1989).

Industrial and Urban Wastes

The disposal of solid and hazardous wastes is an increasing problem worldwide. Although there is nothing distinctively "Mediterranean" about this, seasonal aridity may contribute in slowing down the rate of decomposition of some of these wastes, and reduce dilution and flushing effects in summer. Further, the "closed" system of the Mediterranean Basin presents a special problem, or simply foreshadows problems that will be experienced in larger seas.

At present, for example, industrial and urban landscapes account for 10% of the land area of Spain. These areas are increasing steadily, producing increasing quantities of solid urban waste. Madrid produced 336 000 t in 1960, which increased to 845 000 t by 1975. National solid waste production is more than 10 000 000 t per year.

Loss of Agricultural Land

Closely related to the above problem is the direct loss of agricultural land resulting from the rapid spread of non-agricultural (mainly urban and industrial) land uses. The loss of previously or potentially productive agricultural land, particularly adjacent to major urban centres, is a form of "degradation" which affects all the Mediterranean zones to varying degrees.

In Israel, for example, cities cover extensive areas – about 500 km^2, or 5% of the Mediterranean area of the country – and are expanding rapidly. Urbanized areas in the USA have a similar proportional extent – 363 000 km^2, or 5% of the total country area excluding Alaska (National Research Council (US) 1993). In the Upper Jordan basin, about 50 km^2 or 3.3% of the total area is paved, including roads and other covered surfaces.

Urbanization will increase in the coming decades, with the world's urban population increasing to 5.1 billion in 2020 out of a total population of eight billion. Along the eastern Mediterranean coast, an almost uninterrupted urbanized area is foreseen from Istanbul to Alexandria. Large cities, such as Tel Aviv and Beirut, have expanded over more than 10 km along the coastline, including the suburban areas (Table 13.6).

The Maghreb (North Africa) also experiences a high rate of urbanization, partly linked to very high demographic growth rates and partly to problems in the countryside which encourage rural dwellers to move to the cities. The effects are manifold, particularly the replacement of lands which have a high agricultural potential by gardens and often irrigated suburbs, usually located on the outskirts of towns.

Much prime land in California has been removed directly from agricultural production by urban growth. Although such growth has long been notable around the major urban areas, widespread car ownership and the construction of inter-state and regional freeway systems over the past 40 years have accelerated the process. For example, the greater Los Angeles area is now spreading across prime farm land in neighbouring counties, including the rich farmlands of the Oxnard Plain. Where urban growth has extended along the coast, for example around San Francisco and Monterey bays and in southern California, coastal land and water degradation has increased. There has been a significant increase in mass movement caused by septic tank drainage,

Table 13.6 *Coastal area populations, Mediterranean countries* (source: Blue Plan 1988)

Country	Urbanized coastal zones (km²)	Population in coastal zones (1000s)	Urban population in coastal zone (%)	Population in coastal zones per km of coastline	Length of coastal zone (km)	Coastal islands
Spain	2794	13 860	81	5372	2500	910
France	1203	5496	88	3227	1703	82
Italy	4981	41 829	67	5260	7953	3766
Malta	13	383	85	2128	180	
Yugoslavia	351	2582	54	422	6116	4024
Albania	52	3050	34	7297	418	
Greece	1315	8862	59	591	15 000	7700
Turkey	371	10 000	53	1926	5191	
Cyprus	20	669	49	855	789	
Syria	17	1155	36	6311	183	
Lebanon	86	1668	80	11 867	225	
Israel	154	2886	90	21 250	200	
Egypt	236	16 511	36	17 300	950	
Libya	85	2284	62	1290	1770	
Tunisia	168	4965	67	3819	1300	
Algeria	276	11 500	48	9583	1200	
Morocco	91	3390	45	6621	512	

landscape watering and street drainage, notably along the Malibu coast, Pacific Palisades, and Palos Verdes Peninsula where prehistoric landslides have been reactivated by changes in the groundwater regime.

Western Australia, with the highest population growth rates (2.5% per annum) of the three Mediterranean states in Australia, is placing increasing pressure on agricultural land, especially that used for vegetable and grape production adjacent to Perth. South Australia and Victoria have also reported losses of prime agricultural and horticultural land to urbanization, accounting for up to 1000 ha per year. Small agricultural establishments, especially hobby farms, also take land out of production; in Victoria, for example, such land occupies at least 4.5% of the state's potentially productive agricultural land (Environmental Protection Council of South Australia 1888; Scott 1991).

Degradation of Coastal Areas

Many cities and industries are located on the coast and this, together with the rapid growth of the tourism industry in Mediterranean regions, due to their particular climatic and coastal attractions, is resulting in serious degradation of the coastal zone.

Coastal erosion is a recent aspect of land degradation in Italy, for example. Beaches along Italian coasts were in equilibrium or exhibiting waxing conditions from medieval times to the first decades of the 1900s. However, coastline retreat has now reached as much as 200 m in some parts of the Tyrrhenian coast (D'Alessandro and Lupia-Palmieri 1981). Along the Ionian coast of Calabria, Niccoli and Procopio (1995) have measured a waxing rate of up to 0.5 m/yr from 1870 to 1949, and a waning rate accelerating from about 0.5 m/yr in 1953, to 3 m/yr in 1989.

In the eastern Mediterranean regions, the coastlines are probably the most fragile environment among the geomorphic systems, considering their narrow geographic extent and the dynamics involved in the sea–land boundary, as well as their history (Chapter 14). Human pressures on coastal landforms are enormous, with one of the highest coastal population densities in the world, increased economic investments and strong land-use competition. Erosion and

Plate 13.2 *(a) House foundations revealed by beach erosion along the Malibu coast, southern California. Photo by A.J. Orme. Such erosion is related to sediment starvation resulting in part from damming coastal rivers and in part from house and highway construction that deprive the beach of natural cliff debris. The houses are in the left centre of (b), which shows the landslide-prone coastal slopes. Failure of pervasively fractured rocks in a complex thrust fault zone, favoured by winter rains and seismic shocks, is further exacerbated by road construction and traffic vibration. Photo by L. Loeher*

degradation on land, and marine pollution, are major processes.

In California, coastal dune degradation, initially by grazing but later by military and recreational activities, has either destroyed coastal dunes, as beneath Los Angeles International Airport, or severely disrupted the dune ecosystem, notably around Monterey Bay and in the Santa Maria Basin. Here, stabilized dunes were widely reactivated by off-road vehicles. In all cases, vegetation clearance or conversion meant loss of native habitat and led to changes in soils, hydrology, erosion and sedimentation.

Before 1960, coastal development was encouraged, or at least tolerated, by public agencies and private interests alike, voices of opposition passing unheeded (Plate 13.2). Government's concern for the coast had long been directed mostly towards hazard mitigation and shore protection, both of which favoured development. Wetlands were reclaimed for urban and industrial use, notably around San Francisco Bay, and coastal dunes were converted to military use, particularly around Monterey Bay and the lower Santa Ynez Valley, or for airport development at Los Angeles. From Santa Barbara to San Diego, much of southern California's coastline was transformed into havens for the wealthy interspersed with small-boat marinas, larger harbours and industrial sites. Only the more rugged, less accessible rocky coasts, mostly in northern California, survived widespread damage. By 1960, some 85% of California's population lived within 10 km of a coast that was administered, poorly, by a mix of conflicting federal, state and local laws and ordinances.

In southern Australia, too, the amenity of many coastal areas has been degraded by uncontrolled coastal development, including subdivision and development of housing on fragile dune and saltmarsh areas. Western Australian examples are presented in DPUD 1994a.

14
The historical development of land degradation in the Mediterranean world

---

Compiled by Arthur Conacher and Maria Sala from contributions on the following regions:

Iberian Peninsula and Balearic Islands	**Maria Sala** and **Celeste Coelho**
South of France and Corsica	**Jean-Louis Ballais**
Southern Italy, Sicily and Sardinia	**Marino Sorriso-Valvo**
Greece	**Constantinos Kosmas, N.G. Danalatos** and **A. Mizara**
The Croatian Adriatic coast	**Jela Bilandžija, Matija Franković, Dražen Kaučić**
Eastern Mediterranean	**Moshe Inbar**
North Africa	**Abdellah Laouina**
California	**Antony** and **Amalie Jo Orme**
Chile	**Consuelo Castro** and **Mauricio Calderon**
Southwestern Cape	**Mike Meadows**
Southern Australia	**Arthur** and **Jeanette Conacher**

"The Mediterranean landscape bears the scars of ancient wounds"
(McNeill 1992, p. 351)

Although there are similarities amongst them, each Mediterranean region has a distinctive history of land degradation and is therefore treated in turn. There is, perhaps surprisingly, little duplication. The more serious difficulty comes in separating the historical development of land degradation from the causes of the problem, because the two are clearly very closely interrelated. In general (with the exception of some of Laouina's North African work which shows changes in erosional forms over time), the attempt has been made to discuss more technical, process-orientated research in Chapters 15–17.

The Iberian Peninsula

In the Iberian Peninsula, the first revolution in the relationship of people with their environment occurred in Neolithic times at the end of the fifth millenium BC. This was the beginning of the substitution of natural ecosystems by anthropic ones through pastoralism and agriculture, and its erosive impact has been documented by sedimentary dating (Peña *et al.* 1993). Roman times, the centuries of reconquest from the Arab invasions, and privatization of church land, were other landmarks in the human impact on the environment. But it was with industrialization that the relationship between society and ecosystems became more intense due to technological developments.

Until the 19th century agriculture was located mostly in the mountains, except for the coastal plains of Murcia and Valencia, because in most plains soils were heavy and thus difficult to work with traditional tools, and most of the property was *latifundia* and dedicated to pasture. In addition, Mediterranean coastal plains were inhospitable because they were unprotected from invaders, suffered from catastrophic flooding, and were subject to malaria.

Vegetation clearance

According to Ceballos (1966), in pre-Neolithic times 96% of the Iberian Peninsula was covered by forest, of which 83% comprised evergreen oaks and beech, 8% conifers, and 4% riparian trees. Today, forests have been reduced to 13% and reforestation has favoured introduced conifers and eucalypts against the indigenous oaks.

In classical times, the Greek geographer, Strabo, had already described the existence of a sharp difference between the northern, densely forested slopes of the Pyrenees and the southern, less vegetated ones. It should be noted that this difference is also due to the lower density of the canopy in the evergreen oaks of the south in contrast with the deciduous trees of the north. To give an idea of the loss of vegetation in the Iberian Peninsula it is traditionally accepted that Strabo wrote that a squirrel could jump from tree to tree from north to south and from east to west. At present trees are located mainly in the mountain areas, and decrease markedly from north to south.

During the Roman period, the most important regression of forests took place in the Baetic (Andalusia) and in the Tarraconense (Catalonia and Valencia) provinces, but also in the Ebro basin. Erosion and sedimentation as a result of deforestation and agricultural practices resulted in the seaward extension of many floodplains, particularly those of the Ebro and Guadalquivir.

Agriculture was abandoned in the coastal Mediterranean plains when Barbarian invasions from the north ended the Roman civilization. The Arab invasion caused the retreat of the Christian groups to the mountains, causing an over-population of those areas with the consequent degeneration of the forests. Deforestation was also used during the reconquest in order to create strategic deserts between the two factions.

During the Middle Ages, the establishment in the Christian states of the houses of pastoralists in Aragon and the Mesta in Castille, resulted in the use and abuse of the territory by transhumance stock. McNeill (1992) noted that erosion increased during this conversion from Moorish

Land Degradation in Mediterranean Environments of the World: Nature and Extent, Causes and Solutions.
Edited by A.J. Conacher and M. Sala.

tree crop/terracing to the Christian period of sheep/pastoralism.

In the 15th and 16th centuries, the systematic extraction of wood took place for the construction of ships for the Armada, together with the making of charcoal for smelting iron and for domestic uses. In relation to that King Felipe II wrote in 1572: "I would like to deal with the conservation of our forests because I fear that we are using the richness of the future generations" (Bielza 1989). With the sequestration of the Church lands in 1837 and 1855, a large area of forests became private property and were consequently subject to speculation, and over-exploited for short-term benefits. It was in this century that scientists and intellectuals like Lucas Mallada, Joaquín Costa and Macías Picavea pointed to the ecological disequilibrium generated by deforestation. Their awareness of the degradation and loss of forests led to the first conservation and reforestation measures. Protection laws were issued for the pinus, oak and beech species, but not for the green oak, the most genuine Mediterranean tree, due to its economic value for wood and coal. In 1918 a law was issued in defence of forests, after which permission was needed to cut trees on private properties. In addition, population pressures caused the establishment of agriculture in marginal lands, mostly on mountain slopes which were later abandoned, being the source of erosion and of irreversible degradation (Plate 14.1).

In Portugal, degradation of *montado*, and consequently also of soils, seems to be relatively old and directly connected with the production system associated with it. Shrubs were systematically eliminated to allow cultivation or to promote a better pasture. In the 20th century, the disequilibrium in the *montado* was caused by three factors: the growing of cereals, pasture establishment and tree production.

With the Law of Cereals in 1898, which subsidized the production of wheat, the area under wheat expanded by breaking up uncultivated land for cereal and intensifying wheat under *montado*. Chemical fertilizers were spread in order to increase production. Poor soils of heath (*charneca*) that were in fallow or uncultivated for many years were then cleared and the land ploughed.

The wheat campaign in 1928–1938 supported the increase in production. Fallow was reduced to a minimum of two years. *Montado* was cut and/or pushed to the weaker soils. The introduction of mechanization led to the felling of trees in order to facilitate tillage. But the expected increase in production was not evident and more chemicals were added to the soil.

In 1974 the agrarian reform which followed the April revolution aimed at further increasing wheat production. Again, marginal lands which had been abandoned were put to agriculture. Simultaneously, fallow was reduced and chemical fertilizer application increased. In the late 1970s two very wet years caused runoff and soil erosion, flooding the best crops. Associated with social problems (the devolution of the land to the *latifundario*), this put an end to the idea of self-sufficiency in cereal production in Alentejo.

Water resources

The need to deal with water problems in semi-arid environments caused the Romans to undertake large hydraulic works to provide water for the cities, such as the aqueducts of Segovia, Albarracín and Tarragona, and to initiate irrigation systems, for example the dams of Muel, Almonacid and Proserpina. From the 8th to the 15th centuries the Arabs achieved a very efficient administration of water in semi-arid environments, superior to that of the Christian kingdoms. Previous irrigation works in the Ebro valley and the Mediterranean coast were enlarged, and bath systems and springs in palaces and urban areas were created. In medieval times irrigation laws, and especially juries, were created for the administration of water, such as those of the Jalón river and of the Valencian kingdom.

Coasts

In many coastal areas of Mediterranean Spain, changes in fluvial sediment supply in relation to the construction of dams, together with the construction of ports interfering with the littoral drift, have affected the erosive and sedimentary processes of coastal areas. Coastal evolution of the Gulf of Valencia in historical times has been documented by Sanjaume *et al.* (1996) through

Plate 14.1 *(a) The upper Vallcebre catchment in the pre-Pyrenees, cleared, farmed and grazed by goats; it is now partly abandoned, with farmers finding work in nearby coal mines. (b) Fairly extensive mass movements occur in the catchment following land abandonment. Photos taken in 1986*

the analysis of archaeological, documentary, cartographic and photographic data. Two sectors north and south of Valencia city with a different evolutionary tendency have been identified; from the Ebro delta to the city of Valencia the coast has a recessive tendency; from Valencia to the Cape of Sant Antonio the tendency is to prograde.

South of France and Corsica

Fire history

The very spectacular forest fires observed in the *Midi méditerranéen* and Corsica since the early 1980s have become one of the main causes of soil degradation. Their story is not well known during previous centuries, despite recent research (Amouric 1992). Before the 19th century, references to fires are sparse, although it is often possible to determine the number of fires during the 19th century. For example, in the Var department, there were 46 fires in 1841, 43 fires in 1877, 10 fires in 1906, 26 in 1913 and 121 in 1943. This department also provides important data concerning the areas affected. During the 18th century, the extent of fires was impressive: perhaps 24 km in 1751, more surely 66 km in 1782. During the 19th and the beginning of the 20th centuries, a single fire burnt up to 3000 ha (Fréjus, August 1854) or even 6590 ha (Collobrières, 1931). But, in the whole department, the burnt areas did not exceed 2797 ha/year between 1896 and 1906. On the other hand, 12 420 ha were burnt in 1913 and 45 140 ha in 1919. The 1981–1985 statistics (42 000 ha burnt/year) suggest a sharp increase in the areas affected. Nevertheless, despite these repeated fires, the extent of forests, *maquis* and *garrigues* in the *Midi méditerranéen* is increasing.

Flooding

The story of flooding in the *Midi méditerranéen* has become increasingly accurate since the pioneer researches of Maurice Parde (1925). Catastrophic floods since 1988 have reinvigorated research into floods, especially about the Catalan (Desailly 1990; Becat and Soutade 1993), Rhône (Provansal 1993), Ouvèze (Arnaud-Fassetta *et al.* 1993) and Paillon (Julian 1994) rivers. These and other, unpublished, works have shown the increase in the number of floods of the Rhône and Durance rivers during the "Little Ice Age". But knowledge of the floods of most of the rivers is usually limited to about a century. It is possible to show that cycles do not exist. On the other hand, recurrences are observed, linked to fluctuations in rainfall.

Soil erosion

There is little reference to soil erosion in historical documents, except for the modern period. Nevertheless it can be identified by geoarchaeological and geomorphological studies. Locally, the presence of colluvial deposits proves the existence of phases of soil erosion following cultivation from the Neolithic onwards. But erosion was infrequent before the classical period in the Comtat Venaissin (Meffre 1990; Ballais and Meffre 1995) or in Lower Provence (Provansal *et al.* 1994). Erosion became more general at the end of the classical period and at the beginning of the Middle Ages (Ballais and Crambes 1992; Ballais *et al.* 1993). A second colluvial accumulation, very probably due to cultivation, has been placed at the end of the Middle Ages (Ballais and Crambes 1992). The modern period was characterized by the maximum extension of cultivation and raising of livestock, even on steep slopes and poor soils (Southern Alps). Nevertheless, the widespread use of cultivation terraces significantly limited soil erosion (Jorda and Provansal 1990). During the 20th century, the spreading of rural decline has resulted in a reduction of soil erosion: the spontaneous recolonization by vegetation of the ancient fields, and the cultivation terraces, have usually resisted erosion (Plate 14.2: see also Plate 2.2).

Water and air pollution

In contrast, water and air pollution are recent phenomena and they do not show any signs of improvement. In fact, there are practically no data on historical pollution. As elsewhere in Europe, there was probably a very long phase of limited, domestic pollution. During the 19th century, industrialization produced a great variety of pollutants which were diffused in the water and in the air. Air pollution thresholds were exceeded during the second part of the 20th century when petrochemical industries were concentrated *en masse* around the Berre lagoon, and when nuclear plants were built in Marcoule and Cadarache. On the other hand, water pollution seems to have decreased since the 1970s because of the installation of many domestic water purification stations and the processing of industrial waters (Pelissier 1980; Joannon and Tirone 1980).

Italy

Land degradation is one of the most important concerns for policy, research and development in

Plate 14.2 *Abandoned, revegetating terraces in Languedoc. In the early 20th century the area was farmed for vines and olives and possibly wheat. Abandonment was due to rural depopulation, not EU policies (R.R. Arnett, pers. comm.)*

Italy. Indeed, some historians maintain that land degradation was one of the main reasons for the decay of the Roman Empire. Clear-cutting was the typical activity of human colonization from the Mesolithic, in order to clear land for agriculture or to gather wood for construction and fuel for domestic fires and pottery manufacture. A crisis of general port siltation was reported in historical chronicles of Greek and Roman times (that is, the 8th century BC to the 5th century AD), when the human pressure on forested land in the Mediterranean region was so great that land degradation was widespread. Increased erosion produced a greater sediment load of streams and rivers, resulting in a waxing phase for coasts. Streams were turned into *fiumaras*, and coastal swamps, lakes and lagoons developed. When Barbarians from the east brought malaria, the swamps and lakes provided an optimal environment for the *Anopheles* mosquito to develop. The diffusion of malaria, together with piracy, forced people to leave the coast and to move inland, to safer places. The coastal zones were abandoned and the land, once cultivated, was left for winter grazing (see also McNeill 1992). Droughts became intense and frequent in the 2nd to 4th centuries AD in southern Italy, impoverishing the richest part of the Roman Empire. It is now acknowledged that the causes of these environmental crises were only partly climatic in nature, being principally of anthropic origin. In fact, there are several comprehensive publications on the human impact on the environment in historic times in the Mediterranean region (including Bruckner 1986; Neboit 1980, 1984; Vita-Finzi 1969).

In subsequent centuries, clear-cutting continued throughout the Italian peninsula and islands with varying intensity. Several wars resulted in clear-cutting for military needs. In times of peace, the mechanization of industrial production required wood as the main energy source in the 18th, 19th and the beginning of the 20th centuries, as charcoal was not readily available in Italy (only in Sardinia were small charcoal mines active until a decade ago), and coal for trains, ships, factories and home heating was nearly all imported.

From this research, it seems that since the Greek–Roman Epoch (8th century BC), human activity determined the alternation of erosion and accumulation along the rivers and coasts and, consequently, the slopes. However, these alternations were also influenced by the high vulnerability of the Mediterranean environment, which arises from tectonic activity producing rock weakening and rapid weathering, easily erodible or unstable terrains, steep slopes, alternating wet and dry seasons, and a high variability of daily and average precipitation and temperature.

These conditions were not only historic, as it is well known that they are typical of Mediterranean zones since early Pleistocene times (Judson 1963; Vita-Finzi 1969; Sorriso-Valvo 1993; and many others).

The Croatian Adriatic Coast

Research on land degradation in the region has not revealed the exact time and conditions of its initiation. Some fragmentary investigations indicate that forest was cleared by the first settlers who inhabited the region before Christ. Cropland was extended by clearing forests and by irregular pastoralism from the coast towards the inland. According to Kosović (Horvat 1957), the remains of Yapig graves (5th century BC) indicate that soil was already thin at that time. Bodies were placed in the very shallow (a few centimetres) topsoil and covered by small, stony mounds. Yapig remnants have never been found below today's ground level. Kosović concluded that the soil was already missing at the time the graves were built. He also found that the foundation of a stone wall constructed in the 13th century on Klačnica hill was placed on barren rock, so he inferred that soil was not present at the time that the wall was built.

There is no knowledge of the extent of such areas, but it is assumed that they were not large. For example, Popović (Horvat 1957) cites the first crusade during which its participant Vilim Tyrski described Dalmatia as "a region with poor agriculture, because there are many mountains, forests and grassland, and cattle-breeding is the main activity of the population". Poparić (Horvat 1957) also wrote about the richness of the forest in the region at that time, stating that 1 200 000 logs had been transported from Dalmatia to Venice as foundation materials for the church of Santa Maria della Salute. The first period of forest devastation by unorganized burning and cutting lasted until the 13th century.

Devastation continued and some efforts were made to control it by regulations. These regulations also serve to show that in the past the Mediterranean region in Croatia was much more extensively covered by forests than at the present time. In the 13th century, the danger of deforestation by unorganized land use was comprehended. For this reason, the statutes of Korčula, Split and Grobnik proscribed the cutting of forests, in areas where there is no native forest vegetation today (Kauders 1939). Clearly, devastation of a higher intensity took place after that time.

In the 19th century the degradation was somewhat reduced. The value of natural vegetation was recognized and organized afforestation commenced. A special inspectorate for supervising afforestation activities was established in Senj in 1878, and the first nurseries for seedling production for karst afforestation were established.

Veseli (1877) came to the conclusion that the land ownership system caused the degradation processes, which originated mainly on public and municipal land. The local population was free to cut the trees and to pasture their cattle without restrictions on these areas (the so-called "problem of the commons"). He also wrote that the problem had become so serious that the Austrian Emperor issued an ordinance forbidding the keeping of goats. This prohibition was ignored by the peasants, but the number of livestock decreased during the 19th century (Veseli 1877) because the barren grasslands provided insufficient fodder for the animals.

Over-pasturing, unorganized pasturing and vegetation cutting have been the main causes of land degradation, especially in areas near settlements. Fire was used as a means of gaining new agricultural and pasture areas. The intensity of their impacts is not known. Ziani (1958) warned that inappropriate transportation networks endanger habitat stability because accessible areas (near villages and towns) are more exploited.

After 1945 the pressure on village surroundings was reduced as a result of rural population decline (Table 14.1). People of working age migrated to towns because of intensive industrialization, and large areas of agricultural land were abandoned and left to self-regeneration.

In 1954 the Law on the Prohibition of Goat Keeping was implemented in an attempt to recover barren areas. A new green landscape was supposed to be the result, but it did not happen.

Table 14.1 *Agricultural population decline in Dalmatia since 1890* (sources: Potocic 1957; *Statistical Year Book 1993*)

Year	Agricultural population as % of total
1890	88.7
1900	87.1
1910	85.5
1931	79.7
1948	71.9
1953	54.5
1992	3.3

The establishment of the Section for Torrent Management in 1898 in Zadar (Dalmatia), indicates that torrents occurred at the same time as the degradation discussed above (torrents are natural channels with a steep gradient, or temporary streams which cause soil erosion). The Law on Torrent Flood Management initiated hydrotechnical works on torrent control in Dalmatia. Torrent control works also commenced in Istria at this time (Godek 1957). During the period 1895–1918, 49 torrents were placed under control (the only systematic work ever done in the region). From that time to 1930, torrent control activities were neglected. In 1930, a law on torrent control was implemented and related activities were intensified. World War II stopped all activities. From 1947 to 1960 some progress was made on torrent control, but from 1960 to 1987 these efforts were minimal (Topić 1987).

Extensive unorganized cattle breeding and unregulated removal of wood and forest litter, together with technological pressures (fertilizers, pesticides, uncontrolled industrial and household waste disposal, waste waters, wet and dry pollutants), continue the degradation of forest vegetation which started a long time ago. It is difficult to quantify the effects of these pressures on the forests.

Greece

From an historical point of view, land degradation in Greece has been documented by some ancient writers. Soil erosion was first reported by Homer in his *Iliad*, written in 854 BC according to Herodotus (cited in Ragavis 1888). Greek hillsides were originally forested and covered by a fertile soil which, however, was rather shallow and vulnerable to erosion. Upland grazing and farming probably began around the middle of the second millennium and was related to initial damage to the forests. Agriculture greatly intensified during the Hellenic period, from 800 BC. Forest clearing took place extensively, as did over-cultivation and over-grazing of lands to satisfy the demands of the increasing population (with wood being used for numerous purposes, such as for fuel, construction and ship building).

Erosion rates in the earliest times can only be inferred from weather records. For example, in the period 1585–1605 there were severe winters accompanied by high rainfalls and also very dry years. Severe weather conditions occurring in October 1590, in Rethimnon, forced most of the inhabitants to abandon their houses. The surrounding land was completely ruined (Grove in press). In Tinos, the winter of 1682 was very hard with strong winds, continuous rain and heavy snowfalls; 480 cows and 3427 small animals (pigs, sheep and goats) died from suffering (Grove in press).

As the soil eroded, cultivation was shifted from food grains to commercial crops such as grapes and olives which could be grown on shallower soils. The fruit was processed into wine and olive oil for export, but the increasing reliance on trade made the Greek economy more and more vulnerable. Solon had already advocated this shift of cultivation on the sloping lands of Attica. This advice was echoed in the 4th century BC by Theophrastus in his horticultural treatises. Plato (4th century BC) says in one of his Dialogues (cited in Hillel 1991):

> "What now remains of the formerly rich land is like the skeleton of a sick man, with all the fat and soft earth having wasted away and only the bare framework remaining. Formerly, many of the mountains were arable. The plains that were full of rich soil are now marshes. Hills that were once covered with forests and produced abundant pasture now produce only food for bees."

The soils of the Mesogia area of Attica are mentioned in Plato's Dialogues as "black earth". Those that remain today are severely eroded, very shallow, red or grey soils with very low organic matter content (<2.2%; unpublished data from C. Kosmas; see also Moustakas *et al.* (1995) for further details on these soils).

By the time of the Macedonian hegemony (338 BC), the land had deteriorated markedly. As the country declined, Greeks transplanted their culture and power to such distant centres as Antioch, Seleukia and particularly Alexandria. During the same period, efforts were made to expand the agricultural land of Greece by draining marshlands and the Biotean lake Kopais. In the Roman period, the erosion process continued at even greater rates (Grove in press).

Eroded materials transported by water and deposited in the lowlands have filled valleys, or extended coastal plains and deltas towards the sea. Historical records make clear that the ancient towns Pella (northern Greece) and Thermopyles (central Greece) were situated along the coast, some 2500 years BP. Today, these cities are some kilometres from the sea due to the aggradation of the coastline. Archaeological findings on the Rizomylos alluvial plain (Messinia, south Greece) have demonstrated that alluvial materials have been deposited at thicknesses ranging from 3.5 to 4.0 m in the central part of the alluvial plain, while at its margins, the depth of the alluvial materials deposited since the Mycenean time reaches 2.5 m (Yassoglou and Nobeli 1972).

Under high population pressures during Minoan and Roman times, several hillsides were terraced in Crete. Also, in response to the pressure of the Ottoman occupation (1457–1913), many Greeks moved to the hilly or mountainous areas. During that time, numerous terraces were constructed to protect the soil from erosion. However, this population shift was also related to extensive forest clearing and farming of the uplands as well as over-grazing. As the productivity of these soils decreased, new areas were cleared and introduced to cultivation. This cycle of forest clearing, land cultivation, degradation and abandonment continued for about four centuries throughout the country. As a result, extensive areas were severely eroded to the extent that bedrock was left behind (refer also the discussion in McNeill 1992).

Shepherds damaged the forest by deliberately setting fires to eradicate the woody vegetation and encourage the growth of grass, which they then over-grazed. Farmers added to the damage by extensively growing food grain (wheat and barley) with insufficient soil protection. Once the land was bare of its vegetative cover and the soil was loosened, the torrential rains of autumn and winter began to wash away the topsoil. Several areas throughout the country were extensively degraded. For example, on Chios island (Aegean Sea), there is a badly degraded area extending over more than 250 km^2, and similar conditions occur in other Aegean islands (such as Lesvos and Creta) as well as in semi-arid parts of the mainland, as in Peleponnesus (Plate 14.3). Today, trees taller than 0.5 m are almost absent due to over-grazing and deforestation under the pressure of increasing population.

Since the 1950s, favourable soil and climatic conditions and the availability of ground or surface water have resulted in intensive farming of the lowlands of Greece. The development of high input agriculture in the plains provided much higher net outputs than those obtained from hilly areas or terraced agriculture. The result has been the continuing abandonment of agricultural lands on sloping terrains, followed by the collapse of traditional forms of management. In some cases this has resulted in continued accelerated erosion and land degradation, but in others land degradation has declined where natural vegetation has regenerated.

The Eastern Mediterranean

The history of land degradation by human impact on the Mediterranean areas of Israel was subdivided by Naveh and Dan (1973) into seven major phases of landscape modification. Le Houerou (1981) and Pignatti (1983) followed the same subdivisions and they may be applied to the entire Mediterranean area.

The *first phase* started at the earliest hominid site in the Levant found in Ubeidiya, in the Jordan Rift Valley, south of Lake Kinneret (Figure 14.1), dated at 1.4 million years BP. The phase extended for more than one million years until 100 000 years BP, covering the Lower Palaeolithic period. The Levant was the main route of biotic and hominid dispersal from Africa to

Plate 14.3 *Badly degraded area in Peleponnesus. Photo by C. Kosmas*

Eurasia (Tchernov *et al.* 1994). During the Lower Palaeolithic, human activity was hunting and food gathering and the impact on the environment was insignificant. The only arboreal fossils found at Ubeidiya are *Pistacia lentiscus* and *Rhus tripertita*, both Mediterranean plants found today in the area (Zohary 1983).

The *second phase* started with the use of fire. The oldest site where the use of fire has been dated is in the Petralonia cave in northern Greece, from findings on a layer about one million years old at its lower level to half a million years in the upper level. In France, the first evidence of charcoal and fire is from a site 400 000 years old near the Mediterranean (Naveh and Vernet 1991). Fire was mastered by the Mousterian Neanderthaloides about 60 000 years ago (Perless 1977), who used it to open dense forests.

The *third phase* started with the beginning of agricultural and pastoral livestock husbandry around 11 000 years ago, and is considered to have been a major revolution in human technological development. The Mediterranean areas in Israel, Lebanon, Syria and Turkey were probably the first sites of domestication and cereal cultivation (see also Zohary and Hoff 1993).

The first cereals were wild wheat, including the large-grained *Triticum dicoccoides*, the direct ancestor of the emmer wheat (*Tridicocum*), the wild, small-grained *T. aegilopoides*, ancestor of einkorn *T. monoccoccus*, and the wild ancestor of domesticated barley (*Hordeum spontaneum*) (Butzer 1964). Their present distribution is the Mediterranean climate zone of Asia Minor and the Levant (Figure 14.2). The first step of domestication was the tilling and removal of the other plants until the agricultural stage of ploughing and seeding was reached. Domesticated wheat (*Triticum turgidum* subsp. *dicoccum*) was found at a site near Damascus, dated 9790–9590 BP (van Zeist and Bakker-Heeres 1979).

The first domesticated animals were the dog (*Canis familiaris*) at an early pre-Neolithic stage, and then the goat (*Capra haraus*), descending from the *Capra aegagrus*; the sheep (*Ovis aries*) descended probably from the urial (*Ovis orientalis*), and cattle (*Bos taurus*) descended from the wild *Bos primigenius*. The domesticated pig, descended from the wild pig (*Sus sarofa*), was probably banned by religion at an early stage of the historic civilizations.

Changes in environmental conditions are reflected in agricultural crops of ancient Mesopotamia: the increase of salinity in ancient southern Mesopotamia affected the growth of wheat, which is salt sensitive, and it was replaced by barley. From grain impressions in pottery dated 3500 years BC, the proportions of wheat and barley were equal; 1000 years later wheat was

Figure 14.1 *Main Lower Palaeolithic sites in the eastern Mediterranean area. The Ubeidiya site – approximately 1.4 million years old – is the oldest known human site outside Africa* (modified after Tchernov *et al.* 1994)

20% and 1500 years later only 2%, being abandoned by the year 1700 BC (Goudie 1981).

One explanation for the agricultural revolution is that over-gathering and over-hunting led to the "prehistoric food crisis" and the need to shift to cultivation and domestication (Cohen 1977). This period had a major impact in the lowlands. The increased extent of clearing by fire and the introduction of the plough led to sizeable increases in the rate of soil erosion.

The *fourth phase* was the period with the greatest development of the agro-pastoral economy in the eastern Mediterranean. It started around 5000 years ago and lasted until the end of the Roman period in the 7th century AD. The phase started with the domestication of fruit trees (Zohary and Spiegel-Roy 1975). Land clearance was at its maximum and affected the mountainous areas, and terraces were built in order to minimize erosion and gain agricultural land. Soil and water conservation methods were applied, and the population reached its largest numbers during the whole historical period until present times.

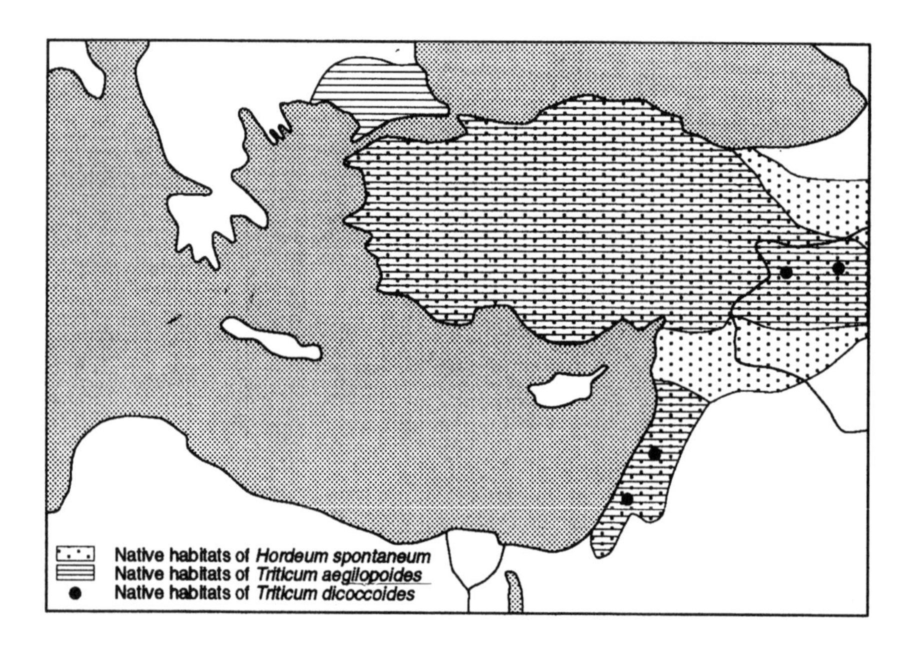

Figure 14.2 *Native habitats of wild cereals (barley and wheat) adopted by humans in the Neolithic period, about 10 000 years BP* (modified after Butzer 1964)

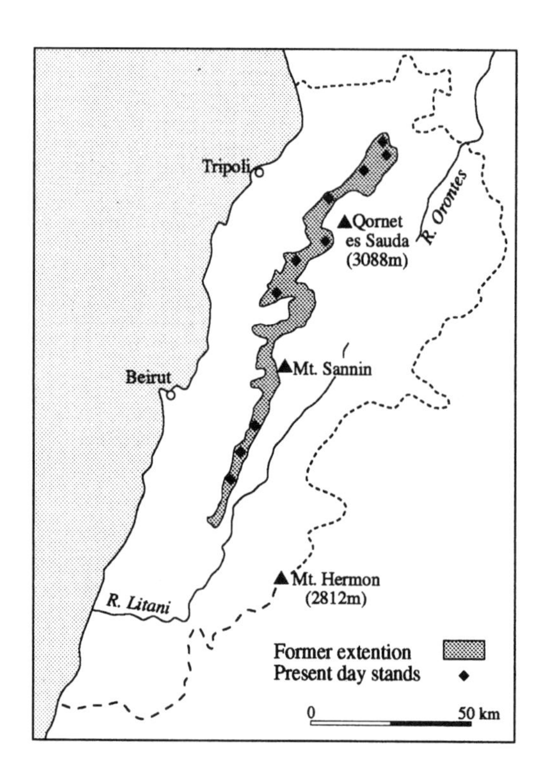

Figure 14.3 *Remnants of cedar stands in Lebanon*

The mountain vegetation included the famous cedars of Lebanon, used in historical times for construction and shipbuilding, and the main source of construction wood for Mesopotamia and Egypt. Egyptian Pharaohs and the Akkadian King Sargon (2300 BC) sent military expeditions to log the Lebanon cedars, and King Salomon reached an agreement with King Hiram of Tyre to obtain timber for building the Great Temple in Jerusalem. The cedars are large but slow-growing trees, and now exist only in scattered stands (Figure 14.3), protected as religious places and forest reserves (Fisher 1978).

Deforestation and the resulting land erosion caused a decline in agricultural resources, and the Phoenicians, who could no longer subsist on their land, had to look for new colonies in the Mediterranean, settling in Carthage on the North African coast, Malta, Sardinia, Sicily and Iberia (Hillel 1994).

The *fifth phase* started with the Moslem conquest of the region and the decline of its economy and agriculture. The pastoral nomadism of the Arab tribes replaced the developed hill lands and irrigation ditches. Increased erosion, accompanied by the loss of soils in the uplands and the creation of swamps in the lower valleys due to siltation of the river channels, was the geomorphic response.

Le Houerou (1981) has strongly criticized the concept that the pastoral and Bedouin civilization had a greater effect on the vegetation and land degradation of the area than the previous development stage. He pointed out that the developed civilization in the Hellenistic–Roman period had a more disastrous impact due to the large population and the pressure on the natural resources of the region. The area of Israel, for example, had a population of five million people according to Flavius Josephus, compared with only 200 000 to 300 000 at the beginning of the 19th century (Naveh and Dan 1973). There is no doubt that a larger population has a larger impact on the natural resources, but under controlled use the impact is minimized. On the other hand, centuries of over-grazing degraded most of the region. The expansion of poor pasture tended to destroy resources in periods of political instability (Aschmann 1973).

The *sixth phase* covered the last one hundred years. Population increased by two to ten times the number at the beginning of the century. Mechanization was introduced in agriculture and irrigation schemes were developed. It is called the technological phase (Pignatti 1983) and it includes major changes, such as the drainage of swamps, land reclamation, monoculture in agriculture and forests, introduction of exotic plants and animals, mechanized work, and the use of pesticides and fertilizers. Classic olive culture gave way to export-orientated citrus orchards.

At the beginning of the 20th century, the development of the Hijaz railway, from Damascus to Mecca, demanded much local timber. Forested areas in the Golan Heights, reported by European visitors in 1880, disappeared during that period. Forests provided charcoal, a manageable small-scale fuel. Villages such as Umm el Fahm ("charcoal" in Arabic) in western Samaria in Israel, received the name from the economic activity of their population, reflecting the intensive use of timber as an energy resource.

Major ecological changes included the effects of opening the Suez Canal, construction of the large Aswan Dam on the Nile, retaining its sediments, and the large water diversions and damming in the Euphrates, Litani, Jordan and Yarmouk basins. The introduction of new species, such as *Eucalyptus* and *Acacia* from Australia and *Pinus radiata* from California, caused major changes in the ecosystem, although plant invasions probably started as a result of anthropic influences and there is evidence that oak (*Quercus calliprinos*) gave way to pine (*Pinus halepensis*) (Naveh and Vernet 1991). Pollens in Lake Kinneret sediments indicate a shift to cultural cultivars, such as olives and grapes, since the Bronze period, about 5000 years BP (Baruch 1987).

Impacts on the environment differed between developed and under-developed regions. In the developed countries, such as Israel in the eastern Mediterranean, accelerated industrialization changed the value of firewood and there has been a decline in the number of grazing animals. The reservation of protected areas increased and reforestation was developed on a large scale. In developing countries, large demographic growth increased the pressure on natural areas, and natural vegetation regressed at a rate of 1% to 2% a year (Le Houerou 1981).

The *seventh phase*, as described in the early 1970s by Naveh and Dan (1973), was one of land degradation. Twenty-five years later, unfortunately we must confirm that the process has been aggravated by population increase, deterioration of water resources, salinization and soil pollution, further impact on natural vegetation by urbanization, and marine and coastal pollution.

The Eastern Mediterranean coasts

Degradation of the coastal zone of the eastern Mediterranean also has a long history. Sand mining started at an early stage, when the Phoenicians commenced glass manufacturing from the siliceous sand. Sand beach mining for construction was well developed in Israel in the 1950s and early 1960s, causing accelerated recession of the coastline. Since 1964 sand mining has been forbidden by law, and alternative quarries inland supply the sand for construction.

Major changes in the coastline include harbour facilities and breakwaters in order to add beach materials and promote tombolo formation. Since

1958, around 20 breakwaters have been built in Israel (Plate 14.4). The sand is deposited between the detached breakwater and the shoreline is enlarged, adding beach area for recreational purposes.

A famous tombolo developed in historical times when Hiram, the Phoenician king in the 11th century BC, connected two islands off the coast of Tyre in southern Lebanon to form a sheltered harbour. Sediment deposition developed a shallow tombolo, which seven centuries later Alexander the Great enlarged to 60 m in width in order to conquer the city of Tyre. Today the tombolo has developed to a 0.8 km wide isthmus.

Another ancient tombolo development occurred in the natural harbour of the old port of Akko (Acre) in northern Israel, which was built 4000 years ago on a coastal ridge 20 m high, close to the coastline (Figure 14.4). The site was opposite an island to the west, which acted as a barrier against storm waves from the open sea. Similar sites were chosen by the early Phoenicians in other Mediterranean coastal sites, for example Tyre, Sidon and Beirut (Emery and George 1963). As sedimentation took place and a tombolo developed, the city and port of Akko became detached from the coastline; the Hellenistic city was built on the former island which became attached to the mainland. Later settlements, like the famous crusade city of Acre, were built on the Hellenistic relicts (Inbar and Sivan 1984).

The filling of harbours by increased sedimentation of rivers between 1000 BC and AD 500 was due to extensive deforestation in the region. Several ports in the eastern Mediterranean, for example Ephesus and Smyrna in Anatolia (Kraft *et al.* 1977) and Abu Hauam near the present port of Haifa in northern Israel, were abandoned for this reason (Galanti *et al.* 1990). Abu Hauam was a port on a broad estuary which filled up and left the site 1.5 km from the shoreline (Plate 14.5). The increased sedimentation in the Near East in this period is known as the "young fill" layer found in many depositional alluvial environments (Vita-Finzi 1969), and this story is repeated throughout the classical Mediterranean (McNeill 1992).

North Africa

The question of the historical development of land degradation and its relation to anthropic

Plate 14.4 *Coastal tombolos formed by detached breakwaters at Tel Aviv beach. Photo by M. Inbar*

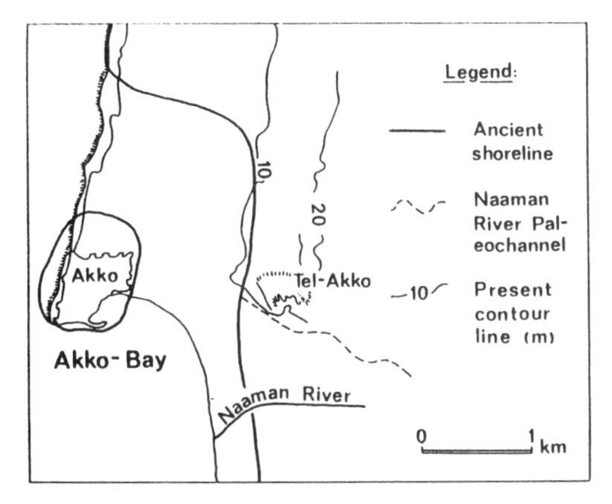

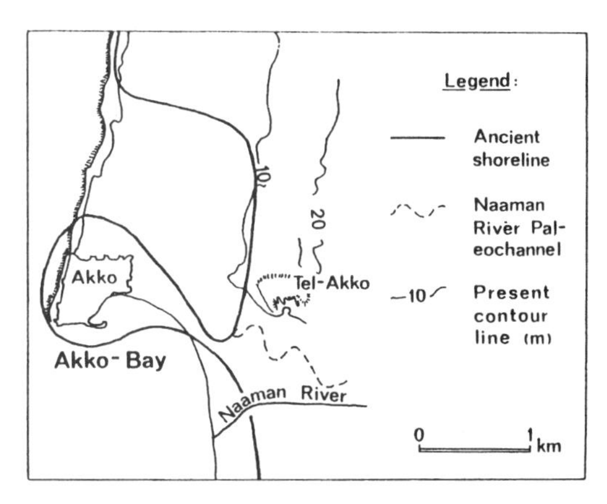

Stage a

Stage b

Figure 14.4 *Development of the shoreline in Akko Bay, Israel, in historical times. The old port of Akko (about 3500 years BP) is now more than 1 km from the sea* (after Inbar and Sivan 1984). *Stage (a): the "old city of Akko", a "kurkar" island opposite Tell Akko; stage (b): a tombolo is formed between the island and the Tell*

Plate 14.5 *Haifa Bay area – a historical silted delta with the Phoenician port of Abu Hauam now 1.5 km inland. Photo by M. Inbar*

factors is posed here. Are we dealing with an ongoing situation or with serious destabilizing crises? And what roles do possible climatic changes play (Neboit 1983; Nafaa *et al.* 1995; Bruckner 1986)?

Several observations militate against the idea of an important climatic change during the last 2000 years:

– the *persistence of springs*, today, that were cited in Roman texts: the conspicuous decline in the level of some groundwater tables is explained by the reduction of recharge;
– *transformations of fauna* (the loss of large, wild mammals and the emergence of the dromedary) are linked to historical circumstances; and

– the *remnants of forests* in the Tell and the Atlas, namely in the vicinity of the marabouts, also provide evidence of the relative stability of the climate since 4000 years BP.

Population pressures

Since 1950 or so, demographic growth has widely asserted itself and locally, populations have more than doubled. Such a concentration of people in somewhat unattractive areas cannot be explained solely by demography. It has something to do with other considerations, since the recent acceleration of rates of degradation and erosion seems also to be linked to attitudes towards the resource. In the Rif, for example, human occupation is very old. One hears about a legendary state whose language was affinitive to the Chleuh of the Masmoudas (Pascon and Van der Wusten 1983). Moreover, numerous ruins, of an ancient and considerable forest population, have been found in areas which are less occupied at the present time. A crisis scenario could have been responsible for the end of that first phase of occupation. It is even possible that the scenario could have occurred several times, with any period of growth being followed by a crisis, possibly due to exceeding a load capacity threshold. Mortality and exodus would have been factors of population regulation and decline (Pascon and Van der Wusten 1983). Under colonization, demographic evolution was complex owing to wars, emigration and the raising of troops for the Spanish Civil War. In addition, there was the disastrous effect of the 1945 famine.

Forest decline

The Tell was described by Roman authors as being abundantly wooded. Agro-sylvicultural exploitation during that period would have caused the first rupture. Arab invasions are often considered responsible for the extensive recession of the forest cover in the Middle Ages. However, fires, and especially the withdrawal of populations from the plains to the mountains before the advance of the Arab tribes, could account for such a state of affairs.

But some travellers of the late 19th century (Mouliéras 1895) have described green, wooded and sparsely inhabited plains which mainly served as rangelands, whereas the mountains were intensively occupied and developed. Still, extensive forest tracts were retained according to descriptions by the first European settlers to arrive in Algeria.

Early in the 20th century, the landscape seemed open, the forest being already widely eroded by sheetwash, with ancient occupation by people (according to the approximate date of the foundation of the villages). The truncation of the soils, then, seems old, but extensive erosion occurred during colonization, including particularly the formation of a number of gullies, as a result of total deforestation (Poncet 1961). Some damage dates back to 1956, just after Independence, when rural populations of the Rif appropriated lands before the government's delimitation of state forests was applied.

The colonial rupture was decisive. The best lands, previously forsaken because they were subject to dispute and thoroughfares, were occupied by settlers and developed intensively (Poncet 1961). The movement of some people to marginal and bordering lands increased the pressure on those difficult environments. With Independence, the three countries underwent both a demographic upheaval and urban growth. Everywhere, a more mobile population and the integration of the monetarized economy with distant regions caused the disintegration of the traditional collective systems (Troin *et al.* 1985).

Excessive exploitation of wood during the two world wars was an important phase of forest decline. It was accounted for by the increase of wood exports to France, on the one hand, and the cessation of timber imports. For example, the production of mine struts from Moroccan forests increased from 240 000 m^3 in 1942 to 1 200 000 m^3 in 1945 and 1 770 000 in 1946 (Boudy 1950). The Algerian war also played an important role in forest loss.

Forest decline in Algeria can be summarized as follows (Plit 1983). The forest covered 5 000 000 ha in 1830. A phase of intense clearing intervened until the implementation of the Forest Code in 1874. By 1887, the forest covered only 3 200 000 ha, and so remained until 1954. In 1967, pessimistic estimates referred to 600 000 ha of forest and 1 800 000 ha of *maquis*. In 1981, the newspaper

El Moujahid presented the following estimations: 1 000 000 ha of productive forest, 1 200 000 ha of degraded forest and 700 000 ha of *maquis*.

A comparative analysis conducted in the Rif region of Morocco by Laouina (1995), using aerial photographs taken at two dates 20 years apart (1962 and 1986), had two objectives: to delimit changes in plant cover in relation to erosion forms in order to identify the anthropic impact on the erosion processes; and to measure existing forms of erosion and to locate the emergence of new ones in order to evaluate the rate of their evolution. The analysis was conducted on 40 stratified samples in three basins, the Loukkos, the Sra and the Leben. The results may be summarized as follows.

Plant cover

The Rif *matorral* undergoes continuous fluctuations in its extent and physiognomy, probably under the effect of clearing and cutting by people. Those fluctuations are both progressive and regressive, indicating that they are dictated by local considerations and that they do not have any general meaning. The usually described process is the following. Some dense and cleared *matorral* is replaced by cultivation until the soil is exhausted; the abandoned land is then colonized by open *matorral*. This can become dense *matorral* again if it is protected. However, that theoretical evolution does not always go in the same direction, since erosion sometimes interferes and scours the land, hindering *matorral* reconstitution. This soil loss can intervene either on cultivated ground or in open *matorral*. Although open *matorral* provides good protection against erosion in normal years, it becomes inefficient in the face of exceptional events which are responsible for the soil loss.

Clearing and the emergence of concentrated erosion forms

Analysis of many sites shows that the clearing of lands which were still covered by vegetation in 1962 did not cause the development of large forms of concentrated erosion. All the big gullies observed in 1986, and all the badland areas, already existed in 1962, apart from some steep slopes directly adjacent to streams, whose erosion was initiated by undermining and destabilization at the base of the slopes. Thus, the rate of emergence of those large erosion processes is slower than the 20 year period between the two sets of aerial photographs. In turn, protected areas, reforestation, and sometimes the cultivation of previously eroded bare lands, makes possible a progressive evolution of plant cover which leads to the stabilization of some forms of concentrated erosion.

The rate of evolution of erosion forms

The forms observed in 1962 have all developed subsequently but at varying rates. Apart from the cases of stabilization referred to above, the developments have often consisted of an increase in the size of the gullies by the extension and the widening of their courses (from some decimetres to some metres per year). Rills are an obvious symptom of accelerated erosion, since they indicate massive loss of materials at a given time. Those rills do not, however, necessarily develop into gullies. But the most notable evolution concerns the hydrographic erosion of the watercourses, with forms of active undermining of the banks and of the intersections of meanders. (Other measurements confirm the acceleration of the speed of erosion. Thus, in the dam of Oued Fodda, in Algeria, 600 000 m^3 of sediments were deposited from 1932 to 1937; between 1937 to 1941 the figure increased to 1 250 000 m^3 and from 1941 to 1944, 3 750 000 m^3; Benchetrit 1972).

Intensification of erosion

In the Maghreb mountains, the historical aggravation of erosion is obvious and indicative of two kinds of processes.

The first model involves an intensification of human occupation. This transforms drainage conditions by suppressing the role of plant cover in regulating flood peaks. About nine million inhabitants are settled in the Tellian mountains, for instance (Maurer 1991). Human occupation, by reducing infiltration, accounts for the most intense functioning of sheetwash, gutters and the rivers. In such a hypothesis, clearing, trampling by animals, over-grazing, or bad management of cultivation exacerbates erosion by substituting concentrated and efficient erosion for the less severe sheetwash.

The second model is the opposite of the first, since it accounts for the aggravation of erosion by the abandonment phenomenon (Dufaure *et al.* 1984). In this case, erosion is associated with the decline of agriculture and the abandonment of works which protected the soils. The process is as follows: a soil, abandoned after its exhaustion, is subject, from the first years, to the development of gutters which hinder vegetal recolonization. A much bigger mass of water then pours out to the rivers whose action is increased (Heusch 1970; Roose 1994). In contrast, good soil management hinders that final land sterilization.

Both models are confirmed on the land at the present time.

In some regions, abandonment of the land is real. Such is the case in the east Algerian Tell, and rural depopulation is intense. Entire lands are forsaken, which has caused local problems of degradation. Yet, during recent years, some return to the land has been noticed (Maurer 1991).

In other regions, agrarian occupation is very important and affects increasingly difficult land towards the highest crests. Some of those marginal occupations end with rapid degradation phenomena.

California

Although land degradation in Greater California is largely a response to human impacts over the past 150 years, a case can be made for certain antecedents. Charcoal found among fluvial, aeolian and marine deposits of Pleistocene age clearly demonstrates the recurrence of fire as a factor in the pre-human landscape (Heusser 1978; Orme 1992b). Blackened hearths and charcoal-rich sediments around prehistoric campsites show that fire was also used by native Americans for cooking, driving game and encouraging new plant growth. Land clearance during the Spanish and Mexican periods was also accomplished in part by fire and it is likely that fire increased in frequency after 1850. Thus along the coast, where indigenous peoples were most concentrated, it cannot be assumed that early European colonists arrived to find uninterrupted expanses of native *chaparral*. For example, the Franciscan diarist Juan Crespi described how, in travelling along the central California coast in 1769, the Portola expedition "set out to the northwest over mesas of good land covered with grass and well supplied with water but without trees" (Crespi, translated in Bolton 1927).

Nevertheless, the arrival of Spanish colonists after 1769 heralded fundamental changes in human attitudes towards the land and its resources which were to have a growing impact on land degradation. Franciscan missions were the initial focus of these changes. Between 1769 and 1823, 21 missions were founded in California on lands which were suitable for raising crops and livestock, and contained sufficient water for irrigation and domestic use. It is likely that much lowland *chaparral* and oak savanna were converted to grassland at this time, although often later abandoned to secondary scrub.

Mexican independence from Spain in 1821 soon led to a souring of relations between the Franciscans and the new government, leading to secularization of the missions in the 1830s, which in turn led to the virtual extinction of native peoples already wracked by introduced diseases. Mission lands were granted to ranchers, including a growing number of Americans. Collectively, Spanish and Mexican land grants accounted for about 3 500 000 ha of California. The uncertain magnitude of soil erosion and land degradation generally during the Spanish and Mexican periods remains a fruitful topic for further investigation.

The growing number of American settlers in Alta California, which led to increased friction with the Mexican authorities in the 1840s, to revolt, independence and in 1850 to statehood, also saw a rapid expansion of cattle ranching along the coast and around the Central Valley. By 1860, there were approximately one million cattle in California. The Gold Rush generated a growing local market for cattle and dairy products, further stimulated after 1869 by transcontinental railroad links and refrigeration. Meanwhile, a boom in wool prices in the 1860s led to expansion of sheep rearing, particularly in foothills above the Central Valley, introducing a large number of Basque shepherds to the southern San Joaquin Valley. Over-production then led to depressed wool prices, countered at first by overstocking which in turn caused over-grazing until, following droughts in the 1870s, sheep rearing declined to more reasonable levels. Meanwhile, over-grazing had caused much soil erosion, making it difficult for trees to become re-established on degraded pastures. Native grasses were replaced by aggressive introduced species, and hay feeding became more important. Stock rearing was thus the dominant feature of the California economy between 1820 and 1870 and, though still important, it is assumed that much of the initial degradation of native vegetation and soils, including the transfer of soil from hillslopes to valley fills, occurred at that time.

Agricultural expansion and diversification were made possible in part by the long growing season, which exceeds 240 days per year throughout the lowlands and locally is unlimited, and partly by the availability of water. Irrigation water came initially from local streams, then from wells tapping into local groundwater reserves and from modest foothill reservoirs, and then, most dramatically, by massive transfers across the state. By 1890, one-quarter of all California farms were irrigated. Today some 35 000 km^2 are irrigated, representing 10% of the Mediterranean region. Reclamation of tule marshes in the delta lands around the confluence of the Sacramento and San Joaquin rivers began in 1851. Half the delta had been poldered by 1900, the remainder by 1930, and the area is now an important source of vegetable and grain crops, but is also plagued by subsidence. It is in the context of irrigation and reclamation that land degradation attributable to the agricultural use and misuse of water emerges.

Agriculture aside, much land degradation in California has been linked historically with mining and logging. Although some mining occurred during Spanish and Mexican times, it was the Gold Rush of 1849 that initiated the first major impact, especially during the period of hydraulic mining of Cenozoic river gravels on the western slopes of the Sierra Nevada between 1852 and 1884 (Plate 14.6). Although the Gold Rush has passed into history, its effects linger on. More recently, mining for petroleum and natural gas has introduced a new series of problems, notably surface degradation and subsidence. Native Americans and early Europeans had long used natural tar seeps for asphaltum with which to waterproof boats and containers. In 1861 an oil well was drilled near Petrolia in Humboldt County, and in 1876 California's first commercial oil refinery was built at Newhall in Los Angeles County. The industry grew slowly until the 1890s when the railroads began using oil rather than coal for fuel. Thereafter, exploitation of oil and natural gas reserves expanded rapidly, bringing with it the problems of land degradation and subsidence.

Logging also expanded rapidly during the Gold Rush as new mining settlements and related cow towns sought wood for buildings. Initially, forests in the Sierra Nevada foothills near the gold workings were the obvious source of timber, felled with little or no regard for erosion. Much of the presently scruffy landscape in those areas reflects the imperfect nature of subsequent revegetation on denuded hillsides. Soon, the growth of San Francisco created a market for nearby timber, generating progressive inroads into the primeval redwood forests along the coast. As railroads expanded, so the market grew for the seemingly limitless timber to be had from northern California hillsides. However, the growth of the timber industry was slow and spasmodic until after World War II. Then, in the 1950s and 1960s, timber harvesting spread beyond readily accessible areas to become more general throughout the forested regions. Problems of erosion and mass movement became significant, leading in turn to growing calls for better forest practices and eventually for conservation of remaining old growth stands.

Chile

At the begining of the Spanish colonization in 1542, chronicles of the time described the Chilean landscape as provided with exuberant vegetation. Afterwards the destruction of natural vegetation began, in order to obtain wood for construction, land for agriculture, fuel for the homes and timber for the mines.

Plate 14.6 *Hydraulic mining in California around 1880. The man in the foreground is directing a monitor (water gun) towards placer gold deposits. Photo courtesy of the Fairbanks Collection, UCLA*

The colonizers' ownership structure, represented by both *latifundia* and *minifundia*, and the rent systems are considered as indirect causes of the vegetation degradation and subsequent soil erosion.

The Spanish colonizers established their villages near the rivers in order to have access to water. Agro-pastoral activities developed at the periphery of the settlements, on the alluvial plains. Later, population growth resulted in the expansion of urban areas on to these good quality soils, thus destroying an important agricultural potential. At present only one-seventh of the country is agriculturally viable.

In 1629 King Philip IV ordered the system to be improved and created the *mayorazgos*, that is, the inheritance of rural properties by the eldest son of a family, in order to avoid the division of lands. This system was abandoned in 1855 with the consequential uncontrolled division of the lands and the increase of *minifundia*. It was only in 1928 that the first Land Colonization Law was issued in Chile.

From 1870 to 1960 soil erosion in the area from Santiago to Lanquihue reached the value of $16 \cdot 10^9$ t. This erosion is attributed to mismanagement of the natural resources. Other factors are international in nature. Due to the destruction of extensive agricultural areas in California, Australia and New Zealand for gold exploration and mining, associated with rapid population growth in those regions, there was strong demand for cereals from these countries and wheat reached very high prices in the international markets. In response to that demand, many areas of Chile (Maule, Ñuble, Concepcíon, Arauco and Malleco) were cultivated and wheat was exported. Production from these virgin forest soils was very high at the beginning, but after a while monoculture gave way to accelerated degradation and erosion and as a consequence productivity declined.

The advance of population towards the south, and especially the German colonization, brought wealth to the country but at the same time a great destruction of forests and soils. In order to clear the vegetation rapidly and economically, there was extensive use of fire. Most of the colonization was on forest soils. By the beginning of the 1960s approximately 6% of the soils of the Coastal Cordillera were subject to some form of sheet erosion, with loss of surficial horizons. In many places erosion was even more severe. In an 18 km^2 area of Region VIII (Tomeco), the number of gullies increased from 420 to 550 between 1943 and 1978, corresponding to a growth of 6.2 to 9.6% (Endlicher 1990).

In relation to vegetation it is estimated that there are 69 species of trees and shrubs at risk (CONAF 1989). Eleven are considered at risk of extinction, 26 vulnerable and 32 rare. The study of the conservation status of plants other than woody species has not yet been undertaken. The country lacks an adequate institution to preserve natural resources.

Chile has suffered from violent earthquakes which have produced landslides and earthflows, which were documented throughout the colonial period. The destruction produced by the 1850 earthquake had such a magnitude that soils were totally ruined in many regions of the country, as in the Coastal Cordillera from Aconcagua to Arauco, in the provinces of Cautín and Malleco, and in most of the Central Depression. Many forests were also destroyed. Another consequence of this catastrophic event was the embankment and formation of bars in many rivers, and the deposition of sands along the coast with the subsequent development of dune fields (Elizalde 1970).

The Southwestern Cape

The recency of European settlement in the New World should make it possible to identify and trace the development of land degradation with much more certainty than is possible in the old. Nevertheless, there is a general paucity of hard data on contemporary land degradation in the southwestern Cape, and tracing the development of degradation problems over time is a difficult task. The only degradation issue for which there is any reliable information at the regional scale concerns exotic invasive organisms, but even in this case the historical picture is far from clear. Nevertheless, the southwestern Cape has a rich archaeological record which, coupled with a well documented historical period, provides some insights as to how environmental problems in the region have evolved through time.

Vegetation clearance

The extent to which early human communities impacted on this aspect of the southwestern Cape landscape is debatable. On one hand, the availability of fire to those people suggests a potentially high level of influence; on the other, their small population numbers point to negligible impact. There is little unequivocal evidence that palaeoanthropogenic fires impacted on *fynbos* communities, although there is a good deal of circumstantial evidence. At Elandsfontein in the western lowlands, there is a suggestion that human-induced fire may have triggered vegetation instability and phases of dune mobility (Deacon 1983).

The activities of later Stone Age hunter-gatherers are more widely accepted as having an impact on surrounding vegetation communities. They used "fire-stick farming" to encourage the growth of geophytes, and their population numbers towards the end of the Holocene may have been close to carrying capacity in the region (Deacon 1992). The introduction of herding, approximately 2000 years ago, heralded further changes and by the beginning of the European contact period in the mid-17th century, there may have been as many as half a million cattle and one million sheep being pastured, mainly on the Cape forelands (Thom 1952). Indeed, the prominence of coastal *renosterveld* shrubland has been attributed to the collapse of a C3 grass-dominated vegetation under heavy grazing pressure (Cowling *et al.* 1988). Elsewhere in the region, palynological evidence indicates significant impacts of both hunter-gatherer and herder populations on vegetation, especially their influence on the abundance of certain *Protea* species in mountain *fynbos*, and the numbers of Clanwilliam cedar trees through adjustments to the fire regime (Meadows and Sugden 1991).

Although, as Deacon (1992, 263) notes "It is too simplistic to equate significant anthropogenic impacts on *fynbos* ecosystems exclusively with modern times", the scale of landscape alteration accelerated markedly following the arrival of European colonists from 1652 onwards. The expansion of the colonists was rapid, and by 1760 the area farmed extended across the entire region (Deacon 1992). Clearance of the local vegetation was especially evident in the Swartland, where a combination of rich soils, adequate winter rainfall and the introduction of Eurasian wheat cultigens eventually produced wholesale landscape alteration. With the opening up of markets in the interior, the establishment of an agricultural economy based initially on grazing was rapidly overtaken by grain production (Talbot 1947). Today there is very little left of the *renosterveld* vegetation (which in itself may have been a product of over-grazing; see above). Vineyards and apple orchards on suitable soils within the southwestern Cape were also established during the 18th and 19th centuries, at the expense of mountain *fynbos* communities.

The consistent demand for large building timbers, coupled with the general scarcity of trees within the *fynbos*, focused forestry attention on the Afromontane vegetation. Only six years after the establishment of the Dutch East India Company base at the Cape in 1652, a proclamation was issued restricting the cutting of yellowwood, *Podocarpus* sp. (Thom 1954). Widespread cutting of the indigenous forests probably reduced their extent in most accessible areas to the east of Cape Town (Deacon 1992). Conventional wisdom holds that these forests were much more abundant before colonial settlement, although Meadows and Linder (1993) have challenged this viewpoint. Much of the transformation probably took place in the immediate post-colonization phase, with negligible rates of recent clearance. Of more than 60 000 ha of indigenous forests in the southern Cape, for example, only 238 ha have been cleared since 1970 (van Daalen and Geldenhuys 1988).

A further impact of the settlers relates to the establishment of "individual ownership", a land tenure system not previously existent in this part of Africa. Deacon (1992) argues that this would have promoted over-stocking, over-grazing and burning at too frequent intervals, all of which must have had serious consequences for the indigenous vegetation.

More recent impacts on vegetation relate to the massive increases in population numbers and the exponential growth of the urban areas of the region. Greater Cape Town has expanded very rapidly, such that the consumption of indigenous vegetation communities has accelerated markedly, especially since the abandonment of influx control legislation in the 1980s (see Chapter 10). McDowell *et al.* (1991, p. 30), in their survey of the area surrounding metropolitan Cape Town, found that the pace of landscape alteration has accelerated dramatically in recent years (Table 14.2). "Strandveld" (coastal vegetation), in particular, has been removed at an increasingly rapid rate (5% in 1975, 10% in 1980 and 37% in 1989).

Invasive exotic organisms

The introduction of alien plant species to southern Africa probably began one to two thousand years ago. Many species were involved, including some which are invasive, for example *Cannabis sativa* (Wells *et al.* 1986). The vast majority of exotics arrived only after European colonization and, as shown in Chapter 12 (Figure 12.13), many of these have become very widespread and are highly invasive. The most important invaders are trees or large shrubs, in particular those belonging to the genera *Acacia*, *Hakea* and *Pinus*, and they have expanded their distribution markedly (Figures 14.5 and 14.6) both through their widespread intentional planting and natural dispersal (Shaughnessy 1986). For a variety of reasons (Chapter 18), many of these species, which have been declared noxious weeds, have been subject to a range of control measures. In some cases this has produced significant reductions in their areas of distribution in recent times (Richardson *et al.* 1992).

Soil erosion by wind and water

Systematic and reliable soil erosion data for the southwestern Cape are few and far between. It is, therefore, difficult to be precise about the

Table 14.2 *Rates of displacement of Cape flats habitats by urbanization* (source: McDowell *et al.* 1991)

Vegetation type	Built % 1983	Built % 1989	% Change	% Remaining
Sand plain *fynbos*	45	56	25	43
Strandveld	10	37	254	73
Whole Cape flats	32	48	51	48

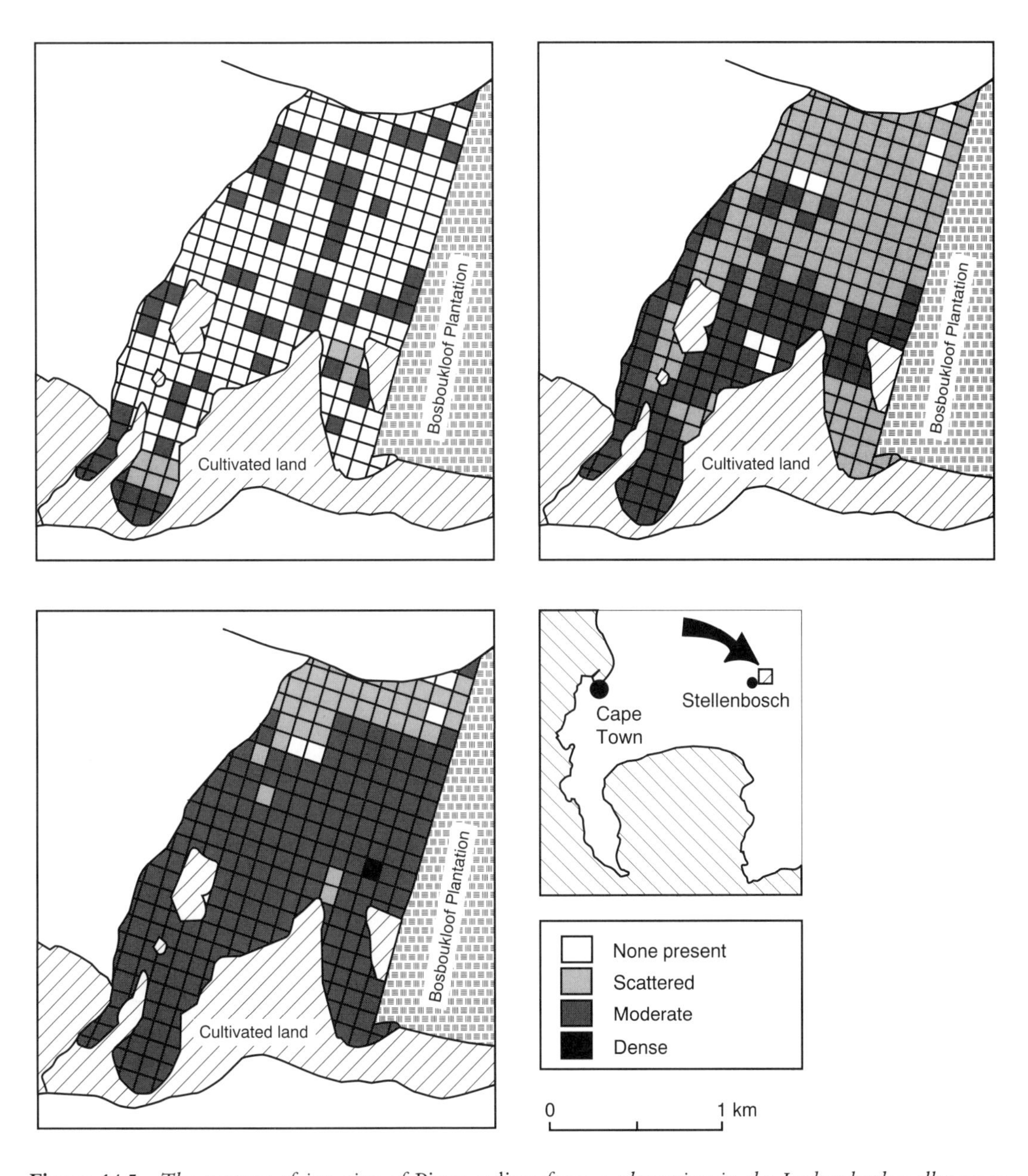

Figure 14.5 *The pattern of invasion of* Pinus radiata *from a plantation in the Jonkershoek valley near Stellenbosch. The top left figure shows the density of* P. radiata *trees in 1 ha blocks in 1953, the top right figure in 1967 and the bottom left figure in 1977* (after Richardson *et al.* 1992)

contemporary situation regarding these problems, let alone reconstruct their evolution over time.

There seems little doubt that the introduction of colonial farming markedly increased the rate of soil erosion, as it clearly did in parts of the western lowlands (Baxter and Meadows 1994). In respect of subsequent developments, however, the question as to whether the southwestern Cape is more or less subject to soil erosion today in comparison with the 1930s and 1940s, remains an open one.

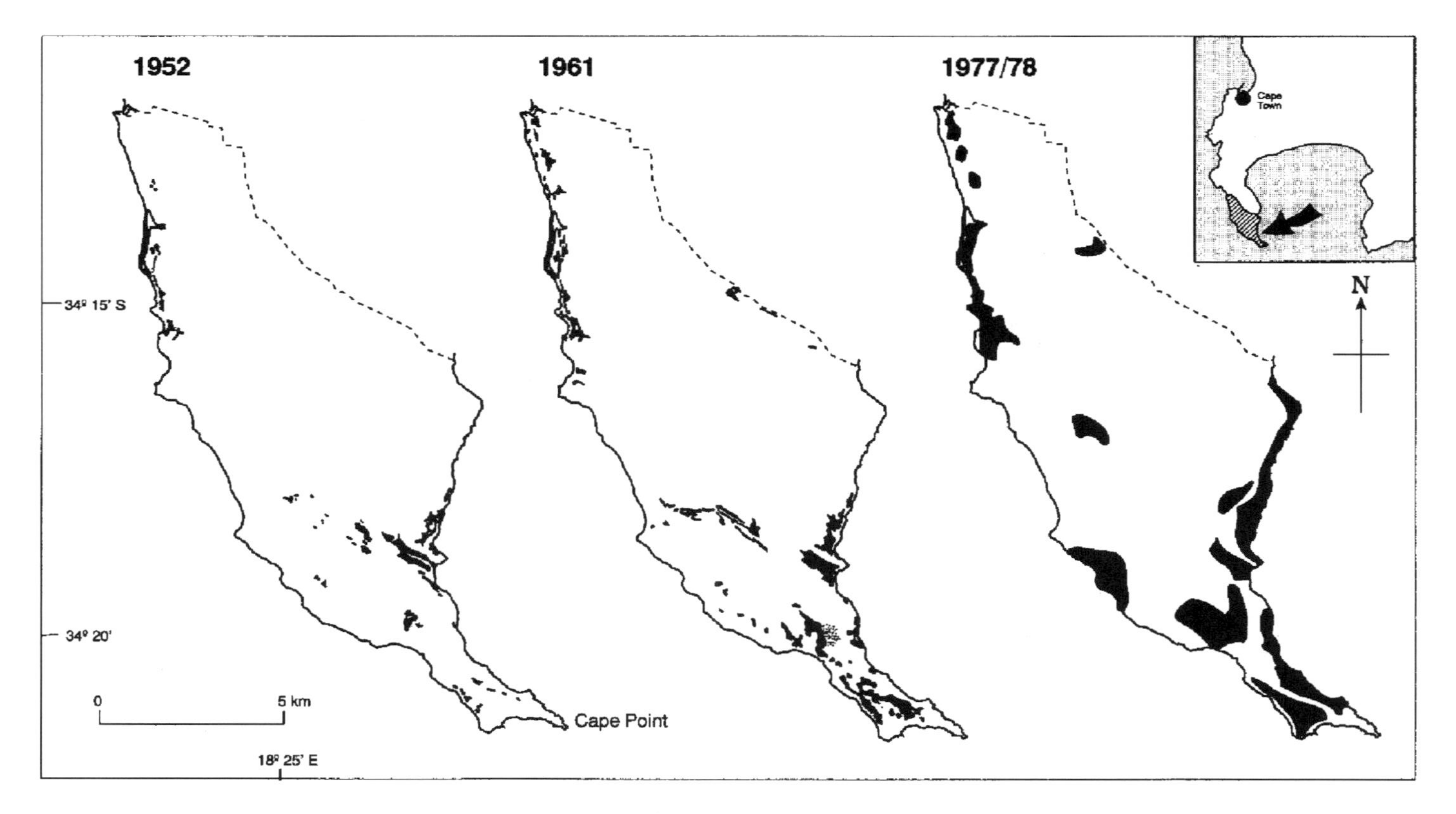

Figure 14.6 *The changing extent of* Acacia cyclops *stands in the Cape of Good Hope Nature Reserve between 1952 and 1978. The implementation of systematic control measures in 1981 has reversed the trend* (after Richardson *et al.* 1992)

Some analyses of soil erosion in the region, for example the study by Talbot (1947), are dated and in themselves provide historical perspectives. Talbot's (1947) study in the Swartland suggested widespread degradation in this wheat farming area. A more recent consideration of the problem (Meadows and Meades in preparation), suggests that the situation has improved considerably (Figure 14.7). Gully drainage density values, for example, declined from more than 7.0/km^2 in 1938 to little more than 1.0/km^2 in 1989. It is assumed that soil conservation and restoration measures have been highly effective in this case in improving the condition of this important agricultural region.

1938

3	6	31	16	0	0
24	37	47	60	52	0
4	13	58	93	96	64
7	19	17	32	45	22

1974

3	18	5	4	0	0
13	53	42	35	22	0
8	42	41	56	30	49
10	21	3	10	41	11

1989

1	1	3	0	0	0
6	10	4	11	0	0
19	5	21	16	22	1
0	0	6	8	15	0

Figure 14.7 *The changing occurrence of gully erosion in the Swartland. The squares represent 1 km · 1 km. Gully density (i.e. frequency of first-order gullies/km^2) is derived from aerial photography for 1938, 1974 and 1989. Reduction in gully density is interpreted to be a result of soil conservation and restoration measures adopted. For location of the area, refer to Figure 12.5*

Long-term data sets on sediment yield do not exist and therefore it is impossible to say whether the reduction in soil erosion in the Swartland is a trend applicable to the southwestern Cape region as a whole. Holmes and Luger (in press), in their study of a coastal dune system on the Cape Peninsula, note that invasion by exotic trees has stabilized the dune forms and reduced wind erosion during the recent past, although it could be argued that the subsequent changes in coastal sediment dynamics are just another form of land degradation.

Southern Australia

Table 14.3 summarizes the main sequence of events relevant to the history of land degradation in Mediterranean Australia, where European settlement took place much more recently than in the southwestern Cape.

Aborigines, vegetation and European settlement

Archaeological and palynological evidence indicates that Aborigines inhabited the Mediterranean regions of Australia for at least 40 000 years, and possibly for more than 100 000 years (Kirkpatrick 1994, pp. 28–29), prior to European settlement in the 19th century. Aborigines used fire extensively (Hallam 1975; Hughes and Sullivan 1986; Latz 1995) for hunting, cooking, warmth, to make tools, for signalling, to provide access and to fell trees. There is disagreement as to whether fire was used deliberately in order to manage vegetation cover so as to encourage certain types of plants and animals. Recalling that these were hunters and gatherers who neither grew crops nor raised stock, tribal groups had a responsibility to "look after" designated areas to ensure the continuing provision of food. Whether modified deliberately or otherwise, it seems very probable that the vegetation which was seen by the first explorers was not the same as that which would have been present in the absence of Aborigines. Indeed, Harris (1986, 30) considers that the impact of sustained firing

Table 14.3 *The main sequence of events in Mediterranean Australia with relevance to the history of land degradation* (compiled by J. and A. Conacher)

Period/event	Activities
19th C/European settlers	Clearing (fell, ring bark, burn) ﬁ pastures and crops
Gold rushes	Trees for pit props, prospecting, growing populations ﬁ land hunger. Succession of Land Acts to allocate land to settlers, break up squatter settlement, Crown land releases
20th C/Closer settement, soldier settlement	Land allocations to new settlers but little back-up support (finance, advice, infrastructure) ﬁ abandonment of land (only half of the 12 635 settlers remain in the northern Victorian mallee)
Consolidation period	"Improvements" to broadcare farming, rotations, between wars ley farming, legume pastures, fertilizers, trace elements, mechanization (clearing, harvesting machinery), government incentives, subsidies
Post-World War II	Soldier settlements more successful, appropriate support and technical advice
1950s, 1960s	Development of marginal lands, large releases
1980s	Large-scale clearing halted

by Aborigines on the composition and structure of plant communities must have been profound. Many botanists now believe that the high fire tolerance, even dependence, of much of the Australian vegetation is the result of a long process of selection initiated by Aboriginal occupation.

But given the climatic fluctuations that occurred during those 40 000+ years, the occurrence of fires lit naturally by lightning, and technological constraints on researching the problem, it is very difficult to separate the effects of Aboriginal actions from environmental changes on the nature of the pre-Aboriginal vegetation in the Mediterranean regions, although palynological research will no doubt throw light on the subject in the near future.

Nevertheless, there can be little doubt that the changes to vegetation wrought by Europeans have been far more fundamental, and have taken place over a much shorter period of time, than those caused by Aboriginal activities, especially in agricultural areas. Here, extensive areas in Western Australia, South Australia and Victoria were cleared of their native vegetation by European farmers and replaced with introduced crop (mainly wheat) and pasture (grasses and legumes) species, mostly during the 20th century (for an example, see Figure 14.8). Large-scale clearing continued until very recently. For example, despite the considerable publicity in recent decades on rural tree decline, estimated at five million trees per year in Victoria's agricultural areas (Campbell *et al.* 1988; cited in Scott 1991, p. 241), clearing continued at high rates, particularly on freehold land and predominantly in the state's western cropping regions, until restrictions were imposed in 1989. In all three states, it is still possible for farmers to obtain permission to clear some of the limited stands of native vegetation remaining on their properties.

The impacts of Europeans on the Aborigines were severe. Scott (1991) observes that the European invasion of Victoria completely destroyed the close relationships the seven major tribal groups of Aborigines had with the land. Fifteen years after settlement, the numbers of European settlers and their stock stood at nearly 80 000 and 7 000 000 respectively. Stock ate out and trampled the small tuberous plants upon which Aborigines depended and destroyed their waterholes and streams. Settlers excluded the Aborigines from their land and commenced wholesale clearing of the natural vegetation. Aborigines were marginalized into reserves and missions, and virtually exterminated in just over three decades by disease, massacre and the disintegration of their culture. Whereas their numbers prior to European settlement were

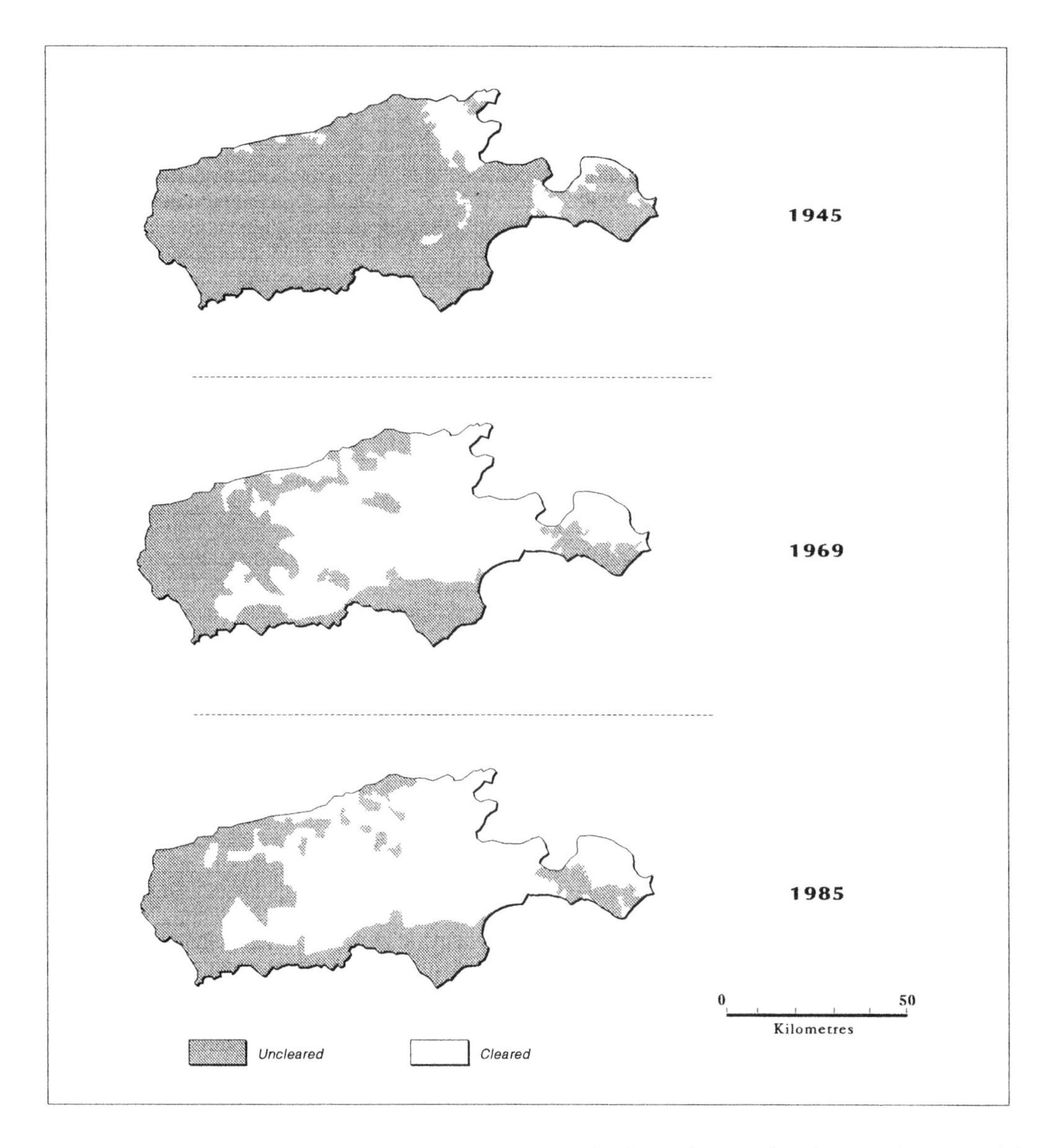

Figure 14.8 *Dates of vegetation clearance on Kangaroo Island, South Australia, showing that most of the clearing took place very recently – between 1945 and 1969* (after Environmental Protection Council of South Australia 1988, figure 4, p. 128). *More generally, most clearing in the Mediterranean regions of Australia occurred after 1920*

variously estimated at 50 000 to 100 000, by the 1870s just over 1000 remained (Scott 1991, p. 120). Further, in view of the attempt to impose European cultivation patterns on the Australian landscape, "... one could hardly suggest we have achieved an agriculture which will be sustainable for at least 40 000 years into the future" (Scott 1991, p. 120).

Salinization

In some cases, evidence of problems was quick to appear. Extensive areas of open grassland and woodland associations on better soils of central and southwestern areas of Victoria were some of the earliest cleared for agriculture during the 1830s and 1840s. As early as 1853, a settler in

Victoria recounted the effect of sheep grazing north of Portland: deep-rooted native grasses were fast disappearing while exotic plants were increasing; cracks, ruts and landslips developed; and springs of salt water were "bursting out in every hollow or water course" (Robertson 1853, cited in Scott 1991, p. 159).

However, it needs to be recognized (as indicated in Chapter 11) that there are extensive areas of primary (naturally occurring) salinity in southern Australia, in the form of saline playas and saline or sodic soils. Early explorers also noted the presence of saline streams. But in Western Australia, the major rivers which drain what is now the wheatbelt were fresh, as were many of the lakes in that region. Evidence of Aboriginal settlement in the region also lends support to the previously healthy state of the rivers and forests of the southwest (O'Connor *et al.* 1989; cited in Olsen and Skitmore 1991).

Unfortunately, there are few early data on the historical development of salinity. Measurements of Western Australia's Blackwood River (the largest in the southwest) at Bridgetown showed that at the turn of the century the river was fresh (Figure 14.9), so much so that its waters were used for steam locomotive boilers. This was also the case at Lake Yealering northeast of the town of Narrogin in the wheatbelt. The waters of the Moore River were used by the Benedictine monks, at the New Norcia monastery 100 km north of Perth, to irrigate their vineyards. In all three cases, the waters are now saline and unfit for human consumption or irrigation.

Secondary soil salinity in Western Australia was first described in the literature in the 1920s (Wood 1924), although its presence had been noted more than two decades earlier. By 1955 there was sufficient concern for the state's Department of Agriculture to conduct a survey of all farmers to ascertain the presence and extent of secondary soil salinity. Further surveys were held periodically – with the most recent one being conducted in 1989 – and, with the exception of the 1984 survey year, the results show a steady increase in the spatial extent of the problem, with a suggestion that the rate of increase may have accelerated (Figure 14.10). Although the data need to be treated with several reservations, which include the ability of farmers to distinguish between areas of primary and secondary salinity or their accuracy in estimating the areas so affected, there can be little doubt about the overall trend. These farmer survey data are

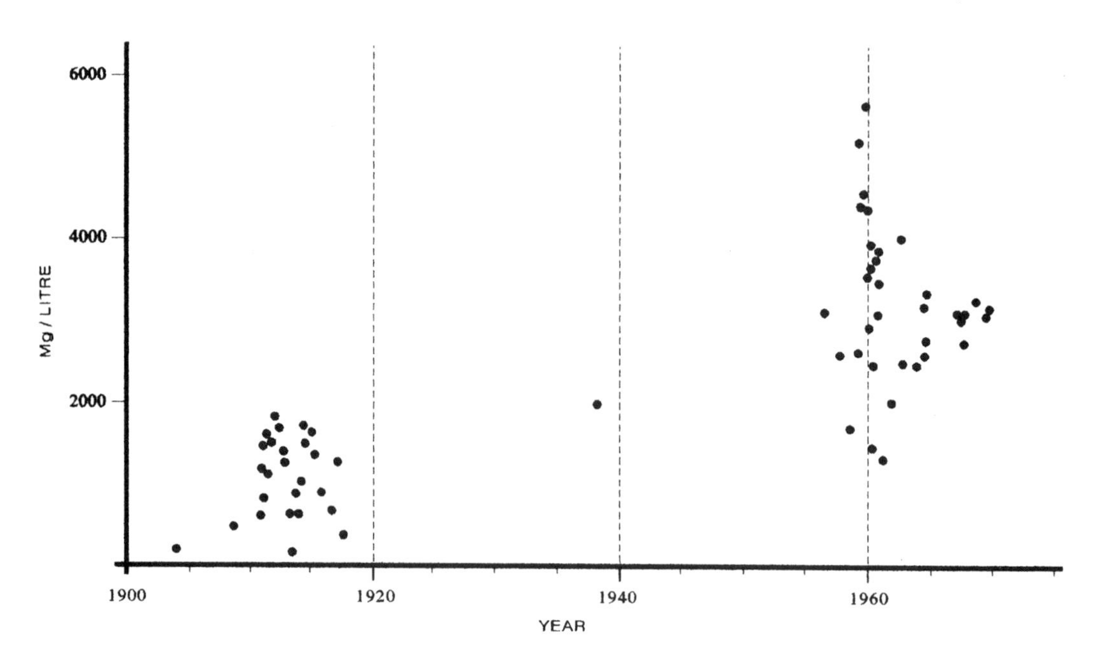

Figure 14.9 *Changing salinity of the Blackwood River, southwestern Australia* (source: Peck 1973, figure 1)

probably the most comprehensive of any form of land degradation in Australia. Since 1989, field-based and remote-sensing surveys, conducted by personnel from the state's agricultural agency, have revealed a much larger area of soils affected by secondary salinization, also shown in Figure 14.10.

Increases in the extent of salt-affected land have also been recorded in South Australia and Victoria. In all parts of the Mediterranean region, consideration of water as well as soil salinization extends the problem far beyond the farm gate; and in South Australia's case the entire population of Adelaide experiences the effects of the salinization of the River Murray.

None of the other forms of land degradation, with the possible exception of vegetation clearance, has equivalent historical data; but it would be safe to suggest that most land degradation problems have paralleled the growth of the secondary salinity problem.

Rabbit plagues

Within decades of first settlement by Europeans, drought, rabbit plagues and seriousious soil erosion had already driven some farmers from the land in different parts of Australia, or had halved farm productivity. One dramatic example from Victoria illustrates the rapid development of the rabbit problem. In 1859, a farmer introduced four hares and 24 wild rabbits for hunting. By 1865, he had killed 20 000 on his property and his irate neighbours were shooting over 14 000 a year, with large sums of money being spent fruitlessly to try and stop the "grey tide" (Scott 1991, p. 313). Following several decades of deliberate introduction, through "acclimatization schemes", by the end of the 19th century some 4 000 000 km² of south-eastern Australia had been invaded by the rabbit (see also, Bolton 1981).

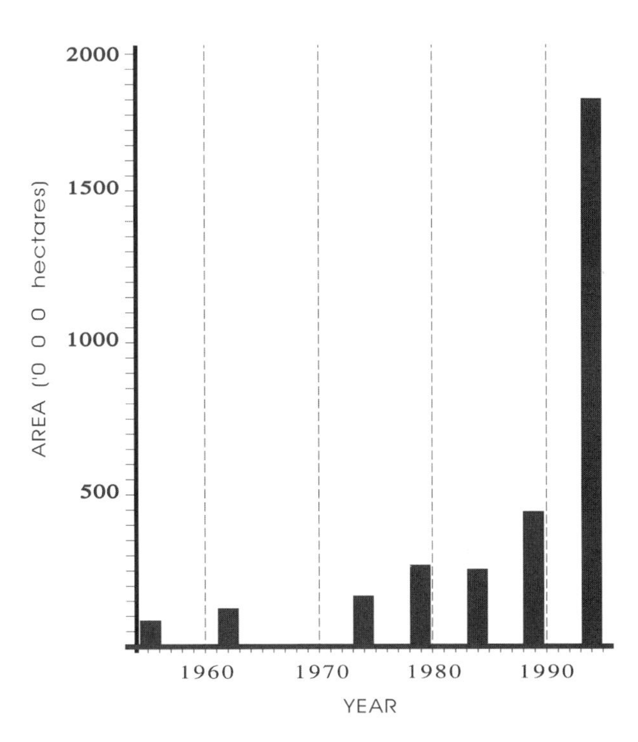

Figure 14.10 *Increasing secondary soil salinity in southwestern Australia* (data from Grant 1992, p. 89; Natural Resources Management Services Unit 1996)

Government schemes

Goyder's line

During the 1860s, South Australia's surveyor-general, George Goyder, recognized that the 400 mm rainfall isohyet and the southern limit of saltbush and bluebush approximated one another. For some years this line was regarded as the northern limit for cereal growing, and settlement was encouraged south of the line only. But a series of good seasons during the 1870s made many believe that the climate was becoming more temperate; settlement was pushed further and further north, with some farmers believing that "rain followed the plough". The inevitable series of subsequent dry seasons and crop failures, severe soil erosion and economic hardship, resulted in abandonment of the land. Ruined homesteads still stand in stark testimony of an unfortunate episode in Australia's history of land settlement (Swanage 1994, p. 309).

Irrigation

In the late 19th century, Victoria (and many other parts of Australia) attempted to "drought proof" land through the encouragement of irrigation schemes. Despite the subsequent clear evidence of rising groundwater tables, increasing salinity, loss of soil structure, waterlogging and economic problems, the government pressed on with major developments during the 1920s and 1930s, and again in the 1950s. The predecessor of CSIRO's (Commonwealth Scientific and Industrial Research Organization) Division of Soils was set up in South Australia in 1927 explicitly "to follow up, by soil survey and profile descriptions and by investigation of the behaviour of the soils under irrigation, the problems of waterlogging and salinity in the Murray Valley . . ." (Wells and Prescott 1983, p. 10). The problems have worsened considerably since that time.

Soil erosion legislation

Following a succession of serious droughts and increasing evidence of severe soil erosion problems, Australia's first Royal Commission on soil erosion was held in 1901, and its first soil conservation agency set up in New South Wales in the 1930s. Western Australia and South Australia introduced sand drift legislation early in the 20th century. These responses took place within a few decades of European settlement and are clear indicators of the severity of land degradation at that time.

Settlement schemes

During the 1920s and 1930s, governments pursued sometimes disastrous "closer settlement" and soldier settler schemes into increasingly marginal lands (Powell 1988). However, problems of isolation, indebtedness, economic depression, rabbits, drought, erosion, salinity, lack of experience, uneconomic farm sizes and inappropriate farming methods forced many to abandon their farms.

The above instances of environmental failure were over-ridden by complex social, economic and political processes. Although concerns were being expressed over the obvious symptoms of land degradation as early as the late 19th century, they continued to be largely ignored by the policy makers with their grandiose political agendas. And despite all the above and many other experiences, many farmers continue to "mine the land". Declining soil fertility is masked by increased inputs of fertilizers and reliance on new plant breeds.

15 The causes of land degradation

1: The nature of the biophysical environment

Compiled by Arthur Conacher and Maria Sala from contributions on the following regions:

Iberian Peninsula and Balearic Islands	**Maria Sala** and **Celeste Coelho**
Southern Italy, Sicily and Sardinia	**Marino Sorriso-Valvo**
Greece	**Constantinos Kosmas, N.G. Danalatos** and **A. Mizara**
Dalmatian coast	**Jela Bilandžija, Matija Franković, Dražen Kaučić**
Eastern Mediterranean	**Moshe Inbar**
North Africa	**Abdellah Laouina**
California	**Antony** and **Amalie Jo Orme**
Chile	**Consuelo Castro** and **Mauricio Calderon**
Southwestern Cape	**Mike Meadows**
Southern Australia	**Arthur** and **Jeanette Conacher**

Introduction

It is very important to identify correctly the causes of land degradation, because unless that is done there is little prospect of finding and implementing workable solutions. But there are numerous difficulties in achieving this objective. First, the topic of land degradation has many feedback loops, and separating cause from effect, and underlying factor from causal process or mechanism, is particularly difficult. Fire, loss of natural vegetation and drought, for example, are problems in their own right and causes of other forms of land degradation. Human actions in themselves do not necessarily carry an explanation of why they are carried out in the way that they are, although historical factors, discussed in Chapter 14, are important.

Second, it is not possible in the confines of one volume to do more than provide an overview and specific examples of research findings from the various regions: the treatment cannot be both comprehensive and detailed for every part of every region. And third, the necessary work has not always been done.

Referring back to Part I of this book and the previous three chapters in Part II, it is evident that the world's Mediterranean-type regions have both similarities and differences in the nature of their biophysical environments, the duration and type of human occupancy, and the nature, extent, severity and history of land degradation. It is not possible to generalize the causes of land degradation for all regions, but on the other hand there are some general themes.

Following introductory paragraphs on some of the regions, this chapter looks at the nature of the Mediterranean-type environments themselves, insofar as they contribute to land degradation, and the following two chapters consider the human influence.

There are both land use and natural causes of land degradation in the Iberian Peninsula. The latter include relief (abundance of mountain areas), climatic conditions (summer drought, torrential rains), and geology (soft sediments with high clay and silt contents). High intensity storms, together with steep relief and changes from rural to urban land uses, are responsible for one of the main environmental problems in Spain: the frequent flooding of the Mediterranean streams.

In the Croatian Adriatic coast, four aspects of the natural environment are particularly important. In relation to the fire risk, the 90 885 ha of inflammable Aleppo pine vegetation in the zone of *Orno-Quercetum ilicis* is of particular significance. The intensity and seasonal distribution of precipitation are particularly relevant to the risk of floods, whereas, ironically, the porous structure of carbonate parent rock in the littoral zone enables the loss of water. Finally, the region has critical slope conditions: owing to the very narrow and elongated Mediterranean climate belt and specific relief forms involving a wide range of slopes (being more than 40 in the Velebit and Biokovo mountain regions), erosion risk is high.

The main land degradation process in Greece is the loss of soil volume capable of supporting a soil-protective vegetative cover. The main factors responsible for land degradation are climate, relief, geological substratum, and the anthropogenic influence.

The causes of land degradation in North Africa can be grouped under six headings: recent orogenesis; the loss of plant cover and increased erosion; lithology and pedology; rainfall concentration and intensity; demographic explosion and human factors; and social aspects.

In addition to the general Mediterranean climatic conditions, in Chile natural land degradation is the result of tectonic activity and the steep terrains of the Andean reliefs. The impacts of human actions concern agricultural and forestry practices. Degradation conditions in Chile are related to topography, thin and erodible soils, clearance of semi-desertic *matorral* and sclerophyll forests, over-exploitation of pastures, shifting agriculture, stubble burning, unprotected fallow, and ploughing up- and down-slope.

In principle, the underlying fundamental cause of all forms of land degradation is injudicious and unsustainable exploitation of natural resources. Thus, in the southwestern Cape, the root of all

Land Degradation in Mediterranean Environments of the World: Nature and Extent, Causes and Solutions.
Edited by A.J. Conacher and M. Sala.

problems of environmental degradation is population growth and the increasing share of the natural resources that are consumed by this population. However, this view needs to be tempered on three counts: first, there is little alternative to exploitation of natural resources, since these are required for survival – it is the unsustainable part of the equation that is problematic; second, population growth *per se* is not really the key issue – rather it is the increased consumption of resources that causes difficulties; and third, neither population distribution nor patterns of resource consumption are evenly distributed across the southwestern Cape landscape. The main causal factors in the southwestern Cape are: agricultural expansion and intensification; population growth and urbanization; alien tree infestation and afforestation, and human-induced changes in the fire regime.

Underlying all of these factors is the prevailing climate which, through summer aridity, may predispose the landscape to disturbance because it represents a time period when the natural vegetation is under moisture stress. Despite this, the word "drought" is not often used in the context of the southwestern Cape, and the causes of environmental problems are more anthropogenic than "natural".

The single most important cause of land degradation in southern Australia – clearing of the native vegetation – is also a major form of degradation in its own right. Related to vegetation clearance is its modification by a wide range of actions. The second most important cause of land degradation is the agricultural practices which followed the removal of the native vegetation.

Climate

Data from three of the regions – Greece, the eastern Mediterranean and North Africa – are used to indicate the fundamental importance of this factor in helping to account for the nature of land degradation in Mediterranean environments.

Greece

Although accelerated rates of erosion and slow rates of vegetation recovery on degraded areas are the main causes of land degradation, rain erosivity and bioclimatic conditions are the relevant parameters.

The contemporary climate of the Mediterranean region is characterized by strong seasonality and spatial variability in rainfall, sudden changes from cold to warm months, and oscillations between minimum and maximum daily temperatures. According to data available from the National Meteorological Service of Greece, rainfall ranges from 1280 mm in the western part of Greece to 370 mm in the eastern part. Average air temperature ranges from 16.5 C to 17.8 C. The most characteristic feature of the climate is the long, dry summer with most of the rainfall occurring during the six to seven cool months. As Figure 15.1 shows, potential evapotranspiration exceeds rainfall for more than six months, creating a large water deficit for plant growth from May to October.

Rainfall amount and distribution are the major determinants of biomass production on hilly lands under Mediterranean conditions. Decreasing amounts of rainfall combined with high rates of evapotranspiration drastically reduce the soil moisture content available for plant growth. Reduced biomass production, in turn, directly affects the organic matter content of the soil and the aggregation and stability of the surface horizon in relation to erosion. A study on the effect of diminishing soil moisture on soil properties and biomass production of rain-fed wheat, in a two year rainfall exclusion experiment, showed that the total above-ground biomass production was reduced in proportion to the amount of rainfall excluded (Kosmas *et al.* 1995c). Reductions in biomass of 90%, 71.4% and 53.4% were measured in the experimental plots in which rainfall was reduced by 65%, 50% and 30%, respectively (total amount of rain falling in the open field during the growing period R = 361 mm).

Diminishing soil moisture negatively affected organic matter content and aggregate stability. The mean organic matter content was decreased to 0.9%, 1.1% and 1.2% from an initial value of 1.4% in plots with 35%, 50% and 70% rain interception in a two year period, respectively.

A rain erosivity index calculated from the product of the Fournier index and the Bagnoulis–Gaussen index (Yassoglou and Kosmas 1988) has

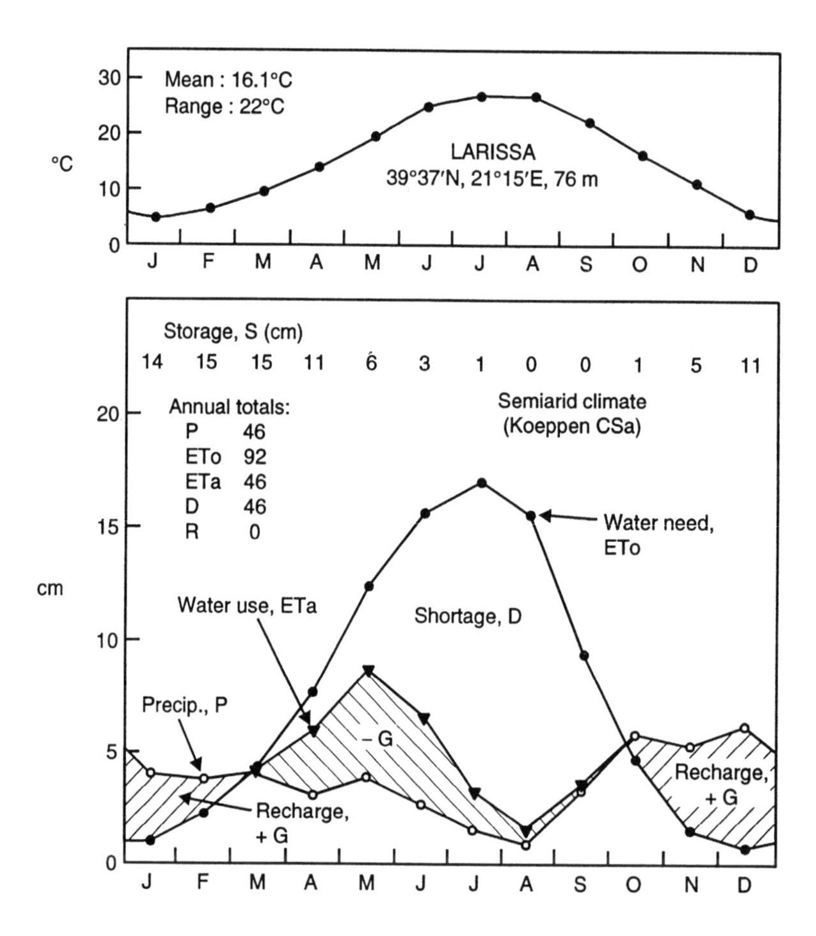

Figure 15.1 *Average monthly rainfall, potential evapotranspiration rate and air temperature of a representative Greek area* (source: Danalatos 1993)

demonstrated that high erosivity is predominant in the western and southwestern part of the mainland and in the eastern Aegean Sea islands (Figure 15.2). Low rain erosivities are mainly confined to the northwestern part of the country, while the medium erosivity class applies mostly to the central and eastern part of the mainland and includes the main agricultural lands.

A study of Mediterranean desertification, conducted within the framework of the EC Project MEDALUS in eight representative experimental sites along the north Mediterranean, showed a tendency of increasing overland flow with decreasing annual rainfall (Kosmas *et al.* in press; refer also to the following section for comparable data from the eastern Mediterranean). Of course there is an inevitable variation among the different sites which is attributed to soil surface properties, slope gradient and length and rainfall intensity and duration.

Vegetation cover is crucial for runoff generation and can be altered readily along the hilly areas of the Mediterranean, depending on climatic conditions and period of the year. In areas with a mean annual precipitation of less than 280 mm (as in Almeria, south Spain) and high evapotranspiration, soil water available to plants is reduced drastically and the soil remains relatively bare, favouring overland flow. Runoff reached values of up to 10% of the total rainfall in Almeria, the driest among the study sites. In the same study it was found that annual sediment loss tends to increase with decreasing precipitation when the latter is less than about 380 mm (Papamichos 1985).

The above findings have clear implications for the effects of possible climate changes. Changing

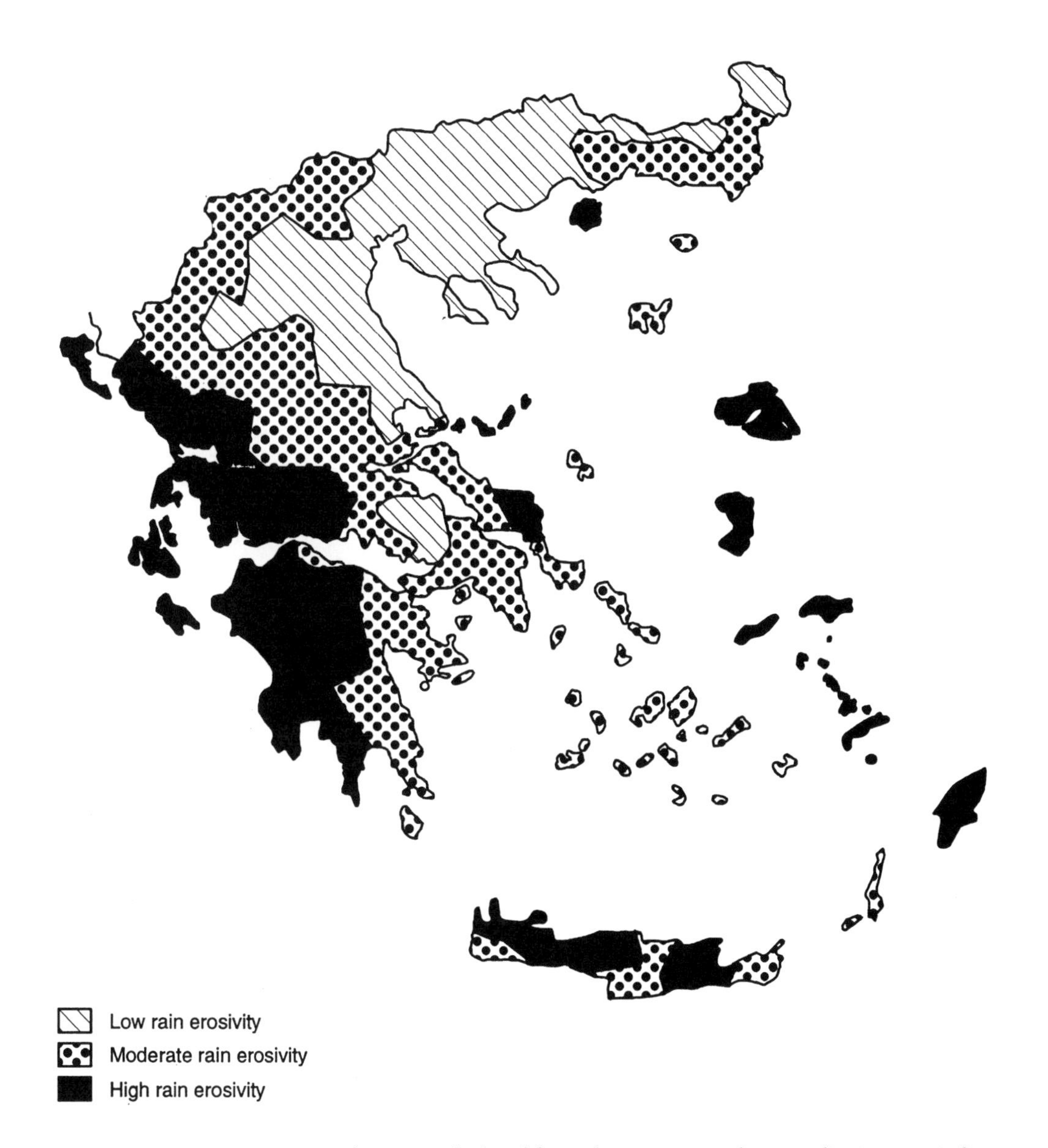

Figure 15.2 *Rain erosivity map of Greece calculated from the Fournier and Bagnoulis–Gaussen indices. In general terms, the classification of an area with "high rain erosivity" means that rainfalls are unevenly distributed during the year (Fournier index high) and occur under dry climatic conditions (high aridity index). In other words, the term "high rain erosivity" includes areas in which short-period intense storms occur during dry seasons* (source: Yassoglou and Kosmas 1988. Full details of the calculations for deriving the erosivity index are given in CORINE 1992)

climates will affect land degradation indirectly by affecting biomass production and therefore plant cover and protection from erosion. Existing growth models (Danalatos 1993) indicate the effect of climate change on biomass production of rain-fed cereals. From this, Yassoglou *et al.* (1994) have estimated the impact of temperature and rainfall change on the relative biomass production of rain-fed cereals in central Greece (Figure 15.3). They calculated that a decrease of rainfall by 30%, without any change in air temperature, may bring about a decrease of biomass

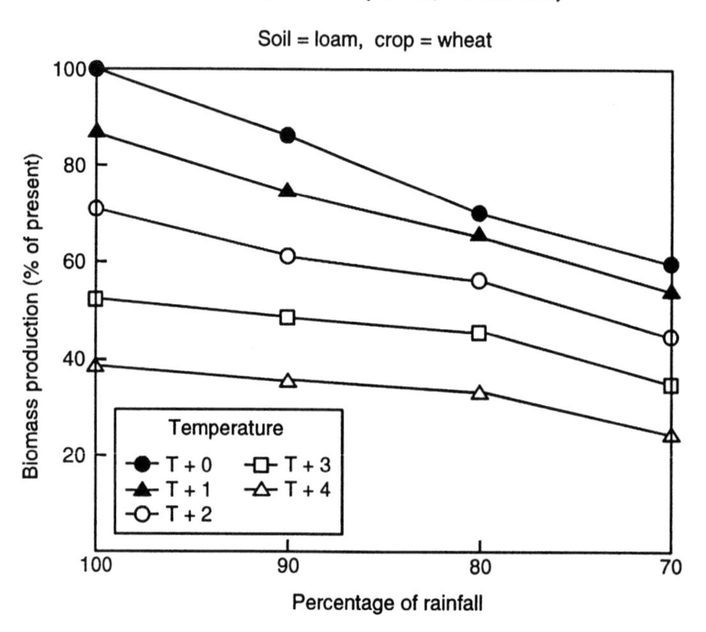

Figure 15.3 *Relative biomass production of rain-fed wheat predicted for various scenarios of reduced precipitation and/or increased air temperature in central Greece* (source: Yassoglou *et al.* 1994)

production by 40%; whilst an increase in air temperature of 2 C accompanied by the same reduction in rainfall would decrease the relative biomass of rain-fed wheat by about 55%.

The eastern Mediterranean

Data from different watersheds in the Mediterranean coastal area of Israel show an increase in sediment yield in the southern basins with a decrease in mean annual precipitation (Table 15.1). The trend is explained by lithological as well as climatic factors: the northern catchments are mainly in karstic areas with low values of sediment yield – 50 to 200 t/km²/yr (or 0.5 to 2.0 t/ha/yr). The largest basin – the Northern Jordan River – has an average sediment yield of 50 t/km²/yr, whereas in the southern loess-covered areas the sediment yield is in the range of 200 to 800 t/km²/yr (Inbar 1992; Table 15.1). The general trend of decrease of sediment yield with increasing precipitation as established by the Langbein and Schumm (1958) curve is nevertheless valid for these Mediterranean areas.

It was found that rates and trends for the eastern Mediterranean are similar to those of Spain, with a sediment peak at 300 mm of annual precipitation and a decreasing rate of 50 t/km²/yr for every 100 mm of increasing annual rainfall. This trend differs from the values found in the Pacific coast Mediterranean areas – California and Chile – where sediment yields peak at 400 mm, with a more rapidly decreasing rate of 100 t/km²/yr for every 100 mm of increasing annual rainfall (Figure 15.4). The values for the Litani and Orontes and the coastal rivers of Lebanon are assumed to be similar to those of northern Israel, that is about 50 t/km²/yr.

Erosion rates depend mainly on runoff if non-climatic factors are constant, but generations of human impacts on the landscape, mainly through vegetational changes, have altered the water and sediment regimes.

General models, such as the Universal Soil Loss Equation (USLE; Wischmeier 1959) bear no relation to erosion data from the Israel basins, as found in a study conducted by the Israel Soil Conservation Service; rather, it was found that erosion values are directly related to runoff (Agassi *et al.* 1986).

Table 15.1 *Sediment yield from Israel's rivers* (source: Negev 1972)

Basin	Area (km^2)	Mean annual precipitation (mm)	Mean annual sediment yield (t/km^2)	Method†
Hillazon	158	680	23	(1)
Netofa	121	600	190	(2)
Qishon	470	480	180	(2)
Qishon	224	650	50	(1)
Snunit	65	620	45	(1)
Alexander	544	650	16	(1)
Ayalon	160	600	117	(2)
Eqron	62	500	185	(1)
Soreq	80	570	47	(2)
Pelugot	200	450	200	(2)
Shiqma	746	390	160	(2)
Adorayim	86	350	165	(2)
Lahav	16	300	840	(2)
Shoval	15	325	200	(2)
Gerar	54	300	310	(2)
Jordan*	1492	820	50	(1)
Meshushim*	160	700	20	(1)

* Lake Kinneret watershed (after Inbar 1982)
† (1) Suspended sediment sampling; (2) sedimentation in reservoirs

North Africa

High rates of surface runoff are related to intense downpours, with the effect of excessive rains being spectacular even if it is limited in time and space.

According to studies of rainfall intensity, the really heavy rains are of a relatively low frequency and the Mediterranean intensities are lower than those recorded in tropical or continentally temperate environments (Couvreur-Laraichi 1972; Amar 1965; Debazac and Robert 1973). In Morocco, the Rif, the Middle Atlas and the high parts of the High Atlas have experienced falls of greater intensity than 100 mm in 24 h, with the highest intensity of 150 mm in 24 h occurring at Jbel Outka-Zoumi-Chaouen in the Rif. In the Atlantic domain, westerly frontal rains, responsible for prolonged precipitation during some days in winter, are recorded. The rains can last just long enough to saturate the soils. But instances of high intensity rainfalls have been noted, namely in a semi-arid region and in the mountains. An example of persistent, high rainfall occurred at Ktama in 1962/63, when 1702 mm was recorded in 17 days, from a total of 3173 mm for the year. In the same year, 10 stations recorded rains of >200 mm in 24 h. Fournier's coefficient p^2/P allows the identification of the regions where rainfalls are really aggressive by their concentration in time and sometimes by their intensity (Table 15.2; Heusch 1982, 1990).

Yet, it is necessary to emphasize the brief duration of periods of effective rainfall. In the basin of Mda, for example, the most dangerous 12 days during three years of measurement recorded 30% of the rains, 50% of the flow rate and 92% of the

Table 15.2 *Fournier's coefficient* p^2/P *for selected regions in North Africa. P is mean annual rainfall, and* p^2 *is the mean wettest month rainfall* (source: Erosion Project, PNUD-FAO)

Coefficient	Regions
30–40	Beni Snassen – Chaouia – Doukkala
40–60	Jerada mountains – Eastern Rif – Basin of Sebou – Basin of Bou Regreg – Middle and W High Atlas – Sous – Haha
>60	Rif and Pre-Rif

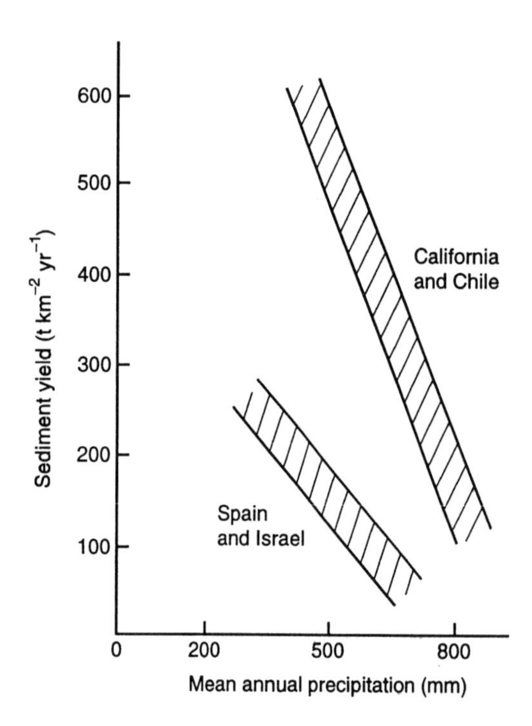

Figure 15.4 *Decreasing rate of sediment yield with increasing precipitation in Spain–Israel and California–Chile* (reproduced from *Catena*, 19(3/4) (Inbar 1992), by kind permission of Elsevier Science – NL, Sara Burgerhartstraat 25, 1055 KV Amsterdam, The Netherlands)

total erosion. Those days were characterized by the intensity of the rains and their continuity, favouring the saturation of the soils (Heusch 1970).

Rainfall intensity is not directly correlated with the pluviometric annual total. All the stations, both in a semi-arid environment and a subhumid one, can receive half, if not more, of the annual total in one month. These situations of high concentration that can happen periodically are responsible for cases of saturation leading to large erosion forms. With regard to normal intensities (with return periods of one or two years), it should be noted that they remain moderate except in the mountains.

There are obvious relationships between mean annual rainfall and land degradation. For example, in eastern Algeria, where mean annual rainfall declines by about 70% from north to south (Bourouba 1993), sediment loss from the subhumid Tell and its southern piedmont is around 400 t/km^2; on the humid littoral, with 900 mm of rain, it is 700 to 900 t/km^2; and on the piedmont of the Aureses, it reaches 1714 t/km^2 with 300 mm of rainfall. The latter rains are concentrated and the soil laid bare. Low annual rainfalls are therefore related to the erosion of large quantities of materials.

At a catchment scale, the runoff coefficient is important and increases according to the amount and intensity of the rainfall. It can reach a value of 52% in small-sized basins, in years of strong concentrations of rainfall. Those are, however, the years which experience record discharges (Heusch 1970; Benchetrit 1972; Bourgou 1995; Bourouba 1993; Debazac and Robert 1973; Dufaure *et al.* 1984; Kalman 1976; Marre 1987). In dry years, runoff coefficients are very low since concentrated runoff, responsible for the flood peaks, becomes less frequent.

Water turbidity does not respond to the same factors as runoff, which is why it is difficult to find a correlation between turbidity and runoff. Continuous rains, responsible for the saturation of the superficial horizon of the soil, account for the highest turbidity values. Heavy and brief rains allow moderate turbidity. Turbidity also varies during a storm event. For example, the movement of saturated soils at particular sites, almost as muddy debris flows, is responsible for the formation of scars. The morphology of the resultant rills means that turbidity rises very rapidly at first, and then more slowly (Kalman 1976); after an initial period of erosion, which is responsible for the rapid increase of turbidity, the rills quickly stabilize and turbidity decreases.

Assessment of sediment sources (Heusch 1970) shows a generally low participation of the slopes compared with gullies and river banks. If the relative contribution by slopes is normally low (from 3 to 19%), it tends, however, to be equal to that of gullies and badlands during extremely wet years (about 20% of the total erosion for each of the two processes), as was the case in the Rif in 1963 and in 1968–1969. It seems that during those particular situations, the slopes start to function under the effect of scars, gutters and solifluction processes (Maurer 1968a; Sari 1977). The river, the focus of the watershed, then receives abundant volumes of water containing high sediment loads. Vertical downcutting does not occur under those circumstances, and sinuous or meandering flow is responsible for undermining the river banks and even slopes underlain by erosive lithologies. Often, soil pipes transform the flow on slopes into a torrential mudflow capable of transporting thick blocks. The load then increases in volume and has the capacity to incorporate large elements. Slope processes are then responsible for more than half of the materials being transported. But it is not certain that the load is transported very far. The sediment-saturated flow is subject to frequent redeposition.

With regard to rainfall intensity and sediment movement (Bourouba 1993), sediment loss reaches its peak in the Saharan Atlas and the Seybouse in autumn, when 70 to 80% of the annual tonnage is removed (Conseil Supérieur de l'Eau 1991a). The material is prepared by the rains of the previous season and then loosened by the heat and stock trampling, and evacuated by the first rains. In contrast, in the mountain resorts the maximum sediment movement occurs in summer, for the rainstorms have high intensities, whereas in the Tellian resorts, the maximum movement occurs in spring (32 to 59%). Here there is an association of mass movement and gullying on ill-protected slopes and the reaction occurs only after saturation.

Indeed, the Mediterranean-type climate is particularly suitable for landsliding, as the maximum supply of water to infiltrate into the ground is provided when deciduous forests are at rest, so that transpiration and canopy protection are minimal. Extensive movements are triggered at the end of the rainy periods, with some being linked to the undermining of channel banks which occurs at the beginning of autumn on account of the first floods.

Topography

In the Mediterranean Sea countries, geologic and Quaternary history and present-day tectonic activity have resulted in a rugged topography, which is an important source of degradation processes, especially taking into account the characteristics of Mediterranean climates as outlined in the preceding section. Alpine mountain uplift created chains of mountains with thousands of metres of altitude located very near the coasts. Post-Alpine movements produced numerous tectonic depressions within the chains. The results are high-energy streams which, during Quaternary climatic changes, had torrential flows producing fluvial incision and accumulation, moulding valleys and building alluvial plains and terraces. In short, high relative relief coupled with valley incision resulted in rugged topographies and steep slopes. The multiplicity of valleys results in a variety of environments, because at these latitudes slope aspect produces important effects on water dynamics and on vegetation. The situation is similar in North and South American Mediterranean environments, and to a lesser extent the southwest Cape. Australia's Mediterranean region coasts are also backed by ranges, but with much lower elevations. Nevertheless, "Mediterranean" is generally synonymous with mountainous terrains and a variety of landscapes.

Despite generally steep topography in Greece, however, in general rill and gully erosion are not evident in forest soils where natural ground covers are undisturbed. But in cases of appreciable disturbance or total destruction (for example, forest fires, mismanagement, logging or overgrazing), resulting in bare soil surfaces, the erosion risk becomes rather high. In those cases, topography has been accepted universally as the dominant factor affecting the rate of soil loss through water erosion. Accelerated erosion favoured by steep slopes combined with improper land use, deforestation and over-grazing has led to the formation of shallow skeletal soils.

Besides slope gradient, aspect also has a great effect on vegetation and land degradation. Soils on south-facing slopes have generally higher temperatures (5–11 C at 5 cm depth), lower moisture contents and greater evaporation rates than soils on north-facing slopes. This may account for the greater loss of vegetation from south-facing slopes in several hilly and mountainous areas of the country.

Earthquakes are a common occurrence all around the Mediterranean Sea, and also in California and Chile. This is a natural hazard which directly triggers land degradation, especially of a catastrophic type.

In Italy, for example, evidence of landsliding triggered by strong earthquakes is widespread in ancient chronicles and scientific reports (De Dolomieu 1785; Cotecchia *et al.* 1969; Agnesi *et al.* 1982). Earthquakes act upon slope stability through different processes. Very briefly, the earthquake imparts an acceleration to the ground which adds to gravity acceleration. In addition, volume waves cause alternate increasing and decreasing of neutral pressure in the ground. Lastly, strong earthquakes may cause surficial modifications (increase of slope height and angle, rock folding and faulting) to which the slope mass reacts by gravitational adjustment.

Over the long term, past and present tectonic activity results in maintaining high relief and slope gradients, and in the jointing and faulting of the rocks, weakening them severely (Plate 15.1).

The tectonic mobility of the North African Mediterranean zone accounts for the continuing uplift (and hence the rate of downcutting by rivers) and the repeated intervention of earthquakes which are responsible for large slumps. In the eastern Tell, the 1985 earthquake caused surface ruptures over 3.8 km along a fault, and slumps were reactivated (Benazzouz 1993). The Moroccan mountains are rising at about 0.5 mm/yr (Weisrock 1980).

Lithology and Pedology

Parent rock and its weathering products create different denudational environments.

Under similar Mediterranean rainfall conditions, differences have been found in erosion rates in relation to lithology in mountainous areas of the northeastern Iberian Peninsula (Table 15.3). On the vegetated phyllite slopes of Montseny mountain, with average annual rainfall ranging from 800 to 1200 mm, 0.003 t/ha/yr of fine materials (<2 mm) were washed from the slope, although the movement of coarse material downslope increased the total rate up to 0.02 t/ha/yr. On weathered granite slopes total material washed from a beech woodland was 0.015 t/ha/yr. But when the comparison is made amongst devegetated areas the role of lithology is still more clear. On marl and clay badland slopes erosion is higher (266 and 454 t/ha/yr respectively), than on burned phyllite and granite slopes (1.1 and 7.3 t/ha/yr respectively).

Southeast Spain is one of the Mediterranean areas recognized as experiencing high erosion rates, particularly in regions underlain by soft lithologies (Plate 15.2). Areas which illustrate the role of marls and clays are Murcia, Granada and

Plate 15.1 *View down a large landslide which has dammed the river at the base. This Calabrian landscape is characterized by tectonic instability, steep slopes and deeply weathered (fissured) rocks*

Table 15.3 *Erosion rates (in tonnes per hectare per year) in relation to lithology, vegetation and rainfall, Spain*

Lithology	Beech	G. oak	Matorral	Felled	Burnt	Bare	Rainfall (mm)	Source
Phyllite		0.003	0.009				500	Sala (1988)
Phyllite		0.006		0.03	0.27		500	Soler and Sala (1992)
Phyllite					0.32		600	Sala *et al.* (1994)
Granite	0.015						1000	Sala (1988)
Granite					3.79		600	Sala *et al.* (1994)
Granite					2.6		550	Cartaña and Sanjaume (1994)
Metasediment					5.44		550	Ubeda and Sala (in press)
Marl						266	1000	Cervera *et al.* (1990)
Clay						454	1000	Cervera *et al.* (1990)

Plate 15.2 *Part of an extensive area of badlands in marls, near Murcia*

Almería (in a more arid zone), which have been studied extensively. While the high erosion rates and the widespread badland landscapes are often related to land use strategies, the marl and clay rock types play a most fundamental role. Literature on badland morphology shows that these landforms can be found in many climatic zones (Bryan and Yair 1988).

All along the Ebro sedimentary basin, for example, except for the alluvial fills, marls are the dominant lithology and with them are rilling, gullying and badland topography. At the lower part of the Pyrenees chain, flysch materials are composed of alternating soft marls and tougher sandstones. Especially when these contrasted sedimentary strata have been tilted, erosion of the marls is severe and with it the overall erosion of the slopes, producing a badland topography and a large amount of sandstone rock fragments.

Eocene conglomerates with limestone cementation all around the Ebro basin are subject to solution processes along cracks, also producing a considerable amount of cobbles released from the conglomerate and from rockfalls.

In relation to soils, Díaz-Fierros and Benito (1996) found that, although the understanding of the erodibility of the soils of the Iberian Peninsula is still incomplete, the two parameters for which some quantitative basis exists are structural stability and the USLE factor K, although representing approximations which explain only isolated aspects of the more general phenomenon of soil degradation (Figure 15.5). For example, in soils rich in calcium carbonate, as is the case of 45% of the soils in the Iberian Peninsula, good structural stability may be associated with high erodibility. Likewise, K values calculated for soils with more than 30 or 40% of free calcium carbonate, as in many Iberian regions, are of doubtful validity. Saline and gypsum soils pose particular problems in terms of evaluating structural stability and erodibility. Soils with a low K value (less than 0.10) are grouped together in the humid zone, where organic matter plays a fundamental role in structural stabilization. At the opposite extreme, with K values in excess of 0.40, are Regosols, Luvisols, Yermosols, certain Fluvisols and, above all, saline soils.

In Italy, the main causes of landsliding in the Mediterranean zone are to be found in the widespread outcropping of low-resistance strata, intense present and past tectonic activity, climate and human activity (Cotecchia and Melidoro 1974).

The occurrence of outcrops of weak formations is illustrated in Figure 15.6. "Scaly shales" (stiff clays with a scaly structure due to tectonics) are the most landslide-prone rock types, as they may reach 100% of landslide incidence.

When data are broken down according to slope inclination (Sorriso-Valvo 1993), it appears that maximum landslide incidence occurs on

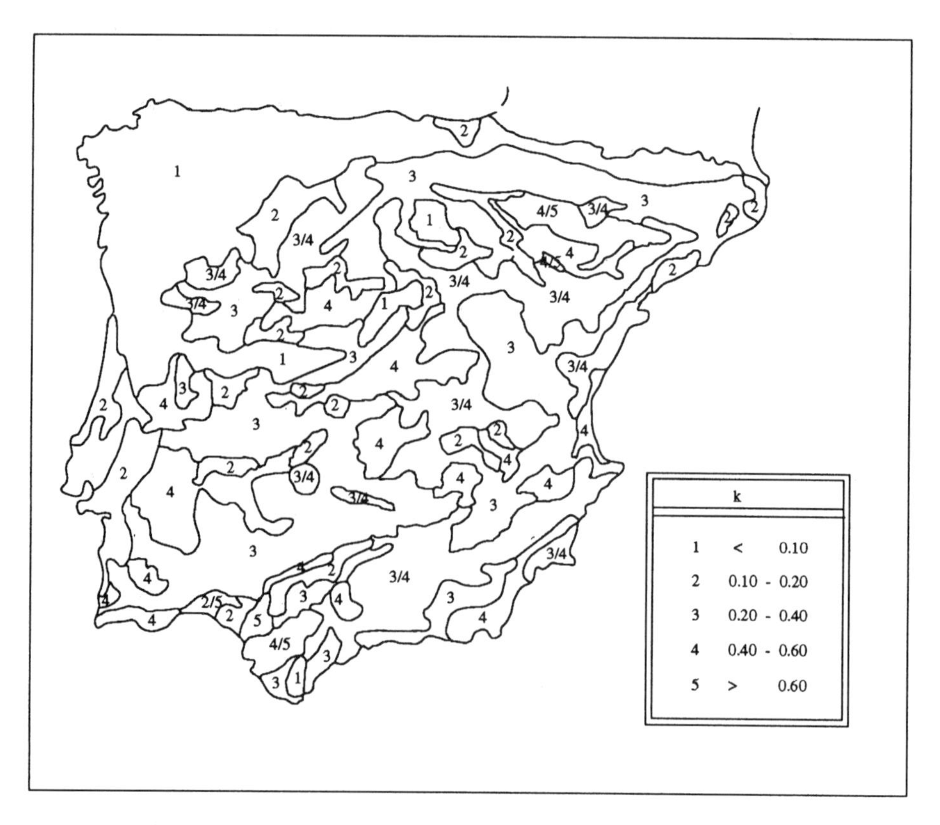

Figure 15.5 *Map of K values in the Iberian Peninsula: rainwash erodibility of Iberian soils* (after Diaz-Fierros and Benito 1996)

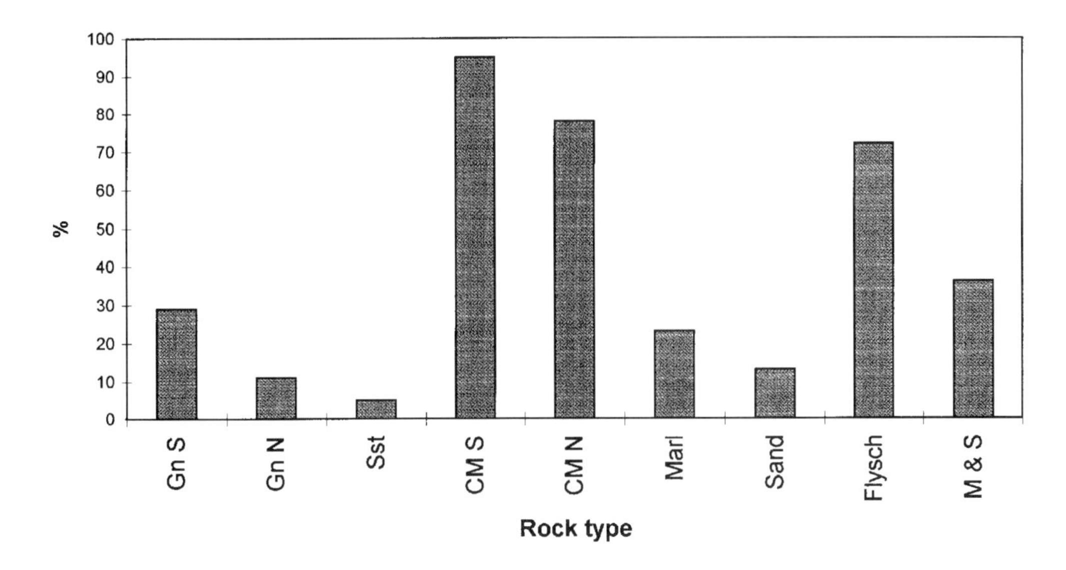

Figure 15.6 *Landslide incidence on different lithologies. Acronyms: Gn S = gneiss of south Calabria; Gn N = gneiss of north Calabria; Sst = sandstone; CM S = clayey melange (scaly shale) of south Calabria; CM N clayey melange of north Calabria; M & S = marls and sands* (after Sorriso-Valvo 1993)

slopes with intermediate inclination, except for hard sandstones and granites (Figure 15.7). This is easily explained by the fact that, in general, the geological land classification is a simplification of the real situation, so that lithostratigraphic formations with non-uniform composition (which is most of the geologic formations) actually include sections with slight differences in geomechanical characteristics. Because of that, in areas with stronger mechanical properties, slopes may rest at higher inclination, and landslides are less frequent. This is typical for cohesive or cohesive-like rocks (such as weathered gneiss). Brittle materials, like sands or hard, unjointed sandstones, behave differently: the rock quality is more regular and landslides occur on the steepest slopes, as expected.

Many Greek soils are formed on marl deposits. Soil data over the last 60 years have shown that a thick, dark, surface horizon with strong structure is formed on these soils (mollic epipedon). In most cases, however, this epipedon has greatly degraded due to the reduction in organic matter content and aggregate stability which occurred after clearing of the natural vegetation followed by cultivation.

Extensive degraded areas are primarily confined to rock formations of Mesozoic limestones and secondarily to acid igneous and metamorphic rocks. Soils formed on limestones usually have moderately fine to fine texture. Drier micro climatic conditions prevail on these areas, reducing the potential for plant growth, and the soils remain bare for long periods, favouring overland flow and erosion. The soils on these areas are very shallow or the parent rock is exposed at the surface.

Soil textures on acid rocks are usually moderately coarse to medium and the aggregate stability is low. Therefore, soil erodibility on these parent rocks is high and land degradation has proceeded over large areas. It is estimated that about 28% of the country is degraded in part due to the presence of these parent materials.

Hilly areas with a substratum of shales/sandstones or flysch exhibit a smaller erosion risk. The soils are moderately to fine textured and permeable, and have a moderate to rich vegetation cover. However, if the natural vegetation is removed (by fires, forest clearance, etc.), the areas on flysch are very susceptible to gully erosion and landslides.

Extensive hilly areas cultivated with rain-fed cereals are underlain by marl, shales/sandstone and conglomerate deposits (Plate 15.3). The importance of lithology on erosion rates has been

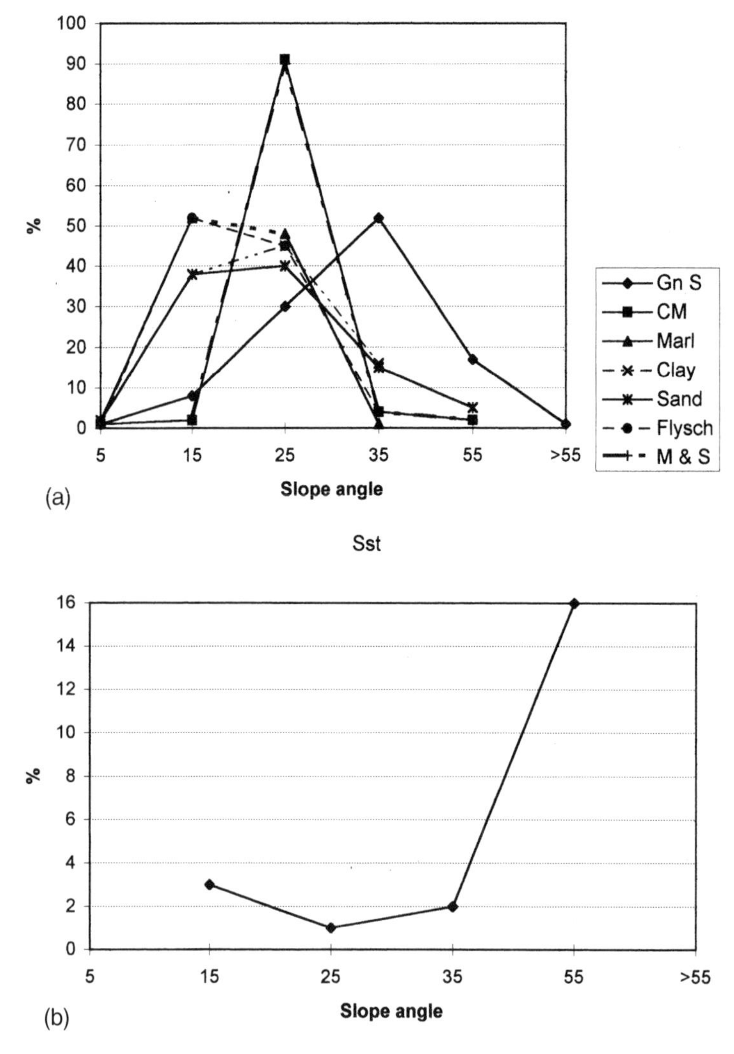

Figure 15.7 *Relationship between landslide incidence and slope angle for the same lithologies as in Figure 15.6, except for Gn N. (a) Data for lithologies which have the maximum incidence of landslides at intermediate slope angles; (b) the same diagram for sandstone, which displays a minimum incidence of landslides for intermediate slope angle* (after Sorriso-Valvo 1993)

demonstrated in several studies (Heusch and Millies-Lacroix 1971; Demmak 1984; Meade and Parker 1985). Studies of erosion rates have been conducted along certain hillslope components and on soils formed on the above parent materials using a rainfall simulator (Kosmas *et al.* 1995c). The results showed that soils on marls appear to have considerably lower (standard) sorptivity, and exhibit higher runoff rates and sediment loss under similar slope gradients and management practices, than soils on shales/sandstone or conglomerates (Figure 15.8). Measured runoff rates varied considerably among bare soils with different parent materials, ranging from 42.5% to 25.8% and 32.5% of the applied rain for soils on marls, shales/sandstones and conglomerates, respectively. Sediment losses also varied considerably, with soils formed on marls losing an average of 491 g/m², compared with soils formed on shales/sandstones and conglomerates losing an average of 21 g/m² and 176 g/m², respectively. The presence of wheat residues on

Plate 15.3 *General view of one of the extensively farmed areas in which Kosmas carried out the experimental work discussed in the text*

the soil surface significantly reduced the runoff and sediment losses on all soils. Crop residues create favourable conditions for protecting the soil aggregates from rainsplash.

Nevertheless, soils formed on shales/sandstones are being abandoned due to land degradation at higher rates than soils formed on marls or conglomerates under the same climatic conditions. Although erosion rates of soils on marls are high, their productivity generally remains at relatively high levels due to the absence of restrictive bedrock layers, such as those present beneath soils on shales/sandstones.

However, studies of biomass production on soils with different parent materials along catenas have demonstrated that productivity varies with climate. Lands on marls are very susceptible to degradation in particularly dry years. The soils cannot support vegetation during such conditions, despite their considerable depth and high productivity in normal and wet years (Kosmas *et al.* 1993a). On the contrary, soils on conglomerates and shales/sandstones, despite their normally low productivity, may supply appreciable amounts of previously stored water to the stressed plants and secure a not negligible biomass production in particularly dry years.

Rock fragments

Soils on shales/sandstones and conglomerates usually contain rock fragments in varying amounts depending on landscape position and degree of erosion, and the mechanical breakage of petrocalcic surface horizons may also create a layer of coarse materials. Further, the washing of fines in a mixed stony soil is a common occurrence in slopes with periglacial deposits. In that sense it should be taken into consideration that cold Quaternary periods around the Mediterranean Basin were characterized by the production of a high amount of debris by gelifraction, especially in interfluves composed of schist and limestone. These fragments moved downslope either by gravity or runoff, producing the well known *grèze littée* deposits, or by solifluction when enough clay-sized materials were available.

The presence of rock fragments, especially on the soil surface, appears very important: (a) in dry years, by conserving appreciable amounts of soil water from evaporation through surface mulching (Kosmas *et al.* 1993b); and (b) by affecting soil erosion rates (depending on the size of the rock fragments) (Moustakas *et al.* 1995).

Rock fragments may have positive or negative effects on soil properties affecting erosion. Large

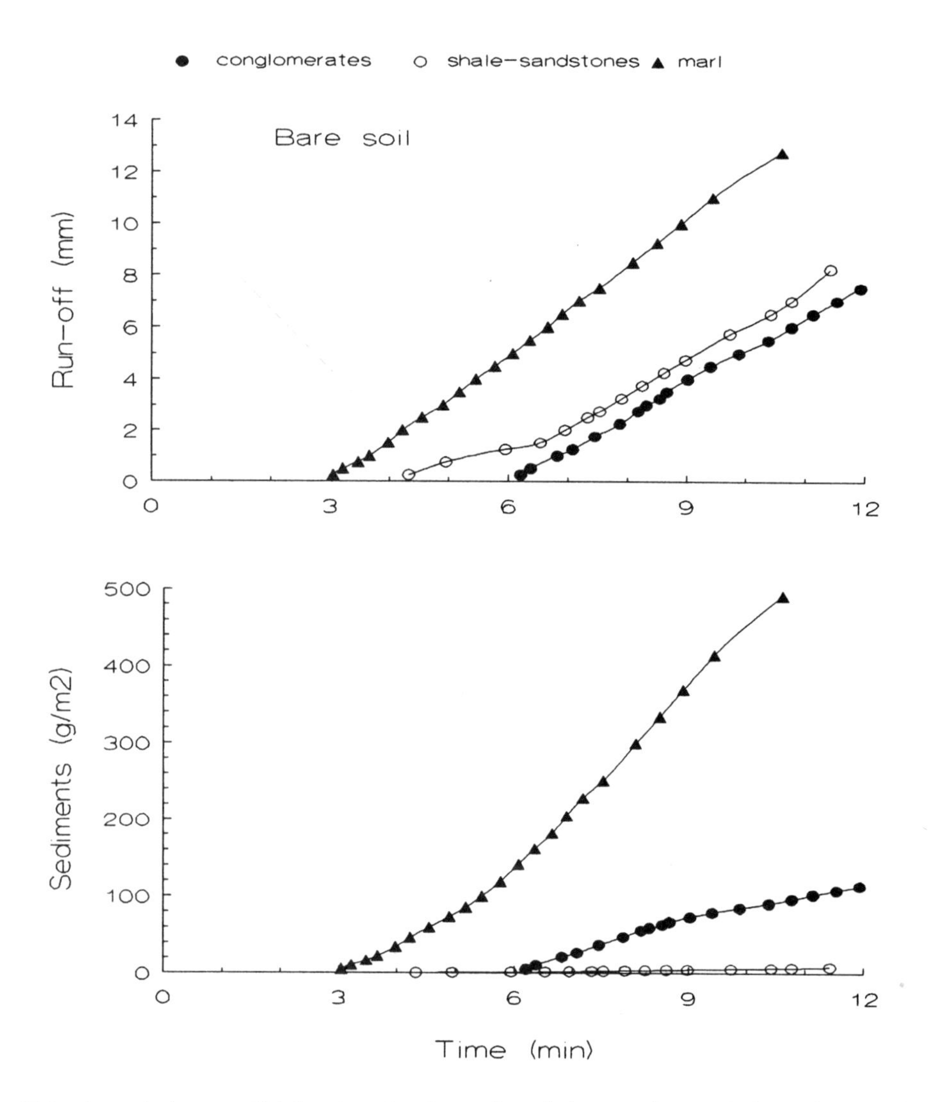

Figure 15.8 *Cumulative runoff (above) and sediment loss (below) on bare soils formed on marls, shales/sandstones and conglomerates measured in a rainfall simulation experiment with a constant rainfall intensity and similar slope* (source: Kosmas *et al.* 1995c)

rock fragments (cobbles) on the soil surface generate greater runoff than smaller fragments (coarse gravel). By increasing rock fragment cover on the soil surface, the water-conducting area decreases and thus the hydraulic conductivity of the surface is decreased. However, soils with appreciable amounts of coarse gravel on the surface generate considerable runoff under rainfalls of low intensity and long duration, but much smaller amounts at greater rainfall intensities (Figure 15.9; Plate 15.4). Sediment loss is more favoured on soils with cobbles than on soils containing coarse gravel, or even from soil free of rock fragments.

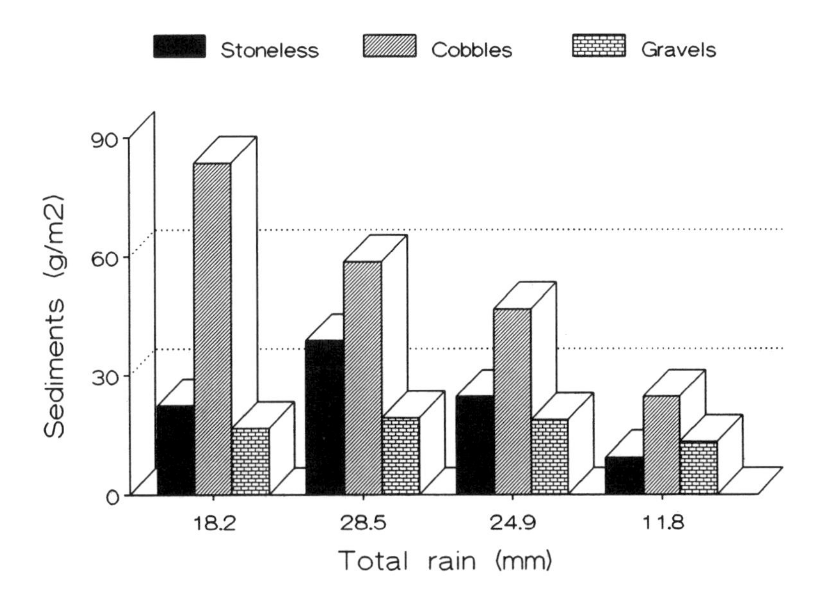

Figure 15.9 *Sediment loss measured in runoff plots containing different amounts and sizes of rock fragments placed on the soil surface (cover 18%)* (source: unpublished data from Kosmas and Danalatos)

Plate 15.4 *Part of the experimental design of the research on the hydrologic effects of rock fragments (amongst a range of other factors) by Kosmas: (a) runoff plots; (b) lysimeters*

16

The causes of land degradation

2: Vegetation clearing and agricultural practices

Compiled by Arthur Conacher and Maria Sala from contributions on the following regions:

Iberian Peninsula and Balearic Islands	**Maria Sala** and **Celeste Coelho**
South of France and Corsica	**Jean-Louis Ballais**
Southern Italy, Sicily and Sardinia	**Marino Sorriso-Valvo**
Greece	**Constantinos Kosmas, N.G. Danalatos** and **A. Mizara**
Dalmatian coast	**Jela Bilandžija, Matija Franković, Dražen Kaučić**
Eastern Mediterranean	**Moshe Inbar**
North Africa	**Abdellah Laouina**
California	**Antony** and **Amalie Jo Orme**
Chile	**Consuelo Castro** and **Mauricio Calderon**
Southern Australia	**Arthur** and **Jeanette Conacher**

Introduction

Chapter 15 considered the nature of the biophysical environment as an important underlying cause of land degradation in Mediterranean environments. But it is clear that without human interference, land degradation as defined in this book would not take place: instead, "normal" geological, geomorphic, pedological, hydrological and biological processes would prevail. Certainly these processes may pose significant elements of risk for humans, but such events are more usually regarded as "natural hazards", not land degradation.

Prominent amongst human actions responsible for land degradation are clearing of the natural vegetation for agriculture, and the agricultural practices themselves. It will be evident that separating discussion of these activities from the historical context in which they took place is impossible. But the focus here is on human actions as a *cause* of degradation, not the historical development of degradation *per se* (which is discussed in Chapter 14). Other important human actions are considered in Chapter 17.

Clearing of Vegetation, Mainly for Agriculture

The Iberian Peninsula

Historical deforestation in the Iberian Peninsula resulted in increased runoff, flooding of lowlands with loss of orchards, and the development of gravel-bed rivers. At present, forests in Mediterranean Spain have become a recreational resource for urban populations, a compensatory breath from the urban atmosphere. Rural settlements have been abandoned.

More recently, and especially in Portugal, clearing of *montado* and *dehesa* systems in order to transform them into solely agricultural areas has become a serious cause of land degradation. The trees of *montado* play an important role in the hydrological cycle and in the recovery and protection of soils, with their root systems improving soil structure and increasing water retention capacity. Concentrations of phosphate, potassium and nitrogen are higher in soils of *montado* than under cereals or pasture (Salgueiro 1976). Other implications include loss of humus, modification of soil structure, reduced soil moisture, and loss of fertility of the superficial layers. Consequently the effects of water erosion and aridity are all aggravated.

Italy and Greece

From a great deal of experimental data, some from Mediterranean Italy (Sorriso-Valvo *et al.* 1995), it is evident that erosion rates are much higher where soil conditions are already degraded and the protective effect of vegetation cover is reduced. The reduced density of root channels reduces infiltration, so runoff, and thus erosion, is increased.

In Greece, the extensive deforestation of hilly areas and intensive cultivation with rain-fed cereals had already led to accelerated erosion and degradation in the 19th century. Soils on hilly Tertiary and Quaternary landforms usually have limiting subsurface layers, such as petrocalcic horizons or bedrock, and under high erosion rates and hot and dry climatic conditions, cultivation of cereals is no longer feasible, leading to land abandonment. Several hilly areas, especially in the eastern part of Greece, are being abandoned at an increasing rate.

The eastern Mediterranean

In mountain areas such as Mt Carmel in the eastern Mediterranean, the *maquis* forest cover is efficient and almost no runoff or sediment yield has been detected in experimental plots (Rosenzweig 1972; Inbar *et al.* 1995). However, the change from natural vegetation to cultivated or grazed lands exposes bare soils to the effects of the winter storms.

Naveh and Dan (1973) have described soil/vegetation degradation in Israel caused by humans. Following the destruction of vegetation, the underlying soils on the coastal plain were

Land Degradation in Mediterranean Environments of the World: Nature and Extent, Causes and Solutions.
Edited by A.J. Conacher and M. Sala.

eroded, with the undulating to rolling slopes being dissected by small gullies. Soil B horizons, and in some areas the underlying sand and sandstones, were exposed by erosion. In the foothills and mountain areas, the effects of erosion processes have been more pronounced, as the soil depth is shallower, and there is both greater exposure of the underlying rock and a greater influence of the lithology on pedogenic processes (Plate 16.1). Eroded areas are colonized by more resistant plants, like *Thymus capitatus*, a shrub component of the low batha formation. More dense areas are covered by *Asphodelus microcarpus* and other perennial grasses. On degraded colluvial slopes, shrubs such as *Pistacia lentiscus* and *Rhammus palaestina* can be found. After cessation of human interference, a *maquis* shrub vegetation develops again over a period of several decades (Naveh 1955), as exemplified in the no-man's-land along the Israel–Lebanon border (Plate 16.2).

Removal of vegetation cover enhances erosion, and major changes occur with a cover ranging between zero and 30% (Nortcliff *et al.* 1990). Erosion declines exponentially with an increase of vegetation cover, as found by Thornes (1990) and shown in Figure 16.1 from study sites in Mt Carmel (Inbar *et al.* 1995).

North Africa

The plant cover of the soil also plays an important role in North Africa, as illustrated by an interesting comparison between wooded sloping catchments and cultivated catchments. Research was carried out on three catchments in Ikaouen, comparable in all respects except soil use (Heusch 1970, 1982). The first catchment, totally wooded, loses only 5.85 t/ha/yr. The second one, partly cleared and cultivated, undergoes a soil loss of 18.45 t/ha/yr. The third catchment, which has been occupied for a long time and is totally cultivated, exhibits much more severe erosion, with soil losses of 93.7 t/ha/yr. Increased human action, then, is expressed in more severe erosion.

In Algeria, the only dam which does not silt up is that of Beni Bahdel in Orania. Its sloping drainage basin is entirely wooded. In comparison, the Oued Fergoug dam, with an entirely cleared catchment, received $23.5 \cdot 10^6$ m^3 of sediment in 22 years, and the Oued Fodda dam was silted up in only 12 years (Benchetrit 1972).

Other research in North Africa has also shown that bare or tilled soil undergoes much more severe erosion than soil with a permanent plant cover. The plantations of corks at Ikaouen, in the central Rif, represent the most stable

Plate 16.1 *Bare rock surfaces in western Galilee: a former terraced slope, now degraded as a result of overgrazing by goats. Photo by M. Inbar*

Plate 16.2 *Re-established* maquis *shrub after the cessation of human interference, in the no-man's land along the Israel–Lebanon border. Photo by M. Inbar*

environment with a runoff coefficient of less than 5% and insignificant soil loss. Soils with wheat crops interspersed with *matorral* are also the less eroded ones (<0.9 t/ha). But slopes with wheat monocultures have soil losses of up to 21 t/ha/yr (Debazac and Robert 1973).

The comparison of several situations is instructive (Heusch 1990):

1. A wooded slope with a close ground cover in the Rif undergoes practically no soil loss. Hilly, wheat-cultivated soil loses on average 2.6 t/ha/yr. On mountain slopes, wheat fields lose an average of 3.2 t/ha/yr.
2. An entirely covered, small, sloping basin loses 5 t/ha/yr.
3. The basin of the Leben river, located in an extensively stripped, hilly zone, undergoes soil loss of 22.5 t/ha/yr.
4. The basin of the Sra river, which drains a densely populated mountain, loses 35 t/ha/yr.
5. Badlands undergo a loss of 150 t/ha/yr in the Pre-Rif and 260 t in the Rif.

In Petite Kabylia, as in the Rif, the extension on the piedmonts of colonial agriculture, particularly vineyards, accounted for the migration of a great number of inhabitants towards the interior of the mountain (Tatar 1995). Thus, in 1946 population densities rose to 100 persons/km^2 in the peripheries and to 40 in the mountains. Such a concentration required the development of new lands at the expense of the natural vegetation cover. Comparing the vegetation maps of 1934 with those of 1960 shows that the forest, originally well preserved, surrounded by fields and glades, "maintains itself only at a third of its former extent" (Benchetrit 1972). The towns, in full growth, influenced the degradation of the plant cover, for energy needs had to be provided for, notably for Tunis and Tetouan. Demands for wood to be used in mines also accounted for the decline of extensive forests.

Clearing rates have undergone different phases, particularly during Independence when the populations wanted to establish property rights in the forest domain which the Spanish colonization had not demarcated. At present, the rate of final clearing has decreased considerably, with the operation taking place increasingly in small patches and surreptitiously. Quantitative estimates by some authors in certain municipalities cannot be generalized to the entire mountain (Benabid 1982, 1992; Ahmadan 1991).

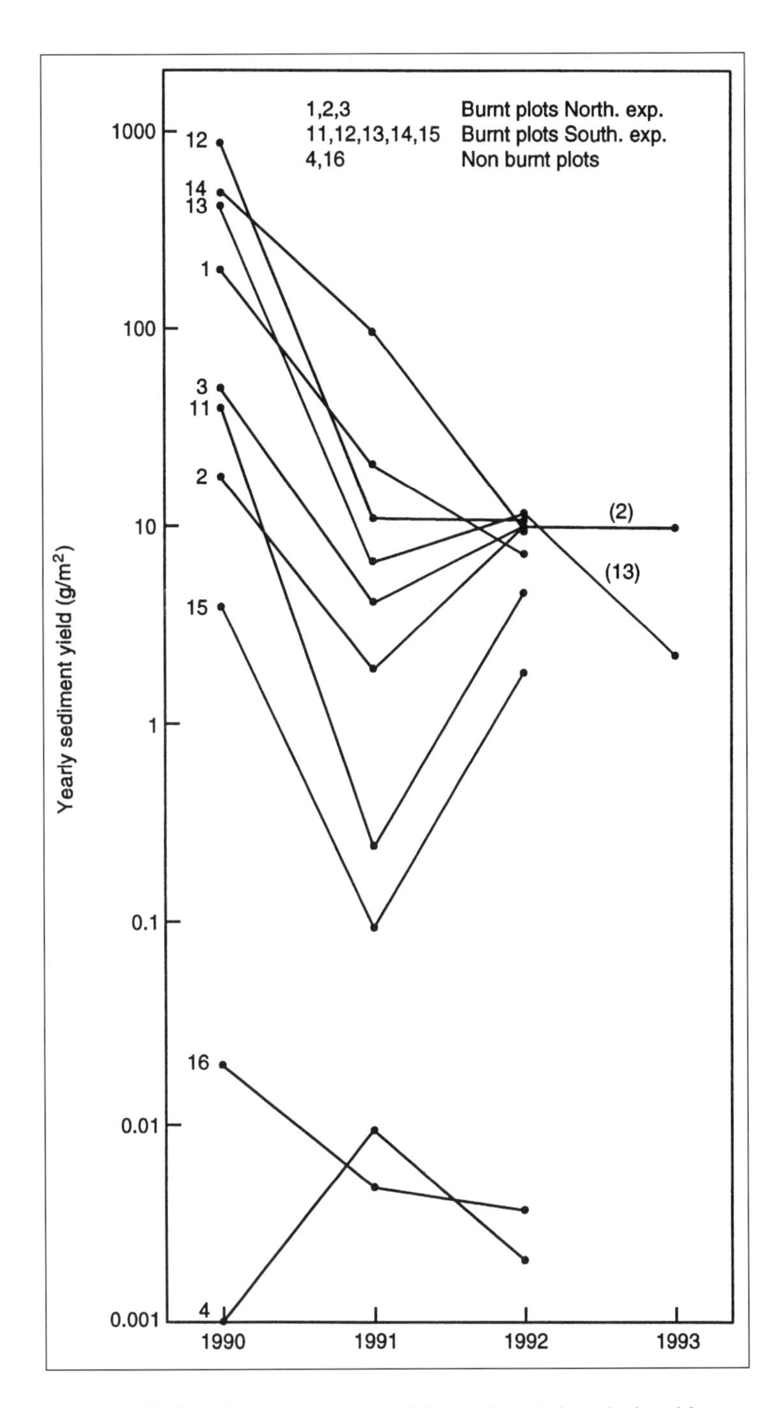

Figure 16.1 *Sediment yield after the 1989 Mt Carmel forest fire. Values declined by one to three orders of magnitude after the first rainy season* (unpublished data by Inbar)

Southern Australia

Indigenous vegetation removal by European settlers (and its replacement with introduced agricultural crop and pasture species) has been directly responsible for the loss of plant cover and habitat for wildlife in southern Australia. The introduction of exotic "pest" species of flora and fauna, and fungal diseases, exacerbated these losses. Fire, by modifying important physical, chemical and biological processes in various vegetation systems, has caused significant species shifts or losses (Gill *et al.* 1981; Adamson and Fox 1982; Kirkpatrick 1994).

Hydrological changes

Less directly, the replacement of natural vegetation with exotic species, which have marked differences in form, structure, canopy and root systems, sets in train a series of hydrological changes. Australian data show that eucalypt forests and mallee woodlands intercept from 10% to as much as 35% of mean annual rainfall (Conacher and Conacher 1995), whereas there are times of the year when agricultural plant cover is sparse or even non-existent. Thus a greater proportion of rain than previously reaches the soil surface directly. Further, the absence of a litter layer (following removal of the indigenous vegetation) means that the soil is unprotected, allowing rainbeat to pack surface soil particles causing surface sealing. The number and richness of soil biota species are also diminished significantly following clearing (Lobry de Bruyn 1993; Hobbs *et al.* 1993), and the consequent loss of soil structure causes a reduction of infiltration capacity (Lobry de Bruyn and Conacher 1994). All the indications, therefore, are that the replacement of indigenous vegetation with introduced crops and pastures leads to: an increase in rainwater reaching the soil surface directly; reduced infiltration into the soil; a consequent increase of overland flow; and, therefore, accelerated erosion and removal of nutrients and biocides as leachates.

The above vegetative changes also lead to increased subsurface water, both in perched aquifers (or "throughflow", which is often ephemeral or seasonal, but occasionally perennial) and in the deeper, permanently saturated groundwater aquifer or aquifers (which are often present in the deep – up to 45 m – intensely weathered soils). The increase in perched and groundwaters occurs because the indigenous vegetation (whether forests, woodlands or heath) has long taproots enabling it to draw on groundwater throughout the year. In contrast, the introduced crop and pasture species have shallow root systems and grow only during spring and early summer, extracting less water from the ground.

From drier (around 300 mm/yr), topographically subdued parts of the Western Australian wheatbelt, George (1992) and George and Conacher (1993) have shown that the rates at which groundwater (in the deeper, regional aquifers at the base of the deeply weathered soils) moves out from catchments are very slow, in the range of 0.05 to 0.3 mm/yr. These discharge or natural drainage rates were presumably in equilibrium with rates of groundwater recharge and water use by plants prior to clearing. With the catchments now cleared for agriculture, recharge to groundwater averages 10 mm/yr (range 6–15 mm/yr, lower than some previous estimates). This large disparity between recharge and drainage of the groundwater aquifers is reflected in rising water levels in deep bores. In areas with <500 mm/yr rainfall, the rate of groundwater table rise is 0.05–0.25 m/yr, and much more rapid in higher rainfall areas (see also Rose 1991).

The rising water table of the saline groundwaters, with salt concentrations (predominantly sodium chloride) ranging from around 10 000 mg/l in upland areas to >100 000 mg/l beneath some broad, flat valley floors, may reach the surface directly, or come within the capillary fringe of the soil surface, or mix with throughflow. All of these processes deliver salts to the surface. Additionally, throughflow alone, when associated with deep aeolian sands, causes mildly saline seeps over much of the eastern wheatbelt of Western Australia.

Varying combinations of these processes are directly responsible for the major problems of waterlogging and secondary, dryland salinization (George and Conacher 1993).

Agricultural Practices, Forestry

Pastoralism and grazing

Pasture is in recession and transhumance has nearly disappeared from the Iberian Peninsula. Since the 1960s, over-grazing has been replaced by subgrazing, mostly in relation to the agricultural policies of the European Union. This has also favoured the recolonization of shrub vegetation in many areas, which favours soil regeneration but increases the risk of fires.

Grazing can be a problem for soil conservation, but the direct effects of grazing, in Italy, have never been particularly serious. Sheep were once the most important livestock in Italy, mostly in the hills and mountains. They are still the most important breed in Sardinia and part of Sicily. A few decades ago, in southern Italy, there were pathways formed of grass-covered strips some 50 to 100 m wide, delimited by stone walls, and tens to hundreds of kilometres in length; these were the *tratturi*, through which livestock were conducted from one pasture to another (this practice was called *transumanza*). Bare paths were instead left for sheep and goats in Sicily. Today, most livestock is kept in fence-delimited ranges. Over-grazing is seldom a problem because the stocking rate of cattle is not very high.

Terracettes are very frequent, and in most cases they are due to trampling by cattle. However, they can also be found in areas where cattle are not present; they affect the part of the slope with steeper gradients, so they are probably due also to the process of micro-slumping or creeping of grass turf, or other phenomena yet to be understood.

In the eastern Mediterranean, on the other hand, over-grazing wrought a strong reduction of primary productivity and standing crop, developing new habitats: *garrigue*, *maquis*, cultivated sites, and an evolutionary process characterized by a reduction of the vegetative apparatus towards a transition to synanthropic habitats (Pignatti 1983).

In North Africa, increased livestock numbers in the 19th century in Algeria and again since Independence, have also constituted a cause of land degradation. From 1970 to 1980, a doubling of the herd in Petite Kabylia was noticed, hence over-grazing, with forest cutting exceeding the resilience of the environment (Tatar 1995).

In California, cattle grazing, which expanded rapidly during the Mexican and early American periods, was an early cause of widespread land degradation, particularly on coastal slopes and foothills around the Central Valley. Over-grazing led to changes in soil chemistry and structure, to soil erosion, and to the replacement of native bunch grasses with European annuals. Despite widespread recognition of the degradation caused by over-grazing, notably by 19th century observers such as Bryant (1848) and Brewer (1864), comparatively few quantitative data presently exist regarding the magnitude of this erosion before 1900. The alluvial fills and *arroyos* found in many valleys within the Coast and Transverse Ranges merit further investigation in this context, notably because they contain palaeosols and charcoal seams which appear to indicate rapid changes in the erosional regime within the historic past.

In Chile, over-grazing of the natural meadows /prairies is considered to have initiated the degradation of the natural resources. Loss of organic matter facilitates soil erosion and desertification, and with them the increasing poverty and out-migration of the rural population.

In diagnosing the factors threatening the flora resource, over-grazing also constitutes an important negative factor in some ecosystems. High densities of animals cause selective pressures on certain plant species or groups of species. After a time, a change in the original composition of the community occurs, tending to alter the prairie condition and diminishing its biodiversity.

Clear-felling and the increased intensity in the use of prairies led to the reduction of cattle numbers until they became insignificant. Sheep numbers then increased, but the final stage of this evolution is the predominance of goats. They are capable of intensely using the prairies, leaving them in a poor condition. Their uncontrolled grazing has led the present ecosystem to a state of "agri-desert" in extensive zones of Regions III and IV.

In addition, the main form of degradation in Regions IV and V is the harm done to the prairies by over-grazing, principally in the prairies of the Upper Cordillera and in the interior dry farming lands. On the other hand, the Coastal Cordillera of Central Chile has suffered a massive environmental impact due to the combined action of dry farming cultures and pastoralism, in association with clear-felling and the harvesting of wood crops.

Cultivation

Once cleared, the land is more often than not used for agriculture. A wide range of practices is involved, all having various ramifications for land degradation.

Traditional agriculture in the Iberian Peninsula was based on rotation and fallow in alternate years (Duero and Ebro basins), or every third or fourth year (western peninsula). Fallow favours wind erosion, especially where the winds are strong as in the Ebro basin, in Extremadura and in La Mancha. At present, soil degradation is related to monoculture, the use of heavy machinery and chemicals. In North Africa, the use of fallow, originally biennial, is decreasing in the rotation of cultivations. Most lands are tilled each year.

In the experimental field of "El Ardal" in the Mula basin, Murcia, the impact of different land uses on runoff and erosion is being studied in areas cleared of natural vegetation (López-Bermúdez *et al.* 1996). Data from the period 1989–1994 clearly demonstrate that maximum erosion occurs under ploughed conditions when the work is done downslope (Table 16.1; average data). The cutting of *matorral* is the second most important cause of erosion. Minimum erosion takes place under fallow conditions in stony soils. When cultivated, wheat is the less protective crop.

Stubble burning and excessive ploughing are considered responsible for the destruction of soil aggregates in Chile. Ploughing down the slope, lack of protection of natural drainage ways and inadequate rotation of crops, are amongst the principal factors contributing to the acceleration of erosion. On the other hand, rains with a high erosive potential occur during the months of June and July, precisely when the seeded lands present the lowest vegetation cover. During the dry summer months, the fallow lands are exposed to significant soil loss by wind erosion, mainly due to the low bulk density of the soil.

In Italy, cultivation practices frequently result in land degradation when insufficient care is taken. First of all, it is well known that the deeper the ploughing the more extensive the damage to the soil structure and composition. Things are worse when ploughing is carried out down-slope instead of along the contours. Studies on the effects of ploughing have been carried out but not within Mediterranean Italy. Instead of general data, some examples give an indication of the present situation.

Table 16.1 *Erosion rates 1989–1994 from the experimental site of El Ardal, Murcia* (source: Lopez-Bermudez *et al.* 1996)

Surface treatment	Soil loss 1989–94 (t/ha)		
	Average	Maximum	Minimum
Matorral	0.142	0.509	0.004
Matorral	0.045	0.124	0.002
Matorral	0.041	0.132	0.002
Cut *matorral*	0.401	2.023	0.004
Cut *mattoral*	0.731	3.501	0.008
Cut *matorral*	0.112	1.034	0.009
Fallow	0.159	0.634	0.011
Fallow with stones	0.028	0.077	0.005
Fallow with stones	0.015	0.021	0.007
Fallow with stones	0.007	0.016	0.001
Ploughed fallow	0.156	0.345	0.017
Ploughed downslope	1.084	2.174	0.031
Barley	0.388	1.003	0.012
Wheat	0.357	1.034	0.009

In the River Mesima valley, south Calabria, slopes are carved out of whitish chalk and sandy chalk of Pliocene age. Here, brown soils were able to develop under natural forest until the beginning of this century. During World Wars I and II, the need for wood for steam energy production resulted in the clear-cutting of forests in this zone, and in its conversion to pasture. More recently, valley bottoms were reclaimed to orchards and orange groves, and most gentle slopes to wheat. Since the 1960s, however, with the increased use of tractor-powered ploughs, steeper slopes are being deep ploughed for vineyards (this is also the case in Greece). The effects are illustrated in Plate 16.3. The scarp is caused by repeated ploughing up to the same height on the slope, resulting in accelerated transport of the soil down-slope and, in most places, exposure of the parent rock. The white-coloured ground is the C horizon of the former brown soil. Scarp heights may reach 4 m or more. In places scarps collapse through rotational sliding, because no attention is paid to the size of the scarp. Considering that these practices were initiated in the 1960s, and that the oldest scarps are 3 m high on average, it can be calculated that the average erosion rate is about 10 cm/yr.

The soil mass moving down-slope accumulates at the base of the slope, the neutral point between erosion and accumulation on the slope profile being from a few metres to a few tens of metres from the scarp, depending on the slope angle. The accumulation rate at the base of the slope is a few millimetres per year (Figure 16.2). The soil undergoes a general deterioration because erosion mixes it with crushed parent rock.

Rill and gully erosion devastating the fields and vineyards is the first and most common consequence of these practices. Often vineyards need to be replanted after a few years. It seems that farmers do not understand the reason for this, as every year they keep ploughing in the same way. But probably there are other reasons, such as the financial support of some selected agricultural activities by the European Union.

The regional distribution of these practices is unknown. Virtually all ploughed fields on hills may be experiencing these conditions. However, in some areas, such as the Aspromonte (south Calabria) high plains, more care was paid in reclaiming land for cultivation. Thus, all valley floors are artificially flattened (Plate 16.4), but the boundaries of the ploughed fields are not marked by a sharp edge or scarp. This may result

Plate 16.3 *Ploughing scarp, south Calabria, Italy. The heavily ploughed field is marked by a ploughing scarp some 3–4 m high formed in less than 10 years. Soil is completely missing; the outcropping lithology is Pliocene silty marl. Photo by M. Sorriso-Valvo*

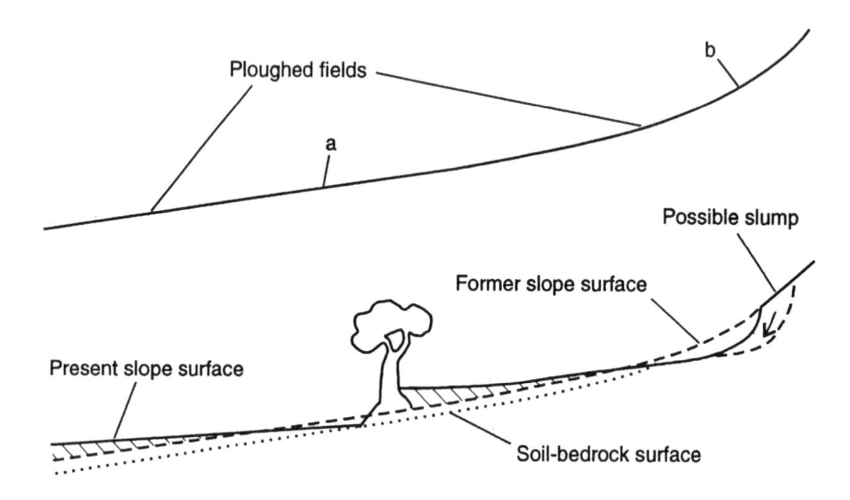

Figure 16.2 *Slope profile modification due to deep ploughing, Italy. The diagrams are not to scale, except for the gradient. Legend: a = property or field boundary; b = upper limit of slope gradient for ploughing*

Plate 16.4 *High mountain flat-floored valley, south Calabria. In this area deep weathering makes the parent rock extremely weak and easy to cut with ploughs, but no scarp is formed because of the continuing collapse of the cut slope, and the vegetative cover ensured by the high humidity. Photo by M. Sorriso-Valvo*

from the local climatic condition, which is more humid, and the presence of thicker soils.

In badland areas of southern Italy, the reshaping of slopes by means of very heavy earth excavation and slope regulation has been a common practice, in association with tree planting. In some cases, the intervention itself is an erosive event with tremendous effects.

Rates of erosion linked with these practices are not known either in general or in detail. They may reach huge values in the short term. The hidden effects of subsurface erosion also need to be considered, and expectations are for rather large values. In general, there is a complete lack of data on this point to date.

In Greece, the particular nature of the Mediterranean relief, with slopes subject to extensive deforestation and intensive cultivation since ancient times, has led to soil erosion and the formation of skeletal soils. Erosion risk is especially high in areas cultivated with rain-fed cereals. For one or two months after sowing winter cereals the land remains almost bare, and rains of high intensity and occasionally long duration occur during that period. The sloping lands of the Thessaly plain, the greatest lowland of Greece, were grazed for centuries, especially in winter by transhumance flocks and herds. Rapid population increase due to immigration in the early 1920s resulted in a sharp increase of the areas which were brought under wheat cultivation. Erosion experiments and estimations from the exposure of tree roots have demonstrated that erosion on these areas has proceeded at rates of 1.2–1.7 cm of soil per year since the introduction of wheat (Kosmas *et al.* 1995b).

In southern Australia, research has shown that there is a range of specific soil properties which are part of the land degradation problem, and which reflect a range of agricultural practices. Each soil property has its own set of specific causes, which often overlap. Soil acidification, for example, is associated with leaching, the use of superphosphate fertilizer, and increased nitrogen both from synthetic fertilizers and legumes (Evans 1991). Various soil nutrient or elemental excesses or depletions are often related to inappropriate fertilization rates and poor field management (McGarity and Storrier 1986), as well as to soil acidification (or alkalization). Water repellency of surface soil materials appears to be related to the use of nitrogenous fertilizers as well as exotic or native legumes (McGhie and Posner 1980). In Greece, too, the use of large amounts of ammonium fertilizers over the previous few decades has resulted in a rapid increase in soil acidity.

In Victoria, soil acidification is especially marked in the pasture areas. Wind and water erosion, on the other hand, is periodically a problem on the poorly structured, sandy soils of the Wimmera and on the fine-textured clay soils associated with the Murray River plains to the north (Scott 1991). In fact, the Mediterranean region of Australia has mostly old, duplex soils (Chapter 11), which are vulnerable to water erosion, waterlogging and structural decline.

The clearing of natural vegetation, cultivation and cropping, grazing, burning of crop stubble, bare fallow practices and irrigation all have major impacts on soil structure (Abbott *et al.* 1979; Clarke 1986; Hobbs *et al.* 1993). Loss of soil structure may be caused by: ploughing of very wet or dry soils; pressure from cloven-hoofed animals; the use of increasingly heavy farm machinery (which can exceed 20 t in weight, exerting hundreds of kilopascals of pressure down soil profiles); and continuous cropping and the loss of restorative pasture phases in longer rotations. These practices reduce soil organic matter (including root systems) and biota, adversely affecting aggregate stability, porosity and nutrient availability. Soil bulk densities and surface sealing are increased; this decreases germination rates and root development, and reduces permeability, which in turn enhances overland flow and soil erosion by water. Soils compact and subsoil hardpans may develop.

The foregoing problems are further aggravated by other forms of degradation such as waterlogging, salinization and the presence of residues from pesticides and fertilizers.

Land abandonment

The abandonment of marginal terraced slopes is a further source of slope degradation in the Iberian Peninsula. Land abandonment in Mediterranean lands generally may result in either positive or negative effects on land conservation (see also the discussion of North Africa in Chapter 14).

Results of land abandonment are positive in areas where geomorphic and climatic conditions are favourable to plant life, while cultural practices were detrimental. In Italy, this condition occurs in abandoned, low-gradient vineyards, dipslope-ploughed fields, grasslands in mountain areas and naturally stable terraced slopes (Plate 16.5: see also Plate 14.2). On the other hand, land abandonment has negative effects in areas where the stability of the soil and of the slope itself was assured by farmers' assistance, as in the ancient, steep, terraced slopes of Liguria, Campania, Calabria and Sicily (Plate 16.6). Slope gradients may be close to 100%, and soil was brought in by hand, filling the back of stone walls. Olive groves and vineyards are the typical crops on these artificial terraces. It is nearly impossible to use agricultural machinery in these lands (in Liguria they use small, toothed rack or funicular railways to transport goods), so the abandonment is not due to the productivity and the quality of products, but to the difficulty of the job. Once abandoned, the collapse of these stepped slopes may be rapid because, by the continuous work of keeping in place the mass that by gravity should have gone down-slope, farmers construct a metastable system whose earth mass, if sustaining work is abandoned, suddenly recovers the energy position which it would have attained if left to evolve naturally.

Plate 16.5 *Abandoned terraced slopes in the Peloritani Mountains, Sicily. The vegetation has been damaged by fire. Photo by M. Sorriso-Valvo*

Plate 16.6 *Abandoned terraced slopes in Aspromonte, Calabria. Photo by M. Sorriso-Valvo*

The occurrence of positive effects was frequent in the 1960s and 1970s in marginal or non-high-profit lands, when people were attracted by the easier life in the cities. But since the mid-1980s there has been a return to the fields, mostly in southern Italy, due to the labour crisis in the industrial cities of northern Italy and central Europe.

Negative effects are currently progressing or stationary, because people find it too difficult to return to these steep lands.

As was the case in Italy, research in Greece has also found that land abandonment may lead to a deteriorating or improving phase of the soils depending on the particular land and climatic conditions of the area, although the processes seem to be a little different. Studies conducted in hilly areas have demonstrated that soils under good conditions of plant cover may improve with time by accumulating organic materials, increasing floral and faunal activity, improving soil structure, increasing infiltration capacity and, therefore, causing a decrease in the erosion potential (Kosmas *et al.* 1995b). In cases of poor plant cover, the erosional processes may be very active and the degradation of these lands may be irreversible. In cases of land partly covered with annual or perennial vegetation, the remaining bare land with soils of low permeability (clays) creates favourable conditions for overland flow, soil erosion and land degradation.

Changes of land use

Changes of land use may also be responsible for land degradation, as has already been implied and as is shown by observations in the south of France. In the rare cases where the area of vineyards is increasing, for example in the Vaucluse department, the extension on to steeper slopes has been made by remoulding the topography, smoothing the slopes and separating parcels of land by enormous banks very different from the traditional cultivation terraces (Figure 16.3). The risks of accelerated soil erosion were quickly indicated (Durbiano 1988). At the same time, the hydrographic network was totally restructured by adapting it to the new parcels. It is clear that the extension of vineyards favoured the acceleration of soil erosion and an increase in the sediment load of overland flow. In particular, a spectacular network of rills and a 1–1.3 m local downcutting appeared in the Sausses valley (close to Vaison-la-Romaine) following the planting of a new vineyard in about 1970–1975 (Arnaud-Fassetta *et al.* 1993). Heavy rains on 22 September 1992 (about 180 mm in 3 h) produced significant runoff and erosion (up to 430 m^3/ha on the steepest marl slopes). On a marly substratum, 10 cm have been stripped from soils over a 40 year period on low-angled slopes in the Violès vineyard (Ballais and Meffre 1995).

The results of detailed research in Greece have produced similar findings. Here, the impacts of agricultural land use change on soil conditions and erosion along a hillslope catena have been investigated by Kosmas *et al.* (1995b) (Plate 16.7). For more than 160 years the major land use of the study area was olive growing. In 1979, all vegetation was removed from part of the area and vines were planted. The data obtained suggested that the change in land use from olives to vines

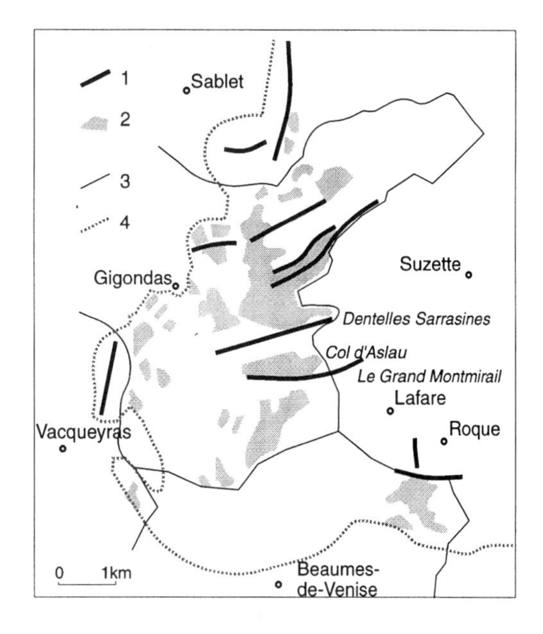

Figure 16.3 *Authorized clearing and recent cultivation terraces in the Dentelles de Montmirail massif* (after Durbiano 1988). *Key: 1 = crests; 2 = clearings and cultivation terraces (some wine growers were permitted to cut the forest and replace it with cultivation terraces planted with vines); 3 = limits of the communes; 4 = limit of the Dentilles de Montmirail massif*

Plate 16.7 *(a) Land under olives (foreground) is well protected by grasses and herbs, and has very low rates of soil loss, whereas the area now under vines (background, previously under olives) has no surface cover between the rows (b) and hence experience high rates of soil loss, associated with other changes to soil properties discussed in the text. This is part of the same experiment conducted by C. Kosmas referred to in Plate 15.4*

had a degrading effect on the size and stability of soil aggregates. The organic carbon content was reduced by about 33% in 12 years, while a tenfold decrease in the size of soil aggregates was determined. The depth of the A horizon was altered as a consequence of the 12 years of cultivation with vines. At least 247 kg/m^2 were estimated to have been eroded from the upper part of the transect under vines. Under existing climatic conditions and cultivation practices, sediment losses under vines have fluctuated between 67 and 460 t/km^2/yr (Kosmas *et al.* unpublished data).

In contrast, very low runoff and sediment loss, and thus soil erosion and land degradation, were found under olives with an understorey vegetation of annual plants (in semi-arid conditions); such a combination occupies an appreciable part of Mediterranean uplands and hilly areas. Under this land use, annual vegetation and plant residues have a high soil surface cover, occasionally up to 90% of the ground, so preventing surface sealing and minimizing the velocity of the runoff water. Based on three years of experimental data collected in Greece, runoff in excess of 5% of total rain and sediment losses greater than 5.3 t/km^2/yr are rare in olive groves (Kosmas *et al.* unpublished data).

"The Lemon Coast"

An excellent example of the consequences of changing land use comes from the so-called "Lemon Coast", a few kilometres north of Catania, along the eastern coast of Sicily. Here some 1000 ha of bare land on lava flows with no soil were reclaimed in the period 1880–1900, by transporting soil from other areas of Mount Etna. The transported soil had a thickness of about 30–40 cm, which was sufficient for vineyards. But when the demand for wine declined in the 1950s, the vineyards were converted into high quality lemon gardens. The limited thickness of soil, however, requires gardens to be irrigated. Water is available in abundance as the lava flows contain groundwater, allowing two harvests per year. But this "forcing" exposes the lemon trees to stresses which reduce their resistance to several types of pests.

It appears that in order to overcome this problem, instead of reducing the harvest to one per year, the local farmers introduced more resistant, but poorer quality species of lemon trees. The result has been that now most of the harvest is not sold on the market, but is stocked for low-profit industrial uses or non-profit controlled destruction, or just left on the trees. As a consequence, some of the lemon gardens are being abandoned as new tourist resorts are being constructed. Thus, the demand for drinkable water increases, unprofitable trees are irrigated less and less and eventually die, and the soil is beginning to be eroded. This phenomenon is at a very early stage, but if no measures are taken, the land will undergo degradation, with heavy consequences for the tourist income of the area, for which the beautiful landscape of the lemon gardens is the most important attraction.

California

Whereas the removal of native vegetation and its replacement by domestic crops and orchards profoundly affect the nature and rate of soil erosion, it is simplistic to suggest that such replacement must lead inevitably to accelerated erosion. This is because the type, location and nurturing of field crops and orchards vary in both a spatial and temporal context. Nevertheless, vegetation removal does expose the land to increased rainsplash and overland flow, and later to the effects of irrigation water on surface flows and subsurface hydrology.

Recent work in Ventura County, southern California, shows that rates of erosion and sediment yield for the period of American occupation exhibited significant geomorphic responses to changes in agricultural practices and the nature of areas under cultivation. For example, the Calleguas Creek basin (888 km^2), which delivers sediment to Mugu Lagoon at the coast, has experienced rapid changes in rates of erosion and sedimentation during the past century. Explanations for these changes lie not only with urbanization and crop diversification, but also with the transformation from early dryland cultivation on the plains to irrigated hillside cultivation.

The 1928 land use survey of Ventura County showed 39 000 ha devoted to dryland grain, hay, beans and apricots, and 14 000 ha to irrigated citrus and walnuts. In large measure the areas under cultivation were restricted to the flat Oxnard Plain and nearby valleys. By 1932, however, lima bean cultivation, the leading field crop in the Calleguas Creek basin, had encroached on to adjacent hillsides, causing very high rates of erosion ($1.45 \cdot 10^6$ t/yr) and sedimentation ($0.54 \cdot 10^6$ t/yr). Significant channel enlargement soon occurred, forming the deep *barrancas (arroyos)* so characteristic of the region. Though agricultural activity continued to expand during the mid-20th century, the nature of the crops changed significantly during the 1960s and 1970s, with less area devoted to lima beans (today only 162 ha remain) and greater production of citrus (lemon, orange and grapefruit), avocado, strawberry and vegetables. More importantly, during the 1980s, cultivation expanded on to hillsides and the total area under irrigation grew to over 37 000 ha – nearly 68% of the total farmland in Ventura County. For example, the dominantly agricultural Sand Canyon basin now experiences a mean annual erosion rate of 12 973 t and a sediment yield of 5443 t. These rates are in sharp contrast to estimates of annual erosion and sediment yields of 1633 t and 544 t respectively during the Native American period (USDA 1995). By comparison, Las Llajas Canyon, which is not under cultivation, experiences only 6350 t of annual erosion and 1360 t of sediment yield, figures very close to estimates for prehistoric time.

Table 16.2 presents erosion and sediment yield data and their source association for the Calleguas Creek watershed. The contrast between new and old orchards is striking. With the clearance of native vegetation or old orchards, erosion yields are high, presumably owing to surface exposure to raindrop impact, overland flow, gully initiation, and increased surface and soil-water volumes from irrigation. Orchard roads also yield substantial volumes of material. For example, in the Sand Canyon basin, orchards and their associated roads annually contribute some 3822 t of debris. However, of the area under direct agricultural use, field crops contribute the largest volumes of eroded soil and relatively limited sediment yield to streams. This can be explained in part as a result of the location of field crops on flatland areas and in part as a result of limited access to debris basins with direct stream conduits. Beyond these direct agricultural contributions to erosion and sediment yield, unstable streambanks and hillslope soil slips yield significant amounts of debris.

Irrigation

Soil degradation is exacerbated by irrigation agriculture. Such degradation commonly occurs as salinization, namely the hyper-concentration of naturally occurring substances such as sodium, selenium, boron and molybdenum, and the accumulation of artificially introduced nutrients associated with fertilizers and pesticides. Saline soils (>2500 mg/l "salt") are endemic to most irrigated Mediterranean regions, though the problems associated with this phenomenon are amplified by limited soil drainage and disposal sites for the salts, poor irrigation practices, substandard water quality, and insufficient water for removing accumulated salts. Augmenting these factors may be the presence of an artificially high water table introduced by irrigation which concentrates salts within the upper 0.5–1.5 m of soil. The most common salts include sodium (Na^+), calcium (Ca_2^+) and magnesium (Mg_2^+) combined with chloride (Cl^-) and sulphate (SO_4^{2-}).

Irrigation of dry lands, for example Monegros in the Ebro basin in the Iberian Peninsula, has

Table 16.2 *Agriculture-related erosion and sediment yield, 1990* (source: USDA 1995)

Sources	Area (ha)	Erosion (t/yr)	Sediment yield* (t/yr)
Field crops	13 000	87 996	12 700
Citrus (new)	4272	47 173	17 236
Citrus (old)	2817	9979	2721
Avocado (new)	1405	27 215	10 886
Avocado (old)	1796	9072	1814
Orchard roads		73 482	34 473
Confined animals	156	3629	907
Pasture	196	907	907
Streambanks		161 478	137 891
Soilslips		170 550	19 958
Gullies		29 030	9979
Total	23 642	620 511	249 472

*Sediment yield is defined as that material which passes through debris basins and is introduced into streams

produced and will produce increased salinization if plans for further irrigation are implemented. On the other hand, intensive irrigated agriculture in glasshouses in southeast Spain, especially in Almeria, is the cause of degradational soil processes, mainly biochemical, and of desertification.

"Biochemical soil degradation" is produced by the increased number of pests and with them the increased number of pesticides needed. Eventually the soil becomes exhausted and the land has to be abandoned. The landscape becomes a succession of patches of active and abandoned glasshouses. "Climatic desertification" is the result of alternating humid and dry conditions. In the humid period agriculture expands beyond the usual limits. In the dry period it is not always possible to retreat to the initial situation. A similar process may occur in the irrigated, arid areas of southeast Spain. The use of subterranean waters for intense irrigation may exhaust the water reserves, leaving the lands totally desertified.

In Italy, 260 000 ha were irrigated in 1988, while 450 000 ha were irrigable. In 1990, due to drought, although more than 502 000 ha of land were irrigable, only 144 000 ha were irrigated (Leone 1994) – that is, less than 0.5% of the nation's land area. However, the area under irrigation doubled in the last decade in southern Italy, with six-fold increments for Puglia and four-fold increments in Basilicata. This trend should remain unchanged in the next decades, as several reservoirs are still under construction in southern Italy. Here, the practice of irrigation for rice crops and fruit gardens is rather recent, thus salinization does not represent a real problem yet. It will become a problem within a few years, especially if very solute-rich water is used. This may be the case in areas where spring water is salt-rich because the aquifers are totally or in part composed of soluble minerals, such as halite or gypsum.

In Ionian Calabria and in central Sicily, irrigation has some limitations because of salt-rich groundwater. In Sicily, surface water may also be of poor quality for irrigation. For instance, there are two rivers which are named *Salso*, which means salty. Carbonate-rich waters are frequently found in Italy, as the Apennine range is constructed mainly of carbonate rocks. Some of these waters have a carbonate content so high that they cannot be used for civil use or to water cattle without mixing with fresh water. This richness in carbonate salt is witnessed by the high frequency of contemporary travertine deposits along many Italian streams.

In Greece, extensive irrigation without effective drainage practices in areas with high moisture deficits may lead to soil salinization. In many cases, rapid development resulted in the over-exploitation of the aquifers for a variety of uses (agricultural, industrial and drinking water), causing a gradual intrusion of seawater (see Figure 16.4). Irrigation using water rich in salts increases the salinity of the soil. Soil salinization is a potential land degradation threat in climatic zones characterized by high xerothermic indices, especially in the eastern Greek provinces (Yassoglou 1989). In addition, irrigation and intensive cultivation of recent alluvial plains or old alluvial terraces with heavy machinery have promoted a decline in soil aggregate stability, an increase in soil bulk density, decreasing porosity and permeability, and the formation of subsoil layers almost impermeable to water, over more than 100 000 ha.

The salinization problem was extensive in Mesopotamia, of course, turning the fertile lands of the Summerians, Akkadians and Babylonians into sterile soils that caused a shifting of cultures toward the north (Hillel 1994; Ghassemi *et al.* 1995). Then, as now, irrigation systems leave salts which accumulate in the soil, and by repeated irrigation to promote leaching, the groundwater table rises; when it reaches within 2 m of the surface, capillary action becomes effective.

In the Jezreel Valley in northern Israel, continuous irrigation of the soils without adequate drainage has caused the salinization of about 300 ha. The reason was that the Qishon River's main drainage basin was transformed into an endoreic basin by damming most of the river tributaries and main channels; the reservoir waters were used for irrigation and there was no leaching of salts as in a natural river system.

In coastal areas salinity problems are created when seawater invades the coastal aquifer. This occurs when the water table of the fresh groundwater is lowered by over-pumping. A reduction of 1 m in the freshwater level brings a rise of 40 m of the saline water according to the Glyben–Herzberg relationship (Figure 16.4: see also Dunne and Leopold 1978, 225–226).

The coastal aquifer of Israel has a total storage capacity of $1 \cdot 10^9$ m^3 and it has been mined for several decades by tens of wells, with a total annual pumping rate of about $400 \cdot 10^6$ m^3. Natural replenishment is about $300 \cdot 10^6$ m^3 each year, with fluctuations according to rainfall variability. The water table has been lowered about 20 m and seawater has intruded in several wells situated close to the shoreline.

The problem of agriculture-related salinity in Greater California is most strongly demonstrated in the San Joaquin Valley, much of which is underlain at shallow depth by impermeable clay. During the early agricultural development of the valley (1870–1915), perched water tables, often within 6 m of the surface, affected over 800 000 ha in Fresno, Stanislaus, King and Kern counties. These restricted the downward percolation of irrigation water and encouraged leached salts to accumulate near the surface, which in turn inhibited plant growth.

At the present time high water tables and salinization persist, particularly in the downstream reaches of the San Joaquin River owing to water diversion and recycling of saline agricultural flows. For example, prior to the construction of the Broadview Water District Drain in Fresno County, recycled agricultural water in 1981 contained as much as 2962 ppm total dissolved solids

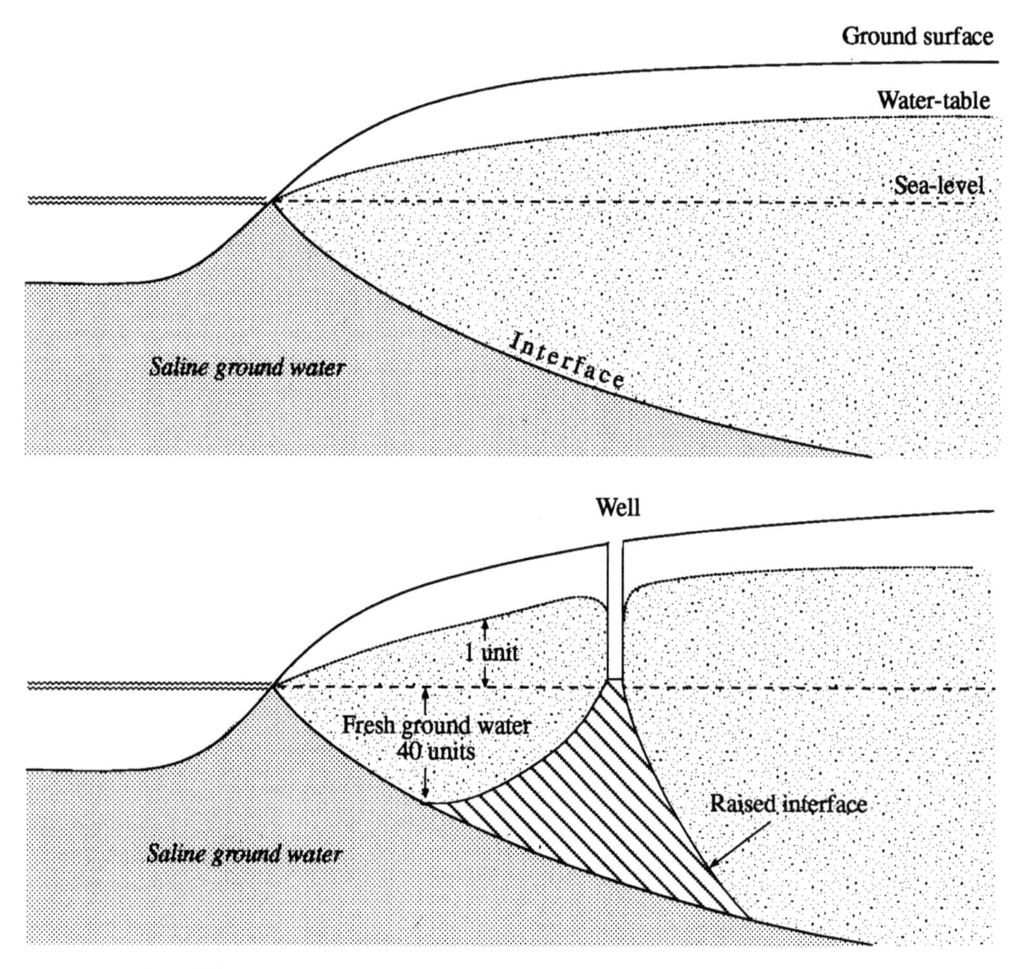

Figure 16.4 *The Glyben–Herzberg relationship for coastal aquifiers. Fresh groundwater decline of one unit raises the saline groundwater by 40 units*

(TDS), with 2158 ppm TDS delivered to local growers. By contrast, concentrations of salts in the nearby freshwater canal were only 346 ppm TDS. Along the western portion of the valley some 60 000 ha have water tables within 3 m of the surface and elevated salt levels. Overall in the San Joaquin Valley, over $2.26 \cdot 10^6$ ha are irrigated, with $0.9 \cdot 10^6$ ha experiencing high salinity and at least $0.6 \cdot 10^6$ ha being characterized by high water tables.

Soils with increased concentrations of salt in the water exhibit less swelling and dispersion of clays, which ultimately decreases the transmissivity of the soil. The presence of high salinity within the root zone encourages the attachment of zoospores to the surface of salt-stressed roots. The zoospores in turn use the sugars and amino acids leaked by root-cell membranes, thus decreasing the ability of the roots to resist disease. With root rot present, especially when associated with a high water table and high salinity, new roots are unable to grow.

Chemical soil degradation in California is not restricted to salt accumulation alone. Selenium is a naturally occurring element in the environment and is commonly associated with shale substrate and alkaline soil. Selenate (SeO_4^{-2}) and selenite (SeO_3^{-2}) dominate selenium concentrations in soils. Passed through the food chain from aquatic organisms to higher vertebrates, selenium deficiency is a more common problem than selenium hyperconcentration. However, in several areas of agricultural activity in California, selenium has been found in unusually high amounts. Concentration occurs by dissolution from selenium-rich bedrock and soils and the accumulation of irrigation water. For example, in the Panoche Creek area of the San Joaquin Valley, concentrations of selenium exceed 4000 ppb in contrast to naturally occurring amounts of 0.2 and 0.1 ppb in fresh and marine water respectively. Toxicity levels are considered to range from 3000 to 5000 ppb, while recommended limits in water for domestic consumption are 10 ppb.

Selenium occurs as both a soluble and an insoluble form in soil. Its mobility depends on the pH, oxidation potential, organic carbon content, calcium carbonate content, and cation exchange capacity of the soil. In its solid phase it occurs as an iron seleno-sulphide within a confining clay layer. In its liquid phase it is associated with perched water tables where it occurs as selenate and selenite. Oxidation by micro-organisms and concentration in soil water encourage its accumulation. Subsequent uptake by soil humus and plants which are disposed towards selenium concentration, such as mustard (*Brassica* spp.) and milkvetch (*Astragalus*), further encourages transmission through the food chain. While relatively little is known about the effects of high concentrations of selenium on wild animals, the toxic effects on humans are expressed by hair and nail loss and skin lesions while the effects on domestic animals with extreme concentrations include embryonic malformation. Some 20% of waterfowl nests in wetlands near Mendota were found to contain deformed embryos and 40% had at least one dead embryo (Tanji *et al.* 1986).

Irrigation agriculture may also cause subsidence. Two processes are involved. First, subsidence may occur as a result of hydro-compaction when loose, moisture-deficient alluvial deposits are wetted for the first time. The destructive effects of hydro-compaction in the San Joaquin Valley were first noticed in 1915, and around Mendota subsidence of 1–5 m eventually came to affect 650 km^2. Second, and far more important, subsidence occurs from compaction of the aquifer system caused by intensive pumping of groundwater. Widespread pumping began in the Central Valley around 1900, increased exponentially to the 1960s, and has since declined. By 1964, $12.3 \cdot 10^9$ m^3 of water were being pumped from 40 000 wells in the San Joaquin Valley, or about one-quarter of all groundwater pumped for irrigation in the United States (Lofgren 1975). Groundwater overdraft was recognized in the 1930s, long before the problem of subsidence was understood.

Subsidence apparently began in centres of overdraft in the 1920s and by 1975 amounted to a total volume of $19.3 \cdot 10^9$ m^3, with a maximum subsidence of 9 m in a 120 km long trough along the west side of the valley. More than 11 000 km^2 of farmland have subsided over 0.3 m. Although the problem is scarcely evident to the casual observer, and the importation of surface water to overdraught areas via canals and aqueducts since the 1950s has returned artesian pressures towards pre-subsidence levels, damage to wells and drainage systems has had a major impact on farming practices in the valley. In the Santa Clara Valley around San Jose, subsidence of up to 4 m was caused by groundwater pumping between 1912 and 1967, initially for irrigated agriculture and later for industrial and urban use. This subsidence has required the construction of levees along stream channels and around southern San Francisco Bay in order to inhibit flooding and salt-water intrusion.

In Chile, one of the most important risks in irrigated agriculture is the use of surficial waters affected by microbiological contaminants. This problem is represented by the impact of microorganisms on the production and exploitation of orchard products near ground level (strawberries) and above all by pathogenic illnesses derived from the ingestion of products irrigated with wastewaters. Although it is estimated that over a certain time span this contamination does not affect the productive potential of the zones of low irrigation, it is evident that there are other negative effects. In the long term, as a consequence of the health requirements of the importing countries, production can be seriously affected. At present the absence of a wastewater treatment system is manifested by the costs borne by farmers in the contaminated areas. They have to absorb the economic consequences derived from this situation, such as the reduction of more valuable crops and the cost of administrative control measures, without receiving compensation.

Limitations related to drainage in the irrigated valleys and oases in the northern regions generally depend on the methods used and soil characteristics. In the rest of the country the conditioning factor is the amount and distribution of rainfall. Many of the problems of insufficient drainage of the soils are related to the type of soil or parent material preventing the free movement of excess water derived from rainfall or irrigation.

Irrigation is also the main factor contributing to the increase in the natural salt content and to the phytotoxic components of irrigated soils. In well drained soils the degradation level depends mainly on the quality of the irrigation water, and also the quantity used.

Forestry and logging

Reforestation for wood production and related management techniques in the Iberian Peninsula induce important changes in soil structure, responsible for major disturbances to soil hydrological properties and soil hydrological and hydrochemical processes. Table 16.3 shows the impact of land management techniques on overland flow and streamflow in the Agueda Basin, central Portugal, during a three year period (Ferreira 1996).

The most deleterious phenomena are rip-ploughing and forest fires, which induce major changes to overland flow (with 46.4% and 11.6% of the incident rainfall respectively) and to streamflow, with more than 40% of the total rainfall amount. The profusion of macropores in soils beneath burned pine compared with eucalyptus probably accounts for the smaller amounts of overland flow when compared with regrowth eucalyptus. The litter layer increase following cutting is responsible for the decrease of overland flow beneath regrowth eucalyptus, in comparison with mature eucalyptus stands. The mature eucalyptus catchment yields streamflow comprising less than 25% of rainfall; unfortunately, equivalent data for the mature pine catchment are not available.

Higher biological activity is probably also responsible for the low nutrient amounts in both

Table 16.3 *Overland flow and streamflow under different land management regimes in central Portugal* (source: Ferreira 1996)

Land use	Overland flow (% of rainfall)	Streamflow (% of rainfall)
Mature pine	0.09	
Mature eucalyptus	2.0	23.7
Burned pine	11.6	48.5
Rip-ploughed eucalyptus	46.4	43.4
Regrowth pine (after fire)	7.8	33.0
Regrowth eucalyptus (after cut)	0.7	

Table 16.4 *Loss of nutrients in overland flow and streamflow under different land management practices in a forested area of central Portugal* (source: Ferreira 1996)

Management	Overland flow (kg/ha)				Streamflow (kg/ha)			
	NO_3	Ca	Mg	K	NO_3	Ca	Mg	K
Mature stands	0.017	0.07	0.03	0.07	0.01	0.27	0.67	0.082
Rip-ploughed eucalyptus	7.4	17.4	14.0	23.7	0.05	1.18	2.2	0.37
Burned pine	0.5	13.8	13.9	11.6	2.5	6.5	8.7	3.1
Regrowth pine (after fire)					0.033	1.5	2.4	0.33
Regrowth eucalyptus (after fire)					0.005	1.1	2.4	0.38

overland flow and streamflow (Table 16.4). In fact, the loss of nutrients from mature stands is only a fraction of the losses from the disturbed forest areas. The stands which had suffered the greatest disruption (rip-ploughing and burning) yielded the highest losses of nutrients at the catchment level. This is particularly the case with the burned pine catchment, where an additional source of nutrients (the ash) is responsible for the mobilization of abnormally high amounts of solutes. Regrowth areas, due to plant regrowth and more favourable hydrological processes, are placed in between the mature and highly disturbed stands (Coelho *et al.* 1995).

Pressure on forests is on the increase in North Africa. The usual, more visible exploitation by the local population at the edges of the forested massifs has been reduced in response to the work of the Forest Services, but it has been replaced by a more concealed exploitation within the massifs, in the more distant sectors. It consists of internal clearing that can affect the most precious forests such as the cedar or fir plantations (Deil 1987, 1988). Thus, the cedar plantation of Ketama lost a third of its wood resource between 1953 and 1976, according to assessments. A renewable resource is, presumably, being used up. One is then approaching a threshold which can lead towards a radical transformation of the vegetated landscape. Yet, it does not seem that this development should be linked to demographic factors, since several other factors intervene, in particular the "free" provision of the forest resource.

Everywhere the needs for firewood represent the first form of excessive cutting. Annually in Morocco, for example, $10 \cdot 10^6$ m^3 of wood are cut (from 22 000 ha), whereas the possibilities of Moroccan reforestation are only $3.3 \cdot 10^6$ m^3 each year. Over-cutting is estimated, in the region of Tensift, at 78%, in the northwest at 61%, in the northern centre at 67% and in the east at 52% (Direction de l'Aménagement du Territoire 1992).

The High Atlas serves as an example of the process of plant degradation (Al Ifriqui 1993). It is a real mosaic of meso-climates with a very mixed vegetation which has a great number of endemic species, including the Atlas cypress (*Cupressus atlantica*), the shrubby brooms (*Genista* spp.) and the argan tree (*Argania spinosa*). The degradation of a leafy, evergreen oak ecosystem takes the following forms. In a subhumid environment, between 1100 and 2100 m above sea level, degradation commences with the opening of the forest canopy resulting in a drying up of the ecosystem. Heliophilous species (those liking sunshine) then penetrate the formation. Because of excessive wood cutting, there is a change from the Mediterranean oak grove with 100% overlapping canopy to a more open canopy with cystus (*Cystus monspeliensius*) and thyme (*Thymus* sp.). The soil, partly stripped, dries up and can disappear from the most vulnerable areas. The recession of the forest is followed by the waste of important genetic potentialities. In other respects, the increased density of small plants growing as secondary vegetation increases the risk of fires.

Commercial logging, or timber harvesting, is a direct form of vegetation change based on the premise that the old-growth timber of primeval woodlands will be replaced by even-aged timber of secondary forest which, growing towards maturity, will again be logged, leading to tertiary forest and so on. Much of the primeval forest of the northern Coast Ranges and the foothills of the Cascade Range and Sierra Nevada in California has been transformed into a patchwork of timber stands in various stages of regeneration, with accompanying changes in geomorphic processes, soil characteristics and habitat. Primeval old-growth forest survives in several national and state parks and in areas difficult to log, but the net effect has been a major change in forest landscape and ecology.

The impact of timber harvesting on forest lands depends in part on environmental factors and in part on the nature of logging operations. Thus such natural factors as frequent heavy rainfall, steep unstable slopes and erodible soils all favour erosion and mass movement when areas are logged. In a curious coincidence, these same factors, together with occasional fire and windthrow, appear to favour the lush coast redwood forests of northern California. It has even been argued by the timber-harvesting industry that logging replicates the natural disturbances which were once common to the ecosystem but have been reduced by fire and flood management. Whether or not that argument can justify logging, the actual nature of logging operations and the degree of care exercised in those operations do have demonstrable impacts on forest land degradation.

Silvicultural practices and yarding methods are particularly important in this context. In the 19th century, loggers were rarely concerned with silvicultural practices, taking such timber as was needed from accessible alluvial flats and nearby slopes with little or no concern for physical and ecological impacts. In the earlier 20th century, however, as logging operations engulfed larger areas and steeper slopes, questions arose as to whether clear-cutting or selective (patch) cutting was more desirable, raising in turn questions of the optimal size of a clearcut or patchcut. Meanwhile, yarding methods – the means whereby felled trees are removed from the forest – were changing from the use of oxen, stationary engines and narrow-gauge railroads, to the use of tractors, cable systems, balloons and helicopters. Of this latter group, tractor skidding became the preferred method in the coast redwood forest, leading to much increased erosion rates. Table 16.5 shows how erosion is favoured by tractor skid trails, logging roads and landings while the intervening forest actually harvested provides much less debris. In recent decades, heated debate has focused on these issues, with preservationists disliking any form of logging, conservationists accepting some cable, balloon or helicopter yarding from patchcuts as a workable compromise, while industry has favoured the most cost-effective silvicultural and yarding practices. All such operations have come under increased scrutiny and government control (see Chapter 17).

In Chile, over-exploitation of the natural forest resource is one of the main environmental problems. The wood resource has considerable importance in the national energy grid, representing more than 25% of the energy used in the country. Between $6 \cdot 10^6$ and $7 \cdot 10^6$ m^3 of wood are consumed each year.

In addition, in recent years pressure on the native forest formations in order to produce wood chips has grown to unsustainable levels (Claro and Wilson 1996). This has also occurred in the southwest of Western Australia (Conacher

Table 16.5 *Surface erosion and mass movement by land use and slope class, northern California coast redwood region, 1975–1977* (source: Hauge *et al.* 1979)

Land use	Slope (degrees) <18	18–26	27–35	>35
Percentage of study area				
Harvested forest	75	71	82	88
Tractor trails	19	21	12	5
Logging roads	3	5	4	3
Landings	3	3	2	4
Percentage of total erosion				
Harvested forest	0	8	1	13
Tractor trails	21	37	24	9
Logging roads	11	45	44	71
Landings	68	10	31	7

1983). External commercialization of woodchips in Chile commenced in 1987 with an export of 452 000 t. In the following year the amount increased to 907 000 t, of which 8% corresponded to chips of native species. By 1990 exports were 2 200 000 t, 52% of which were indigenous woodchips.

These production forestry activities are responsible for soil degradation due to the management of the adult plantations and of the native forests, especially the cropping operations by total tree cutting, which initiate or accelerate erosive processes or other forms of soil degradation. The main factors or activities within this context are the following:

– long periods of unprotected soil exposure producing erosive processes after logging;
– the use of heavy machinery and/or inadequate systems of wood extraction leading to soil compaction, loss of structure and reduced aeration of the soil;
– extraction of nutrients from the soil by fast-growing tree species;
– use of clear-cutting and residue burning over extensive areas irrespective of slope or the season;
– agricultural use, over-grazing and over-cutting of trees on soils suitable for forestry; and
– intensive construction of logging roads.

17

The causes of land degradation

3: Other human actions

Compiled by Arthur Conacher and Maria Sala from contributions on the following regions:

Iberian Peninsula and Balearic Islands	**Maria Sala** and **Celeste Coelho**
South of France and Corsica	**Jean-Louis Ballais**
Southern Italy, Sicily and Sardinia	**Marino Sorriso-Valvo**
Greece	**Constantinos Kosmas, N.G. Danalatos** and **A. Mizara**
Dalmatian coast	**Jela Bilandžija, Matija Franković, Dražen Kaučić**
Eastern Mediterranean	**Moshe Inbar**
North Africa	**Abdellah Laouina**
California	**Antony** and **Amalie Jo Orme**
Chile	**Consuelo Castro** and **Mauricio Calderon**
Southwestern Cape	**Mike Meadows**
Southern Australia	**Arthur** and **Jeanette Conacher**

There is a wide range of human actions other than those concerned primarily with agriculture. They include activities in the coastal zone, the introduction of exotic biota, plant diseases and pests, the use of fire, urbanization, industrialization and tourism, various polluting practices, and a range of social and political factors.

Human Actions and Coastal Erosion

Coastal erosion is a recent aspect of land degradation in some areas. On the Spanish Mediterranean coast, recession of the Ebro delta due to the increasing damming of the river, and especially since the construction of the Mequinensa dam in 1946, has occurred as a consequence of reduced sediment reaching the north–south littoral drift. In addition, in this sector there are only ephemeral streams with small drainage areas, thus resulting in scarce and sporadic sediment supply. Evidence of coastal recession can be seen in the destruction of several works, including home gardens along the urbanized coastline, the road from Benicarló to Peñíscola, the littoral road that has to be continuously rebuilt towards the interior, the coastguard houses built around 1920, and the watchtowers constructed in the 16th century found drowned 160 m from the coastline in 1953 when constructing the port of Castelló (Sanjaume *et al.* 1996).

In contrast, south of Valencia city the coast has prograded 2 km since Roman times, although the last 1000 m accumulated during the last two centuries. In this sector, drainage systems are larger and provide more sediment. Accumulation is also favoured by the development of the dune fields of the Devesa del Saler (Sanjaume *et al.* 1996).

Another phenomenon observed in this sector is related to the construction of ports and the subsequent interruption of the littoral drift. Accumulation occurs in the northern part of the ports and erosion in the southern parts. The port of Borriane is an example of rapid and recent coastal changes, with progradation rates of about 1.26 m/yr, involving a surface accumulation of 5.4 ha, and a considerably higher rate of loss south of the port, at 3.8 m/yr over an area of 33 ha. Another example is the port of Valencia, where progradation of more than 1000 m north of the breakwaters took place between 1812 and 1971. In the south of the port erosion has been estimated at 3 m/yr since 1900, although losses of 10 m/yr between 1926 and 1933 have occurred (Pardo Pascual 1991).

Beaches along Italian coasts were in equilibrium or exhibiting waxing conditions from medieval times until the first decades of the 1900s.

In the 1960s and '70s, it was thought that the coastal erosion problem of the 20th century was a consequence of the excavation of gravel from riverbeds and beaches. But this activity nearly ceased from the early 1980s, whereas beach erosion continues at nearly the same rate. In the early 1950s and in 1972/73, two major storms occurred in southern Italy, triggering several landslides. This replenished the beaches, but it was a transient phenomenon. The only factor likely to have reduced the debris budget of the streams, ironically, has been the widespread corrective measures against soil erosion in mountain areas since the 1950s. Check dams keep a large part of debris along the tributaries, thus the lower reaches of the rivers have been brought into a degradation phase. Consequently the equilibrium or waxing conditions on the beaches were disturbed, resulting in the general retreat of the coasts (Plate 17.1), except for a few locations corresponding to the mouths of major rivers

Beach erosion has been accelerated along the Tyrrhenian coast because of inappropriate coastal works, such as T-groynes and parallel reefs, constructed without any preliminary study of the dynamics of the coastal system (Plate 17.2). In response to the drawbacks of countermeasures, beach replenishment has been undertaken, but using earth-rich material, while the natural beach materials were pebbles. The earth gives some cohesion to the material being deposited along the coast; but whilst this may prove useful against low-energy summer waves, it will not resist the high-energy winter breakers. A second drawback will be the pollution of the sea floor with muddy deposits.

Land Degradation in Mediterranean Environments of the World: Nature and Extent, Causes and Solutions.
Edited by A.J. Conacher and M. Sala.

Plate 17.1 *Example of coastal erosion along the north Adriatic coast, Italy. The eroded rocks are Pleistocene sands. Photo by M. Sorriso-Valvo*

Plate 17.2 *T-groynes off the west Calabrian coast*

The coastlines of the eastern Mediterranean are probably the most fragile environment among the geomorphic systems of the region, considering their narrow geographic extent and the dynamics involved in the sea–land boundary. Human pressures on coastal landforms are enormous, with one of the highest coastal population densities in the world, increased economic investments and strong land use competition.

The Mediterranean sea currents carry waste materials dumped by cities and ships, eventually reaching the shorelines. Oil pollution caused by leakages, or emptying the ballast water of oil tankers, causes the formation of tar, which particularly affects bathing beaches.

Other major impacts of human activities on the eastern Mediterranean coast include:

(a) the opening of the Suez Canal which allowed the introduction of Indian Ocean species into the Mediterranean Sea: about 20% of the marine fauna in the eastern Mediterranean have their origins in the Indian Ocean; and
(b) the High Aswan Dam, which retains most of the sediments – about $80 \cdot 10^6$ t annually – that the Nile previously carried to the Mediterranean. The sand fraction was deposited in the delta area and drifted toward the Sinai and Israeli coasts; the lack of supply has promoted accelerated erosion in the Nile delta area and eventually the supply to the northern coasts will be reduced, triggering erosion processes.

An important demographic feature of southern Australia, as in the eastern Mediterranean, is that over 70% of the population lives at or near the coast. Since 1975, population growth in non-metropolitan coastal areas has been most rapid of all Australian statistical areas in the Perth region (600%), followed by the non-Mediterranean Sunshine Coast in Queensland (250%; Zann 1995). Enormous pressures are placed on coastal environments as a result.

As in other Mediterranean environments, climate, coastal scenery and amenities attract increasing numbers of tourists, with the industry now accounting for 5% of Australia's gross domestic product and employing a similar proportion of the total workforce (as against a decline in the agricultural sector). But the rapid growth of the tourist industry has been accompanied by over-use of major attractions, causing problems of beach and coastal dune erosion, decline in natural habitats, loss of water quality, increasing demands on water supply and loss of amenity values (noise, crime, aesthetics, congestion).

The impacts of different kinds of land use (agriculture, forestry, urban, industrial and recreational) on southern Australian coastal and marine environments are considerable and have been outlined in greater detail by Zann (1995). The causes include:

– *clearing of river catchments*, resulting in land erosion and thus sedimentation of estuaries and coastlines; although not in the Mediterranean region, in Queensland it has been estimated that four times more sediment is entering the sea than during pre-European times;
– *nutrients* from agriculture and urban sewage being transported in sediments and runoff water, contributing to eutrophication of coastal waters;
– *damming*, flood control, diversions and infilling, which affect the hydrodynamics of estuaries and the supply of sand to beaches;
– contributions of *toxic materials* (pesticides, heavy metals and organic compounds) from agricultural, urban, forestry and industrial sources, including rubbish dumps, marinas, industrial discharges, urban runoff and ships' bilge tanks;
– *coastal engineering*, including walls, groynes and dredging, affecting current circulation and sediment movements; and
– *other activities* such as mineral sand mining.

Introduction of Exotic Plants

It is interesting that of all the world's Mediterranean regions, only in the southwest Cape is the introduction of exotic, invasive plants considered to be a major land degradation problem, although introduced plants, diseases and pests are also causing problems in southern Australia and, no doubt, elsewhere.

The spread of exotic trees is at once both a form of land degradation and a causal factor. The form and nature of the problem have been outlined in Chapter 12 and its historical development in Chapter 14, so it remains to attempt to understand why this factor has become such an important cause of degradation in the southwestern Cape.

It has been suggested that the region is especially susceptible to invasion by exotic trees, and there are indeed strong indications that the region is subject to more extensive infestations of exotic trees and shrubs than other parts of South Africa (Macdonald 1984). The fact of there being rather few trees *per se* within the *fynbos* has led to the suggestion that this renders it more vulnerable to invasion under the so-called "vacant tree niche" hypothesis (Moll *et al.* 1980). Macdonald *et al.* (1986), on the other hand, can find no statistical evidence to support this contention and suggest rather that it may be the longer period over which invasions have been occurring in the southwestern Cape that has resulted in its greater degree of infestation. Notwithstanding this, it is obvious that the scarcity of trees in the region led the early colonial settlers to introduce exotic varieties in order to supply the expanding supply base at Cape Town (Shaughnessy 1986). Thus, for the purposes of shade, timber and sand-binding, many species of trees were introduced and cultivated (see Chapter 12).

The nature and extent of the initial introductions were, however, only part of the problem, since it has been the subsequent expansion and invasion of adjacent areas which have determined the intensity of the resultant degradational impacts (Richardson *et al.* 1992). Two of the most widespread invasives in the mountains, *Hakea sericea* and *Pinus pinaster*, illustrate the kinds of biological features which have favoured rapid range expansion. Both are dispersed by wind, and both have winged seeds which are held in fire-resistant cones. The seeds are released and germinate quickly after fire such that the opportunity for successful seedling establishment is maximized (Shaughnessy 1986). In addition, the fact of their being alien in the first place advantages them over their indigenous competitors, because in having no "natural enemies" they may be less prone to seed predation and herbivory.

Exotic trees have been ecologically successful, then, because they were introduced in large numbers over extensive areas of the southwestern Cape, respond favourably to the fire regime, have a high reproductive success rate and are not impacted heavily by local herbivore populations. There is also some evidence which suggests that alien trees selectively invade ecosystems which have already been subject to some form of disturbance. This implies that one form of land degradation (agriculture or urbanization) may pave the way for another. Further, since the areas of both agriculture and urban land in the region are predicted to increase in response to population growth, the alien infestation cause of degradation may intensify over the next few years, assuming that efforts to intervene and control the problem are not successful. A related underlying cause is that of afforestation. Besides the invasion into adjacent areas of certain tree species commonly used in commercial forestry plantations (in particular *Pinus pinaster*), there are off-site impacts of this process on, for example, surface and groundwater resources.

A closely related problem concerns the introduction of plant diseases and pests.

Introduced Plant Diseases and Pests

Chile's forests are being affected by various harmful agents which, due to the forest's endemism, are systematically degrading the resource. In the case of exotic plants, the monoculture of *Pinus radiata* has a fragility which makes it especially susceptible to damaging agents. Amongst the insects which attack the pines are a moth (*Rhyacimia buoliana*), a lepidoptera which is widely distributed in the plantations of the *Pinus* genera introduced from Argentina. In addition there is the harm caused by rabbits (*Oryctolagos cuniculus*) which can affect as much as 75% of recent forest plantations. The *Eucalyptus* genera are affected by *Phoracantha semipunctata*, which produces mortality rates after two or more years of successive attacks. Poor sites are the most affected, that is in arid zones where plantations grow under unfavourable agroclimatic conditions, or areas susceptible to drought.

Plant diseases in southern Australia have been caused by the disturbance of the natural ecosystems as distinct from their complete replacement. "Jarrah dieback" is the prime example, affecting nearly 20% of Western Australia's forests (see Plate 11.2) and much of the heathlands and woodlands of the southwest of the state. But the name is a complete misnomer. The disease kills at least 900 species of both native (mostly) and introduced plants in Western Australia, not only the eucalypt jarrah (*Eucalyptus marginata*) (Grant 1992, p. 40). It also occurs in South Australia and Victoria, where there is no jarrah.

The disease is caused by a microscopic soil-borne fungus, *Phytophthora cinnamomi.* The fungus attacks the roots of the vegetation, reversing the osmotic pressure differential across the outer membranes of the roots. The roots become unable to absorb soil water from the soil and the plant starts to "die back" from the tips of the branches, as the plant becomes starved of nutrients and moisture.

The *Phytophthora* fungi – there are several species in addition to *cinnamomi* – apparently were introduced to Western Australia from Indonesia in the 1920s. Nevertheless, for many years they did not seem to cause a problem. In undisturbed conditions, they live and move slowly in soil water. But with increasing human activities – not only intensive logging and clear-felling of the forests, but also the building of roads and power lines, quarrying, dam construction and recreational activities – the fungus was spread rapidly as soil containing the spores was transported physically in road gravels and on vehicles from diseased to healthy vegetation. These processes continue in the late 1990s.

There were, and are, also more subtle processes at work. The prescribed burning programme of the Western Australian (then) Forests Department, now part of the Department of Conservation and Land Management (CALM), has also helped to spread the fungus in at least two ways. Note that prescribed burning is carried out in all native vegetative communities, not only forests. One effect was the removal of the litter layer. Research by Springett (1976) showed that the fungus is inhibited by the presence of a good litter layer. In contrast, its spread is more rapid where litter layers are absent or poorly developed. The second effect of burning was the observation that prickly moses (*Acacia pulchella*), whose root systems inhibit the fungus, was being depleted by the prescribed burning programme. In contrast, the banksias, whose root systems harbour the fungus, were being encouraged by the same fire regime.

"Dieback" is now considered to be the single biggest threat, after clearing, to the long-term survival of Western Australia's native plant communities in the Mediterranean region, and it is also a major problem in plant communities in South Australia and Victoria. It should be pointed out that pathogens other than *Phytophthora* also attack native vegetation (Grant 1992, 40–41; Wills 1993; Withers *et al.* 1994).

In southern Australia, feral pests compete with native animals and birds for food and water, act as disease vectors and foul watering points. Small birds and animals are particularly vulnerable to predation by introduced foxes and feral cats. According to an Adelaide zoologist, one feral cat preys on 64 species of native mammals and can be responsible for as many as 1000 kills each year (Szabo 1995). Foxes (introduced by English settlers for the "sport" of fox hunting) are considered to have an even more devastating effect than feral cats on Australia's wildlife.

Rabbits over-graze and displace small herbivores, and cause erosion. They are becoming increasingly resistant to the myxomatosis virus, which was introduced as a control in the 1950s. Rabbits are now being seen in increasing numbers in agricultural areas and even in towns (as are foxes). (An accidental escape of *Calcivirus* in 1995 had dramatic, albeit localized, impacts on rabbit populations in South Australia; subsequently, in 1996, the virus was spread deliberately throughout Australia – a practice which was accompanied by scientific controversy.) Feral goats and pigs are increasing in numbers, with large herds of goats reported in the Victorian mallee and pigs in Western Australia's forests. Weeds, including many Mediterranean species such as wild oats (*Avena* sp.), barley (*Critesion* sp.) and brome grasses (*Branus* sp.), are a serious and growing problem in many areas (Commonwealth Department of Primary Industries and Energy, undated). They compete for moisture, nutrients and light and can reduce crop yields by 10–50%. Contamination of wool and wheat with unwanted materials can lead to buyer resistance and even rejection. Numerous introduced insect pests and diseases, such as spider mites, weevils, locusts, root fungi and mildews, cause significant crop losses.

Fire

Forest fires are a natural hazard in Mediterranean environments due to summer drought, thunderstorms and the combustibility of the vegetation. In addition, *Pinus* and *Eucalyptus*, the main reforestation species in the Mediterranean Basin, are highly combustible in comparison with oaks, especially cork oak. But the increase in fire extent and frequency is also related to human causes, including conflicts of land use, such as the replacement of pasture lands with forests, tourist developments especially in coastal mountains and forests, the increasing use of forest roads, and human negligence and arson.

Fire appears to have served as a most beneficial tool in the skilled hands of our Mediterranean ancestors, but it has been abused greatly by later generations. It is now being neglected and rejected in some areas (Naveh and Lieberman 1984), while in others it is being used as an integral part of forest management. Fire, like erosion, can also be seen as just another element in the natural evolution of Mediterranean ecosystems with soils and vegetation which have adapted to it (Kozlowsky and Ahlgren 1974; Naveh 1974). People have traditionally used the positive effects of fire for renewal of pastures, pest and species control, and fertilization. The problem, especially today, is the way the balance is being upset by the increase in the number and extent of uncontrolled fires.

In Spain, the increase in forest fires is considered to be related to human-induced causes, either accidental or deliberate. In Catalonia, the spatial distribution of fire initiation is closely related to the road net and the location of second residence settlements, implying negligence. In former times an important cause of fires was sparks from coal-fired railway locomotives.

In the case of deliberate lighting of fires the causes differ in relation to the socio-economic problems of the region. In Galicia the government policy of reforestation of pasture lands is often related to fire. In the Catalan Mediterranean coast, most of the areas burnt during the 1970s were transformed from rural to urban uses such as second residences and hotels. It could also be related in some cases to the urban mentality of new landowners.

At least up to the middle of the 19th century, the main cause of forest fires in the south of France was *écobuage* or *culture sur brûlis*, a 1000 year old agricultural practice which allows the expansion of the cultivated area. In particular, during the 18th and 19th centuries, population increase required the exploitation of very large territories, even far from the houses, on the steep slopes and the less favoured soils. The *écobuage* allowed this extension but, when it was poorly controlled, considerable areas of forests were burnt.

Accurate statistics concerning the causes of forest fires in the south of France have become available only recently. They show that only 30% of the causes are known and that most of them are due to carelessness and lack of knowledge about danger. Sources include rubbish dumps, farm works, machinery used in forestry, hazardous installations, shepherds, hunters, firebugs, children or smokers' carelessness, heedless tourists and reckless farmers. Natural or fortuitous causes such as lightning are probably rare (Entente Interdepartementale 1988). Many studies have shown that forest fires burn mainly during the dry summer season, when temperatures are high, when strong winds blow and, above all, when soil-water storages are very low (Godde 1976; Douguedroit 1991; Prosper-Laget 1994).

In Italy, as elsewhere, fire has been the means of clearing forested areas in order to obtain land for grazing and agriculture. This has been done everywhere since the transformation of human culture from hunting and gathering to pastoralism and agriculture. Even today, in the Mediterranean lands, shepherds and cow-hands light illegal fires in order to put the land to pasture. Southern Italy, and Sardinia in particular, are no exception, even though a national law forbids the use of burnt lands for any use other than new forest plantations. But while the trees are planted and grow, the land is used illegally for pasture. This is only one example to introduce the concept that fires, today, are essentially of anthropic origin.

As in France, the causes of fires in Italy are natural in only a minority of cases: most (70–75%) fires are lit deliberately. Back-burning, preparing land for grazing, property speculation (insurance, land use), vandalism and ignorance are the most common reasons for such acts. Finally, and not only in Italy, there are rumours about workers and firms appointed every year for fire control, ensuring the need for their employment.

In Greece and Chile, most fires can also be attributed to people's carelessness, with the majority occurring in areas with high xerothermic indices and moisture deficits. Soil dryness and wind speed are the principal factors of fire evolution.

The abandonment of vineyards that were not profitable causes the accumulation of fuels which burn easily, especially in summer, in the Mediterranean region of Croatia. Research on identifying fuel types, quantities and accumulation rates was started in 1991 and it is not possible to draw conclusions yet. Data on fires in the karst region for the 1952–1955 and 1986–1990 periods are shown in Table 17.1. According to forest fire data for the last 20 years it can be stated that 99.9% are due to human activities. About 50% of them are ascribed to unknown or doubtful causes (Bilandzija 1993). There is probably a significant proportion of arson fires in the category "cause unknown", but it is very difficult to prove due to very complicated political relations in recent years. There are no quantified data on the impact of fire on the state of the environment (landscape, hydrology, increased erosion), because no systematic monitoring or problem-solving activities have been carried out.

Fires sparked by lightning are rare in the eastern Mediterranean and there are almost no natural forest fires. Intentional burning, as in other Mediterranean countries, is for agricultural purposes, or to "clean" the greenbelt for construction and development; it may be used by shepherds to increase pasture, and by arsonists for political reasons. At present, intense fires are encouraged by the accumulation of debris as fuel in the forest layers, and the high cost of labour for cleaning or prescribed burning.

Forest fires are an important cause of soil degradation in North Africa (Zitan 1987). During the Algerian war, fire was used by both sides.

Table 17.1 *Burnt areas and number of forest fires in the Karst region of Dalmatia* (sources: Krpan 1957; Bilandzija 1993)

Year	Burnt area (ha)	No. of fires
1952	1563	142
1953	156	75
1954	114	133
1955	223	64
Total	2056	414
1986	3178	91
1987	3409	81
1988	14 821	161
1989	3849	70
1990	21 937	266
Total	47 194	669

The cedar forests of Djurdjura were devastated by napalm, and in Petit Kabylia, 48% of the plant cover was destroyed by fire during the war.

A consequence of the expanding population of the southwestern Cape and the fragmentation of its natural vegetation has been a manipulation, intentional or otherwise, of the long-term fire regime with associated degradational impacts. Fire frequency, season, intensity and size have all altered under human influence. Fuel load and fuel type of transformed ecosystems are different from those of the indigenous vegetation, and people change the fire potential significantly by adding to the sources of ignition. A prescribed burning programme is followed on all land under state control and on privately owned land considered as being part of a mountain catchment area; season of burn is usually late summer (van Wilgen *et al.* 1992). There is reason to suppose that fires now occur more frequently than in precolonization times, and that where there are significant levels of alien tree infestation, fire intensity is somewhat higher, with attendant impacts on plant community structure and diversity (van Wilgen *et al.* 1992). Global warming scenarios for the region point to a further intensification of this effect in the future (van Wilgen *et al.* 1992).

As in other Mediterranean-type regions, several factors predispose an area to wildfires in southern Australia: extreme weather conditions (prolonged drought, high temperatures, low humidity and strong winds); high fuel loads (especially dry litter and grasses); proneness to fire (the presence of *Eucalyptus* and other combustible species); and an ignition source (lightning, broken glass, accidental fire escape, carelessness or arson).

While Western Australia has had no major wildfires since 1961, in recent years there has been an increasing number of deliberately lit fires around the Perth metropolitan area. Additionally, the state's Department of Conservation and Land Management carries out prescribed fuel reduction burns each spring and autumn throughout the state forests and reserves. Fire frequency is 5–10 years, with the objective of preventing wildfires by reducing the fuel load. Unfortunately, these "cool" burns encourage fire-prone shrubs, open up the canopy and understorey, and allow the germination of invasive grasses and other weeds, thereby increasing the fire hazard. Furthermore, a varying proportion of wildfires are caused by escapes from prescribed burns.

Urbanization, Industrialization and Tourism

Twenty countries surround the Mediterranean Basin, supporting a population of 300 million and visited each year by 120 million tourists. Tourist numbers are expected to treble by the year 2025 (ICCOPS 1995).

Progressive industrialization and increasing tourism are the main causes of economic progress in many Mediterranean countries and at the same time the sources of degradation. In Croatia, for example, during the summer season (end of June to end of August) tourists increase the total population by more than three times, causing enormous pressure on the environment (1 360 000 inhabitants versus 5 064 000 tourists; Statistical Year Book 1991, 1993).

After World War II, improved social and economic conditions in Italy also brought the development of seaside tourism, which has a heavy drawback in causing the urbanization of coastal zones. A large part of Mediterranean Italian coasts is only a narrow strip, unable to accommodate strong urban pressures (see Plate 3.1). This notwithstanding, in several parts of south Italy urbanization is uncontrolled. Here holiday houses are built where once there were dunes, or directly on the beach. In some places, the strong coastline retreat of recent decades is resulting in the complete destruction of the rear-beach environment. Building on coasts blocks the supply of material from the land towards the beach, which in consequence is more intensely eroded. Coasts in these conditions represent at least 30% of the Italian Mediterranean coast. Better control is exerted by local authorities in northern Mediterranean regions, and in very high-value zones in Sardinia, parts of Sicily, the Amalfi–Sorrento coast, Capri, and a few other islands. The same occurs for some tracts of coast difficult to reach from the mainland because of the lack of roads.

In the Mediterranean coast of Spain, tourism has brought wealth and with it increased urbanization (see Plate 1.5) and industrialization which in turn have led to an increase of impervious surfaces. Since the early 1970s, there has also been an increase of single family houses built within the forests (mostly *Pinus*), and accordingly an increase of road building. This is a way of living similar to the northern, more humid countries, whereas the Mediterranean people were more accustomed to being grouped in cities.

The results have been an increase in the incidence of forest fires (too many people in the inflammable *Pinus* forests) and higher and more frequent floods (due to the increase of impervious surfaces in the catchments). Streamflow regimes have been affected in total runoff, peak discharges and lag times between rainfall and runoff (Sala and Inbar 1992). The cumulative curve of annual precipitation shows a steady increase over the years, thus indicating no change in rainfall. In contrast, the cumulative curve of annual runoff shows one or two breaks corresponding to an increase in runoff at the end of the 1950s and in the 1960s. These periods correspond to pulses of industrial and urban development in their watersheds. As a result, floods have become a common feature in recent years in downstream urban areas, especially in the coastal *ramblas*.

The economic costs and losses of human life caused by floods result from disregard of these simple hydrologic facts and the belief that the responsibility is to be charged politically to the party in power. Because the floodplains are dry for most of the year and extreme floods occur only every five or ten years, local governments allow the construction of large buildings adjacent to them and also their use for recreation activities or parking, further increasing the incidence of runoff as well as contributing to the hazard.

Research in the south of France has also shown that many buildings have been constructed on floodplains over recent decades, exacerbating the effects of floods, or the flood hazard. As early as 1982, Gabert and Nicod drew attention to the particular characteristics of the Huveaune and Arc floods in 1978, which were a response to the urbanization of Marseille and Aix-en-Provence (Figures 17.1 and 17.2). More recently, urban floods (in Nîmes in 1988, and in Aix-en-Provence in 1993) have shown that the huge increase of impervious surfaces in the towns over

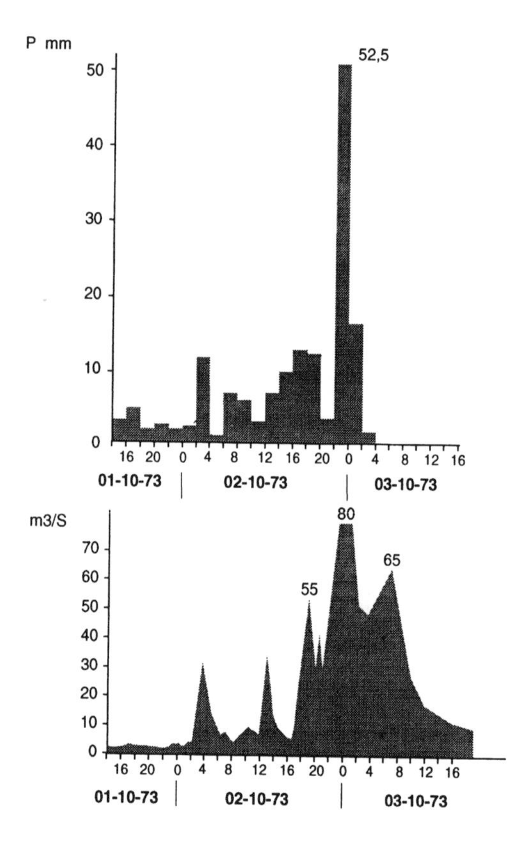

Figure 17.1 *Pluviogram and hydrograph of the Huveaune River flood in Marseille, 2–3 October 1973* (after Gabert and Nicod 1982)

the past few years has produced a new type of catastrophic flood.

A temporal comparison of peak discharges of two neighbouring watersheds in southern California (Grimes Canyon, 16.6 km^2, and Gabbert Canyon, 21.8 km^2) also reveals the differences between a natural system and an urbanized system (Table 17.2). Clearly the return period of the 100 year discharge is less affected than that of smaller events such as the two year discharge. With paving of the ground surface and the introduction of storm-drain systems, the likelihood of frequent increased water volumes is apparent.

Data from Italy are also of interest in this context. The road density in Italy is very close to 1 km/km^2. Considering an average width of 8 m, this means that 0.8% of the territory is occupied by roads. In addition to this 0.8%, areas occupied by cities, industrial zones and airports must be added, which means that 2 to 3% of territory is invaded by impermeable settlements and services. This figure is extremely high, if compared with the 3.7% of territory occupied by natural parks. The percentage of impermeable surfaces is much higher on the plains, where it may reach over 10%; but it is also high in valleys, where the negative effects of reduced infiltration and increased runoff contribute to the direct effect of subtraction of good lands to agriculture, with heavy consequences for the flood hazard.

One example is given from recent flash-floods which struck the city of Genova (Liguria) in the winters of 1993/94 and 1994/95, causing a great loss of goods and some casualties. The steep slopes behind the city, which is built along the coast, have been invaded by houses and roads, with the result that intense storms of non-exceptional magnitude regularly cause flash-floods downtown. In this region, however, the expansion of areas occupied by greenhouses for flower cultivation must also be considered, as the areas affected may be greater than the usual forms of urbanization.

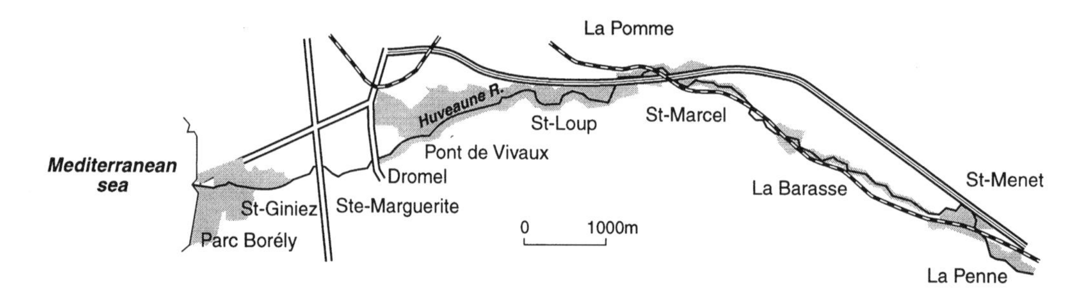

Figure 17.2 *Flooded areas (shaded) in Marseilles and la Penne, 16–17 January 1978* (after Gabert and Nicod 1982)

Table 17.2 *Peak discharges (m^3/s) in Grimes Canyon (undeveloped) and Gabbert Canyon (urbanized) over time* (source: USDA 1995)

Return period (years)	Native American	Spanish-Mexican	1932	1990	2010
Grimes Canyon					
2	25	55	170	170	170
5	60	135	280	280	280
10	90	190	360	360	360
25	210	345	535	535	535
50	240	380	570	570	570
100	330	500	690	690	690
Gabbert Canyon					
2	35	90	175	250	290
5	85	210	330	425	460
10	140	295	430	525	560
25	305	510	670	765	805
50	345	565	720	820	860
100	465	720	875	975	1020

A similar case is the city of Catania (east coast of Sicily). The lower end of the main road of the city was located on the bed of a small stream on whose banks the primordial city was built. The city developed considerably in recent decades, practically rendering waterproof the entire watershed of this stream. Major storms in the 1970s resulted in repeated flooding of this road. Countermeasures have improved the superficial drainage and now the problem seems to have been overcome. As a further example, in 1994 a flash-flood killed a woman who was walking down a road built along a former steep canyon in Enna, a recently expanding city built on the top of a mesa-like mountain in the centre of Sicily. The small (<1 km^2) watershed of the canyon, once occupied by orchards, had been completely urbanized in the previous decade.

An important effect of road building in mountain areas or in landslide-prone zones is the triggering of landsliding along the sides of road cuts. Data on this topic are unavailable for Italy as a whole. However, it is difficult for a road not to cross landslide-prone or landslide-affected areas in many parts of the country. The effect is, of course, the activation or reactivation of landslides. This effect can be delayed. For example, the 80 km State Way No. 18, in north Calabria, was built in the early 1970s. It crosses several landslides. Some moved during the construction of the road, and corrective measures were undertaken. Several others instead had a long-lasting creeping stage of deformation, and eventually moved in December 1980, some 15 days after the earthquake of 23 November in south Italy, whose epicentre was some 200 to 250 km north of the landslides. Although the landslide occurred after a period of intense rain, that event was not an absolute maximum in the rain time series of the area. Damage was extensive to both cropland and buildings, and of course to the road itself. Although the earthquake and rain were the triggering factors, the construction of the road had a determinant long-term role.

Rapid urban/industrial growth of the major cities over the past few decades in Greece has resulted in a series of processes which collectively reduce the physical potential of the land and restrict its agricultural area and productivity. For example, the populations of Patras and Heraklion, two major Greek cities, increased by 7.3% and 11.9% respectively between 1980 and 1990 (data from the National Statistical Service), resulting in a considerable decrease of surrounding agricultural areas. Furthermore, serious air pollution and traffic problems in cities such as Athens have induced the expansion of the city into the surrounding area, reducing dramatically the existing vegetation cover and further aggravating the problems of air pollution, soil erosion and flooding of low-lying urban areas. Industrial zones have also been located in most of the major cities, further restricting the availability of high quality agricultural land.

Cities have affected landscapes since their first development in the eastern Mediterranean. In Jerusalem, for example, the present square pavement near the Western Wall is more than 15 m above the street of the Herodian City, built 2000 years ago. The ancient mounds, or "tells", of the Middle East are the sites of ancient settlements which changed the original topography by tens of metres. Tell Hazor, an Early Bronze Age site in Upper Galilee, founded about 4000 years BP, has an area of 70 ha reaching an elevation of 40 m above the surrounding area (Nir 1983).

Modern urbanization affects all aspects of the ecosystem: there is a large increase in runoff, especially in karstic areas where the surface runoff/rainfall ratio is only about 1–2% under natural conditions. Changing the surface to an impervious one may increase the ratio by more than 50%. Flooding in the lower parts of the Carmel hills in the city of Haifa is a common feature nowadays due to the urbanized area of the karstic hills.

Sediment discharge also increases in urbanized catchments, yielding concentrations 10 times higher than from non-built areas. Flows from urbanized areas have a large content of dissolved and suspended sediments, increasing the pollution potential of rivers and underground water.

Changes induced to the hydrological cycle by urbanization are: (1) the import of water from other watersheds, for Tel Aviv, Jerusalem, Amman, Istanbul and almost all big cities in the area; (2) export of water as sewage; (3) decrease of infiltration to groundwater and increase of flooding in the downstream area; and (4) depletion of aquifers by intensive exploitation (Lindh 1972).

Urban growth, together with the industries and transport networks associated with such growth, has undoubtedly played a major role in land degradation in Greater California. Just as much of the natural landscape was converted to agricultural use, so much original agricultural land has been subsumed by urban development, especially around the major metropolitan areas. Further, several natural environments have been converted directly to urban or industrial use. For example, about 78 km^2 of wetlands in southern San Francisco Bay have been enclosed to provide evaporation ponds for salt production.

Loss of arable soils due to the advance of cities and villages, and to changes of land use from rural to industrial, is one of the major problems in relation to the soil resource in Chile. On the other hand, soils of low agricultural capacity are not occupied by urban areas. Soils of higher quality and productivity potential are concentrated in the plains of the Central Zone, in Regions V, VI and the Metropolitan Region, precisely the areas where numerous cities and population centres are concentrated. In relation to the surface occupied by urban centres consolidated during the last 10 to 12 years, it is estimated that the average rate of urban growth is of the order of 800–1000 ha/yr (CONAMA–MINAGRI 1994).

Under present legislation, which allows the subdivision of lands down to 0.5 ha, soils of excellent productivity have been used for residential purposes, a situation which particularly affects the Metropolitan Region and Regions V and VI (MINAGRI 1994).

Urbanization has a range of other impacts on the nature of a landscape. This is because, apart from the physical space that urbanized land occupies, the degree of transformation is usually very high and the activities associated with urbanized areas, for example industry and waste disposal, are also prominent degradation factors. The major urbanized area in the southwest Cape, Cape Town, has been subject to rapid and extensive population and spatial growth (Figure 17.3). Having taken 290 years to reach a population of 600 000 inhabitants in 1945, it took only a further 30 years to double in size to 1 200 000 by the early 1970s and only another 15 years to reach the three million mark (Palmer Development Group 1990). During the process, the physical expansion of the urbanized area has often been indiscriminate and environmentally uninformed, such that the loss of natural resources has been very high (Western Cape Regional Services Council 1994).

A major problem with respect to the growth of Cape Town is that it has been characterized by poor form and structure. On the one hand there are large areas of low density development,

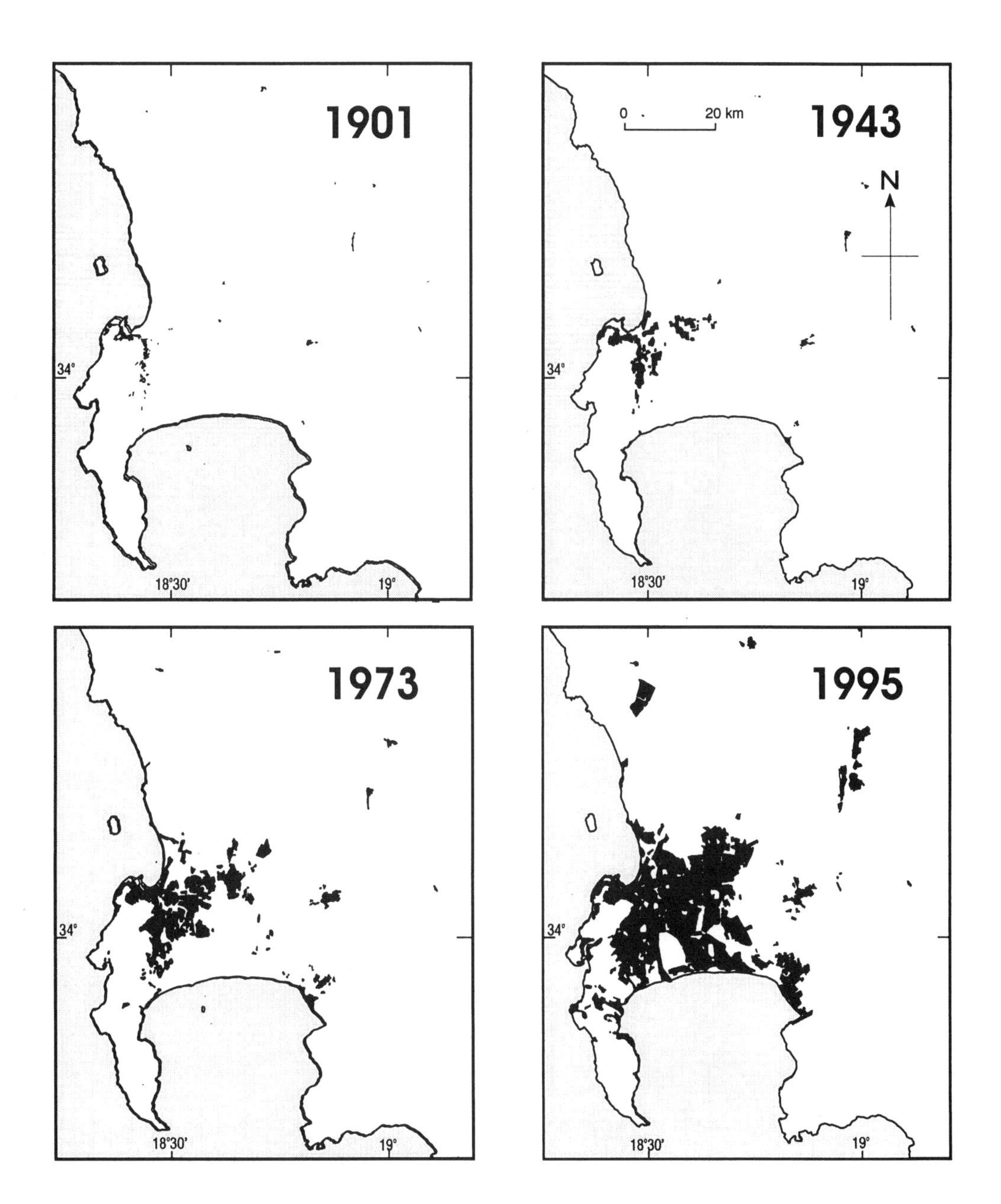

Figure 17.3 *The physical growth of Metropolitan Cape Town: urban sprawl 1901 to 1995* (source: Gasson in preparation)

which ranges from nine to 20 dwelling units per hectare (Western Cape Regional Services Council 1994). On the other hand, much of the urbanization has been informal, with relatively little planning control; the resultant high density shack settlements frequently lack services such as sewerage and electricity, so that there is an associated high degree of environmental degradation and environmental health problems (Plate 17.3). The lack of electricity may have secondary environmental impacts because the principal source of heat and light is the combustion of fuelwoods

Plate 17.3 *Part of the Crossways shack settlement near Cape Town, built partly on a rubbish dump and with no services. Accurate population figures are not available but total numbers are thought to exceed 500 000. Photo taken in January 1994*

which produces high concentrations of particulate carbon and associated forms of atmospheric pollution, especially in winter months.

Urban sprawl clearly means that the consumption of land for the increasing population has been inefficiently high and that the resulting ecological and biogeographical impacts have been great (McDowell *et al.* 1991). There are indications that the situation may worsen in the future. Estimates of the quantity of land which is required to meet the current and projected future housing backlog vary from 18 000 ha to as much as 50 000 ha (Western Cape Regional Services Council 1994), an area almost half the size of the contemporary inner metropolitan area.

It is not only "natural" vegetation which is consumed in the process. Often it is prime agricultural land that becomes the focus for housing development; in this way urbanization could be considered to intensify degradation which is already underway. During the last 10 years, for example, 1560 ha (representing 23% of the total urbanized area) of valuable agricultural land have been developed into middle and high income suburbs in northeastern Cape Town around Durbanville, Kraaifontein and Kuilsriver (Western Cape Regional Services Council 1994). Gasson (in preparation) notes that during the 1960s and 1970s, farmland was over-run at a rate approaching one hectare per day and that this consumption rate doubled during the 1980s.

Associated impacts include the disturbance of recharge of the Cape Flats aquifer, which has been largely covered by low-cost housing. Other components of degradation also have urbanization as their underlying cause. This is particularly true of water and air pollution. The despoiling of the wetlands of the Cape Flats through inputs of sewage and other chemicals and particulates, for example, has been described as the worst case of water pollution in South Africa (Gasson, in preparation).

Adverse effects of urbanization have also been identified in southern Australia. As in the other Mediterranean regions, expanding cities cause the direct loss of often highly productive agricultural land, and change the hydrology as well as the air and water quality of the landscape. In addition, cities place large demands on water, with the average Australian household using 400 000 l/yr, of which 30–80% is used on the suburban garden. Agriculture, however, uses most water (80%), mainly for irrigation, followed by urban uses (10%) and industry (7%) (Australian Academy of Science 1994, p. 349).

Cities pollute surface and groundwaters with their sewage, toxic wastes, urban runoff, illegal waste disposal and legal rubbish dumping (Commonwealth of Australia 1982; Environmental Protection Council of South Australia 1988; Grant 1992). They place demands on the surrounding countryside not only for water but for building materials, timber, recreational areas and hobby farms. Urban arsonists light bushfires every summer around Perth and Adelaide, and the unwanted pets of urban residents escape or are left in the bush to prey on wildlife. The recreational activities of city dwellers are having considerable impacts on adjacent rural areas, including soil degradation, habitat destruction, vandalism, littering, pollution of water supplies by campers, and over-use of wood in popular camping areas.

Quarrying and mining

Quarrying is a particular form of land use whereby materials are directly "eroded" from the landscape by human actions. Italy is a case in point, and one interesting location where quarrying has been practised since Roman times is the famous Carrara marble quarries in Tuscany. Long-term effects are evident over an area of several tens of square kilometres. The extraction rate of marble today is nearly equal to the maximum reached during the Roman period, but it was much lower in the intervening period, with rates increasing since 1830. About 48% of the more than 55 · 10^6 m^3 has been extracted in the last 40 years. The average "erosion" rate for the 2000 year period is approximately 0.5 mm/yr. In addition to excavation, the spoil from the quarries occupies (so far) a volume of more than 100 · 10^6 m^3 (Colli and Grandini 1994).

For a second example, data from a 100 km^2 area in north Calabria, given in Table 17.3, show the amount of erosion over a time span of about 30 years, due to different human activities. These data indicate an average erosion rate of approximately 0.33 mm/yr.

In California, hydraulic mining had a devastating impact on the environment. The Stewart surface mine in Placer County, for example, was 5 km long, 2 km wide and 120 m deep, and yielded $6 million worth of gold between 1865 and 1878 (Clark 1979). Furthermore, these deposits had long been stored in the foothills but were now liberated and flushed downstream, causing widespread sedimentation and flooding in lowland areas. As the Yuba River bed rose at a rate of 0.3 m/yr, the city of Marysville built levees for protection, but it was still flooded in 1875 and 1876. By 1880, 20 000 ha of the Sacramento Valley had been inundated by gold tailings; by 1884, 990 · 10^6 m^3 of debris had been washed from the Sierra Nevada foothills in just 35 years.

Table 17.3 *Volume of materials removed for various extraction purposes over a 30 year period, over an area of 100 km², in north Calabria* (source: Marzulio 1986)

Extraction activity	Erosion (m^3)
Clay for concrete and bricks	1 600 000
Aggregate for concrete	8 000 000
Soil for gardens	200 000
Total	9 800 000

This situation led to litigation between upstream miners and downstream farmers which in 1884 was decided in favour of the latter. Thereafter, hydraulic mining declined, many pits were abandoned to secondary scrub, and dredging became more important. In recent years, however, these stream gravels have become an important source of sand and gravel for construction purposes. But the damage had been done, notably in San Francisco Bay which began to fill rapidly with sediment from the gold fields. In his classic work on hydraulic mining debris in the Sierra Nevada, Grove Karl Gilbert (1917) suggested that it would take until 1950 for the debris to be flushed from the system. As Gilbert's biographer has stated, "hydraulicking was strip mining at its most rapacious" (Pyne 1980).

California contains many other minerals whose exploitation has caused local degradation. Most important among these is petroleum, because its recovery has led not only to the compelling bleakness of the industrial oil field, but to extensive subsidence associated with its removal. Petroleum is recovered mostly from folded Cenozoic marine sediments beneath the southern San Joaquin Valley, from the Santa Maria, Ventura and Los Angeles basins along the coast, and more recently from the continental borderland offshore. In the Midway–Sunset oil field at the southern end of the San Joaquin Valley, petroleum production began in 1894 and peaked in 1914. The subsidence measurement of between 0.4 and 0.6 m between 1935 and 1965 represents only a small part of the total subsidence likely to have occurred over the past hundred years (Lofgren 1975). In the Wilmington–Long Beach oil field in the Los Angeles Basin, subsidence associated with petroleum production has caused port facilities at Wilmington to subside below sea level.

In Australia, mining is an important element in the industrialization and urbanization processes, and it produces its own range of consequences for land degradation. In addition to the direct disturbance of land by excavation, mining produces tailings (which may be toxic), storage dams, leachates, dust and noise. In Western Australia's Darling Ranges southeast of Perth, every tonne of bauxite mined produces two tonnes of highly alkaline "red mud" contaminated by caustic soda, which is stored in large (200 ha) containment areas (Australian Academy of Science 1994, p. 363; plate 17.4). It has also been claimed that every hectare mined for bauxite leads to three hectares infected by *Phytophthora cinnamomi*, the dieback fungus.

Polluting Activities

Activities which produce pollution are related to economic development in agricultural and industrial fields and the associated development of cities. In general terms, it could be said that highly productive agriculture and certain industrial activities are responsible for most of the water pollution, whilst air pollution (not considered further here) is related primarily to industry.

The deterioration of water quality, from which all developed countries suffer and which is increasing rapidly in developing countries, has clear examples in Mediterranean environments. The cause is partly an erroneous attitude in the sense that all pollution risks are considered to be solved in a natural way by the auto-purifying power of running water in rivers in the case of the biodegradation of organic matter, and by the dilution capacity of the aquatic environment in relation to nutrients and toxic products. The disequilibria inherent in the growing eutrophication of many lakes and the accumulation of several products in aquatic organisms (heavy metals, organochlorides and radioactive compounds), amongst other phenomena, have demonstrated

Plate 17.4 *Land "quarantined" by caustic "red mud" tailing dams associated with the Pinjarra alumina refinery south of Perth, one of four alumina refineries in the southwest of Western Australia. The toxic wastes will need to be managed long after the refineries have been abandoned, in order to avoid contamination of adjacent land and groundwaters*

the illusory protection offered by the natural processes of purification and solution (Ortiz-Casas 1995).

In Spain, water pollution has become one of the most important environmental problems, far more serious than soil erosion. It is not only rivers which are contaminated in certain areas but also groundwater. At a first glance, it is industry which is mainly responsible for this situation because its impact on rivers is easily seen (and smelt!).

At the end of the 1970s no more than 10% of the Spanish population had the benefit of treatment of residual waters, while by 1993 this percentage had reached 59%. Despite this, the present situation shows one-third of the total length of rivers with various pollution problems, 70% of the dams with signs of eutrophication more or less advanced, and nearly 50% of the groundwaters with excessive concentrations of nitrates or high salinity.

A very typical Mediterranean water pollution problem is related to the industrial production of olive oil (Cabrera-Capitán 1995). Olive oil is obtained by discontinuous press or by continuous centrifuging solid–liquid processes in three production phases: olive oil (20%), solid residue (30%) and aqueous liquor (50%). The solid residue (*orujo* in Spanish), which is made of the pressed olive husks and stones, oil and water, is used for the extraction of oil and, when exhausted (*orujillo*), is used as a fuel. Other uses of the olive cake are animal feeding and the production of organic compost, active carbon and furfural. The aqueous liquor composed of the vegetation water and soft tissues of olives, and the water used in the olive processing, is the olive mill wastewater (OMW; *alpechín* in Spanish). The quantity of OMW produced in the process ranges from 0.5 to 1.5 l/kg of olive. OMW is a dark liquid comprising 83–94% water and 4–16% mineral salts (carbonates, phosphates, K, Na), with a high pollution potential (BOD 35–100 g/l, COD 45–130 g/l, EC 8–22 dS/m). The disposal of OMW is becoming a critical problem in Mediterranean countries.

Andalucía produces 80% of the Spanish olive oil (Spain is the world's third largest olive oil producer) and its annual mean OMW production amounts to $2 \cdot 10^6$ m^3, which represents an equivalent pollution to that produced by 16 000 000 inhabitants during the short period of the milling campaign (*c.* 100 days). Because of the polyphenol content, treatment of OMW by conventional methods is difficult and expensive. These methods only achieve a reduction of BOD down to 3000 mg/l, at a cost of $0.03–0.06 per kilogram of oil. Alternatives are land treatment (infiltrating the sludge into soils) and the co-composition of OMW and composting of OMW sludge with agricultural residues, in which some of the OMW components are recycled. A new solution to the OMW problem is the extraction of oil by a two-phase (oil and cake) process which reduces water consumption and OMW to a minimum. The new pressed cake contains 55–60% water and most components of the OMW, being available for the same uses as the solid residue derived from traditional processing.

Agricultural areas devoted to irrigation practices have increased three-fold in the last 25 years in Spain and, as a result, N-fertilizer use has increased from 11 kg/ha (1950) to 54.5 kg/ha. Therefore, extensive areas have been affected by nitrate pollution. The problem is especially acute for groundwater aquifers, for example those associated with the Guadiana, Guadalquivir, Segura and Júcar rivers, and some restricted areas in the Tajo and Ebro watersheds. The same can be said about other pollutants derived from agricultural practices, such as phosphorus and synthetic pesticides.

Water pollution due to salt water intrusions in coastal aquifers has been discussed elsewhere. The cause of this degradation of water in the Spanish Mediterranean coast is uncontrolled and excessive water pumping, in the first place due to the increased demand for irrigation, and in the second place for urban/tourism needs.

In the Ebro River basin, Tertiary evaporite strata (mainly gypsum) occupy 22% of the total area, and therefore natural salts are a main component of the water. Nevertheless, a steady increase in the amount of salt transported along the Ebro River course has been observed and related to agricultural practices and runoff decrease (Navas and Machín 1995). A total of $6.95 \cdot 10^6$ t of salt are supplied annually to the Mediterranean Sea through Tortosa by the Ebro. The main components of the salinity are sulphates, chloride, calcium and sodium.

Water pollution is an important component of land degradation in Italy. In general, the problem is a consequence of bad practices in urban and industrial waste disposal (Plate 17.5). For instance, only 10 to 20% of waste disposal sites have been approved regularly by authorities. More recently, the effects of heavy use of fertilizers have been pointed out. Studies in Italy have been published for the River Po valley (which is out of the Mediterranean zone), where the effects are becoming serious. Studies are under way in the Puglie region, where aquifers are composed of carbonatic rocks. Here, too, pollution by heavy metals and other industrial products is at dangerous levels (Cotecchia pers. comm.), but available data need to be analysed further before publication. In this case, the effects on land degradation may result from the use of the polluted water for irrigation without prior treatment.

In coastal zones, water pollution, which may cause heavy damage to vegetation, can be due to a variety of reasons: dune erosion, salt-water wedge intrusion to the root zone, over-grazing, acid rain, contamination by air pollution (chemical compounds of industrial origin, such as SO_2, NO, ozone, HF), and anionic surfactants. In Tuscany (Bussotti *et al.* 1984) and Latium (Gisotti and De Rossi 1980), it was found that the main reason for vegetation degradation in the coastal zone is the presence of anionic surfactants. Their concentration in rainwater reached 37 mg/l before a national law enforced the use of anionic surfactants with at least 80% of biological degradability. Afterwards, the concentration of anionic surfactants was reduced, but subsequently it increased again to 32 mg/l because of the increasing urbanization of the coastal zone of Tuscany.

For the effects on vegetation, a combination of two factors is important: (i) the surfactants make a very thin veneer floating on the seawater; and (ii) strong winds spray the salt water containing surfactants on to the coastal vegetation. The destruction of the sand dunes for urbanization makes it possible for the polluted spray to reach further and further inland. Indeed, the damaged area extends over a strip up to 5 km wide.

In Croatia, discharges of wastewaters are mainly uncontrolled because of poorly maintained or non-existent sewage treatment plants. That situation will be changed by implementation of the inventory of polluters – identifying pollutants by sources, types and quantities – which is in the process of adoption.

Deteriorating water quality is a major land degradation problem in southern Australia. Some

Plate 17.5 *Rubbish dumped in a southern Calabrian stream bed awaiting the next flood for its disposal to the Mediterranean Sea. Photo taken in 1990, with apologies to Italian colleagues*

examples have been presented in Chapter 13, and the processes responsible for secondary salinization in dryland agricultural areas have been referred to in Chapter 16. Eutrophication is caused by the runoff of excess phosphatic and nitrogenous fertilizers from farms and, in specific locations, by effluent from piggeries, abattoirs, and poorly treated sewage from country towns. Phosphates from fertilizers are removed in bonded form on soil particles, and to a lesser degree in solution, and cause problems of nutrient enrichment in surface waters. Nitrogenous fertilizers, use of which has increased dramatically since the 1970s, move readily in solution particularly through poorly buffered sandy soils, exacerbated by heavy rains and irrigation practices. The increasing accumulation of nitrates in ground and surface waters is a direct result (Lawrence 1983; Environmental Protection Council of South Australia 1988; Scott 1991; Grant 1992).

Other chemical pollutants in water which originate from agricultural activities have been recorded in limited surveys in Western Australia, Victoria and South Australia (Olsen and Skitmore 1991; Scott 1991; Environmental Protection Council of South Australia 1988). They include metals such as cadmium, zinc and lead which are present as impurities in fertilizers, pesticide residues (especially persistent organochlorines and some herbicides) and veterinary chemicals. They may contaminate water bodies from spray drift, leachates or accumulations in sediments, or as a result of careless disposal or accidental spillage of the chemicals.

Some problems of water pollution are caused by poorly sited and managed industry, not agricultural practices. Part of the problem of Lake Bonney in South Australia, for example, is drainage for agriculture; but the main culprits are industrial sources (Chapter 13). Groundwater has been polluted with pesticide residues and other contaminants (sulphates, sodium hydroxide and hydrocarbons) beneath the Kwinana industrial area south of Perth in Western Australia. Plumes of contaminated groundwater have been reported to be entering Cockburn Sound with adverse effects on sea grasses and fish breeding grounds (Department of Conservation and Environment 1979; Hirshberg 1988, in Grant 1992, p. 148).

Social and Policy Factors Responsible for Land Degradation

The more we understand about interactions between people and nature and the roles played by institutions and property rights in fundamental socio-economic and ecological management systems, the more capable we will be of designing policy tools for sustainable land management.

Spain and the EU

Ever since Spain joined the European Union (EU) in 1986, farmers have had to adopt the policies of the Common Agricultural Policy. A policy which has been under constant debate is the "set-aside policy" created in 1988. The purpose of this policy is to reduce the production of herbaceous crops (cereals, oilseeds, protein crops) by setting aside 20% of productive lands in order to reduce the costly surpluses of these crops in the EU. Since September 1995, the set-aside percentage of land has been reduced to 10%. According to Mata-Porras (1993), this particular policy will influence the process of soil erosion negatively because the affected lands need to be kept in fallow, leaving some of the agricultural land bare all year round and thus unprotected from wind and water erosion.

A study by Campos Palacín (1992) has indicated that livestock grazing in the southwest of Spain, in the *dehesa* regions, representing about 25% of the Spanish land, causes serious soil erosion and land degradation due to excessive grazing by cattle, sheep and game animals. Uncontrolled grazing destroys the regenerative capacity of the covering vegetation and turns the land into desert-like conditions. The underlying reason for over-grazing is the lack of enclosures. Here the problems of ill-defined property rights arise. Property rights do exist in this region but seem to be weak and the incentives to defend these rights also seem to be relatively few. The situation is reminiscent of open access, where public authorities have abandoned the management of public lands, allowing animals to graze uncontrolled.

A similar situation is that of excessive grazing activities in the Pyrenees and Pre-Pyrenees (García-Ruíz *et al.* 1991). Here, over-grazing seems to occur when small-holders abandon their private estates and thus turn some areas into open access (see Plate 14.1a). Since 1980, more than $3 \cdot 10^6$ ha have been abandoned and the farming population has fallen to 50% (Valladares 1993; Sandstrom 1996). The extent of the areas used as open access is not known.

About 30% of Spanish forests are owned by local councils, with central and regional governments in fact owning very limited land. However, regional governments, introduced in 1980, control and manage most of these forests. More than 60% of the forests are in the hands of private landowners, with an average forest holding of 3 ha (Campos Palacín 1992). Modern Spanish forestry is characterized by fast-growing plantations and clear-cutting practices. Clear-cutting will inevitably lead to erosion, because tree roots bind the soil, and the underlying bushes protect the ground from rainsplash.

The European Common Agricultural Policy influences silvicultural practices by encouraging farmers to transform some of their agricultural land into forest, providing subsidies for fast-growing plantations. These plantations have led to an increased number of forest fires, owing to their highly combustible nature (Sandstrom 1996).

"Uncertainty" is a good word for describing Spanish land management institutions. The areas of responsibility amongst different governing levels (local, regional, national and European) have not yet been clearly defined (Mata Porras 1993). There also seems to be a great variety of contradictory policies facing Spanish farmers and other people directly concerned with land management. On the one hand, there is environmental legislation which provides opportunities for sustainable land management, but on the other hand some European and Spanish governmental policies seem to indirectly produce devastating, long-term soil problems for short-term gains (Valladares 1993). A highly diversified decision-making system influences land management at various scales. The public sector has appreciably increased its capacity for control of natural resources since the mid-1970s. This capacity, however, is limited by difficulties of coordinating the numerous federal institutions operating in land management. The costs of coordination and monitoring seem to be very high.

The transition of land management power from local to regional and finally to European institutions since the 1970s has reduced the ability of local communities and individuals to make decisions concerning their land resources. Distant decisions, such as the "set-aside policy" created by the European Union, are not always suited to fragile Spanish ecosystems.

In addition, land management policies based on a single-factor model, involving solely economic or solely ecological parameters, are unable to create a model for sustainable land management. The set-aside policy is derived from such a single-factor model, with the strictly economic purpose of reducing the expensive surpluses of herbaceous crops, without any awareness of the ecological consequences it produces.

The Croatian Adriatic coast

Elsewhere in Mediterranean regions, ever-increasing population numbers reduce the space per capita, changing relationships amongst rural communities, landowners and the aristocracy. In Dalmatia, the northern Adriatic coast and Istria, a colonial system was established during the Roman occupation (peasants were formally free but subordinated to the landowner by contract) and remained practically until World War II (Sabadi 1994).

According to Veseli (1877), this early establishment of land ownership was the main cause of land degradation. In 1877, it was easy to differentiate private property from the public ones – it seemed like an oasis in the midst of stony desert. In 1958, 74% of land was so-called "social property" (today, public or state property). Exploitation of public and "social" properties (during the socialist era) continued in the second half of the century but the intensity of the degradation still has not been measured (Plate 17.6).

Although the number of inhabitants in the Dalmatian region increased more than sevenfold, from 100 000 in the 16th century to 714 200 in 1953, their attitudes and methods of land use remained the same (Potočić 1957). In 1992 there

Plate 17.6 *State forest of pubescent oak near Knin in Dalmatia, used for fodder and litter. Photo by B. Vrbek*

were 863 752 inhabitants in Dalmatia; 28 537 of them (3.3%) were employed in agriculture (Statistical Year Book 1993). There is still a lack of awareness of environmental matters but recently it has been improving slowly.

Croatia is a country with an economy in transition, with a series of legal changes concerning business relations and financing. Since 1990 Croatia has been involved in the liberation war, and part of the Mediterranean region is occupied. The few years of independent state sovereignty and of developing democracy constitute too short a period to be able to draw conclusions on the importance and influence of governmental organizations at any level, or to obtain data on the roles of agri-business and multi-nationals.

Except for the arson fires mentioned previously, it is not possible to assess the impact of military activities on the environment, although an example was the threat by occupying forces to destroy the Peruca dam in Dalmatia.

Greece and the EU

In Greece, land mismanagement is strongly connected with farm fragmentation: the average farm covers only 1.8 ha. On small, hilly fields it is inevitable that the land will be cultivated perpendicular to the contours. This is one of the main reasons for the rapid occurrence of accelerated erosion and land degradation.

More generally, the new reforms (1993–1996) of the Common Agricultural Policy of the EU will have a strong impact on land degradation/protection. The main points of the reforms are: (a) price cuts for key products such as cereals; (b) the withdrawal of land from production; and (c) accompanying measures for agri-environmental protection, afforestation of agricultural land and early retirement of the farmers (Commission of the European Communities 1993).

The above reforms will affect extensive areas cultivated with cereals in Greece. Price reductions of 29% over three years are expected, together with the removal of land from production and compensatory payments according to land area and not production. Declining cereal prices, which are not offset by compensatory payments, will decrease the intensification of agriculture (reduced use of fertilizers and pesticides, and decreased ploughing), resulting in the protection of land and the environment. Land abandonment, as discussed previously, will have both positive and negative effects: positive through increasing plant cover and hence reducing the erosion risk; negative through increased exposure to the effects of climate change and more intensive grazing pressures on the land. Under drought conditions, as in 1989, the land remains bare and at risk of both water and wind erosion. Abandoned lands undergo intensive grazing which dramatically increases the rate of land degradation.

North Africa

Unlike many of the depopulating mountain areas of other parts of the Mediterranean Basin (McNeill 1992), the population of the Maghreb has grown strongly. In Algeria there were 2 000 000 inhabitants in 1830. By 1876 the numbers had increased to 2 800 000, by 1926 to 6 000 000, and in 1981 they had reached 19 100 000. Similarly, in Morocco the population was 5 000 000 in 1911, 21 700 000 in 1981 and 26 500 000 in 1994 (Plit 1983; Tatar 1995).

In Petite Kabylia and in the Rif, population densities greater than 100 persons/km^2 are common. The demographic evolution of Beni Chougranes can be recounted since the census of 1886 (Benchetrit 1972). In 1886, the density might have been about five persons/km^2 and the wooded massifs were practically uninhabited. In 1926, the population had doubled in the mountains (10 persons/km^2), and if the inhabitants of the newly created centres of colonization at the mountain limit are added, that density would be 40 persons/km^2. The extension on the piedmonts of colonial agriculture, particularly vineyards, accounts for the migration of a great number of inhabitants towards the interior of the mountain. Thus, in 1946 population densities rose to 100 persons/km^2 in the peripheries and to 40 in the mountains.

The population of the Rif alone and in the zone of the massive chains of the Sanhaja, Rhomara and nearby massifs, increased from 77 000 to 163 000 inhabitants from 1950 to 1980. The rate of growth seems to have accelerated, since it increased from 1.6% between 1930 and 1950 to 2.5% subsequently. Such rates are not found in the other Moroccan mountains, and are possible only in developed rural areas, such as irrigated plains.

Population movements, linked to the European colonization of the best lands and political disturbances, are the most frequent explanation of the present state of land degradation. The mountain refuge notion was often used by authors who noticed the hasty and extensive clearings by the migrated populations, which caused damage (Poncet 1961).

The destruction of the anterior social organization has often been depicted as being responsible for the acceleration of resource degradation. In the High Atlas of Azilal, the different ecological stages were used in a complementary fashion, thanks to co-operation amongst the tribes (Herzenni 1993). In summer, several tribes used collective altitudinal rangelands, requiring an inter-tribal agreement, especially upon the opening and closing dates of the pastures in question. Forests closer to the villages were exploited according to internal understandings in the village (with a quota of wood cutting per family), and that use was rationed during snowy periods. Cultivated lands could not be extended without the collective's decision; on the other hand, private lands were managed by the private status. However, irrigation as well as large works were carried out in an arranged manner (Rachik 1993).

The present tendency is to accommodate populations in smaller areas, to reduce the transhumance ranges (which accounts for the over-grazing of the *thurifer* juniper), to extend cultivation and to appropriate lands. In the collective pastures, customary regulations are no longer implemented, resulting in over-exploitation. The breaking up of private property and fear of expropriation of the lands developed at the expense of the forest and rangelands, are other underlying causes of degradation. Indeed, the lack of guarantee as to the future is one of the reasons for the exploitative behaviour of the people (Fay 1979, 1984).

In Algeria, in the region of Oran, the transition from the socialist economy towards a liberal one was implemented by division of the big state domains into small concerns by drawing lots and a definite appropriation of ownership. Because of conflicts between beneficiaries, many lands remained uncultivated and were abandoned. Big rural houses were built within the subdivided land, then agreements were made with the municipalities and thus those lands became urbanized (Benjelid 1993).

The present social mutations in North Africa are deep. Many of the youth who studied, and are now unemployed, long to emigrate. That category of youngsters has lost the knowledge of former generations. Traditional techniques are seen as archaic, old-fashioned and useless. For example, few youngsters are capable of building the dry stone walls properly and according to the inherited norms.

Southern Australia

In Australia, the Conachers (1995) have discussed a range of underlying social and economic factors responsible for land degradation. It is important for these to be identified, because often the processes and mechanisms responsible for land degradation are reasonably well known, and yet inappropriate practices persist. These socioeconomic factors include: economic pressures on farmers, especially the cost/price squeeze; people's perceptions of and attitudes towards the environment (namely the perception that it is limitless and will withstand abuse, and the pioneering attitude that it is a resource to be used, not an environment to be nurtured); farmer education (generally poorer than their urban counterparts), conservatism and resistance to change; related to that, the ageing of the farm population; people's lack of awareness of the problems of degradation, especially those aspects which are not readily visible, or which are apparently ephemeral; and the not inconsiderable roles of government agencies and policies. Drought relief, for example, may encourage farmers to retain stock during dry years, thereby causing considerable damage to vegetation and soils which would not have occurred had the farmers destocked during the drought.

California

In California, there are essentially two groups of underlying causes of human use and misuse of land: those related to human attitudes towards the land, and those dealing with the principles and policies of land stewardship.

Human attitudes

In terms of human impacts on the land, North America is a relatively young continent and, within that context, the Mediterranean lands of Greater California are even younger. While recognizing an antecedent scatter of Native American peoples and mostly coastal Spanish and Mexican colonists, it may be argued that massive human impact began essentially around 1850 with the coming of California statehood. At that time, the region was perceived as blessed with an abundance of natural resources – gold and timber certainly but above all, land. To this region, the new, mostly American immigrants brought what Zelinski (1973) would call an intense, almost anarchistic individualism, a high valuation on mobility and change, a mechanistic vision of the earth, and perhaps a messianic perfectionism. Simply translated, this implied that resources were limitless, there to be used unhindered by government. Attitudes such as this help to explain much of the region's early development, how resources were exploited and land used and abused before people either moved on or began to have doubts about this morality. But to where does one move on in California? The Pacific Ocean is something of a barrier and, perhaps for that reason alone, the increasing clutter of people along the Pacific coast began to have doubts and to reassess its options.

It is in the above context that one can recognize five principal phases in human attitudes towards the land of California: an Exploitation Phase before 1890; a Progressive Phase between 1890 and 1930; a Government Phase from 1930 to 1945; a Post-war Expansion Phase from 1945 to 1970; and a Conservation Phase since 1970. There is of course much temporal and spatial overlap. Exploitation persists as individual property rights are guarded against government encroachment, and conservation efforts score some victories but many failures. But the above phases do embrace a spectrum of attitudes from 1850 to the present that has seen decreasing individualism, increasing corporate investment and expanding government control with respect to the land and its resources. Not that individualism is dead or government paramount, just that the individual cannot get away with as much now as in the past. There is accountability and an increasing concern for responsible stewardship of the land.

The *Exploitation Phase* began in dramatic style with the Gold Rush and the ancillary activities it spawned and which later developed their own momentum, specifically logging and ranching. Simple exploitation faltered when hydraulic gold mining collapsed under legislative edict in 1884, but the concept of limitless resources continued for many decades. The *Progressive Phase*, ushered in by the creation of Yosemite and Sequoia National Parks in 1890 and by the Forest Reserve Act of 1891, recognized that marginal lands could be conserved, ideally by government stewardship, while much of the rest could continue to be exploited. The *Government Phase* from 1930 to 1945 was really a government response to decades of over-grazing and soil exhaustion, and to individuals brought low by the Great Depression and its aftermath, including World War II. It saw broad government intervention in human welfare and resource use, captured respectively in such organizations as the Civilian Conservation Corps and the Soil Conservation Service. The *Post-war Expansion Phase* saw renewed economic growth and accelerated urban growth as demobilized service personnel, perhaps having first tasted California in the war years, sought a better life in "lotus land". Growing concern for environmental issues, however, led in the 1960s to a plethora of government legislation for the more orderly use of the land, ushering in the present *Conservation Phase* from around 1970. In reality, the period since 1970 has seen a continuing battle between conservation and development interests, the former supported by some of the people and some politicians, the latter protected by American respect for property rights.

An unkind observer would opine that the exploitation of resources is driven by greed and compounded by ignorance – greed for gold, greed for black gold (oil), greed for timber. A kinder observer would suggest that resource use is a legitimate means of achieving a good life for individual and family and ultimately for society. If so, we are still left with ignorance. Many early resources were exploited in ignorance of on-site and off-site impacts. Gold miners cared nothing for the worthless tailings they shed downstream, at least not until they were educated about downstream flooding and sedimentation. Early cattle graziers thought little about the problems of range deterioration and erosion, at least until stock values declined. Ignorance of the repercussions of resource use, rather than simple greed, largely explains many of the problems of land degradation that have beset California during historic times. It is in this context, through education, that sustainable resource management has often been achieved, leading for example to the replacement of logging by timber harvesting.

Human stewardship of the land

In California, perhaps more than in other Mediterranean regions, an understanding of who controls the land goes far towards explaining how it has been used and misused. Native Americans maintained the land and its resources for the common good and, because their numbers were so few, issues of land ownership did not apply. Such notions soon died, along with most of the native peoples, when Europeans arrived. Missions and presidios were established under the Laws of the Indies, ultimately controlled by the King of Spain. After 1821, Mexico continued the Spanish tradition of land ownership with embellishment. Before 1850, some 3 500 000 ha of prime grazing and crop land, mostly in coastal areas, had been transferred to private ownership.

When California joined the Union in 1850, a board of commissioners was established to segregate private land claims from the public lands of the United States. In this way, the land was defined according to the origin of title, and this helps to explain how the land came to be used in later years. First, land titles derived from Spanish or Mexican authority were mostly confirmed in private ownership. These were *rancho* and *pueblo* lands containing some of the best pasture and crop areas, mostly along the coast. Second, lands unconveyed under Spanish and Mexican authority, though once vested in the King of Spain or the Mexican nation, passed directly to the United States as so-called *public lands*. The United States in turn began conveying much of this public domain to individuals or corporate bodies under American laws governing preemption, homestead, desert, timber and mining claims, military bounty warrants, federal townsites, railroad titles, and Native American reservations (Robinson 1948). In addition, the United States soon granted to California from the public domain the so-called *State lands*, to be used primarily for raising funds for education and reclamation. With little thought for conservation or development, or future income, California largely disposed of these areas within 20 years of becoming a State. Between 1848 and 1965, some 20 000 000 ha of public land were transferred to private and state ownership. State lands also included tidelands extending "three English miles" (5 km) from mean high water, a zone of much contention in later years. Land beyond the "three-mile limit" remained part of the United States, at least subject to international agreement, and also proved contentious later because of the federal government's offer of leases for petroleum exploration and recovery on the outer continental shelf.

The relevance of land stewardship soon becomes evident in the context of land degradation. Today, 46% of California and 49% of Oregon remain "public land" though, because of large federal holdings in the desert, the actual percentages within the Mediterranean region are less. Some of this land is strictly controlled within national parks, but the remainder is subject to a variety of federal jurisdictions and variable control. For example, the Bureau of Land Management, which controls extensive mountain and desert lands, has been criticized for mismanaging several critical areas, leading to degradation from over-grazing, mining and off-road vehicles. The Department of Agriculture, through its control of the National Forest System, has been criticized for abusing its stewardship of forest lands in the Coast Ranges and Sierra Nevada. Nevertheless, degradation of federal lands is less problematic, if only because they are open to public scrutiny, whereas private lands are not. About 2% of California remains in state ownership.

Confirmation of Mexican *rancho* titles and the subsequent transfer of federal lands to private ownership have created numerous instances of land degradation which have gone unheeded and uncontrolled until the land has been irreparably damaged. These problems are compounded by an American reverence for private property rights and feelings towards government ranging from muted intolerance to outright hostility. Soil erosion in the state correlates well with the distribution of private lands subject to prolonged and unrestricted grazing. The 52% of California in private ownership includes nearly all the land suitable for intensive agriculture and urban/industrial development.

18
Some broader implications of land degradation

Compiled by Arthur Conacher and Maria Sala from contributions on the following regions:

Iberian Peninsula and Balearic Islands — **Maria Sala** and **Celeste Coelho**
South of France and Corsica — **Jean-Louis Ballais**
Italy, Sicily and Sardinia — **Marino Sorriso-Valvo**
Greece — **Constantinos Kosmas, N.G. Danalatos** and **A. Mizara**
Dalmatian coast — **Jela Bilandžija, Matija Franković, Dražen Kaučić**
Eastern Mediterranean — **Moshe Inbar**
North Africa — **Abdellah Laouina**
California — **Antony** and **Amalie Jo Orme**
Southwestern Cape — **Mike Meadows**
Southern Australia — **Arthur** and **Jeanette Conacher**

Introduction

Since many forms of land degradation (removal of natural vegetation, fire, accelerated erosion) are also causes of other forms (accelerated erosion, change of species composition and sedimentation, respectively, for example), a number of implications of land degradation have already been discussed as "Problems" or "Causes" in the previous chapters. This chapter, therefore, looks at a selection of the broader implications, whilst recognizing that some of them – notably the loss of species – are themselves aspects of land degradation in their own right.

The most important implications identified by the contributors are ecological, economic and perhaps social, with some interesting additional thoughts contributed from some regions.

Ecological Implications of Land Degradation

Vegetative implications of seasonal drought

Seasonal drought has been discussed in Chapter 13 as a problem; whether it should be regarded as an aspect of "land degradation" is doubtful. Nevertheless, in natural vegetation it produces interesting responses, some of which are probably unique to Mediterranean-type climates.

In the Mediterranean-type climate there are two ways in which vegetation can survive a prolonged dry and hot summer (Fisher 1978). The first is by completing the life cycle during the wet winter period, for example grass and cereals (wheat, barley), some of which are indigenous to the region. The second way is by special structural adaptations, such as deep or extensive roots, or dormancy during the dry season, as leaf shedders, or by maintaining nutriment in bulbs or tubers, such as members of the liliaceae family. Other plants are active during the summer but develop a thick outer layer, in order to reduce transpiration and heat stress, on the trunk or on leaves, as in the oak, laurel and olive, where leaf surface and size are also reduced, even to spines or spiky leaves as in the tamarisk and heath forms. These adaptations are typical of sclerophyllous vegetation. In the Mediterranean Basin, adaptation to these conditions developed the vegetational *maquis* formation, usually a dense forest with oak (*Quercus calliprinos*) and pine (*Pinus halepensis*), and *garrigue* in the less moist areas, with bushes up to 1 m high. Analogues of these forms are found in southwestern Africa and southern Australia.

Implications of fire

Fire has been prominent in Mediterranean forest ecosystems for thousands of years. The burning of the vegetation cover affects water and sediment regimes, nutrient cycling and the vegetation composition of the damaged areas, causing an adjustment of the ecosystem to the new conditions (Bourdeau *et al.* 1987; Gill *et al.* 1981; Kutiel and Naveh 1987). Fires alter the physical and chemical properties of the soil and partly or totally destroy the biomass. Their effects on vegetation include species loss, reductions in or alterations to species abundance and diversity, and structural and functional changes. Species diversity may increase after fire as a result of the creation of patches and landscape heterogeneity (Kutiel 1994). Fire also reduces the resistance of the soil to erosion and decreases infiltration, affecting surface erosion processes, increasing runoff and sediment yield. The type and magnitude of geomorphic effects depend on two factors; fire regime and geomorphic sensitivity. Intense fires in areas of high erosion potential will dominate sediment production and cause major alterations to the ecosystem, while no major effects may be noticed on soil after a light burning (Swanson 1981).

In Spain, experiments have shown that the increase in erosion in the first months after a fire can be 40 times greater than the normal loss of soil under a forest canopy (Soler and Sala 1992), and runoff can be 80 times greater (Díaz-Fierros *et al.* 1982; Vega *et al.* 1983). For this reason, it is probable that if the impacts of burnt surfaces

Land Degradation in Mediterranean Environments of the World: Nature and Extent, Causes and Solutions,
Edited by A.J. Conacher and M. Sala. 1998 John Wiley & Sons Ltd.

were considered when estimating erosion rates for various Spanish provinces (MOPU 1989a), there would be a notable increase in the figures for Galicia, Valencia and Catalonia.

In the wetter areas of central Portugal (Agueda basin), recovery after fire is rapid in pine forests due to the development of a thick layer of *maquis* which completely covers the soil surface, inhibiting the detachment of soil particles by splash. However, during the first year after fire, runoff values can reach 90% of rainfall, decreasing in subsequent months to less than 40%.

Post-fire overland flow responses are higher and decline more slowly with time after fire in eucalyptus than pine in the north-central Portuguese environment. These differences appear to be linked to: (a) more severe hydrophobicity of post-fire (and pre-fire) eucalyptus forest soils; (b) higher erosion rates and associated stone layer development on the less litter-covered post-fire eucalyptus soils in the immediate post-burn period prior to timber clearance, compared with the needle-covered burnt pine soils; and (c) fewer routeways for water to penetrate the hydrophobic soil because of slower undergrowth regeneration on burnt eucalyptus soils.

A dual model of overland flow generation has been found to apply to all post-burn plots in Portugal. Overland flow responses were highest after drought (when soils were highly hydrophobic) and after very wet antecedent weather (when soils were saturated or near-saturated), but were muted in moderately wet conditions (when soils were neither hydrophobic nor saturated). Post-fire, overland flow responses tend to be higher on shallower soils because they are more easily saturated. This tends to produce high erosion rates even when soils have become very shallow (Coelho *et al.* 1995).

Burning enhances solutional losses of Mg, Ca, K, NO_3 and SO_4, reflecting the natural mobility of the substances. The increased losses are due to increased runoff and elevated solute concentrations at the burnt sites. Burning converts large stores of nutrients held in vegetation into readily soluble and mobile elements in ash.

In France, it appears that the Sainte-Baume mountain forests (Figures 18.1 and 18.2) have diminished considerably since 1939, at the same time as *Pinus halepensis* has substituted itself for the oaks as a result of fire (Juramy and Montfort 1982). However, provided it is not disturbed, recolonization by the *Quercus coccifera garrigue* takes only two to three years on argillaceous and calcareous (Sainte Victoire Mountain; Baccouche 1990) or metamorphic (Maures massif) substrata. In addition, it seems that the floristic composition of the vegetation is influenced by the number of fires which have burnt it (Ballais *et al.* in press).

As in Spain, the semi-natural geosystems are also marked by the considerable increase in erosion rates, which can be multiplied by 10 or even 100 during the first hydrologic year after the fire

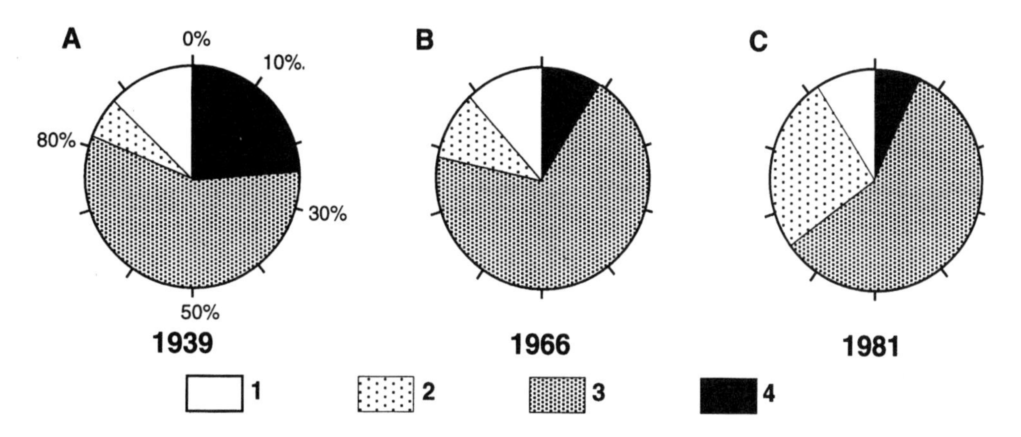

Figure 18.1 *Composition of the vegetation cover on the western slopes of Sainte-Baume mountain: evolution from 1939 to 1981. Key: 1 = forest; 2 =* garrigues; *3 = "fallow lands" or grass; 4 = cultivated areas* (after Juramy and Montfort 1982)

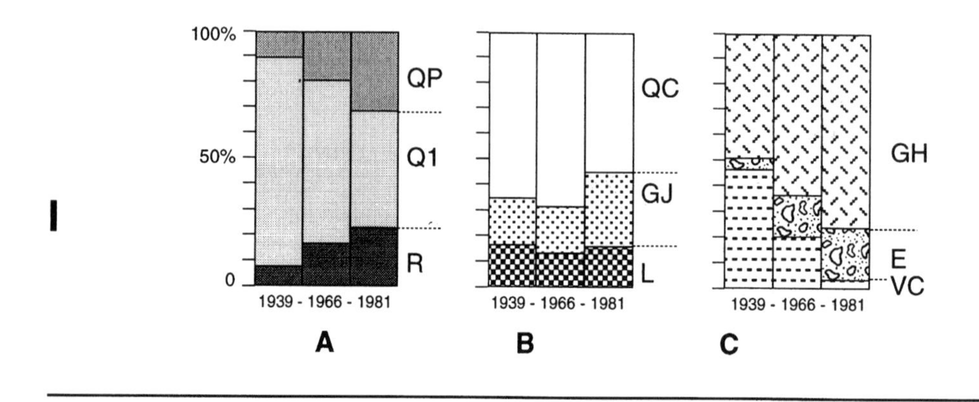

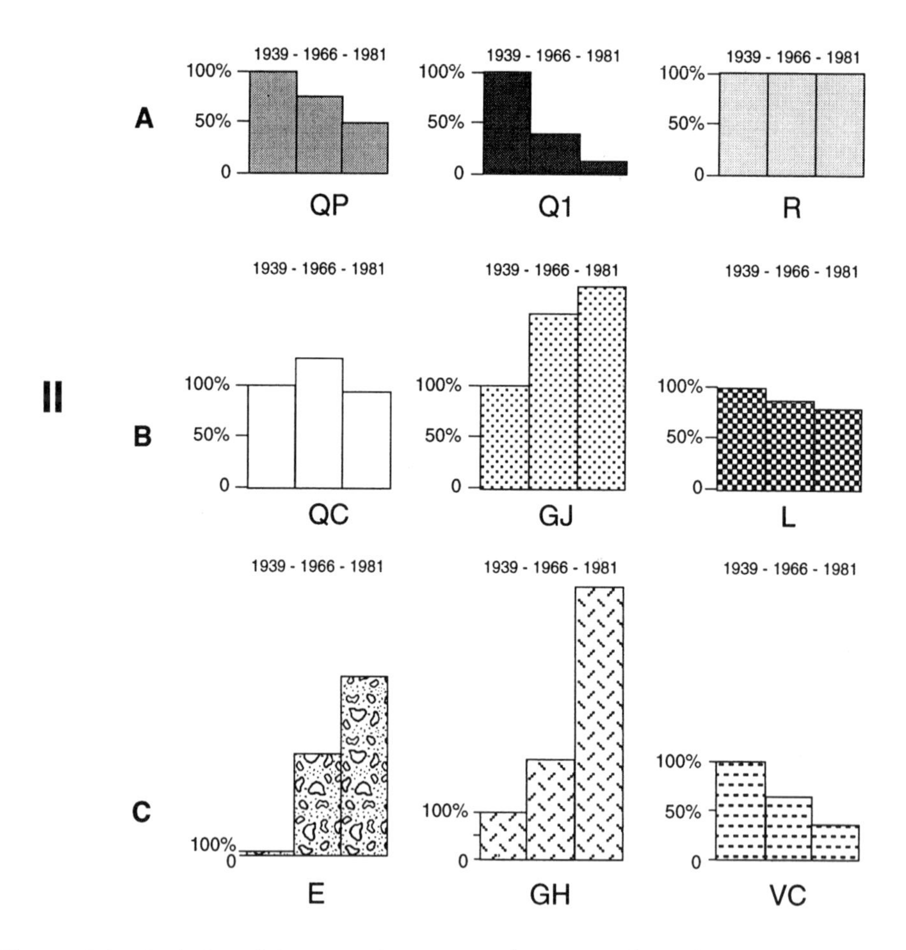

Figure 18.2 *Evolution of the vegetation cover on the western slope of Sainte-Baume mountain from 1939 to 1981. (I) Changes in the vegetative species composition: A, forest; B, garrigue; C, fallow land, screes. (II) Quantitative changes in each vegetation species: QP,* Quercus pubescens; *Q1,* Quercus ilex; *R, riparian forest; QC,* Quercus mirbecki; *GJ,* Juniperus *and* Juniperus *forests; L, lavender formation; E, screes; GH, herbaceous grouping; VC, vegetation on the crests. 100% is the 1939 value for the whole of part II* (after Juramy and Montfort 1982)

(Beguin 1992; Ballais 1993; Martin *et al.* 1993). Then, very quickly, after two to three years according to the pluviometric characteristics (annual total and rainfall intensity/erosivity), erosion rates return to their previous levels. This result is explained by the quick recolonization by *garrigue*, which rapidly achieves about the same amount of cover as was provided by the forest before the fire. So, in soil erosion terms, what is important is the rapidity with which the soil surface becomes protected by a vegetative cover.

Furthermore, fires have produced many stone flakes on the calcareous slopes of the Sainte Victoire mountain. These angular, very flat and thin flakes have been termed "ignifracts". In some cases, 1 m^2 of calcareous substratum produces as much as 22 500 g of ignifracts, as during the fire of 28 August 1989 (Ballais and Bosc 1994). This amount, produced in a few minutes, can be compared with 181 $g/m^2/yr$ for the rate of gelifraction on limestones in lower Provence. It means that today, and perhaps during the historical period, ignifraction is the main mechanical fragmentation process on limestones in the south of France.

There are few systematic studies on naturally fire-stricken areas in Italy. From direct observation, it can be stated that fire is not a big problem for the *macchia*, which is capable of self-regeneration with a resilience time of one year in the case of light fires, to three or four years in the case of heavy fires. Results from experimental fires confirm this observation (Lucchesi *et al.* 1994). Complications may arise if the burnt area is affected by severe erosion (Plate 18.1).

Plate 18.1 *Gully erosion on a fire-effected slope, Italy. Photo by M. Sorriso-Valvo*

Recovery times for other types of woodland vary considerably, and normally new plantings are required, except for riparian vegetation and pines in the mountain areas, where they find their climax. If burned vegetation was already stressed, steppe followed by *macchia* replaces other types of vegetation.

By means of fire experiments on *maquis* in Tuscany, it has been observed that light fires, that is fires that raise ground temperature to no more than 180–200 C, do not alter substantially the chemical properties of the soil but promote the solubility of various cations, with beneficial effects. Thus they allow for a rapid recovery, with new vegetation reaching 2 m in height after 12 months. In contrast, in the case of severe fires where ground temperature may exceed 450 C, the structure of the soil is altered: porosity is decreased and runoff increased. A longer period is necessary for vegetation recovery, so that after one year the tallest plants are only 1.5 m in height (Lucchesi *et al.* 1994).

In the eastern Mediterranean, Kutiel and Inbar (1993) investigated fire impact on soil nutrients and soil erosion in a pine plantation, with *P. halepensis* and *P. brutia* trees over 30 years old, after a light wildfire moved through the understorey and litter layer without causing major damage to the overstorey tree canopies. A closed, unburnt site was used as a control. An increase of NH_4–N was noted in the soil and NO_3–N concentrations determined nitrification processes after the fire (Figure 18.3). A 15-fold increase in available P was found at the burnt site one month

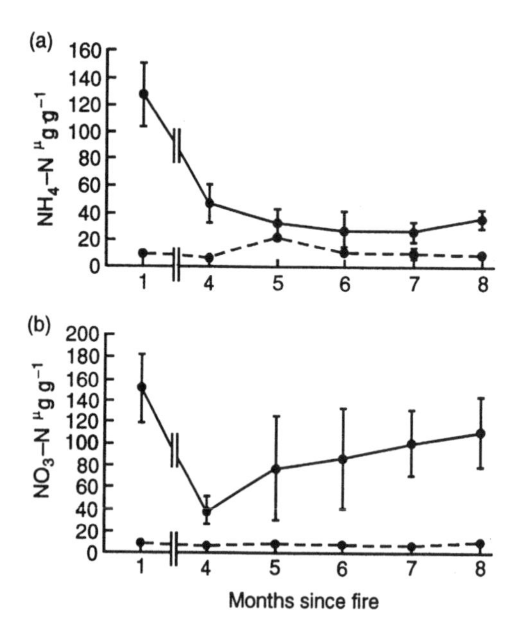

Figure 18.3 *Effects of a moderate forest fire in a mature pine plantation on NH_4-N (a) and NO_3-N (b) concentrations. Solid line = burnt soil; dotted line = unburnt soil. Vertical lines indicate standard deviations* (reproduced from *Catena*, 20 (Kutiel and Inbar 1993), by kind permission of Elsevier Science – NL, Sara Burgerhartstraat 25, 1055 KV Amsterdam, The Netherlands)

after the fire. Runoff and sediment values were very low, reaching a maximum of 0.12 l/m^2 and 0.02 g/m^2, respectively, for the largest storm in the season with 84 mm. Total runoff was higher in the unburnt site than in the burnt site, due to the light fire promoting a higher infiltration capacity.

In contrast, after one of the most severe fires in the country in September 1989 on Mt Carmel (Plate 18.2), erosion and runoff rates during the first year were found to be five and four orders of magnitude higher, respectively, in the burnt areas than in the unburnt control areas (Figure 16.1). The rates declined rapidly during subsequent years after vegetation became re-established.

In the southwestern Cape, Boelhouwers *et al.* (in press) have documented the development of water-repellent soil and consequent accelerated erosion following a hot fire in an area of Table Mountain. The situation has been described elsewhere in southwestern Cape catchments occupied by both forest and *fynbos* (Scott and van Wyk 1990). There are indications that inappropriate burning practices, particularly in areas heavily infested with alien species, can produce marked changes in soil wettability and, therefore, susceptibility to erosion by surface runoff.

Plate 18.2 *Fire on Mt Carmel. Photo by M. Inbar*

In countries with very old civilizations such as the Mediterranean Basin countries, continuous human use of fire (natural fires are rare in the eastern Mediterranean) has created a floristic composition which is adapted to fire. The plants of sclerophyllous communities in the Mediterranean region – such as the *maquis* and *garrigue* in southern France, *macchia* in Italy, *xerovuni* and *phrygana* in Greece, *choresh* and *batha* in Israel, *gatha nabati* in Syria and Lebanon, and *tomillares* and *matorral* in Spain – are adapted to survive after fires, especially medium to light fires (Trabaud 1981). There are various adaptations to fire: by strong vegetative subterranean stems, such as *Q. calliprinos* in the eastern Mediterranean (Naveh 1974); by disseminating seeds, for example, *P. halepensis* and *P. brutia*; or by increasing their seedlings after fire (*Q. ilex*, *Q. calliprinos* and *Arbutus*).

Trabaud (1981) has also described several types of adaptation to fire by the pyrophytic ecosystem of the *garrigue* and *maquis*: (1) rhizomes, bulbs or other subterranean parts of the plant; (2) buds, which are kept in the soil, and develop into trunk or branches of the tree; (3) thick bark that prevents penetration of heat; and (4) production of great quantities of seeds. These responses are very similar to the adaptations to fire of the natural vegetation in Australia's Mediterranean climate regions, where tens of thousands of years of Aboriginal burning in all likelihood have also been responsible for a pyrophytic ecosystem (Gill 1981).

In short, the floristic composition of the Mediterranean vegetation, as well as the Mediterranean Basin's soils and even its geomorphology, are probably very strongly influenced by a fire environment induced by people over a long period of time.

Other ecological implications of land degradation

In Italy, Brandmayr and Pizzolotto (1994) found that in environments disturbed by strong human intervention, such as high mountain pasture, the number of animal and vegetal species increases as common species colonize the area. As a counterpart, however, the specialized forms of life are under stress and in the long term some may disappear. In particular, some coleopterans (beetles), which are good indicators of the degree of conservation of the soil and of the environment in general, show a reduction of variety of species and individuals from west to east. The influence of human activity has been stronger in the eastern part of southern Mediterranean Italy because it is drier (Figure 3.1) and thus more susceptible to disturbance. The state of degradation of the environment is reflected also in the lack of great ungulates and predators (only a few wolves remain), eliminated by humans in recent centuries.

At present, in the lowlands (up to 600 m above sea level), where human pressure has been higher, the environment of sclerophylls has been destroyed by up to 95% of the area, and that of deciduous forest by up to 80%. In mountain areas, the proportion of preserved environment is higher but in the best conditions the degraded area is not less than 60% (Bransmayr and Pizzolotto 1994).

Unfortunately, there is only general knowledge of the presence of different biotope types and/or their status in Croatia. The mapping of biotopes in Croatia is now in an initial phase. A good basis for biotope mapping activity within the "Information system for environment and landscape management" will be the Vegetation Map of Croatia (scale 1:100 000) when completed, with 40% of the sectional maps having been printed and released in mid-1995. Nevertheless, according to the Red Book of the plant species of Croatia (Ministry of Civil Engineering 1994), 62 species are under threat due to the destruction of habitat by human activities (directly or by indirect changing of habitat conditions), for example by constructing roads, dams and industrial zones, and by pollution. There appear to be no data here or elsewhere in the Mediterranean regions on the loss of local control of plant genetic resources through plant variety rights (patenting).

Greece presents a positive feedback loop whereby land degradation leads to the intensification of agriculture which in turn has ecological and degradation consequences. Undoubtedly a similar argument could be constructed in the other regions.

In particular, it is argued that the rate of restoration of natural vegetation on sloping, degraded land with shallow soils and dry climatic conditions is low. Consequences of such disturbance include increased flooding, landslides and the siltation of rivers and dams. Large dams have been constructed throughout the country in recent decades, resulting in the intensification of agriculture with the consequences of soil structure deterioration and soil contamination with agro-chemicals, and the creation of favourable conditions for salinization on recent floodplains.

One of the main effects of land degradation in Greece is the reduction of soil resources. As the soils become increasingly scarce, agriculture becomes concentrated in areas with rich soils. This aggravates the problem of land degradation by concentrating higher and higher inputs into restricted areas, overloading the land system and creating further problems of land degradation.

The reduction in the thickness of hill soils beyond a critical point drastically reduces the potential for biomass production of any value. The effect of effective soil depth on total above-ground biomass production is depicted in Figure 18.4 for soils formed on shales/sandstones, which are highly variable in depth to bedrock and in gravel and stone contents, according to their landscape position. It can be observed that total above-ground biomass production increased logarithmically with soil depth, pointing to the impact of the latter on water availability (especially in dry years) as well as soil fertility status (especially in wet years) (Kosmas *et al.* 1993a).

The impact of humans on vegetation in the Levant areas of the eastern Mediterranean has been summarized by Zohary (1983) as having three main effects.

1. *Extinction* – hundreds and probably thousands of species have disappeared due to human impacts, over-grazing and fire. Over the past 100 years, hundreds of endemic plants previously collected and described by 19th century botanists are no longer found in Turkey and Iran (Zohary 1983).
2. *Antipastoralism* – non-palatable plants have invaded areas which were deprived of their palatable plants. Examples include *Peganum harmala*, *Zygophyllum* species, non-palatable species of *Artemisia*, and many other thorny species; thorniness is a naturally selected property for high survival in heavily grazed regions. This change affected mainly the marginal Mediterranean areas, with less than 300 mm precipitation.
3. *Antipyrism* (fire adaptation) – plants have acquired the ability to regenerate and survive after fire. This category includes almost all the trees and shrubs of the Mediterranean *maquis* species of *Quercus*, *Arbutus* and *Pistacia*, and

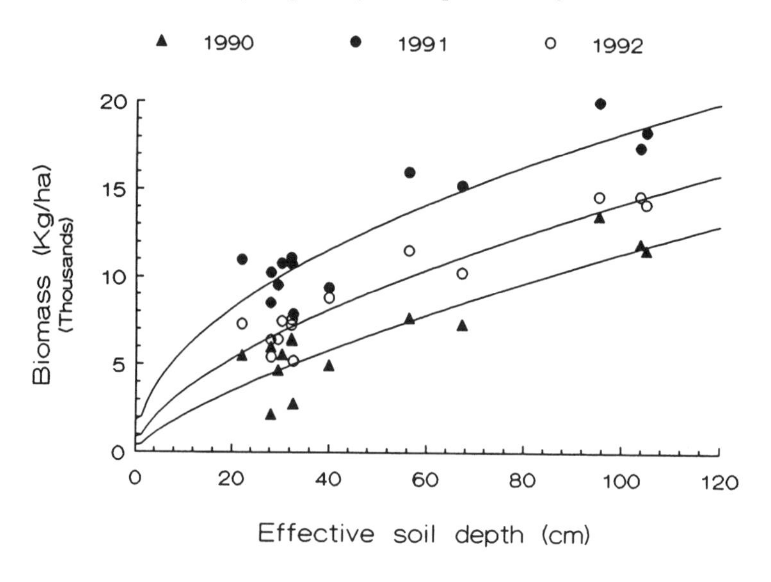

Figure 18.4 *Relation between the total above-ground biomass production of rain-fed wheat and effective soil depth of soils formed on shales/sandstones for three successive years* (source: Kosmas *et al.* 1993a)

many succulent plants such as *Artemisia herba-alba* and *Anabasis syriaca*, which are inedible and incombustible. A typical shrub is *Quercus calliprinos*, branching from its base probably as a means of survival after fire. The arboreal form, with a well developed stem, is found around burial places which were protected from burning and grazing.

Here, as throughout the Mediterranean Basin, agriculture has introduced a large number of species from different geographical regions over the centuries, to the extent that it is not always easy to identify unambiguously what is truly "native" vegetation.

There are also important ecological implications of land degradation in North Africa, where the present rate of afforestation is less than 10% and is much lower than the international average of 20%.

Real and precious ecosystems have disappeared.

(a) The *pistachios* of the semi-arid plateaux of eastern Morocco and of Orania were the habitat of a wild fauna, rich in gazelles, lions and hyenas. The use of *pistachio* wood to make soap accounts for the disappearance.
(b) *Oleaster* plantations (*Olea europaca*) (wild olive tree) were a main element of a climatic vegetation of the totally cleared Atlantic plains.
(c) Red juniper plantations (*Juniperus phonicea*) used to cover the southern slope of the Atlases. Those slopes are now completely deforested. They also used to be a natural barrier facing the desert. The mountains provided plant refuges during the dry periods, then residual remnants from the extension of the degradation from the lower lands, which alone maintain ecosystems which are otherwise vanishing
(d) The Moroccan fir tree (*Abies pinsapo*), an endemic species of the Rif, now covers only 5000 ha from an original area of 15 000 ha. The fir tree provides shelter for more than 30 species of nesting birds.
(e) Cedar (*Cedrus atlantica*) has widely receded. Its present extent in Morocco is 115 000 ha, compared with a potential area of more than 200 000 ha. Yet, its vascular flora reach about 1000 species (trees, shrubs, annual and perennial herbaceous). These contain pastoral, aromatic, medicinal, ornamental and other useful plants. The example of the cedar cemeteries of the eastern High Atlas is often quoted.
(f) The *thurifer* juniper (*Juniperus thurifera*), known as a "living fossil", is the only tree of the Atlantic summits between 2000 and 3000 m above sea level. It is a very hardy tree and its economic role is fundamental because it ensures the highlanders' living (heating, cooking, leaves for cattle feed), and it is an irreplaceable formation which is receding rapidly.
(g) The *orophil* flora (steppe of high mountains) is very rich (150 endemic species) but is undergoing rapid degradation.

Many animal species have disappeared since the beginning of the century, either through destruction of their habitats or through hunting. They include the lions of the Atlas, panthers and hyenas, as well as reptiles and birds which are widely threatened. Pollution also represents a serious threat to fauna which are linked to aquatic habitats.

Considerable pressures have been brought to bear on natural habitat and native species by human encroachment over the past 150 years in California. As Table 18.1 shows, the losses incurred by certain specialized habitats are exceptionally severe. California's biologically productive coastal wetlands, which offer valuable habitat to many bird and fish species and serve as natural water purifiers, have been reduced by 80%. The interior wetlands of the Central Valley, once the home to grizzly bear, pronghorn antelope (*Antilocapra americana*), Tule elk (*Cervus*

Table 18.1 *Habitat losses in California since 1850* (source: Kreissman 1991)

Habitat	Area lost	
	ha	%
Coastal wetland	82 000	80
Interior wetland	1 520 000	96
Riparian woodland	330 000	89
Valley grassland	8 895 000	99
Vernal pools	1 121 000	66

nannodes), mule deer (*Odocoileus hemionus*) and others, and the nesting, breeding and foraging sites of millions of migrating water fowl, now cover but 4% of their former range. Riparian woodlands, which provided similarly valuable habitat in and around the Central Valley, have been reduced by 89%. Oak woodlands, especially those comprising valley oak or *roble (Quercus lobata)*, have been thinned and isolated. Native valley grasslands, which covered about one-quarter of the state a hundred years ago, have almost vanished. Some two-thirds of California's vernal pools, so important to wildlife in a region of summer drought, have disappeared and attempts to recreate these unusual ecosystems have met with little success.

Loss of such habitat has severely impacted the region's wildlife. Certainly, in earlier times Native Americans hunted animals for food and other purposes but with little long-term impact. However, the prolonged fight between bears and Portola's soldiers in Los Osos Valley, near San Luis Obispo, in 1769 heralded a rapid decline in wildlife which brought many species to extinction. Grizzly bear, grey wolf (*Canis lupus*), white-tailed deer (*Odocoileus virginianus*), and long-eared kit fox have all disappeared from California. Of the native birds, San Clemente Bewick's wren, Santa Barbara song sparrow and sharp-tailed grouse have gone. Of the fish, bull trout, Clear Lake splittail, Shoshone pupfish, Tecopa pupfish and thick-tailed chub have similarly disappeared, as have scores of invertebrates, notably those associated with dune, lake and wetland habitats.

Although some species now extinct in California still survive elsewhere in North America, human impacts have decimated much of the remaining fauna. Mountain lion (*Felis concolor*), the largest North American member of the cat family, was once widespread throughout the United States but is now restricted to the western states. When given protected status in 1985, only 600 survived in California, mostly in mountain habitats but some close enough to expanding suburbia to present management problems. Beaver (*Castor canadensis*) has been more resilient. Trapped nearly to extinction for its fur in the 19th century, initially by American trappers during the Mexican period, it is now re-established over most of its former range. In contrast, the giant California condor (*Gymnogyps californianus*) struggles to survive. Brought close to extinction by hunting and pesticide ingestion, the remaining species in the Transverse Ranges were brought to the Los Angeles and San Diego zoos in the 1980s for breeding, a programme that has led more recently to the reintroduction of selected breeding pairs to the wild, where their survival is now being monitored.

The loss of natural habitat has been partly offset by the introduction of new habitat, especially in and around cities, to which some native animals have adapted. Though naturally shy, grey fox (*Urocyon cinereoargenteus*) and coyote (*Canis latrans*) have both proved adaptable to urban life, living off scraps and domestic pets on the margins of many cities. Though rarely seen, mountain lion (*Felis concolor*) and bobcat (*Felis rufus*) have shown similar adaptability.

"The world's richest flora at risk"

The floral wealth of the southwestern Cape is regarded as being one of the "Mediterranean" world's great biogeographical assets. Unfortunately, the nature and extent of land degradation in the region have threatened a substantial proportion of these flora with extinction. Many of the plant species naturally have small populations which, combined with the loss of habitat, have resulted in the *fynbos* region having a particularly high concentration of Red Data Book species. Critically rare, endangered, vulnerable or extinct species account for approximately 15% (more than 1400) of the region's 8500 plant species. Twenty-nine percent of the mammals, 7% of the birds, 18% of the amphibians, 43% of the freshwater fish and 23% of the butterflies also qualify for Worldwide Fund for Nature (WWF) Red Data status (Cowling and Richardson 1995).

As in other Mediterranean regions, the threatened species are not evenly distributed across the area, as Figure 18.5 shows, and the southwestern extremity around Cape Town and its hinterland is especially rich in rare and threatened plants. Indeed, even the remnant communities within the city boundaries are loaded with species threatened with extinction. McDowell *et al.* (1991), in a survey of just 484 ha of acid sand plain *fynbos* within the metropolitan area of Cape Town, recorded 74 Red Data plant taxa, representing a ratio of more than 15 species/km^2. One site of just 56 ha, in the centre-field area of a Cape Town racecourse, has 18 Red Data plant species, arguably the highest concentration of any equivalent area in the world (McDowell *et al.* 1991). Such small areas emphasize their extreme vulnerability to any kind of disturbance.

Loss of indigenous species in southern Australia

Clearing of native vegetation, draining of wetlands and a range of land-use practices have been responsible directly for species loss or decline in southern Australia. Most broadacre crop areas of the region are characterized by fragmented stands of native vegetation. Any remnants of less than 5 ha in extent are considered to be ecologically non-viable and many are considerably degraded. Without adequate protection, most of these remnants will not survive. They are subjected to external pressures from the surrounding "hostile" agricultural landscape – such as grazing by stock, nutrient enrichment, chemical drift, fire, predation and rising water tables. Individual trees scattered across farms stand an even smaller chance of survival. Loss of native vegetation also means loss of habitat for native fauna. "Mallee has been largely replaced by wheatfields; eucalypt forest by orchards or pasture; mangroves and coastal vegetation by suburbs, factories and holiday shacks" (Aslin 1986, p. 57; see also Breckwoldt 1986; Davidson and Davidson 1992; Hobbs and Saunders 1993). For these and other reasons, a number of species have become extinct, rare or endangered.

Western Australia has by far the largest number of "presumed extinct" (70), endangered (91), vulnerable (363) and poorly known (573) species of plants of all the Australian states (Australian

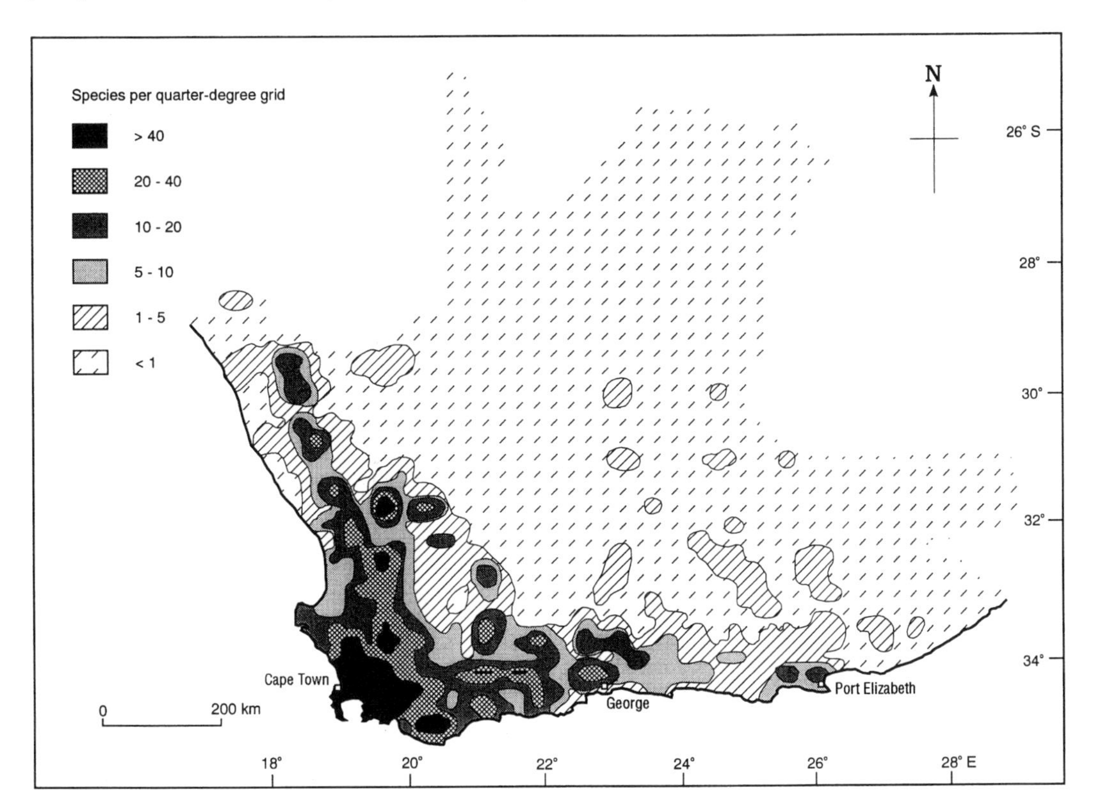

Figure 18.5 *Distribution intensity map of rare and threatened plant taxa in the southwestern Cape* (source: Hilton-Taylor and Le Roux 1989)

Bureau of Statistics 1992). Most of Western Australia's 1024 rare or endangered species of native plants and animals are located in the wheatbelt, in the Mediterranean region, where the highest number of plant and animal extinctions has occurred. Approximately 60% of the species of medium-sized mammals have disappeared from the wheatbelt since European settlement (Select Committee into Land Conservation 1991).

Of the more than 3000 vascular plant taxa known in South Australia, four are extinct and 200 endangered. And of more than 600 land vertebrate species known in the state, 28 species of mammals and birds are extinct and 36 species rare, endangered or vulnerable. The areas at greatest risk include the southeast, the mallee and the Mt Lofty ranges (Department of Environment and Land Management 1993). In Victoria, 35 (and possibly as many as 100) species of native plants are now extinct, with 110 species endangered and 315 vulnerable. Less than 0.5% of the original lowland grasslands of Victoria remain intact (Scott 1991). Nearly a quarter (125) of all known species of *Eucalyptus* are at some degree of risk throughout Australia, especially in the east and southwest of the continent (Department of Arts, Heritage and Environment 1986). Kennedy (1990, p. 17) has summarized the situation neatly:

"The problem of disappearing species represents not only a conservation crisis on a scale never faced before, but also one of the greater environmental challenges we are ever likely to confront".

Implications of the spread of exotics

Mediterranean regions, because of their natural features and long history of human settlement, are extremely susceptible to biological invasions. The impacts are profound in terms of disruption to ecosystems, as well as for agriculture and human welfare (see di Castri *et al.* 1990; Groves and di Castri 1991).

The most significant impact of the introduction of new species in California saw aggressive sod-building European grasses expand at the expense of native bunch grasses, particularly as a result of sheep and cattle raising from Spanish times onward. Native bunch grass (*Aristida, Festuca* and *Poa* spp.) and needlegrass (*Achnatherum, Nassella* spp.) have been largely replaced in most pastures by European species of *Avena, Bromus, Festuca* and *Lolium*. The effect of this change from bunch grass to sod grass on erosion and mass movement has yet to be quantified. However, during heavy rains those hillslopes covered in sod grass such as European wild oats (*Avena fatua*) are more prone to soil slippage than slopes vegetated with bunch grasses, because their sod restricts downward infiltration and increases shear stresses at the base of the saturated sod.

Current attempts at revegetating selected areas with native bunch grasses have raised interesting questions regarding the relative importance of annual and perennial grasses in the native ecosystem. In the 19th century, it was widely accepted that California's native grasslands had been dominated by annual plants, because of the stress imposed upon perennials by protracted summer drought. In the earlier 20th century, influenced by Clementsian concepts of plant succession, it came to be believed that purple needlegrass (*Nasella pulchra*) and other perennial bunch grasses were the natural dominants. This view now has little empirical support and recent research suggests that the native grasslands contained abundant annual grasses (for example, *Vulpia microstachys, Orcuttia* spp.). It was these annuals as well as the perennial bunch grasses which were replaced by European annuals, especially on favourable mesic sites (Blumler 1995). Native annuals were thus relegated to marginal xeric or hydric sites of low productivity. The introduction of fast-growing European annual grasses to oak savannas formerly dominated by perennial bunch grasses has also had an impact on fire frequency and oak regeneration.

In a study of threats posed to plant species in an especially species-rich area of the southwestern Cape, Hall and Veldhuis (1984) identified invasive exotic plants and agriculture as the principal culprits. This is mainly because of the competition for habitat and relates to the aggressive and ecologically successful nature of the invasive trees, as discussed in Chapter 17. In a study of the impacts of invasion of mountain *fynbos* by *Pinus radiata*, Richardson and van Wilgen (1986) reported a 64% decline in indigenous species diversity over a 40 year period.

There are, in addition, several other significant impacts of the spread of exotics, including effects on coastal sediment dynamics, streambank erosion, nutrient cycling, fire regime and, most importantly, catchment hydrology.

The reduction of sand movement by wind in coastal dunes which have been planted with *Acacia cyclops* has produced dramatic (intentional) reductions in the occurrence of bare sand, and profound effects (unintentional) on the distribution and extent of sandy beaches (Macdonald and Richardson 1986). River bank erosion is accelerated following the invasion of riparian vegetation by exotic trees because they replace the mats of the indigenous sedge (*Prionium serratum*) which otherwise protects these banks from disturbance during winter floods. Soil erosion is definitely a problem following the clear-felling of long-established pine plantations, especially on steeper slopes (Macdonald and Richardson 1986). Dense infestations of the leguminous and nitrogen-fixing, alien *Acacia* species produce marked changes in soil nutrient status, and because *fynbos* plants are generally adapted to oligotrophic soils, the eutrophication process has significant effects on the indigenous flora, even following removal of the invasives. Fuel loads are much greater in stands of introduced vegetation and this has important consequences for fire intensity and fire size. Such impacts may permanently degrade *fynbos* ecosystems if there are subsequent changes in soil physical structure (soils subjected to extremely hot fires may become water repellent, for example (Boelhouwers *et al.*) (in press) or increased soil erosion.

Dense stands of exotic woody species, because of greater overall biomass and transpiring leaf area, are now known to have a very marked impact on catchment hydrology, in particular on water yield (Richardson and Van Wilgen 1986). Van Wilgen *et al.* (in press) have compared the effects on hydrological output of two hypothetical mountain *fynbos* catchments, one with exotic tree removal as a specific management strategy, and concluded that the reduction in biomass results in a 30% improvement in runoff in the case of controlling alien tree infestation. Despite the initial costs of clearing the catchment and the ongoing management costs of maintaining the slopes, there is considerable financial benefit. Indeed, the unit cost of providing water is reduced by almost 20% over the non-management option. Taking other ecosystem services into account, Higgins *et al.* (in preparation) have concluded that proactive catchment management has potentially huge economic advantages. Clearly, the conservation of biodiversity which results from establishing and maintaining an alien-free *fynbos* environment pays for itself in several other ways, not least in providing considerably greater quantities of water for the expanding population of the region.

Some Economic and Social Implications of Land Degradation

The importance of land degradation does not end with the deterioration of environmental ecosystems, significant though that is. It also has serious ramifications for people's economic and social well-being, which, in the view of some, is the main justification for the increasing effort being made to understand and solve the problems. For example, although the Mediterranean forest is considered to have no economic value because the timber, especially that of *Pinus halepensis*, is not really appreciated, this forest is a place for walking, relaxation and recreation for the local urban populations and for tourists. From this viewpoint, which is convergent with the ecologists' one, the disappearance of trees is regarded as a serious loss. This section samples these broader implications, since it is impossible to provide a detailed evaluation of the topic here.

The south of France

With regard to floods, the ecological implications are not very important compared with the economic and human consequences. In the south of France, for example, the October 1940 floods probably killed 48 people in the Tech valley alone (Soutade 1993), the flood of 3 October 1988 caused nine deaths and about $1 billion worth of damage in Nîmes (Leoussoff 1993), and the flood of 22 September 1992 resulted in 37 deaths and five missing persons in the Vaucluse department (Mennessier 1992). Road infrastructures were destroyed by floods (six bridges in the Ouvèze watershed in September 1992), as were houses, dykes and railways. Riparian forests and harvests were devastated. Floodplains are often deeply remoulded, as in the Tech valley where 10 to 20 million tonnes of alluvium were accumulated in October 1940 (Parde 1941). Restoration of communications sometimes takes several years. Such catastrophic events lead one to question hydrologic and planning specialists as to the efficacy of their methods.

Italy

For the whole of Italy, the cost of damage and repair for all natural geomorphologic catastrophes has been estimated at about $4 million *per day*; and land degradation must be considered one of the principal causes of the delay in the development of south Italy, that is, one half of Mediterranean Italy.

There are feedback effects between the economy and land degradation, and south Italy entered this circle since the Roman Empire (2nd century AD) due to drought and heavy land use. A specific example is coastal erosion, which has not only destroyed the beach and the rear beach, but also the coastal plain, some 2 to 4 m in elevation, which was the site of orchards and wheat fields. Thus, there was a substantial, non-recoverable loss of cultivated land. The extent of this loss has not been estimated to date, as cultivated land is attacked only along limited parts, and countermeasures have been undertaken in places. The retreat of the coastline is being blocked by massive intervention along some 60 km of coast. It is evident that in future years the problem will be felt in those parts of the coast down-drift from the southern end of the intervention zone, as the dominant drift is southwards.

Another example concerns the economic and social implications of landsliding. Table 18.2 displays information obtained by two agencies of the Cosenza province (north Calabria), as they appear in Nossin (1972), with the costs of treatment updated to 1994. The two agencies care for an area which is estimated to cover about 65% of the area of the province.

The records of ANAS, the national agency for road construction and maintenance, show that in

Table 18.2 *Extent of landsliding and costs of treatment in north Calabria* (source: Nossin 1972, with costs of treatment updated to 1994)

Year	Number of landslides	Surface (ha)	Damage (US$ million)
1966	49	196	1.9
1967	48	201	1.9
1968	33	230	1.6

1971 in the River Crati basin alone, which extends over 2577 km^2, there were 34 active landslides involving the ANAS road network, covering a surface of 7.6 km^2. Thus, the data for the 1966–1968 period in Table 18.2 are only partial, as stressed by Nossin (1972).

Attempts to obtain similar data for other regions have been unsuccessful. The province of Cosenza, however, can be considered representative of all Mediterranean areas in Italy except Puglia and Sardinia, where landslide incidence is much lower due to different geological and morphological conditions.

Flooding also has serious social and economic consequences. As a result of the 1951 flood in Calabria discussed in Chapter 12, 64 people died and 5000 had to abandon their houses. Because of collapsed bridges, 30 major roads and 29 aqueducts were cut. The total cost was estimated at more than $1 billion, a massive amount in one of the poorest regions of Italy, only a few years after World War II. Another major event occurred in October 1953, involving a larger area than in 1951 (8750 km^2). Rain intensity was higher, at least locally, reaching 138 mm/hr. Casualties (more than 100) and damage were more serious than in 1951, and included the destruction of at least 2000 ha of orange gardens and olive groves. The November 1966 flood in Florence is remembered because of the damage it caused to the cultural heritage, but soils also suffered severe damage.

Fire also has economic implications. In the most recent year (1993) for which data are available, fires in Italy caused damage estimated to cost $41 million. In Sardinia, the region most affected by fire, the damage from 1970 to 1990 amounted to several tens of millions of US dollars. According to an environment-orientated Italian party, in two years (1993–1994) fires caused damage valued at about $400 million.

At present, the various forms of land degradation extend their effects over nearly 30% of Sicily, Calabria, Lucania and Molise, 20% of Sardinia and Puglie, and 5 to 10% of remaining Mediterranean Italy. This has been a steady-state condition since the 18th century at least (De Tavel 1812); but when the new nation's government introduced an unfair tax on some agricultural and agriculture-derived products of south Italy (wheat, cotton, silk), marginal lands were abandoned and migration, first to America and then to central Europe, took place to such an extent that the Italian-derived population living abroad today is nearly the same as the number of Italians remaining in Italy. From this inadequate policy, for which land degradation provided the substratum but not the excuse for every mistake, the "Meridional Question" (the economic disparity between north and south Italy) arose at the turn of the last century, and is still far from solution at the end of the current one.

There is no doubt, however, that the solution of land degradation in Mediterranean Italy would constitute 90% of the solution of the Meridional Question.

Agricultural and forest statistics reflect the different conditions of the eastern Mediterranean. Turkey, Cyprus and Israel are closer to the north Euro-Mediterranean countries, which have a trend of farmland abandonment in marginal areas, and an increase in reforestation and urbanization, whereas in the southern Mediterranean Basin, forest land has receded and stock numbers increased, with a dependence on imported, concentrated food (Le Houerou 1992). The number of sheep and goats exceeds the carrying capacity of the fields and exerts an even greater pressure on marginal lands, as there is no rational management. Countries which were net food exporters now import food needs (FAO 1987).

North Africa

In North Africa, the degradation of the natural environment first means the lowering of forest production, despite the growth of plantations. It also means the lowering of the tourist value of the devastated regions, and the silting of dams, which are a basis of the modern agricultural economy and of urban development. There is also damage to infrastructure.

The growing danger of flooding is illustrated by the example, in 1984, of the Rhumel, where a bridge was destroyed claiming 1700 victims. Part of an industrial area of Rhumel was also destroyed. The most famous case is the floods of the lower plain of the Rharb.

One of the major implications of flooding is sedimentation, which in turn has a range of economic consequences. Thus, in the Rif and littoral rivers, river beds aggrade, fields are drowned and trees covered (Maurer 1968a, b). The sterilization of good plains soils by massive detrital materials occurred over 400 ha at the outlet of the streams of the Sbiba watershed in Tunisia (Boujarra 1993). Those gravel cones were all formed in 1969 and covered extensive agricultural lands.

A more serious aspect, however, is the silting up of the dams. An asset of the agricultural policy of Morocco and Tunisia and a fundamental basis for the maintenance of irrigation, the dams are threatened by the risk of both long- and short-term siltation. Annual alluviation is estimated at $50 \cdot 10^6$ m^3 in Morocco (Conseil Supérieur de l'Eau 1991a).

Most importantly, water resources of the Maghreb are limited, and damage to infrastructure and reduced capacities of dams as a result of sedimentation are serious. Taking the example of Mohammed V dam on the Moulouya River, from a total capacity of $726 \cdot 10^6$ m^3, the lost capacity is $260 \cdot 10^6$ m^3. Thus, the dam will be totally silted up by the year 2030 (Conseil Supérieur de l'Eau 1991b). Such a silting up has subsidiary effects (using Morocco for the examples):

– a reduction of electricity generation by $60 \cdot 10^6$ kWh by the year 2000 and of $300 \cdot 10^6$ kWh by the year 2030;
– a reduction of drinking and industrial water by $40 \cdot 10^6$ m^3 around the year 2000 and by more than $200 \cdot 10^6$ m^3 in 2030;
– a reduction of the available water for irrigation by $50 \cdot 10^6$ m^3 in the year 2000 and by $120 \cdot 10^6$ m^3/yr around the year 2030; and
– a threat to 100 000 irrigated hectares at a cost of more than $100 million and 50 000 jobs. The cost of investments in hydraulic fittings should be added.

The hilly dams of Orania in Algeria silt up at a particularly rapid rate. Of 30 dams, the majority have silted up by about 60% and some to 100% (Taabani and Kouti 1987; Oulad Chrif 1984).

Land degradation in the form of mass movement also has economic consequences. For example, in Algeria, 1543 houses were evacuated and 1543 families had to be relodged in 1972 after extensive slumping in the vicinity of Constantine (Benazzouz 1993).

Accelerated soil erosion also has economic implications for agriculture. Although studies seem to show that total yields continue to increase (but at a very low rate), the increased use of fertilizers should be taken into account. A calculation carried out for the watershed of Sidi Salah in the Rif (Merzouk and Backe 1991) shows that water erosion of soils causes the annual loss of 41 kg/ha of nitrogen, 16 kg/ha of phosphorous and 20 kg/ha of potassium, the replacement cost of which has been estimated at nearly $100/ha.

It has been noted in Chapter 15 that maximum erosion occurs during the years of high rainfall. These are, however, the years which provide record yields. Maintaining those yields at the same level in degraded regions, despite the use of chemical fertilizers and the selection of grains, testifies to a soil loss which is not negligible.

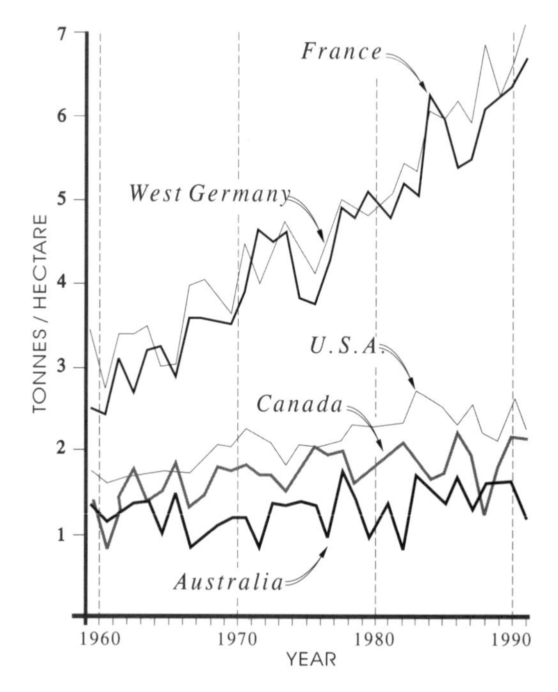

Figure 18.6 *Average wheat yields per unit area 1960–1994: comparative data for Australia and selected countries* (plotted by J. Conacher from data in *FAO Production Yearbooks*, Rome)

Southern Australia

Farm productivity and costs of repair

Australia's major crop is wheat, and Western Australia's Mediterranean region is the biggest state producer. Since 1960, there has been considerable research expenditure on improving the yields of wheat and other crops – by improved cultivars, fertilizers, pesticides, farm machinery and cultivation practices (notably minimum tillage). But average yields of wheat per unit area have not improved in that time (Figure 18.6) (the marked variations from year to year are of interest, reflecting mainly seasonal variations in the amount and timing of rainfall). There is a strong argument that Australia's failure to improve wheat yields is due to soil degradation. In other words, the increasing inputs into farming are needed purely to maintain yields: without increased inputs, productivity would have declined.

Counter-arguments are not particularly convincing. Perhaps the strongest is the suggestion that the static trend reflects the opening up of increasingly marginal land, where lower yields depress the averages. But McWilliam (1981) showed that neither average commercial yields nor record yields (as measured by field crop competitions) showed any significant change between 1960 and 1980 (even though about 60% of plant breeding research funds were spent on wheat). In other words, the best farmers were not increasing their productivity either.

For the whole of Western Australia, the Department of Agriculture estimated the annual cost of production losses due to land degradation at US$480 million in 1989 (Select Committee into Land Conservation 1991). However, this and the following data need to be treated with considerable caution: essentially, they provide only an indication of the economic implications of land degradation.

The major dollar losses occurred in the agricultural areas, in the Mediterranean zone: the data have been presented in Chapter 12, Table 12.7. Alarmingly, the annual losses of gross production due to secondary salinity alone were predicted to exceed the present total from all causes – to US$760 million – within the next 30–50 years. The data in Table 12.7 do not include the costs of repairing the damage (a 1994 estimate for Australia as a whole exceeded US$1.6 billion (Conacher and Conacher 1995); Yapp *et al.* suggested over US$2.4 billion in 1992), or production losses caused by insects and weeds, or inappropriate use of farm chemicals, or the economic consequences of air pollution, the enhanced greenhouse effect or the consequences of increasing ultraviolet radiation caused by the depletion of ozone in the upper atmosphere. Off-farm costs such as water salinization and eutrophication are also excluded from Table 12.7. Table, 18.3 indicates how just one problem, salinity, can impact on a wide range of uses.

It also needs to be emphasized that declines in productivity are generally due to a range of factors rather than any one cause. For example, work in the Western Australian wheatbelt by Marsh and Carter (1983) showed that productivity could be halved due to soil losses caused by specific farm practices. But these and other experimental studies assessed the effects of one

Table 18.3 *Tolerance of selected uses to various concentrations of soluble salts in water or soil water* (source: Conacher 1982, table 9.2)

Concentration of total soluble salts (mg/l)	Use of water
200	pulp and paper manufacture
250	high grade steel production
450	clover*
500	human consumption – WHO standard
600	stone fruits, citrus*
1050	vines*
1500	human consumption – WHO maximum permissible limit
2000	beans†
2500	carrots, onions, clovers†
3000	lettuce†, poultry
3800	potato, sweet potato, corn, flax, broadbean†
4400	pigs
5000	lucerne, broccoli, tomato, spinach, rice†
5800	soybean†
7000	horses
7300	dairy cattle
7500	beets, sorghum†
9000	safflower, wheat†
10 200	sugarbeet, cotton†, beef cattle, lambs, weaners, ewes in milk
11 200	barley†
13 200	adult sheep on dry feed
18 000	adult sheep on green grass
36 000	mean salinity of seawater in Cockburn Sound, Western Australia
55 000	camels

* The water quality criterion for clover, stone fruits, citrus and vines refers to the quality of irrigation water used and should not be confused with the soil-water quality criterion in the second footnote. The latter criterion is often more meaningful (although it, too, varies with soil properties amongst other variables), because the effects of irrigation can vary widely depending on the method of application (drip, sprinkler, flood), the duration of irrigation and the effectiveness of soil drainage.

† These are approximate values, from saturated soil extracts during the period of rapid plant growth and maturation, at which 50% yield reductions can be expected.

form of soil degradation (in this instance, surface soil loss) in isolation from others, such as soil structural decline or acidification, which are often taking place at the same time. As another example, lack of nitrogen is a common agronomic diagnosis of poor crop performance and yellowing of wheat crops (Hamblin and Kyneur 1993). But a decade earlier, Ellington *et al.* (1981) had demonstrated that several factors, including root defects, soil hardpans, poor rotation and acidic soils, were responsible for these symptoms. This is an important point which has a major bearing on the selection of appropriate solutions to land degradation problems (Part III).

In South Australia, it has been estimated that soil structure decline and reduced rainfall infiltration cause a loss in production of approximately US$48 million each year. Under experimental conditions, wheat yields declined by 120 kg/ha for every centimetre of soil lost by erosion, with yield reductions of up to 40% depending on the depth of soil lost. In another experiment, the equivalent of 54 kg of soil per hectare were lost in one minute during the course of a simulated 75 km/h wind. This represented a loss of 0.17 kg/ha of nitrogen and 0.35 kg/ha of phosphorus every minute. If such an event lasted for one hour, it would cost approximately US$9.60/ha to replace the lost nitrogen and phosphorus with fertilizer. The report points out that the loss of organic matter was of much greater importance, and that its replacement, which is vital to nutrient levels and soil structure, is much more difficult to achieve and takes considerable time (Community Education and Policy Development Group 1993).

In 1993, annual losses of production in South Australia attributed to secondary salinity were estimated at US$21 million, to water repellence US$1.6 million, and to soil acidification US$8 million. In the Adelaide Hills the average cost of purchasing and spreading the recommended rate of lime is US$44–60/ha and may be required every 10–15 years (Community Education and Policy Development Group 1993). Other costs of rehabilitation, such as contouring, fencing and tree planting, can be considerable. Costs of tree planting can range from hundreds to thousands of dollars per hectare depending on the methods used.

Victoria also experiences significant economic costs. For salinity alone (both irrigated and dryland), up to US$24 million of production are lost each year, and more than US$126 million were spent between 1984 and 1991 on remedial and preventative measures (Scott 1991, p. 470). Indicatively, across the border in New South Wales (north of the Mediterranean region), production losses due to secondary salinity constitute only 18% of all losses in the Murray/Darling basin. There, as is probably also true for Victoria, soil structure decline accounts for by far the greatest percentage (68%) of production losses (Murray/Darling Basin Ministerial Council 1987; cited in Edwards 1993, p. 148).

These problems are compounded or intensified during droughts and by external market factors and increasing costs – the cost/price squeeze. It becomes very difficult to separate the economic effects of land degradation on farm productivity from all the other factors which affect economic productivity (see Conacher and Conacher 1995). The 1983–1988 period was one such period of rural crisis, for which research by Smailes (1989) in the district of Cleve, normally a highly productive district in the Eyre Peninsula, has provided some interesting findings. They mirror the processes which have occurred throughout the dryland crop areas of South Australia, Western Australia and Victoria.

Impacts of rural crisis: Cleve, South Australia Loss of farm production was considerable over the 1983–1988 period. The losses were associated with reduced income and a relatively slight increase in debt, and significant cost cutting. Investment-type expenditure on land, buildings, installations, fences, farm plant and breeding stock shrank from US$1.36 million to less than $0.4 million for an aggregate of 26 farms. Such deferral of expenditure has obvious implications for land degradation and has also been described elsewhere for Australian farmers (Farley 1991; Knopke and Harris 1991).

The level of investment expenditure in the local Cleve economy fell from 77% to 59%. About 30 of 291 families quit farming, 18% of habitable houses were empty at June 1987, and the number of people per farm dropped from 6.2 to 4.8 between 1984 and 1987.

Although the number of businesses in Cleve remained more or less constant (eight firms closed and six opened), aggregate turnover dropped by 9.5% in nominal terms and by 26.8% when adjusted for inflation. Employment declined by 9.6%. Credit available to customers was reduced. More than half the households in the towns said they had been adversely affected economically by the rural crisis. Those on the lowest incomes were affected the most.

Nearly all businessmen expressed the view that their economic advantage was much better served by a large number of small farms in the district, even if not affluent, than by a smaller number of large and presumably wealthy farms. The number of families in the district was seen as crucial to maintain turnover and staffing levels. This was also the finding of an earlier study by Smailes (1979).

Rural population decline, social consequences and health

Rural population decline and its social consequences are perhaps a somewhat tenuous consequence of land degradation, but we consider the connection to be real. Rural population has been declining in Australia, and in parts of most Mediterranean countries, for many years. There are numerous causes, including mechanization of agriculture (reducing the need for a rural labour force), social changes (young people prefer to work in cities), smaller families, ageing populations, and increasing farm sizes. It is the latter factor which links with land degradation.

The argument is that yields per unit area have not increased (for example, Figure 18.6). Farm costs are rising much faster than prices received by farmers (Figure 18.7). Therefore, farmers have had to increase the area farmed (by purchasing other farms) in order to increase their total production in an attempt to remain solvent. Average farm sizes in the Western Australian wheatbelt have increased from around 800 ha in the 1950s to about 2000 ha in the 1990s. As discussed above, the failure to improve productivity per unit area is probably a consequence of land degradation. Larger farms have meant that the number of farms in any given area has decreased. Since farmers cannot afford to take on additional labour (relying increasingly on unpaid family help), due to the cost/price squeeze, the area worked by each farm worker has increased, and the total population in the hinterland of country towns has declined. In turn, this has resulted in the decline of country towns, particularly the smaller ones with populations of under 2000; larger central places, and towns with a broader economic base (including in particular mining, tourism and recreation, and retirement centres), have maintained their populations or even increased in size (Conacher and Conacher 1995).

The decline of smaller country towns has been exacerbated by a range of other actions, often by government. Normally, public sector employment has a stabilizing effect on towns through the supply of services and thus jobs and regular income. But in recent years many of these services in small towns have been closed, such as branch railway lines, post offices, police stations, hospitals and schools, which in turn has forced the closure of other town businesses (Bowie and Smailes 1988).

From the study of the Cleve district, discussed above, Smailes (1989) found that the rural crisis of 1984–1989 had increased the normal amount of rural out-migration by 25–50% over the period – mostly of young or working age people. Only 4% were retired farmers. These trends continued into the 1990s and apply throughout the extensive crop and pasture areas of Mediterranean Australia.

The social consequences have been severe – in terms of lifestyle, increasing distance from neighbours, schools and essential services, and loss of amenities and sense of community. Another consequence of increased out-migration of young people is a developing imbalance between the farm labour force and the work to be done, leading to stress, over-work, increasing health problems and suicides, increasing age, and postponement of retirement (Smailes 1989; Conacher and Conacher 1995). As Campbell (1992, p. 38) has commented poignantly: "the rural sector is ageing, declining, stressed and going broke".

Other health-related aspects of land degradation

Many land degradation problems tend to become recognized only when they pose a threat to

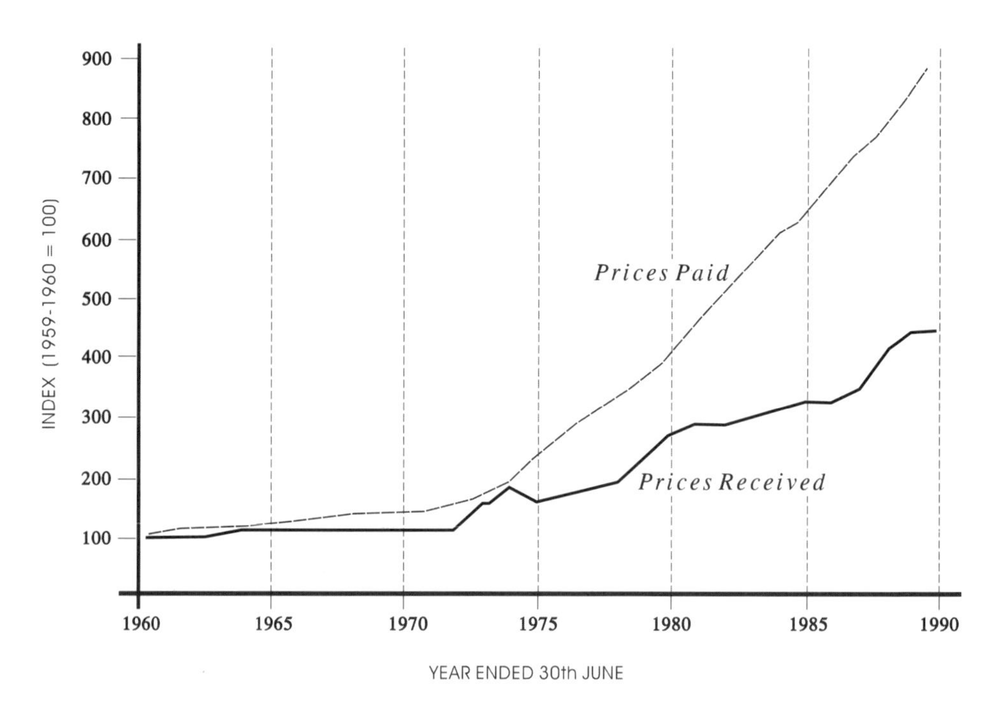

Figure 18.7 *Prices received in relation to costs paid by Australian farmers, 1960–1990* (after Cribb 1991, figure 4)

human health through the intake of contaminated air, water or food. In Victoria, for example, the fertilizer contaminant, cadmium, has been detected in food samples at higher than permissible concentrations (Scott 1991, p. 444). Of 650 vegetable and fruit samples tested since 1989 in South Australia, 22 contained a pesticide residue in excess of prescribed limits (Community Education and Policy Development Group 1993). The periodic discovery of pesticide residues in meat exported from Australia to the United States and Japan (caused by contaminated fodder) highlights the serious economic implications of the problem as well as the chain of cause and effect. Some marine organisms continue to accumulate heavy metals. For example, data from the Environmental Protection Council of South Australia (1988, table 4.14) show that significant percentages of sharks and rays have concentrations of mercury (14% of samples tested) and selenium (8%) above National Health and Medical Research Council (NHMRC) standards. The other marine organisms which exhibited some accumulation of heavy metal residues were crustacea (where 2% of the sample exceeded the NHMRC standard for copper) and mollusca (3% for lead and 5% for selenium).

Thus, other health-related aspects of land degradation include the toxic effects of blue-green algae on stock and humans (Main 1994), the accumulation of heavy metals in the body (Fergusson 1990), and the consequences of the increasing use of pesticides (Scott 1991, chapter 28). Blue-green algae (cyanobacterial) blooms, in their toxic form, may cause skin, respiratory and gastrointestinal problems. High lead levels in children's blood appear to affect detrimentally their IQ. Heavy metals also affect organisms other than humans (Shepherd 1986). Pesticides may cause acute illness from direct contact or long-term problems from chronic exposure through occupational use or food and water intake.

The huge increase in the use of pesticides, especially herbicides, has caused concern throughout the farming communities of southern Australia. Conacher and Conacher (1986) have reviewed the literature on this topic in some detail and provide more recent information in their 1995 publication. Often the people most directly affected are the operators – in most cases, the farmers. Much of the poisoning probably arises from improper or careless use of pesticides, such as lack of care in mixing the chemicals, spraying on windy days, especially if driving tractors with unsealed cabins, and failure to use protective clothing. A Western Australian survey in 1993 found that 36% of 920 farmers using synthetic agricultural chemicals experienced ill effects, of which more than half were attributed to herbicides. Symptoms (from most to least frequent) included headaches, nosebleed, nausea, skin rashes and burning, breathing difficulties, vomiting and diarrhoea, and eye problems (Kondinin Group 1993).

Some Other Considerations

The political crises of water

As discussed in Chapters 13 and 15, the nature of the Mediterranean environment, with its summer drought, limited water resources, "inconvenient" locations of water supplies and burgeoning populations, has led to problems of water shortages and deterioration of water quality, with a serious potential for trans-national conflicts over water use. Who "owns" the water, who pollutes the resource, and who pays? Of all the Mediterranean regions, these questions are perhaps most urgent in the eastern Mediterranean. They are discussed more fully in Chapter 13 and again in Part III, in the context of "Solutions".

Implications of a possible rise in sea level

Coastal processes in the eastern Mediterranean region, as elsewhere in the Basin, are characterized by a low tidal range and winter storms with exceptional waves up to 10 m. Catastrophic events, such as tsunamis and tectonic activities, have affected the morphology of the present coastline, with accelerated sedimentation of river estuaries (Pirazzoli 1986).

Most of the coastline is structurally controlled and will not be subjected to flooding by a rise in sea level of 10 to 50 cm in the next century. However, impacts on port infrastructures and beach facilities would be considerable, and have particular implications for the region's economy. The following effects are expected (Jelgersma and Sestini 1992):

1. lowlands and wetlands will be invaded by the sea, aggravating coastal flooding;
2. the salinity of the coastal aquifer and estuaries will increase;
3. there will be a negative effect on fisheries;
4. beach widths will be reduced by 20 to 60 m with a sea level rise of 30 cm. Since most beaches are less than 100 m in width this will have a serious impact; and
5. coastal floods will increase and flood protection structures will be needed.

Urbanization processes, involving industrial, commercial and communication centres, are concentrated in the coastal zone, as are most of the new hotels and other recreational structures. They are therefore most vulnerable to sea level rise. Future planning will need to prevent the worst scenarios posed by this threat.

Geoscientists' training

Geoscientists have an important role to play in dealing with environmental problems. In the past, they were pivotal to economic growth, by finding and making available mineral, energy and water resources. The main future task will be the conservation of resources, especially water and land; and scientific excellence and an adequate number of well trained earth scientists will be required.

The number of geoscientists in the United States in 1989 was about 71 000 or 300 per million inhabitants (American Geological Institute 1989), with 58% of them employed in finding and developing oil and gas (Figure 18.8). In Israel, there are an estimated 700 geoscientists, or 140 per million inhabitants, with a low percentage in the oil industry and about 10% in universities. As an illustration, there were about 40 Masters graduates in geomorphology in Israel in the 1980s decade, which is more than the total graduated in the previous 30 years; most of them are active in environmental issues.

Environmental alterations by anthropogenic activities demand new efforts from earth scientists in order to deal with these problems. It is this new trend that many geoscientists are dealing

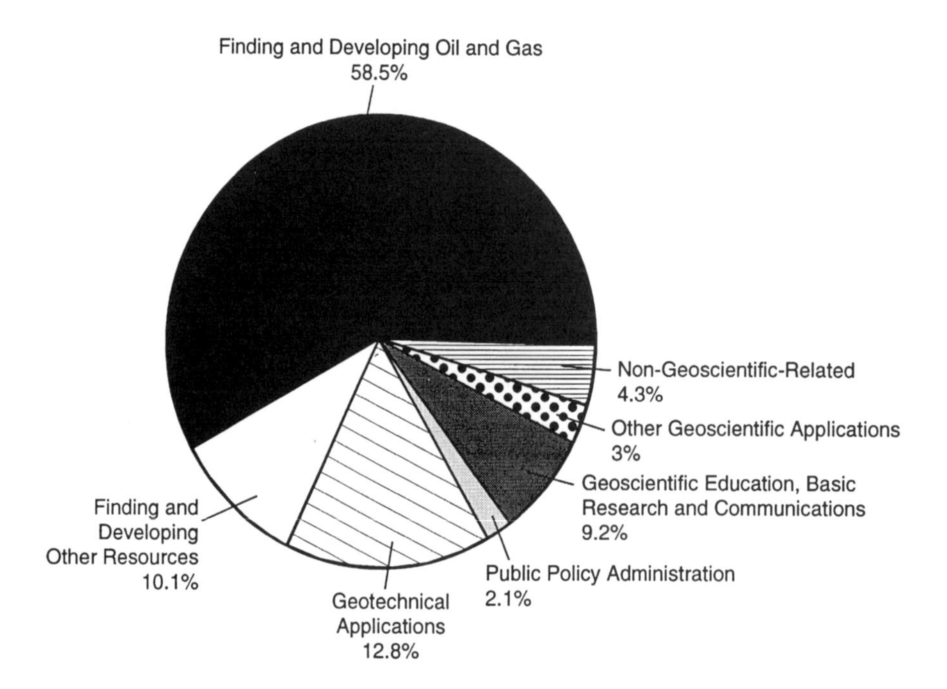

Figure 18.8 *Occupational objectives of employed solid-earth scientists in the US* (after American Geological Institute 1989)

with today, and it will be the focus of research in the future. Gathering and interpreting data, and integrating the effects of population pressures on the environment, are and will be important tasks for geoscientists.

Global collaboration and co-operative earth sciences programmes

The final implication of land degradation discussed here is the need for global collaboration and co-operative earth science programmes to understand how earth systems function. Current research in regions of Mediterranean-type climates is a good example of international co-operation. One such international study group is MEDALUS (Mediterranean Desertification and Land Use), which includes about 20 experimental groups in several Mediterranean countries and is financed by the European Union. Most European Mediterranean countries have study centres for Mediterranean topics, mainly on environmental issues. In the IGU (International Geographical Union) there was a Study Group on Mediterranean Erosion and Desertification (MED); this included 300 members of the world's scientific community drawn mainly from the five Mediterranean parts of the world, and was responsible for the production of this book. The Study Group has subsequently been upgraded to a Commission on Global Land Degradation and Desertification, and retains a Task Force on Mediterranean Land Degradation. An inter-governmental institution – MAP (Mediterranean Action Plan), based in Athens – carries out permanent work. It has issued a Blue Plan for practical activities, and a Priority Action Programme (PAP) is implementing 11 working plans in the 1990s. The major efforts are concentrated in the coastal areas.

PART III
Solutions to land degradation

Introduction

Arthur Conacher

Part I of this book described the nature of the world's Mediterranean environments, with an emphasis on those aspects which are most relevant to the widespread problem of land degradation. Whilst an overall definition of "Mediterranean-type climate" was adopted, based on the FAO classification, there are regional variations. In southern Australia, for example, the extensive crop/grazing regions are relatively uniform, but have a very marked boundary with the adjacent, uncleared, pastoral areas at their semi-arid margins. It makes no sense to place an artificial climatic boundary across an essentially homogeneous region when drawing the line a little further to the north coincides with marked changes on the ground.

Nevertheless, all the world's regions classified here as "Mediterranean" have characteristic features in common, of which the hot summer drought and cold season rainfall maximum are both definitive and the most distinctive. Many characteristics, both biophysical and human, of the Mediterranean regions flow from these essential climatic features; and they, in turn, account for some of the common problems of land degradation.

Another feature common to most Mediterranean climate regions is their proximity to more arid lands, especially in the eastern Mediterranean and North African regions, the southwestern United States, Chile, southwestern South Africa, and southern Australia. One of the consequences of this geographical fact is an acute awareness amongst the agricultural scientists in those regions of the concept of "desertification" and the manifold environmental, land use and socio-economic implications of possible global warming.

The problems of land degradation were described and explained in Part II, and their further implications discussed. Whilst there are similarities amongst the regions, particularly in relation to soil degradation, vegetation clearance and fire, there are also some marked differences. Some of these differences reflect historical factors, especially the duration of intensive land management in the Old World in relation to the New. Others reflect characteristics, such as the nature of the lithology, which are unrelated to the Mediterranean climate (although extensive limestone lithologies reflect common maritime and tectonic origins amongst most Mediterranean Basin regions). Some of the most severely eroding areas – which in turn generate a wide range of off-farm and socio-economic consequences – are underlain by soft marls and mudstones. But even there, the detailed nature of the processes still reflects the seasonality of the rainfall regime.

It is also apparent from Part II that there is a very wide range of problems of "land" degradation, extending well beyond soil erosion. Many specific soil properties, often not evident to the casual observer or even the farmer, are deteriorating, with adverse effects on yields. Terrestrial and aquatic biotic communities are becoming increasingly impoverished, with many species either extinct or endangered. Water quality and water shortages (the two are related), either existing or pending, are two of the major problems

Land Degradation in Mediterranean Environments of the World: Nature and Extent, Causes and Solutions.
Edited by A.J. Conacher and M. Sala.

here, and indeed in other environments. Sedimentation, salinization and accumulation of nutrients and toxic wastes in streams, rivers, dams, estuaries and groundwaters, mean that increasing proportions of previously freshwater resources are no longer suitable for irrigation or for human or animal consumption. Many of the various aspects of land degradation interact with one another, compounding the problems.

One of the major by-products of land degradation has been rural depopulation, although other important social, economic and political factors are also responsible. Rural depopulation and the associated decline of country towns and villages in turn have further implications for land degradation: lands which are no longer managed carefully, or which have even been abandoned, are perhaps a luxury which a world characterized by rapid population growth and increasing demands for food cannot afford. On the other hand, intensification of agriculture in the more restricted but more fertile areas, carries with it a range of actual and potential environmental costs.

Numerous causes of the complex problems of land degradation have been discussed. Direct causes are usually more obvious, including factors such as the removal of natural vegetation, fire, the deliberate introduction of exotic species of plants and animals, cultivation practices, tourism and recreational pressures on the land, the dumping of rubbish, the disposal of sewage, and the construction of dams. The less immediately obvious underlying causes include the nature of the environment itself, as referred to briefly above, economic pressures on farmers, people's perceptions of and attitudes towards the environment, lack of knowledge, land ownership and tenure, the important roles of government policies and agency actions, and, of course, the historical context. Not surprisingly, these causes vary, sometimes considerably, from one region to another; but again there are some constants. Human actions, perceptions and attitudes, and governmental failure to act effectively to deal with the problems, appear to be common to all regions.

Nevertheless, many positive actions are being taken throughout the world's Mediterranean regions to deal with the problem of land degradation, and these form the subject of Part III. The intention here is to identify and discuss those actions which are proving to be effective, and also to identify actions which may be effective. A major intention, which was an underlying reason for undertaking the writing of this book, is for scientists within the various regions to learn from one another about both successes and failures. The extent to which that has been achieved is considered in the concluding chapter.

Part III is structured in such a way that each chapter deals with a particular range of solutions. This is not always effective or easy to do, because in many, if not most, cases, degraded land is characterized by more than one form of land degradation, requiring more than one solution. But since it is not possible to discuss all problems and their solutions simultaneously, a structure had to be imposed.

Thus, Chapter 19 deals with solutions which concern stock (stocking rates, the kinds of animals being raised and grazing rotations) and cultivation and horticultural practices (including rotations between crop and pasture, strip farming, minimum tillage and cultivation implements). Solutions of an engineering type are considered in Chapter 20, such as terracing, contour banks, gully control structures, dams and coastal groynes. In contrast, solutions focusing on vegetation, or "ecological" type responses, are described in Chapter 21. These include "integrated" or whole farm or catchment solutions as well as more specific practices such as windbreaks and shelter belts, changing crop types, and agro-forestry. Recognizing the importance of some of the underlying causes, economic, social and policy solutions are discussed in the final two chapters, with Chapter 22 focusing on the Mediterranean Basin and Chapter 23 on the New World. These chapters consider changes such as incentives and penalties to persuade land managers to change their ways: direct subsidies, tax incentives and fines; education; information and communication; the responsibilities of various government agencies, or the need to create new bureaucracies; and the need (if appropriate) for new or changed government policies, legislation and regulations. How do we legislate to achieve environmentally sustainable agriculture using appropriate technologies?

19
Solutions dealing with animals, cultivation and horticultural practices

---

Compiled by Arthur Conacher and Maria Sala from contributions on the following regions:

Iberian Peninsula and Balearic Islands	**Maria Sala** and **Celeste Coelho**
South of France and Corsica	**Jean-Louis Ballais**
Italy, Sicily and Sardinia	**Marino Sorriso-Valvo**
Greece	**Constantinos Kosmas, N.G. Danalatos** and **A. Mizara**
Dalmatian coast	**Jela Bilandžija, Matija Franković, Dražen Kaučić**
North Africa	**Abdellah Laouina**
California	**Antony** and **Amalie Jo Orme**
Chile	**Consuelo Castro** and **Mauricio Calderon**
Southwestern Cape	**Mike Meadows**
Southern Australia	**Arthur** and **Jeanette Conacher**

Solutions dealing with animals – stocking rates, types of animals and grazing rotations – are considered in this chapter. The control of animal pests is also discussed. The chapter also considers how cultivation and horticultural practices may be varied to solve problems of land degradation in rural areas. Topics discussed include cultivation implements and practices such as crop rotations, strip cropping, minimum tillage and orcharding. The latter topic has some degree of overlap with agro-forestry, which is dealt with in Chapter 21.

Solutions Dealing with Animals

The Iberian Peninsula

The disorganization of traditional Mediterranean biological systems and landscapes, and the disintegration of rural societies in relation to the agricultural policy of the European Union (EC), is counteracted by the EU's promotion of studies of the viability of pastoral land use in the marginal, low productivity agricultural lands. Pasture with autochthonous animals, which have a low demand on the environment, can produce valued food such as industrial milk and lean meat (kid goat), and generate permanent jobs. It has been found that goats are a very appropriate animal for this use, and that pasture in arid Mediterranean environments produces an increase of biomass and a decreased vulnerability to drought (Boza *et al.* 1985; Le Houerou 1989; Montserrat 1990).

South of France and Corsica

The total number of grazing animals is decreasing as part of the general trend of rural decline. This is especially the case for sheep. Taken as a whole, the flocks play only a marginal role in the degradation of Mediterranean geosystems. In fact, their role is complex because, on the one hand, people have created high mountain pastures at the expense of the forests but, on the other hand, the use of high mountain pasture by sheep has also had positive consequences (Soutade 1980). Today they play an effective role in relation to the problem of forest fires. The best example is in Corsica, where burning of the *maquis*, and sometimes the forest, is used to favour the growth of palatable herbs (Dubost 1991). Cows and sheep were reintroduced into the Maures massif forest after the great fire of 1989, limiting the occurrence of new fires by eating the very inflammable herbs. This clearing role could be systematically developed, particularly in the extensive firebreaks.

The Dalmatian coast

Solutions dealing with animals, stocking rates, types of stock and grazing rotations are in an initial research phase. There are only some preliminary results at present. The projects "Integral use of natural vegetation resources in the region of Zadar" (Čižek and Varga 1988) and "Sustainable development of the island of Unije" (Dumančić 1994), could serve as pilot projects on the possibilities of revitalizing the Mediterranean region, especially the islands. According to preliminary results, the Croatian Adriatic region is considered to be particularly well suited to producing ecologically clean food of plant and animal origins, aromatic and pharmaceutical plants, and for apiculture and forestry purposes (Plate 19.1).

North Africa

The policy of pasture conservation aims to delimit collective groups and protection areas as well as to prevent the cultivation of some rangelands. For example, there are 14 projects of pasture improvement spread over 250 000 ha in the Atlas, while another 19 pastoral improvement programmes have allowed the constitution of 42 co-operatives over 77 000 ha. New forms of farming have appeared, notably ranching over considerably improved and fenced, extensive rangelands. In the mountains, actions have also been taken to improve conditions for the cattle, such as the provision of shelters and feeding centres at higher altitudes, and the elimination of

Land Degradation in Mediterranean Environments of the World: Nature and Extent, Causes and Solutions.
Edited by A.J. Conacher and M. Sala.

Plate 19.1 *An example of sound land use on the island of Hvar in Dalmatia. Vineyards are situated in the karst field, and aspic (*Lavandula spica*) and Aleppo pine (*Pinus halepensis*) are grown on the slopes. Photo by B. Vrbek*

goats from some forests (Bencherifa and Johnson 1993).

In Algeria, pastoral co-operatives, bordered by Aleppo pine trees, aim to reduce wood exploitation by using gas-ring equipment for cooking, but often, over-grazing has not been avoided.

Chile

As a solution to the desertification process it is suggested that stocking rates be reduced to 0.05 animal units per hectare (UA/ha) in the community of Acarquindaño (Region IV) and to 0.07 UA/ha in Yerba Loca (Region IV), and to restrict the use of the grasslands to goats. This would mean ending the grazing of sheep, which have higher requirements for fodder, particularly if the predominant breed in the zone is Hampshire Down, a crossbreed with the Australian Merino.

In order to undertake this rational management it is necessary to identify the appropriate terrains for pasture, considering the requirements of fodder in relation to soil type and fertility, and physiographic conditions. The number of paddocks should be at least five, used in autumn, winter, spring/summer and for special purposes, and/or a reserve when the lack of fodder is critical.

Annual crops should be relocated to the plains sectors of the valley floors. Irrigated and semi-irrigated terrains, dedicated to the production of introduced fodder species such as *Medicago sativa* L. and *Trifolium subterraneum*, are also needed. They can be used for strategic purposes, such as providing supplementary fodder for pregnant and young animals.

With regard to the management of animal births, these should be concentrated during June and July. This would mean a semi-intensive production which should not have a duration greater than two months, considering three ruts in the females in the months of January and February, during the summer pasture in the upper zones of the Andean Cordillera.

The percentage of kids should be 4%. Mothers and kids should be managed separately from the rest of the herd, considering that they have major nutritional requirements which differ from the other animals. In this way fodder needs will be concentrated when the grassland has initiated its growing period. This will allow weaning the small goats earlier, when they weigh 5–7 kg, and shifting them to irrigated or semi-irrigated grasslands where they can increase their weight by 200–250 g per day. This would make it possible to put them to market when demand and prices are higher. In this way a rapid recovery of the meadows will take place, permitting a major availability of fodder during periods of lower rainfall.

The necessity of having a better selection of stock is also an important issue; this involves finding stock which fulfils a double purpose (such as milk and meat) amongst the native animals. Females of indigenous breeds are highly productive, because of their adaptation to a desertified environment.

Enclosures for keeping the animals and for milking should be considered, with a paved area in order to provide more hygienic conditions.

The practice of regenerating degraded grasslands was initiated during the 1960s, and it permitted the recovery of grassland without labour or cultivation, with consequent low seeding costs. The establishment of dry farming grasslands, either natural, ameliorated or artificial, with introduced fodder species has also been of great conservation value, at the same time achieving good land rental values. Amongst the dry farming fodder species are *Phalaris*, "ballica" and subterranean *Trifolium* species.

The use of electric fencing has made direct grazing easier and more efficient, and is a powerful tool in the management of grasslands and their herds.

The southwestern Cape

The Department of Agriculture consults with all stock-farmers in the region in respect of farm planning and grazing rotations. In particular, Extension Officers, allocated to each district within the region, offer advice on appropriate stocking levels, farm division and grazing area management. Proper planning of all watering points, with respect to both numbers and location, is given special attention (H. Germishuys, pers. comm.). Other measures include the use of a fallow period in which the cultivation of pastures is replaced by a crop of legumes, but this practice is limited in extent. Gullies are usually fenced off to discourage stock from further disturbing them.

In the Mediterranean-climate regions *per se* of the region, measures currently undertaken by landowners practising mixed farming, usually of wheat and cattle, appear to be sufficient to ensure adequate soil conservation.

Southern Australia

There are perhaps two main problems of land degradation in southern Australia which are associated with farm animals. One is the effect of high stocking rates, particularly over summer, and the other concerns the leaching or disposal of effluent into water bodies, especially from intensive operations such as piggeries, dairies, poultry sheds, beef feedlots, stock-holding yards, and also from abattoirs. Additionally, farm animals play a major role in complementing extensive grain cropping. The alternation of crop with pasture, usually including subterranean clover, is particularly important for resting the land, breaking pest and weed cycles, and adding nitrogen to the nutrient-poor soils. It also spreads the economic risk for the farmer.

Summer grazing

Wind erosion is due mainly to the lack of surface vegetation and indirectly to the lack of soil moisture. Additionally, soil aggregates break down rapidly if dry soils are cultivated or trampled by stock, making them very susceptible to erosion by wind and water. Western Australian research suggests that trampling can loosen 0.3 to 0.7 t of soil per animal each week (Environmental Protection Council of South Australia 1988, p. 170). Accelerated wind erosion from summer pastures can be reduced, therefore, by decreasing stocking rates, and avoiding where possible grazing on land where the soil is particularly vulnerable to erosion. But for economic and agronomic reasons this is not easy to achieve. Some stock can be agisted, and a significant proportion of herds are culled in summer, leaving the main breeding stock.

Other options include farming animals which are less destructive in their grazing habits (that is, which do not crop close to the soil or, in the case of goats, even pull the plants out of the ground, including their roots), and raising soft-footed animals instead of the introduced, hard-hoofed sheep or cattle. Soft-footed animals such as kangaroos and even emus do not break up the soil to the same extent as sheep, and the soil surface is therefore less vulnerable to wind erosion.

In Australia, the latter suggestion is still a somewhat sensitive issue. However, kangaroo meat is now served in restaurants, and it has long been a staple for pet meat. Commercial emu farms have also been established and offer several products, such as oil, leather and meat. Conservationists need to recognize that the controlled farming of a species will ensure its survival.

Controlling effluent from intensive farming operations

Intensive farming is here to stay. It will inevitably increase in importance in southern Australia, as it has for many years in other parts of the world (notably the west coast of North America), as people come to recognize its role in increasing global food production. The problem of effluent disposal is very real, however, and will have to be tackled in a sound, systematic way.

For example, in 1990 Western Australia had 881 piggeries, with a total of more than 270 000 animals. Each pig daily produces nearly 5.5 kg of odorous, turbid waste with a high concentration of nutrients (principally nitrogen and phosphorus) and microbes. Even when diluted with up to 60 litres of wash water per pig per day to maintain hygiene, piggery wastes contain 20–50 times more degradable organic matter per unit volume than municipal sewage waste. Problems which arise are associated with nutrient enrichment, nitrate pollution and bacterial and organic matter contamination of water bodies, odours, noise, toxic gases and risks of disease (Grant 1991, p. 127).

Australian methods of dealing with such problems are essentially a variation on the methods used to treat sewage. The effluent is passed through a series of aerated, effluent disposal ponds in which bacteria break down the solid matter. The ponds must be lined with an impermeable material to prevent the leaching of wastes into groundwater. In the final pond, the quality of the effluent should be sufficient for it to be released into streams, or sprayed on pastures and crops. The dried, solid matter is also used as a manure (although care is needed in relation to heavy metal residues if vegetables are being grown). However, the high nutrient loads of some streams in Western Australia (such as Ellen Brook and The Serpentine on the coastal plain) and in South Australia, as well as continuing problems of unpleasant odour, indicate that treatment methods need to be upgraded. The authorities appear to have been reluctant to enforce effective remedial measures.

However, in Western Australia an Environmental Code of Practice is being developed on a co-operative basis between industry groups, the Environmental Protection Authority and the Department of Agriculture. Twenty-two of the largest piggeries have to operate under licence in that state.

Effluent from intensive feedlots may also include residues of growth hormones and antibiotics from commercially prepared feed. Measures such as those outlined above do not appear to have been fully adopted as yet in intensive chicken farming, dairying, or for rural abattoirs. Recommendations for further studies of alternative treatment measures have been made by Klingberg and Schrale (1985).

There are also problems of nutrient enrichment of water bodies from the dung produced by grazing animals. In Western Australia's Avon River, dead aquatic species have been observed in pools fouled with sheep manure after summer storms. The problem seems to be most severe where stock can graze along the river and stream banks (Olsen and Skitmore 1991, p. 62). Such problems are dealt with by increasing the protecting fringing vegetation along waterways, and conservation works on farms (contour banks, revegetation, stubble mulching and controlled grazing: see the following chapters).

Solutions Dealing with Cultivation and Horticultural Practices

The Iberian Peninsula

Three types of agriculture can be distinguished and in each case specific recommendations should be taken into account:

– *highly productive* farming, located in small favourable areas, with intensive monoculture, often under glasshouses, with Mediterranean products (fruits and early season vegetables): irrigation with modern techniques;
– *productive in extensive, hilly areas* (olives, vineyards): suppression of cultivation on slopes except with terracing, and use of no-till techniques;
– *poorly productive* but more conservative ecologically and in terms of landscape, over extensive areas (cereal polyculture, *dehesa* (sylvo-pastoralism)): fallow should be reduced, because it leaves the soil without protection,

and replaced with fodder crops. Straw should be used as mulch instead of burning it.

Agricultural techniques

The use of heavy machinery has had a big impact on soil erosion in Mediterranean areas. It is evident, therefore, that it is necessary to reach a compromise in soil management which permits its exploitation and at the same time preserves the resource. Good soil management in semi-arid environments includes ploughing the soils, as has been stated by Henderson (1979).

In Spain, given the wide range of existing agricultural systems due to the varied soil, climate, relief and crop types, it is difficult to simplify conservation methods to a few recommendations. At this point one should go back to Hugh Bennett's suggestion: the best conservation method is good agricultural practice.

According to Giráldez (1996), the most common traditional system of dry farming tillage consists of: (a) burning the straw or stubble after harvest; (b) mouldboard ploughing to break the hardened soil surface; and (c) successive discing and harrowing to reduce soil clod size. After sowing, several harrow passes may further reduce soil crusts and eliminate weeds growing between plant rows. This system has at present many disadvantages due to the high energy consumption, time required to establish the crop, and increased erosion risk. But if seedbed is prepared with reduced tillage, weed competition for water and light would limit crop growth. Developments by the chemical industry have allowed the successful implementation of reduced and zero tillage practices, resulting in what is called "conservation tillage". The main characteristics of conservation tillage consist in keeping the surface covered by stubble, application of herbicides, and fewer tillage passes.

Conservation tillage

Conservation tillage is being adopted increasingly in the many different agricultural regions of Spain. As with all new technologies, adoption depends on the socio-economic advantages which can be achieved (Fereres-Castiel 1990). In semi-arid areas of Spain, especially in Andalucía where soil deterioration is the worst in the country, particular efforts are being made, with minimum tillage and direct seeding being applied to the clayish soils of the lowlands (Giráldez 1996). Work has also been undertaken in Castilla-León and several other regions, often by the Technical Service of Monsanto.

Numerous comparative experiments have shown that it is possible to obtain similar production under conservation tillage as under traditional cultivation. Experimental work done in Tomejil (Sevilla) by Giráldez *et al.* (1989) found that in dry years direct seeding is more productive, while conventional tillage is more productive in wetter years.

During a workshop organized by the General Direction for Agrarian Research and Extension of the Andalucían Autonomous Community, the effects of conservation tillage were reviewed under different crops: under annuals (Fereres-Castiel 1990), under woody plants including olives, almonds and vines (Pastor Muñoz-Cobo 1990), and under pasture (Terceño Ramos 1990). The effects of minimum tillage on production costs were reviewed by Hernanz Martos (1990). According to Valera (1990), a researcher from Monsanto España, acceptance of the technologies by farmers is increasing.

Conservation tillage needs the correct use of herbicides, and farmers who intend using the technique need to understand the problems associated with their use. These problems include the development of species resistant to the residual herbicide, the presence of species not controlled by the herbicide usually used and, in the case of humid and temperate winters, the development of a rapid, microbial degradation of the herbicide so that seeds can germinate at the end of winter.

Italy

Reducing land degradation by adapting cultivation practices to the susceptibility of land to erosion is an ancient practice in Mediterranean areas. This may seem controversial, as the cultivation itself has generally negative effects on soil conservation. Soil only develops and is preserved under forest cover. However, adopting some cultivation techniques may substantially reduce soil erosion.

First of all, the depth of ploughing must be considered. Old farmers in south Italy say "land wants caresses". Experiments confirm this assumption. Second, the orientation of furrows has a variety of effects. In central Italy, if plough furrows have a true slope direction (Figure 19.1A), then tillage is called *a rittochino* (straight down-slope). If instead the direction of furrows follows contour lines (Figure 19.1B, it is called *a girapoggio* (round the hill). If furrows keep a constant direction whatever the slope direction (Figure 19.1C), it is called *a cavalcapoggio* (hill-riding). *Girapoggio* tillage reduces runoff and superficial erosion and might therefore seem the best method to preserve the soil from erosion. But it increases infiltration, which may have a negative effect in cohesive, poorly developed soils, by enhancing soil slip and deep-seated landsliding. There is little evidence of proper use of these techniques in Italy, as on steep slopes tillage is much easier down-slope (*a rittochino*). In south Italy, the fragmentation of fields works against the adoption of the best tillage technique, so that farmers till using the most convenient way.

Another technique for reducing soil erosion was the rotation of different crops on the same field, with a period of four to five years. For example, after one or two years of wheat, a legume was grown for one or two subsequent years, and for one year the field was left uncultivated. This also avoids biological problems, such as the development of pests resistant to pesticides, linked to monocultural practices. In addition, manure was used primarily, and chemical fertilizers were used with care. Yields were less than at present, but the soil was not exhausted, as is happening today in all Mediterranean lands. In the 1960s and 1970s, the massive use of deep ploughing, chemical fertilizers and pesticides has resulted in heavy pollution of groundwater and a substantial worsening of the quality of the produce. Thus today the early, intensive techniques are being reconsidered and the massive use of chemical fertilizers and chemical pesticides is being reduced, while biological pesticides are being widely introduced.

Greece

There is no doubt that soil erosion is the major land degradation process in Greece. In most cases it is controlled not by a single practice but by a system composed of a number of actions such as conservation tillage, contour farming, terracing, grassed waterways, and maintaining a rich vegetation cover.

Conservation tillage has been shown to reduce rill and inter-rill erosion in hilly areas cultivated with cereals. Unfortunately, the incorporation of plant residues into the soil after harvesting is restricted due to the very dry soil conditions during summer. Farmers prefer to burn wheat, corn or cotton residues (partly as a means of

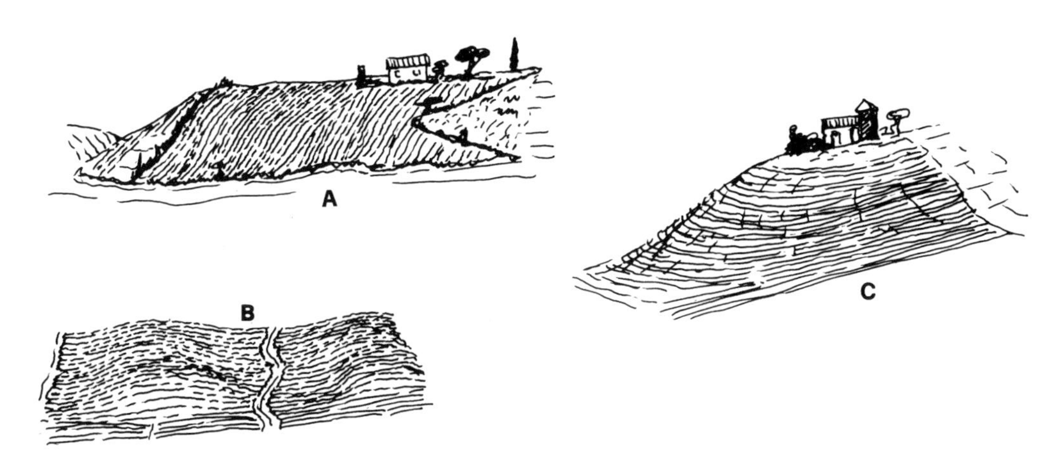

Figure 19.1 *Three ploughing styles typical in the hills of central Italy: (A)* rittochino *(or* ritocchinoro*); (B)* girapoggio*; (C)* cavalcapoggio

controlling weeds and diseases), than to incorporate them into the soil, so drastically reducing organic matter content and aggregate stability.

Contour farming has been applied recently to large areas cultivated with winter crops. Farmers were accustomed to cultivating in straight rows up- and down-hill, or in oblique lines. This type of cultivation was preferred for safety reasons and because of the small size of the fields belonging to each farmer. However, this type of ploughing led to accelerated erosion of the soil and deposition in the bottom land. Following contour farming, each furrow acts as a reservoir to receive and retain the runoff water. Data on water runoff from contoured land, compared with up and down-hill farming, show that runoff and sediment loss are greatly restricted (Figure 19.2). However, on long slopes and under heavy and prolonged rainfall events, contoured rows cause considerable rilling or gullying and soil erosion may actually be increased. Although contouring may be an attractive economic alternative for land protection, it cannot alone control erosion. Furthermore, large farm machinery cannot follow contours exactly, leading to portions of the fields that are not contoured.

North Africa

Agriculture in marginal, degraded regions is changing rapidly. Sundry examples in mountainous zones in North Africa demonstrate this (Maurer 1993).

Cereal cropping prevails both on the good lands and in more marginal lands. It is increasingly accompanied by leguminous plants, which accounts for the limitation of fallow to less than 30% of the cultivated area.

There has been an important extension of market gardening on irrigated parcels. Income generated by market gardens can be up to 20 times that from a good harvest of cereals. Those cultivations are cared for and receive an important contribution in manure.

The cultivation of cannabis (*kif*) is increasing in the Rif mountains (Ahmadan 1991). It is a plant which is adapted to the Rif environment owing to its short cycle, and which does not suffer from either intense winter rains or summer drought. It can grow with or without irrigation. Today, 200 000 people directly or indirectly live by the cultivation of cannabis, which has introduced a real upheaval in the economy and in the Riffian society. The mountain has become a

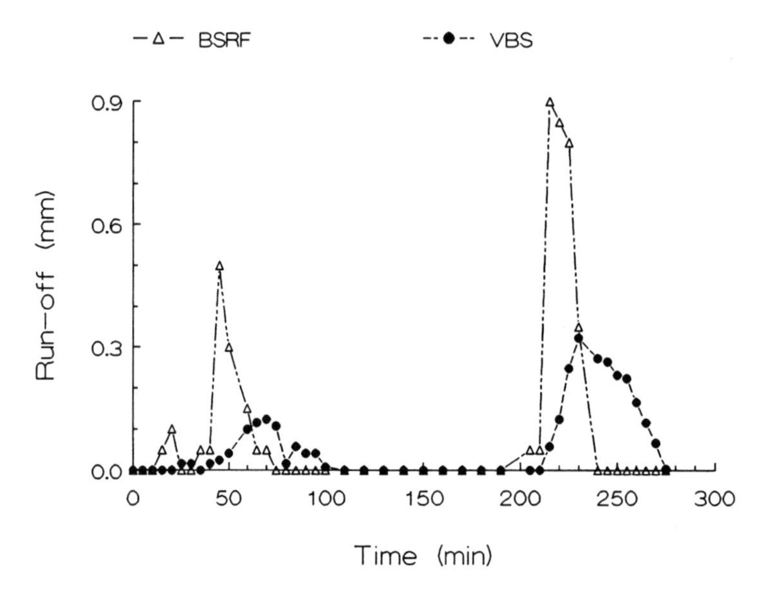

Figure 19.2 *Runoff change over time in plots with similar slope gradients but under different cultivation practices: VBS = contour ploughing, BSRF = ploughing perpendicular to the contours* (source: Kosmas *et al.* unpublished data)

focus for in-migration, mainly of workers. Housing is improving and consumption is becoming increasingly varied. The cultivation of cannabis is an example of speculation in marginal lands without causing too much soil degradation, because precautions were taken to control surface runoff; but in many cases it replaces precious forests and must be considered as a factor of vegetation degradation. Today, cannabis yields seem to have increased (by 50% in 10 years), despite the efforts being made for its limitation.

Orcharding is an old tradition, mainly with the olive tree which is increasing in many sectors; but other tree species can be considered, for the mountainous Mediterranean environment is highly favourable to trees. Trees have always been foremost in development proposals. The olive tree or the almond were suggested to rural dwellers with the aim of converting them into nurserymen. For example, the goal of the second "green dam" or "fruit dam", which was initiated in 1977 in Algeria, was to change from a subsistence food production system to a modern commercial mode (Plit 1983). In Morocco, mountain areas which have been planted to orchards extend over 240 000 ha.

Within the framework of soil conservation (DRS – Defense et Restauration des Sols – in Morocco and Algeria; CES – conservation of water and soils – in Tunisia), fruit plantations have played an important role. Conservation has been practised mainly on private lands. Its objectives are to ensure that the value of mountain lands is improved and production diversified. The olive tree, the most common fruit tree, plays only a secondary protecting role. Recommendations included: the abandonment of olive trees from the wetter areas, where it is threatened by water surplus; the restriction of the tree to the most favourable fruit-producing lands, and the association of the tree with runoff control banks. However, these recommendations have not been implemented (Laouina 1995).

Indeed, the trees have been planted systematically regardless of local environmental conditions. This accounts for the big gaps, ecologically speaking, between the planted sites. Some areas are favourable to the tree by virtue of their soil or their climate and show obvious examples of success of arboriculture with, as a corollary, the disappearance of stock farming. Other zones, however, are unfavourable to the olive tree; in such areas the trees are widely spaced, and have a high mortality, low vigour and low productivity.

Plantation densities are generally between 80 and 120 trees/ha, that is, four to six rows of 20 trees. Low density is a rule, and differs from the high densities which are observed in traditional orchards, located near the villages. The reason for this choice lies in the intention of the agricultural engineers responsible for the development to keep a productive space for cereals for the farmers. Yet, the low density plantation cannot be maintained by the same methods as would a dense orchard, located in the most favourable site. The trees, often ill-maintained, are mostly small, with thin trunks and low productivity.

The tree increases land values (in some Rif villages, for example, the price of the land increased from about US$120 to $1200/ha after the plantation), but efforts made by the farmers to improve the condition of those plantations and their productivity seem to have decreased. That is why, nowadays, many plantations have higher densities without respecting the linear disposition. Presumably, this kind of orchard will be better maintained.

California

The concepts of the US Natural Resources Conservation Service (NRCS, previously the Soil Conservation Service) are well known and have been applied widely throughout the world, such that only some aspects of its work are detailed here. Many of these methods also overlap with technical and vegetation-related solutions, but are discussed here primarily for convenience.

Conservation tillage, namely those practices which cover at least 30% of the soil surface with crop residues, is particularly important in areas subject to overland flow erosion and wind deflation. About 3 000 000 ha of California land were planted to conservation tillage in 1987 (CTIC 1988). There are five principal approaches:

1. *No-till* – soil is left undisturbed prior to planting which is completed in a narrow seedbed or slot; weeds are controlled by herbicides;

2. *Ridge-till* – soil is left undisturbed prior to planting which is completed using one-third of the soil surface to prepare a seedbed on ridges 10–15 cm high; weed control combines herbicides and cultivation;
3. *Strip-till* – soil is left undisturbed prior to planting which is then completed in rows over one-third of the soil surface using rototillers, in-row chisels and row cleaners; weed control combines herbicides and cultivation;
4. *Mulch-till* – the soil surface is tilled by chisels, field cultivators, discs, sweeps or blades prior to planting; weed control combines herbicides and cultivation;
5. *Reduced-till* – any other tillage and planting system that meets the 30% residue requirement.

Conservation tillage not only reduces soil erosion and compaction, but also saves time, fuel and equipment costs. It also reduces the runoff of farm chemicals and may reduce the need for chemical fertilizers. However, more herbicides and pesticides may be needed to combat weeds and pests respectively, raising concerns for other elements in the ecosystem, notably birds. That conservation tillage presently represents less than 20% of the total area planted in California each year may be due in part to the lack of a perceived need, especially in the irrigated flatlands, and in part to ignorance and traditional conservatism among farmers.

In addition to soil conservation, the NRCS periodically reviews conservation practice standards over a wide range of other operations, from the construction of dykes and firebreaks to the treatment of landslides, subsidence and mine closure. In a series of codes for each practice, the NRCS addresses applicable conditions, environmental considerations, and construction and maintenance techniques.

The US Department of Agriculture now provides several hundred conservation practice standards and specifications designed to mitigate environmental deterioration. These practices, augmented by locally adapted methods, range in scope from control and remediation of accelerated erosion and sedimentation to water quality improvement in agricultural and range lands. At the present time, California has implemented 20 "interim practice standards" in addition to USDA recommendations, to address specific problems associated with agriculture.

Erosion control employs over 60 practices with varying degrees of effectiveness. For example, *filter strips* are applied at the lower margins of fields, above terraces or diversions, and on fields adjacent to streams and lakes. A filter strip typically consists of a uniform grass area of at least 3 m for slopes less than 2 and at least 7.6 m for 17 slopes, with an effectiveness of 40–85%. By contrast, *contour orchard* and *contour farming* practices may yield only 15–20% effectiveness. These methods are used where there are continuous slopes with a nearly uniform grade and non-tilled planting of up to 10% of the slope for a maximum distance of 122 m. *Cover and green manure cropping*, including "between row" crops, and erosion-control planting yield 10–35% effectiveness. Grade *stabilization structures*, designed to prevent the development or stop the enlargement of gullies, use rock drops, sills and fords with 5–45% effectiveness. Steel sheets, rock riprap, straw bales, sandbags and silt fences provide 20–40% effectiveness. Beyond these practices, *underground outlets,* such as culverts, drop inlet boxes, and storm drains are considered 15–65% effective in erosion control, while *surface systems*, including sprinkler and trickle irrigation, lined ditches, pipelines and tailwater recovery yield 15–90% effectiveness. Sediment control, while employing many of the above mitigation measures, also uses *runoff management systems* including sediment basins, infiltration trenches and sediment traps which are 5–100% effective.

A recent study in Ventura County focused on 20 mitigation measures to address problems of erosion, sedimentation and excessive runoff in part of Calleguas Creek watershed under orchard cultivation. The study recognized the difference between new and old orchards, as well as orchards with variable slope and fruit type (Table 19.1).

It is estimated that implementation of these controls within the top 11 eroding basins in the Calleguas Creek catchment will involve over 10 150 ha of agricultural land, 190 km of roads, and 30 km of streams. Within streams requiring treatment, it is projected that six sediment basins and 291 structures will be needed. Beyond this, at

Table 19.1 *Erosion, sedimentation and runoff controls, Calleguas Creek catchment, Ventura County, California. Controls or practices are listed which are used to mitigate erosion, sedimentation and runoff, but not all of them apply to all land uses. Thus, contour farming as an erosion control practice can be used to control E (erosion), S (sedimentation) and R (runoff) from new citrus and avocado orchards but cannot be applied to old orchards after they have been planted, nor is it applicable to unpaved roads* (source: USDA 1995)

Control	Citrus (new)	Citrus (old)	Avocado (new)	Avocado (old)	Unpaved road (>3)
Access road	E, R	E, R	E, R	E, R	E, R
Conservation tillage	E, S, R	E, S, R	E, S, R	E, S, R	
Contour farming	E, S, R		E, S, R		
Cover crop	E, S, R	E, S, R	E, S, R	E, S, R	
Critical area planting	E, S	E, S	E, S	E, S	E, S
Crop residue use	E, S, R	E, S, R	E, S, R	E, S, R	
Diversion	E, S, R	E, S, R	E, S, R	E, S, R	
Filter strip	E, S, R, Q	E, S, R, Q	E, S, R, Q	E, S, R, Q	
Grade stabilization structure	E, S	E, S	E, S	E, S	E, S
Grassed waterway	E, S, R, Q	E, S, R, Q	E, S, R, Q	E, S, R, Q	
Hillside bench terrace*	E, S, R	E, S, R	E, S, R	E, S, R	E, S, R
Irrigation (trickle, drip, sprinkler)	E, S, R, Q	E, S, R, Q	E, S, R, Q	E, S, R, Q	
Irrigation water management	E, S, R, Q	E, S, R, Q	E, S, R, Q	E, S, R, Q	
Irrigation water conveyance (pipeline)	E, S	E, S	E, S	E, S	
Runoff management	E, S, R, Q	E, S, R, Q	E, S, R, Q	E, S, R, Q	E, S, R, Q
Sediment basin	E, S	E, S	E, S	E, S	
Streambank protection	E, S	E, S	E, S	E, S	
Stream corridor improvement*	E, S	E, S	E, S	E, S	
Underground outlet	E, S	E, S	E, S	E, S	E, S
Wildlife upland habitat management	E, Q	E, Q	E, Q	E, Q	

E = erosion, S = sediment, R = excess runoff water, Q = water quality
* Used on slopes >2

least 90% of orchards less than seven years old should be planted with cover crops and filter strips, until litter from the trees can provide row cover.

The proposed mitigation measures within the Calleguas Creek watershed (890 km²) involve some 175 km². Of these measures, sediment basins (whose practice life is around 50 years) within seven "priority" subwatersheds are projected to cost $3 180 000 with average annual maintenance costs of $614 000. It is estimated that these basins would control 1 105 943 t/yr of erosion and 324 861 t/yr of sediment. Other proposed mitigation methods include: *orchard cover crops* (Blando Brome), with a practice life of 20 years at $857 600 for installation and annual maintenance costs of $481 100; *bank protection* (including shaping and planting with purple needle grass, giant wild rye, California buckwheat, willow, alder and cottonwood), with a practice life of 20 years at $64 100 for installation and annual maintenance cost of $1500; and *grade stabilization structures* with practice lives of 50 years at $6 357 000 for installation and average annual maintenance of $153 000. These combined mitigation measures are designed to provide erosion control of 1 016 586 t/yr and sediment control of 334 840 t/yr.

Chile

The Institute of Agro-pecuarian (*pecuarian* = stock breeding) Research in Chile (INIA) has investigated and developed conservation cultivation practices based on the establishment of crops with stubble on the surface. The INIA Experimental Station of Quilamapu has worked in the interior dry farming area of Region VIII where the traditional practices of cultivation, such as stubble burning and ploughing with machinery, have initiated erosive processes.

Minimum tillage

At present it is considered appropriate to plough the soil at a minimum, that is, only to the extent necessary to reduce the scrub and favour the conditions for the growth of the cultivated plants. With the principle of minimum tillage, equipment for tractors has been developed, in particular the chisel plough which loosens the soil without mixing or inverting the different layers of the soil profile. This has been termed "vertical ploughing".

To a certain extent these ploughing principles have been used with animal traction, common in the country, as a gradual modification of the implements used by primitive man since remote times and present in American indigenous cultivation.

Ploughing with implements

According to the findings obtained from evaluations of appropriate instruments to undertake vertical ploughing, the use of animal-drawn implements named "vibration cultivators", which are made of very strong steel and are highly resistant to breakdown, has been recommended. Its design ensures a large number of vibrations both lateral and longitudinal.

The work capacity of the equipment permits the reduction, or in some cases the elimination, of fallow, one of the most negative practices in relation to soil erosion because it leaves the soil without the protection of vegetation. Thanks to increased labour efficiency, it is possible to break the soil surface and seed it in the same year. In order to eliminate compaction, generally caused by traditional cultivation and located between 15 and 20 cm depth, a subsoil animal-drawn plough has been developed. It can reach a depth of 20 cm with an average traction of 200 kg. This demand can be met by a pair of oxen or the use of a horse collar.

In order to eliminate the excessive mulch accumulated at the surface, a mesh rake is used. The raking and seeding is done after the second significant rainfall, using a heavy rake with nails.

Direct seeding

Direct seeding involves the complete elimination of soil cultivation and the use of a seeder which permits the introduction of the seed through the mulch or the superficial vegetation already chemically controlled.

From research undertaken in Cauquenes (Region VII), where the effects of three systems of cultivation on the production of wheat and lentils has been analysed, notable savings of time and money have been obtained when applying conservative cultivation systems, as can be observed in Table 19.2. Table 19.3 reflects the yields obtained under each system for the most important dry farming products.

Crop rotations

The rotation of crops has been widely used in Chile, especially in the central and southern zones and in the dry sectors of the coastal areas. In these areas agricultural production is based on the establishment of traditional crops on soils which are highly erodible due to the rugged topography.

Table 19.2 *Comparison of three cultivation systems in the interior drylands, Cauquenes, Region VII, Chile* (source: Raggi 1994)

Activity	Conventional		Minimum tillage		Direct seeding
	MWP	AWP	MWP	AWP	MWP
Fallow	3.9	10.3	1.0	2.7	0
Dragging	0.6	0.8	0.6	0.8	0
Herbicides	0	0	0	0	0.06
Seeding	2.2	2.0	1.3	0.8	1.5
Total	6.7	13.1	2.9	4.3	1.6

MWP, man work power; AWP, animal work power

Table 19.3 *Yields obtained in the interior drylands, Cauquenes, Region VII, Chile* (source: Raggi 1994)

Crop	Conventional (t/ha)	Minimum tillage (t/ha)	Direct seeding (t/ha)
Wheat	3.06	3.49	3.17
Lentils	1.28	1.58	1.02

The use of protective crops between the lines of fruit trees has allowed the protection of the soil from raindrop impact. In this sense legumes have considerably ameliorated soil characteristics in the zone of Concepción (Region VIII).

The application of green manures has permitted the replacement of lost humus and nitrogen in the soils of the province of Cauquenes, although this practice is not commonly used by the farmers of Regions VII and VIII.

A model of crop rotation considers the weeded crop as the beginning of the rotation (papa, kidney bean, beetroot), followed by cereals (mainly wheat), and ending with two to three years of pink trefoil.

The Southwestern Cape

The Department of Agriculture is involved in thorough planning of farms, especially in terms of, for example, the positioning of roads, footpaths and even railways in relation to the cultivated fields. The Department of Agriculture allows no indiscriminate cutting of natural vegetation and, indeed, the Conservation of Agricultural Resources Act of 1989 (see Chapter 23) stipulates that a permit be granted before any clearance of virgin land.

Commonly used conservation measures

Soil conservation measures are widely practised in the agricultural areas of the southwestern Cape. Farms are generally planned with water runoff control as an important management priority (H. Germishuys, pers. comm.) and cultivation methods are generally assumed to take this into consideration. The history of land use in the region (Chapter 14) has seen increased intensity of farming operations over time, including a reduction (even complete removal) of the fallow period that was once common (Talbot 1947). Notwithstanding this trend, as has been pointed out in Chapter 12, apparently the problem of soil erosion is not a very serious one, an observation which suggests that land use planning conducted under the auspices of the Department of Agriculture is effective in reducing land degradation by soil erosion.

Other measures include strip cropping, although this is confined mostly to the "Sandveld" regions of the western coastal platform (Plate 19.2). Here, land use is mainly low intensity cattle and sheep grazing and supplementary crops of winter wheat are planted in thin strips to minimize soil loss through wind erosion. Minimum tillage is not widely applied in the southwestern Cape.

What is needed?

The most obvious problem in seeking solutions to land degradation which relate to cultivation

Plate 19.2 *Strip cropping north of Saldanha Bay, as protection against wind erosion*

practices is the lack of useful data on the cultivation practices themselves. Although the Department of Agriculture collects census data on such matters from every farm in the region, the data are not summarized or collated in such a way as to enable comparisons to be made between areas or between land uses. There is clearly a need for the effective summarizing of such information to facilitate broader scale management decisions on practices which are considered to promote, or mitigate against, forms of degradation.

Southern Australia

Fallow and ley farming

As discussed by the Environmental Protection Council of South Australia (1988) and Scott (1991, pp. 143–145), before the introduction of improved farming methods, the soils of southern Australia had low organic matter contents. The long, dry summers and the widespread use of fallow did not favour the build-up of organic matter. The use of a long fallow involved leaving the land bare for periods of 9–12 months – up to 15 months in Victoria – before sowing a crop. The land was ploughed during winter and, to conserve moisture, weed growth was destroyed by cultivation during the following summer. Organic matter fertility and crop yields declined, and the soil surface was exposed to erosion by water and wind. A similar situation arose in orchards and vineyards.

Since the 1940s, the recognized method of maintaining good soil structure has been to grow vigorous pastures between cropping years (the "ley farming" system). "Conservation farming methods" were adopted, involving the use of crop rotations, residue retention, manures and legume pastures. Subclovers on neutral-acid soils, medics on alkaline soils, rye grasses and other pastures were used to build up soil fertility between crops. However, many farmers continued to cultivate their soils excessively during the cropping phase, resulting in a loss of nitrogen from the soil. Cropping became heavily mechanized, resulting in an increase in the area cropped and the speed and efficiency of cultivation, sowing and harvesting. Fallows are still used in some drier areas (<400 mm mean annual rainfall) but are of much shorter duration – as little as three months.

The ability of shallow-rooted annual pastures to restore soil structure in the short period of the pasture phase is questionable. Further, the effectiveness of legumes in replacing soil nitrogen was greatly reduced by the low number of farmers inoculating their clovers and medic seed with *Rhizobium* bacteria before sowing. Moreover, trials showed that organic carbon levels still decline with a pasture–wheat rotation using conventional cultivation. Both fallow–wheat and continuous cereal rotations lost about 14% of initial fertility, and grain yields and protein content continued to decline, even with large inputs of up to 80 kg/ha of nitrogen.

Reduced tillage, stubble retention and herbicides

Soil erosion problems in the cereal/sheep regions continued, despite the ley farming system. But from the 1970s the use of the herbicides Spray.Seed (a paraquat/diquat mix) and Roundup (glyphosate), in conjunction with suitable tillage equipment to incorporate crop and pasture stubbles into the soil, made it possible to reduce soil cultivation. This was also assisted by the availability of a number of other herbicides, particularly Hoegrass (diclofop-methyl) for both wild oats and annual rye grass control. But by 1980, a strain of annual rye grass which is resistant to Hoegrass had appeared. Other problems also emerged, some of which have been discussed in Part II.

By 1989, most (89%) of the extensive cropping farmers in Victoria had tried herbicides, compared with 78% in 1984. But there was little net change in the number of farmers undertaking direct drilling. Non-adopters in 1984 had taken up the practice in 1989 and *vice versa*, and it is evident that farmers were concerned about health problems and herbicide residues in soil and crops, as was also the case in Western Australia (Gorddard 1991). In South Australia, about 80% of landholders in northern cereal districts have adopted stubble retention and reduced tillage systems for cropping in areas prone to water erosion. Of interest is that state's development of a "herbicide economic threshold model" for wheat and barley crops, with the clear intention of reducing herbicide spraying (Department of Environment and Land Management 1993, pp. 82, 193).

The use of grain legumes in crop rotations can be important for controlling grasses such as brome, barley and silver, which are difficult to control in reduced tillage systems with cereal growing. Stubble burning is also being reduced (but not eliminated), improving organic matter contents and reducing nutrient loss and the risk of soil erosion. This has been made possible by the introduction of machinery which can handle stubble and incorporate it into the soil. Cutting of stubble soon after harvest has also been an important development. South Australian research has shown that an extra tonne of stubble is needed to counter the decline in organic carbon from a single cultivation. Increasing organic carbon by as little as 0.1% would boost grain yield by 150 kg/ha, as well as increasing grain protein content and therefore the price received for the wheat crop (Department of Environment and Land Management 1993, pp. 82). Western Australian research at Wongan Hills showed that organic matter was 6% higher on ungrazed plots where all stubble was retained (Leonard 1993).

Scott (1991, pp. 326–327) cites Budd (1987), who reported tillage research results on soils similar to those of Victorian mallee farmland. He found that traditional farming systems employing three-year rotations and aggressive tillage systems based on stubble incorporation result in unacceptable levels of erosion. To achieve erosion control, Budd considered that stubble cover must be maintained at >0.6 t/ha. That requires less aggressive fallow tillage systems and a two-year rotation. Management to retain adequate stubble is critical. Trash retention farming leaves some 90% of residues on the surface, compared with about 50% when working with traditional equipment. In Western Australia, Leonard's (1993) data are comparable. She recommended that between 30% and 60% of anchored cover is needed to control erosion, with 50% being a safe compromise. These percentages convert to 1–3 t/ha of stubble, depending on crop type.

Budd also commented that a purely chemical system is ideal for erosion control, but impediments include cost, considerable farmer resistance to the use of residual chemicals, and farmers' dislike of the untidiness of unploughed fields.

Leys (1990) also found that the stubble component is the key to wind erosion control. He identified an appropriate integrated farming system which balances soil conservation, agronomic factors and production requirements for sandy soils in a low rainfall environment. The system includes:

– harvesting at 30 cm and spreading the straw;
– light stocking to ensure retention of standing stubble;
– two-year cropping rotation;
– effective annual grass control at fallow opening;
– use of trash-retention tillage equipment and herbicides where necessary to kill grasses before sowing; and
– ensuring that stubble does not fall below 0.6 t/ha for sandy loams, and approaches 2 t/ha.

Note, however, that many years previously, Woodruff *et al.* (1966) observed from North America's Great Plains that there are no simple rules for the amount of stubble to be retained on the surface. The quantities needed to control wind and water erosion vary according to the type of stubble, whether it is loose or fixed, and factors such as the degree of slope, soil type, rainfall intensity and type of implements used.

Further, Cooke *et al.* (1985) found that after five years of trials in northern Victoria's wheat country, grain yields were about 0.5 t/ha greater on land tilled and scarified, than on land which was left untilled and subject to herbicide treatment. Similarly, reducing the length of fallow reduces soil erosion by wind but also reduces crop yields by diminishing moisture conservation and cereal disease control (Kent 1987; Semple and Budd 1988; Leys 1990).

Diversification

Some components of the agricultural sector have both diversified and specialized since the 1970s. Some cereal/sheep farmers have diversified into oilseeds and livestock such as Angora goats, deer and emus. Though spring wheats are traditionally grown in Australia, in Victoria winter wheats (with a longer growth period and winter hardiness) are being developed to suit the higher rainfall and cooler areas of the southwest of that state. Cropping in this region should be of less risk to soil erosion and structural decline. In horticulture, however, there has been a shift towards monocultural practices, namely crop specialization and a reduced number of cultivars associated with high chemical inputs and irrigation. This may lead to increased disease, pest and other problems in the future. Yet, ironically, the greatest diversity of recent times has occurred within the horticultural industries, though not on individual farms or individual fields.

Australian agriculture in general is diversifying, although not all of the following are taking place in the Mediterranean areas (Cribb 1989). To traditional exports of wheat, barley, rice and sorghum, Australia has added lupins, peas and hay, and may add other grain legumes and oilseeds. Demand for specialist protein grains is likely to lead to a further development of industries producing triticale, chick peas, fababeans, guar, lentils, various millets, cow and pigeon peas, adzuki and mung beans. To this range are being added new crops such as kenaf, sesame, grain amaranth, fenugreek, linola, guayule, jojoba, buckwheat and indian mustard, all of which were being experimented with by Australian farmers and researchers in 1989. The scope for mixed and inter-cropping in order to maintain and improve soil fertility is enhanced if farmers have a wide range of crops, with differing nutrient requirements and susceptibilities to diseases and pests, to draw upon.

20
Technical solutions

Compiled by Arthur Conacher and Maria Sala from contributions on the following regions:

Iberian Peninsula and Balearic Islands	**Maria Sala** and **Celeste Coelho**
South of France and Corsica	**Jean-Louis Ballais**
Italy, Sicily and Sardinia	**Marino Sorriso-Valvo**
Greece	**Constantinos Kosmas, N.G. Danalatos** and **A. Mizara**
Eastern Mediterranean	**Moshe Inbar**
North Africa	**Abdellah Laouina**
California	**Antony** and **Amalie Jo Orme**
Chile	**Consuelo Castro** and **Mauricio Calderon**
Southwestern Cape	**Mike Meadows**
Southern Australia	**Arthur** and **Jeanette Conacher**

This chapter considers technical, engineering-type solutions to the problems of land degradation. The practices depend on the nature of the problem, and include terracing, contour ditches, banks or interceptors, gully control structures, dams, and coastal groynes.

The Iberian Peninsula

Problems posed by seasonal and annual rainfall irregularities are being ameliorated by the construction of additional dams, even though they cause regression of the deltas and coastal plains. Other actions include the transference of water from basins with a surplus to those with deficits, not only at a national level but also at a European level (such as the transference of water from the Rhône to Catalonia). In Spain, a water plan under discussion is causing tensions amongst regions and with Portugal. Water from the Ebro should go to the highly developed Catalan coastal areas, and water from the Tajo to the SE Mediterranean coastal plains (some is already going to the Segura river).

Desalinization of seawater is another action the Spanish Government is undertaking, hopefully with support from the European Union (EU). The project includes desalinization plants in Mallorca, Menorca, Alicante, Cartagena, Almería, Málaga and Sevilla.

More efficient use of water is needed, such as avoiding leakages in the distribution networks, improving irrigation practices (drip irrigation is being encouraged) and convincing the population that water is scarce.

South of France and Corsica

The use of mechanical, sometimes massive, techniques to stabilize the channel walls of rivers, or to straighten the channel or construct artificial levees, does not take into account the natural behaviour of the river. Armouring river banks with boulders generally creates slopes which are steeper than equilibrium slopes. It cannot be a durable solution: undermining by the current displaces the blocks which eventually collapse in the riverbed, making it necessary to replace them. Channel straightening and smoothing are done to accelerate the velocity of flow. From this viewpoint, they are locally efficient because they limit the maximum height of the flood, but the problem is transferred downstream. Levees are locally able to protect people and infrastructures, but their effect is to increase flood heights, to decrease infiltration on floodplains and thus to exacerbate the downstream situation.

Mechanical clearing is used to prevent forest fires. While it is certainly the most rapid method it prevents the germination of young trees and destroys regenerating saplings. Moreover, in general (and this has been shown in relation to soil erosion in the south of France), machinery must be used with considerable care. For example, studies carried out on the consequences of forest fires on erosion, in the Mont Lozère as well as in the Maures massif, have shown that erosion produced by forestry works is more severe than the erosion caused by the fire itself (Bernard-Allee *et al.* 1991; Beguin 1992). In particular, the ruts created by heavy machines form preferred sites for water movement during heavy rains, favouring the concentration of drainage and erosion.

Italy

Slope terracing with trellis-work, stone walls and gabions on land, building check dams along mountain torrents, and groynes and levees along streams, are the most common technical practices for soil conservation in Italy.

Slope terracing is very ancient (Plates 16.5 and 16.6), as it is throughout the Mediterranean Basin, and has been used since the time of Greek colonization. It is also very effective in keeping the soil in its place, but requires continuous maintenance work. In Italy, it is used on all slopes cultivated with highly productive cultures (olive groves, vineyards, orchards, fruit trees) with gradients of more than 30%. Stone walls are the most frequently used retaining structures. In

Land Degradation in Mediterranean Environments of the World: Nature and Extent, Causes and Solutions.
Edited by A.J. Conacher and M. Sala.

certain regions (Liguria, Calabria, west Sicily) terracing may represent more than 10% of the hill and low mountain zones.

Badland slopes with up to 100% gradient have been reshaped step-wise and their height reduced either by cutting the tops or filling the gullies. New plantations of Aleppo pines or eucalyptus trees were then established with the aim of reducing erosion and producing wood for the paper industry. This type of intervention has been widespread all over Italy, mostly in the 1960s and '70s (Avolio *et al.* 1980; Sorriso-Valvo *et al.* 1995), and particularly in Calabria and Sicily. They have proved to be of limited effect, however, as following an initial phase of reduced soil loss, a resurgence of erosion normally occurs, even on gentle slopes, and severe piping develops (Sorriso-Valvo *et al.* 1992, 1995).

Check dams have been used since Roman times, but since the 1950s their use has become widespread. All streams which appeared to be undergoing even potential erosive action, have been filled with check dams. In some streams of Calabria and Liguria, where natural streams are eroding rapidly, it is possible to see up to four generations of check dams. The oldest were normally made with gabions. They work well, but being permeable, normally collapse because of undermining, especially in non-cohesive or deeply weathered rocks (cf. Plate 20.8). New generations of check dams may be quite large (more than 10 m high and 50 m long) and are constructed of concrete. In streams where debris flows carry large blocks, selective dams are built. These are rather frequent in the Alps, but are still rare in south Italy. It is strange, however, that check dams are also used on clay or clay-rich rocks. Here the solid load is transported as suspended load, and the siltation behind the check dam is nil or insignificant.

The general effectiveness of these structures in reducing sediment transport does not mean that erosion is also reduced. Due to siltation, erosion is reduced upstream but it is increased downstream as the clear waters have a higher erosive capability. In order to reduce erosion, each check dam should stay within the siltation basin of the next check dam downstream. This situation is realized in some cases, but it is rare.

There may be heavy drawbacks to using concrete for building check dams, principally in terms of adverse effects on the environments for natural flora and fauna.

Levees and groynes along the banks of streams are widely adopted (Plate 20.1). Levees were used by the Romans, and their use increased after medieval times, when the cities were expanding rapidly. Several large cities were located on the banks of rivers, and clear-cutting in the forest to reclaim lands for cultivation caused an increment of flood frequency and magnitude, as the regulating effects of the forests were reduced. Engineers reacted by constructing larger and larger levees. This is the case of Florence and Pisa along the Arno, for instance. Of course, aggrading stream beds are a long-term consequence of these practices, and the probability of catastrophic floods is enhanced by the intensive use of territory adjacent to the levees, or between two or more levee sets, as is the case for major rivers like the Arno and Tevere.

These corrective measures have been adopted along all streams in Mediterranean Italy, even the smallest ones (Plate 20.2), except those that are well entrenched naturally. But the latter happens only to streams that reach the sea at a high cliff, and these may represent only 5 to 10% of Italian coasts.

We have already seen the consequences of river erosion control works on beaches. This occurs because check dams block the coarse material in mountain zones, and beach deposits are made of sand and gravel. Technical reduction of land degradation along beaches consists of two measures, to be used contemporaneously: the construction of groynes and the replenishment of beach material.

Italians love great engineering works, so that constructing groynes is the most widespread practice along coasts where something useful needs to be implemented against sea erosion. There are several types of groynes, each suitable for different conditions of beach dynamics. By observing the works so far adopted, it seems that the choice of the type of groynes has been based on rather randomly assorted criteria. For instance, T-shaped groynes have been constructed along a 5 km tract of Tyrrhenian coast (Plate 17.2). The effect has been a rapid replenishment

Plate 20.1 *Torrenta Turbulo, a concrete channelized stream in Calabria; the catchment area is 50 km², with annual rainfall ranging from 350 mm at the coast to 2000 mm in the mountains (to 1200 m above sea level)*

Plate 20.2 *Cemented stone wall, designed to protect this southern Calabrian village from debris avalanches initiated in the mountains in the background*

of inter-groyne tracts, and a catastrophic beach retreat down-drift (2 to 10 m/yr in two years). In order to avoid the complete destruction of houses along the 2 km of coast subject to this rapid retreat, beach replenishment has been undertaken. The effectiveness of this work has not been tested, since it has been implemented only recently.

Greece

Terracing is an important method for water conservation and erosion control, which decreases slope length and reduces damage by surface runoff. As in Italy and elsewhere, it is a very old practice, having been used for centuries. Many uplands have been terraced for cultivating cereals, vines, olives and other crops throughout the country (Plate 20.3). In many cases, terraces which have been constructed with stones are hundreds or even thousands of years old.

For example, in Lesvos, where large areas are cultivated with olives, crescentic terraces have been carefully constructed for individual trees. Also, in several hilly areas of Epirus, Peloponnesus and the Aegean islands, terraces have been constructed with stone walls. Soil was removed from elsewhere to fill these terraces. This conservation management requires high labour inputs to maintain the terraces. In recent decades, the value of such terraces has declined markedly because they are difficult to access and because they cannot be cultivated easily with tractors. At present, most of these areas have been abandoned and the terraces have collapsed, causing the rapid removal of soil by runoff, apart from some cases where the stone walls are protected by the roots of fast-growing shrubs and trees (Plate 20.4). Maintaining such terraces appears to be a very expensive practice compared with most other alternatives for soil erosion control. Considering that such terraces protect very valuable soil which in turn preserves the natural vegetation, these agricultural structures should be maintained with the aid of national consolidation schemes, particularly in environmentally sensitive areas.

The Eastern Mediterranean

In the eastern Mediterranean, too, terracing is one of the soil and water conservation practices which has been used since historical periods (Plate 20.5). In the hilly and mountainous areas of the Levant, terracing was the solution to soil erosion and for water retention. Natural slopes with shallow soils were transformed into a developed system of agricultural terraces. After the terrace is built, soil accumulates within a few years (Inbar *et al.* 1995). Terraces require permanent maintenance in order to repair and protect the stone walls. However, the walls are breached frequently by intensive storms, grazing animals that dislodge the stones or by the pressure of saturated soil behind the wall. Therefore, terrace abandonment has a negative effect on land degradation through the processes of rilling, gullying, piping and fan deposition, and an overall increase of sediment yield. The cost of labour implied in terrace conservation will probably increase the area of abandoned terraces.

Salinity problems are treated in two technical ways.

Plate 20.3 *Sloping land with stone terraces and cultivated with vines. Photo by C. Kosmas*

Plate 20.4 *Abandoned terraces on steep, severely eroded slopes. Photo by C. Kosmas*

Plate 20.5 *Terraced hills near Bethlehem*

1. *Eradication by drainage or soil leaching.* Drainage, either open or underground, is very expensive, although it was adopted in the Jezreel valley in northern Israel, where around 1000 ha suffered from salinization. Underground plastic pipes were introduced mechanically into the subsoil on parallel lines every 100 m and saline groundwater was flushed to the drainage channel.
2. *Irrigation methods*, such as drip irrigation, which reduce salinity by providing less salts to the soil and keep the plant area leached of salts during the growing period. Many of the irrigated fields in the Jordan rift valley are using this system with water containing up to 1000 mg/l chlorides (Plate 20.6).

Marine pollutants are severely prohibited in the Mediterranean Sea, being an almost closed

Plate 20.6 *Drip irrigation, Jordan valley: an efficient irrigation and fertilizing system, saving about 70% of conventional water use. Photo by M. Inbar*

inner sea. Non-polluting waste, for example ash from the burnt coal in thermo-power stations, is dropped on to the sea floor at depths greater than 500 m, in order not to pollute the beaches. Strict regulations have been imposed on oil tankers prohibiting the disposal of ballast water into the open sea. Tar pollution of the beaches has been reduced, but still affects the attractive beaches of the eastern Mediterranean.

About 90% of wastewater in Israel is treated and a large proportion – about $200 \cdot 10^6$ m^3 in 1995 – is used for irrigation, mainly for cotton crops. With population growth and increasing demand for domestic use, less water will be available for agricultural purposes. In order to maintain the coastal aquifer of Israel in a non-deteriorating state, about 300 wells monitor the subsurface level and water quality, and a centralized system controls the pumping. Besides natural recharge by rains, artificial recharge of the aquifer is performed during winter by diverting floodwater from coastal streams and water from the National Water Carrier (pumping water from Lake Kinneret) into sand areas. Recently, wastewater from the Tel Aviv conurbation is being treated and percolated into the aquifer and it is envisaged that 80% of domestic water will be recycled and used for irrigation or to replenish the coastal aquifer. Indeed, the eastern Mediterranean's major problem, water shortages, requires large-scale engineering solutions. However, their implementation carries with them important social and political implications. Moreover, water resources use and development schemes affect all components of the environment and in many cases are irreversible processes.

Water development projects in Turkey have caused international problems with the two downstream countries, Syria and Iraq. Further irrigation development schemes in these countries may aggravate the present situation. A probable global climate change in the near future may reduce rainfall and bring the situation to a political or even military crisis. On the other hand, land development may bring welfare to an underdeveloped region in east Anatolia, which has a large community of the Kurdish minority. One of the political aims may be the dilution of the Kurdish population by bringing in more Turkish people (Hillel 1994).

The water crisis in Syria can be mitigated by improving current irrigation technology, which demands 17 000 m^3/yr (Wakil 1993), by using water-saving technologies such as drip irrigation, which can more than halve present demands.

Transferring water from the Tigris River basin to the Euphrates River via canals may be another possibility. These solutions demand large investments from countries which spend a large part of their economic and human resources on their armies.

Although relatively small, the Jordan River basin epitomizes many of the problems of water resources in the area. The constraints of a highly intensive water management system, and the necessity for an optimal utilization of natural resources, are compounded in the Jordan River basin by the particular geo-political structure of the region to which Israel is bound by the heritage of its past and by recent history. The Jordan River-Lake Kinneret (Lake of Galilee) basin supplies an annual average of 650 · 10^6 m^3 or about 35% of Israel's total water resources (Figure 20.1).

The water development scheme for the Jordan basin as proposed by Lowdermilk (1944) was probably one of the first comprehensive river development schemes on a regional and multinational basis. Patterned along the lines of the Tennessee Valley Authority, it encompassed the development of agriculture, power and manufacturing for the socio-economic progress of the people living in the region. Lowdermilk envisaged the diversion of sweet water from the Upper Jordan and the Yarmuk River – its major tributary entering the Jordan River south of Lake Kinneret – into open canals or closed conduits running around the slopes of the Jordan valley.

He also pointed out the difference in altitudes between the Jordan valley and the Mediterranean Sea, which offered a splendid opportunity for a combined power and irrigation scheme. This proposal and other projects embodying similar ideas were the forerunners of a unified plan for the integrated development of water resources in the Jordan valley. This was prepared in 1953 at the request of the United Nations and is known as the Johnston Plan. The final plan regarded Israel, Jordan and Syria as co-riparian states to the Jordan and Yarmouk rivers, while Lebanon was considered a riparian state to the Jordan basin.

The Johnston Plan was endorsed by experts from all sides as a logical and equitable approach, but was rejected by the Arab League Council on political grounds. Israel and Jordan proceeded thereupon individually, and with USA support, on their water development schemes: the Israel National Water Carrier (Plate 20.7) and the Jordanian Great Yarmouk Project respectively. The Israel National Water Carrier extends from Lake Kinneret through nearly the full length of the country, interconnecting with existing branch networks, storage and irrigation systems. It supplies water from the northern water surplus areas to the southern areas, with their extensive lands but water deficiencies.

Lake Kinneret serves as a yearly reservoir, receiving the winter floods, with a storage capacity of nearly 700 · 10^6 m^3, less than twice the yearly supply of 400 · 10^6 m^3. The lake is a critical component of the Jordan River integrated and centrally controlled water system, and to date it has afforded considerable flexibility and regulative capacity for meeting fluctuating demands. In 1991, after four consecutive dry years, the water level was lowered by one additional metre, reaching the –213 m level. The exceptionally wet years which followed again filled the lake to its storage capacity. Such a system must depend increasingly on a delicate balance in terms of quantity and quality. High lake levels affect some low coastal areas and the town of Tiberias; while low levels have negative impacts on nutrient accumulation, and beach resorts and religious sites along the coast find themselves several hundreds of metres from the shoreline.

North Africa

Traditional techniques

The adaptation of inhabitants to difficult conditions was the origin of a number of soil conservation techniques and of water management, with the objective of improving production. They can be found in regions with ancient installations and in recently developed regions. Yet, soil conservation techniques tend to be associated with regions of high densities of population and housing. The CES (conservation of water and soils) works appeared spontaneously and often reflected the need to manage the difficult environment.

The aims of the techniques are manifold: the improvement of lands by clearing stones and

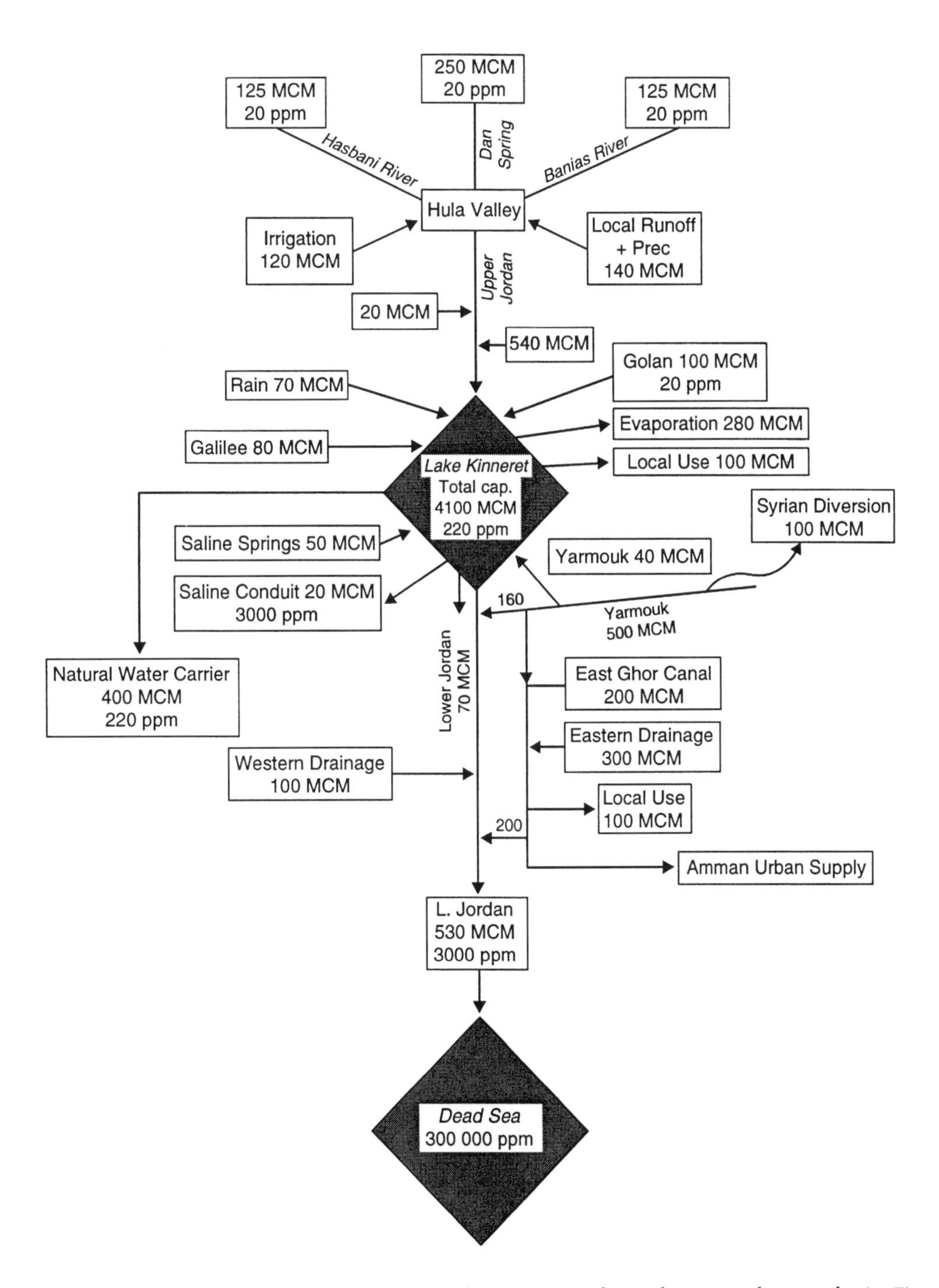

Figure 20.1 *Schematic chart of water resources and water use in the Jordan River drainage basin. Figures are approximate yearly flows in millions of cubic metres (MCM) and salinity in parts per million (ppm)* (original diagram by M. Inbar)

Plate 20.7 *The Israel National Water Carrier before it enters a 2.5 m diameter closed pipe system. Photographer not known*

establishing geometrical fields; the need to conserve soil; and the wish to possess land parcels.

Dry stone walls

This kind of development concerns the slopes with stony, detrital materials. The walls are built perpendicular to the gradient, blocking the transportation of surficial materials whether by surface runoff, slumping or under the effect of tillage. There is a tendency to build higher walls on steeper slopes, with the morphology of the slope guiding the construction which is adapted to the minor contours. These constructions allowed the natural creation of small terraces behind the old walls, now totally fossilized on the up-slope side. The small terraces have a double advantage, partly because they are composed of a lower gradient than the primary slope, which facilitates their tilling and reduces their erodibility, and partly because they are freed from a large number of the biggest rocks which previously covered the surface, thus improving their productivity (Aït Hamza *et al.* 1995).

Dry stone walls constructed along the gradient delimit parcels of land and are also simply disposal heaps for the rocks which have been removed from the soils. They are often lower and less well constructed, but they have an indirect effect on the soil's stability since they hinder the movement of stock into the fields.

Stone heaps coincide with rock outcrops which could not be dug up or reshaped, and which are used as a focus for stone disposal. Their spacing is related to stone-throwing distance; cultivation surrounds these heaps of stones which also play a role of fixing the whole land parcel.

Stones and larger rocks are thrown into gullies, which provide a useful place to dispose of stones removed from neighbouring fields. The piling up of loose stones acts to prevent any further downcutting of the gullies.

Several conclusions can be drawn.

1. Traditional, erosion control works are well maintained and are rebuilt after each event which damages them.
2. Recent works are constructed on newly developed parcels of land, perhaps in the hope of making them a parcel of permanent and intensive cutivation. In the semi-arid piedmonts, small dams and dry stone walls are constructed to hold the soil and water (Plate 20.8). Actually, the resort to CES techniques in the semi-arid piedmonts aims particularly at reducing newly appeared forms of erosion. On the other hand, CES works are necessary in order to store some of the rainfall in the soil.
3. Constructing or maintaining CES works is increasingly being carried out individually by members of the family, whereas previously they were carried out with the help of other inhabitants of the village.
4. All the lands in which CES traditional techniques are carried out are private lands. The concern for soil protection is much less evident in the collective lands.
5. The cost of constructing the traditional CES techniques is high, if the time for the necessary work is counted. However, they are constantly maintained and extended. Those works would never be carried out if the farmers tried to account for the invested efforts.

At the present time in some regions in Morocco, those techniques are being extended, giving rise to some questions. Is it a revival of interest in agriculture as a result of the failure of other development options? Is it a simple recognition of the need for soil conservation? Or can it be a response to information campaigns about the need for soil conservation practices and about the development projects?

Modern erosion control banks with lines of fruit trees

In North Africa, erosion control banks have always been preferred in association with biological methods, with multi-purpose objectives. Economic transformation has resulted in increased mechanical efficiency.

There are many techniques, yet one technique is adopted in the same project whereas conditions may vary from one site to another. The adopted technique does not always comply with recommendations. Thus, many banks have been constructed in wetter areas such as the High Loukkos. Attempts were made to adopt a less unfavourable technique – that of the diversion bank – but it is perhaps the most delicate technique to manage, particularly in terms of controlling linear runoff adjacent to the troughs, and of designing the drainage outlets which are liable to develop into gullies (Laouina 1993a).

The choice of construction density can be expressed in linear metres per hectare. It has several effects: first, on the efficiency of erosion control, for the steeper the gradient the closer the spacing between the banks; then on the loss of productive area; and finally, on the cost of the project. In reality, the chosen density is often standard or based on considerations other than technical ones, with the spacing remaining the same over long distances regardless of physical conditions. Sometimes, densities seem excessive.

In some cases, there have been negative effects on the slope of the embankment, namely the emergence of erosion forms (rills and mass movement, among others) which have degraded the scree of the embankment. This means that the embankment is incapable of preventing the processes. There are even cases where slopes without banks have less erosion than those with banks. Physical factors can account for the degradation of banks, and can even result in their disappearance. Indeed, although banks can intercept all the runoff from normal rains of medium intensity, they may be unable to intercept unusual, highly aggressive runoff (Heusch 1995). The silting up of the Fodda River channel in Algeria, for example, does not seem to have been reduced after the construction of banks in a large part of the watershed.

An evaluation of banks carried out in Tunisia, between 1962 and 1975, showed that 17% of the network had disappeared and that 16% was partly degraded (Heusch 1982).

The lack of maintenance seems to be the main reason for the degradation of erosion control structures. Crumbling of dry stone walls and the emergence of scars in inter-bank areas are due to drainage or consolidation shortcomings. The

Plate 20.8 *(a) Check dam in a gully incised in a steeply sloping alluvial fan (b). The porosity of the rock-constructed, wire-encased dam is resulting in loss of accumulated sediment by sapping through the wall*

emergence of gullies or gutters derives from poor management of diversion banks and of the outlets. Thus, a situation favourable to erosion involves processes which thorough maintenance could have avoided. This lack of maintenance means that the population's agreement with the development choice was not obtained.

Deliberate rupture is another cause of degradation. This fact poses the problem of the real efficiency of costly developments which may disappear within a relatively short period of time. Deliberate ruptures can be noticed everywhere except in the case of dry stone walls. These ruptures, caused by tilling the trough and the embankment, aim, above all, at reconquering lost ground (Laouina 1993a).

There are two means of giving the banks a chance of survival. One is to construct the banks adequately, with a deep trough which is difficult to fill up and a more massive embankment which is difficult to undermine. The other is to encourage people to care for the structures. Presumably, farmers' agreement with developments would provide some assurance of the development's durability.

The following conclusions can be drawn in relation to development techniques which would be helpful.

(a) Techniques need to be reconsidered in relation to efficiency and durability, which cannot be guaranteed without good cooperation from the farmers.
(b) The techniques to be used must be the most profitable ones possible with minimal outlays.
(c) The techniques must be able to be easily reproduced by farmers over extensive areas.
(d) It is only where serious problems occur that engineering will intervene with sufficient funds to prevent the damage.

California

In relation to the problems associated with hydraulic mining, ambitious engineering works, including debris basins and river levees, were built to alleviate the problems, accelerate debris discharge and help agriculture. The Sacramento thus became the third American river, after the Mississippi and Missouri, to be managed by engineers (Pyne 1980).

The recovery of petroleum and natural gas has introduced problems of degradation through replacement of a natural landscape with an oilscape dominated visually by producing and abandoned wells, storage tanks and pipelines, ecologically by waste that is toxic to plant and animal life, and structurally by widespread subsidence. There is little that can be done to combat visual degradation, although producing wells at Long Beach have been disguised as tropical islands. Nor are the effects of toxic groundwater and soil contamination easily rectified. However, subsidence in the Long Beach–Wilmington oil field has been partly offset by replacing the petroleum with water.

Irrigation makes California's Central Valley one of the world's most productive regions but also introduces some of irrigation's worst problems: high concentrations of salt and toxic chemicals such as selenium. To counter this problem, the Bureau of Reclamation began building the San Luis Drain in 1968 to remove contaminated waters to the Sacramento–San Joaquin delta, but various environmental and health concerns ended this project short of its objectives. In 1984, the San Joaquin Valley Interagency Drainage Program began evaluating various options including groundwater management, use of salt-tolerant crops, idling of croplands, water treatment, reuse and disposal, water conservation, and wildlife habitat restoration and protection.

Accelerated erosion caused by the irrigation of crop lands and orchards has been seriously addressed in recent decades, notably by the University of California's Cooperative Extension Service, an advisory system for farmers. Erosion problems, often linked to general field irrigation involving furrows and sprinkler systems, began to be countered by the drip-trickle method of irrigation. In the 1960s, drip-trickle irrigation was used only in greenhouses but by 1977 it had been introduced to nearly 40 000 ha of California and has since spread widely (Hall 1978). Basically, the system consists of a flexible plastic pipe with small diameter emitters attached to serve water and fertilizer at designated intervals, foreign matter being screened out by filters. Though emitters and filters must be cleaned periodically, the system has been shown to save sufficient water to cover initial installation costs and yield a profit from increased yields of row crops, orchard fruit and vines. Above all, the system has greatly reduced the deleterious erosional impact of excessive runoff.

Irrigation-related subsidence is being addressed by various measures. Problems caused by hydro-compaction are countered in the fields by replacing drainage ditches with sprinkler systems, and for structures by pre-wetting and pre-consolidation of susceptible deposits. Some 65 km of the California Aqueduct, designed to cross hydro-compacted areas along the west side of the San Joaquin Valley, required expensive pre-consolidation at a cost of $620 000/km prior to construction. The more subtle but extensive problems of overdraught and subsidence caused by groundwater withdrawal began to be addressed in the 1950s when San Joaquin River waters were diverted into the Friant–Kern Canal along the valley's eastern margin, and canal imports via the Delta–Mendota Canal were directed along the western margin. By the 1970s, subsidence trends were levelling out and artesian pressures were recovering to pre-subsidence levels, but the areas remained depressed (Lofgren 1975).

Solid waste management became an important environmental issue in the 1960s, leading to the elimination of open dumping and burning, to improved sanitary landfill practices, and to reduced incinerator emissions. By 1988, there were 720 landfills in California, mostly for urban solid waste. Voluntary recycling was recommended but largely ignored in California until, faced with shortages of landfill space, mandatory regulations were imposed in the early 1990s. In Los Angeles, the proscription of incineration in the 1960s led to a remarkable growth of sanitary landfills whereby whole canyons were filled with garbage, then adapted for methane production, and finally landscaped for suburban development. As of 1996, the problems of solid waste disposal have yet to be adequately resolved, as Los Angeles seeks to remove its waste to create a mountainous landfill in the Mojave Desert.

Chile

In Chile, owing to the lack of an official soil conservation policy, conservation techniques are not applied systematically. Nevertheless, concern to reach a solution to these problems has always existed, especially in the agricultural organizations. The participation of these institutions in Regions IV and VIII is worthy of mention. These dry farming lands of the Coastal Cordillera correspond to the Mediterranean zone, where a series of conservation practices are being implemented in order to attenuate the effects of soil degradation.

The main objective of the conservation techniques is to prevent or diminish the concentration of runoff waters flowing down the slopes, to reduce the velocity of the flow, to trap and accumulate transported sediment down-slope, and to prevent the progressive deterioration of the soils when submitted to inadequate exploitation techniques. Topographic conditions, the productive and land use capacity, and the severity of the damage are the issues taken into account when designing control techniques.

In Region IV for instance, in response to the severity of the problem and the presence of a continuing desertification process, it has been necessary to establish, in an integrated form, physical structures and mechanical conservation practices. In Region VIII the physical structures for conservation are channels for deviating runoff, infiltration ditches, channel protection works, gully control (cement and mesh), talus control, terracing and water harvesting.

The most important soil conservation techniques in Chile are the following.

1. *Mixed dykes in brooks.* The most common dykes are made of a mixture of wood and stone, and they are capable of retaining erosive discharges in the drainage nets of the micro-basins. The water can be stored and used for irrigation. The benefits of these structures are the consolidation of slopes and fluvial beds.
2. *Evacuation channels.* These are located on alluvial fans and cones at the base of the "*quebradas*" (gorge, ravine, brook), about 500 m apart. The objective is to catch the water which overflows from the infiltration ditches (below) and direct it to the natural drainage channels.
3. *Infiltration ditches* are constructed at 15 m intervals, on slopes which range from 10 to 20%. Their objective is to catch the water flows produced over the alluvial fans and cones. The construction is made in relation to the natural infiltration capacity of the soil, which depends on its texture.
4. *Retention dykes.* In the Yerba Loca community, 12 dry masonry dykes have been constructed, equivalent to 1250 m^3 of total construction. Each retention dyke has a base 3 m wide and a total height of 2 m. The gauging structure is 0.5 m high and 3 m wide.
5. *Ridges between furrows in slopes.* These works have an identical objective to the dykes, with the difference being that retention of runoff takes place on the slopes. These ridges are 0.5 m high and 0.5 m wide and are complemented by two lateral arms. They are made of stone and earth.

6. *Reservoirs* are aimed at rainfall water harvesting. The overall structure is similar to a small drainage basin; that is, with a wide reception zone and a channel zone leading towards the reservoir.

Gully control works have been undertaken in the localities of Los Vilos (Region IV); Peñuelas (Region V), Pichilemu (Region VII) and Laguna Torca (Region VII), usually involving a combination of structures and revegetation (with species such as *Pinus radiata*, *Lupinus* and *Chusquea*).

The Southwestern Cape

Planning for water runoff control appears to be central to the advice provided to landowners by the Soil Conservation Control technicians and this frequently involves the construction of a range of physical measures to reduce or combat soil erosion (H. Germishuys, pers. comm.).

Commonly used technical and engineering solutions to land degradation

A wide range of technical and engineering practices is employed throughout the region. Terracing is employed in deciduous fruit production areas on the steep slopes around Grabouw, where the enterprise is considered sufficiently profitable to maintain such an expensive measure. In the wheatlands of the Swartland, on the other hand, contouring, including the construction of contour bunds, is the most widely utilized practice as neither the slopes nor the profit margin warrant terracing. Water runoff management, particularly in the form of wide, shallow ditches lined with either an indigenous perennial grass ("kweek") or, increasingly, concrete, is widespread (Plate 20.9). According to the Soil Conservation Department of the Department of Agriculture for the Winter Rainfall Region, this is the main reason why the scenario of severe land degradation and gullying, which prevailed in the Swartland in the 1930s and 1940s (Talbot 1947), has been reversed and replaced by a landscape which is broadly speaking operating sustainably (H. Germishuys, pers. comm.).

In addition to these measures which are clearly aimed at discouraging the development of gullies, or *dongas*, several practices, such as the construction of weirs and small dams, are aimed at controlling the development of existing erosion features (Table 20.1). The Winter Rainfall Region has expended considerable effort and finance (Chapter 23) to encourage the use of such technical solutions, especially in the deciduous fruit, vine and wheat-growing areas of the southwestern Cape.

For the future, the current situation in the region suggests that apart from the need for more personnel and financial investment by the Department of Agriculture in promoting the construction of technical solutions to soil erosion, the most significant improvements could be made by enforcing the provisions of the relevant legislation (Chapter 23) and educating more landowners as to the erosion potential of the region.

Southern Australia

Water erosion tends to lend itself to engineering-type solutions, and in South Australia's agricultural region, about one-third of the area at risk from water erosion has control structures such as contour banks and waterways (Department of Environment and Land Management 1993, p. 82). According to Scott (1991), contour banks in Victoria, designed to slow water movement down the slope, offer some protection against soil loss but do not prevent the movement of soil between the contour banks if the soil is finely cultivated or if the vegetation cover is insufficient. In times of severe erosion, contour banks may overflow and breach.

The problem of eutrophication in Western Australia's Peel/Harvey estuary, south of Perth, has been tackled by a combination of methods, including the use of slow-release fertilizers in the intensely farmed dairying country in the estuary's catchment. Agriculture was responsible for an estimated 80% of the phosphorus load in the estuary; thus the first steps were to reduce phosphorus inputs by agronomic means (Department of Conservation and Environment 1984). In addition, this area is perhaps unique in

Plate 20.9 *Concrete-lined erosion control ditch, southwestern Cape*

Table 20.1 *Conservation works built and subsidized by the Department of Agriculture in the Winter Rainfall Region of the southwestern Cape* (source: Department of Agriculture Annual Reports for the Winter Rainfall Region)

	1992/93		1993/94		1994/95	
	Number	Subsidy (US$)	Number	Subsidy (US$)	Number	Subsidy (US$)
Contours (km)	1330	396 970	711	185 203	820	230 289
Waterways	93	350 912	16	64 114	29	81 028
Subsurface drainage works	219	271 475	59	79 086	72	99 558
Pasture improvement works	67	15 448	27	10 047	72	35 594
Small-scale erosion structures	46	14 254	32	14 754	36	15 830
Total		1 049 059		353 204		462 299

Australia in the other major approach to dealing with the problem. A massive engineering project (the Dawesville cut) was completed in 1994 in order to improve the flushing regime of the severely eutrophic estuary by providing a second, broad entrance to the ocean. Land sales helped to provide funds for the project. Prior studies involved extensive use of dyes to trace current movements in the estuary, as well as integrated measurements of salinity, temperature and other variables. Mathematical modelling was used intensively to predict the effects of the cut. Early indications are that the cut is achieving its intended objective; but one unanticipated side-effect, an increase in the number of mosquitos caused by the exposure of larger areas of estuarine muds due to the increased tidal range in the estuary, has become apparent. In turn this has led to concern over the spread of mosquito-borne diseases in the area, especially the Ross River virus which has a debilitating effect on humans. The estuary is the focus of a major and fast-growing tourist, recreation and retirement centre which extends north and south from the town of Mandurah.

An earlier, major engineering project was the construction of the barrage at the mouth of the Murray River in South Australia, although it is debatable as to whether this was done to solve a problem of degradation. Its intention was (and is) to prevent the incursion of salty seawater, which would render much of South Australia's drinking water unpotable. Other South Australian projects, such as the draining of the extensive wetland systems in the southeast (as with major wetland drainage schemes south of Perth), also do not come under the heading of solutions to problems of land degradation. On the contrary, they have resulted in serious losses of wildlife habitat, despite providing useful agricultural land.

Interceptors and secondary salinization

One of the most interesting and controversial instances of an engineering-type solution to land degradation has been the initiation, testing, development and adoption of systems of interceptor banks designed to solve the problem of secondary salinization in the extensive, dryland grains/sheep areas. Developed in Western Australia, farmers on Kangaroo Island (South Australia) and near Swan Hill (Victoria) have also adopted the system. Of particular interest is the fact that the system was developed and implemented by farmers.

In the 1950s, a Mr Harry Whittington, farming in the Brookton area of the Western Australian wheatbelt, was experiencing an increasing problem of soil (and stream water) salinization. He drilled numerous holes on his property and observed the movement of water over and through his soils; he also wrote to the then Director of the United States Department of Agriculture, Mr Hugh Bennett, for advice.

In brief, he concluded that the cause of his problem was the development of a perched water table at a shallow depth beneath his salt-affected land. That water was initially fresh, causing a waterlogging problem, but then became increasingly saline over time. Whittington considered that the perched groundwater was supplied by what geomorphologists term throughflow, moving down the valley sides and accumulating above a relatively impermeable hardpan about 1 m below the surface in the low-lying, increasingly waterlogged and salt-affected land.

The solution was to prevent the throughflow from reaching the low-lying areas. To do this, a system of interceptor banks was constructed, first on an experimental basis on a small part of the property, and later, over many years, extending over the entire property. Each interceptor was constructed on the contour, with the first interceptor excavated as high as possible on the slope. The earliest interceptors were surveyed with the assistance of State Department of Agriculture personnel, and then excavated by bulldozer, with the blade cutting down through the (approximately) 0.5 m of sand-textured topsoil into the relatively high clay content, B horizon materials. The excavated materials formed a barrier down-slope of the excavation, with the clayey materials pushed up against the up-slope side of the bank. The purpose was to seal the interceptor when it filled with water. Subsequent interceptors were constructed progressively down-slope. Because they are built on the contour, each must be capable of retaining the water which falls in the catchment between adjacent interceptors. Unlike drains, however, they have the advantage that gaps can be left to allow the movement of farm machinery up and down the slopes (Plate 20.10).

Western Australia's agricultural agency was opposed to this system, because it considered that the major cause of secondary salinization was the rising of deeper groundwater tables towards the surface, and not the throughflow/perched water mechanism outlined above. If the rising groundwater table mechanism is responsible, then holding water on the slopes with interceptors will, if anything, exacerbate the problem by increasing recharge to groundwater (refer to Chapter 16, "Southern Australia, Hydrological changes" for a discussion of the processes responsible for secondary salinization in dryland agricultural areas). However, Whittington claimed significant success with his system and, over time, many farmers came to adopt his methods. A farmer group named WISALTS (Whittington Interceptor Salt Affected Land Treatment Society) was formed in 1978 to promote and implement the method: farmers were trained to "read the landscape" and to survey appropriate systems for other farmers. Variations on the original scheme were developed to deal with problems such as greater depths to the impermeable layer, or situations where farmers did

Plate 20.10 *Interceptors designed to prevent throughflow from reaching valley floors and causing waterlogging and secondary salinization. (a) The mode of construction; (b) a system of interceptors holding water on a valley side in the Western Australian wheatbelt*

not have control over the catchments of their salt-affected land. By 1980 there were nearly 1000 members of WISALTS, and it is still an active group in the 1990s. It is not known how many farmers have constructed interceptor systems, but there must be several hundred, not only in Western Australia, as indicated above.

Research has been carried out to test the effectiveness of the system. This has included questionnaire surveys of the success claimed by the earliest farmers to implement the system, as well as of all farmers in one local government area who had implemented the system at that time (Conacher *et al.* 1983a). Related research on the hydrological processes responsible for the problem was carried out on one salt-affected property on which interceptors had been constructed (Conacher *et al.* 1983b). In brief, it was found that interceptors succeeded well on some properties, failed on others, and achieved some measure of success on the majority of the 43 farms investigated. Reasons for disappointing results fell into two groups. First, interceptors often failed to intercept water effectively because they were not comprehensive enough over the catchment, and/or were poorly constructed, allowing leakages below the base of the excavation. Second, in many (not all) locations throughflow was not the sole or prime cause of the problem, and even a well designed and constructed system would be only partly effective under such conditions.

Subsequent, detailed field-based measurements from many locations in the Western Australian wheatbelt have reinforced those findings and enabled improved quantification of the various mechanisms of water and soluble salt movements in the landscape (George 1992; George and Conacher 1993). There is still a role for technical solutions: interceptors, to control overland flow and throughflow; drains, to divert water coming onto an area from a neighbouring farm; and pumps, to lower groundwater tables in situations where the saline water can be disposed of without causing further damage downstream. But in most if not all situations, such measures should be combined with vegetative options (Chapter 21).

21
Vegetation-related solutions

Compiled by Arthur Conacher and Maria Sala from contributions on the following regions:

Iberian Peninsula and Balearic Islands	**Maria Sala** and **Celeste Coelho**
South of France and Corsica	**Jean-Louis Ballais**
Italy, Sicily and Sardinia	**Marino Sorriso-Valvo**
Greece	**Constantinos Kosmas, N.G. Danalatos** and **A. Mizara**
Dalmatian coast	**Jela Bilandžija, Matija Franković, Dražen Kaučić**
Eastern Mediterranean	**Moshe Inbar**
North Africa	**Abdellah Laouina**
California	**Antony** and **Amalie Jo Orme**
Chile	**Consuelo Castro** and **Mauricio Calderon**
Southwestern Cape	**Mike Meadows**
Southern Australia	**Arthur** and **Jeanette Conacher**

This chapter deals with biological or ecological solutions to land degradation. Included in the discussion are: establishing reserves; reintroducing native species and replacing habitats; the use of exotics (for windbreaks, erosion control, stock shelter); changing crop types; agro-forestry (or sylvo-pastoralism); and whole farm planning.

The Iberian Peninsula

Preserving natural habitats

As stated previously, the Iberian Peninsula still has a very rich flora and fauna. This richness is being preserved in national parks and protected areas (Figure 21.1). Wetlands are also being protected (Figure 21.2) due to their importance for migratory birds.

Reforestation and revegetation

Both Spain and Portugal have increased their forested areas but for different reasons. In Spain, wooded areas have expanded due to rural outmigration, with the consequent abandonment of forestry practices; additionally, reforestation has been implemented mainly to reduce dam siltation. In Portugal, forest land use has trebled this century from about 1 000 000 to 3 200 000 ha at present, occupying over one-third of the total extent of the country. It is expected that forestry will continue to expand in the entire peninsula by afforestation of marginal lands due to the "set-aside" policy of the European Union. In relation to this policy in Spanish Mediterranean mountains, forestry practices are being encouraged and reorientated towards long-lived species, but also towards leisure uses.

The first Law of Reforestation in Spain dates from 1887, although it was not very effective. It was in 1936, with the establishment of the State Forest Patrimony, that the General Reforestation Plan of Ceballos was issued, consisting of a two-stage development, each stage lasting 50 years: at the end of the stages, 6 000 000 ha should have been reforested. The first stage has been accomplished and 3 000 000 ha have been reforested. But there has been no agreement on the selection of the species used for reforestation. Species of rapid growth such as *Pinus* and *Eucalyptus* have been favoured, and often heavy machinery used. From 1940 to 1981, 25 219 km^2 were reforested with pines, 4165 km^2 with *Eucalyptus*, 203 km^2 with several conifers, and only 704 km^2 with indigenous oaks. Nevertheless, in many cases reforestation with pines and *Eucalyptus* is the only possibility on slopes with poor soils.

Unfortunately, the rate of reforestation was offset by the destruction of holm and cork oak woodlands by felling and by forest fires. During the decade 1976–1985 the total area affected by forest fires (972 790 ha) exceeded that subjected to afforestation schemes (830 222 ha). In addition, the type of plantation undertaken has significantly modified the species composition. Eucalyptus plantations have displaced Mediterranean *matorral*, and have even degraded holm and cork oak woodlands.

Although the Spanish forest authorities recognize that the optimum use of forest land in dry areas should combine trees and shrubs suitable for forage purposes (simultaneously feeding extensive livestock holdings and conserving soils), opportunities for livestock/dairy farming in forest lands have not been given the attention they deserve by those same authorities. In the case of cork there has been a reduction of annual production, which in the 1980s stood at only 70% of production in the second half of the 1960s.

South of France and Corsica

Biological or ecological solutions are recommended in most cases. This is not because solutions should in some way be "natural", in conformity with nature. Neither do they refer to the "good old days", or to a Golden Age which, in reality, never existed. Ecological solutions simply are the most efficient. If one undertakes a serious economic analysis, taking into account not only immediate costs but also the costs in

Land Degradation in Mediterranean Environments of the World: Nature and Extent, Causes and Solutions,
Edited by A.J. Conacher and M. Sala.

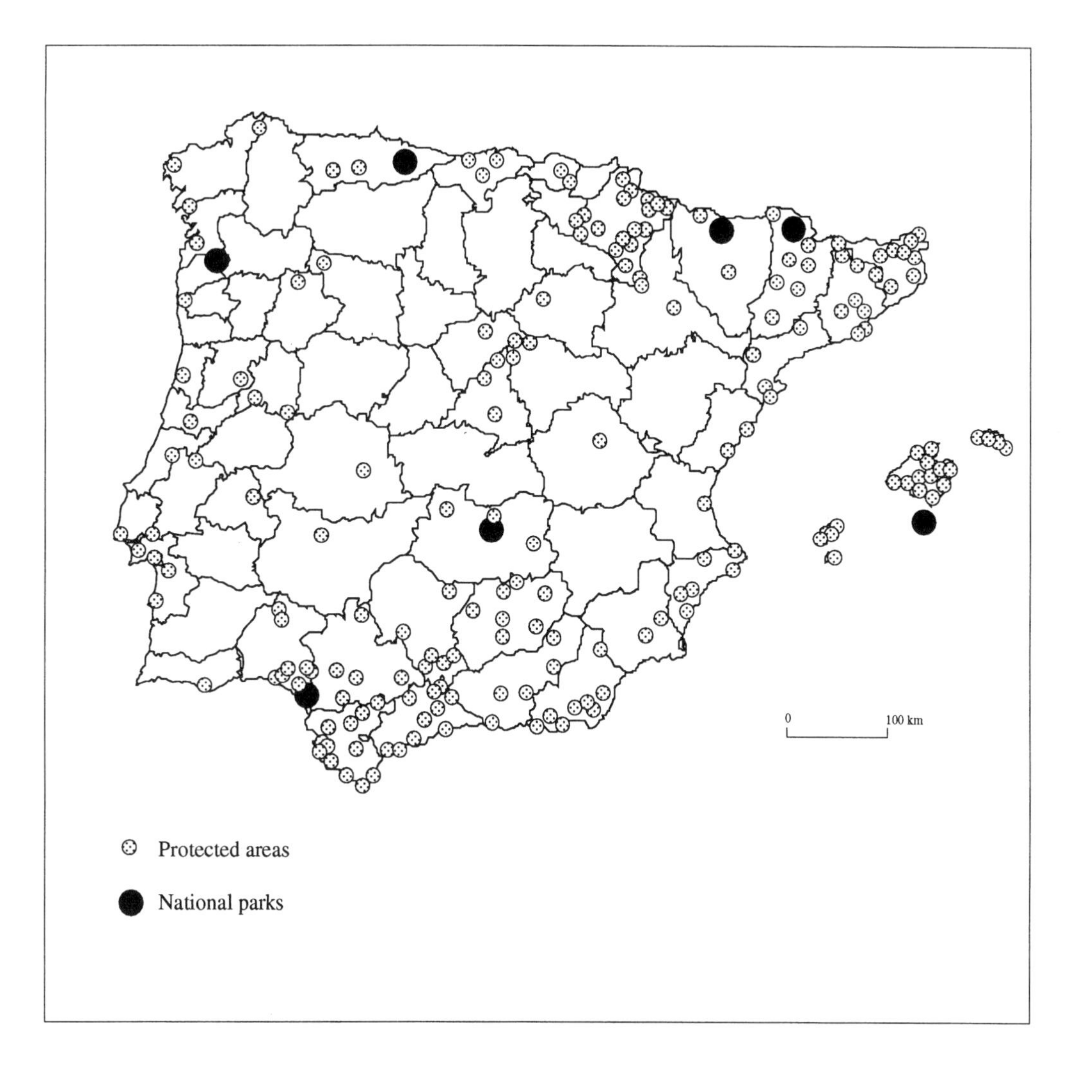

Figure 21.1 *National parks and protected areas, Iberian Peninsula* (source: Atlas Nacional de España 1996)

pedological and ecological terms, they are also the most inexpensive methods.

Some ecological methods have been used for many years. For example, reforestation in the badland zones of the southern Alps began more than a century ago, by planting *Pinus nigra* sp. *austriaca*. The results were particularly spectacular: in the 87% reforested Brusquet watershed (108 ha), mean annual soil loss (1985–1990) is 3.6 m^3, whereas in the 32% reforested Laval watershed (86 ha), mean annual soil loss (1985–1990) is 1801 m^3 (Meunier and Mathys 1993). Reforestation is often an efficient means of stabilizing slopes, providing conditions to encourage pedogenesis and decreasing runoff.

Spontaneous reforestation is also possible and is as efficient as the anthropogenic method, but is cheaper. This is particularly so on the ancient cultivation terraces where *Pinus halepensis* recolonizes swiftly. Nevertheless, forest does not constitute an efficient means of combating Mediterranean-type floods. In fact, runoff tends to concentrate under forest. Renewed downcutting is favoured in the thalwegs and this channelization allows a more rapid evacuation of the flood flows (Arnaud-Fassetta *et al.* 1993).

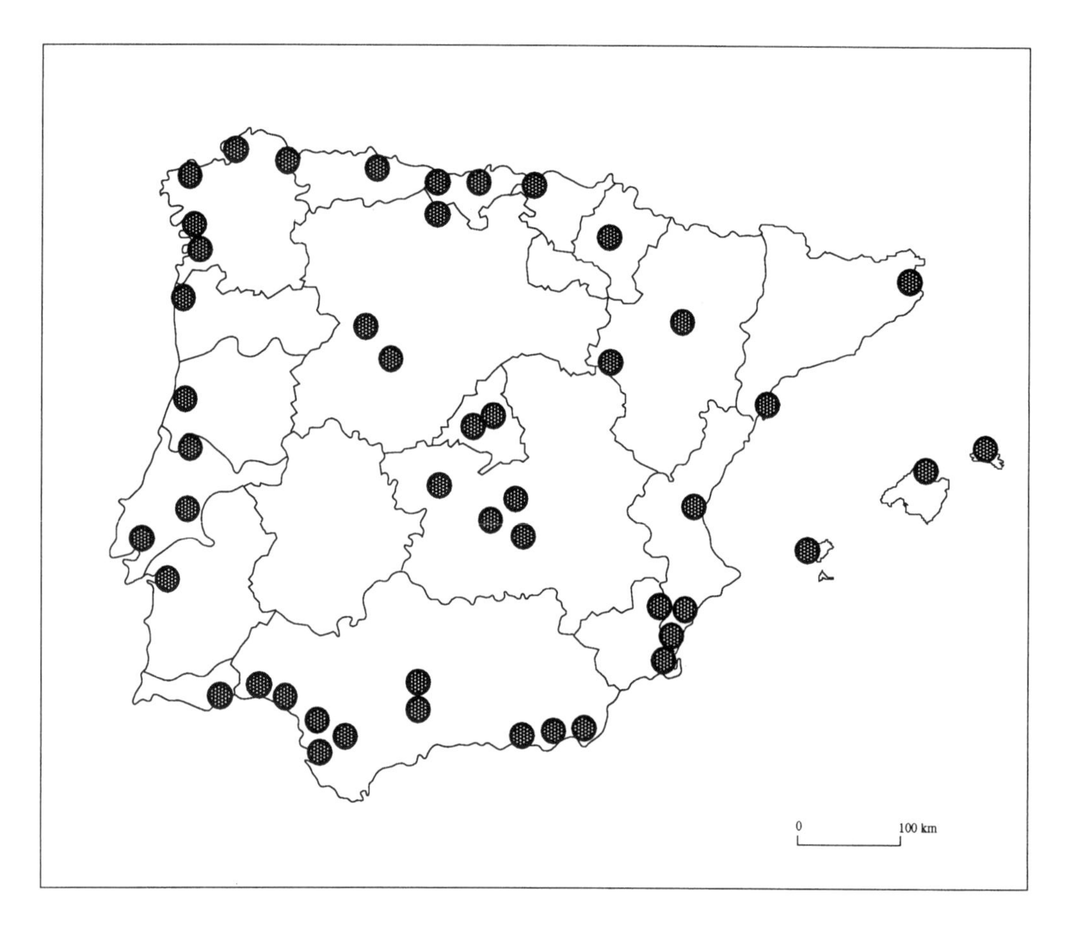

Figure 21.2 *Protected wetland areas, Iberian Peninsula* (source: Atlas Nacional de España 1996)

On the other hand, the fight against floods requires integrated ecological solutions (Masson 1993; Ballais *et al.* in press). Generally, the watershed and hydro-system scales must be used to understand what the consequences will be downstream if an action is performed upstream. In particular, instead of calibrating and rectifying the minor bed upstream, thus causing aggravated flooding downstream, it is necessary to systematically encourage infiltration during a flood. This decreases the severity of the flood downstream and the phreatic and deep aquifers are recharged (and can be used for agricultural, industrial or domestic purposes). Several experiments have been conducted on the Gardon d'Anduze (Languedoc) (Masson 1993) and on the Lez (Provence). In detail, it is necessary to conserve the three "beds" of the river: the minor one, the middle one and the major one (all part of the floodplain shown in Figure 21.3), even through towns. They can then form a pleasant "green stream" with places to walk and light recreational facilities able to support periodic inundation.

This same concern for integrated planning is at the forefront of the Sainte Victoire mountain rehabilitation project following the 1989 fire. This project envisages the preservation of agriculture with some cultivated fields and with herds that will limit the regrowth of shrubs.

Finally, as discussed previously, the preservation of a herbaceous layer in the vineyards is a very efficient means of preventing soil erosion.

Italy

Biological practices against land degradation consist essentially of new plantations, restoration of

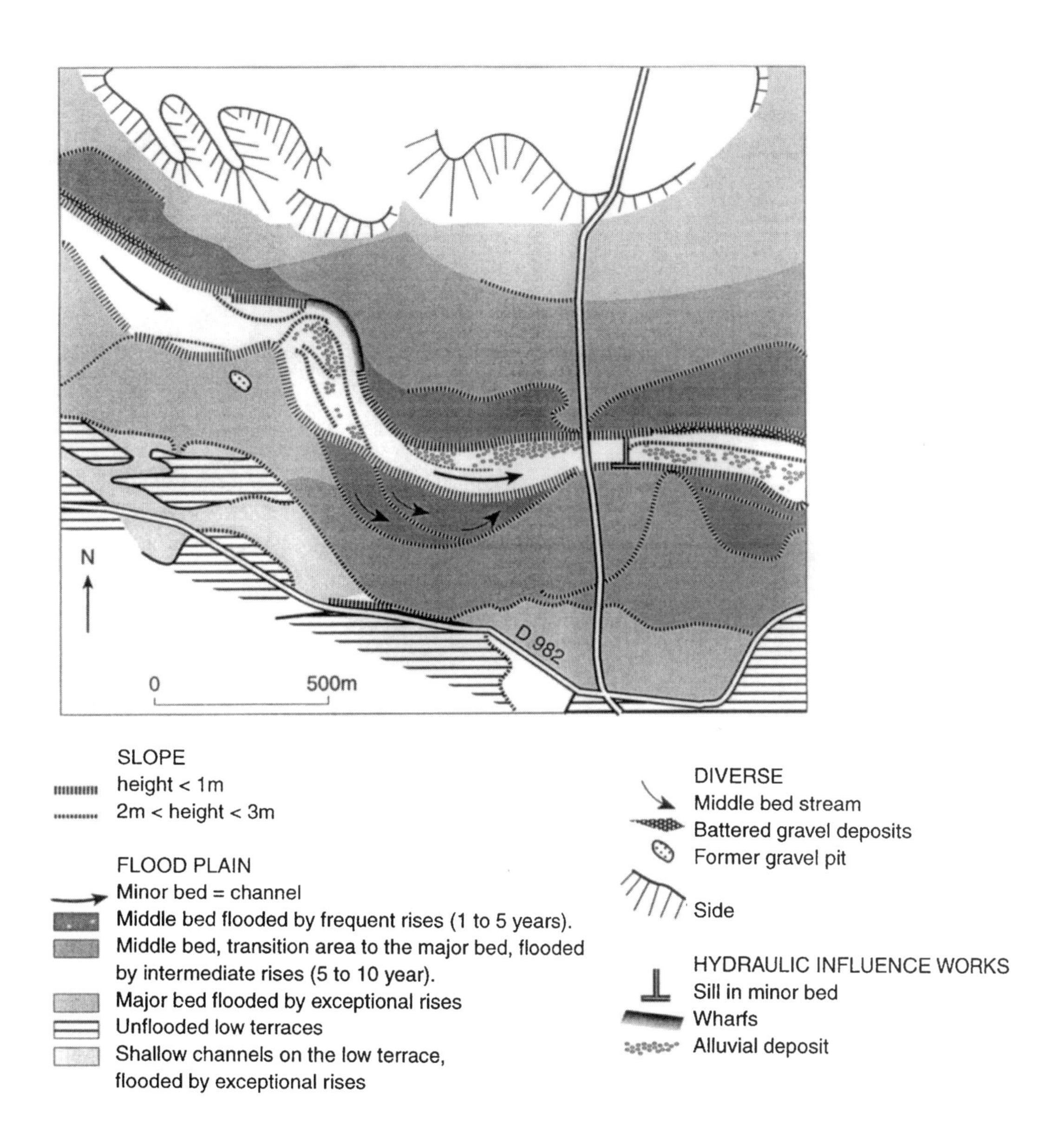

Figure 21.3 *Floodplain geomorphological mapping of Garden d'Anduze (Gard)* (after Masson 1993)

the biological variety of forests, and improvement of vegetation on grazed lands. These practices are undertaken by public enterprises, as it is usually the public forests which experience various attacks on their integrity (essentially clear-cutting and fires). Private forests are normally well kept. They represent about 50% of forest ownership, but data on private forests are not available.

If degradation involves the *macchia*, this is normally due only to fires. In this case, corrective measures are undertaken only in the case of complete or nearly complete destruction of the vegetation cover, in order to reduce soil erosion; otherwise badlands may develop within a few years. The common practice consists of new plantations of pine or eucalyptus trees, according to the kind of parent rock and morphological conditions.

As discussed in the previous chapter, since the 1950s public agencies have been attempting to recover badland zones by means of new plantations and reshaping of the steep slopes into stepwise forms. This practice has some drawbacks as piping erosion is increased, particularly in chalk and marly siltstones. Piping is initiated along rock joints and along the contact surface between *in situ* rock and earth fill used for step-wise fashioning (Sorriso-Valvo *et al.* 1992).

Reforestation efforts may be considerable, but in less developed zones they may also represent a source of employment. For example, after the widespread clear-cutting of public forest in the 1940s in Calabria, from 1951 to 1991 the main public forestry company provided a total financial budget of $330 million (Maiolo 1993a). Between 700 and 4100 men were employed daily for a total of 16 300 000 working days. Reforestation of barren slopes and improvement of vegetation involved 34 800 ha. New plantations as a direct measure against erosion involved 2000 ha. Corrective measures have been undertaken also on 200 ha of pasture land. Check dams (normally with gabions) required a total volume of 400 000 m^3 of material.

Such an effort required other works, such as the construction of non-metalled roads (11 425 km, including 218 km with asphalt pavement), 12 forest refuges, 29 forestry barracks and service buildings, 242 km of aqueducts and 46 km of sewage systems. In addition, complementary works such as picnic areas, sports facilities and parks have been constructed.

With regard to the reduction of land degradation, it is very difficult to obtain exact data concerning reclaimed lands and reduced rates of erosion. Data are limited to a few locations where results have been either extraordinarily effective or completely unsuccessful. An indirect indication of the general effectiveness of the intervention in river catchments is the strongly increased beach erosion, as already mentioned.

The Croatian Adriatic Coast

There are no data on this subject, only the 1949 report by Beltram on the use of narrow-crowned cypress trees as windbreaks. He wrote that in 1933/34 the peasants on the island of Brac planted 300 000 cypress trees to protect the soil from winds. Beltram reported that the effort was successful despite its unsystematic organization (because of many small family holdings), but he did not cite any related data.

Greece

Vegetated areas constitute an effective sediment filter, usable in agricultural and other lands. Many fields cannot be cropped efficiently or, if cropped, are extremely susceptible to erosion. Irregularly shaped and unproductive dry areas, such as those occurring in eastern Greece, can be kept under natural vegetation for controlling runoff and sediment loss.

Lands with a high actual erosion risk are located on hilly terrains with shallow soils, since they have undergone significant erosion in the past. These lands have high "climatic erosivity" indices (Yassoglou 1989). Their stabilization requires the maintenance of large amounts of vegetation, which the soil and the bioclimatic conditions can no longer adequately provide. These conditions are aggravated by fires which have intensified recently. Therefore, under existing conditions, poorly vegetated areas may remain bare, enhancing runoff and sediment loss.

The Eastern Mediterranean

Over-grazing is still a major problem in marginal areas, especially in mountains. Erosion effects are notable and thousands of hectares are being lost. The deterioration of rangelands prompted Jordan to develop an integrated programme which took into account the ecological, human, legislative and technical aspects (Ministry of Agriculture 1986), namely:

1. establishment of protected areas;
2. improvement of rangelands;
3. improvement of grazing conditions in irrigated areas and ample rainfall areas;
4. settlement of nomads;

5. establishment of national parks and conservation of wildlife; and
6. application of the Law on Rangelands. Application of the Law had many difficulties, attributed more to human – such as settlement of nomad Bedouins – than to technological reasons.

Rangeland improvement was undertaken by artificial seeding of forage crops: *Poa sinaica*, *Festuca pratensis*, *Phalaris tuberosa*, *Oryzopsis holciformis*, *Oryzopsis miliacea* and *Colutea istaria*. Forage shrubs included the following species: *Atriplex nummularia*, *Atriplex canescens*, *Prosopis specigera* and *Salsola vermiculata*. The improvement of the rangelands resulted in an increased grazing capacity from 1 sheep/10 ha to 1 sheep/2 ha.

The settlement of nomad Bedouins was intended, among other social and economic purposes, to reduce pressure on natural rangelands to allow for their regeneration. The Agricultural Law No. 20 of 1973 empowers the Ministry of Agriculture to adopt measures to improve pasture and forage, to regulate pastoral rotation, to provide water by drilling wells and building dams, and to improve forest resources and reverse the deterioration process.

Forest management policy consists of massive logging of forests and planting single species of trees (plantations) in clear-cut areas. This has a considerable effect on stream morphology and chemistry, and on landslides, erosion and water availability. Maintaining biodiversity is important geomorphologically as well as for the health and stability of the ecosystem. In the eastern Mediterranean, the forest industry is quite well developed in Turkey and to a lesser extent in Cyprus and certain areas of Lebanon and Israel, and greater awareness is needed by managers and developers of long-term ecological effects.

The present trend in agriculture in the eastern Mediterranean countries is to a more specialized agriculture based on vegetables and subtropical fruits. Greenhouses cover extensive areas of the coastal zone and the rift valley. Reforestation by public policy has converted extensive areas (200 000 ha, or about 20% of the total Mediterranean areas in Israel) to forest plantations.

North Africa

Reforestation

In Morocco, 153 500 ha were reforested to 1939. In the first 20 years of Independence, 370 000 ha were reforested, but by 1977, only 500 000 ha of forests were planted. At the present time, 1 000 000 ha of forests have been planted, of which 100 000 ha are protected. One should note the exclusivity of two species in those actions: the eucalyptus for the plain lands or hills and the pine for mountain lands. Research on the use of local species is not very advanced.

In Algeria, only 8200 ha were reforested from 1851 to 1930 (Plit 1983). After Independence, reforestation was pursued vigorously, but the large scale of the programmes was beyond the capabilities of the seed banks; hence there was inefficiency and many abandonments. With the agrarian revolution, the works in the Tell have stagnated to some extent and, in turn, there has been an intensification in the steppe (belts and "green dams" against the Sahara). During the period 1963–1981 in Algeria, 560 000 ha were reforested, of which 300 000 ha were planted by volunteers.

The anti-erosive effect of those plantings is diminished because the trees are planted following clearing of the *matorral*, which, in the first years, leaves the soil totally bare and at the mercy of runoff. Extensive damage can occur if serious rain events take place during that period. However, with the growth of the trees and the accumulation of a litter of pine needles on the ground, protection becomes very good, particularly where the trees are planted at high densities (Laouina 1995).

However, at a practical level, it should be noted that pastoral use is totally excluded from the pine forests. The reforestation, then, no longer has a function of production, other than as a side benefit to the anti-erosive effect. Further, the high density of the trees, implemented for soil protection reasons, is unfavourable for good sylvicultural production. Moreover, with protection being the primary objective, forest maintenance and thinning, fundamental for satisfactory production, have not always been carried out.

Reforestation for erosion control, then, seems to have the necessary efficiency against erosion but it remains unsatisfactory on the production level, particularly if the cost of establishing the plantations is taken into account. It would be desirable to avoid the confusion which is detrimental for sylvicultural production, either by opting for soil protection based on improvement of the natural *matorral*, or by choosing to establish really productive forests. The first choice would be interesting both in terms of cost and efficiency, while avoiding the risks of soil stripping. Such a forest could also retain its pastoral function.

Reserves

With regard to national parks, biological and hunting reserves, Morocco has two national parks, Toubkal and Tazekka. Toubkal National Park is the oldest; it was established in 1942 over 36 000 ha, covering the highest summits of the High Atlas – country with many endemic species of vegetation and with a rare animal, *Ammotragus lervia*.

Many other reserves have been established for various purposes, particularly to conserve precious species of vegetation and to protect some rare animals. The Talassemtane Reserve is located in the Rif to protect *Abies moroccana*. The Souss-Massa Park is a reserve for a rare kind of bird, the "Ibis chauve". Lake Ichkeul in the north of Tunisia was reserved to protect migratory birds, and the coastal reserve of Al Hoceima in the northeast Rif has the objective of preserving *Monachus monachus*.

Sylvo-pastoral developments

The sylvo-pastoral arrangement, experimented with in Morocco, is complex. It aims to produce wood, protect the soil and produce improved fodder units in relation to the primary rangeland. In the Rif, herbs and numerous fodder shrubs such as *Acacia cyanophylla* were planted. In the valleys of the Atlas, such as the Azzaden, recommendations insisted on the necessity of modifying the structure of the forest by planting fodder trees (Troin *et al.* 1985).

In the Loukkos, experiments relating to grazing by goats were carried out, based on a sylvo-pastoral development, but without real success. The sylvo-pastoral actions were limited to private or collective and very degraded lands, for example the slopes of the High Atlas, which are covered by a shrubby and secondary formation of *Cystus* on south-facing slopes and *Oleaster*, *Lentiscus* and *Arbutus* on northerly aspects. In the most degraded environments, which supply a scanty yield of fodder during the harsh periods of the year, plantations of *Acacia cyanophylla* were established in order to increase the density of the plant cover and to contribute fodder. After a ranking of the lands and a sociological investigation aimed at determining the beneficiaries, a programme was set up. In Dhar el Oued, for example, the plantation was established by companies in 1977 with "replanting" in 1978 and 1980, with a density of 1650 plants/ha and at a cost of about $500–600 per hectare (plants and management included over two years). The success was estimated at 80% and the protection lasted four years but was not total (Laouina 1994, 1995).

The anti-erosive effect works in two ways. On the one hand, the protection of the plantations obliged the herders to reduce the size of their flocks, which had a beneficial effect on the planted sites and other lands. On the other hand, the increased density of the plant cover allowed the rehabilitation of a certain number of erosion forms, especially gutters.

Yet the plantations of acacias are not well appreciated by the rural populations, for several reasons. First, due to the problem of land security, any plantation is perceived as the beginning of the state's seizure of the lands. This poses the problem of trust and shows the importance of engaging in a permanent dialogue with the population. Second, there is the problem of managing the reduced area left to the rural populations. The protection of the plantations, which was able to last, in some cases, for more than four years, concentrated the pressure of the herd on the remaining areas, causing excessive degradation at some sites. Fortunately this was corrected by reducing the size of the herds. Finally, there are problems relating to productivity and economic interest. Economically, the plantations of acacias (*Acacia cyanophylla*) were not sufficiently profitable despite their undeniable protecting effect. Other fodder shrubs may have been more beneficial. At any rate, it seems necessary, in the future, to recommend the introduction of a wider range of species in these difficult mountainous areas (Laouina 1995).

Permanent meadows

Permanent meadows were established to improve the degraded collective rangelands in hilly areas. The treatment consisted of clearing the brushwood, tilling and fertilizing, and sowing. The work was carried out by a private company and cost between $450 and $500 per hectare (today, the same operation would cost more than $600/ha). Maintenance included the weeding of the meadow.

By paying a rental of $3.00/yr per head of cattle and $0.6/yr per sheep, and by the management and fertilization being taken over by the group, the refunding of 70% of the initial expenses, on the collective group's behalf, was to be carried out gradually, presumably over a 20 year period. In fact, the repayments were really symbolic, the collective group having paid, for example, the sum of $1500 for the year 1987/88. Then the rental, fixed by the representatives of the collective group, declined (Laouina 1995).

The meadow seems to have had a good anti-erosive effect, since the emergence of new erosion forms has not been noticed and previous forms were covered over. The meadow has therefore really improved the land, even if its economic success is questionable. However, over-stocking effects, entailing a degradation of the meadow, were conspicuous. The lack of maintenance was also responsible for these degradation effects.

Although this costly development was not a real success on the collective land, it represented for the rural dwellers a pattern which they tend to reproduce on their own lands, today. Indeed, some of them tend to replace their cereal parcels with permanent meadows, which will have obvious effects on the reduction of erosion on the clayey calcareous lands of the hills and, undoubtedly, also appreciable economical effects.

California

Of the commercial timberland in California and Oregon, 50% is part of the National Forest, a further 5% is state-owned, 16% is owned by the forest products industry, and the remaining 29% is owned by farmers and other private interests (US Forest Service 1988). This implies that different solutions to problems of degradation may be applied depending on ownership (Chapter 23). In reality the situation is more complex. The National Forest contains timber which is harvested on a sustained basis with appropriate constraints, together with commercial stands which remain unharvested because they are of low quality or located in remote areas. Furthermore, commercial holdings continue to decline as old-growth timber is transferred to park and wilderness status. The problems of timber harvesting thus focus mainly on land owned by the forest products industry, and on the different approaches to timber harvesting practised by the logging companies.

The transfer of land from active development to passive use has a long and controversial history in California, controversial because it defies the tenets upon which the state's resources have long been exploited. Even the transfer of land from unrestricted development to managed development or "wise use", explicit in the creation of the national forest system and federal grazing legislation, was greeted with alarm by commercial interests. But the creation of national parks at Yosemite and Sequoia in 1890 introduced a more radical approach to land stewardship, specifically the preservation of land for scenic and recreational purposes. Creation of these parks, far removed from centres of population and more accessible resources, was not widely perceived as a limitation to growth at the time. Indeed, some park proponents sought to limit logging only in order to protect water supplies for irrigation purposes downstream.

California's system of state parks may trace its origins to federal action in 1864 which transferred Yosemite Valley to the state, only to be reconveyed to the fledgling national park system some decades later. For many decades, logging and other commercial interests sought to prevent the creation of a formal park authority but, in 1902, some 1000 ha of redwoods near Santa Cruz were afforded state protection, and a state park system was eventually set up in 1926. Today, the California Department of Parks and Recreation manages 570 000 ha of parks and reserves in nearly 275 units, including the nation's first marine reserve at Point Lobos (1960) and several World Heritage Sites in the coast redwood forest. Among its activities are programmes of prescribed burning, designed to inhibit large destructive fires in areas where fire suppression was previously dominant, control of non-native plant species, reserves for endangered wildlife such as the California least tern, and habitat restoration in important fish streams.

Chile

Introduction of vegetation species

The Government of Chile initiated the conservation of soil resources in 1951 with the creation of the Directorate for the Protection of Natural Resources (DIPROREN). This organization provides technical assistance to farmers in relation to the establishment of prairies, amelioration and management of natural pasturelands, and the construction of tanks for irrigation, amongst other activities.

At present there is no plan specifically directed to the protection of agricultural soils. On the other hand, the National Forestry Corporation (CONAF) maintains technical assistance and provides subsidies for activities which protect the soil resource, but which are evidently orientated towards forestry production. In this respect CONAF already has experience in sylvicultural practices such as the afforestation and reforestation of the arid and semi-arid zones of Chile. It is estimated that 24 000 000 ha have been reforested in seven regions of the country (Regions I to VII).

There have been 90 Chilean experiments to introduce new species, located in the seven regions, with 2509 species or varieties tested. The most frequent genera are shown in Table 21.1. The most outstanding species in relation to their adaptability to the conditions of the different pedo-climatic units have been *Acacia cyanophylla*, *Acacia cyclops*, *Atriplex canescens*, *Atriplex glauca*, *Atriplex repanda*, *Atriplex halimus*, *Eucalyptus camaldulensis*, *Eucalyptus globulus* spp. bicostat, *Eucalyptus globulus* spp. globulus, *Eucalyptus globulus* spp. maidenil, and *Eucalyptus sideroxylon*.

The available information from these experiments and from additional observations permits the recommendation of the *Eucalyptus* species listed in Table 21.2.

In the semi-arid zone of Chile, new experiments have been carried out in relation to the introduction of new species at altitudes ranging from 85 m to 1700 m above sea level: 240 experimental plots have been set up with the 36 forest species listed in Table 21.3.

Large-scale reforestation

In Chile nearly 250 ha of *Simmondsia chinnensis* have been planted between Region I and the Metropolitan Region. The biggest plantation is located to the south of La Serena city (Region IV),

Table 21.1 *Plant genera planted in the arid and semi-arid regions of Chile* (source: Latorre 1990)

Genus	No. of species or varieties
Eucalyptus	60
Acacia	28
Atriplex	27
Pinus	25
Cupresus	5
Kochia	5
Populus	5

Table 21.2 *Species of* Eucalyptus *tried with success in the arid zones of Chile* (source: Latorre 1990)

Mean annual rainfall (mm)	*Eucalyptus* species
<350	*citriodora*
	estradingeus
	cladocalix
	diversicolor
	resinifera
	sideroxylon
350–400	*camaldulensis*
>400	*delegatensis*
	fastigata
	globulus
	bicostata
	maideninitens
	regnans

Table 21.3 *Forestry genera trialled in the semi-arid regions of Chile* (source: Latorre 1990)

Genus	No. of species
Acacia	9
Callistris	1
Casuarina	1
Cupresus	4
Eucalyptus	14
Pinus	5
Quercus	1
Robinia	1

with an area of 120 ha and an average density of plantation of 1370 plants/ha.

From the experiments with *Atriplex*, replanting with *Atriplex numularia* in Coquimbo (Region IV) is outstanding; it is the second most extensively planted species in Chile after *Pinus radiata*. At present, more than 30 904 ha have been planted with *Atriplex* in this Region, of which more than 98% are *Atriplex numularia*. This extent of planting is the result of a series of conditions which can be summarised as follows:

- the need to find supplementary alternatives to the feeding of livestock (goats and sheep), which provide the dietary base of most of the rural population;
- the presence of an abundant soil resource which is not appropriate for agriculture due to the arid and semi-arid conditions; and
- the existence of a legal body which favours afforestation including forage shrubs.

Prosopis represents another very important genus in the development of the agro-silvo-pastoral development of drylands, with areas extending over more than 50 000 ha. Of this area, 26 000 ha are planted with *Prosopis chilensis* and 24 000 ha with *Prosopis tamarugo*. The latter grows in marginal soils with high contents of salts of Na, Ca, Mg and K, and the existence of a saline hardpan often at the surface with a thickness exceeding 500 mm. Under mixed populations of *Prosopis chilensis* and *Prosopis tamarugo*, production of honey by bees has been developed with yields of 14 kg of honey for each honeycomb. Firewood production is 14 m^3/ha and stocking density 1 sheep/ha/yr. The adapted species of sheep are Karakul, Suffolk Down and Australian Merino. The adapted goat species is Angora.

Reintroduction of species

Forestry enterprises of Region VIII, under the guidance of CONAF and INFOR, have undertaken studies on the reintroduction of species which had traditionally been planted for ornamental purposes, for shade, shelter and wind protection. This is the case of the Australian acacia (*Acacia metanoxylon*), which is also a rapidly growing species used for dune stabilization. At present its dispersion is heterogeneous between Regions VI and X. In these areas it has been able to develop in marshy locations, forming nearly pure stands of good quality trees. *Ricinus communis* is an oleaginous shrub also well adapted in Region IV. Its potentialities are multiple, including the lubricant industry, pharmacy, paints, varnish and explosives.

Since 1987, the National Forestry Corporation, together with Japanese experts, has initiated an ambitious Project of Technical Cooperation in the conservation and reforestation of drainage basins in the semi-arid zone of Chile. The project aims at re-establishing the vegetative communities of San Pedro (Metropolitan Region), Las Condes (Metropolitan Region) and Illapel (Region IV).

The Southwestern Cape

Given that the most serious environmental problems in the southwestern Cape region appear to be those of an ecological and biological nature arising from the destruction of natural vegetation communities (Chapter 12), it is obvious that solutions of a biophysical and ecological nature are most prevalent. Indeed, the problems of loss of indigenous plant communities, more particularly the forests, and the growth of alien vegetation infestation have been recognized for some considerable time. Solutions to such problems were sought as early as 1652 (Verster *et al.* 1992). Perhaps one of the most obvious solutions has been the establishment of numerous areas set aside specifically for conservation of the indigenous plant and animal communities, although as

Rebelo (1992) points out, the reserved area network is far from comprehensive and many regard it as inadequate for the task of ensuring the continued existence of the region's prolific biodiversity.

Alien plant and animal control

"The control of invasive trees and shrubs is the largest single task facing managers of most natural areas in the biome" (Van Wilgen *et al.* 1992, p. 354). There are, in essence, three control options, namely biological, chemical and mechanical control.

Biological control is a relatively recent development which has proved most successful in restricting the spread of several of the Australian acacia tree types (Kluge *et al.* 1986), although it has not been without controversy. Biological control methods exist in a variety of forms (see Van Wilgen *et al.* (1992) for a complete list), including a gall rust which reduces seed production in *Acacia saligna* and a gall wasp which attacks the buds of *Acacia longifolia*, both of which have been imported from Australia and released in a controlled manner. Chemical controls include the spraying of post-fire regrowth with systemic herbicides, and mechanical methods (the most widely applied) rely basically on the "hacking" of juvenile plants. The latter two methods are labour intensive and frequently unaffordable, particularly by private landowners. Without doubt, the most effective methods are those which utilize a combination of more than one control option, so-called integrated control (Kluge *et al.* 1986).

While calls have long been made by conservationists in the region for the removal of alien vegetation on the grounds of conserving biological diversity, more recently it has been recognized that alien infestation also has a major impact on the available water resources. Water resource-based arguments for the widespread reduction in alien tree infestation have proved considerably more successful in persuading the appropriate authorities to take the problem more seriously. The Minister for Water Affairs and Forestry announced, towards the end of 1995, a major new initiative, funded by the Reconstruction and Development Programme (see Chapter 23), to effect alien tree removal in large parts of the southwestern Cape mountains. Crucial to the implementation of this initiative was the recognition (a) that alien trees consume vastly greater quantities of water than the indigenous *fynbos*, and (b) that the labour-intensive nature of the commonly used forms of control represented significant job-creation potential.

Not all calls for removal of aliens are greeted with such enthusiasm. Plans by the Cape Town Municipality to remove dense pine tree plantations from what is known as Newlands Forest on the eastern slopes of Table Mountain, were greeted with considerable hostility by the public (P. Britten, pers. comm.), and alternative plans which left some of the pine areas intact were put forward as an alternative.

Integrated management

There have been frequent calls for a more holistic and integrated approach to managing the southwestern Cape environment, although complications regarding land ownership and the complexity of the legislation (see Chapter 23) have mitigated against such methods. The concept of integrated catchment management has received much publicity, but its principles are almost impossible to apply in the context of privately owned land with so many different stakeholders and so many apparently disparate objectives. Both the Mountain Catchment Areas Act and the Water Act (Chapter 23) essentially embody the principles of integrated management, but their implementation is hampered by ownership issues and the difficulty of actually enforcing the law (Fuggle and Rabie 1992).

Southern Australia

Protecting wildlife habitat

For Australia as a whole, 10% of the landmass has been set aside for conservation land use (national parks and various kinds of reserves). However, by far the largest reserve areas are in the arid zones; the proportions set aside in the Mediterranean region are small, scattered and often of dubious ecological value. South

Australia, for example, has a reserve system covering more than 20 000 000 ha. But 96% of these national and conservation parks and regional reserves are in the arid zone (Department of Environment and Land Management 1993, 129). Thirty-four plant communities had an improved conservation status rating in 1992 compared with 1982; but 26% of all plant communities in the state are still considered to be poorly conserved or not conserved at all – although this is an encouraging improvement on the 1982 figure of 37% (Department of Environment and Land Management 1993, pp. 116–117). Western Australia's wheatbelt has over 630 nature reserves, covering about 7% of the region. Most reserves are small, with the average size being about 114 ha.

An important development in South Australia was the introduction of the Native Vegetation Act in 1991, under which no further large-scale clearing of vegetation is envisaged. The Act also has the potential to control habitat degradation caused by wood cutting and brush harvesting. Western Australia, too, has ceased the opening up of new agricultural lands in the southwest of the state, although clearing continues on individual farms. Permits have to be obtained, and are usually granted unless the land is considered to be susceptible to degradation. When clearing is approved, farmers may be required to plant trees on an equivalent area on another part of their property.

A significant development in Western Australia's state forests was the introduction in 1994 of an extensive fox-baiting programme. Foxes were introduced in the 19th century to provide sport, and have become the major cause (after clearing) of the loss of wildlife, especially mammals. Many endangered, indigenous species are now found only on offshore islands. The baiting programme uses the 1080 poison (sodium fluoroacetate), which occurs naturally in some of the native plants, and to which native animals are resistant (but not dogs or cats). Once the fox populations have been depleted, native animals are reintroduced. Although there have been some early successes in some forested areas, notably with the numbat, in other places the reintroduced animals simply provided easy meals for the resident populations of foxes and feral cats. The need to provide continued management and control of the introduced predators is a major difficulty. The areas involved are far too extensive for fencing to be a practical proposition. Fox-proof fencing needs a close mesh and has to extend at least 50 cm into the ground: it is very expensive to construct.

Retention and re-establishment of native vegetation on farms

In Western Australia, the Commissioner for Soil and Land Conservation has guidelines for the area of native vegetation which should be left in a subcatchment: 30% in high rainfall zones (700–1100 mm mean annual rainfall); 25% in medium rainfall zones (500–700 mm); and 20% in low rainfall zones (<500 mm). However, in central wheatbelt shires clearing already exceeds 90%. To assist in achieving these guidelines, in 1988 the state government introduced a scheme to subsidize the cost of fencing remnants of native vegetation on farms. In the three years to 1991, more than US$880 000 were provided in 470 grants to fence 26 000 ha. In 1990, a survey by the Western Australian Farmers Federation found that 57% of the farmers surveyed protected an average of 53 ha of native vegetation on their farms. However, only 25% of the vegetation was protected from grazing by stock (Grant 1992, p. 126).

South Australia and Victoria also have schemes which aim to encourage the retention and re-establishment of native vegetation on farms – to a large extent because it is considered that native vegetation is an integral component of sustainable agriculture. Thus, the Victorian *State of the Environment Report* (Scott 1991, p. 26) identified six major, related instances of positive change. These are:

- increased awareness of the need for integrated catchment management (ICM);
- establishment of Landcare groups throughout the state;
- increased interest in and adoption of whole farm planning;
- establishment of Tree Victoria, in response to the federal One Billion Trees programme;
- planning regulations controlling clearing native vegetation on freehold land; and

– the development of the Decade of Landcare Plan for Victoria.

Agroforestry

Originally, in Australia, agro-forestry was a method of introducing some grazing beneath rows of trees planted for timber – usually pine trees. In some quarters this is still the case. But the increasing recognition that trees play a much wider role than merely supplying wood for the timber and paper pulp industries, has extended the agro-forestry concept (Reid and Wilson 1985).

Agro-forestry attempts to provide farmers with a combination of regular, annual (or seasonal) income from normal farming practices with the much longer term (and higher risk) income from tree crops. The two systems are fully integrated within individual farm paddocks. In an increasing number of instances, the purely economic objectives are being combined with other, long-term objectives of maintaining and improving the quality of soils and water and, increasingly, the purely environmental objectives of providing habitat for the rapidly diminishing wildlife. Thus, the role of agro-forestry is now seen to include the following objectives (modified from Conacher and Conacher 1995, table 7.2, and not in order of importance):

- producing a wide range of shrub and tree crop products (including timber, posts, firewood, flowers, honey, oils, fruit, nuts, cork, tannin, camphor, resins, seeds, medicines, pesticides)
- reducing wind erosion
- providing shelterbelts for stock
- conserving flora and fauna
- controlling water erosion
- protecting water catchments
- reducing the impact of droughts
- controlling salinity
- providing cover on salt-affected land
- improving aesthetics (including wider community benefits)
- modifying micro-climates
- reducing greenhouse gases
- increasing nutrient cycling
- adding value to the property.

There are numerous management options to assist farmers to attain these objectives. They include: protection from grazing by fencing; replanting, including understorey species; providing linking corridors; managing drift from chemicals; controlling exotic weeds and feral animals, and reintroducing local fauna (Breckwoldt 1983; Davidson and Davidson 1992; Carne 1993; Hussey and Wallace 1993). As some of these authors have shown, the beneficial effects of these approaches can be spectacular.

Nevertheless, Ryan (1990) has noted that trees can reduce yields by competing for nutrients, light and water, flow regimes in streams may be diminished, and some native species are not suited to soils which have received applications of synthetic fertilizers. It is important, therefore, to conduct careful planning and management in order to balance productivity needs against ecological objectives; and the Davidsons (1992) have stressed that the two objectives can be achieved if the schemes are properly designed. It is also clear that a fully comprehensive agro-forestry system merges with whole farm planning and integrated catchment management.

Whole farm planning

Whole farm planning (WFP) is a particularly relevant concept which incorporates the full range of land degradation solutions on a particular property, and which is also very applicable under the National Landcare and Land Conservation Districts programmes (see Chapter 23). WFP aims to integrate management of a farm's soil and water resources, through mapping, contouring, drainage and the use of the land, in order to optimize sustainable production over the long term. It has the potential to include ecological factors as well, such as revegetation to combat tree decline and fencing for farm-based conservation of native vegetation and species habitat. Scott (1991) points out that the concept is not new, having been promoted by the Victorian Soil Conservation Authority in the 1950s.

Figure 21.4 is a schematic diagram indicating how vegetation retention and revegetation can be incorporated in a farm plan in order to achieve several objectives: in this instance to combat secondary salinization, to provide stock shelterbelts

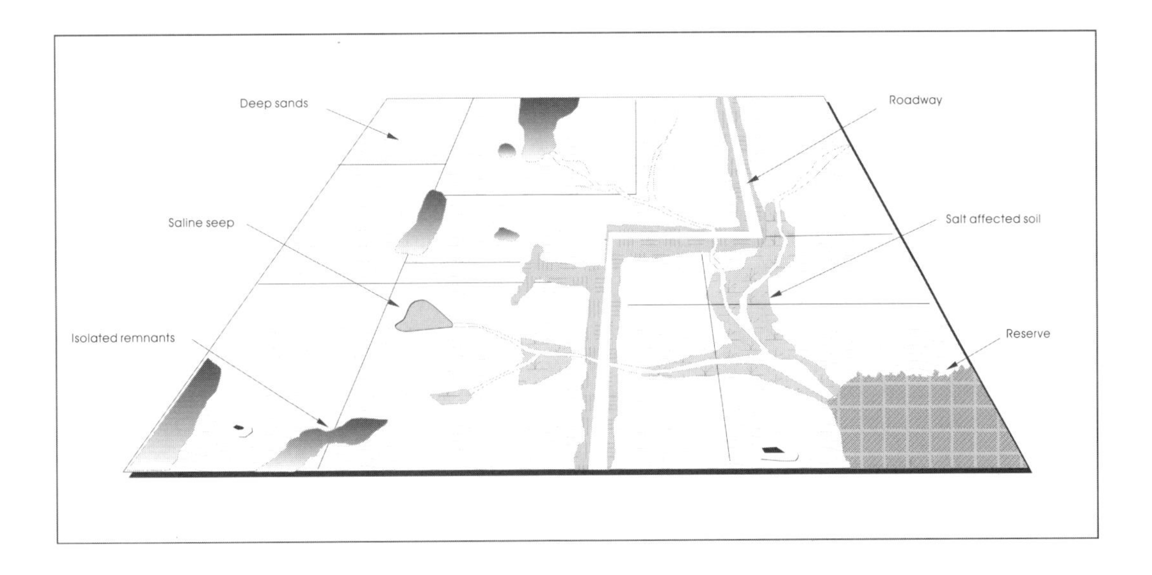

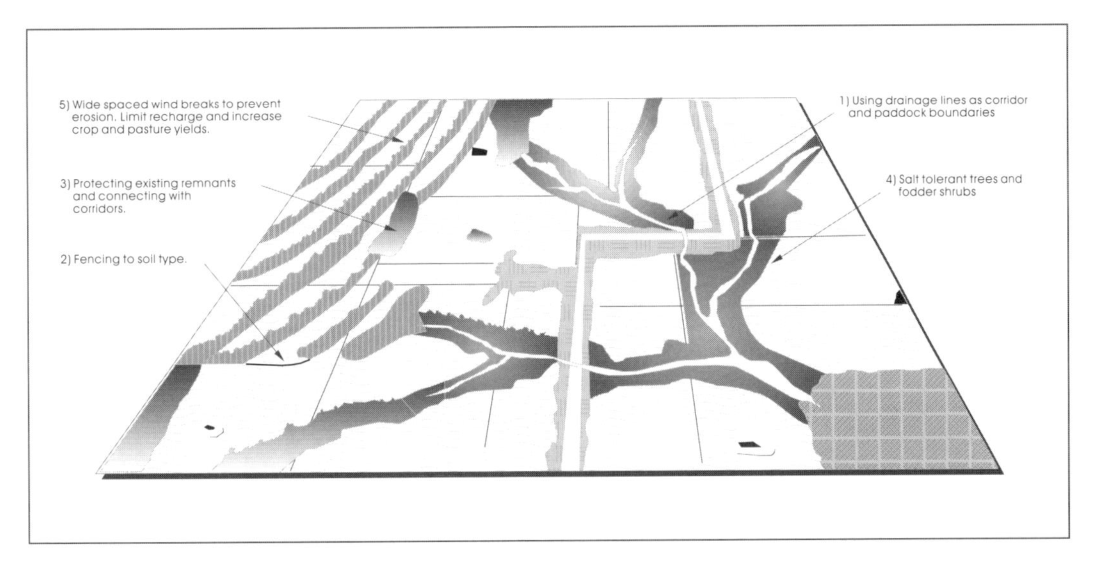

Figure 21.4 *Schematic diagram showing how vegetation retention and revegetation can be integrated into a farm plan* (after Hobbs and Saunders 1993). *Compare with Figure 12.8, which provides an indication of the scope for imaginative revegetation, and Plate 21.1*

and windbreaks, and corridors for the movement of wildlife.

The Watkins Farm

A specific example of whole farm planning is the work of Ron Watkins, a farmer in the Frankland District of the Western Australian wheatbelt. The farm is relatively small for the district, covering an area of only some 700 ha. This permits the farm to be managed much more intensively than is generally possible in the wheatbelt, where farm sizes average around 2000 ha.

A system of WISALTS-type interceptors (see Chapter 20, although Watkins has excavated them on a slight gradient rather than on the contour) has been constructed to control surface and subsurface water movements, using Yeoman's "keyline" principles whereby water is diverted from the wetter hollows to the drier spurs (Yeomans 1954). Part of the scheme involved the construction of farm dams. Water from one dam is used to irrigate lucerne (alfalfa), and Watkins intends to introduce aquaculture. The property is now drought-proof.

Down-slope from the interceptors Watkins has planted rows of trees, mostly four rows wide, again along the contour (Plate 21.1). He is experimenting with a range of species, including native eucalypts and acacias, salt-tolerant breeds which have been genetically engineered for growing in saline soils, the Chinese "miracle tree" (*Pawlonia*), poplars, carobs and tree lucerne (tagasaste) which is harvested annually and fed to the sheep. The planted trees, and several blocks of remnant native vegetation, are protected from stock with electric fencing (Plate 21.2). Watkins claims that this is more effective and less expensive than the normal strained and barbed fences.

The tree belts are multi-purpose. They provide shelter and food for stock (sheep and beef cattle), act as windbreaks for the between-trees contoured strips of crop (canola) and pasture, and attract an increasing variety of bird species which in turn control insect populations on the property. As a result, no synthetic pesticides are used except some herbicides on firebreaks. The trees are also intended to draw on water from the deeper aquifer in the deep-weathered soils, thereby controlling a small but developing salinity problem. This whole farm approach is further supplemented by the application of non-synthetic nutrients to the soil in the form of dolomite, gypsum and judicious crop and pasture rotations, including the retention of crop stubble on the soil surface and its later incorporation into the soil, together with direct drilling and occasional deep ripping. The farmer is currently avoiding the use of synthetic fertilizers and pesticides and is essentially practising ecological farming.

Watkins was awarded a Churchill Fellowship, using it to travel to North America, and then a local soil conservation award in 1991 (Plate 21.3). In 1995 he won a United Nations Environment Programme award for his innovative farming practices and the positive influence he is having on other farmers in the district.

Plate 21.1 *Interceptor and rows of trees planted on the Watkins property in the Frankland District, Western Australian wheatbelt. Soil in the foreground has been deep ripped to break up a subsoil hardpan. The farm is planned and managed explicitly on a "whole farm" basis by the Watkins family*

Plate 21.2 *Electric fencing used on the Watkins farm, Frankland. The relatively small size of the farm makes it possible to check and repair the fencing on a regular basis. Within a few years, indigenous ground covers and shrubs in stands of remnant trees protected by the fencing from grazing by stock were regenerating; protection from stock is essential for the rows of planted vegetation*

Plate 21.3 *Ron Watkins and his soil conservation award*

22

Economic, social, agency and policy solutions

1. The Mediterranean Basin

Compiled by Arthur Conacher and Maria Sala from contributions on the following regions:

Iberian Peninsula and Balearic Islands — **Maria Sala** and **Celeste Coelho**
South of France and Corsica — **Jean-Louis Ballais**
Italy, Sicily and Sardinia — **Marino Sorriso-Valvo**
Greece — **Constantinos Kosmas, N.G. Danalatos** and **A. Mizara**
Dalmatian coast — **Jela Bilandžija, Matija Franković, Dražen Kaučić**
Eastern Mediterranean — **Moshe Inbar**
North Africa — **Abdellah Laouina**

Introduction to Chapters 22 and 23

Economic solutions to the problem of land degradation include the provision of incentives, such as direct subsidies and tax concessions to encourage good management practices, or the imposition of financial penalties to punish harmful practices. Social measures may include the relocation of people from badly degraded areas, and education and improved communication to change attitudes or to improve landholders' skills. Such measures are implemented at a level higher than that of the individual landholder, and merge with political, or policy, and agency solutions. These are discussed in this and the following chapter, with the material arbitrarily split between that dealing with the Mediterranean Basin (this Chapter) and the New World (Chapter 23).

With regard to land degradation, there is a need to balance short-term economic and production needs against long-term environmental goals. Increasingly, governments are attempting to reconcile these imperatives through the development of sustainable development policies (WCED 1989; OECD 1994).

In purely economic terms, farming is a business where farmers respond to market forces driven by consumer demand and preferences. However, this process is distorted by government policies such as subsidies, trade and marketing practices, and taxation or funding

Land Degradation in Mediterranean Environments of the World: Nature and Extent, Causes and Solutions.
Edited by A.J. Conacher and M. Sala.

requirements. These in turn may influence the ways in which farmers use their land or make decisions to invest in land conservation. In reality, even in good years farmers do not receive a profitable return on the real costs of production. It follows that in an economically recessionary environment, funding priorities will not be directed to land conservation. In fact, decisions to make such expenditures tend to be based on ethical, ecological and long-term considerations rather than short-term production or monetary goals.

The involvement of government and non-government agencies and organizations in the management of land degradation is also discussed. Problems associated with agency structures may include the question of which agency has the responsibility for soil conservation, and whether those responsibilities are separate from those concerning the growing of crops. Who deals with the revegetation of farmland if there are separate agricultural, forestry, and flora and fauna protection agencies?

Finally, the chapters discuss political solutions, considering government policies, legislation and regulations in relation to land degradation. Arguably, these underlie all actions: are there policies to encourage sustainable agriculture or the use of appropriate technologies?

The Iberian Peninsula and the European Union

Forestry and fires

To date, Spanish forestry policy has been dominated by fast-growing plantations which have led inevitably to increased numbers of forest fires. These trends are now being exacerbated by the devolution of forest management to regional governments and the inadequate nature of European Union (EU) forestry policy and practice (Rojas 1987). However, the Common Agricultural Policy is currently being reformed to permit a more rational use of natural resources, and to include explicitly the improvement of environmental goods and services, and the maintenance and development of the rural way of life.

Until recently, agricultural price protection led to the expansion of croplands at the expense of forest areas. But agricultural policy has begun to include the evaluation of non-economic functions in forest areas. Nevertheless, contradictory tendencies still exist between private interests which attempt to maximize economic income, and growing social demands for environmental goods and services. In principle, the latter, especially soil and water conservation, should differentiate Spanish forest lands, which are mainly Mediterranean in nature (agri-sylvo-pastoral), from many sylvicultural practices and policies in central and northern Europe.

Over two-thirds of Spain's woodlands are in the hands of private landowners. Given that private business cannot (or is unwilling to) undertake required woodland renovation and soil conservation, public intervention is needed to guarantee more rational exploitation of the environmental and economic goods and services of Iberian forests. This is presently done under the auspices of the EU's Agrarian Policy, which funds the management of cork forest lands, which in turn also includes measures to prevent forest fires.

Structural measures to reduce the fire hazard should include diversification of land use, management of recreation and illegal dumping, introduction and enforcement of appropriate legal instruments, and environmental education (Sala and Rubio 1994). The number and intensity of forest fires can be reduced by co-ordinating concerted action by forest managers, technical and scientific groups, and administrative departments. Other recommended measures are to accelerate the development and application of new techniques for detecting and extinguishing fires.

Prevention, reforestation and research projects are being carried out in Spain, especially in the worst affected of its Autonomous Communities, to confront the serious consequences of forest fires. The research has been grouped by Vega-Hidalgo (1986) into four main areas, dealing with ecosystem evolution and conditions before and after a fire: (1) the role of vegetation; (2) soil physico-chemical properties; (3) soil biology; and (4) soil erosion. Prevention is orientated along three main lines: (1) development of techniques to control combustibility such as prescribed firing, and pasture and brush clearing; (2) determination of the calorific power of the vegetation; and (3) development of indices of danger which combine a meteorological indicator, a spatio-temporal risk, causation and inflammability.

In view of the dramatic increase in the occurrence of forest fires, in 1986 the Council of the EU adopted a Regulation which established a Community action plan for the protection of forests. The Regulation comprises the fostering of prevention by sylvicultural operations, the construction of forest roads and firebreaks, and vigil (monitoring) points. On the other hand, the Spanish forest policy, with its programme of protection and improvement of the natural environment, has as its objectives the protection and restoration of soil and vegetation, the equilibrium of natural ecosystems, and defence against forest fires and erosion. As a specific example, the Catalan Autonomous Government has recently (1995) limited the free circulation of people and vehicles in the forests, and also discourages the building of villas in the forest. The construction of firebreaks around urbanized areas is compulsory.

Erosion

The Spanish public administrations are confronting this problem by means of Co-operation Agreements for the hydrologic restoration of drainage basins, formalized between the

Autonomous Communities and the Ministry of Agriculture through ICONA (Instituto para la Conservación de la Naturaleza).

The administration, through ICONA, has also initiated the LUCDEME programme (Lucha Contra la Desertifición en el Mediterráneo – Fight against Desertification in the Mediterranean). The long-term objective of the programme is the control of desertification in the southeast of Spain by analysing the resources and areas affected, determining systems and techniques to combat it, initiating restoration plans for vegetation and soil, and correcting the effects of torrential rivers.

The LUCDEME programme has fostered the installation of erosion plots and experimental basins at several points along the Mediterranean coast and has generated a great deal of interesting data. It has also fostered research on vegetation responses to environmental conditions, and on how root development could be improved in order to repair soils more effectively and more rapidly. At present, the LUCDEME programme has been substituted by the RESEL programme (Red de Estaciones de Seguimiento y Evaluación de la Erosión Lucdeme – Net of Lucdeme Pilot Sites for Monitoring and Evaluation of Erosion), with the objective of maintaining the experimental sites created during the LUCDEME project 1996.

Soil pollution

The problem of soil pollution is related to dysfunctionalities, concerning the environment, in the cycles of productive activities in the territory. In that sense degradation and contamination can be caused by direct actions (mineral extractions, public works and forest fires) and inadequate management of wastes. It is understood that control and protection of soil pollution has to be partly focused on the polluted soils themselves and partly on the global context of prevention measures.

During 1989 the ELIS project was initiated with the following objectives:

- define and characterize the different types of soils already polluted
- characterize the forms of superficial pollution
- characterize the different origins of pollution
- analyse damage and evaluate the risks produced by polluted soils
- obtain quantitative estimates of the existing contaminated soils
- estimate the cost of sanitation and rehabilitation techniques.

The development of the project has two main thrusts: an internal one which integrates all Autonomous Communities, and an external one of co-ordination with EU directives within which specific actions are undertaken through the programme "Community Actions for the Environment".

The techniques used are innovative and the EU provides financing in order to demonstrate and implement them. In 1989 Spain presented four projects dealing with: (1) the establishment of a measurement net of soil water pollution parameters in western Andalucía; (2) a monitoring system for forest fires; (3) continuous control of coal wastes in a rehabilitation process; and (4) extraction and washing of polluted soils and waters and treatment by oxidation/precipitation of organic compounds and heavy metals.

Inventories of polluted soils have been commenced in many of the Autonomous Communities with similar methodologies, and also estimates of the cost of amelioration (Table 22.1). The unitary costs applied are $3.5/m^3, which seems extremely low if compared with the standard used in Europe (three hours of an engineer's time, that is $120/m^3). Nevertheless, these estimates permit decision making at a legislative level and the provision of the necessary funds.

European resources and instruments for environmental information

The EU CORINE programme is defined as an experimental project for the collection and coordination of information concerning the environment and natural resources in the European Union. It was adopted by Decision 85/388 of the European Council in 1985 for a development period of four years.

The specific objectives of the programme were explicitly formulated by the Decision. They are to:

Table 22.1 *Estimated cost of "sanitizing" polluted soils in Spain* (source: MOPU 1995)

Source of pollution	Cost (10^3 US$)
Industrial wastes	1 676 600
Spilling from tanks	
combustibles	59 000
other	2300
Accidents	
railway transport	1500
road transport	6800
pipes	250
Production	208
Storages	250
Industrial zones	175 000
Sediments in rivers and ports	50 000
Total	1 971 908

1. collect reports on the environmental situation for a certain number of priority applications;
2. co-ordinate initiatives of the member states or of international bodies; and
3. provide coherence of codes, rules and definitions in order to facilitate data comparability.

Spain has participated actively in the development of the CORINE projects corresponding to biotypes, emissions to the atmosphere, erosion risks and soil resources, water resources and water quality, coastal erosion, land uses and boundary regions.

A list of norms on environmental issues has been summarized in MOPU (1989b), classified by European Union, Spanish State, Autonomous Communities, and in relation to the different topics. Public money spent on the control and restoration of the environment and the main results obtained are also reported. MOPU (1989b) has also provided data in relation to: (a) environmental planning; (b) the different systems for evaluation of environmental impact; and (c) environmental research and education.

Control of water quality

In Spain, water quality problems have been considered in merely qualitative terms until very recently. General concern about water deterioration did not arise until a few decades ago. This can be illustrated by the lack of infrastructure for urban sewage disposal and treatment, which has recently been subject to a policy of expansion and upgrading. It is the Spanish Government's goal to attain a complete coverage for towns with >2000 inhabitants by the year 2005, as required by the European National Sanitary and Sewage Treatment Plan, the Regularization and Control of Wastewater Discharges, the SAICA Project, and others (Alvarez Cobelas and Cabrera Capitán 1995).

At the same time, Spanish investigators in environmental agencies, universities and scientific institutions have increased efforts through the development of national limnological surveys, studies on biotic indices, thematic mapping, model calibration, pollution sources assessment and reservoir typology. In summary, prospects for water research in Spain appear to be truly optimistic, bearing in mind a higher public and political awareness, tremendous technological progress and promising economic development.

The European Union has developed criteria for water quality control in rivers and aquifers, suggesting water quality levels for swimming, fish life and water consumption by humans. In addition, the Spanish government has considered criteria for other purposes, such as irrigation, recreation and industrial uses. In order to obtain automatically data on water quality that can be useful for comparing objectives and preventing users from violating such criteria, the General Directorate of Water Quality from the Spanish Ministry of Public Works is implementing a network of sampling stations in Spanish rivers, comprising 1000 conventional stations and 115 automatic stations, the so-called SAICA Project (Sistema Automático para la Información de la Calidad del Agua – Automatic System for Information on Water Quality). In the conventional stations water will be sampled for further laboratory analyses. In the automatic sampling stations water variables will be measured continuously (discharge, pH, electrical conductivity, dissolved oxygen, ammonia plus ammonium, total/dissolved organic carbon, turbidity, soluble reactive phosphorus, nitrates, lead, cadmium and chromium). Each automatic station will send data to the HISPOSAT satellite from which data will be sent to earth stations through a Geographic Information System. Up to 70 microcomputers and 10 UNIX servers distributed in 10 centres throughout Spain (based in the network of Spanish Water Authorities) will provide hardware for automatic operations.

Marine waters

In accordance with the directions of the Council of the EU, member states have to communicate regularly the quality of waters used for swimming. This order has been included in Spanish legislation since 1988. Since 1986, the Ministry of Health and Consumption, in collaboration with the Sanitary Administrations of the Autonomous Communities, has published regular reports for swimming waters synthesizing data from the vigil network established at several representative sites. In 1988, there were 1020 sampling sites with a total of 14 075 samples analysed.

The "Blue Flag" campaign was an initiative of the European Year of the Environment in 1987, and is promoted by the Commission of the EU and the European Foundation for Environmental Education. The objective is the protection of the coastal marine environment, including beaches. The awarding of a Blue Flag to a port or a beach indicates a high water quality level, and the availability of essential equipment, education and information on environmental issues. In Spain, 567 beaches and 126 recreation ports have obtained a Blue Flag since 1987.

In 1975, the governments of 16 Mediterranean countries approved the Action Plan for the Mediterranean. At a second meeting in 1976, the governments agreed to take appropiate measures to "prevent, reduce and fight pollution . . . and protect and improve the marine environment". Several protocols have been added to that agreement in relation to pollution induced by waste discharges from ships, emergency situations due to discharges of hydrocarbons, terrestrial pollution, and in relation to specially protected zones, the exploitation of the continental shelf and the sea floor.

Several coastal areas have been declared as protected zones. In Spain there are 124 protected areas with an area of 583 041 ha and extending 1373 km along the coast.

South of France and Corsica

The fight against forest fires must call on the responsibility of the citizens: information about the causes of fires is easy to elaborate and to communicate. The cost of the fight against fire is so high that it should be less expensive to pay shepherds to move "cleaning" flocks through areas at risk than to renew indefinitely a fleet of specialized helicopters and planes.

With reference to pollution, it seems fair that those who pollute should be those who pay. But this would cause increased production costs which could result in refusal to take on additional labour, or to pay employees adequately, in a period of high unemployment.

With regard to floods, the main problem is the thousands of houses which have been built on the floodplains of Mediterranean rivers over the past few decades. Trading and industrial zones, as well as camping grounds and caravan parks, have also been built in such locations. In Vaison-la-Romaine, following the flood of 22 September 1992, the town council decided to destroy all the remaining houses on the floodplain and to forbid reconstruction. But this decision was taken after the catastrophe and not before. On the other hand, the economic cost of this type of decision is so large that it is very improbable that similar decisions will be made elsewhere. Basic information must be given to the actual and future inhabitants about the dangers of living in a house built on the floodplain of a Mediterranean river.

Agency and political solutions comprise two parts: regulations and knowledge.

With reference to the first part, the problem is to apply and extend existing regulations: it is forbidden to build a house in the floodplain of a river; the owner of a Mediterranean forest must clean it; it is forbidden to walk in the Mediterranean forests during most of the summer; the European pollution norms must be applied. This objective can be achieved only by a long process which must combine information (and education) and repression.

Referring to the second part, two main orientations seem to be necessary. The first is to apply systems analysis to knowledge of the varied effects of human actions on natural systems when formulating specialist programmes, from the technical level upwards. The second is to recognize the particular characteristics of the Mediterranean geosystems, so different from temperate ones, when formulating such programmes. In particular, one can show the Mediterranean climate characteristics and their possible consequences on floods (slow in the temperate zone, brutal in the Mediterranean domain), on forest fires (recurrent and dangerous in the Mediterranean domain, rare and only occasionally dangerous in the temperate zone) and on soil erosion.

The Croatian Adriatic Coast

According to the Law on Forests (adopted in 1990), 0.07% of gross national product is being assigned to the forest protection fund (part of which is used for karst afforestation). There is a list of prohibited activities in a forest, such as wood stealing and setting an open fire. These prohibitions have not been effective because hyper-inflation has decreased the value of the fines, while legal proceedings are very slow. New forest regulations are being issued.

The Republic of Croatia has been independent since 1990. Many Acts and Regulations have been or are being issued. In 1991, the Ministry of Civil Engineering and Environmental Protection was established, and in 1994 it was separated into the Ministry of Physical Planning, Building and Housing, and the State Directorate for Environment. The Ministry of Agriculture and Forestry is included in the Environment Directorate.

Active non-governmental organizations in Croatia are Green Action Zagreb, Lijepa Nasa, and professional societies such as the Croatian Ecological Society, Croatian Biological Society and the Croatian Forestry Society. It is not possible to evaluate the extent of their influence.

The Law on Environmental Protection came into force in 1994. It serves as a basis from which a series of regulations are to be derived, such as the Book of Rules on Monitoring the Global Environment. It is based on an integrated approach of monitoring the whole ecosystem, not separating biological, soil, air and water resources. Monitoring will be based on legally prescribed methods. Accurate and precise data on the state of the environment will be available after a proper inventory of the natural resources and environmental processes has been carried out.

Greece

An effective measure which might combat further land degradation is the development of alternative, integrated, environmentally sound and socially acceptable land use schemes. Such schemes should be implemented by zoning in each threatened area (Yassoglou 1989). Slope gradient and elevation should be the main criteria for the delineation of each management zone. Soil maps, bioclimatic maps, erosion risk maps and erosion control modelling, accompanied by further research, are required for the implementation of zoning. Social and educational work is also required for the local population to accept the consequences of zoning.

Agricultural production has become more and more concentrated in the most promising lands. Old agricultural systems based on human and animal power and serving local or regional needs could not survive in areas with high wages and increasing expectations. They were therefore replaced by more capital- and energy-intensive systems depending on national and international markets.

Generally speaking, Greek agriculture is affected by European Union policies. According to the Commission's Regulation 797/85, Article 19, member states are explicitly permitted to introduce their own national aid schemes to support farming practices which preserve or improve the environment (Commission of the European Communities 1985). An annual payment per hectare may be provided to farmers in sensitive areas who agree to adhere to appropriate farming methods for protecting the environment. Such practices might involve restricting the intensity of livestock farming, limiting the use of water for irrigation, or taking measures to reduce soil erosion, for example by the conversion of arable land to wildlife reserve.

Regulation 1760/87 of the European Commission encouraged the member states to define areas of sympathetic agricultural practices. They were able to claim a quarter of the cost of such schemes, up to a limit of 100 ECU (approximately $50) per hectare. The Commission allocated 319 million ECUs (about $160 million) to the EU to protect soils from erosion, for biotope management and selective reforestation. In the decade 1982–1992, the strategy was modified progressively to allow greater support for conservation of the landscape and its component parts.

In Greece, the Ministry of Agriculture's Directorate of Environmental Protection is responsible for supporting practices which preserve or improve the environment. For example, a programme was organized in 1995 by the Ministry of Agriculture, supported by the EU, to reduce the amount of nitrogenous fertilizers applied to cotton by reimbursing farmers for lost production. An amount of 2.3 million ECUs (about $1.15 million) was distributed to the farmers participating in the programme in the agricultural areas of Thessaly, where farms covering an area of 10 000 ha were required to reduce the amount of fertilizer applied to cotton fields to 100 kg of nitrogen per hectare. The Institute of Soil Survey and Classification in Larissa is responsible for executing and monitoring the programme, by measuring the amount of inorganic nitrogen in soils from randomly selected fields during the entire growing period of the crop.

According to Regulation 2078/92 of the European Union, the Greek Ministry of Agriculture is required to take measures for applying erosion control cultivation measures in abandoned areas. A new programme was initiated in 1996 to protect 25 000 ha of abandoned land over 20 years by planting trees and repairing terraces. Additionally, areas of national parks considered important for wildlife protection, such as certain wetlands, are incorporated in a programme of environmental protection by reimbursing adjacent farmers for reducing applications of fertilizers and pesticides.

The Maastricht Treaty (1992) recognized that the EU must promote measures at an international level to deal with environmental problems and ensure "sustainable growth respecting the environment" (Corrie 1991; Baldock and Beaufoy 1992a,b). It is planned for the coming years that 15% of the land cultivated with crops such as cereals and sugar beets is to be put in fallow (the farmers have to exclude 15% of their land from any cultivation). Extending the fallow land could mean an improvement in soil fertility.

The unemployment problem which is now faced by the EU has environmental and land use implications. It has been stated that the Community intends to maintain rural populations in environmentally sensitive areas so as to ensure their proper management.

Whilst local agricultural agencies play an important role in agricultural issues and environmental protection, there is a continuous effort to modify regulations in accordance with EU policies. Additionally, there are numerous initiatives at the local scale for environmental protection, such as the construction of dams or small reservoirs along waterways in hilly areas for irrigation as well as to protect the land from erosion, to control flooding in low-lying areas, and to conserve water by increasing infiltration to groundwater aquifers. Trees are planted in abandoned areas at a small scale, mainly for recreation, and by the Forestry Department after forest fires.

The Eastern Mediterranean

Rules and laws protecting the natural resources of water, soil and forests have existed since the earliest periods of history. In the famous Code of Hammurabi, King of Babylon from 1728 to 1686 BC, there are laws concerning water management and the proper use of water in order not to damage neighbouring fields; and Adrian, ruler of Rome from AD 117 to 138, decreed rules restricting the exploitation of forests in Lebanon, Syria and Palestine.

The need for environmental impact assessments to be undertaken prior to development projects was accepted by the early 1970s in many countries of the world, following the initiative of the US Congress which passed the National Environment Policy Act in 1969. In Israel, similar legislation was passed in the early 1970s but the Ministry of Environment was created only in 1988. There is a broad consensus about ecological problems and many laws cover different aspects of economic life. For an example, legislation has stopped the quarrying of coastal sand, which was accelerating erosion of the coastline. However, the implementation of laws and their economic costs are a permanent cause of conflict. For instance, in 1991 the High Commissioner for Water in Israel reduced water allocations for agricultural purposes by raising the water price by 30%. The measure found strong opposition from the powerful agricultural lobby. Following the change of government in 1992, the High Commissioner, Professor Dan Zaslavsky, an eminent water hydraulics scientist, was dismissed.

In Cyprus, a joint soil and water conservation project was carried out by the government with the assistance of the UN during 1968–1979. The plan consisted of soil conservation measures such as terracing and the construction of stone walls and earth banks, and reforestation. About 20% of the area of Cyprus is government forest land, and measures have been taken to protect and reforest the areas (Table 22.2). Tree planting as part of the World Food Programme project increased fivefold from previous years. Water projects were also developed, involving the construction of dams and distribution systems, drilling to underground water, and replacing traditional wasteful water systems with drip systems. Government personnel, with the assistance of the UN, were crucial in the project's development (Savvides 1986).

North Africa

Techniques of soil protection and water management cannot be set up and improved unless the following conditions are fulfilled. The loss of rural areas, especially in the mountainous regions, must be reversed to allow the commercialization of production, taking advantage of the rapid and inexpensive communications with urban centres. Production and yields must be improved by the use of selected seeds and fertilizers, and the development of specialized and high-value products. All these developments require farmer education and training. At the present time, farmers increasingly prefer leasehold cultivations, growing fruit trees such as the almond, fig, walnut, apricot, peach and olive – species which are grown everywhere. Seedlings are distributed in all areas, with the apple, quince and almond taking precedence. The potato is also making progress in the valleys but problems of lack of transport, processing and storage facilities, and price instability, currently hinder beneficial developments. Finally, it is necessary to choose the most appropriate and suitable techniques of water and soil conservation in relation to the natural environment and the socio-economic conditions of the people. Methods for managing soils and water should draw inspiration from traditional techniques, but should be improved and, above all, kept simple (saving time and maintenance efforts).

Table 22.2 *Measures to protect and reforest government lands in Cyprus* (source: Savvides 1986)

Type of work	Works completed (m^2)	Man-days worked
Dry stone walls	2 499 655	1 662 553
Masonry and concrete structures	351 548	1 134 986
Bench terracing by machinery	95 710 368	987 975
Bench terracing by hand	505 726	36 725
Levelling	30 926 454	126 845
Earth banks (linear m)	924 695	60 679
Tree planting	1 032 122	749
Tree planting and maintenance	88 390 431	642 047
Ripping and rock raking, clearance and dry cultivation, uprooting and deep cultivation	98 531 134	637 079
Roads	628 369 m	137 516
Total		5 427 154

The organization of user groups

Popularizing techniques for protecting the land, and the participation of the population in protective land management, have been fundamental goals of integrated operations in Morocco. However, there are two contradictory refusals and rejections. Rural dwellers reject proposals from the state due to their lack of confidence. On the other hand, the authorities reject all that is traditional. Yet, development must be consensual.

In the Atlas of Azilal, the forsaking of customary regulations largely accounts for the degradation of forests and the altitudinal meadows (Herzenni 1993). Preservation and better productivity could be ensured if the system of management was carried out at organizational levels of smaller size. User groups would take shape at the village scale but also according to the variety of ecosystems, and would be the device for discussion with the technical services.

Organizing the beneficiaries of the group pastures, in order to involve them still further in the development of land management programmes, their implementation and follow-up processes, is a goal which has still not been achieved. Such organizations should also cater for the people who use the forests through the pastoral groups for regular exploitation and supplementary incomes.

Judicial regulations and their effects

As early as 1838, the first Act which forbade the burning of forests in North Africa was promulgated. At that moment the organization of the forest service began, and the Forest Code was issued in 1874. Fines were levied, as a result of which the rate of degradation decreased and the wooded area increased.

In Morocco, the forest service was created in 1913 and the legal Act dates back to 1917.

The approach

There are fundamental contradictions between clauses which were inserted to aid the struggle against erosion, and those which had, as their objective, the improvement of production. Indeed, attempts to make farmers adhere to the DRS (Defense et Restauration des Sols) or the CES (conservation of water and soils) failed. Yet compensation offered to farmers in return for works on their lands, or taking land out of production, are very costly for the state's budget – all the more so because those works are financed by loans in strong foreign currencies. Repayment costs, then, are very high for a precarious profitability. Approaches which aimed at improving fertility and water management, hence production, were more rare. The effective participation of the farmers was never really guaranteed.

In one specific instance, in order to reduce the use of wood, gas cooking rings were distributed at the beginning of 1975. However, the programme was later abandoned because of a lack of funds.

The present approach is one of integrated arrangement. It aims both at the downstream objective of reducing sedimentation in order to guarantee the water supply and the upstream objective of increasing production and improving incomes, as well as the social context, with the intention of setting a limit to rural depopulation.

The management projects

The DERRO project was launched in the Riffian provinces as early as the 1960s for a period of 20 years, and became operational in 1966. Planning was carried out at the level of the Riffian chain, which is divided into subregions which group two to five rural communes and are called Unities of Regional Development. These in turn are further subdivided into Perimeters of Intervention spread over 2000 to 6000 ha. The programme of interventions was based on sociological investigations and a ranking of the lands:

- ploughable lands on slopes less than 25%;
- non-ploughable agricultural lands which serve to produce fodder, thanks to the conservation of strips of natural vegetation on slopes ranging between 25 and 50%; and
- non-agricultural lands on steeper slopes.

Two principles underlie the interventions: on the one hand, firmness, and on the other, the constant concern of convincing the farmers and settling real contracts with them. The proposed techniques were very varied and were based on a general orientation: to transform the agricultural lands and the often intensively exploited Riffian runs into highly productive orchards and intensive fodder production zones. For such a policy of modernization to succeed, a programme of popularization and training of cadres had to be established.

The integrated projects

The watersheds of Loukkos, Nekkor and Tleta in Morocco were decreed Perimeters of National Interest, which justified any intervention thought necessary, regardless of the land's ownership. The Loukkos has, besides, benefited from funding thanks to a loan from the World Bank which commenced in 1981.

As far as the organization of the Loukkos scheme is concerned, nationally the programme was under the direction of the Forestry Administration, whereas management was the responsibility of a committee comprising various representatives from the Ministries of Agriculture, Public Works, Interior and National Education. Regionally, the Provincial Commission of Agriculture, assisted by a committee of co-ordination trained by the provincial authority and the people's representatives, was responsible for managing the scheme.

Conclusion

It is necessary to distinguish the following:

(a) the big investments of protection carried out by the states, the effects of which remain partly precarious but which aim for a global benefit for the whole community – this implies choices, sometimes draconian ones, but carried out with the knowledge of the different economic and social components; and
(b) rural development actions, with directly measurable gains, which must be within the competence of the local initiative, but which imply a real support of the population. With the aim of obtaining the best investment benefits, those actions must be selected from amongst a range of competing choices.

The institutions must be integrated:

– *nationally*, a National Arrangement Plan must be adopted by an inter-ministerial institution responsible for setting priorities;
– *regionally*, a scheme for an integrated arrangement must incorporate the struggle against the degradation processes, rural development, and precise options in the domain of infrastructures; and
– *locally*, a scheme council must regroup the involved population (elected people, groups of agricultural producers, stock breeders and natural resources users), local authorities, and the technicians responsible for supporting those groups in order to properly define the action projects.

Multiple schemes have been launched to initiate the development of marginal regions but there are few successful cases. Reasons include the schemes themselves (their duration and problems with finance), the techniques used – namely the fact of having ignored traditional techniques, and having over-used maladjusted techniques in relation to the physical and social context – and failure to ensure the effective participation of the people. Other, structural reasons include the judicial status of the lands and property rights. The population continued to rely on the practice of burning *matorral* in order to develop extra lands. In addition, farmers had the right to collect wood and to graze their stock in the forests.

Forest legislation considered timber as belonging to the state and submitted the forests to a regime of management and supervision by the state. The tribes would keep a regulated right of use, with clearing and wood cutting being outlawed. Immediate effects included rapid clearing to assert claims over property and a real over-exploitation while waiting for the delimitation of the forest domain. The lack of means put at the forest services' disposal hindered the carrying out of a balanced and sustained protection policy of the resources. The fines soon proved to be ineffective in the face of the various forms of wood gathering, strongly influenced by the "free" nature of the resource (resulting in excessive wood exploitation for the urban centres, the ovens and the *hammams* (traditional baths), in particular). For the farmers, the problem of land ownership will remain of prime importance as long as the forest domain has not been entirely delimited and as long as property security, with reference to the marginal lands illegally acquired by clearing, has not been resolved.

Integrated interventions have always had as proclaimed aims the protection and development of the resources of national interest, such as water, on the one hand, and development of the living standard of the populations as well as maintaining them on the land, on the other. In fact, only technical aspects were implemented without initiating any real mountain policy.

In order to succeed, the arrangements require improved research and a better co-ordination with the population. In that respect, all of a series of conditions must be fulfilled:

1. intensifying agricultural production in the mountains and improving profitability in order to interest the farmers in the arrangements. Profitability implies reducing the competition of the necessarily less costly plains products. To comply with such a demand, special quality products and producers' organizations are required;
2. creating new activities (tourism/handcrafts) and research for new resources;
3. taking into account traditional strategies in choosing methods of intervention;
4. improving the organization and management of schemes by avoiding a top-heavy administrative apparatus and by creating local cells capable of intervening rapidly and taking appropriate decisions;
5. searching for new relationships with the population, by setting up real contracts making the farmers responsible, and by identifying exactly the share they are entitled to in every arrangement effort; and
6. providing national financing for the various actions, with a capability of quickly supplying necessary funds and releasing emergency funds to comply with particular situations, where necessary.

23
Economic, social, agency and policy solutions

2. The New World

Compiled by Arthur Conacher and Maria Sala from contributions on the following regions:

California	**Antony** and **Amalie Jo Orme**
Chile	**Consuelo Castro** and **Mauricio Calderon**
Southwestern Cape	**Mike Meadows**
Southern Australia	**Arthur** and **Jeanette Conacher**

California

If one sought to define events which formed a watershed between widespread exploitation and a more responsible use of the land and its resources, these would probably be the passage of the National Environmental Policy Act in 1969 and, for California, the California Environmental Quality Act of 1970. Despite many attempts to curb excesses prior to 1969/70, notably through mining legislation and soil conservation, 19th century attitudes towards resource use persisted well into the mid-20th century, leading in the 1960s to a growing clamour for reform. What the above Acts did was to impose an accountability in dealing with the environment, codified through a requirement for environmental impact reports prior to development. These Acts did not entirely resolve the problems of environmental degradation, for they created a subindustry of report writing that often generated more heat than light on environmental problems, and relegated many contentious issues to politicians and lawyers who were poorly qualified to make wise decisions. Nevertheless, these Acts changed the ground rules by which the human use of the land was henceforth to be evaluated.

The following discussion examines the solutions to land degradation more or less in chronological order with respect to the problems they addressed. Mining legislation came first, followed by the creation of national parks and the national forest system, by initiatives dealing with over-grazing and soil conservation, and then more recently by legislation dealing with urban planning and coastal zone management. Legislative solutions, proposals from non-government groups, and private initiatives do of course continue to address each of these areas, but it is argued that mining, at least onshore, is far less of a problem today than continuing environmental threats posed by urban growth.

Solutions to problems caused by mining

The formation by the US Congress in 1893 of the California Debris Commission to regulate the hydraulic mining industry (refer Chapter 17), was important in California as the first co-ordinated attempt by one group of interests (farmers), supported by government, to combat degradation caused by other interests (miners). Although the General Mining Law of 1872 still permits prospectors to stake claims and obtain patents on valuable minerals on federal lands, in reality access to large areas is now denied by government action.

The problematic development of offshore oil resources, through the Outer Continental Shelf Oil and Gas Leasing Program administered by the Department of the Interior, acknowledges conservation needs through various restrictions, bans and lease-pricing policies.

Solutions to problems of timber harvesting

Federal involvement with forest lands began with the Forest Reserve Act of 1891, which authorized the General Land Office to establish forest reserves, and the Forest Management Act of 1897 which placed general restrictions on timber harvesting. In 1905, management of the national forests was transferred to the Department of Agriculture's new Forest Service headed by Gifford Pinchot, a leading advocate of conservation and the multiple use of forests. Today, the Forest Service provides guidelines which restrict or prohibit clear-cutting, such that the landscape becomes a mosaic of patchcuts of various ages, but these guidelines apply only to federal lands.

Privately owned forests have been subject to much less control. Certainly, there are constraints applied indirectly via federal water quality regulations, but most control over forest practices has fallen to the state, which often finds itself in the middle of a continuing battle between preservationists on the one hand and forest landowners on the other. Most forest owners are indeed concerned with the long-term vitality of their resources, but some are less so. California created the nation's first State Board of Forestry in 1885, but the battle lines for future conflict were soon confirmed when the California Forest Protective Association was organized by forest

Land Degradation in Mediterranean Environments of the World: Nature and Extent, Causes and Solutions.
Edited by A.J. Conacher and M. Sala.

industries in 1909 and the Save-The-Redwoods League was formed in 1918. In 1945, passage of California's Forest Practice Act led to reorganization of the Board of Forestry but, during the following two decades of widespread clear-cutting, legislative intent was considered unsatisfactory by many.

In response to the continuing clamour for conservation, and following massive floods in 1964 and 1972 which caused widespread damage within and downstream from northern California's logging areas, a revised Forest Practice Act was passed in 1973. This Act imposed requirements for state review and prior approval of timber harvesting plans involving evaluation not only of sylvicultural systems, road and landing designs and yarding methods, but of likely impacts on erosion hazards, water quality, fish and game, and other resources. Over the past two decades, the controversy has subsided somewhat with implementation of the 1973 Act, but the conservationist lobby remains vigilant, ever ready to seek further restraints on the logging industry.

Solutions for soil conservation

Prior to the 1930s, there was little attempt to address problems of soil deterioration and erosion caused by excessive grazing and unwise cropping. Although Gifford Pinchot, together with President Theodore Roosevelt and Interior Secretary James Garfield, encouraged a leasing system for federal grazing lands outside the national forest, the era of "free-range" did not end until the Taylor Grazing Act of 1934. Under this Act, grazing districts were established on lands considered most suitable for grazing. The Act's broader purposes were to restrict over-grazing and associated soil degradation, to end large-scale disposal of federal lands, and to increase the federal government's role as steward of the land (Mason and Mattson 1990).

Continuing concern for over-grazing and ecological damage on federal rangelands has led more recently to passage of the Federal Land Policy and Management Act of 1976 (the Organic Act), which requires multiple use and sustained yield management of federal grazing areas, and the Public Rangelands Improvement Act of 1978 which provides incentives to ranchers for range improvement. By the 1980s, rangeland quality had improved somewhat, notably in California where 48% of federal rangeland was considered good to excellent in 1987, that is, with over 50% of potential natural vegetation (BLM 1987).

Growing concern for cropland soils led the Department of Agriculture to establish the Soil Erosion Service in 1933 which, in response to the Dust Bowl disaster on the Great Plains, was renamed the Soil Conservation Service in 1935. Over the next 30 years, 3000 soil conservation districts were established across the United States, bringing farmers advice on soil capabilities and conservation techniques. Several districts were established in California, notably in coastal and foothill areas with significant soil erosion. This service has recently been renamed the Natural Resources Conservation Service.

Passage of the 1985 Food Security Act marked a new beginning in agricultural conservation (Mason and Mattson 1990). The Act seeks: (1) to remove highly erodible lands from cultivation through a Conservation Reserve Program which pays farmers rents and subsidies; (2) to promote better management of the remaining croplands through a Conservation Compliance Policy involving specific conservation plans, failure to implement making farmers ineligible for federal loans, insurance and price supports; and (3) to stem the conversion of wetlands and highly erodible lands to cropland. By mid-1988, between 10% and 24% of eligible land in California had contracted into the Conservation Reserve Program.

Solutions to problems caused by irrigation and water use

Because water is so important to life and livelihood in Greater California, any significant deterioration in its quality is cause for concern. The region grew by manipulating its surface and groundwater resources, especially by massive water transfers into southern California, and its future must be bound intimately with the continued use of these waters and, perhaps, with seawater desalinization. Some 64% of the problems of water pollution in California's rivers are attributable to non-point sources, primarily agricultural pesticides and nitrates in the Central Valley, a further 16% to municipal sources including feedlots, especially around San Francisco Bay, with much of the remaining 20% coming from industrial activity leading to heavy metal concentrations also around the bay (EPA 1989).

These problems of water quality have been addressed primarily by a series of federal Water Pollution Control Acts in 1948, 1956, 1966, 1972, 1977 and 1987, and by the Safe Drinking Water Act of 1974 and subsequent amendments. The earlier legislation sought to enforce water quality standards for inter-state, navigable and coastal waters where pollution threatened human health or welfare, and also gave grants for municipal sewage and wastewater treatment. The 1972 amendments, the Clean Water Act, decreed that all waters should be "fishable and swimmable" by 1983 and zero discharge of pollutants should be achieved by 1985, objectives which have been approached but not yet wholly achieved. The 1977 and 1987 amendments expanded regulation of toxic substances, directed more emphasis to the states, and replaced the technology-based approach of the 1972 Act with a more pragmatic water quality approach ("do what is needed"). Amendments in 1986 and 1995 to the Safe Drinking Water Act of 1974 set standards for such contaminants as arsenic and radon. On a related issue, the Federal Insecticide, Fungicide and Rodenticide Act (FIFRA) of 1972 banned the use of the pesticide DDT but, together with dieldrin and chlordane, much remains locked in stream and lake sediments and much long-term damage had already been done, notably through ingestion by the California condor.

Solutions for solid waste

Toxic or hazardous wastes, differing from normal household waste only by degree, pose problems which have been addressed since the 1970s by legislation designed to regulate the production, use and disposal of toxic chemicals. In 1987, California ranked first among the United States in terms of the Toxics Release Inventory (EPA 1989). Some 1662 facilities produced $2650 \cdot 10^9$ g of toxic emissions in that year, 65.7% to surface waters, 26.2% to underground injection wells, 4.2% to public sewers, 1.4% to the air, and 2.5% to on-site landfills and off-site facilities. Reflecting the Comprehensive Environmental Response, Compensation and Liability Act of 1980 and the Superfund Amendments and Reauthorization Act of 1986, "superfund sites" throughout the state, but especially in and around urban centres, have been designated for clean-up by federal and state agencies and by polluters. Continued funding of these clean-up efforts is a serious political issue.

Solutions to coastal degradation

As early as 1931, a legislative report had recommended the formation of a single state agency to control coastal activity, but nothing was done. Then, within two decades, all that changed. First, in 1960, a Save-the-Bay Association was organized by citizens opposed to further infilling of San Francisco Bay, leading in 1965 to the San Francisco Bay Conservation and Development Commission. This became a model for initiatives at the federal level, leading via the Marine Resources and Engineering Development Act of 1966 and the Estuary Protection Act of 1968, to the Coastal Zone Management Act of 1972. Such is the nature of the US political structure, however, that this Act provided a framework designed to assist states rather than a mandate for coastal management. The onus thus remained with the states.

Meanwhile, building on the work of San Francisco Bay's citizenry, California appointed an Advisory Commission on Ocean Resources in 1964, and in 1968 produced a Comprehensive Ocean Area Plan. In the late 1960s, the California Assembly sought three times to create a statewide authority over the coast, but the bill was killed each time in the Senate where more conservative interests prevailed. Then, in one of the passive citizen revolts for which California is noteworthy, an alliance of concerned citizens, frustrated by legislative inaction, obtained sufficient signatures to place Proposition 20, the "Coastal Initiative" on the ballot for the November 1972 General Election. Proposition 20 passed by a margin of 55% to 45% of the voters, in part because inland support came out to block coastal development interests. Proposition 20 created a California Coastal Commission with six regional commissions whose goal was to develop a coastal plan within four years and meantime to regulate development within 914 m (1000 yards) of mean high water. A feature of California's constitution is that a law passed by voter initiative cannot be changed except by another initiative.

In 1976, the required plan was accepted by the legislature and became the California Coastal Act. This provided for orderly conservation and development of a coastal zone extending 5 km seaward and 8 km landward, excluding San Francisco Bay which had already been addressed. The Act, which required local authorities (15 counties and 52 cities with coastal lands) to generate appropriate coastal plans, was, however, a legislative enactment and not an initiative, and thus could be changed by political action – and was. By 1980, popular activism had begun to wane even as development interests and their legislative bedmates were reorganizing. Since 1980, as regional commission offices have closed and federal and state support for coastal planning has declined, so coastal management has once again become a political football, subject to legislative whim. The California Coastal Commission struggles on, achieving some successes such as the provision of improved coastal access, but frustrated by lengthy public hearings, legal challenges and declining citizen involvement. The coast has again become a battleground on which private property rights and development interests are ranged against the supposed public good, but at least coastal development is now more accountable.

Protection of wildlife

As discussed in Chapter 18, California has lost many species over the past 150 years, some forever, while some are present in neighbouring states. Other species have seen their habitat contract and have become endangered in one way or another by introduced land uses. At present, wildlife is protected by a variety of federal and state legislation, including the California Wildlife Protection Act of 1990 that resulted from Proposition 117. However, the threat continues and there are frequent attempts to limit or abandon such protection. For example, in response to an alleged threat to suburban communities by the state's 600 or so remaining mountain lions (*Felis concolor*), a thinly veiled attempt to legalize their hunting was placed on the March 1996 ballot in California as Proposition 197, designed to repeal the protected status of mountain lions and direct wildlife authorities to manage them according to a code which lists hunting as a management tool. The proposition was heavily defeated but it did emphasize the need for continuing vigilance in the protection of wildlife.

Other animals once close to extinction have also been favoured by protected status or relocation. These include the tule elk (*Cervus nannodes*), reduced from more than 500 000 to only 28 individuals during the 19th century but now relocated and numbering around 3000. Along the coast, the elephant seal (*Mirounga angustirostris*) once ranged from Alaska to Mexico but, hunted for its oil, had been reduced to a few score by 1892, including nine individuals on Guadelupe Island off Baja California Norte, seven of which were removed as specimens! However, protection was afforded by Mexico in 1911 and later by the United States, and there are now more than 50 000 along the coast south of San Francisco (Kreismann 1991). Similarly hunted for its fur, the California sea otter (*Enhydra lutris nereis*) was almost extinct when afforded protection in 1938, but its numbers have since expanded hesitantly to around 2000. The giant California condor (*Gymnogyps californianus*) had also been reduced to near extinction by hunting and poisonous pesticides when the remaining 27 birds were removed from the wild in the 1980s and nurtured in San Diego and Los Angeles zoos. A number of breeding pairs were returned to nature in the Transverse Ranges in the 1990s but their future remains uncertain.

The ultimate political solution: removing land from active development

The case for preservation, made by the Sierra Club founded by John Muir in 1892, was pursued reasonably quietly for many decades but, by the 1960s, had begun to conflict more forcefully with both commercial interests and the recreational needs of a rapidly growing urban populace. Thereafter the problem of land stewardship became a triangular contest, with development interests in one corner, preservationists – or more realistically, conservationists – in another corner, and the federal, state and local governments in the third corner, at times as umpires but increasingly as active participants seeking the middle ground but pleasing nobody.

The Wilderness Act of 1964 established preserves on federal lands where human impacts were to be kept to a minimum. Large areas of national forest land in California were designated wilderness to be managed by the Forest Service, particularly in the Sierra Nevada high country and the forests of the Coast and Transverse Ranges. The Endangered Species Acts of 1966 and 1973, the Wild and Scenic Rivers Act and the National Trails System Act of 1968, the Marine Mammal Protection Act and the Marine Protection, Research and Sanctuaries Act of 1972, and the National Parks and Recreation Act of 1978 were among a large body of federal legislation which sought to protect or at least conserve important natural resources. Additions to the national park system included the Golden Gate National Recreation Area, the Point Reyes National Seashore, and the Santa Monica Mountains National Recreation Area, all important for serving the recreational needs of nearby urban communities. In the high mountains, Kings Canyon National Park (1940) and Lassen Volcanic National Park had been added. Meanwhile, the state parks and recreation system was setting aside much land, especially along the coast. In Baja California, with but modest development pressures beyond Tijuana and the west coast highway, Constitucíon National Park was established in the Sierra Juarez and the Sierra San Pedro Martir National Park further south in the 1970s.

In addition, land trusts were being established by private non-profit organizations, mostly urban in origin, to acquire by donation or purchase large areas of land for reasons of ecological protection and passive recreation. The National Audubon Society, the Nature Conservancy and the National Parks and Conservation Association have been prominent in acquiring land trusts. These organizations, together with the Sierra Club and others, have been particularly successful in generating support among the large, often frustrated urban populations of California's major cities, combining youthful idealism with middle-class romanticism to fashion a now considerable environmental lobby in local, state and federal corridors of power. Well educated, financially comfortable white citizens are particularly well represented in these groups.

Inevitably, designation of wilderness and park lands led to massive and continuing conflict between the various parties. Most contentious of all was the extension of Redwood National Park in the northern Coast Ranges, because this cut directly to the heart of the timber harvesting industry. Around 1970, a spectrum of environmental concerns was ranged against the timber interests, combining traditionalists from the Sierra Club with more radical but highly effective elements from the Environmental Defense Fund, founded in 1967, the Natural Resources Defense Council, founded in 1970, and the confrontational Friends of the Earth. In some respects, the issues were simple. On the one hand, the timber industry did not wish to surrender valuable stands of old-growth redwood to an expanded national park. On the other hand, environmental interests supported the proposed park, prodding government representatives with barbs appropriate to their individual perspectives – some muted, some sharp. Both sides over-stated their hand.

Nevertheless, there have been some notable expressions of compromise. For example, the US Bureau of Land Management has joined with other agencies and oil companies to create the Carrizo Plain Natural Area, a 75 000 ha reserve southwest of the San Joaquin Valley where tule elk (*Cervus nannodes*) and pronghorn antelope (*Antilocapra americana*) have been reintroduced and the endangered San Joaquin kit fox (*Vulpes macrotis*) is now protected. The authority which supplies drinking water from the Sierra Nevada to 1.2 million people around eastern San Francisco Bay recently began a voluntary programme to improve aquatic life in the Lower Mokelumne River. The programme has significantly raised Chinook salmon production from local hatcheries. Recently, the Kern County Water Agency returned 300 ha in the southern San Joaquin Valley from agricultural use to wetland, leading to a return of waterfowl and shorebirds, even the rare American white pelican (*Pelecanus erythrorhynchos*). In Riverside County, southern

California, the Metropolitan Water District has entered into public/private partnerships to finance and manage a 4000 ha reserve near Lake Skinner, a 2000 ha reserve at Lake Matthews, and 3000 ha of oak woodland and native grassland on the Santa Rosa Plateau near Murrieta.

Chile

Instruments for the encouragement of conservation practices

At present Chile does not have a national environmental policy, although there are numerous laws which directly or indirectly impinge on environmental problems. The encouragement of soil conservation is stated in Organic Law No. 19283 of the Agricultural and Stock Breeding Service (SAG). The Law gives this body the responsibility for regulating and administering the provision of incentives for incorporating conservation practices in the land uses of Chile. In parallel, Act No. 701 provides incentives for reforestation by providing loans when a management plan of the land is prepared by the owners. Having a management plan gives the owner the right to receive an allowance equivalent to 75% of the net costs of the project, to be exempt from territorial and other taxes.

Law No. 18450 has the objectives of increasing the irrigated area of the country, of reducing water distribution, and of draining waterlogged agricultural soils. The Law promotes private investment in irrigation and drainage works by making available an allowance which can reach 75% of the construction and repair costs of these works, and of the investment in equipment for mechanical irrigation. However, the Law does not promote standards at research and training levels, nor rational and sustainable agricultural and pastoral land use.

Education

Education at the various levels has clear deficiencies in soil conservation topics. Although programmes at basic and middle education levels are relatively well structured, training of the teachers who have to implement these programmes does not incorporate the subject adequately. It can be concluded that the trainers are unable to provide the necessary ideas to the students. None of the Chilean universities offers a postgraduate programme in soils or soil conservation.

Development strategies

Under the current dry farming circumstances, the development of the area, in terms of conservation of natural resources, tries to focus on the problems of small farmers from two simultaneous angles. It is based on the premises that the farmers operate in a fragile environment with a continuous decrease of productivity due to soil degradation, and that low income levels dictate that the main objective is to ensure basic subsistence needs. Thus, land use is not always compatible with land capability, and overexploitation of natural resources makes the task greater than acceptable.

On the one hand, there are measures to alleviate poverty by providing complementary sources of income to agriculture, thus making possible an improved economic situation and with it the possibility of introducing an exploitation system compatible with the soil's capability. On the other hand, part of the actions are orientated towards improved management of natural resources, providing technological assistance adapted to the resources, which also leads to an increase in income, this time on a sustainable base.

Government institutional structures

Within the Ministry of Ariculture there is a Subsecretary of Agriculture which provides technical, judicial and administrative support to the Minister. Attached to this Subsecretary are the following organizations which have a direct responsibility for aspects of land degradation.

CONAF (Corporación Nacional Forestal – National Forestry Corporation) has objectives orientated towards the conservation, protection, development, management and exploitation of the forestry resources of the country. The emphasis is on increasing productivity compatible with conservation of resources and environment. CONAF also administers SNAPE (Sistema

Nacional de Areas Silvestres Protegidas del Estado – National System of Protected Areas of the State) manages the national parks and cultural monuments in the country, and is in charge of the protection of the flora and fauna.

CONAMA (Comisión Nacional del Medio Çambiente – National Commission for the Environment) is an inter-ministerial commission, created in 1990 with the objective of working on the investigation, proposal, analysis and evaluation of all matters related to the protection and conservation of the environment.

Non-government organisations (NGOs) dealing with soils

The organization CIAL (Comisión de Investigación en Agricultura Alternativa – Commission on Research for Alternative Agriculture) deals with soil problems through an agro-ecological approach based on the peasants and farmers who work in organic agriculture, dealing with the consequences of the indiscriminate use of synthetic fertilizers and pesticides.

CLADES (Consorcio Latinoamericano sobre Agroecología y Desarrollo – Latin American Consortium on Agro-ecology and Development) is an institutional axis to promote research, training and information on the necessary agro-ecological basis for the sustainable development of peasant agriculture in Latin America.

CET (Centro de Estudios y Tecnología – Centre of Studies and Technology) has developed a series of exhibitions with an emphasis on crop diversification, recycling of organic matter and soil conservation, amongst other environmental practices.

The Ecology and Sustainable Development Programme established by Centro El Canelo de Nos (Canelo de Nos Centre) has two complementary thematic areas.

(a) *Agro-ecology*. In the Metropolitan Region, El Canelo has demonstration experimental plots where experiments, investigations, training and diffusion activities on sustainable development take place. Based on agro-ecology, the production system is structured around crop rotations, the use of waste materials and biological pest control. It is aimed at the combination of these factors to control weeds, insects and diseases while maintaining soil fertility and productivity.

(b) *Alternative techniques*. The centre has compiled and developed a series of alternative technologies for the rural environment, such as alternative energy systems, animal-drawn farming implements, prefabricated houses and the design of rural buildings.

The Southwestern Cape

Soil erosion and land degradation have been recognized as an environmental problem in the southwestern Cape at least since the arrival of European settlers. A form of conservation policy was in place as early as 1652, just a few years after the initial permanent settlement of the Cape by the Dutch colonists, although the ensuing legislation was aimed mainly at controlling fires and protecting vegetation (Verster *et al.* 1992). There was, however, no formal legislation dealing with the problem of soil erosion *per se* until the mid-20th century and it is implicit that society as a whole did not deem conservation measures worthy of much investment.

Social solutions to the problems: reconstruction and development

Although it does not deal specifically with the issue of environmental degradation, the ANC's Reconstruction and Development Programme (RDP) has been formulated to address the fundamental socio-economic and environmental problems in South Africa (African National Congress 1993). By implication, it is designed to help solve significant inequalities in access to resources in such a way as to ensure the sustainability of those resources. It is a long-term, five to ten year action plan to address, *inter alia*, the problem of political and criminal violence, shortage of housing, shortage of employment opportunities, inadequate education and health care, and lack of democracy in state and private structures. Above all, the RDP aims to boost an ailing economy in such a way as to increase the access to wealth and prosperity for all South African citizens.

From the point of view of the South African environment in general, and of the southwestern Cape environment in particular, the programme may be regarded as having some serious shortcomings, for although the concept of "sustainability" is alluded to, it is clear that economic growth is the fundamental engine of the RDP and, since this automatically involves exploitation of natural resources, it is difficult to imagine how this can be achieved without at least some negative environmental consequences. The environmental consequences of the RDP have been reviewed by Schreiner (1995), who concludes that its environmental implications have been given insufficient attention and that, at the very least, environmental education should be a priority.

Progress with the RDP appears to have been very slow. However, there has been some success in the establishment of small community projects, and this has been especially marked in the hundreds of small communities which now have access to clean water as a result of RDP-funded schemes. Some aspects of the RDP do have a positive environmental component: for example, the alien vegetation eradication programme was recently announced by the Minister of Water Affairs and Forestry which will have the multiple advantages of removing exotic vegetation, increasing water supply to dams, promoting the conservation of the indigenous flora and creating several thousand jobs. More time is needed before the real successes of this socio-economically innovative programme can be judged.

Economic incentives and penalties

The provisions of the Conservation of Agricultural Resources Act of 1983 allow for punitive measures to be taken against landowners who do not comply with its strictures. This means that resource inspectors can impose fines on landowners who do not adequately protect their land from degradation, in particular erosion. Unfortunately, the number of prosecutions under this Act, and other related pieces of legislation (see below) have been few, mainly because of a shortage of personnel to enforce the measures. Verster *et al.* (1992) have noted that it is not so much the inadequacy of the law *per se* which is in question, rather the logistical difficulties involved in implementing it. Aside from fines for non-compliance, there are incentives in the form of subsidies to individual farmers for conservation works of various types. The subsidies are made available through the Department of Agriculture. They represent a tangible and substantial action on the part of the state to encourage conservation practice and, although the amount of funding fluctuates from year to year, they have been widely applied to conservation works such as the establishment of contours and the construction of waterways (Table 20.2). Annual subsidies for the various conservation measures have varied between US$320 000 and more than $1 million during the last few years (Department of Agriculture Annual Reports); these amounts are supplemented by approximately 150% for works conducted entirely under the finances of the landowners themselves.

Agency and legislative solutions

As noted above, undoubtedly the principal player in land degradation issues in the southwestern Cape is the state, through the Department of Agriculture and the Department of Water Affairs and Forestry. State involvement in dealing with the problem of land degradation in the region must be seen in the context of prevailing environmental awareness, or lack of it, in South Africa as a whole for much of its post-colonial history.

Soil conservation legislation in South Africa

In 1944, the South African Government invited Dr Hugh Bennett, Chief of the Soil Conservation Service of the United States Department of Agriculture, to visit the country. His subsequent critique (Bennett 1945) appears to have been instrumental in persuading the South African Government to introduce legislation. The Forest and Veld Conservation Act 13 of 1941, which merely gave the state powers to expropriate privately owned land that was considered in need of reclamation and had little impact on the promotion of soil conservation as such, was replaced by the Soil Conservation Act 54 of 1946 (Verster *et al.* 1992). The Act operated through the establishment of soil conservation district committees operated by local landowners and was a model of democratic and co-operative involvement. The committees put forward soil conservation schemes and, assuming their approval by the Minister, they were implemented in such a way that non-compliance by a farmer could result in punitive action.

An analysis of the situation by Verster *et al.* (1992) pointed to the Act's serious shortcomings, not least of which was the fact that the establishment and implementation of a conservation scheme rested on the initiative of the farmers themselves and that, in most cases, these landowners were unwilling to take action against non-complying colleagues. Thus, the progressively democratic nature of the 1946 Act resulted in its ineffectiveness: only 21 prosecutions were ever promulgated and, of these, only 14 were successful (Verster *et al.* 1992). An additional problem was that the Act was only applicable to "white"-owned land in South Africa; the tribal trustlands were specifically excluded from its provisions (Verster *et al.* 1992), although admittedly this had no implications for the south-western Cape as such (there being no "homelands" within the region).

The weaknesses of the 1946 Act were to some extent addressed by its successor, the Soil Conservation Act 76 of 1969 (Verster *et al.* 1992). In this instance, executive power to ensure that the Act was effective was vested in the Division of Soil Protection of the Department of Agriculture (rather than with the farmers themselves). This meant that the landowners could be directed to construct and maintain soil conservation works, and grants and subsidies were available to assist in such actions. This Act was also repealed, however, and replaced by the more wide-ranging Conservation of Agricultural Resources Act 43 of 1983. In essence this recognized the failure of previous initiatives to actually enforce a soil conservation ethos on the farming community. The Act operates through the aegis of a Conservation Advisory Board, which advises the Minister of Agriculture who has ultimate jurisdiction over soil conservation matters. The Board works through a hierarchy of Conservation Committees (regional and district) which are appointed by the Minister.

The so-called Winter Rainfall Region, as defined by the Department of Agriculture (Figure 23.1), is divided into five agricultural extension and soil conservation subregions, which are in turn subdivided into a number of districts. Each of these administrative units is supported by an infrastructure involving an Extension Officer, whose primary duties are to liaise with landowners in regard to cultivation and stocking principles and practice. Each of the subdivisions also has access to at least one Soil Conservation Control Technician, whose responsibility (and budget) is to oversee aspects of the constituent farms relating specifically to soil conservation matters. Over and above these personnel is the nationally administered Resource Inspectorate, which in essence "polices" the implementation of, in particular, the Conservation of Agricultural Resources Act and imposes the penalties provided for non-compliance.

Subsidies are available to assist with the construction of soil conservation works or the restoration of eroded, degraded or damaged land (Verster *et al.* 1992). Arguably, the most significant problems remain the lack of public consciousness in regard to the problem of land degradation in South Africa, and a shortage of personnel to ensure that the spirit of the Act is complied with.

Other environmental legislation

Environmental law in South Africa has been reviewed comprehensively in Fuggle and Rabie (1992), and it is apparent that relevant legislation is highly diffuse. There is a single Act which attempts to embrace conservation of the environment in a more general sense, namely The Environment Conservation Act 73 of 1989 which, according to Rabie (1992), strives for effective protection under a non-utilitarian, ecocentric perspective. This Act makes provision for the establishment of two statutory bodies, the advisory Council for the Environment and the Committee for Environmental Management, with powers to ensure that the Act is complied with. Essentially, the Act deals with controlling activities or developments which may have a detrimental effect on the environment, and limiting such activities or developments in certain designated areas.

In principle at least, the Act ensures that actions do not result in developments in areas

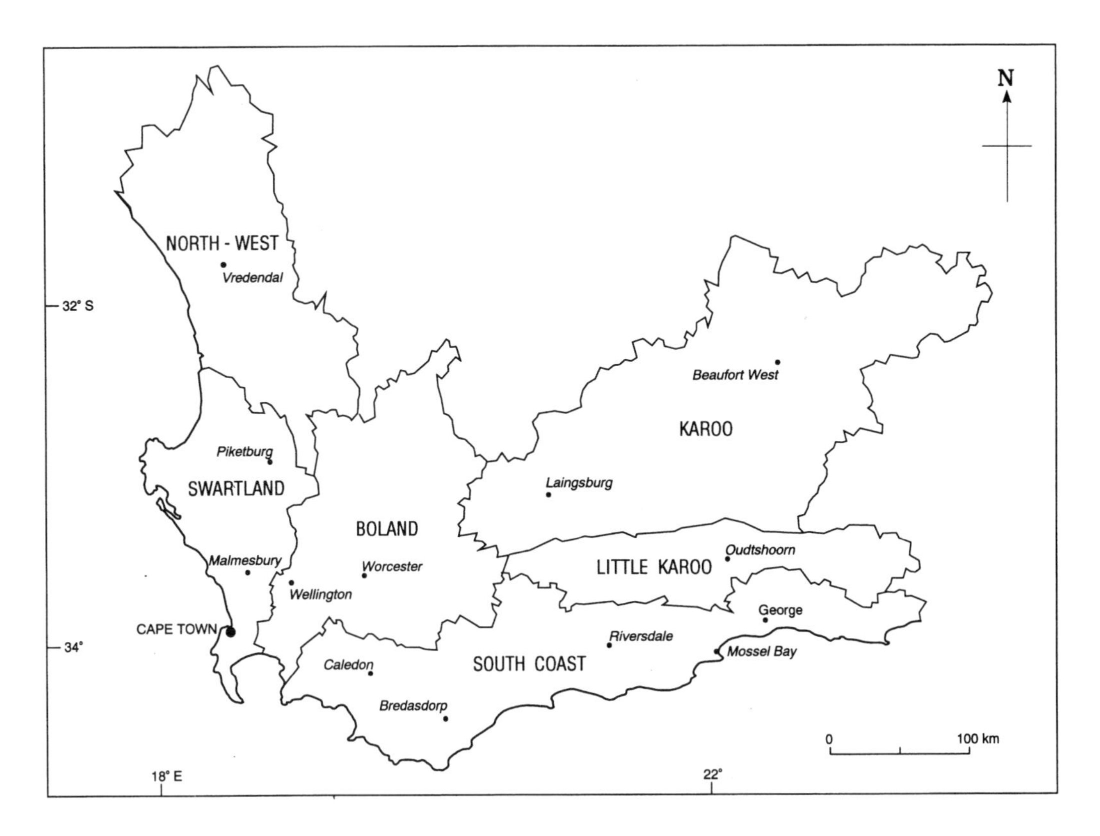

Figure 23.1 *Agricultural subregions and districts of the Winter Rainfall Region, South Africa* (source: Department of Agriculture)

regarded as important to the future survival of endangered plants or animals, and in so doing establishes several categories of "protected area" (Rabie 1992). The obvious difficulty of dealing with problems which arise out of the fact that numerous different administrative bodies have accountability for various elements of the environment, is partly dealt with by making the Minister of Environment Affairs responsible for the determination of an environmental policy. But as Rabie (1992) points out, this in itself does not avoid the difficulty of achieving the concurrence of several other ministers who also have "environmental" concerns, for example the Ministers of Agriculture and of Water Affairs and Forestry.

Several other significant pieces of environmental legislation exist on the South African statutes. Outside protected areas, for example on privately owned farmland, the task of ensuring the conservation of indigenous plant species is vested in various provincial conservation ordnances which, in the case of the southwestern Cape, means the Cape Nature and Environment Conservation Ordnance 19 of 1974. This ordnance protects some designated plant species by restricting picking to licence holders only, and it is forbidden to pick any indigenous species along road verges. The ordnance does not, however, protect indigenous plants in general, and a member of the public only requires the permission of the landowner concerned in order to pick non-designated plants on private land.

The issue of noxious weeds is dealt with in terms of the Agricultural Resources Act mentioned above. The Mountain Catchment Areas Act 63 of 1970 has provisions dealing specifically with the conservation of soil, vegetation and water as well as with afforestation, invader plants and fire. Other relevant pieces of legislation include the Forest Act 72 of 1968 and the Water Act 54 of 1956, both of which contain conservation-orientated provisions (Fuggle and Rabie 1992).

In summary, the legislative situation as it pertains to the environment in South Africa may be deemed barely adequate to the task of ensuring the proper protection of its environment. Not only is environmental law itself fragmented, but it has proved difficult to enforce. The law appears to be applied only in selective instances and it has proved ineffective because of under-staffing and under-resourcing in appropriate administrative areas. Moreover, South African environmental law is practised against a background of continued and widespread environmental ignorance and apathy.

NGO involvement

A wide range of non-governmental organizations (NGOs) is involved in the issue of land degradation, more particularly in the rural areas. A comprehensive listing is given in Ramphele and McDowell (1991). They range from the more "establishment" organizations such as the Wildlife Society of South Africa and the Botanical Society, to environmental pressure groups such as the Green Action Forum and Friends of the Earth, to more direct "hands-on" bodies such as Abalimi Bezekhaya. This is a Cape Town service organization encouraging township dwellers to grow their own food and protect the environment.

What is still needed?

It is worth reiterating the need for a change in the mind-set of South Africans as a whole who appear to have viewed the environment as an expendable resource. A concerted and nationally administered policy of more effective environmental education for all South African citizens must be considered a priority. Additionally, it is clear that the legislation, in itself being reasonably comprehensive (if fragmented), is ineffectively implemented and does not adequately protect the environment.

The crucial challenge remains for the southwestern Cape to maintain, indeed increase, its economic productivity while allowing improved access to land and prosperity historically unavailable to a substantial majority of the population. Elsewhere in South Africa, the policies of apartheid were instrumental in promoting widespread and possibly irreversible land transformation and degradation (Wilson 1991). The hope is that the new dispensation which now prevails does not repeat those mistakes in the southwestern Cape which, until now at least, has escaped the worst ravages of environmental degradation by supporting (admittedly on a highly selective basis) a policy of protecting the land.

Southern Australia

Economic solutions

Not surprisingly, there is some argument amongst economists concerning the costs and benefits of land repair (Kirby and Blyth 1987; Yapp *et al.* 1992; Chisholm 1992; Grivas *et al.* 1995). Some argue that investment in expensive and specific works for, say, serious gullying is not justified in relation to the benefit obtained. Others suggest that funding land repair, and more particularly degradation prevention, yields long-term benefits which justify the investment. Others again argue this is a favoured option purely in terms of the precautionary principle.

Some of the economic policies dealing with land degradation are discussed below.

Tax concessions

Tax incentives for conservation works are the usual means of encouraging better management. Under Sections 75B and D of the Income Tax Assessment Act 1936, farmers may claim a deduction for expenditure on fencing, weed control, tree planting, habitat protection and land rehabilitation. In reality, few farmers appear to make these claims (or they are unaware of the provisions) and in any event the allowances are insufficient to make an impact (Nelson and Mues 1993; Mues *et al.* 1994). Furthermore, the provisions do not benefit farmers below the tax threshold. Over 20% of Australian broadacre farmers fell into this category during the period of recession in the late 1980s and early 1990s.

Often these are the farms which are most in need of some conservation management as land runs down due to the lack of capital investment or is placed under increased production pressure to generate further income. Thus a considerable number of farmers are being penalized – and land conservation works are not being undertaken.

Grants, loans, rebates

Direct payments or compensation to farmers or communities can be made for costs of land repair or conservation. One of the main sources of funding in recent years in this category has been the National Landcare Programme. While these funds undoubtedly have fostered the rapid development of many Landcare groups around Australia, there is increasing concern over the way funds are being disbursed. According to Donaldson (1995), of the US$11.3 million expended on community-based projects in 1993/94, only US$3.76 million were spent on ground works, and even these are intended to be demonstration programmes, meaning they would have very little impact on a catchment-wide basis. The land use director for Western Australian pastoralists and graziers has stated that government bureaucracy soaked up 90% of the money earmarked for Landcare in Western Australia in 1994/95, with only about US$1.17 million reaching the landholders (*Farmers' Weekly* 25 May 1995, p. 3). Nevertheless, it was found that small government grants can have a stimulatory effect on private expenditure and inputs from other research organizations and funding bodies.

Other assistance has been available to farmers in the form of low interest or interest-free loans, and various rural assistance schemes. However, funding institutions require security and, given the high level of indebtedness of the rural sector, this option is not a favoured one. Again, those who need the assistance most are least likely to obtain it.

A United States scheme, operating under the Farm Act, has some merit and may be considered for Australia. A farmer removes degraded land from production for 10 years in return for a rental fee. The farmer is required to implement land conservation measures. But Bradsden (1991) asks: What happens at the end of the designated period? What is to prevent the farmer from once again degrading the land? And does the removal of land from production mean that pressures on the remaining land are increased? The other point that needs to be made in this connection is that such schemes are often introduced primarily for economic reasons – namely as a response to overproduction – and not as a rehabilitative measure.

Incentives, awards

Rewards for farmers or other individuals who promote or implement good land management have proved useful. Examples have included the Australian Broadcasting Commission Tree Awards, the Sir William McKell Medal for outstanding work in soil and water conservation, and National Landcare Awards. However, innovative farmers are not always recognized: they take the risks (in effect a private risk for a public benefit) and often are subjected to criticism from government agencies and other farmers. When farmers are recognized, the awards in effect represent some compensation for unpaid research and for risks taken (Donaldson 1995). There is also the notion that farmers will only undertake conservation works when a subsidy is available.

Other possible incentives can include a system of trade-offs. In the United Kingdom, for example, a 1989 nitrates pilot scheme permits farmers with land in sensitive, designated areas to enter a five-year management agreement in return for payments to cover the cost of meeting scheme obligations beyond good agricultural practice. In 1995, the Western Australian government announced US$720 000 per year funding for the remnant vegetation protection scheme, and a subsidy of up to US$480/km for farmers to fence off stands of remnant natural vegetation which the landholder is not permitted to use for 30 years (*The West Australian* 11 September 1995). An area of 50 000 ha of remnant vegetation was targeted for fencing in the five-year period 1995–2000 (*Farmers' Weekly* 25 May 1995).

Subsidies

Until the 1980s, Australian farmers had the benefit of subsidized fertilizer, fuel, agricultural chemicals, water and fodder under a system of price subsidies, as well as various commodity price support schemes. It has been argued successfully that the effect of these policies was to encourage wasteful use of inputs (or to encourage greater output), often with harmful environmental consequences (Kirby and Blyth 1987; Chisholm 1992). As a result, most of these subsidies were removed during the 1980s. For example, the previous drought relief policy which subsidized fodder during periods of drought encouraged farmers to retain stock rather than removing stock from the land to avoid serious damage to the soil. Current policies require farmers to destock and adopt drought risk management strategies (DPIE/ASCC 1988).

Penalties

Through fines, regulations are generally used as a last resort for farmers who do not voluntarily repair or prevent serious land damage, or control noxious weeds or feral animals. Soil conservation agencies have the power to issue directives or levy fines, and even to take over the land and manage it directly (for example in the Western Australian Soil Conservation Act). For obvious reasons such powers are used reluctantly in Australia: agency personnel rely on good personal relationships with farmers in order to do their jobs. However, increasing action has been taken against farmers in Western Australia. In 1994/95, the Commissioner for Soils issued 21 Soil Conservation Notices relating to land clearing and drainage. There were three successful court prosecutions and a further 10 cases were pending in 1995 (*The West Australian* 11 September 1995).

Quarantine powers have been used to isolate farms on which contagious stock diseases have been discovered, and (in recent years) where beef containing pesticide residues has been identified. Such punitive systems have been seen as inflexible, are difficult and costly to administer, and lead to considerable ill-feeling.

Agency and policy solutions

As mentioned previously, one of the major difficulties in Australia has been the separation of government agencies responsible for trees and agriculture. As a result, those promoting the combined use of the two, in agro-forestry (broadly defined), find there is no single agency which can assist. Forestry agencies are concerned generally with growing trees for commercial timber production, usually in higher rainfall areas. Agricultural agencies are concerned primarily with animals and crops. Even commercially useful species such as those which produce fruit, nuts and oils are dealt with by the horticulture branch of the agriculture agency (although some forestry agencies have become aware of the value of some species for producing oils), but are not generally seen as being relevant to agro-forestry.

Despite this and other problems, however, there have been many positive responses to the problem of land degradation in southern Australia.

Legislative policies/responses to land degradation

Early concerns over land degradation in Australia are reflected in the various soil conservation Acts and programmes which were introduced in the 1930s and 40s. By the 1970s, the increased understanding of land degradation problems and their management, and a greater community awareness, saw the emergence of a range of federal and state legislation to deal with rural and urban land use concerns (see the list in Environmental Protection Council of South Australia (1988, p. 6), and the list of programmes relevant to sustainable agriculture in Australian Agricultural Council (1991).

The principal impetus for legislation directed at land degradation arose from the National Conservation Strategy (Department of Home Affairs and Environment 1984), which in turn was a response to the World Conservation Strategy (IUCN/WWF 1980). The National Strategy was quickly followed by equivalents in the various Australian states and territories. In 1988, the National Soil Conservation Strategy (ASCC 1988) listed a set of land management goals, and funding was made available under the National Soil Conservation Programme for research and community projects to be administered in conjunction with the states. Some developed their own soil conservation strategies. Also around this time, the federal government embarked on a series of sectoral Ecologically Sustainable Development (ESD) investigations, including one on agriculture (ESD 1991). Land degradation problems were addressed and policy recommendations made. But it seems that the ESD programme as such was abandoned in the early 1990s by the re-elected Labour Government, and its role largely taken over by the National Landcare Programme (NLP) and the Land and Water Resources Research and Development Corporation, established in 1990 as a national statutory authority to support the sustainable use of Australia's land, water and vegetation.

In 1989, the Prime Minister's Statement on the Environment (Hawke 1989) was something of a landmark in land conservation in Australia. He announced the Decade of Landcare, which was to focus on and implement ecologically sustainable land use by the year 2000. A number of important initiatives were established and there is little doubt that these provided a considerable impetus to improving awareness, research and action on land degradation problems around Australia.

The Prime Minister's Statement established major revegetation and replanting programmes, which included: the One Billion Trees and remnant vegetation Save the Bush programmes (with particular encouragement given to the formation of Landcare groups, operating at the community level, with an emphasis on education and community projects (see further below)); a National Reforestation Programme, encouraging hardwood plantations on private land to relieve logging pressure on native forests; the establishment of a National Resources and Research Development Corporation which was to examine soil, water and forestry issues and to promote an integrated approach to land and water research; and additional funding for the Murray/Darling Basin Natural Resources Management Strategy, to implement land, water and vegetation programmes. Agricultural chemicals legislation was to be reviewed. Total funding for all the programmes was to exceed US$240 million over 10 years, later upgraded by Prime Minister Keating (Keating 1992). Seven years after the 1989 Statement, many of its programmes and objectives were being implemented.

State of the Environment reporting has provided valuable background assessments and a perspective on the condition of the biophysical environment nationally and in different states. Victoria (Scott 1991), South Australia (Department of Environment and Land Management 1993) and Western Australia (Grant 1992) have produced such reports, which have been referred to in previous chapters. The reports are intended to be produced at regular intervals to assess changes and trends.

Landcare

The Landcare concept was developed in the late 1980s, also as a federal programme (Campbell 1994). This seems to have merged conceptually with the Land Conservation Districts (see below), but generally Landcare groups are smaller, covering subcatchments. There are also other community groups in Western Australia associated with the Waterways Commission. They are working on management requirements for a number of degraded estuaries and their catchments.

Landcare is not restricted to farmers. Country towns provide members, and many urban and rural schools are involved in the Kids for Landcare programme (for example, Saltwatch, Wormwatch, Ribbons of Blue). Numerous groups are involved in tree planting, vegetation retention and enhancement, and wetland conservation. Targets of the Landcare plan include: incorporation of revegetation and vegetation management into district and individual property management planning; and increased revegetation (including agro-forestry) to complement and improve agriculture production.

The plan puts the contribution of revegetation into context at the farm and district level. In 1992, a central office was established which enables groups seeking community grants from the National Landcare Programme, Natural Resources Management Strategy, the One Billion Trees Programme or Save the Bush to apply using a standardized application process.

The timing of national and state initiatives on soil conservation and land use could not have been better in terms of capturing the mood of rural communities, which were increasingly aware of serious problems in their areas. Within four years of setting up Landcare programmes, more than 2000 Landcare groups had formed around the country (Mues *et al.* 1994; Campbell 1994). Significantly, membership has been greatest among broadacre farmers, especially in the Mediterranean zones. In 1992/93 about one-third of all broadacre farmers belonged to a Landcare group. Regional differences were notable, from a high of 44% in Western Australia to a low of 12% in Tasmania. The comparable figures for Victoria and South Australia were 38% and 21% respectively (Mues *et al.* 1994). However, as indicated earlier in this chapter, some serious problems relating to management and funding of Landcare programmes have emerged in some regions.

Soil conservation legislation and land conservation districts

The three Mediterranean states have made important amendments to soil conservation legislation in recent years, with Victoria being discussed in the following section under "Catchment planning".

In Western Australia, the Soil and Land Conservation Act (1945–88) includes a facility for establishing Land Conservation Districts (LCDs) (formerly Soil Conservation Districts, established in 1982 (Select Committee into Land Conservation 1991, p. 130)). LCDs now cover 85% of the agricultural area. Their primary role is to develop co-operation amongst land users, local governments and agricultural agencies in order to implement sustainable land management systems and to solve local and regional land degradation problems.

Some LCDs have been successful, benefiting from good leadership, well defined goals, and the energetic involvement of farmers and community groups. However, discussions with agency personnel indicate that a large proportion of the LCDs are effectively "defunct", only going through the motions of holding committee meetings and recording proceedings. Failures have arisen for a variety of reasons, including lack of focus and consensus, disillusionment, anti-bureaucratic attitudes, the inability of some farmers to recognize the existence of the less visible degradation problems, and difficulties in accessing information, technical assistance or funds (see Select Committee into Land Conservation 1991, pp. 132–133). Moreover, funding for projects which have an impact on the ground is now directed through Landcare and the other programmes discussed above, and this is clearly having the effect of downgrading the perceived importance of the LCDs. This is reinforced by the establishment of staffed Landcare centres (offices) in several country towns, providing a valuable, tangible focus for that initiative.

In South Australia, the equivalent of the Western Australian Land Conservation District Committees are District Soil Conservation Boards, with 26 (in 1993) covering the state's agricultural areas except Kangaroo Island and the southeast. The Boards are also based on broad community membership. Their role is to promote the use of land within its capability and to conserve and rehabilitate degraded land. Under South Australia's Native Vegetation Act 1991, the Boards must be consulted about the development of management guidelines for heritage agreements and about applications for minor clearance of vegetation. Under South Australia's Soil Conservation and Landcare Act 1989, the Boards have to develop district plans by 1995. District plans identify (Department of Environment and Land Management 1993, p. 93):

- classes into which land falls
- capability and preferred uses of the land
- land uses
- nature, causes, extent and severity of degradation
- measures that should be taken to rehabilitate each type of degradation
- land management practices best suited to preventing degradation of the various classes of land.

Catchment planning

Increasingly, the focus in land management is away from the individual farm to integrated catchment or regional scales to help address land degradation problems more effectively. Several Australian states now have formal catchment management policies, including Western Australia, South Australia and Victoria in the Mediterranean zone.

For example, Victoria's Catchment and Land Protection Act 1994 addresses land degradation issues and embraces the concept of sustainability. A framework for integrated catchment management, which encourages community participation, has been established. Regional boards are being set up to correspond with areas covered by existing Landcare groups. Landowners are required to take reasonable steps to conserve soil, protect water and prevent the spread of feral animals and noxious weeds.

In Western Australia, under the Waterways Conservation Act 1976, the Waterways Commission (now part of the Water and Rivers Commission) has powers to conserve and manage the state's waterways in declared management areas.

Two important developments have been the formation of authorities to plan and manage the drainage basins of the Avon and Blackwood rivers, the two largest – and most degraded – rivers in the state's southwest. The authorities operate through Land Conservation Districts, catchment groups, state government agencies and local governments to help improve land management in the drainage basins through sustainable agricultural methods and care of the river systems. Farmers are advised of impacts of their practices and river water quality is monitored to assess the effects of the management initiatives (Avon River Management Authority 1993).

The basic elements of integrated catchment management are:

- identify the problems
- involve land holders, community and government agencies
- set goals and objectives
- collect base data and maps
- identify vulnerable areas, natural features, land use
- delimit management units
- incorporate risk management strategies (for example, droughts, commodity price fluctuations)
- prepare plans
- make plans available for public comment
- implement plans
- monitor and review.

Incorporating land degradation solutions into regional planning

Australia's Mediterranean states have been active in developing regional plans which incorporate a range of sound land use practices.

Ideally, objectives need to ensure that land is kept in good condition, harmful practices are avoided, rural functions and landscape values are retained, and the community is actively involved in the planning process.

A fine example of a regional management programme for a large area concerns the Murray/Darling basin. This major drainage system extends over one-seventh of the area of the Australian continent, and experiences a range of trans-state land management and environmental problems. The basin is especially significant in containing half of Australia's irrigated land and growing one-third of the country's agricultural produce (valued at about US$8 billion each year). In 1988, the Murray/Darling Basin Commission was established as an autonomous, intergovernmental body to deal with the development and administration of land, water and environmental strategies for the basin. State bodies have been superseded, and long-term sustainable agricultural and ecological issues are being addressed at catchment and regional scales (Murray/Darling Basin Commission 1989).

South Australia's Planning Act (1982–85) includes the provision that where a development is likely to have a major social, economic or environmental impact, the minister can call for an environmental impact assessment. More recently, comprehensive regional planning has been undertaken in order to combine land use planning with environmental protection and the prevention of land degradation. Two such plans had been approved by 1995, with the Mount Lofty Ranges Regional Strategy also involving the creation of a regional authority to ensure the plan's implementation (Department of Housing and Urban Development 1993).

A quiet revolution has been taking place in Western Australian rural policy and planning, and is discussed in some detail here as it is considered that the developments may have broader application throughout the Mediterranean zone. Since 1988, and more particularly in the 1990s, the Department of Planning and Urban Development (DPUD; since renamed the Ministry for Planning) has been preparing regional plans, or strategies, which have a major rural component. These plans, or strategies, have also explicitly been seen as vehicles for integrating environmental protection with land use planning.

In 1985, the State Planning Commission Act made the State Planning Commission (SPC) responsible for planning at the state level. Twelve regions based on ward boundaries of the Country Shire Councils Associations were identified (Figure 23.2). However, regional plans produced to date are for realistically smaller regions than these 12 SPC regions (Figure 23.3).

Even though the regions are often very different from one another, there is a recurring pattern. The regional plans discussed here are the Bunbury–Wellington Region Plan (DPUD 1993a), the Central Coast Regional Strategy (DPUD 1994b) and the Albany Regional Strategy (SPC 1994).

Figure 23.2 *The 12 major Western Australian planning regions* (source: DPUD 1993a, figure 1)

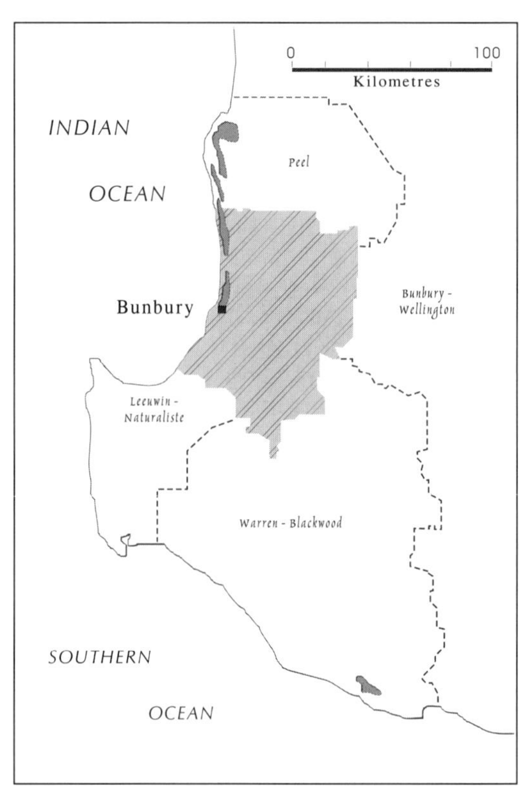

Figure 23.3 *Planning regions in the southwest of Western Australia* (source: DPUD 1993a, figure 1). *Shaded area indicates the Bunbury–Wellington planning region*

Methodology In general terms, the work was carried out by a Project Team, comprising representatives from each of the local government authorities in the region, and from each of the relevant government agencies. All the regional planning studies involved a programme of extensive community consultation. This took the form of a series of public workshops, making documentation available to the community and inviting written submissions. The Central Coast public workshops addressed an Issues and Opportunities Paper and the results of the workshops were published as reports.

All the regional plans were preceded by the release of a number of documents, with those of a non-technical nature being made available for public comment. The draft Albany Regional Rural Strategy (DPUD 1991), for example, was published in October 1991. In a foreword, the Minister for Planning invited community response to be forwarded to DPUD's offices in Albany or Perth by February 1992; and the final page of the document contains a tear-out form for comment, with the addresses to which it should be posted. Comparing the draft Albany Region Plan (DPUD 1993b), which contained an equivalent invitation and timeframe for response, with the final strategy (SPC 1994), it is evident that changes were made.

Each region was subdivided into catchments (some of them are rather artificial, especially on the coastal plain), and these in turn were further subdivided into Planning Units. The basis of these Planning Units is similar in all three cases. In the Central Coast Regional Strategy, for example, they are based on (DPUD, 1994a, p. 53):

- surface water catchments
- physiographic zones
- land systems and land capability analysis
- public and private land tenure
- the existing pattern of land use
- functional and administrative geographical boundaries.

The studies developed extensive spatial databases. In the most recent plans or strategies (Central Coast and Albany), the data were stored on a geographic information system (GIS), using Arc/Info software, for subsequent analysis and data manipulation. The data relate to the natural and modified environments, and (in the case of the Central Coast Regional Strategy) include (DPUD 1994a, p. 49):

- physiography
- geology
- wetland systems and catchments
- vegetation systems
- remnant vegetation
- coastal dune stability
- marine habitat (to 5 km offshore)
- mining and petroleum tenements and resource areas
- reserves
- vacant Crown land and unvested reserves
- townsites and squatter settlements
- general land use.

With regard to the planning process, all reports comment to the effect (DPUD 1994b, p. 49) that:

"the process used to prepare the land use plan included elements of intuitive decision making and is not necessarily based on scientific analysis. The values and opinions of the community, the Study Team and Steering Committee input, have contributed to the outcome".

Contents of the reports All reports are similar in their structure and contents, which are summarized below. Nevertheless there are differences. The Bunbury–Wellington and Albany regions each focus on an urban area, whereas the Central Coast region has no such focus. Specific problems and issues reflect the characteristics of the regions. Thus, the rural parts of the Albany region are mostly agricultural, with some tourism; Bunbury–Wellington has a large state forest component as well as mining (coal, mineral sands) and secondary industry which is to some extent decentralized outside the city of Bunbury; and the Central Coast region has a sparse population of a little over 4000 – yet with a significant problem of more than 1000 squatter shacks, whose inhabitants are not counted in the census – and also sand mining, an important marine component, and relatively large areas of unallocated state-owned land.

Underlying principles Each plan sets out a number of underlying principles. All state that a fundamental objective is "to provide a link between State and local planning which is based on a balance of economic, social and environmental considerations" (DPUD 1994b, p. 13).

Turning to the final Albany Regional Strategy, eight broad issues or principles (all with their counterparts in the other reports) are identified. The regional strategy sets out to (SPC 1994, p. 4):

- provide for long-term sustainability of land use
- ensure the quality of soils and water including marine waters are maintained or improved;
- protect land uses which contribute to the economy of the region and the state
- protect the natural, human and environmental heritage

– enable land uses to respond to changes in the economy
– guide the expansion of urban areas to avoid adverse impact on valuable rural land and waterways
– ensure an adequate level of government services and facilities
– plan for development which enhances, or at least minimizes, any adverse impacts on the values of places of cultural heritage significance.

It will be noted that environmental principles are high on the list.

Strategies The Albany Regional Strategy outlines "Strategies" under five headings:

- natural and human environment
- rural land
- housing and residential land
- industry, transport and infrastructure
- commercial and other economic activity.

Under each of these headings, the Strategy contains a brief discussion and then lists separately Issues, Objectives and Key Actions. For the Natural and Human Environment, 21 Issues are identified, of which the first four are:

- land degradation
- water degradation
- sustainability
- loss of and need for retention of natural vegetation.

Thus it is clear that environmental issues are seen as being paramount.

Nine Objectives follow, and then the crucial 24 Key Actions. The Objectives and the Key Actions clearly match the identified Issues; and importantly, each Key Action identifies the agency (or agencies) responsible for implementation. The first four Key Actions are (SPC 1994, p. 29):

1. Establish the location, nature and extent of land and water degradation, and implement actions for their improvement.
2. Prepare Integrated Catchment Management Plans for each catchment, to guide the formulation of subcatchment/farm plans, to address land and water degradation and ensure long-term productivity of the rural environment.
3. Investigate the need for an Environmental Protection Policy to clarify requirements for land and water conservation and sustainable rural land use.
4. Support and expand current programmes encouraging the retention of remnant vegetation and the reforestation of private landholdings.

Ten Issues, four Objectives and 12 Key Actions are identified under the Rural Land heading, with some overlap with the Natural and Human Environment section. The second Key Action refers explicitly to the preparation of local rural strategies and rural planning guidelines, as part of integrated catchment management plans. The local rural strategies and rural planning guidelines are to be reflected in local authority planning schemes, produced by Local Government Authorities (LGAs) with the assistance of the State Planning Commission, the Office of Catchment Management and Land Conservation District Committees. These "location-specific strategies" refer to the Planning Units referred to previously – the smaller spatial units nested within the catchments.

Implementation Each report concludes with a section headed "Implementation". The Central Coast and Albany region reports identify the specific agencies responsible for implementing their recommended Actions; presumably the final Bunbury–Wellington Plan will do likewise.

However, it is interesting to note that the draft Bunbury–Wellington Region Plan and the draft Albany Region Plan both recommended that adoption of the plan as a whole should have statutory backing. Despite this, the final Albany Regional Strategy notes that the State Planning Commission preferred to adopt the Strategy as policy. Reasons are not given. Similarly, the four LGAs "will also have regard to" the strategy in their various planning processes, as distinct from "will be required to".

The SPC (1994, p. 51) concludes by noting that the LGAs which comprise the Albany Regional Strategy area have formed a voluntary regional organization of councils, known as the Rainbow Coast Regional Council, and comments that "this organisation could provide an excellent forum for the Councils of the region to implement many of the actions of the strategy". Similar voluntary associations of LGAs are also appropriate for the other regions.

Conclusion The reports briefly reviewed here are particularly encouraging in a number of respects. These include: the integration of environmental protection with land use planning; the considerable effort taken to involve the community in the preparation of the plans or strategies; the improving methodology, routinely incorporating land capability evaluation and GIS; the recognition of Aboriginal values, especially in the Bunbury–Wellington Plan, and the explicit nesting of the regional plans within state planning policies.

Of course, there are also some areas where improvements could be made. The attention given to Aboriginal requirements in the Bunbury–Wellington Plan should be carried through in all regional planning; actions need to be more precise and detailed (although it is recognized that these are intended as guidelines only for LGAs and other agencies, the more specific the better); the GIS could be used more effectively than merely as a means of storing and overlaying spatial data; and the lack of a statutory underpinning for the adopted Albany Regional Strategy is a matter for some concern – hopefully unwarranted.

It is still far too early to determine the "success" of these plans/strategies in terms of identifying improvements on the ground – although, in a number of instances, the strategies are formalizing actions which are already being undertaken. But two major areas are cause for congratulation: the commitment to meaningful public participation, and the high priority given to environmental considerations.

24

Summary and conclusions

Arthur Conacher

In the Introduction, the objectives of this book were stated as being to bring together information on the major problem of land degradation in the world's Mediterranean environments. The information, provided by experts from the various regions, would include the cultural and biophysical contexts of the problems as well as a description of their nature and extent, and would also include solutions. A further objective was to ascertain whether the seasonal nature of Mediterranean climates produces distinctive forms of land degradation.

The Introduction continued with a discussion of the characteristics of these regions, noting that they are distinctive in many respects: their sensual environments, their peoples and their history, as well as their biophysical environments. Whilst being distinctively "Mediterranean", however, the regions are also very different from one another and are often further characterized by complex intra-regional and local diversity (described by some as conferring a "mosaic" quality on Mediterranean landscapes). There are differences between the northern and southern hemispheres, between Old and New World, between north and south and east and west within the Mediterranean Basin, and in all regions between coastal zone and hinterland, between plains, interior valleys and mountains, and between coastal cities and inland towns and villages. These many aspects of the Mediterranean regions were elaborated in Part I, which provided a context and framework for the discussion of land degradation in Parts II and III.

The Problems of Land Degradation

There are distinctive, regional differences in land degradation. Desertification is highlighted in the Iberian Peninsula, water shortages in the eastern Mediterranean (and also in North Africa, Spain and Greece), forestry problems in Croatia, soil deterioration and erosion in Chile, plant invasions in the southwest Cape (but actually widespread historically: what is truly "indigenous"?), a range of problems associated with population pressures on rural land in North Africa and Greece, landslides in Italy, North Africa and California, subsidence in Italy, replacement of indigenous vegetation with introduced crops and pastures in southern Australia (with a broad, resultant array of land degradation problems – also in the eastern Mediterranean, Spain and Greece), wind erosion in parts of Greece and especially some Aegean islands and in southern Australia, and severe secondary soil salinization in parts of California and southern Australia. Water pollution (both surface and subsurface) is also a problem in several regions, and is particularly referred to in Spain and the south of France (industry), North Africa (nitrates in groundwater), and the eastern Mediterranean, California and southern Australia (salinity and agricultural pollutants).

There are also common problems amongst the regions. Aperiodic and unpredictable streamflow regimes, sometimes resulting in serious flooding, are ubiquitous, reflecting the seasonal and episodic nature of the climate (amongst other factors). In contrast, annual summer droughts are also characteristic of Mediterranean environ-

Land Degradation in Mediterranean Environments of the World: Nature and Extent, Causes and Solutions.
Edited by A.J. Conacher and M. Sala.

ments, and are particularly severe where the regions border on more arid environments – as in the eastern Mediterranean, North Africa, southern California, northern Chile, the north of the southwest Cape, and southern Australia. Accelerated soil erosion by water occurs everywhere and is a long-standing problem in the Old World. As a result, sedimentation is also ubiquitous, with historical aggradation of coastlines and river valleys in many parts of the Mediterranean Basin, and more recent silting of dams, particularly in North Africa. The problem is less severe in southern Australia (due largely to the absence of mountains) and in the southwest Cape (perhaps due to the retention of a good vegetation cover in catchments, allied with relatively minor pressures on the land).

Fire is ubiquitous and is considerably enhanced by the occurrence of hot, dry seasons and strong winds, and highly inflammable vegetation. In addition to losses of human life and infrastructure, fires also cause changes in the physical and chemical properties of soils and contribute to water repellency and hence accelerated erosion. Vegetation loss and degradation is widespread: in southern France and Corsica to the end of the 19th century, and in Croatia, Greece, North Africa, the southwest Cape and southern Australia.

There are increasing pressures on the coastal environments of all Mediterranean regions as a result of booming tourist numbers attracted by the climate and coastal amenities. In contrast, many inland areas are suffering from rural depopulation, with the notable exception of parts of North Africa (and parts of the eastern Mediterranean). This has serious social and economic consequences, and both beneficial and adverse effects on land degradation – beneficial through increased plant cover, and adverse as terraces and stone walls are abandoned. Most areas are being subjected to increasing pollution associated with urbanization and industrialization, with adverse effects on vegetation, soils and water.

The Historical Development of Land Degradation in Mediterranean Regions

In Spain, the Neolithic period at the end of the 5th century BC saw the beginning of the substitution of natural ecosystems with anthropic ones through pastoralism and agriculture. At that time 96% of the Iberian Peninsula was forested, compared with 13% today.

The Roman period saw the regression of Iberian forests, increased erosion and sedimentation and the seaward extension of many floodplains. Agriculture was abandoned with the Barbarian invasions from the north, and the later Arab invasions caused the retreat of Christians to the mountains, resulting in over-population and further forest destruction. During the Middle Ages, Moorish tree crops and terraces were converted to a Christian period of sheep pastoralism, once again causing increased erosion. The 15th and 16th centuries were a period of increased use of the forests for timber, charcoal and shipbuilding, while the sequestering of church lands in the mid-19th century to private ownership resulted in speculation and over-exploitation. This generated the first conservation and reforestation measures. Agriculture was practised mainly in the mountains until the 19th century, except on the coastal plains of Murcia and Valencia. Here, as elsewhere in the Mediterranean Basin, the coasts were inhospitable, malarial, and unprotected from invaders and catastrophic flooding.

Soil erosion is also traced back to the Neolithic period in the south of France, but it was generally infrequent before the classical period. Erosion then became more general, and also at the beginning and end of the Middle Ages. The modern period was characterized by the maximum extension of cultivation and raising of livestock, even on steep slopes and poor soils, although the widespread use of cultivation terraces limited soil erosion. During the 20th century rural depopulation has resulted in vegetation recolonization and reduced erosion. Detailed knowledge of the history of fires and floods extends back only to the mid- to late 18th century (fires) and the last century (floods).

In Croatia, soils were already severely eroded by the 5th century BC, associated with extensive deforestation. Regulations in the 13th century AD prohibited the cutting of forests in areas where today there is no forest – clearly an ineffectual response! Organized reforestation commenced in the latter part of the 19th century.

Soil erosion was recognized in Homer's *Iliad* (854 BC) in Greece. Upland farming and grazing probably started around the middle of the second millennium BC, and were related to the initial damage to the forests. Agriculture greatly intensified during the Hellenic period, from 800 BC, associated with extensive forest cutting, over-cultivation and over-grazing. As soil eroded, cultivation shifted from food grains to commercial crops such as grapes and olives which could be grown in shallower soils. These eroded lands were described eloquently by Plato nearly 2500 years ago. By the time of the Macedonian hegemony (338 BC), the land had deteriorated markedly, and erosion continued at an even greater rate during the Roman period.

During the Ottoman occupation of 1457 to 1913, many Greeks moved to the hills and mountains, and constructed numerous terraces to control erosion. But they also cleared forests and over-grazed pastures (cf. Spain and North Africa). The cycle of forest clearing, cultivation, degradation and abandonment continued for about 400 years, causing severe erosion and the exposure of bedrock in some areas.

Intensive farming of the Greek lowlands has developed since 1950, accompanied by abandonment of agricultural lands on sloping terrains and the collapse of traditional forms of management. This rural depopulation has sometimes been associated with accelerated erosion due to the abandonment of terraces, whereas in other areas decreased erosion has followed the regeneration of natural vegetation.

Agriculture and pastoral livestock husbandry began around 11 000 years ago in the eastern Mediterranean, with wheat, barley, goats, sheep and cattle. By 2500 BC, increasing salinity in ancient Mesopotamia was affecting grain cultivation. There was a major development of the agro-pastoral economy in the eastern Mediterranean starting around 5000 years ago, which continued until the end of the Roman period in the 7th century AD. This period was associated with the domestication of fruit trees, construction of terraces, the use of soil and water conservation measures and increased population. The Moslem conquest of the region was associated with declining agriculture and increased pastoral nomadism, resulting in increased erosion, loss of soils in uplands and the creation of swamps in lower valleys. Over the last 100 years there has been a huge increase in population, mechanization of agriculture, the development of irrigation schemes, drainage of swamps, introduction of monocultures and exotic species (pines, eucalypts and acacias) leading to changed ecosystems, and increased use of fertilizers and pesticides. Classical olive cultivation has been replaced by export-orientated citrus, and forests have been felled for timber and energy. The current period is one of land degradation, aggravated by continued population increases and characterized by deteriorating water resources, salinization and soil pollution, further impacts on natural vegetation by urbanization, and increased marine and coastal pollution.

The North African Tell was abundantly wooded in Roman times, when agro-silvicultural exploitation commenced. Arab invasions were responsible for some forest recession in the Middle Ages, and the use of fire. However, extensive forest tracts were still present at the arrival of the first European settlers, and the main forest decline took place after the 1830s. By the 20th century the landscape was open and widely eroded by sheetwash. Extensive erosion occurred during the French and Spanish colonial period, including gully development resulting from deforestation; excessive exploitation of wood occurred during the two world wars. With Independence, Algeria, Tunisia and Morocco underwent demographic upheavals and urban growth, and the disintegration of the traditional collective system. Both the intensification of human occupation (particularly in the Rif) and the abandonment of land (in eastern Algeria) have aggravated erosion problems.

Although fire was used by California's indigenous inhabitants, and some soil erosion and land degradation was associated with the Spanish and Mexican periods from 1769 to 1850, there is little doubt that land degradation is largely a product of the last 150 years. This was associated initially with the expansion of cattle ranching (in turn driven by the Gold Rush and the development of railway networks), and booming wool prices. An array of problems followed, associated with the massive expansion of areas under irrigation and, somewhat ironically, the draining of marshes and deltas. Other forms of land degradation were linked historically with mining – particularly hydraulic mining of river gravels and, later, oil extraction – and logging.

Land degradation in Chile dates from Spanish colonization in 1542, when the destruction of native vegetation commenced in order to obtain wood for construction, land for agriculture, fuel for homes and timber for the mines. World demand for cereals initiated during the gold exploration and mining period of the 19th century saw the extensive cultivation of wheat, which led eventually to accelerated erosion and land degradation and declining productivity. Population growth and shifts towards the southern parts of Chile resulted in continued deforestation and increasing use of fire; 69 species of trees and shrubs are now at risk. In the mountainous areas, landslides are a continuing hazard, often associated with earthquakes.

In the southwest Cape, Stone Age hunters used "fire stick" farming but their numbers were small. Cattle herding was introduced about 2000 years ago and may have been responsible for the collapse of grass-dominated vegetation and its replacement with *renosterveld* shrubland. The European occupation from the mid-17th century resulted in rapid population growth, clearing of natural vegetation and the development of agriculture, and the introduction of invasive, non-indigenous species of the genera *Acacia*, *Hakea* and *Pinus*. Wind and gully erosion developed but appear to have been managed successfully.

Aborigines have occupied Australia for possibly 100 000 years and used fire extensively, resulting (as in all Mediterranean regions) in a fire-adapted vegetation. Europeans occupied the land from the early 19th century, and extensive clearing of the natural vegetation for agriculture took place mostly in the 20th century. Adverse effects on the land were described as early as 1853, including problems of soil and water salinization, wind erosion and rabbit plagues. Southern Australia experiences the full range of land degradation problems, which are often severe, but accelerated erosion and sedimentation are less pronounced than in other regions owing to the more subdued topography.

Causes of Land Degradation Problems

Causes of land degradation were grouped into those relating to the nature of the biophysical environment, and human actions. The climatic factor is common to all Mediterranean regions: strong seasonality, intensity and spatial variability of rainfall, with sudden changes from cold and wet to hot and dry months (and *vice versa*). Most regions have steep topography backing coastal plains – partial exceptions are southern Australia and parts of the eastern Mediterranean – and several have dry, interior valleys or basins, often irrigated (with an associated array of land degradation problems). Reflecting the other environmental factors, all regions have poor soils except in localized areas such as floodplains. Soft lithologies are a major factor in some areas, notably parts of Spain, Italy and California, as is tectonic activity in triggering landslides. Inflammable native vegetation is common to all Mediterranean regions.

Human factors can be further grouped into direct and indirect causes. Direct causes include inappropriate land management practices – excessive removal of indigenous vegetation, grazing, cultivation, burning, introduction of exotic species, irrigation, industrialization, urbanization and tourism. Indirect causes include factors such as attitudes towards and perceptions of the biophysical environment, population growth and movements (including rural depopulation), historical changes in social organization and land tenure (especially the loss of traditional social structures and associated land management methods, notably in the Old World), and actions and policies of agencies and governments. All regions have been subjected to invasions by foreign cultures – repeatedly in the Mediterranean Basin – and in all regions the occupations were accompanied by episodes of accelerated erosion and sedimentation as the occupiers introduced new land management practices (or abandoned old ones), or indigenous populations retreated to mountains or deserts, exerting increased pressures on the environment. Historically, wars were often associated with deforestation as a result of increased demand for timber, notably for shipbuilding.

Experimental results have demonstrated the relationships between vegetation removal and loss of biomass with loss of soil moisture and organic matter content, increased runoff and increased erosion. Rainfall is an important factor, with data showing increased erosion with decreased mean annual rainfall. However, in some areas this trend is attributed to limestone lithology in wetter areas, not higher biomass. All experimental findings show that soil loss is essentially correlated with runoff, indicating the need for water management on the slopes. But processes change during intense rainfall events, from stream bed and bank erosion (primary sediment sources) to slope rilling and gullying and even mass movement as the duration of a high intensity rainfall event increases. Once formed, gullies may remain relatively stable for long periods. Sediment preparation is important, with surface soils loosened by heat and stock trampling during prior dry periods; and experimental data show that erosion rates are much higher where soil conditions are already degraded and the protective effect of vegetative cover is reduced. Major changes occur when plant cover is reduced to between 0% and 30%.

Stoniness has variable effects on runoff generation, depending on stone size and percentage cover. Plant stubble and mulches decrease runoff, sediment and nutrient loss, and protect soil aggregates from rainsplash. Soils with wheat crops interspersed with *matorral* have much lower erosion rates than slopes with wheat monocultures.

In southern Australia, vegetation removal has changed landscape hydrology in such a way as to raise saline groundwater tables closer to the surface and increase the water content of perched aquifers over extensive areas of both dryland (especially) and irrigated agriculture, causing waterlogging and salinization of soils and increasing salinization of streams, rivers, wetlands and dams. Soils here and elsewhere are responding to cultivation practices in a number of adverse ways, including loss of structure and increasing acidification.

Land abandonment may cause degradation. Although it may result in increased vegetation cover and therefore reduced erosion, it also means that traditional forms of soil conservation, especially terraces and stone walls, are no longer maintained. The crumbling of these structures may result in serious erosion problems. Although in some areas there has been a return of people to the fields as a result of urban unemployment, often the traditional skills or the will to undertake the hard work have been lost.

Changes from olive growing to vineyards also result in increased erosion, as previously stable, vegetation-covered slopes under olives (with an understorey of annual plants) are ploughed. Organic matter content and soil aggregate sizes decrease markedly, and A horizons thin significantly in only 12 years.

Ironically, in some areas (notably Italy), erosion control measures in river catchments have decreased sediment transport to river mouths to the extent that some coastlines are now eroding. Elsewhere, inappropriate coastal structures such as T-groynes and parallel reefs, constructed without preliminary study of coastal dynamics, have caused beach erosion. In the eastern Mediterranean, the opening of the Suez Canal allowed the introduction of Indian Ocean marine species to the Mediterranean, and the trapping of sediments by the High Aswan Dam on the Nile has resulted in serious coastal erosion.

Fires are regarded as being significantly more serious in Mediterranean regions than elsewhere, due to the climate and the nature of the vegetation itself. Fires are lit either deliberately – to expand pastures into forests, to express political views (notably in Greece), or simply as acts of arson – or accidentally, through negligence associated with increased human activities in forests (including recreation and increasing tourism), agricultural areas and rural/urban fringes. Severe fires are associated with prolonged, desiccating high temperatures and strong winds, low soil water content and accumulations of fuel (such as leaf litter and dried grass). Fires affect water and sediment regimes, nutrient cycling and the species composition of indigenous biota. Recovery rates vary with fire frequencies, vegetation associations, the severity of erosion, and post-fire climatic conditions.

The above problems and their causes are not necessarily unique to Mediterranean regions; rather, they are exacerbated by the nature of Mediterranean-type environments. In combination, however, they are distinctive to the regions. Other land degradation problems are not peculiarly "Mediterranean": for example, introduced plant diseases and pests, urbanization and industrialization producing increased flooding, quarrying and mining, water and air pollution and over-fishing.

Implications of Land Degradation

Ecologically, the Mediterranean regions exhibit a range of vegetative adaptations to seasonal drought, and the floristic composition and structure of Mediterranean vegetation has also probably been strongly influenced by fires induced by people over tens of thousands of years.

Soil loss can be considered as being the loss of ecological capital. The restoration of natural vegetation on sloping, degraded land with shallow soils and dry climates is slow. Consequences include increased flooding, landslides and siltation of rivers and dams. New dams, associated with the intensification of agriculture, unfortunately also contribute to soil structure deterioration and soil contamination with agro-chemicals, and the creation of conditions favourable to salinization on recent floodplains. As fertile soils become increasingly scarce, agriculture becomes more and more concentrated on those soils, in turn aggravating the problem of land degradation by concentrating higher and higher inputs into diminishing areas. On the hills, the progressive reduction of soil thickness beyond a critical point is drastically reducing the potential for biomass production.

In the Levant, the impact of humans (including over-grazing and fire) on vegetation has resulted in the extinction of hundreds and probably thousands of plant and animal species, and invasions of non-palatable plants. Indeed, throughout the Mediterranean Basin, agriculture has introduced a large number of species from different regions over the centuries, so that it is not always easy to identify unambiguously what is truly "indigenous" vegetation in any particular region. Australia – especially the Mediterranean areas of Western Australia – has recorded exceptionally high rates of species extinctions over the past 100 years as a result of clearing, draining of wetlands and various land use practices.

In social and economic terms, the loss of forests results in a significant loss of amenity for urban populations, the loss of an important resource, and increased floods with their attendant economic costs (destruction of bridges and transport routes, and buildings) and losses of human life. Fires, too, cause the loss of property and life as well as accelerated erosion. Increased sedimentation and silting of dams results in the loss of hydro-electric power, and reduced water for human consumption, industry and irrigation.

Water shortages arising through increased abstraction for agriculture and other uses, associated with increasing population, and through increasing pollution, will have extremely serious international consequences in some regions.

Large proportions of land in various regions are affected by land degradation, estimated at up to 30% in Sicily, and 44% in Western Australia, for example. Failures to improve crop yields – noted in North Africa and southern Australia – and huge losses of production have been attributed to land degradation. There is a growing reliance on the importation of agricultural products and food commodities in some Mediterranean Basin areas.

Increasing farm size (to increase total production, or associated with land reform) is one of a number of causes of rural depopulation, which in turn leads to the decline of country towns and villages and loss of social services and amenities in most Mediterranean regions. Adverse effects on human health have been identified both as a result of these social changes and through the impacts of agro-chemical residues.

Other important implications of land degradation which have been identified include the nature of geoscientists' training which must adjust in order to deal with the problems, and the need for global collaboration and co-operative earth sciences programmes.

Solutions

Animals

The European Union is promoting research into the viability of pastoral land use in marginal, low productivity agricultural lands. Pasture with indigenous animals, which place low demands on the environment, can produce valued food and generate permanent jobs. Similar arguments would also be appropriate in Australia. Stock also have a "cleaning" role in open forest or big firebreaks, by eating the highly inflammable herbs.

Jordan has an integrated programme involving the establishment of protected areas, improvement of grazing conditions in rangelands, the development of irrigated areas, and the settlement of nomads. In North Africa, pasture conservation aims to delimit protection areas, establish co-operatives, improve pastures and establish permanent meadows, reduce herd sizes, improve conditions for stock (for example, providing shelters at high altitudes), and prevent the cultivation of some rangelands. This may be integrated with wood production and prevention of over-cutting (for example, by introducing gas stoves) and planting fodder trees within the forests. However, plantations are often perceived by local populations as precursors to takeover of the forests by the state, and the new meadows are often over-stocked and poorly maintained, with consequent degradation. Electric fencing has been an important means of improved pasture management in Chile.

Controlling effluent – which may contain high nutrient loads, toxic metals, pathological organisms, growth hormones and antibiotics – from intensive animal farming is a major need in California, southern Australia and elsewhere. Methods used are essentially variants on the treatment of human sewage and are often inadequate.

Cultivation and horticulture

The reduction of fallow and the introduction of zero or minimum tillage techniques are the most common means of reducing the degrading effects of traditional cultivation practices. Allied with these techniques are rotational cropping (noting that crop rotations assist in breaking pest, weed and disease cycles and building up nutrient levels), inter-cropping, and the retention of strips of natural vegetation between crops, and rebuilding terraces and stone walls. Astonishingly, it is still necessary to discourage ploughing across the contour. Terracing is an ancient and effective means of achieving this, but terraces are very labour intensive and costly to maintain. It also needs to be noted that under some conditions contour ploughing and terracing may enhance erosion by favouring subsurface processes (piping and tunnel gully erosion) and mass movement. Rilling or gullying may also occur on contoured land under prolonged rainfall events. Therefore, clear identification of specific erosive processes must be undertaken before appropriate remedial measures can be introduced at any particular site, and numerous specific techniques have been developed, particularly by the United States Department of Agriculture. It further needs to be noted that small farm and field sizes often militate against introducing more effective cultivation practices and in such cases co-operative measures amongst farmers are necessary, although not always readily obtained.

Incorporating crop residues into the soil, or retaining standing stubble, are encouraged but are resisted by farmers in many regions. They prefer to burn, partly to control weeds and diseases, but also because stubble gets caught up in farm machinery, harbours pests and temporarily robs the soil of nitrogen.

Fruit orchards play an important soil conservation role in North Africa, sometimes in association with runoff control banks. Difficulties are associated with determining appropriate planting densities, cropping between trees and orchard maintenance. There are also problems associated with orchard monocultures (in Australia and elsewhere), whereas other forms of agriculture are diversifying.

Engineering/technical approaches

Dams are widely perceived as being the main solution to the increasing problem of water shortages – and more are planned. But dams lead to a wide array of land degradation and associated problems, including waterlogging and salinization of irrigated areas, the introduction of diseases through the changed ecology associated with the artificially created lakes, coastal erosion and, in some areas, major regional and international disputes over water ownership and use.

River "training" is also widely recommended by engineers, including bank stabilization, dredging, construction of levees and channel straightening. These may be locally efficient but often merely result in a transference of the problem downstream. Check dams are in widespread use and have been built since Roman times. Erosion is reduced upstream due to siltation but is increased downstream. To solve this problem, each check dam should be located in the siltation basin of the next check dam.

Terraces, stone walls and banks are used throughout the Mediterranean Basin, and terraces are used in association with deciduous orchards in the southwest Cape: their effectiveness and associated problems have been referred to above. In North Africa, erosion control banks have always been preferred in association with biological methods, usually fruit trees, with multi-purpose objectives. In the southwest Cape, water runoff is managed by means of wide, shallow ditches lined with an indigenous, perennial grass or, increasingly, concrete. Several engineering structures are used extensively for soil conservation in Chile. Grade banks and grassed waterways are widely used for erosion control in southern Australia, while deeper interceptors, developed in relation to the control of secondary salinization, mainly in Western Australia, are perhaps unique in being designed to intercept throughflow as well as overland flow. As in North Africa, the interceptors are increasingly being constructed in association with tree planting.

In California, a variety of major engineering approaches are used to deal with subsidence resulting from oil extraction and pumping of groundwater. This region, with Israel, pioneered a range of drip-sprinkler methods of delivering irrigation water more efficiently to plants.

Biological approaches

The setting aside of naturally vegetated areas in reserves or national parks is a widely used means of protecting flora and fauna. However, this is not a substitute for careful management of the great bulk of land which is not so protected.

Reforestation is implemented in all regions, sometimes in association with terracing, but usually with introduced species, not natives (*Pinus*, *Acacia* and *Eucalyptus* in the Mediterranean Basin and elsewhere, including Australia, but even in the latter region, plantation eucalypts are often not local species). The trees are intended to stabilize slopes, encourage pedogenesis and control surface and subsurface water movements; but in order to succeed, a shrub layer and ground covers (and a herbaceous layer in vineyards and orchards) are also necessary, both for soil protection and to encourage soil biota and bioturbation. A catchment approach to reforestation is particularly appropriate, preferably integrated with agriculture.

Agro-forestry is a related development, whereby farmers are provided with a combination of regular, annual (or seasonal) income from normal farming practices with much longer term and higher risk income from tree crops. Trees may be grown for a variety of uses, of which timber is only one. In its most successful forms, agroforestry combines the objectives of maintaining and improving soil and water quality with providing habitats for diminishing wildlife.

Whole farm planning is being encouraged in southern Australia, involving the integrated management of a farm's water and soil resources through mapping, contouring, drainage, fencing and appropriate land use to optimize sustainable production. This approach may incorporate and encourage the retention and re-establishment of native vegetation on up to a third of farm areas; and in parts of Western Australia there is an extensive fox-baiting programme allied with attempts to reintroduce rare and endangered native mammals.

Economic and social solutions

In relation to the problem of forest fires, solutions include diversifying land use, managing recreation and illegal dumping, using stock to consume fuel, introducing and enforcing legal instruments, and improving environmental education. The Catalan government limits the movement of people and vehicles in forests, discourages the building of villas in forests and requires the construction of firebreaks around urban areas.

Flood "management" may include removing houses from floodplains and forbidding reconstruction. If the population is unwilling to comply, then the authorities need to ensure that current and future inhabitants have good information concerning the danger of living on the floodplain of a Mediterranean river.

Rural depopulation, especially in mountains, may be reversed by commercializing production and improving marketing, production and yields by using selected seeds and fertilizers, and developing specialized high-value products. Education and training are necessary. Soil and water conservation techniques need improvement, which can be done by modifying traditional methods and keeping the approach as straightforward as possible. Economic activities need to be diversified and tourist infrastructure improved. Local populations may reject state initiatives because of a lack of confidence, and state agencies often reject traditional methods. Therefore, development must be consensual and not imposed "from above". User groups need to be organized co-operatively at the village/country town level.

Numerous economic measures have been applied to achieve various land management objectives, including the control of land degradation. The measures include tax incentives, grants, loans, rebates, awards, subsidies and penalties. All hold promise, but all have or give rise to a number of difficulties and distortions.

Policies

The European Union promotes measures to ensure "sustainable growth protecting the environment", and most if not all Mediterranean countries have environmental protection legislation, largely modelled on the United States National Environmental Policy Act (NEPA) of 1969. In some areas this legislation has been introduced only recently and there has been insufficient time for it to have had an impact. On the other hand there have been laws for nearly four millennia in the eastern Mediterranean concerning the proper use of water so as not to damage neighbouring fields, and rules for 2000 years restricting forest exploitation. The efficacy of such laws is clearly dubious.

There are policies concerning integrated approaches to land management in North Africa which are of particular interest. They have downstream objectives of reducing dam sedimentation in order to guarantee water supply, and upstream objectives of increasing production and improving incomes (with highly productive orchards and intensive fodder production zones), and limiting rural depopulation. Unfortunately there are few successful schemes, due to problems with funding, their duration, land tenure, failure to modify traditional techniques and failure to ensure the effective participation of the people.

Legislative/policy solutions to land degradation in California commenced with mining legislation, followed by the creation of national parks and the national forest system, initiatives dealing with over-grazing and soil conservation, and more recently legislation dealing with urban planning and coastal zone management. Onshore mining is considered to be far less a problem today than continuing environmental threats posed by urban growth.

Chile, too, has numerous laws which impinge on environmental problems. A particular difficulty is the potential conflict between policies designed to increase the incomes of essentially subsistence farmers in dry farming areas, and policies designed to manage natural resources on a sustainable basis.

In southern Australia there have been difficulties concerning the jurisdiction of agricultural and forestry agencies in relation to agro-forestry schemes. There has also been a progression of federal and state policy developments over the past two decades, from the National Soil Conservation Programme through Land Conservation Districts and District Soil Conservation Boards, to Ecologically Sustainable Development, One Billion Trees, Save the Bush, National Reforestation, Landcare, Integrated Catchment Management and Regional Planning programmes. Commendable efforts are being made to incorporate land degradation solutions into integrated, regional land use planning. In terms of planning and management policies, this approach incorporates many of the necessary requirements for effective action to combat land degradation. Whether it will succeed remains to be seen.

References

Aalouane, N., 1995: La dynamique actuelle des milieux forstiers dans le Rif central, *Publ. de l'Association des Géographes Africains*, Rabat, 21–26.

Abbott, I., Parker, C.A. and Sills, I.D., 1979: Changes in abundance of large soil animals and physical properties of soils following cultivation, *Australian Journal of Soil Research*, 17, 343–353.

Adamson, D.A. and Fox, M.D., 1982: Change in Australasian vegetation since European settlement, in J.M.B. Smith (ed.), *A history of Australian vegetation*, McGraw-Hill, Sydney, 109–146.

African National Congress, 1993: *The reconstruction and development programme*, ANC, Johannesburg.

Agassi, M., Benyamini, Y., Morin, J., Marish, S. and Henkin, E., 1986: *The Israeli concept for runoff and erosion control in semi-arid and arid zones in the Mediterranean basin*, Miscellaneous Report 30, Soil Erosion Research Station, Emek-Hefer, Israel.

Agnesi, V., Carrara, A., Macaluso, T., Monteleone, S., Pipitone, G. and Sorriso-Valvo, M., 1982: Elementi tipologici e morfologici dei fenomeni di intabilità dei versanti indotti dal sisma del 1980 (alta Valel del Sele), *Geologia Applicata e Idrogeologia*, XVIII, 309–341.

Ahmadan, A., 1991: *L'évolution récente d'un espace rural périphérique marocain, le pays Rhomara*, Thèse de Doctorat, Univ. de Tours, 2 vols.

Aït Hamza, M., Chaker, M., El Abbassi, H. *et al.*, 1995: *Analyse des techniques traditionnelles de conservation des sols et de gestion de l'eau*, Centre for Development Cooperation Services, Amsterdam.

Alexandris, S., 1989: The impacts of fires on the forestry and natural environment in *The protection of the environment and the agricultural production*, Geotechnical Society of Greece, Thessaloniki, 353–364.

Al Ifriqui, M., 1993: La dégradation du couvert végétal dans le Haut Atlas de Marrakech, in A. Bencherifa (ed.), *Montagnes et hauts pays d'afrique*, Publ. Fac. Lettres, Rabat, sér. Colloques et séminaires, 29, 319–332.

Alvarez Cobelas, M. and Cabrera Capitán, F. (eds), 1995: La calidad de las aguas continentales españolas. Estado actual e investigación (Spanish Inland Water Quality. Current State and Research). Geoforma Ediciones, Logroño, 307.

Amar, M., 1965: *Intensité-durée des précipitations au Maroc*, Météorologie Nationale, Casablanca.

American Geological Institute, 1989: *Geoscience employment and hiring survey*, American Geological Institute, Alexandria, Virginia, USA.

Amouric, H. 1992: *Le feu à l'épreuve du temps*, Narration, Aix-en-Provence.

Anon, 1994: *Crvena knjiga životinjskih svojti Republike Hrvatske – sisavci*, Ministarstvo graditeljstva i zaštite okoliša, Zavod za zaštitu prirode, Zagreb.

Arabi, M. and Roose, E., 1989: Influence du système de production et du sol sur l'érosion, *Bulletin du Réseau Érosion*, ORSTOM, Montpellier, 9, 39–51.

Archibold, O.W., 1995: *Ecology of world vegetation*, Chapman & Hall, London.

Arianoutsou, M. and Groves, R.H. (eds), 1994: *Plant and animal interactions in Mediterranean-type ecosystems*, Kluwer Academic Publishers, Dordrecht.

Arnáez-Vadillo, J., Larrea-Saenz, V. and Ortigosa Izquierdo, L., 1991: Environmental and topographical controls in geomorphological evolution of hill-roads (Iberian System, La Rioja, Spain), in M. Sala *et al.* (eds), *Soil erosion studies in Spain*, Geoforma Ediciones, Logroño, 27–39.

Arnaud-Fassetta, G., Ballais, J.-L., Beghin, E. *et al.*, 1993: La crue de l'Ouvèze à Vaison-la-Romaine (22 septembre 1992). Ses effets morphodynamiques, sa place dans le fonctionnement d'un géosystème anthropisé, *Revue de Géomorphologie Dynamique*, 42, 34–48.

ASCC, 1988: *National soil conservation strategy*, Australian Soil Conservation Council, Australian Government Publishing Service, Canberra.

Aschmann, H., 1973: Man's impact on the several regions with Mediterranean climates, in F. di Castri and H.A. Mooney (eds), *Mediterranean type ecosystems*, Springer-Verlag, New York, 363–371.

Aslin, H.J., 1986: Native animals, in C. Nance and D.L. Speight (eds), *A land transformed: environmental change in South Australia*, Longman Cheshire, Melbourne, 55–77.

Atlas Aguila, 1961: Aguilar, Madrid.

Atlas du Maroc, 1962: Précipitations naturelles, au 1/2 000 000, Sheet 4a.

Atlas du Maroc, 1963: Etages bioclimatiques, au 1/2 000 000, Sheet 6b.

Atlas du Maroc, 1984: La population rurale, au 1/2 000 000, Sheet 31.3.

Atlas Nacional de España, 1996: Ministerio de Obras Públicas y Transportes & Dirección General del Instituto Geográfico Nacional, Madrid.

Atlas of Australian Resources, Division of National Mapping, Canberra. 6 vols.

Audurier-Cros, A., 1976: *Contribution à l'etude de l'environnement de la Zone de Fos et de l'Etang de Berre. La Pollution Atmospherique. Ses implications, ses aspects*, Thèse 3ème cycle, Université d'Aix-Marseille III, 404 pp.

Audurier-Cros, A., 1982: Environnement industriel et qualité de l'air: le cas de Gardanne, *Méditerranée*, Aix-en-Provence, 3, 95–103.

AUSLIG, 1988: *Atlas of Australian resources, third series, volume 5. Geology and minerals*, Australian Surveying and Land Information Group, Department of Administrative Services, Canberra.

Australian Academy of Science, 1994: *Environmental science*, Australian Academy of Science, Canberra.

Australian Agricultural Council, 1991: *Sustainable agriculture*, Report by Working Group, CSIRO, Melbourne.

Australian Bureau of Statistics, 1992: *Australia's environment: issues and facts*, Catalogue No. 4140.0, Commonwealth of Australia, Canberra.

Avolio, S., Ciancio, O., Grinovero, C. *et al.*, 1980: Effetti del tipo di bosco sull'entità dell'erosione in unità idrologiche della Calabria – Modelli erosivi, *Annali dell'Ist. Sperimentale per la Selvicoltura*, XI, 45–131.

Avon River Management Authority, 1993: *Avon river: report to the community*, Western Australian Waterways Commission, Perth.

Baccouche, N., 1990: La forêt du massif de Sainte Victoire et les incendies: le cas de l'incendie de 1989, Mémoire D.E.A., Université d'Aix-Marseille II.

Bailey, R.G., 1995: *Description of the ecoregions of the United States*, Forest Service, US Department of Agriculture, Publication 1391, Washington DC.

Baldock, D. and Beaufoy, G., 1992a: *Plough on! An environmental appraisal of the reformed CAP (Common Agricultural Policy)*, A report to the World Wildlife Fund from the Institute for European Environmental Policy, London.

Baldock, D. and Beaufoy G., 1992b: *Green or mean? Assessing the environmental value of the CAP reform "accompanying measures"*, A report to the CPRE from the Institute for European Environmental Policy, London.

Baleste, M., Boyer, J.-C., Gras, J., Montagné-Villette, S. and Vareille, C., 1993: *La France: 22 régions de programme*, Masson, Paris.

Ballais, J.-L., 1993: L'érosion consécutive à l'incendie d'août 1989 sur la Montagne Sainte-Victoire: trois années d'observation (1989–1992), *Bulletin de l'Association de Géographes Français*, Paris, 5, 423–437.

Ballais, J.-L. and Bosc, M.-C., 1994: The ignifracts of the Sainte-Victoire mountain (Lower Provence, France), in M. Sala and J. Rubio (eds), *Soil erosion and degradation as a consequence of forest fires*, Geoforma Ediciones, Logroño, 217–227.

Ballais, J.-L. and Crambes, A., 1992: Morphogenèse Holocène, géosystèmes et anthropisation sur la montagne Sainte-Victoire, *Méditerranée*, Aix-en-Provence, 1, 29–41.

Ballais, J.-L. and Meffre, J.-C., 1995: Vaison et ses campagnes dans l'Antiquité et le Haut Moyen Age (Haut Comtat Venaissin, Vaucluse): Archéologie de l'espace rural, in S.E. Van Der Leeuw (ed.), *Understanding the natural and anthropogenic causes of soil degradation and desertification in the Mediterranean basin*, report for the European Community, Cambridge, vol. 3, part I, 37–53.

Ballais, J.-L., Jorda, M., Provansal, M. and Covo, J., 1993: Morphogenèse Holocène sur le périmètre des Alpilles, *Travaux du Centre Camille Jullian*, Aix-en-Provence, 14, 515–547.

Ballais, J.-L., Garry, G. and Masson, M., in press: *Prévention des risques d'inondation*, Ministère de l'Equipement, Paris.

Baric, A. and Gasparovic, F., 1992: Implications of climatic change on the socio-economic activities in the Mediterranean coastal zones, in L. Jeftic, J.D. Milliman and G. Sestini (eds), *Climatic change and the Mediterranean*, Edward Arnold, London, 129–174.

Baruch, V., 1987: The Late Holocene vegetational history of Lake Kinneret (Sea of Galilee), Israel, *Paleorient*, 12, 37–48.

Battesti, A.. 1991: Les incendies en Corse, *Méditerranée*, Aix-en-Provence, 1, 39–42.

Batušić, N., 1978: *Povijest hrvatskog kazališta*, Školska knjiga, Zagreb.

Baxter, A.J. and Meadows, M.E., 1994: Palynological evidence for the impact of colonial settlement within lowland *fynbos*: a high resolution study from the Verlorenvlei, southwestern Cape, South Africa, *Historical Biology*, 9, 61–70.

Beaudet, G., 1969: *Le plateau central Marocain et ses bordures. Étude géomorphologique*, Thesis, Imprimeries Françaises et Marocaines, Rabat.

Becat J. and Soutade, G. (eds), 1993: *L'aiguat del 40*, Generalitat de Catalunya, Barcelona.

Beguin, E., 1992: *Erosion mécanique après l'incendie de forêt du massif des Maures de 1990: exemple du bassin versant du Rimbaud*, Mémoire de Diplome d'Etudes Approfondies, Université de Provence.

Benabid, A., 1982: *Etude phytoécologique, biogéographique et dynamique des associations et séries sylvatiques du Rif occidental, Thèse de Doctorat, Univ. d'Aix-Marseille.*

Benabid, A., 1992: Dégradation des écosystèmes forestiers marocains, in *Environnement-Développement*, Publ. Comité Environnement, Rabat, 79–84.

Benazzouz, M.T., 1993: La dégradation des terres dans les montagnes telliennes orientales Algériennes, in A. Bencherifa (ed.), *Montagnes et hauts pays d'afrique*, Publ. Fac. Lettres, Rabat, sér. Colloques et séminaires, no. 29, 257–266.

Bencherifa, A. and Johnson, D.L., 1993: Environment, population pressure and resource use strategies in the Middle Atlas mountains of Morocco, in A. Bencherifa (ed.), *Montagnes et hauts pays d'afrique*, Publ. Fac. Lettres, Rabat, no. 29, 101–122.

Benchetrit, M., 1972: *L'érosion actuelle et ses conséquences sur l'amenagément de l'Algérie*, Publ. Univ. Poitiers, vol. 11, Presses Universitaires de France, Paris.

Benito, G., Gutiérrez, M. and Sancho, C., 1991: Erosion patterns in rill and interrill areas in badlands zones of the middle Ebro basin (NE Spain), in M. Sala *et al.* (eds), *Soil erosion studies in Spain*, Geoforma Ediciones, Logroño, 41–54.

Benjelid, A., 1993: Processus récents de dégradation de l'environnement agricole autour des villes algériennes, in *Communication au 2ème Congrès des Géographes Africains*, Rabat.

Bennett, H.H., 1945: *Soil erosion and land use in the Union of South Africa*, Department of Agriculture, Pretoria.

Bennett, H.H., 1960: Soil erosion in Spain, *Geographical Review*, 10, 59–72.

Bensaad, A., 1993: Climat et potentiel hydrologique en Algérie, *Travaux de l'institut de Géographie de Reims*, 85/86, 5–14.

Benzarti, Z., 1987: La pluviométrie indice de sécheresse, tendances pluriannuelles, report for UNESCO, abstract published in *Communication au 2ème Congrès des Géographes Africains*, Rabat et Agadir, avril 1993, 274.

Bernard-Allee, P., Valadas, B., Cosandey, C. *et al.*, 1991: Forest harvest geomorphic effects in a submediterranean granitic middle mountain, *Zeitschrift für Geomorphologie*, Supplementband, 83, 1–8.

Biasini, A., D'Alessandro, L. and De Marco, R., 1993: Geomorphological stability in Italy, in E. Lupia-Palmieri (coordinator), *Report on the state of the environment in Italy*, Ministry of the Environment, Instituto Poligrafico e Zecca della Stato, Roma.

Bielza, V., 1989: La degradación del cuadro natural, in V. Bielza (ed.), *Territorio y sociedad en España I*, Taurus, Madrid, 403–431.

Bilandzija, J., 1993: Forest fires in Croatia – a permanent danger, *International Forest Fire News*, no. 9, UN Economic Commission for Europe and FAO, Geneva, 13–15.

Bilello, A., 1994: Emergenza idrica in agricoltura, in G. Rossi and G. Margariota (eds), *La siccità in Italia, 1988–1990*, Presidenza del Consiglio dei Ministri, Diparimento della Protezione Civile, 143–162.

Birot, P., 1964: *La Méditerranée et le Moyen orient*, tome 1, Presses Universitaires de France, Coll. Orbis, Paris.

BLM, 1987: *Public land statistics 1987*, Bureau of Land Management, Washington, DC.

Blue Plan, 1988: Futures of the Mediterranean Basin: Environmental Development 2000–2025, Sophia Antipolis.

Blumler, M.A., 1995: Invasion and transformation of California's valley grassland: a Mediterranean analogue system, in R. Butlin and N. Roberts (eds), *Human impact and adaptation: ecological relations in historical times*, Blackwell, Oxford, 308–332.

Boelhouwers, J.C., de Graaf, P.J. and Samsodied, M., in press: The influence of wildfire on soil properties and hydrological response at Devil's Peak, Cape Town, South Africa, *Zeitschrift für Geomorphologie*, Supplementband, 7.

Bolton, G., 1981: *Spoils and spoilers: Australians make their environment 1788–1980*, Allen & Unwin, Sydney.

Bolton, H.E., 1927: *Fray Juan Crespi, missionary explorer of the Pacific Coast, 1769–1774*, University of California Press, Berkeley and Los Angeles.

Bonazountas, M. and Katsaiti, A., 1995: *Selected topics on management of the environment*, Goulandri Museum of Natural History, Greece.

Bond, P. and Goldblatt, P., 1984: Plants of the Cape flora. A descriptive catalogue, *Journal of South African Botany*, supplementary volume 13, 1–455.

Borgel, R., 1983: *Geomorfología de Chile*, Coleccion Geografia de Chile, Tomo II, Instituto Geográfico Militar, Santiago.

Borgel, R., 1988: *Geomorfología*, Colección Geografía de Chile, Ed. Instituto Geográfico Militar, vol. 2, Santiago, Chile.

Borzan, Ž., Lovrić, A.Ž. and Rac, M., 1994: Hrvatski biljni endemi, *Šume u Hrvatskoj*, Grafički zavod Hrvatske, Zagreb, 223–236.

Boudy, P., 1950: *Economie forestière nord-Africaine*, Editions Larose, Paris, vols 1–4.

Boujarra, M., 1993: La dynamique des milieux dans la moitié est du bassin-versant de l'oued Sbiba, Tunisie centrale, in A. Bencherifa (ed.), *Montagnes et hauts pays d'afrique*, Publ. Fac. Lettres, Rabat, sér. Colloques et séminaires, 29, 301–318.

Bourdeau, P. Rolando, C. and Teller, A. (eds), 1987: Influence of fire on the stability of Mediterranean forest ecosystems, *Ecologia Mediterranea*, Marseille, vol. 13, special issue.

Bourgou, M., 1995: Le bassin-versant du Kébir-Miliane, Tunisie nord-orientale, *Publ. de l'Association des Géographes Africains*, Rabat, 69–86.

Bourman, R.P. and Harvey, N., 1986: Landforms, in C. Nance and D.L. Speight (eds), *A land transformed: environmental change in South Australia*, Longman Cheshire, Melbourne, 78–125.

Bourouba, M., 1993: Erosion dans les bassins-versants d'Algérie orientale, *Trav. Inst. Géogr. Reims*, 85/86, 15–24.

Bourriau, J., 1993: *Le catastrofi*, Edizioni Dedalo, Bari.

Bowie, I.J.S. and Smailes, P.J., 1988: The country town, in R.L. Heathcote (ed.), *The Australian experience*, Longman Cheshire, Melbourne, 233–256.

Boyatzoglou, B., 1983: *Progressive inventory of the requirements of Greek agriculture with special reference to agronomic research*, Commission of European Communities, Report EUR 863 EN, FR.

Boza, J., Silva, J. and Azocar, P., 1985: Recursos alimenticios en zonas áridas, *Simposium Int. Explotación caprina en zonas áridas*, Fuerteventura, 191–223.

Bradsden, J., 1991: Perspectives on land conservation, *Environmental and Planning Law Journal*, 8, 16–40.

Brandmayr, P. and Pizzolotto, R., 1994: Ecosistemi terrestri in Calabria e loro stato di conservazione, *Proceedings, 1st Convegno "Energia, Clima e Ambiente"*, Cittadella del Capo, 6–8 Maggio 1994, 155–158.

Brandt, C.J. and Thornes, J.B. (eds), 1996: *Mediterranean desertification and land use*, John Wiley & Sons, Chichester.

Breckwoldt, R., 1983: *Wildlife in the home paddock: nature conservation for Australian farmers*, Angus and Robertson, Sydney.

Breckwoldt, R., 1986: *The last stand: managing Australia's remnant forests and woodlands*, Australian Government Publishing Service, Canberra.

Brewer, W.H., 1864: *Up and down California in 1860–1864*, ed. F.P. Farquhar, 3rd edition, 1966, University of California Press, Berkeley and Los Angeles.

Bruckner, H., 1986: Man's impact on the evolution of the physical environment in the Mediterranean region in historical times, *GeoJournal*, 13, 7–17.

Bryan, R. and Yair, A. (eds), 1988: *Badland geomorphology and piping*, Geo Books, Norwich.

Bryant, E., 1848: *What I saw in California . . . in the years 1846, 1847*, Appleton, New York.

Bryant, L., 1992: Social aspects of the farm financial crisis, in G. Lawrence *et al.* (eds), *Agriculture, environment and society: contemporary issues for Australia*, Macmillan, Melbourne, 157–172.

Budd, G., 1987: Tillage research update and findings, in A.W. McGufficke (ed.), *Wind erosion and its control on the aeolian soils of south eastern Australia: Proceedings of the inter-state wind erosion workshop and research update held in Mildura on the 23rd and 24th September 1987*, National Soil Conservation Programme.

Buondonno, C., Di Giaimo, A., Leone, A.P. *et al.*, 1993: Carta dell'erosione potenziale massima della Comunità Montana "Fortore Beneventano", *Annali Facoltà Scienze Agrarie Univ. di Napoli*, Serie IV, vol. XXVII, 20–33.

Bussotti, F., Rinallo, C., Grossoni, P. *et al.*, 1984: La morìa della vegetazione costiera causata dall'inquinamento idrico, *Monti e Boschi*, 6, 47–55.

Butzer, K.W., 1964: *Environment and archeology*, Aldine Publishing Co., Chicago.

Cabrera-Capitán, F., 1995: El alpechín: un problema mediterráneo, in M. Alvarez-Cobelas and F. Cabrera-Capitán (eds), *La calidad de las aguas continentales españolas: Estado actual e investigación*, Geoforma Ediciones, Logroño, 141–154.

Caelleigh, A.S., 1983: Middle East water: vital resource, conflict and cooperation, in R.S. Joyce and A.S. Calleigh (eds), *A shared destiny*, Praeger, New York.

Caloiero, D. and Mercuri, T., 1982: *Le alluvioni in Basilicata dal 1921 al 1980*, Geodata, 16, 67pp.

Campbell, A., 1994: *Landcare: communities shaping the land and the future*, Allen & Unwin, St Leonards, New South Wales.

Campos Palacín, P., 1992: Spain, in T. Jones and S. Wibe (eds), *Forests: market and intervention failures, five case studies*, Earthscan, London, 165–200.

Carne, R.J., 1993: Agroforestry land use: the concept and practice, *Australian Geographical Studies*, 31(1), 79–90.

Cartaña, X.U. and Sanjaume, M.S., 1994: Erosion as a consequence of rains immediately following a forest fire, *Proc. 2nd Int. Conf. Forest Fire Research*, Coimbra, vol. II, D32, 1139–1148.

Ceballos, L., 1966: *Mapa forestal de España*, Ministerio de Agricultura, Madrid.

Cervera, M., Clotet, N. and Sala, M., 1990: *Runoff and sediment production on small badland basins in the Upper Llobregat catchment (submediterranean environment)*, Internal Report, LUCDEME Project.

Chaker, M., 1995: La dégradation actuelle du couvert végétal dans le massif de Boukhouali, processus et impact sur l'éqilibre du milieu, *Publ. de l'Association des Géographes Africains*, Rabat, 87–102.

Chisholm, A.H., 1992: Australian agriculture: a sustainability story, *Australian Journal of Agricultural Economics*, 36, 1–29.

Ćirić, M., 1984: *Pedologija*, I izdanje, SOUR "Svjetlost", OOUR Zavod za udžbenike i nastavna sredstva, Sarajevo.

Čižek, J. and Varga, B., 1988: *Tehničko tehnološke mogućnosti razvoja Zadarske regije, razvojni pravci primarne poljoprivredne proizvodnje, razvoj oranične i travnjačko-pašnjačke proizvodnje*, Projekt, Agronomski fakultet, Zagreb.
Clark, W.B., 1979: Fossil river beds of the Sierra Nevada, *California Geology*, 32, 143–149.
Clarke, A.L., 1986: The impact of agricultural practices on Australian soils: cultivation, in J.S. Russell and R.F. Isbell (eds), *Australian soils: the human impact*, Queensland University Press, St Lucia, 273–303.
Claro, E. and Wilson, G.A., 1996: Trans-Pacific wood chip exports: the rise of Chile, *Australian Geographical Studies*, 34, 185–199.
Cody, M.L. and Mooney, H.M., 1978: Convergence versus nonconvergence in Mediterranean-climate ecosystems, *Annual Review of Ecology and Systematics*, 9, 265–321.
Coelho, C.O.A., Shakesby, R.A. and Walsh, R.P.D., 1995: *Effects of Forest Fires and Post-Fire Land Management Practice on Soil Erosion and Stream Dynamics*. Agueda Basin, Portugal. European Commission – Soil and Groundwater Research Report V, EUR 15689 EN, p. 91.
Cohen, M.N., 1977: *The Food Crisis in Prehistory*, Yale University Press, Yale.
Cole, M.M., 1961: *South Africa*, Methuen, London.
Colli, M. and Grandini, G., 1994: Evoluzione e compatibilità ambientale dell'attività estrattiva del marmo di Carrara, *Geoengineering Environment and Mining*, 83(2–3), 111–116.
Commission of the European Communities, 1985: *Situation of agriculture in the Community. Annual report: 1985*, Office for Official Publications of the European Communities, L-2985, Luxembourg.
Commission of the European Communities, 1988: *Travelling in the EEC*, EEC XI/E-5, Luxembourg.
Commission of the European Communities, 1992: *CORINE – Soil erosion risk and land resources in the southern regions of the European Community*, EUR 13233, Office for the Publications of the European Communities.
Commission of the European Communities, 1993: *The agricultural situation in the Community*, 1992 report, Luxembourg.
Commonwealth Department of Primary Industries and Energy, undated (November 1992): *Towards a national weeds strategy*, Canberra.
Commonwealth of Australia, 1982: *Hazardous chemical wastes: storage, transport and disposal*, Report of the House of Representatives Standing Committee on Environment and Conservation, First Report on the Inquiry into Hazardous Chemicals, Parliament of the Commonwealth of Australia, Australian Government Publishing Service, Canberra.
Community Education and Policy Development Group, 1993: *The state of the environment report for South Australia 1993*, Department of Environment and Land Management, Adelaide.
Conacher, A.J., 1982: Dryland agriculture and secondary salinity, in W. Hanley and M. Cooper (eds), *Man and the Australian environment*, McGraw-Hill, Sydney, 113–125.
Conacher, A.J., 1983: Environmental management implications of intensive forestry practices in an indigenous forest ecosystem: a case study from southwestern Australia, in T. O'Riordan and R.K. Turner (eds), *Progress in Resource Management and Environmental Planning*, vol. 4, Wiley, Chichester, 117–151.
Conacher, A.J., 1995: Definition of Mediterranean climates, *GeoJournal*, 36, 298.
Conacher, A.J. and Conacher, J.L., 1995: *Rural land degradation in Australia*, Oxford University Press, Melbourne.
Conacher, A.J. and Murray, I.D., 1973: Implications and causes of salinity problems in the Western Australian wheatbelt: the York-Mawson area, *Australian Geographical Studies*, 11(1), 40–61.
Conacher, A.J., Combes, P.L., Smith, P.A. and McLellan, R.C., 1983a: Evaluation of throughflow interceptors for controlling secondary soil and water salinity in dryland agricultural areas of southwestern Australia: I. Questionnaire surveys, *Applied Geography*, 3(1), 29–44.
Conacher, A.J., Neville, S.D. and King, P.D., 1983b: Evaluation of throughflow interceptors for controlling secondary soil and water salinity in dryland agricultural areas of southwestern Australia: II. Hydrological study, *Applied Geography*, 3(2), 115–132.
Conacher, J.L. and Conacher, A.J., 1986: *Herbicides in agriculture: minimum tillage, science and society*, Geowest no. 22, Dept. Geography, University of Western Australia.
CONAF, 1989: *Red list of Chilean terresterial flora*, ed. I. Benoit, Chilean Forest Service, Santiago, Chile.
CONAMA-MINAGRI, 1994: *Propuesta para un Plan Nacional de Conservación de Suelos*, Comision Nacional del Medio Ambiente, Ministerio de Agriculture, Ducumento Técnico.
Conseil Supérieur de l'Eau, 1991a: *Aménagement des bassins-versants et protection des barrages contre l'envasement*, Rabat, 5éme session.
Conseil Supérieur de l'Eau, 1991b: *Plan-Directeur de développement des resources en eau du bassin de la Moulouya*, Rabat, 5éme session.
Conseil Supérieur de l'Eau, 1991c: *Préservation du patrimoine hydraulique, protection de la qualité de l'eau contre la pollution*, Rabat, 5éme session, Rabat.
Conseil Supérieur de l'Eau, 1993: *Plan-Directeur intégré d'aménagement des eaux des bassins du Loukkos, du Tangérois et des Côtiers Méditerranéens*, Rabat, 7ème session.

Conte, M., 1994: Situazioni meteorologiche del Mediterraneo ed andamento delle precipitazioni sul territorio Italiano, in G. Rossi and G. Margariota (eds), *La siccità in Italia, 1988–1990*, Presidenzas del Consiglio dei Ministri, Dipartimento della Protezione Civile, 13–40.

Cooke, J.W., Ford, G.W., Dumsday, R.G. and Willatt, S.T., 1985: Effects of fallowing practices on the growth and yield of wheat in south-eastern Australia, *Australian Journal of Experimental Agriculture*, 25, 614–627.

CORINE, 1992: *Soil erosion risk and important land resources*, Commission of the European Communities, Directorate-General, Environment, Consumer Protection and Nuclear Safety, B-1049, 97 pp.

Corrie, H., 1991: *Reforming the EC Common Agricultural Policy*, World Wildlife Fund International CAP (Common Agricultural Policy) Discussion Paper, Brussels.

Corte, M., 1995: *Il libro dei fatti 1995*, Adnkronos, Roma.

Cote, A. and Legras, J., 1966: La variabilité pluviométrique interannuelle au Maroc, *Revue de Géographie du Maroc*, 10, 19–30.

Cote, M., 1983: La population de l'Algérie, *Méditerranée*, 4, 95–100.

Cote, M., 1996: *L'Algérie*, Masson-Armand Colin.

Cotecchia, V. and Melidoro, G., 1974: Some principal geological aspects of the landslides of Southern Italy, *Bulletin International Association Engineering Geology*, 9, 23–32.

Cotecchia, V., Travaglini, G. and Melidoro, G., 1969: I movimenti franosi e gli sconvolgimenti della rete idrografica prodotti in Calabria dal terremoto del 1783, *Geologia Applicata e Idrogeologia*, 4, 1–24.

Couvreur-Laraichi, F., 1972: Les précipitations dans quelques stations de la mer d'Alboran, *Revue de Géographie du Maroc*, 21, 85–104.

Cowan, G.I. (ed.), 1995: *Wetlands of South Africa*, Department of Environmental Affairs and Tourism, Pretoria.

Cowling, R.M. (ed.), 1992: *The ecology of fynbos: nutrients, fire and diversity*, Oxford University Press, Cape Town.

Cowling, R.M. and Richardson, D.J., 1995: *Fynbos: South Africa's unique floral kingdom*, Fernwood Press, Vlaeberg, Cape Town.

Cowling, R.M., Campbell, B.M., Mustart, P. *et al.*, 1988: Vegetation classification in a floristically complex area: the Agulhas Plain, *South African Journal of Botany*, 54, 290–300.

Cowling, R.M., Holmes, P.M. and Rebelo, A.G., 1992: Plant diversity and endemism, in R.M. Cowling (ed.), *The ecology of fynbos: nutrients, fire and diversity*, Oxford University Press, Cape Town, 62–112.

Cribb, J., 1989: Agriculture in the Australian economy, in J. Cribb (ed.), *Australian agriculture: the complete reference on rural industry*, vol. 2, Morescope, Camberwell, Victoria, 11–48.

Cribb, J. (ed.), 1991: *Australian agriculture: the complete reference on rural industry*, 3rd edition, Morescope, Camberwell, Victoria.

CTIC, 1988: *1987 national survey: conservation tillage practices*, Conservation Technology Information Center, West Lafayette, Indiana.

CWA, 1995: *Survey of rural poverty and its effects on rural and remote communities*, Country Women's Association of Australia, National Office, Queensland.

D'Alessandro, L. and Lupia-Palmieri, E., 1981: Lineamenti morfologici ed evoluzione della spiaggia emersa, in *Primi risultati delle indagini di geografia fisica, sedimentologia e idraulica marittima sul litorale del golfo di S. Eufemia, Progetto Finalizzato "Conservazione del Suolo", Sottoprogetto "Dinamica dei Litorali"*, CNR, publication no. 127, 25–36.

Danalatos, N.G., 1993: *Quantified analysis of selected land use systems in the Larissa region, Greece*, PhD Thesis, Agricultural University of Wageningen, Wageningen, 370 pp.

Daveau, S., 1982: Les températures au Portugal et en Espagne d'aprés les satellites Météosat et HCMM, *Finisterra*, 33, 53–96.

Davidson, R. and Davidson, S., 1992: *Bushland on farms: do you have a choice?* Australian Government Publishing Service, Canberra.

Davies, B.R., O'Keeffe, J.H. and Snaddon, C.D., 1993: *A synthesis of the ecological functioning, conservation and management of South African river ecosystems*, Water Research Commission Report TT62/93, Pretoria.

Deacon, H.J., 1983: The peopling of the fynbos region, in H.J. Deacon *et al.* (eds), *Fynbos palaeoecology: a preliminary synthesis*, South African National Scientific Programmes Report 75, 183–204.

Deacon, H.J., 1992: Human settlement, in R.M. Cowling (ed.), *The ecology of fynbos: nutrients, fire and diversity*, Oxford University Press, Cape Town, 260–270.

Deacon, H.J., Jury, M.R. and Ellis, F., 1992: Selective regime and time, in R.M. Cowling (ed.), *The ecology of fynbos: nutrients, fire and diversity*, Oxford University Press, Cape Town, 6–22.

Dean, W.R.J., Hoffman, M.T., Meadows, M.E. and Milton, S.J., 1995: Desertification in the semi-arid Karoo, South Africa, review and reassessment, *Journal of Arid Environments*, 30, 247–264.

Debazac, E. and Robert, P., 1973: *Recherches relatives à la quantification de l'érosion*, Document no. 4, Publications du projet érosion, FAO, Rabat.

De Dolomieu, D., 1785: *Memoria sopra i tremuoti della Calabria*, Librai Francesci rimpetta S. Angelo a Nido.

Deil, U., 1987: La végétation actuelle et l'occupation des terres dans la région du Jbel Arz, *Etudes méditerranéennes*, 11, 241–256.

Deil, U., 1988: La distribution actuelle et potentielle du cèdre dans le Haut Rif central, *Revue de Géographie du Maroc*, 12/1, nouvelle série, 17–32.

Delannoy, H., 1971: Aspects du climat de la région de Marrakech, *Revue de Géographie du Maroc*, 20, 69–106.

Demmak, A., 1984: Recherche d'une relation empirique entre opports solides specifiques et parametres physico-climatiques des bassin: application au cas Algérien, in International Association of Hydrological Sciences Publication, 144, 403–414.

Department of Agriculture, 1994: *Agriculture in South Africa*, van Rensburg Publications, Johannesburg.

Department of Arts, Heritage and Environment, 1986: *State of the environment in Australia source book*, Australian Government Publishing Service, Canberra.

Department of Conservation and Environment, 1979: *Cockburn Sound environmental study 1976–1979*, Department of Conservation and Environment, Perth.

Department of Conservation and Environment, 1984: *Management of Peel Inlet and Harvey Estuary. Report of research findings and options for management*, Department of Conservation and Environment Bulletin 170, Prepared by Kinhill Stearns, Perth.

Department of Environment and Land Management, 1993: *State of the environment report for South Australia*, Department of Environment and Land Management, Adelaide.

Department of Environment, Housing and Community Development, 1978: *A basis for soil conservation policy in Australia: Commonwealth and State Government collaborative soil conservation study 1975–77. Report 1*, Australian Government Publishing Service, Canberra.

Department of Home Affairs and Environment, 1984: *A national conservation strategy for Australia*, Australian Government Publishing Service, Canberra.

Department of Housing and Urban Development, 1993: *Mount Lofty Ranges regional strategy plan*, Department of Housing and Urban Development, Adelaide.

Department of Primary Industries and Energy, 1990: *Public land fire management*, Standing Committee, Australian Forestry Council, Australian Government Publishing Service, Canberra.

Department of Water Affairs, 1994: *Explanation of the Cape Town (3317) Hydrogeological Map, 1:50 000*, DWA, Cape Town.

De Ploey, J., 1994: Introduction and conclusions, in M. Sala and J.L. Rubio (eds), *Soil erosion as a consequence of forest fires*, Geoforma Ediciones, Logroño, 13–15; 275.

Desailly, B., 1990: *Crues et inondations en Roussillon. Le risque et l'aménagement – Fin du XVIIème siècle – milieu du XXème siècle*, Thèse, Université Paris X.

Despois, J. and Raynal, R., 1967: *Géographie de l'Afrique du Nord-Ouest*, Payot, Paris.

Development Bank of Southern Africa, 1994: *South Africa's nine Provinces: a human development perspective*, DBSA, Halfway House.

De Tavel, D. 1812: *Lettere della Calabria* (transl. in Italian 1985), Rubettino, Reggio Calabria.

DGQA (Direcção Geral da Qualidade do Ambiente), 1992: Data in Ministério do Planeamento e Administração do Território & Ministério do Ambiente e Recursos Naturais, *Relatório do Estado do Ambiente e Ordenamento do Território*, Lisbao.

Díaz-Fierros, F. and Benito, E., 1996: Rainwash erodibility of Spanish soils, in J.L. Rubio and A. Calvo (eds), *Soil degradation and desertification in Mediterranean environments*, Geoforma Ediciones, Logroño, 91–104.

Días-Fierros, F., Gil, F., Cabaneiro, A., Carballas, T., Leiras, M.C. and Villar, M.C., 1982: *Effectos erosivos de los incendios forestales en suelos de Galícia*, Anales Edafología y Agrobiología, 41, 3–4, 627–639.

Di Castri, F., 1991: An ecological overview of the five regions of the world with a Mediterranean climate, in R.H. Groves and F. di Castri (eds), *Biogeography of Mediterranean invasions*, Cambridge University Press, Cambridge, 3–16.

Di Castri, F., Goodall, D.W. and Specht, R. (eds), 1981: *Mediterranean type shrublands*, Elsevier, Amsterdam.

Di Castri, F. and Hajek, E., 1977: *Bioclimatología de Chile*, Vicerrectoría Académica de la Universidad Católica de Chile, Santiago.

Di Castri, F., Hansen, A.J. and Debussche, M. (eds), 1990: *Biological invasions in Europe and the Mediterranean basin*, Kluwer Academic Publishers, Dordrecht.

Di Castri, F. and Mooney, H.A. (eds), 1973: *Mediterranean type ecosystems: origin and structure*, Chapman and Hall, London.

Dimase. A.C. and Iovino, F., 1988: *Carta dei suoli dei bacini idrografici del Trionto, Nicà e torrenti limitrofi (Calabria)*, CNR, Istituto di Ecologia e Idrologia Forestale, Cosenza.

Direction de l'Aménagement du Territoire, 1992: *Schéma National d'Aménagement du Territoire*, Dossier Ressources Naturelles, Rabat.

Division of National Mapping, 1980: *Atlas of Australian resources, third series, volume 1. Soils and land use*, Commonwealth of Australia, Canberra.

Division of National Mapping, 1982: *Atlas of Australian resources, third series, volume 3. Agriculture*, Commonwealth of Australia, Canberra.

Donaldson, S., 1995: Farm level perspective of constraints to dealing with Landcare, *Outlook 95 Conference*, Australian Bureau of Agriculture and Resource Economics, 190–197.

Douguédroit, A., 1991: Influence of a global warming on the risk of forest fires in the French Mediterranean area, in K. Takouchi and M. Yoshiro (eds), *The global environment*, Springer-Verlag, Berlin, 85–96.

Douguédroit, A. and Zimina, R.-P., 1987a: Le climat méditerranéen en France et en U.R.S.S., *Méditerranée*, Aix-en-Provence, 2, 75–84.

Douguédroit, A. and Zimina, R.-P., 1987b: La végétation méditerranéenne en France et en U.R.S.S., *Méditerranée*, Aix-en-Provence, 2, 71–72.

Downes, R.G., 1956: Conservation problems on solodic soils in the State of Victoria, *Journal of Soil and Water Conservation*, 11, 228–232.

DPIE/ASCC, 1988: *Report of working party on effects of drought assistance measures and policies on land degradation*, Department of Primary Industries and Energy/Australian Soil Conservation Council, Australian Government Publishing Service, Canberra.

DPUD, 1991: *Albany regional planning study regional rural strategy*, Department of Planning and Urban Development, Perth.

DPUD, 1993a: *Bunbury–Wellington region plan*, Department of Planning and Urban Development, Perth.

DPUD, 1993b: *Albany regional planning study Albany region plan*, Department of Planning and Urban Development, Perth.

DPUD, 1994a: *Central coast regional profile, incorporating parts of the Shires of Irwin, Carnamah, Coorow, Dandaragan and Gingin*, Department of Planning and Urban Development, Perth.

DPUD, 1994b: *Central coast regional strategy*, prepared for the Central Coast Planning Study Steering Committee by the Department of Planning and Urban Development, Perth.

Dubost, M., 1991: Pastoralisme et feux en Corse. Recherche de synthèses: pour en sortir, *Méditerranée*, Aix-en-Provence, 1, 33–38.

Dufaure, J-J., Guérémy, B., Dumas, R. *et al.*, 1984: *La mobilité des paysages méditerranéens, Hommage à P. Birot*, Rev. Géogr. Pyrénées et Sud Ouest, Travaux II.

Dumančič, J., 1994: Island of Unije-sustainable agricultural development, Bulletin no. 2, Croatian-American Society, Zagreb, 5–19.

Dunne, T. and Leopold, L.B., 1978: *Water in environmental planning*, Freeman, San Francisco.

du Plessis, M.C.F., 1986: Grondagteruitgang, *Die Suid-Afrikaanse Tydskrif vir Natuurwetenskap en Tegnologie*, 5, 126–137.

Durbiano, C., 1988: L'expansion du vignoble des Côtes du Rhône méridionales, *Méditerranée*, Aix-en-Provence, 3, 3–11.

Edwards, K., 1988: How much soil loss is acceptable?, *Search*, 19(3), 134–140.

Edwards, K., 1993: Soil erosion and conservation in Australia, in D. Pimentel (ed.), *World soil erosion and conservation*, Cambridge University Press, Cambridge, 147–169.

El Gharbaoui, A., 1982: Géographie et typologie des sols "tirs" au Maroc, *Revue de Géographie du Maroc*, 6, nouvelle série, 81–93.

Elizalde, M.R., 1970: *La sobrevivencia de Chile*, Ministerio de Agricultura, Servicio Agrícola y Ganadero, Santiago, Chile.

Ellenic Soil Science Society, 1996: *Sixth Ellenic symposium on soils*, held in Nauplio 29/5–1/6 1996, National Soil Science Society, Thessaloniki, Greece.

Ellington, T., Reeves, T.G. and Peverill, K.I., 1981: Chlorosis and stunted growth of wheat crops in N.E. Victoria, *Proceedings of the Soil Management Conference*, Australian Society of Soil Science, Dookie, Victoria, 91–109.

El Moujahid, no. 5130, 26 dec. 1981, Alger, 4p.

Emery, K.O. and George, C.J., 1963: *The shores of Lebanon*, Woods Hole Oceanographic Institute Contribution 1385.

Emsley, J., 1992: Weedy gauge of ozone pollution, *New Scientist*, 136, 15.

Endlicher, W., 1990: Landscape damage in central Chile, *Applied Geography and Development*, Tubingen, Germany, 35, 45–62.

Entente Interdepartementale en vue de la Protection de la Foret Contre l'Incendie éd, 1988: *La forêt méditerranéenne*, 3rd edition.

Environmental Protection Council of South Australia, 1988: *The state of the environment report for South Australia*, Department of Environment and Planning, Adelaide.

EPA, 1989: *The toxics release inventory: a national perspective, 1987*, US Environmental Protection Agency, Washington, DC.

Ergenzinger, P., 1988: Regional erosion: rates and scale problems in the Buonamico basin, Calabria, *Catena*, Supplement 13, 97–107.

Errazuriz, A.M. *et al.*, 1992: *Manual de Geografía de Chile*, Andrés Bello, Santiago.

ESD, 1991: *Ecologically Sustainable Development Working Group. Final report. Agriculture*, November 1991, Australian Government Publishing Service, Canberra.

Evans, D.G., 1991: *Acid soils in Australia: the issues for government*, Bureau of Rural Resources, Canberra.

Fadloullah, A., 1987: L'évolution récente de la population dans le Haut Rif central, *Etudes Méditerranéennes*, 11, 463–482.

FAO, 1987: *Commodity review and outlook, 1986–1987*, Food and Agriculture Organisation of the United Nations, Rome.

FAO–UNESCO, 1989: *Soil map of the world*, revised legend, World Resources Report 60, FAO, Rome. Reprinted as Technical Paper 20, ISRIC (International Soil Reference and Information Centre), Wageningen.

Farley, R., 1991: President's message, *Australian Journal of Soil and Water Conservation*, 4(2), 1.
Fassoulas, A. and Fotiades, N., 1966: *The adaptation of plants to extensive cultivation in Greece*, Aristotle University of Thessaloniki, Thessaloniki (in Greek).
Faucher, D. (ed.), 1951: *La France. Géographie-tourisme*, Larousse, Paris, 2 vols.
Fay, G., 1979: L'évolution d'une paysannerie montagnarde, les jbalas du Sud Rifain, *Méditerranée*, 1/2, 81–92.
Fay, G., 1984: Un projet agro-sylvo-pastoral pour le Rif occidental, *Revue de Géographie du Maroc*, 8, nouvelle série, 3–22.
Fereres-Castiel, D., 1990: Agronomía del laboreo de conservación en cultivos anuales, *Ponencia Jornadas Técnicas sobre El Agua y el Suelo: Laboreo de Conservación*, Junta de Andalucía, Sevilla, 1–19.
Fergusson, J.E., 1990: *The heavy elements: chemistry, environmental impact and health effects*, Pergamon, London.
Ferrari, E., Gabriele, S., Rossi, F., Villani, P. and Versace, P., 1990: *La valutazione delle piene in Calabria. Aspetti metodologici di una analisi a scala regionale*, Atti del XXII Convegno di Idraulica e Costruzioni Idrauliche, Cosenza, 4–7 Ottobre 1990, Editoriale Bios, 511–534.
Ferreira, A.J.D., 1996: *Processos hidrologicos e hidroquimicos em povoamentos de Eucalyptus globulus* Labill. e *inus pinaster* Aiton., PhD thesis, Universidade de Alveiro.
Fey, M.V., Manson, A.D. and Schulze, R., 1990: Acidification of the pedosphere, *South African Journal of Science*, 86, 403–406.
Fisher, W.B., 1978: *The Middle East*, Methuen, London.
Fisher, W.B., 1993a: Syria – physical and social geography, *The Middle East and North Africa*, 786–805.
Fisher, W.B., 1993b: Turkey – physical and social geography, *The Middle East and North Africa*, 850–870.
Floriani, B. (ed.), 1991: *Popis posebno zaštićenih objekata prirode*, Ministarstvo zaštite okoliša, prostornog uredenja i stambeno-komunalne djelatnosti Republike Hrvatske, Zagreb.
Font-Tullol, J., 1983: *Climatología de España y Portugal*, Instituto Nacional de Meteorología, Madrid.
Fox, R., 1991: *The inner sea: the Mediterranean and its people*, Sinclair Stevenson, London.
Fuggle, R.F. and Ashton, E.R., 1979: Climate, in J. Day *et al.*, (eds), *Fynbos ecology: a preliminary synthesis*, South African National Scientific Programmes Report 40, 7–15.
Fuggle, R.F. and Rabie, M.A. (eds), 1992: *Environmental management in South Africa*, Juta, Cape Town.
Gabert, P. and Nicod, J., 1982: Inondations et urbanisation en milieu méditerranéen. L'exemple des crues récentes de l'Arc et de l'Huveaune, *Méditerranée*, Aix-en-Provence, 3, 11–24.
Gabriele, S., Govi, M. and Petrucci, O., 1994: Individuazione delle aree soggette a rischio di inondazione: il caso delle fiumare calabre, *IV Geoengineering International Congress – Soil and Groundwater Protection*, Torino, 10–11 Marzo 1994, 133–142.
Galanti, Y., Inbar, M. and Raban, A., 1990: The development of the Qishon river delta in Holocene times, *Ofakim beGeographia*, 31, 133–146 (in Hebrew).
García-Ruíz, J.M., Lasanta, T. and Martínez, R., 1991: Erosion in abandoned fields, what is the problem?, in M. Sala *et al.* (eds), *Soil erosion studies in Spain*, Geoforma Ediciones, Logroño.
Garfunkel, Z., 1981: Internal structure of the Dead Sea leaky transform (rift) in relation to plate kinematics, *Tectonophysics*, 80, 81–108.
Gasson, B., in preparation: Environmental inventory of the Cape Town Metropolitan Area.
Gaussen, H., 1958: *Précipitations annuelles, carte de l'Atlas du Maroc au 1/2.000.000, planche 4a, avec notice*, Publ. du Comité National de Géographie du Maroc.
Gaži-Baskova, V. and Bedalov, M., 1983: Biljni pokrov Kornatskog otočja, *Povremena izdanja Muzeja grada Šibenika*, Šibenik, 10, 455–462.
Gentilli, J.G. (ed.), 1971: *Climates of Australia and New Zealand*, vol. 13 of *World Survey of Climatology*, Elsevier, Amsterdam.
George, R.J., 1992: *Interactions between perched and deeper groundwater systems in relation to secondary, dryland salinity in the Western Australian wheatbelt: processes and management options*, PhD Thesis in Geography, University of Western Australia.
George, R.J. and Conacher, A.J., 1993: The hydrology of shallow and deep aquifers in relation to secondary soil salinisation in southwestern Australia, *Geografia Fisica e Dinamica Quaternaria*, 16, 47–64.
GEOTE, 1996: *Land reclamation, management of water resources, and agricultural mechanisation*, second Ellenic symposium, held in Larisa, 24–27 April, 1996, Greece, 2 vols.
Ghassemi, F., Jakeman, A.J. and Nix, H.A., 1995: *Salinisation of land and water resources: human causes, extent, management and case studies*, CAB International, Wallingford, England.
Giangrossi, L., 1973: Nubifragi ed alluvioni in Calabria nel periodo 20 dicembre 1972 – 2 gennaio 1973, in Instituto di Ricerca per la Protezione Idrogeologica, CNR (ed.), *Il dissesto idrogeologico in Calabria*, Regione Calabria, Arti Grafiche Rubettino, 1990, Soveria Mannelli (CZ), 107–132.
Gilbert, G.K., 1917: *Hydraulic mining debris in the Sierra Nevada*, US Geological Survey Professional Paper 105.

Gill, A.M., 1981: Adaptive responses of Australian vascular plant species to fire, in Gill, A.M., Groves, R.H. and Noble, I.R. (eds), *Fire and the Australian biota*, Australian Academy of Science, Canberra.

Gill, A.M., Groves, R.H. and Noble, I.R. (eds), 1981: *Fire and the Australian biota*, Australian Academy of Science, Canberra.

Gil Olcina (ed.), 1983: *Lluvias torrenciales e inundaciones en Alicante*, Instituto Universitario de Geografía, Alicante.

Ginnivan, D. and Lees, J., 1991: *Moving on: farm families in transition from agriculture*, Rural Development Centre, University of New England, Armidale, New South Wales.

Giráldez, J.V., Laguan, A. and González, P., 1989: Soil conservation under minimum tillage techniques in Mediterranean dry farming, *Soil Technology*, 1, 139–147.

Giráldez, J.V., 1996: Reduced tillage as soil and water conservation practice, In J.L. Rubio and A. Calvo (eds), *Soil degradation and desertification in Mediterranean environments*, Geoforma Ediciones, Logroño, 251–264.

Girgis, M.S., 1987: *Mediterranean Africa*, University Press America, Lantham, Maryland.

Gisotti, G. and De Rossi, C., 1980: Il depauperamento della vegetazione litoranea nell'ambito del degrado delle coste Italiane, *Ingegneria e Architettura*, 5/6, 2–14.

Glavaš, M., Harapin, M. and Hrašovec, B., 1994: Zaštita šuma, *Šume u Hrvatskoj*, Grafički zavod Hrvatske, Zagreb, 171–179.

Godde, S., 1976: Données climatiques et risques d'incendie de forêts en Provence, *Méditerranée*, Aix-en-Provence, 1, 19–33.

Godek, I., 1957: Bujičaustuo i problem erozije tla kraškom prodručjn Hruatske, Savezno saújetowanje o kršu, *Proceedings*, Split, 221–259.

Goldblatt, P., 1978: An analysis of the flora of southern Africa: its characteristics, relationships and origins, *Annals of the Missouri Botanical Garden*, 65, 369–436.

Goldhammer, J.G. and Jenkins, M.J. (eds), 1990: *Fire in ecosystem dynamics: Mediterranean and northern perspectives*, SPB Academic Publishers, The Hague.

González-Nicolás, J. and López-Asio, C., 1996: The future of water quality in Spain: the SAICA project, in M. Alvarez-Cobelas and F. Cabrera-Capitan (eds), *La calidad de las aguas continentales españolas. Estado actual e investigación*, Geoforma Ediciones, Logroño, 23–30.

Good, R., 1964: *The geography of the flowering plants*, 3rd edition, Longman, London.

Gorddard, B., 1991: The adoption of minimum tillage in the Western Australian wheatbelt, Paper to *35th Annual Conference of the Australian Agricultural Economics Society*, Armidale, New South Wales.

Goudie, A., 1981: *The human impact*, The MIT Press, Cambridge, MA.

Grant, R. (ed.), 1992: *State of the environment report*, Government of Western Australia, Perth.

Grivas, J., Moon, L., Mues, C. *et al.*, 1995: The Landcare taxation provisions – some issues, *Outlook 95*, Australian Bureau of Agricultural Resource Economics, Canberra, 157–169.

Grove, A. (ed.), in press: *Mediterranean Europe: desertification, sustainability, land and water management*,

Groves, R.H. and di Castri, F. (eds), 1991: *Biogeography of Mediterranean invasions*, Cambridge University Press, Cambridge.

Guerricchio, A., Melidoro, G. and Tazioli, S., 1976: *Lineamenti idrogeologici e subsidenza dei terreni olocenici della Piana di Sibari*, Sviluppo, Numero Speciale Atti 68 Convegno Società Geologica Italiana, Praia a Mare, 77–80.

Gutiérrez-Elorza, M. (coord.), 1994: *Geomorfología de España*, Rueda, Madrid.

Hakim, B., 1985: *Recherches hydrologiques et hydrochimiques sur quelques karsts méditerranéens, Liban, Syrie, Maroc*, Publ. Univ. Libanaise, Beyrouth, 2 vols.

Hall, A.V. and Veldhuis, H.A., 1984: *South African Red Data Book: plants – fynbos and karoo biomes*, South African National Scientific Programme Reports No. 117, Pretoria.

Hall, B.J., 1978: Drip-trickle irrigation in California, *California Geology*, 31, 266–267.

Hallam, S.J., 1975: *Fire and hearth*, Australian Institute of Aboriginal Studies, Canberra.

Hamblin, A. and Kyneur, G., 1993: *Trends in wheat yields and soil fertility in Australia*, Department of Primary Industries and Energy, Bureau of Resource Sciences, Australian Government Publishing Service, Canberra.

Hamza, A., 1955: Les conséquences géomorphologiques des inondations de janvier 1990 en Tunisie centrale et mériodiionale, *Publ. de l'Association des Géographes Africains*, Rabat, 113–130.

Harris, C.R., 1986: Native vegetation, in C. Nance and D.L. Speight (eds), *A land transformed: environmental change in South Australia*, Longman Cheshire, Melbourne, 29–54.

Haub, C. and Yanagishita, M., 1994: *World population chart*, Population Reference Bureau, Washington DC.

Hauge, C.J., Furniss, M.J. and Euphrat, F.D., 1979: Soil erosion in California's Coast Forest District, *California Geology*, 32, 120–129.

Hawke, R.J.L., 1989: *Our country, our future: Prime Minister's statement on the environment*, 2nd edition, Australian Government Publishing Service, Canberra.

Healy, D.T., Jarrett, F.G. and McKay, J.M., 1985: *The economics of bushfires: the South Australian experience*, Oxford University Press, Melbourne.

Heathcote, R.L., 1992: Settlement advance and retreat: a century of experience on the Eyre Peninsula of South Australia, Paper presented to the *International Conference on the Impacts of Climate Variations and Sustainable Development in Semi-arid Regions*, Fortaleza, Brazil, 27 January – 1 February 1992.

Heizer, R.F., 1978: *California – handbook of North American Indians*, vol. 8, Smithsonian Institution, Washington DC.

Henderson, D.W., 1979: Soil management in semiarid environments, In A.E. Hall (ed.), *Crop productivity in arid and semiarid environments*, Ecological Studies No. 35, Springer Verlag, Amsterdam, 224–237.

Hernanz Martos, J.L., 1990: Repercusión de las nuevas técnicas de laboreo de conservación en los costos de producción, *Ponencia Jornadas Técnicas sobre El Agua y el Suelo: Laboreo de Conservación*, Junta de Andalucía, Sevilla.

Herzenni, A., 1993: Gestion des ressources et conditions du développement local dans la haute montagne de la province d'Azilal, in A. Bencherifa (ed.), *Montagnes et hauts pays d'afrique*, Publ. Fac. Lettres, Rabat, sér. Colloques et séminaires, 29, 333–346.

Heusch, B., 1970: L'érosion dans le Prérif: une étude quantitative de l'érosion hydraulique dans les collines marneuses du Prérif occidental, *Annales des recherches forestières*, 12, 9–176.

Heusch, B., 1982: *Etude de l'érosion et des transports solides en zone semi-aride*, Recherches bibliographiques sur l'Afrique du Nord, Projet PNUFD, Rab/80/04, roneo.

Heusch, B., 1990: *Définition d'un programme de recherche pour l'aménagement des bassins-versants*, rapport préliminaire, Banque Mondiale, Rabat.

Heusch, B., 1995: Pourquoi la banquette CES diminue les rendements et augmente l'érosion, *Bull. Réseau Erosion*, 15, ORSTOM, 317–325.

Heusch, B. and Millies-Lacroix, A., 1971: *Une methode pour estimer l'eccoulement et l'erosion dans un basin*, Mines et Géologie, Rabat, no 33.

Heusser, L., 1978: Pollen in Santa Barbara Basin, California: a 12 000-yr record, *Geological Society of America, Bulletin*, 89, 673–678.

Hickman, J.C. (ed.), 1993: *The Jepson Manual – higher plants of California*, University of California Press, Berkeley.

Higgins, S.I., Turpie, J.K., Costanza, R. *et al.* (in preparation): An ecological economic simulation model of mountain *fynbos* ecosystems: dynamics, valuation and management.

Hillel, D., 1991: *Out of the earth: civilisation and the life of the soil*, University of California Press, Berkeley.

Hillel, D., 1994: *Rivers of Eden*, Oxford University Press, New York.

Hilton-Taylor, C. and le Roux, A., 1989: Conservation status of the fynbos and karoo biomes, in B.J. Huntley (ed.), *Biotic diversity in southern Africa*, Oxford University Press, Cape Town, 202–223.

Hirshberg, K-J., 1988: *Groundwater contamination in the Perth metropolitan region*, Geological Survey of Western Australia, Perth.

Hobbs, R.J. and Saunders, D.A. (eds), 1993: *Reintegrating fragmented landscapes: towards sustainable production and nature conservation*, Springer Verlag, New York.

Hobbs, R., Saunders, D., Lobry de Bruyn, L. and Main, A., 1993: Changes in biota, in R.J. Hobbs and D. Saunders (eds), *Reintegrating fragmented landscapes: towards sustainable production and nature conservation*, Springer Verlag, New York, 65–106.

Holmes, P.J. and Luger, A., in press: Geomorphic implications of a headland bypass dune system in the Cape Peninsula, South Africa, *Zeitschrift fur Geomorphologie*, Supplementband.

Horvat, A., 1957: Històrìjski razvoj devastacije i degradacije i degradacije krša, Savejno savjetovanje o kršu, *Proceedings*, Split, 185–194.

Horvat, I., 1962: Die Grenze der mediteranen und mitteleuropaeischen Vegetation in Suedosteuropa in Lichte neuer pflanzensoziologischer Forschungen, *Berichte der Deutschen Botanischen Gesellschaft*, Stuttgart, 75, 94–104.

Horvatić, S., 1958: Tipološko raščlanjenje primorske vegetacije gariga i borovih šuma, *Acta Botanica Croatica*, Zagreb, 17, 7–98.

Horvatić, S., 1967: *Flora Analytica Iugoslaviae*, SV. 1, no. 1, Institut za botaniku Sveučilišta u Zagrebu, Zagreb, 15–61.

House of Representatives, 1989: *The effectiveness of land degradation policies and programmes*, Report of the House of Representatives, Standing Committee on Environment, Recreation and Arts, Commonwealth of Australia, Canberra.

Hughes, P.J. and Sullivan, M.E., 1986: Aboriginal landscapes, in J.S. Russell and R.F. Isbell (eds), *Australian soils: the human impact*, Queensland University Press, Brisbane.

Hussey, B.M.J. and Wallace, K.J., 1993: *Managing your bushland*, Department of Conservation and Land Management, South Perth.

Husson, A., 1985: Télédétection des incendies de forêts en Corse entre 1973 et 1980, *Méditerranée*, Aix-en-Provence, 1/2, 53–59.

ICCOPS, 1995: *The Mediterranean exercise: pursuing the sustainable development of the Mediterranean*, Report on International Workshop, International Centre for Coastal and Ocean Policy, Genoa, 19–22 April 1995, 9–11.

Imeson, A.C. and Emmer, I.M., 1992: Implications of climatic change on land degradation in the Mediterranean, in L. Jeftic, J.D. Milliman and G. Sestini (eds), *Climatic change and the Mediterranean*, Edward Arnold, London, 95–128.

Imeson, A.C. and Sala, M. (eds), 1988: *Geomorphic processes in environments with strong seasonal contrasts: Vol. 1, Hillslope processes; Vol. 2, Geomorphic systems*, Regional Conference on Mediterranean Countries, Barcelona, Catena Verlag, West Germany.

Inbar, M., 1982: Measurement of fluvial sediment transport compared with lacustrine sedimentation rates: the flow of the River Jordan into Lake Kinneret, *Hydrological Science Journal*, 4, 439–449.

Inbar, M., 1987: Effects of a high magnitude flood in a Mediterranean climate. A case study in the Jordan river basin, in L. Mayer and D.B. Nash (eds), *Catastrophic flooding*, Allen and Unwin, London, 333–354.

Inbar, M., 1992: Rates of fluvial erosion in basins with a Mediterranean type climate, *Catena*, 19, 393–409.

Inbar, M. and Maos, J.O., 1984: Water resource planning and development in the northern Jordan valley, *Water International*, 9, 18–25.

Inbar, M. and Sivan, D., 1984: Paleo-urban development and Late Quaternary environmental change in the Akko area, *Paleorient*, 9/2, 85–91.

Inbar, M., Tamir, M. and Wittenberg, L., 1998: Runoff processes and sediment yield after a Mediterranean forest fire – Mount Carmel, Israel, *Geomorphology* (forthcoming).

INSEE Corse, 1994: *Tableaux de l'Economie Corse 94*, Paris.

Instituto Geográfico Militar, 1993: *Atlas de la República de Chile*, IGM, Santiago.

Instituto Nacional de Estadística, 1994: *Compendio Estadístico 1994*, INE, Santiago.

Iovino, F., Menguzzato, G. and Veltri, A., 1988: Studio sulle condizioni termoigrometriche dell'aria e del suolo nelle abetine di Serra San Bruno, *CNR*, Istituto di Ecologia e Idrologia Forestale, Cosenza, Publ., 3, 1–44.

IUCN/WWF, 1980: *World conservation strategy*, International Union for the Conservation of Nature and the World Wildlife Fund, United Nations Environment Program.

Jackson, R.B., 1983: Pesticide residues in soils, in *Soils: an Australian viewpoint*, CSIRO, Melbourne/ Academic Press, London, 825–842.

Jakić, Ž., 1993: Rimljani u Hrvatskim zemljama, *Povijest staroga vijeka*, Školska knjiga, Zagreb, 170–174.

Jakšić, P., 1988: *Provisional distribution maps of the butterflies of Yugoslavia*, Entomological Society of Yugoslavia, Special Edition No. 1, Zagreb.

Jalžić, B. and Pretner, E., 1977: Prilog poznavanju faune koleoptera pećina i jama Hrvatske, *Krš Jugoslavije*, Zagreb, 9/5, 239–274.

Jelgersma, S. and Sestini, G., 1992: Implications of a future rise in sea level on the coastal lands of the Mediterranean, in L. Jeftic, J.D. Milliman and G. Sestini (eds), *Climatic change and the Mediterranean*, Edward Arnold, London, 282–303.

Jensen, A., 1986: Inland waters, in C. Nance and D.L. Speight (eds), *A land transformed: environmental change in South Australia*, Longman Cheshire, Melbourne, 148–181.

Joannon, M. and Tirone, L., 1980: L'alimentation en eau des villes de Provence-Alpes-Côte d'Azur: le cas de l'agglomération Marseillaise, *Méditerranée*, Aix-en-Provence, 2, 87–102.

Joly, F., 1952: *Quelques phénomènes d'écoulement sur la bordure du Sahara et leurs conséquences géomorphologiques*, Congr. Géol. International.

Jorda, M. and Provansal, M., 1990: Les terrasses de culture dans le bilan érosif méditerranéen: l'exemple du Vallat de Monsieur (Basse-Provence), *Méditerranée*, Aix-en-Provence, 3, 55–61.

Jovanović, B., 1986: Prirodne karakteristike jugoslavenskih šuma, *Šume i prerada drveta Jugoslavije*, Savez inženjera i tehničara šumarstva i industrije za preradu drveta Jugoslavije, 15–33, Beograd.

Judson, S., 1963: Erosion and deposition of Italian stream valleys during historic times, *Science*, 140, 21–27.

Julian, M., 1994: Un torrent dans la ville: le Paillon à Nice (Alpes-Maritimes), in J. Riser (ed.), *Aménagement et gestion des grandes rivières méditerranéennes*, Etudes Vauclusiennes, Avignon, 151–157.

Julivert, M. and Fontbote, J.M., 1974: *Mapa tectónico de la Península Ibérica y Baleares*, escala 1:1 000 000, I.G.M.E., Madrid.

Juramy, S. and Monfort, I., 1982: Dégradation d'un milieu naturel sur le versant Ouest de la Sainte-Baume, *Méditerranée*, Aix-en-Provence, 3, 77–85.

Jury, M., Tegen, A., Ngeleza, E. and du Toit, M., 1990: Winter air pollution episodes over Cape Town, *Boundary-Layer Meteorology*, 53, 1–20.

Kahrl, W.L. (ed.), 1979: *The California water atlas*, State of California, Sacramento.

Kalin Arroyo, M.T., Zedler, P.H. and Fox, M.D. (eds), 1994: *Ecology and biogeography of Mediterranean ecosystems in Chile, California and Australia*, Springer Verlag, New York.

Kalman, R., 1976: Etude expérimentale de l'érosion par griffes, *Revue de Géographie Physique et de Géologie Dynamique*, 13(5), 395–406.

Kassab, A. and Sethom, H., 1980: *Géographie de la Tunisie, le pays et les hommes*, Publ. Univ. Tunis.

Kauders, A., 1939: Prilog povijesti brvatskog šumarstva štumarski list, Februav–Mart, Hrvatsko šumarsko društuo, Zagreb, 205–216.

Keating, P., 1992: *Australia's environment: a national asset. Prime Minister's statement on the environment*, Australian Government Publishing Service, Canberra.

Kennedy, M. (ed.), 1990: *Australia's endangered species: the extinction dilemma*, Simon and Schuster, Brookvale, New South Wales.

Kent, T. (ed.), 1987: *Cultivation in the northern Wimmera*, Agdex 100/510, Department of Agriculture, Victoria.

Kirby, M.G. and Blyth, M.J., 1987: Economic aspects of land degradation in Australia, *Australian Journal of Agricultural Economics*, 31, 154–174.

Kirkpatrick, J., 1994: *A continent transformed: human impact on the natural vegetation of Australia*, Oxford University Press, Melbourne.

Klingberg, C.M. and Schrale, G., 1985: *Guidelines for land application of effluent from intensive animal enterprises and related industries*, Department of Agriculture, Adelaide.

Kluge, R.L., Zimmerman, H.G., Cilliers, C.J. and Harding, G.B., 1986: Integrated control of alien weeds, in I.A.W. Macdonald *et al.* (eds), *The ecology and management of biological invasions in southern Africa*, Oxford University Press, Cape Town, 295–304.

Knopke, P. and Harris, J., 1991: Changes in input use on Australian farms, *Agriculture and Resources Quarterly*, 3(2), 230–240.

Kondinin Group, 1993: Personal protection equipment: cover-up tactics for farmers, *Farming ahead with the Kondinin Group*, Perth, 19, 27–30.

Korenčić, M., 1979: *Naselja i stanovništvo SR Hrvatske 1857–1971*, Edicija i djela Jugoslavenske akademije za znanost i umjetnost, knjiga 54, JAZU, Zagreb.

Kosmas, C.S., Danalatos, N.G., Moustakas, N. *et al.* 1993a: The impacts of drought on the wheat biomass production along catenas in the semi-arid zone of Greece, *Soil Technology*, 6, 337–349.

Kosmas, C.S., Moustakas, N., Danalatos, N.G. and Yassoglou, N., 1993b: The effect of rock fragments on wheat biomass production under highly variable moisture conditions in Mediterranean environments, *Catena*, 23, 191–198.

Kosmas, C.S., Danalatos, N.G., Moustakas, N. *et al.* 1995a: A methodology for mapping desertification units in Mediterranean landscapes, in C. Kosmas *et al.* (eds), *Red mediterranean soils. Third International Meeting*, Chalkidiki, Greece, May 1995, 120–124.

Kosmas, C.S., Moustakas, N., Danalatos, N.G. and Yassoglou, N., 1995b: The effect of land use change on soil properties and erosion along a catena, in J. Thornes and J. Brandt (eds), *Mediterranean desertification and land use*, J. Wiley and Sons, Chichester.

Kosmas, C.S., Yassoglou, N.G., Danalatos, N. *et al.* 1995c: *The Spata field site*, MEDALUS II final report, EEC project no. EV5V-CT92–012V.

Kosmas, C.S., Cammaraat, E., Chabart, M. *et al.*, in press: The effect of land use on soil erosion rates and land degradation under Mediterranean conditions, *Catena*,

Kozlowski, T.T. and Ahlgren, C.E. (eds), 1974: *Fire and ecosystems*, Academic Press, New York.

Kraft, J.C., Kayan, I. and Erol, O., 1977: Paleogeographic reconstructions of coastal Aegean archaeological sites, *Science*, 195, 41–47.

Kreismann, B., 1991: *California: an environmental atlas and guide*, Bear Klaw Press, Davis, California.

Krpan, R., 1957: Obnova šuma na kršu Hrvatske, Savezno savjetovanje o kršu, *Proceedings*, Split, 195–212.

Kutiel, P., 1994: Fire and ecosystem heterogeneity: a Mediterranean case study, *Earth Surface Processes and Landforms*, 19, 187–194.

Kutiel, P. and Inbar, M., 1993: Fire impact on soil nutrients and soil erosion in a Mediterranean pine forest plantation, *Catena*, 20, 129–139.

Kutiel, P. and Naveh, Z., 1987: The effect of fire on nutrients in a pine forest soil, *Plant and Soil*, 104, 269–274.

Labhar, M., 1995: Dynamique actuelle du milieu forstier dans le Moyen Atlas occidental, *Publ. de l'Association des Géographes Africains*, Rabat, 175–190.

Lambrechts, J.J.N., 1979: Geology, geomorphology and soils, in J. Day *et al.* (eds), *Fynbos ecology: a preliminary synthesis*, South African National Scientific Programmes Report 40, 16–26.

Langbein, W.B. and Schumm, S.A., 1958: Yield of sediment in relation to mean annual precipitation, *American Geophysical Union Transactions*, 39, 1076–1084.

Laouina, A., 1982: La sécheresse au Maroc et dans les pays riverains du Sahara, *Revue de Géographie du Maroc*, 6, nouvelle série, 13–36.

Laouina, A., 1984: Les croûtes calcaires et leur contexte géomorphologique en région semi-aride, *Revue Géographie du Maroc*, 8, nouvelle série, 23–62.

Laouina, A., 1987: *Dégradation du milieu et action anthropique, conséquences de l'exploitation agricole du sol dans le secteur Triffa-façade nord des Bni Snassen, Maroc oriental*, Centre National de la Recherche Scientifique et Technique, Rabat.

Laouina, A., 1990: *Le Maroc oriental, reliefs, modelés et dynamique du calcaire*, Publ. du Rectorat de l'Univ. Mohammed Ier, Oujda.

Laouina, A., 1992: L'aménagement des montagnes dans une prespective de protection de l'environnement, Publ. Comité Environnement, Rabat, 95–104.

Laouina, A., 1993a: Evaluation des périmètres de DRS fruitière, le cas du Projet Loukkos, *Bull. de Réseau Erosion*, 14, 271–278.

Laouina, A., 1993b: L'impact humain sur la morphodynamique en milieu montagnard Méditerranéen Marocain, in A. Bencherifa (ed.), *Montagnes et hauts pays d'Afrique*, Publ. Fac. Lettres, Rabat, sér. Colloques et séminaires, no. 29, 201–214.

Laouina, A., 1994: Démographie et dégradation de l'environnement, le cas de la montagne rifaine, in *Le Maroc méditerranéen, quels enjeux écologiques?*, Publ. du Groupement d'Etudes et de Recherches sur la Méditerranée, Rabat, 19–46.

Laouina, A., 1995: L'érosion en milieu méditerranéen, une crise environnementale? *Publ. de l'Association des Géographes Africains*, Rabat, 191–220.

Laouina, A. and Watfeh, A., 1993: Le littoral de Salé et de la Mamora, les héritages et la morphodynamique, in *Aménagement littoral et évolution des côtes*, Publ. Comité National de Géographie du Maroc, Rabat, 53–64.

Laouina, A., Chaker, M., Naciri, R. and Nafaa, R., 1993: L'érosion anthropique en pays méditerranéen, le cas du Maroc septentrional, *Bulletin de l'Association de Géographie Français*, Paris, 384–398.

Latorre, J., 1990: Reforestation of arid and semiarid zones in Chile, *Agriculture, Ecosystems, Environment*, 33, 111–127.

Latz, P., 1995: *Bushfires and bush tucker*, IAD Press, Australia.

Lawrence, C.R., 1983: *Nitrate-rich groundwaters of Australia*, Australian Water Resources Council Technical Paper 79, Australian Government Publishing Service, Canberra.

Leece, D.R. (ed.), 1974: Fertilisers and the environment, *Proceedings of the symposium on ecological effects of fertiliser technology and use*, Australian Institute of Agricultural Science (New South Wales), University of Sydney.

Lefroy, T. and Hobbs, R., 1992: Ecological indicators for sustainable agriculture, *Australian Journal of Soil and Water Conservation*, 5, 22–28.

Le Houerou, H.N., 1981: Impact of man and his animals on Mediterranean vegetation, in F. di Castri, D.W. Goodall and R.L. Specht (eds), *Mediterranean-type shrublands*, Ecosystems of the World 11, Elsevier, Amsterdam, 479–521.

Le Houerou, H.N., 1987: Vegetation wild fires in the Mediterranean basin: evolution and trends, *Ecologia Mediterranean*, Marseille, 13, 13–24.

Le Houerou, H.N., 1989: Agrosilvicultura y silvopastoralismo para combatir la degradación del suelo en la cuenca mediterránea, in: MOPU, *Degradación de zonas áridas del entorno mediterráneo*, Monografía Dirección General del Medio Ambiente, Madrid, 105–116.

Le Houerou, H.N., 1992: Vegetation and land use in the Mediterranean basin by the year 2050: a prospective study, in L. Jeftic, J.D. Milliman and G. Sestini (eds), *Climatic change and the Mediterranean*, Edward Arnold, London, 175–232.

Le Maitre, D.C., Scott, D.F. and Fairbanks, D.H.K., 1995: *Handy reference manual on the impacts of timber plantations on runoff in South Africa*, unpublished contract report, FOR-DEA 914, to the Department of Water Affairs and Forestry, CSIR Division of Forest Science and Technology, Pretoria.

Leonard, L., 1993: *Managing for stubble retention*, Bulletin 4271, Agdex 579, Western Australian Department of Agriculture, South Perth.

Leone, G., 1994: Effetti della siccità sulle irrigazioni nel mezzogiorno d'Italia, in G. Rossi and G. Margariota (eds): *La siccità in Italia, 1988–1990*, Presidenza del Consiglio dei Minstri, Dipartimento della Protezione Civile, 137–142.

Leoussoff, J., 1993: L'inondation catastrophique de Nîmes, 3 Octobre 1988, in J. Becat and G. Soutade (eds), *L'aiguat del 40*, 129–133.

Leys, J.F., 1990: Blow or grow? A soil conservationist's view to cropping mallee soils, *Proceedings, national mallee conference*, Adelaide, April 1989, CSIRO.

Linder, H.P., Meadows, M.E. and Cowling, R.M., 1992: History of the Cape flora, in R.M. Cowling (ed.), *The ecology of fynbos: nutrients, fire and diversity*, Oxford University Press, Cape Town, 113–134.

Lindh, G. 1972: Urbanization, a hydrological headache, *Ambio*, 1, 185–201.

Lobry de Bruyn, L., 1993: Ant composition and activity in naturally vegetated and farmland environments on contrasting soils at Kellerberrin, Western Australia, *Soil Biology and Biochemistry*, 25(8), 1043–1056.

Lobry de Bruyn, L.A. and Conacher, A.J., 1994: The influence of ant biopores on water infiltration in soils in undisturbed bushland and in farmland in a semi-arid environment, *Pedobiologia*, 38, 193–207.

Lofgren, B.E., 1975: *Land subsidence due to groundwater withdrawal, Arvin-Maricopa area, California*, US Geological Survey Professional Paper 437D.

López-Bermúdez, F., Romero-Díaz, M.A. and Martínez-Fernández, J., 1996: The El Ardal field site: soil and vegetation cover, in C.J. Brandt and J.B. Thornes (eds), *Mediterranean desertification and land use*, Wiley, Chichester, 169–188.

Lopez-Ontiveros, A., 1984: Actividad agraria y medio ambiente, in *MOPU, Geografía y Medio Ambiente*, Ministerio de Obras Públicas y Urbanismo, Madrid, 213–253.

Loup, J., 1962: L'Oum Rbia, études sur une grande rivère des montagnes marocaines, *Review de Géographie Alpine*, 50, 519–554.

Lowdermilk, W.C., 1944: *Palestine, land of promise*, Victor Gollancz Ltd., London.

Lucchesi, S., Ansaldi, M. and Giovannini, G., 1994: Regeneration of Mediterranean Maquis after the passage of an experimental fire, in M. Sala and J.L. Rubio (eds), *Soil erosion and degradation as a consequence of forest fires*, Geoforma Ediciones, Logroño, 177–183.

LUCDEME, 1996: *Red de estaciónes experimentales de Seguimiento y evaluación de al erosión y desertificación, RESEL*. Catálogo de estaciones, Dirección General de Conservación de la Naturaleza, Ministerio de Medio Ambiente, p. 121.

Macdonald, I.A.W., 1984: Is the fynbos biome especially susceptible to invasion by alien plants? A reanalysis of available data, *South African Journal of Science*, 80, 369–377.

Macdonald, I.A.W., 1989: Man's role in changing the face of southern Africa, in B.J. Huntley (ed.), *Biotic diversity in Southern Africa: concepts and conservation*, Oxford University Press, Cape Town, 51–78.

Macdonald, I.A.W., Powrie, F.J. and Siegfried, W.R., 1986: The differential invasion of southern Africa's biomes and ecosystems by alien plants and animals, in I.A.W. Macdonald, F.J. Kruger and A.A. Ferrar (eds), *The ecology and management of biological invasions in Southern Africa*, Oxford University Press, Cape Town, 209–228.

Macdonald, I.A.W. and Richardson, D.M., 1986: Alien species in terrestrial ecosystems of the fynbos biome, in I.A.W. Macdonald, F.J. Kruger and A.A. Ferrar (eds), *The ecology and management of biological invasions*, Oxford University Press, Cape Town, 77–92.

Maclear, L.G.A., 1994: *A groundwater hydrocensus and water quality investigation of the Verlorevlei primary aquifer – Elands Bay*, Department of Water Affairs and Forestry Technical Report Gh 3835, Cape Town.

Maclear, L.G.A., 1995: *Cape Town needs groundwater. A note on the potential of the Cape Flats aquifer unit to supply groundwater for domestic use in the Cape Town Metropolitan Area*, Department of Water Affairs and Forestry Technical Report Gh 3868, Cape Town.

Main, D.C., 1994: *Toxic algal blooms*, Agdex 582, No. 43/94, Western Australian Department of Agriculture, Perth.

Mainguet, M., 1991: *Desertification. Natural background and human management*, Springer Verlag, Berlin.

Maiolo, G., 1993a: Il Gran Bosco d'Italia, *Calabria Verde*, 4, 7–29.

Maiolo, G., 1993b: Il Piano Forestale, *Calabria Verde*, 7, 30–76.

Mancini, F., 1966: *Breve commento alla carta dei suoli d'Italia*, Comitato per la Carta dei Suoli, Tipografia Coppini, Firenze.

Marković-Gospodarić, Lj., 1966: Die verbreitesten Pflanzengesallschaften der Ruderalvegetation Kroatiens, *Angewandten Pflanzensoziologie*, Wien, 18/19, 205–209.

Marre, A., 1987: *Etude géomorphologique du Tell oriental algérien, de Collo à la frontière tunisienne*, Thèse d'Etat, Univ. Aix-Marseille, 2 vols.

Marsh, B. and Carter, D., 1983: Wind erosion, *Western Australian Journal of Agriculture*, 24, 54–57.

Martin, C., Bernard-Allee, P., Beguin, E., Levant, M. and Quillard, J., 1993: Conséquences de l'incendie de forêt de l'été 1990 sur l'érosion mécanique des sols dans le massif des Maures, *Bulletin de l'Association de Géographes Français*, Paris, 5, 438–447.

Martin, J., 1981: *Le Moyen Atlas central, étude géomorphologique*, Notes et Mem. Serv. Géolog. Rabat, no. 258.

Martinović, J., 1987: Odnos tia i Šumskin požara, Oznove zaštite šumo od požara, CIP, Zagreb, 97–109.

Martinović, J. and Valanović, A., 1990: Pedološka Karta Hrvatske, 1: 1 750 000, Urbanističk, Instituti.

Martín-Penela, A.J., 1994: Pipe and gully systems development in the Almanzora Basin (Southeast Spain), *Zeitschrift für Geomorphologie*, 38, 207–222.

Marzulio, M., 1986: *Attivita estrattive per materiali da costruzione*. Thesis, Dip. Scienze della Terra, Universita della Calabria, Arcavacata.

Mason, R.J. and Mattson, M.T., 1990: *Atlas of United States environmental issues*, Macmillan, New York.

Massa, B. and Mingozzi, A., 1991: Considerazioni sulla conservazione e la gestione della fauna in Italia, in E. Parnzini and G. Voldrè (eds), *La gestione dei parchi e delle aree protette*, Edizioni delle Autonomie, Roma, 82–103.

Masson, M., 1993: Après Vaison-la-Romaine. Pour une approche pluridisciplinaire de la prévision et de la planification, *Revue de Géomorphologie Dynamique*, Paris, 42, 73–77.

Mata-Porras, M., 1993: Influence of Common Agriculture Policy of the European Community on socio-economy and land degradation, in R. Fantechi *et al.* (eds), *Desertification in a European context: physical and socio-economic aspects*, Official Publications of the European Community, Brussels, 509–521.

Matheson, W.E., 1986: Soils, in C. Nance and D.L. Speight (eds), *A land transformed: environmental change in South Australia*, Longman Cheshire, Melbourne, 126–147.

Maurer, G., 1962: L'évolution des versants dans le Rif occidental, *Revenu de Géographie du Maroc*, 1/2, 63–66.

Maurer, G., 1968a: *Les montagnes du Rif central, étude géomorphologique*, Thèse de Doctorat d'Etat, Paris-Sorbonne.

Maurer, G., 1968b: Les paysans du Haut Rif central, *Revue de Géographie du Maroc*, 14, 3–70.

Maurer, G., 1975: Les mouvements de masse dans l'évolution des versants des régions rifaines et telliennes d'Afrique du Nord, *Actes Symp. Versants en pays méditerranéens*, Aix en Provence, CEGERM, 5, 133–137.

Maurer, G., 1990: Le Rif occidental et central, montagne méditerranéenne à influence atlantique, *Travaux du Centre de Géographie humaine et sociale*, Poitiers, 17, 443–455.

Maurer, G., 1991: Les dynamiques agraives dans les montagnes vifaines et telliennes du Maghreb. Bulletin de l'Association des Géographie Français.

Maurer, G., 1993: L'agriculture de montagne dans les pays rifains et telliens au Maghreb, in *African mountains and highlands (2)*, Publ. Fac. Lettres et Sciences Humaines de Rabat, Série Colloques, no. 29, 35–54.

McDowell, C.R., 1988: *Factors affecting the conservation of* Renosterveld *by private landowners*, PhD thesis, University of Cape Town.

McDowell, C.R., Low, A.B. and McKenzie, B., 1991: Natural remnants and corridors in Greater Cape Town: their role in threatened plant conservation, in D.A. Saunders and R.J. Hobbs (eds), *Nature conservation 2: the role of corridors*, Beattie and Sons, Surrey, 27–39.

McFarlane, D., undated: *Water erosion on potato lands during the 1983 growing season*, Donnybrook, Soil Conservation Branch, Technical Report no. 26, WA Department of Agriculture, South Perth.

McGarity, J.W. and Storrier, R.R., 1986: Fertilisers, in J.S. Russell and R.F. Isbell (eds), *Australian soils: the human impact*, University of Queensland Press, St Lucia, 304–333.

McGhie, D.A. and Posner, A.M., 1980: Water repellence of a heavy-textured Western Australian surface soil, *Australian Journal of Soil Research*, 18, 209–223.

McNeill, J.R., 1992: *The mountains of the Mediterranean world: an environmental history*, Cambridge University Press, Cambridge.

McWilliam, J.R., 1981: Research and development to service a sustainable agriculture, *Search*, 12, 15–21.

Meade, R.H. and Parker, R.S., 1985: Sediment in rivers of the United States, in United States Geological Survey Water Supply Paper, 2275, 49–60.

Meadows, M.E. and Linder, H.P., 1993: A palaeoecological perspective on the origin of Afromontane grasslands, *Journal of Biogeography*, 20, 345–355.

Meadows, M.E. and Meades, A. (in preparation): Soil erosion by gullying in the Swartland, southwestern Cape, a reassessment.

Meadows, M.E. and Sugden, J.M., 1991: A vegetation history of the last 14 500 years on the Cederberg, SW Cape, South Africa, *South African Journal of Science*, 87, 34–43.

Meffre, J.C., 1990: Habitats augustéens et aménagements de versants. Séguret (Vaucluse), *Méditerranée*, Aix-en-Provence, 3, 17–21.

Meiggs, R., 1982: *Trees and timber in the ancient Mediterranean world*, Clarendon, Oxford.

Mekati, S. and Bellatreche, A., 1995: Mouvements de terrain, cas de la région d'Azazga, Grande Kabylie, Algérie, *Publ. de l'Association des Géographes Africains*, Rabat, 233–240.

Menessier, M., 1992: Vaison: un torrent de négligences, *Sciences et Vie*, Paris, 902, 96–103.

Merzouk, A. and Backe, G.R., 1991: Estimation of interrill erodability of Moroccan soil, *Catena*, 18, 537–550.

Meunier, M. and Mathys, N., 1993: *Panorama synthétique des mesures d'érosion effectuées sur trois bassins du site expérimental de Draix (Alpes-de-Haute-Provence, France)*, CEMAGREF (Research Centre of the French Ministry of Agriculture), Grenoble.

Michard, A., 1976: *Eléments de géologie marocaine*, Notes et Mém. Serv. Géolog. Rabat, no. 252.

Middleton, N.J., 1984: Dust storms in Australia: frequency distribution and seasonality, *Search*, 15, 46–47.

Midgley, D.C., Pitman, W.V. and Middleton, B.J., 1994: Surface water resources of South Africa 1990, Report No. 298/4 1/94, Water Research Commission, Pretoria.

MINAGRI, 1994: *Marco general de la politíca ambiental*, SMASS, Ministerio de Agricultura, Santiago, Chile.

Ministerio de Obras Publicas, 1989: *Estadisticz cas del Departamento de Vialidad*, MOP, Santiago, Chile.

Ministry of Agriculture, Government of the Hashemite Kingdom of Jordan, 1986: Jordan, in J.K. Jain (ed.), *Combating desertification in developing countries*, Scientific Publications, Johpur, India.

Ministry of Civil Engineering, 1994: *Crvena knjiga biljnih vrsta Republike Hrvatske*, Ministry of Civil Engineering and Environmental Protection, Department of Nature Protection, Zagreb.

Minnich, R.A., 1983: Fire mosaics in southern California and northern Baja California, *Science*, 219, 1287–1294.

Moll, E.J. and Bossi, L., 1984: Assessment of the extent of the natural vegetation of the fynbos biome of South Africa, *South African Journal of Science*, 80, 355–358.

Moll, E.J., McKenzie, B. and McLachlan, D., 1980: A possible explanation for the lack of trees in the fynbos, Cape Province, South Africa, *Biological Conservation*, 17, 221–228.

Montserrat, P., 1990: Pastoralismo y desertificación, in *Strategies to combat desertification in Mediterranean Europe*, Report EUR 11175, Luxemburg, 85–103.

Mooney, H.A., 1977: *Convergent evolution in Chile and California Mediterranean type ecosystems*, Dowden, Hutchinson and Ross, Stroudsberg, PA.

MOPU, 1984–1987: *Medio Ambiente en España, Dirección General del Medio Ambiente*, Ministerio de Obras Públicas (MOPU) y Urbanismo, Madrid.

MOPU, 1989a: *Degradación de zonas áridas del entorno mediterráneo*, Monografía Dirección General del Medio Ambiente, Madrid.

MOPU, 1989b: *Medio ambiente en España*, Monografías de la Secretaría General del Medio Ambiente, Madrid.

MOPU, 1995: *The Spanish agrofood sector facts and figures*, Secretaria General Tecnica, Ministerio de Agricultura, Pesca y Alimentacion, Madrid.

Morales, E. and Cañon, R., 1985: *Geografía del Mar de Chile*, Coleccion Geografia de Chile, Tomo IX, Instituto Geografico Militar, Santiago.

Mouliéras, A., 1895: *Le Maroc inconnu*, 2 vols, Oran-Paris, Librairie coloniale et africaine.

Moustakas, N., Kosmas, C., Danalatos, N. and Yassoglou, N., 1995: Rock fragments I. Their effect on runoff, erosion and soil properties under field conditions, *Soil Use and Management Journal*, 11, 115–120.

Mrakovčić, M., Mišetić, S. and Povž, M., 1995: Status of freshwater fish in Croatian Adriatic river systems, *Biological Conservation*, Elsevier, 72, 179–185.

Mriouah, D. and Messaoudi, O., 1987: Les crues de l'Oriental de novembre 1986, *Eau et Développement*, 5, 23–26.

Mues, C., Roper, H. and Ockerby, J., 1994: *Survey of Landcare and land management practices 1992–93*, ABARE Research Report 94.6, Australian Bureau of Agricultural and Resource Economics, Canberra.

Murray/Darling Basin Commission, 1989: *Murray/Darling natural resources management strategy*, Working Group, Background paper 89/1.

Naciri, M., 1988: Calamités naturelles et fatalité historique, in *Drought, water management and food production*, Conference Proceedings, Agadir, 21–24 Nov. 1985, 83–102.

Nafaa, R., Watfeh, A. and Evin, J., 1995: Indices de dégradation de l'environnement depuis l'Holocène dans la région de la Mamora, *Publ. de l'Association des Géographes Africains*, Rabat, 241–252.

Naff, Th., 1993: Water, "that peculiar substance", *Research and Exploration*, 9, 6–17.

Naff, Th. and Matson, R.C., 1984: *Water in the Middle East*, Westview Press, Boulder and London.

Nagel, J.F., 1956: Fog precipitation on Table Mountain, *Quarterly Journal of the Royal Meteorological Society*, 82, 452–460.

Nance, C. and Speight, D.L. (eds), 1986: *A land transformed: environmental change in South Australia*, Longman Cheshire, Melbourne.

National Research Council (US), 1993: *Solid earth sciences and society*, National Academy Press, Washington DC.

Natural Resources Management Services Unit, 1996: *Profile of the catchment hydrology discipline group, group profile as at June 18, 1996*, Agriculture Western Australia.

Navas, A., 1991: Application of simulated rainfall for studying runoff yield and erosive behaviour of gypsiferous soils, in M. Sala *et al.* (eds), *Soil erosion studies in Spain*, Geoforma Ediciones, Logroño, 181–189.

Navas, A. and Machín, J., 1995: Salinidad en las aguas superficiales de la cuenca del Ebro, in M. Alvarez-Cobelas and F. Cabrera-Capitán (eds), *La calidad de las aguas continentales españolas: estados actual e investigación*, Geoforma Ediciones, Logroño, 223–235.

Naveh, Z. 1955: Some aspects of range improvement in a Mediterranean environment, *Journal of Range Management*, 8, 265–270.

Naveh, Z., 1974: Effects of fire in the Mediterranean region, in T.T. Kozlowski and C.E. Ahlgren (eds), *Fire and ecosystems*, Academic Press, New York, 4011–4434.

Naveh, Z., 1979: A model of multiple use management strategies of marginal and untillable Mediterranean upland ecosystems, in G.P. Patil and W.E. Waters (eds), *Environmental biomonitoring assessment, prediction and management*, International Cooperative Publishing House, Maryland, 269–286.

Naveh, Z., 1984: Dynamic conservation management of Mediterranean landscapes, in Z. Naveh and A.S. Liberman (eds), *Landscape ecology*, Springer-Verlag, New York, 256–338.

Naveh, Z. and Dan, J., 1973: The human degradation of Mediterranean landscapes in Israel, in F. di Castri and H.A. Mooney (eds), *Mediterranean type ecosystems*, Springer Verlag, New York, 373–390.

Naveh, Z. and Lieberman, A.S., 1984: *Landscape ecology: theory and application*, Springer-Verlag, New York.

Naveh, Z. and Vernet, J.L., 1991: The paleohistory of the Mediterranean biota, in: R.H. Groves and F. di Castri (eds), *Biogeography of Mediterranean invasions*, Cambridge University Press, Cambridge, 19–32.

Neboit, R., 1980: Morphogenèse et occupation humaine dans l'Antiquité, *Bulletin Association Géographique Française*, 466, 21–27.

Neboit, R., 1983: *L'homme et l'érosion*, Publ. Fac. Lettres de Clermont-Ferrand.

Neboit, R., 1984: Erosion des sols et colonisation en Sicile et en Grande Grèce, *Bulletin Association Géographique Française*, 499, 5–13.

Negev, M., 1972: *Suspended sediment discharge in western watersheds of Israel*, Hydrology Paper no. 14, Hydrological Service, Jerusalem, Israel.

Nelson, R.A. and Mues, C., 1993: A survey of Landcare in Australia, *Agriculture and Resources Quarterly*, 5, 400–410.

Newell, F.H., 1894: *Report on agriculture by irrigation in the western part of the United States at the Eleventh Census: 1890*, US Bureau of the Census, Washington DC.

Niccoli, R. and Procopio, F., 1995: Primi risultati delle indagini sull'evoluzione del tratto costiero compreso tra atanzaro Lido e Soverato (Mar Ionio), *Geologia Tecnica & Ambientale*, 1/95, 33–43.

Nicod, J., 1980: Les réserves en eau de la région Provence-Alpes-Côte d'Azur. Importance et rôle des réserves souterraines, *Méditerranée*, Aix-en-Provence, 2/3, 23–34.

Nicod, J., 1993: Hydrologie et érosion dans quelques bassins-versants de l'Algérie orientale et du Maroc septentrional, in *Hommage à P. Gabert,* Centre National de la Recherche Scientifique, Caen, 141–153.

Niemeyer, H. and Cereceda, P., 1984: *Hidrografía,* Coleccion Geografia de Chile, Tomo VIII, Instituto Geografico Militar, Santiago.

Nir, D., 1983: *Man, a geomorphological agent,* Keter Publishing House, Jerusalem.

NOAA, 1973: *Precipitation-frequency atlas of the western United States, volume 11, California,* US National Weather Service, Washington DC.

Nortcliff, S., Ross, S.M. and Thornes, J. 1990: Soil moisture, runoff and sediment yield from differentially cleared tropical rainforest plots, in J. Thornes (ed.), *Vegetation and erosion,* John Wiley & Sons, Chichester, 419–436.

Nossin, J.J., 1972: Landslides in the Crati Basin, Calabria, Italy, *Geologie en Mijnbouw,* 51, 137–141.

Novak, P., 1970: Rezultati istraživanja kornjaša našeg otočja, *Acta Biologica,* Zagreb, 6 (38), 5–57.

Odendaal, D. and Horwood, P., 1995: The Western Cape, *African Panorama,* 40, 18–28.

OECD, 1989: *Agricultural and environmental policies: opportunities for integration,* Organisation for Economic Co-operation and Development, Paris.

OECD, 1994: *Towards sustainable agricultural production: cleaner technologies,* Organisation for Economic Co-operation and Development, Paris.

Oficina Meteorologica de Chile, 1995: *Compendio estadistico 1995,* OMC, Santiago, Chile.

Ogniben, L., Parotto, M. and Praturlon, A. (eds), 1975: *Structural model of Italy,* Quaderni de "La Ricerca Scientifica", CNR, Uffico Publicazioni, Roma.

Olsen, G. and Skitmore, E., 1991: *State of the rivers of the South West Drainage Division,* Western Australian Water Resources Council, Leederville, Perth.

Orme, A.R., 1985: California, in E.C.F. Bird and M.L. Schwartz (eds), *The world's coastline,* Van Nostrand Reinhold, New York, 27–36.

Orme, A.R., 1992a: The San Andreas Fault, in D.G. Jannelle (ed.), *Geographical snapshots of North America,* 27th International Geographical Congress, Guilford Press, New York, 143–149.

Orme, A.R., 1992b: Late Quaternary deposits near Point Sal, south-central California: a timeframe for coastal-dune emplacement, in C.H. Fletcher and J.F. Wehmiller (eds), *Quaternary coasts of the United States: marine and lacustrine systems,* Society for Sedimentary Geology, 48, 309–315.

Orme, A. R. and Bailey, R.G., 1970: The effect of vegetation conversion and flood discharge on stream-channel geometry: the case of southern California watersheds, *Proceedings of the Association of American Geographers,* 2, 101-106.

Orme, A.R. and Bailey, R.G., 1971: Vegetation conversion and channel geometry in Monroe Canyon, southern California, *Yearbook, Association of Pacific Coast Geographers,* 33, 65–82.

Ornduff, R., 1974: *Introduction to California plant life,* University of California Press, Berkeley and Los Angeles.

Orse, R.J., 1974: *A list of references for the history of agriculture in California,* University of California Agricultural History Center, Davis.

Ortiz-Casas, J.L., 1995: Problemática de la calidad de aguas en España, in M. Alvarez-Cobelas and F. Cabrera-Capitán (eds), 1995: *La calidad de las aguas continentales españolas: Estado actual e investigación,* Geoforma Ediciones, Logroño, 23–30.

Oulad Chrif, B., 1984: Les lacs collinaires au Maroc, *Hommes, Terres et Eaux,* 14, 59–64.

Palmer Development Group, 1990: *A definition of the present structure of the Western Cape economy,* Palmer Development Group, Cape Town.

Pantani, F., 1983: Il problema delle piogge acide e la relativa situazione in Italia, *L'Italia Forestale e Montana,* 38, 10–23.

Papamichos, N., 1985: *The forest soils, formation, properties and behaviour,* Aristotle University of Thessaloniki Press, Thessaloniki.

Parde, M., 1925: *Le régime du Rhône,* Masson, Paris.

Parde, M., 1941: La formidable crue d'Octobre 1940 dans les Pyrénées-Orientales, *Revue géographique des Pyrénées et du Sud-Ouest,* Toulouse, 12, 237–279.

Pardo Pascual, J.E., 1991: *La erosión antrópica en el litoral valenciano,* Conselleria d'Obre Públiques, Urbanisme i Transports.

Pascon, P. and Van der Wusten, H., 1983: *Les Beni Bou Frah, essai d'écologie sociale d'une vallée rifaine,* Institut Universitaire de la Recherche Scientifique, Rabat.

Pastor Muñoz-Cobo, M., 1990: Agronomía del laboreo de conservación en cultivos leñosos, *Ponencia Jornadas Técnicas sobre El Agua y el Suelo: Laboreo de Conservación,* Junta de Andalucía, Sevilla.

Pate, J.S. and Beard, J.S. (eds), 1984: *Kwongan: plant life of the sandplain,* University of Western Australia Press, Perth.

Peck, A.J., 1973: Salinity of soils and streams: putting numbers on the problems in Western Australia, Paper presented at the *ANZAAS Congress,* Perth.

Pelissier, F., 1980: La lutte contre la pollution de l'eau et sa prévention, *Méditerranée,* Aix-en-Provence, 2, 65–76.

Peña, J.L., Echeverria, M., Petit-Maire, N. and Lafont, R. 1993: Cronología e interpretación de las acumulaciones holocenas de la Val de las Lenas (Depresión del Ebro, Zaragoza), *Geographicalia,* 30, 321–332.

Pérennes, J.J., 1993: *L'eau et les hommes au Maghreb, contribution à une politique de l'eau en Méditerránee,* Editions Karthala, Paris.

Perless, C., 1977: *Prehistoire du feu*, Masson, Paris.
Phillips, J.D., 1993: Biophysical feedbacks and the risk of desertification, *Annals of the Association of American Geographers*, 83, 630–640.
Pignatti, S., 1982: *Flora d'Italia*, Edagricole, Bologna.
Pignatti, S., 1983: Human impact on the vegetation of the Mediterranean basin, in W. Holzner, M.J.A. Werger and I. Ikusima (eds), *Man's impact on vegetation*, Dr. W. Junk Publishers, The Hague, 151–161.
Pirazzoli, P.A. 1986: The Early Byzantine tectonic paroxism, *Zeitschrift für Geomorphologie*, Supplementband, 62, 31–49.
Pitman, W.V., Potgieter, D.J., Middleton, B.J. and Midgley, D.C., 1982: *Surface water resources of South Africa*, volume 4, Water Research Commission Report 13/81, Pretoria.
Plit, F., 1983: La dégradation de la végétation, l'érosion et la lutte pour proteger le milieu naturel en Algérie et au Maroc, *Méditerranée*, 3, 79–88.
Poncet, J., 1961: *La colonisation et l'Agriculture européenne en Tunisie depuis 1881*, Editions Mouton, Paris-La Haye.
Potočić, Z., 1957: Ekonomski problemi krša I. Iruatske, Savenzo savjetovanje o kršu, *Proceedings*, Split, 325-346.
Powell, J.M., 1988: Patrimony of the people: the role of government in land settlement, in R.L. Heathcote (ed.), *The Australian experience*, Longman Cheshire, Melbourne, 14–24.
Preston-Whyte, R.A. and Tyson, P.D., 1988: *The atmosphere and weather of Southern Africa*, Oxford University Press, Cape Town.
Principe, I. and Sica, P., 1967: L'inondazione di Firenze del 4 Novembre 1966, *L'Universo*, 2, 1–32.
Prosper-Laget, V., 1994: *Sécheresse et risque d'incendies de forêt en région méditerranéenne française: évaluation par imagerie satellitaire NOAA-AVHRR*, Thèse, Université de Provence.
Provansal, M. (ed.), 1993: La Camargue et le Rhône, hommes et milieux, *Méditerranée*, Aix-en-Provence, 3, 116 pp.
Provansal, M., Bertucchi, L. and Pelissier, M., 1994: Les milieux palustres de Provence occidentale, indicateurs de la morphogénèse Holocène, *Zeitschrift für Geomorphologie*, 38, 185–205.
Prpić, B., Seletković, Z. and Ivkov, M., 1991: Propadanje šuma u Hrvatskoj i odnos pojave prema biotskim i abiotskim činiteljima danas i u prošlosti, *Šumarski list*, 3–5, Savez društava inženjera i tehničara šumarstva i drvne industrije, Zagreb, 107–129.
Pyne, S.J., 1980: *Grove Karl Gilbert: a great engine of research*, University of Texas Press, Austin.
Quezel, P., 1981: Floristic composition and phytosociological structure of sclerophyllous matorral around the Mediterranean, in F. di Castri, D.W. Goodall and R.L. Specht (eds), *Mediterranean-type shrublands*, Elsevier, Amsterdam, 107–120.
Quintanilla, V., 1983: Biogeografia, in *Coleccion Geografia de Chile*, IGM, Santiago, Chile.
Quirantes, J., Barahona, E. and Iriarte, A., 1991: Soil degradation and erosion in southeastern Spain. Contributions of the Zaidín Experimental Station C.S.I.C. (Granada, Spain), in M. Sala *et al.* (eds), *Soil Erosion Studies in Spain*, Geoforma Ediciones, Logroño, 211–217.
Rabie, M.A., 1992: Environment Conservation Act, in R.F. Fuggle and M.A. Rabie (eds), *Environmental management in South Africa*, Juta, Cape Town, 99–119.
Rachik, H., 1993: Espace pastoral et conflits de gestion collective dans une vallée du Haut Atlas atlantique, in A. Bencherifa (ed.), *Montagnes et hauts pays d'afrique*, Publ. Fac. Lettres, Rabat, no. 29, 181–200.
Ragavis, A., 1888: *Dictionary of the Ancient Greek language*, vol. II, Athens (in Greek).
Raggi, R., 1994: Prácticas conservacionistas en Chile, in *Memorias del taller sobre Planificación Participativa de Conservación de suelos y aguas*, Proyecto Regional GCP/RLA/07/JPN, 261–283.
Ramphele, M. and McDowell, C. (eds), 1991: *Restoring the land*, Panos, London.
Rapetti, F. and Vittorini, S., 1994a: *Carta climatica della Toscana centro-settentrionale*, CNR – Centro di Studio per la Geologia Strutturale e Dinamica dell'Appennino, Pisa.
Rapetti, F. and Vittorini, S., 1994b: *Carta climatica della Toscana centro-meridionale ed insulare*, CNR – Centro di Studio per la Geologia Strutturale e Dinamica dell'Appennino, Pisa.
Raus, D., 1976: Povijest šuma otoka Raba, Ekološko valobriziranje primorskoy krša, *Proceedings*, Split, 127–128.
Rebelo, A.G., 1992: Preservation of biotic diversity, in R.M. Cowling (ed.), *The ecology of fynbos: nutrients, fire, diversity*, Oxford University Press, Cape Town, 309–344.
Regional Development Advisory Committee, 1994: *South Africa's Western Cape in focus*, RDAC, Old Oak.
Reichenbach, P., Acevedo, W., Mark, R.K. and Pike, R.J., 1993: *Landforms of Italy* (shaded 1:1 500 000 scale DEM), Special Project AVI, GNDCI, CNR, Roma.
Reid, R. and Wilson, G., 1985: *Agroforestry in Australia and New Zealand*, Goddard and Dobson, Victoria.
Rice, R.M., Corbett, E.S. and Bailey, R.G., 1969: Soil slips related to vegetation, topography and soil in southern California, *Water Resources Research*, 5, 637–659.

Richardson, D.M. and van Wilgen, B.W., 1986: Effects of thirty-five years of afforestation with *Pinus radiata* on the composition of mesic mountain *fynbos* near Stellenbosch, *South African Journal of Botany*, 52, 309–315.

Richardson, D.M., Macdonald, I.A.W., Holmes, P.M. and Cowling, R.M., 1992: Plant and animal invasions, in R.M. Cowling (ed.), *The ecology of fynbos: nutrients, fire, diversity*, Oxford University Press, Cape Town, 271–308.

Robinson, W.W., 1948:, *Land in California*, University of California Press, Berkeley.

Rocha, J.S., 1993: Indicadores do Estado de Erosão do Solo, in *Estudo Preparatório para Definição de Projectos de Ambiente Elegiveis no Contexto do Fundo de Coesão*, Laboratório Nacional de Engenharia Civil, Lisboa.

Roglić, J., 1975: Geološka osnova i reljef, *Geografija SR Hrvatske*, knjiga 5, Školska knjiga, Zagreb, 9–18.

Rojas, E., 1987: El desinterés hacia el bosque en la política española y comunitaria, *El Campo*, 104.

Roose, E., 1994: *Introduction à la gestion conservatoire de l'eau, de la biomasse et de la fertilité des sols (GCES)*, Bull. Pédol. FAO, no. 70.

Rosa, C.A.S., 1980: Uso dos talhöes experimentais de erosäo. Alguns resultados do Centro Experimental de Vale Formoso, *Actas das Jornadas de Drenagem e Conservaçäo do solo para a Agricultura de Sequeiro Alentejana*, DGHEA, M.12.80, Lisboa, 33–42.

Rose, S., 1991: Landcare news: West Hume Landcare Group, *Australian Journal of Soil and Water Conservation*, 4(1), 60–61.

Rosenzweig, D., 1972: *Study of difference in effects of forest and other vegetative covers on water yield*, Research report 33, Volcani Center, Bet Dagan, Israel, 6–17.

Rovira, A., 1984: *Geografía de los Suelos*, Coleccion Geografia de Chile, Instituto Geografico Militar, Santiago.

Rubio, J.M., 1989: Vegetación y fauna, in V. Bielza de Ory (coord.), *Territorio y Sociedad en España I, Geografía Física*, Taurus, Madrid, 315–328.

Ruellan, A., 1971: *Les sols à profil calcaire différencié des plaines de la Basse Moulouya (Maroc oriental)*, Mém. Office de la Recherche Scientifique et Technique d'Outre Mer (ORSTOM), no. 54, Paris.

Ryan, P.J., 1990: Role of trees in controlling land degradation, *Proceedings of the Land Degradation Conference*, Geographical Society of New South Wales, Sydney, 47–57.

Sabadi, R., 1994: *Review of forestry and forest industries sector in Republic of Croatia*, Ministry of Agriculture and Forestry, Public Corporation 'Hrvatske šume', Zagreb.

Sala, M., 1984: Iberian Massif, Baetic Cordillera and Guadalquivir Basin, in C. Embleton (ed.), *The geomorphology of Europe*, Macmillan Press, London, vol. 12, 294–322; vol. 13, 323–340.

Sala, M., 1988: Slope runoff and sediment production in two Mediterranean mountain environments, in A. Imeson and M. Sala (eds), *Geomorphic processes in environments with strong seasonal contrasts, vol. 1, hillslope processes*, *Catena* Supplement 12, 13–30.

Sala, M., 1989: Las aguas continentales, in V. Bielza de Ory (coord.), *Territorio y Sociedad en España*, Taurus, Madrid, 257–314.

Sala, M. and García Ruíz, J.M., 1984: Pyrenees and Ebro Basin Complex, in C. Embleton (ed.), *The geomorphology of Europe*, Macmillan Press, London, vol. 11, 268–293.

Sala, M. and Inbar, M., 1992: Some hydrologic effects of urbanization in Catalan rivers, *Catena*, 19, 363–378.

Sala, M. and Rubio, J.L., 1994: Preface, in M. Sala and J.L. Rubio, *Soil erosion as a consequence of forest fires*, Geoforma Ediciones, Logroño, 9–11.

Sala, M., Rubio, J.L. and García-Ruíz, J.M. (eds), 1991: *Soil erosion studies in Spain*, Geoforma Ediciones, Logroño.

Sala, M., Soler, M. and Pradas, M., 1994: Temporal and spatial variations in runoff and erosion in burnt soils, *Proceedings of 2nd International Conference on Forest Fire Research*, Coimbra, vol. II, D34, 1123–1134.

Salgueiro, T.A., 1976: *O coste dos montados*, vida rural no 3.

Sandstrom, E., 1996: *Property rights and soil erosion in Spain: an ecological and socio-economic approach*, ERASMUS report, Department of Geography, University of Barcelona.

Sanjaume, E., Rosselló, V.M., Pardo, J.E. *et al.* 1996: Recent coastal changes in the Gulf of Valencia (Spain), *Zeitschrift für Geomorphologie*, Sup. 102, 95–118.

Sanroque, P. and Rubio, J.L., 1982: El suelo y los incendios forestales, Diputació de Valencia, Valencia.

Sari, Dj., 1977: *L'homme et l'érosion dans l'Ouarsenis*, Société Nationale D'édition, Alger.

Sauvage, Ch., 1963: *Etages bioclimatiques, carte de l'Atlas du Maroc au 1/2.000.000, planche 6b, avec notice*, Publ. du Comité National de Géographie du Maroc.

Savvides, K.L., 1986: Cypress, in J.K. Jain (ed.), *Combating desertification in developing countries*, Scientific Publications, Johpur, India.

Schick, A.P. and Inbar, M. 1972: Some effects of man on the hydrological cycle of the Upper Jordan-Lake Kinneret watershed, *Freiburger Geografische Hefte*, 12, 27–43.

Schoeman, J.L. and Scotney, D.M., 1987: Agricultural potential as determined by soil, terrain and climate, *South African Journal of Science*, 83, 260–268.

Schreiner, B., 1995: *A popular guide to environment reconstruction and development*, International Development Research Centre, Johannesburg.

Schulze, B.R., 1965: *Climate of South Africa, part eight: general survey*, Weather Bureau, Pretoria.

Scotney, D.M. and Dijkhuis, F.J., 1990: Changes in the fertility status of South African soils, *South African Journal of Science*, 86, 395–403.

Scott, D., 1991: *1991 State of the environment report. Agriculture and Victoria's environment: resource report*, Office of the Commissioner for the Environment, Government of Victoria, Melbourne.

Scott, D.F., 1993: The hydrological effects of fire in South African mountain catchments, *Journal of Hydrology*, 150, 409–432.

Scott, D.F. and van Wyk, D.B., 1990: The effects of wildfire on soil wettability and hydrological behaviour of an afforested catchment, *Journal of Hydrology*, 121, 239–256.

Segura, F, 1990: Procesos fluviales en lechos con materiales gruesos, *Cuadernos de Geografía*, 16, 123–138.

Select Committee into Land Conservation, 1991: *Final report*, Legislative Assembly, Perth.

Seletković, Z. and Katušin, Z., 1994: Klima Hrvatske, *Šume u Hrvatskoj*, Grafički zavod Hrvatske, Zagreb, 13–18.

Semple, B. and Budd, G.R., 1988: Wind erosion and some options for its control on non-irrigated cropland on aeolian landscapes in far south-western NSW, *Australian Journal of Soil and Water Conservation*, 1, 24–32.

Serva, I. and Brunamonte, F., 1994: I fenomeni di abbassamento deò suolo nella ree sottoposte a bonifica idraulica: l'esempio della Pianura Pontina, in L. Serva (ed.), *Monografia: prima*, CNR & Consiglio Nazionale Difesa del Suolo, 149–188.

Shaughnessy, G.L., 1986: A case study of some woody plant introductions to the Cape Town area, in I.A.W. Macdonald, F.J. Kruger and A.A. Ferrar (eds), *The ecology and management of biological invasions in Southern Africa*, Oxford University Press, Cape Town, 37–46.

Shepherd, S.A., 1986: Coastal waters, in C. Nance and D.L. Speight (eds.), *A land transformed: environmental change in South Australia*, Longman Cheshire, Melbourne, 182–199.

Siegfried, W.R., 1989: Preservation of species in southern African nature reserves, in B.J. Huntley (ed.), *Biotic diversity in southern Africa: concepts and conservation*, Oxford University Press, Cape Town, 186–201.

Škorić, A., 1977: *Tipovi naših tala*, Sveučilišna naklada Liber, Zagreb.

Smailes, P.J., 1979: The effects of changes in agriculture upon the service sector in South Australian country towns, 1945–1974, *Norsk Geografisk Tidsskrift*, 33, 125–142.

Smailes, P.J., 1989: The impact of the rural crisis of 1985–89 on a rural local government area, and its wider implications: Cleve, Eyre Peninsula, S.A., Paper presented to the *Institute of Australian Geographers Conference*, Adelaide, 13–16 February 1989.

Soffer, A., 1992a: *Changes in the geography of the Middle East*, Am Oved Publishers, Tel Aviv (in Hebrew).

Soffer, A., 1992b: *Rivers of fire – the conflict of water in the Middle East*, Am Oved Publishers, Tel Aviv (in Hebrew).

Soler, M. and Sala, M., 1992: Effects of fire and of clearing in a Mediterranean *Quercus ilex* woodland: an experimental approach, *Catena*, 19, 321–332.

Sorriso-Valvo, M., 1993: The geomorphology of Calabria – a sketch, *Geografia Fisica e Dinamica Quaternaria*, 16, 75–80.

Sorriso-Valvo, M., 1994: Movimenti di massa e modificazioni climatiche, *Atti del 1 Convegno "Energia, Clima ed Ambiente"*, Cittadella del Capo (CS), 6–8 Maggio 1994, 137–142.

Sorriso-Valvo, M., Bryan, R., Yair, A., *et al.*, 1995: Impact of afforestation on hydrological response and sediment production in a small Calabrian catchment, *Catena*, 25, 89–104.

Sorriso-Valvo, M., Antronico, L. and Borelli, A., 1992: Recent evolution of badland-type erosion in southern Calabria (Italy), *Geoöko plus*, 3, 69–82.

Sorriso-Valvo, M. and Sylvester, A., 1993: The relationship between geology and landforms along a coastal mountain front, northern Calabria, Italy, *Earth Surface Processes and Landforms*, 18, 257–273.

Soutade, G., 1980: *Modelé et dynamique actuelle des versants supra-forestiers des Pyrénées orientales*, Imprimerie cooperative de Sud Ouest, Albi.

Soutade, G., 1993: Les inondations catastrophiques d'Octobre 1940 en Catalogne Nord: le pourquoi d'une commémoration, in J. Becat and G. Soutade (eds), *L'aiguat del 40*, 55–64.

South Australian Year Book, 1994: No. 28, Australian Bureau of Statistics, South Australian Office, Government Printer South Australia, Adelaide.

SPC, 1994: *Albany regional strategy*, adopted by the State Planning Commission as a basis for coordination of local planning, control of subdivision and advice to the Hon. Minister and other Government Agencies, Perth.

SPC and Department of Conservation and Land Management, 1988: *Shark Bay region plan*, State Planning Commission, Perth.

Spiga, Y., 1993: Aménagement de la montagne en Algérie, *Traveaux de l'Institut de Geógraphie de Reims*, 85/86, 49–58.

Springett, J.A., 1976: The effect of prescribed burning on the soil fauna and on litter decomposition in Western Australian forests, *Australian Journal of Ecology*, 1, 77–82.

Stannage, C.T. (ed.), 1981: *A new history of Western Australia. Part 1: first settlers and white settlers*, University of Western Australia Press, 3–178.

State Directorate for Environment of Republic of Croatia, 1994: *National report on environment and development*, Zagreb.

Statistical Year Book, 1991, State Bureau of Statistics, Zagreb.

Statistical Year Book, 1993, State Bureau of Statistics, Zagreb.

Statistical Year Book of Croatian Counties, 1993, State Bureau of Statistics, Zagreb.

Statistical Year Book of SFR Yugoslavia, 1970, SR Croatia, State Bureau of Statistics, Beograd.

Stockton, C.W., 1988: Current research progress toward understanding drought, in *Drought, water management and food production*, Conf. Proceedings, Agadir, 21–24 Nov. 1985, 21–36.

St Quentin, D., 1944: Die Libellenfauna Dalmatiens, *Verhandlungen der Zoologisch-botanischen Gesellschaft*, Wien, 90/91, 66–76.

Swanage, M. (ed.), 1994: *Environmental science*, Australian Academy of Science, Canberra.

Swanson, F.J., 1981: Fires and geomorphic processes, in *Proceedings – Fire regimes and ecosystems*, Honolulu 1979, Gen. Tech. Rep. WO-26 USDA, Washington, DC, 401–420.

Swarbrick, J.T., 1986: Pesticides, in J.S. Russell and R.F. Isbell (eds), *Australian soils: the human impact*, Queensland University Press, St Lucia, 357–373.

Szabo, M., 1995: Australia's marsupials – going, going, gone? *New Scientist*, 145(1962), 30–35.

Taabani, M. and Kouti, A., 1987: Etude de l'aménagement du milieu naturel en Algérie, ex. d'un bassinversant des Bni Chougrane, Tell occidental, *Etudes Méditerranéennes*, 11, Poitiers, 291–306.

Talbot, W.J., 1947: *Swartland and Sandveld: a survey of the land utilization and soil erosion in the Western Lowland of the Cape Province*, Oxford University Press, Cape Town.

Tamames, R., 1994: *Introducción a la economía española*, Alianza Editorial, Madrid.

Tanji, K., Lauchli, A. and Meyer, J., 1986: Selenium in the San Joaquin Valley: a challenge to western irrigation, *Environment*, 28, 8–16.

Tatar, H., 1995: La dégradation des couverts végétaux et la lutte pour la protection du milieu (Petite Kabylie, Algérie), *Publ. de l'Association des Géographes Africains*, Rabat, 261–272.

Taylor, H.C., 1978: Capensis, in M.J.A. Werger (ed.), *The biogeography and ecology of Southern Africa*, Junk, The Hague, 171–229.

Tchernov, E., Horwitz, L.K., Ronen, A. and Lister, A., 1994: The faunal remains from Evron quarry in relation to other Paleolithic hominid sites in the Southern Levant, *Quaternary Review*, 42, 328–339.

Terceño Ramos, J., 1990: El laboreo de conservación en el manejo de praderas y pastos, *Ponencia Jornadas Técnicas sobre El Agua y el Suelo: Laboreo de Conservación*, Junta de Andalucía, Sevilla.

Thom, H.B., 1952: *Journal of Jan van Riebeck*, volume 1, Balkema, Cape Town.

Thom, H.B., 1954: *Journal of Jan van Riebeck*, volume 2, Balkema, Cape Town.

Thornes, J.B., 1990: The interaction of erosional and vegetational dynamics in land degradation: spatial outcomes, in J.B. Thornes (ed.), *Vegetation and erosion*, John Wiley, Chichester, 41–53.

Tolba, M.K., El-Kholy, O.A., El-Hinnawi, E. *et al.* 1992: *The world environment 1972–1992: two decades of challenge*, Chapman and Hall on behalf of the United Nations Environment Program, London.

Toledo, X. and Zapater, E., 1989: *Geografía General y Regional de Chile*, Ed. Universitaria.

Tomanbay, M., 1993: Sharing the Euphrates – Turkey, *Research and Exploration*, 9, 53–61.

Topić, V. 1987: Lirozija i bujica na kraškom prodručju Dalmacije, Pruojargoslavens ko savjetovanje o croziji i inredenju bujica, *Proceedings*, Le penski viv, 29–38.

Topić, V., in press: Ekološka obilježja mediteranskih područja Republika Hrvatske, Savjetovanje "100 godina Instituta za jadranske kulture u Splitu".

Tosato, G., 1994: La convenzione quadro sui cambiamenti climatici e le politiche energetiche di risposta, *Proceedings 1st Convegno "Energia Clima e Ambiente"*, Cittadella del Capo 6–8/5/94, Tipograf SrL, Roma, 9–17.

Trabaud,L., 1981: Man and fire: impacts on Mediterranean vegetation, in F. di Castri, D.W. Goodall and R.L. Specht (eds) *Mediterranean-type shrublands*, Ecosystems of the World, 11, Elsevier, Amsterdam, 523–537.

Trinajstić, I., 1986: Fitogeografsko raščlanjenje šumske vegetacije istočnojadranskog, sredozemnog područja – polazna osnovica u organizaciji gospodarenja mediteranskim šumama, *Annales pro experimentis forestìcis*, Zagreb, editio peculiaris II (Separatum), 53–67.

Troin, J.F., Brulé, J.C., Escallier, R. *et al.* 1985: *Le Maghreb, hommes et espaces*, Armand Colin, Paris.

Ubeda, X. and Sala, M., in press: Variation in runoff and erosion in three areas with different fire intensities, *Geoöko-dynamik*.

UNESCO–FAO, 1963: *Bioclimatic map of the Mediterranean zone*, UNESCO, Paris.

US Bureau of the Census, 1990: US Department of Commerce, Government Printing Office, Washington D.C.

US Bureau of the Census, 1994: US Department of Commerce, Government Printing Office, Washington D.C.

USDA, 1995: *Calleguas Creek watershed erosion and sediment control plan for Mugu Lagoon, Ventura and Los Angeles counties, California*, US Department of Agriculture, Natural Resources Conservation Service, Davis, California.

US Forest Service, 1988: *Area of timberland of the United States by ownership and State, 1987*, US Forest Service, Washington DC.
Valera, A., 1990: Aceptación de las nuevas técnicos de laboreo por los agricultores, Jornados Tecnicas sobre el Agua y el Svelo, Laboreo de Conservación, Sevilla, Quinta Ponencia.
Valladares, M.A. 1993: Effects of EC policy on natural Spanish habitats, *Science of the Total Environment*, 129, 71–72.
Van Daalen, J.C. and Geldenhuys, C.J., 1988: Southern African forests, in I.A.W. Macdonald and R.J.M. Crawford (eds), *Long term data series relating to southern Africa's renewable natural resources*, South African National Scientific Programmes Report 157, Pretoria, 237–248.
Van Wilgen, B., Bond, W.J. and Richardson, D.M., 1992: Ecosystem management, in R.M. Cowling (ed.), *The ecology of fynbos: nutrients, fire, diversity*, Oxford University Press, Cape Town, 345–371.
Van Wilgen, B.W., Cowling, R.M. and Burgers, C.J., in press: Valuation of ecosystem services: a case study from South African *fynbos*, *Bioscience*.
Van Zeist, W. and Bakker-Heeres, J.A.H., 1979: Some economic and ecological aspects of the plant husbandry of Tell Aswad, *Paleorient*, 5, 161–169.
Vaudour, J., Covo, J. and Martin, C., 1991: Un cycle de mesures hydrochimiques sur le Bayon en année sèche et après l'incendie de la Sainte Victoire du 28 août 1989, *Etudes de Géographie Physique*, Aix-en-Provence, 20, 37–46.
Vega, J.A., Bara, S., Villanueva, M.A. and Alonso, M., 1983: *Erosion despues de un incendio forestal*, Centro de Investigaciones Forestales, Lourizan, Pontevedra.
Vega-Hidalgo, J.A., 1986: La Investigacion sobre Incendios Forestales en España: Revision bibliografica, *Ponencias presentadas en Jornades sobre bases ecologiques per la gestio ambiental*, Diputacio de Barcelona, Barcelona, 17–24.
Versace, P., Ferrari, E., Gabriele, S. and Rossi, F., 1989: *Valutazione delle piene in Calabria*, CNR-IRPI, Geodata, 30.
Verster, E., du Plessis, W., Schloms, B.H.A. and Fuggle, R.F., 1992: Soil, in R.F. Fuggle and M.A. Rabie (eds), *Environmental management in South Africa*, Juta, Cape Town, 181–211.
Veseli, J., 1877: Kras hrvatske krajine i kako da se spasi, za tiem kraško pitanje uploške, *Šumarski list*, 2, c. kr. gl. zapovjedničtvo u Zagrebu k jednu zemalska upravna oblast hrvatsko-slavonske vojničke krajine, 57–118.
Victorian Year Book, 1993: Australian Bureau of Statistics, Victorian Office, Government Printer, Melbourne.
Viscomi, P., 1994: *Indagine sulla salinità dei suoli in Calabria*, Thesis, Univ. of Firenze, Fac. di Agraria, Dipart. di Scienza del Suolo e Nutrizione della Pianta.
Vita-Finzi, C., 1969: *The Mediterranean valleys*, Cambridge University Press, Cambridge.
WA Greenhouse, 1989: *Addressing the greenhouse effect*, Western Australian Greenhouse Coordination Council, Perth.
Wakil, M., 1993: Sharing the Euphrates – Syria, *Research and Exploration*, 9, 63–71.
Walling, D.E., 1986: Sediment yields and sediment delivery dynamics in Arab countries: some problems and research needs, *Journal of Water Resources*, 5, 775–798.
WAWRC, 1992: *The state of the rivers of the south west*, Western Australian Water Resources Council, publication no. WRC 2/92, WAWRC, Perth.
WCED, 1989: *Our common future (the Brundtland report)*, World Commission on Environment and Development, Oxford University Press, Oxford.
Webster, J., 1986: *The complete Australian bushfire book*, Thomas Nelson, Melbourne.
Weisrock, A., 1980: *Géomorphologie et paléoenvironnements de l'Atlas Atlantique, Maroc*, Thèse, Université de Paris.
Wellington, J.H., 1955: *Southern Africa: a geographical study*, Cambridge University Press, Cambridge.
Wells, C.B. and Prescott, J.A., 1983: The origins and early development of soil science in Australia, in *Soils: an Australian viewpoint*, CSIRO, Melbourne/ Academic Press, London, 3–12.
Wells, M.J., Poynton, R.J., Balsinhas, A.A. *et al.*, 1986: The history of introduction of invasive alien plants to southern Africa, in I.A.W. Macdonald, F.J. Kruger and A.A. Ferrar (eds), *The ecology and management of biological invasions in Southern Africa*, Oxford University Press, Cape Town, 21–36.
Wesgro, 1994: *Investment guide to the Western Cape*, Wesgro, Cape Town.
Wesgro, 1995: *South Africa's Western Cape: a corporate location report*, Wesgro, Cape Town.
Western Australian Year Book, 1992: No. 29, Australian Bureau of Statistics, Western Australian Office, Advance Press, Perth.
Western Cape Regional Services Council, 1994: *Metropolitan development framework: urban edge study*, Western Cape Regional Services Council, Cape Town.
Wildi, W., 1983: La chaîne tello-rifaine (Algérie, Maroc, Tunisie): structure, stratigraphie et évolution du Trias au Miocène, *Revue de Géologie Dynamique et de Géographie Physique*, 24, 201–297.
Wills, R.T., 1993: The ecological impact of *Phytophthora cinnamomi* in the Stirling Range National Park, Western Australia, *Australian Journal of Ecology*, 18, 145–159.
Wilson, F.M., 1991: A land out of balance, in M. Ramphele and C. McDowell (eds), *Restoring the land*, Panos, London, 27–38.

Wischmeier, W.H., 1959: A rainfall erosion index for a universal soil loss equation, *Soil Science Society of America Proceedings*, 23, 243–249.

Withers, P.C., Cowling, W.A. and Wills, R.T. (eds), 1994: *Plant diseases in ecosystems: threats and impacts in south-western Australia*, Proceedings of a Symposium, *Journal of the Royal Society of Western Australia* (special issue), 77(4).

Wood, W.E., 1924: Increase of salt in soil and streams following the destruction of the native vegetation, *Journal and Proceedings of the Royal Society of Western Australia*, 10, 35–47.

Woodruff, N.P., Fenster, C.R., Harris, W.W. and Lundquist, M., 1966: Stubble mulch tillage and planting in crop residue on the Great Plains, *Transactions of the American Society of Agricultural Engineering*, 9, 849–853.

Yapp, T.P., Walpole, S.C. and Sinden, J.A., 1992: Soil conservation and land degradation: finding the balance, *Search*, 23, 308–310.

Yassoglou, N., 1989: Desertification in Greece, in J.L. Rubio and R.J. Ricon (eds), *Strategies to combat desertification in Mediterranean Europe*, Commission of European Communities Report, EYR 11175 E/ES, 148–162.

Yassoglou, N. and Kosmas, C., 1988: *Potential erosion and desertification risks in Greece*, Second Meeting of the Balkan Initiative Committee for the Balkan Scientific Conference on Environmental Protection in the Balkans, Sofia, 17–18 May 1988.

Yassoglou, N. and Nobeli, C., 1972: Soil studies, in W. McDonald and G. Rapp Jr (eds), *The Minnesota Messenia expedition: reconstructing a bronze age regional environment*, University of Minnesota Press, Minneapolis, 171–176.

Yassoglou, N., Kosmas, C. and Danalatos, N., 1994: Global change and land degradation, in *International Symposium on Eastern Europe and Global Change*, Chalkidiki, Greece, 3–10 October 1994.

Year Book Australia, 1994: No. 76, Australian Bureau of Statistics, Commonwealth of Australia, Canberra.

Yeomans, P.A., 1954: *The keyline plan*, Keyline, Sydney.

Zann, L.P., 1995: *Our sea, our future*, Department of Environment, Sport and Territories, Australian Government Publishing Service, Canberra.

Zelinski, W., 1973: *The cultural geography of the United States*, Prentice-Hall, Englewood Cliffs, NJ.

Ziani, P., 1958: Opéi program kompleksne melioracije krša Jugoslavije, Institut za somavska i lovna istraživanja NRH, Zagreb.

Zitan, A., 1987: Les feux de forêt au Maroc, *Revue de Géographie du Maroc*, 11/1, nouvelle serie, 83–101.

Zohary, M., 1973: *Geobotanical foundations of the Middle East*, G. Fischer, Stuttgart.

Zohary, M., 1983: Man and vegetation in the Middle East, in W. Holzner, M.J.A. Werger and I. Ikusima (eds), *Man's impact on vegetation*, Dr. W. Junk Publishers, The Hague, 151–161.

Zohary, M. and Hoff, H., 1993: *Domestication of plants and animals in the old world: the origin and spread of cultivated plants in West Asia, Europe and the Nile Valley*, 2nd edition, Clarendon Press, Oxford.

Zohary, M. and Spiegel-Roy, P., 1975: Beginning of fruit growing in the Old World, *Science*, 187, 319–327.

Index

Note: **bold** page number indicates a major entry for the particular topic; f denotes a Figure; p denotes a Plate and t denotes a Table.